SOUTHEASTERN VEGETABLE EXTENSION WORKERS

2026 SOUTHEASTERN U.S. VEGETABLE CROP HANDBOOK

HANDBOOK SENIOR EDITOR:

A.L. Wszelaki
University of Tennessee

ASSOCIATE EDITORS:

K. Albornoz
Clemson University, Postharvest

M.B. Bertucci
University of Arkansas, Weed Science

T.R. Bilbo
Clemson University, Entomology

A.J. Cato
University of Arkansas, Entomology

I.M. Meadows
North Carolina State University, Plant Pathology

R.A. Melanson
Mississippi State University, Plant Pathology

C. Rodrigues
Auburn University, Food Safety

R.E. Rudolph
University of Kentucky, Horticulture

H.E. Wright-Smith
University of Tennessee, Weed Science

PHOTO CREDITS: *(from top to bottom)*

Squash bug laying eggs.
 Credit: P. Cremonez, Auburn University

'Mountain Gem' fresh market, determinate tomato.
 Credit: I. Meadows, North Carolina State University

Tomato spotted wilt virus on Tabasco pepper.
 Credit: R. Singh, Louisiana State University

Yellow nutsedge emerging through white-on-black polyethylene mulch.
 Credit: A. Wszelaki, University of Tennessee

EVERYTHING YOU NEED ON THE DASHBOARD OF YOUR TRUCK

SOUTHEASTERN VEGETABLE EXTENSION WORKERS
2026 SOUTHEASTERN U.S. VEGETABLE CROP HANDBOOK

The Southeastern Vegetable Extension Workers (SEVEW) are proud to present the 27th edition of the *2026 Southeastern U.S. Vegetable Crop Handbook*, a comprehensive resource for growers, advisers, educators, Extension agents, and specialists across the region.

This handbook is a joint effort among Extension specialists and researchers from 15 land-grant universities who work in the area of vegetable production. These specialists and researchers represent a wide array of disciplines: agricultural engineering, agribusiness, entomology, food safety, horticulture (vegetable production), plant pathology, postharvest physiology, soil science, and weed science.

This handbook comprises up-to-the-minute information developed from research and Extension projects conducted throughout the southeastern U.S. The purpose of this handbook is to provide you with a practical resource that conveniently fits on your dashboard. It contains the information that you need to manage your vegetable crops, including which varieties to plant, planting dates, fertilizer recommendations, cover crop selection and conservation tillage options, pesticide selection, alternative pest management tools, grafting, fertigation, plasticulture, postharvest handling, and food safety considerations as well as many other topics.

An electronic version is available at: *www.vegcrophandbook.com*

In addition to developing this handbook, the SEVEW focus on strengthening and supporting vegetable production programs around the region, identifying emerging issues, and providing a forum for multistate programming that will benefit growers in the southeastern U.S. Vegetable production in this region faces many challenges. Members of the SEVEW have combined their knowledge and experience to develop approaches and answers that enable growers to optimize their production practices and increase the sustainability of their operations.

We gratefully acknowledge the generous contributions of multiple funding sources, whose support enabled the development and dissemination of this handbook.

This material is based upon work supported, in part, by the National Institute of Food and Agriculture, United States Department of Agriculture, award numbers 2024-70006-43504, 2024-70006-43496, SC-1700674 and SC-1700666, as well as partially funded by the Southern IPM Center (Project S25-062) as part of USDA National Institute of Food and Agriculture *Crop Protection and Pest Management Regional Coordination Program* (Agreement No. 2022-70006-38002).

Any opinions, findings, conclusions, or recommendations expressed in this publication are those of the author(s) and should not be construed to represent any official USDA or U.S. government determination or policy.

We hope you enjoy this handbook!

Sincerely,
SEVEW Group

*The 2026 handbook was prepared and reviewed by the following authors at their respective institutions.
We also wish to thank all of the past authors and participants that have helped to refine and continually improve this handbook.*

ALABAMA A & M UNIVERSITY
Food Safety – A. Jackson-Davis

AUBURN UNIVERSITY
Entomology – P. Cremonez
Food Safety – C. Rodrigues
Horticulture – A. Da Silva*
Plant Pathology – E.J. Sikora
Postharvest – M. Trandel-Hayse

CLEMSON UNIVERSITY
Agribusiness – K. Burkett
Entomology – J. Ballew*, T. Bilbo, and M. Tayal
Food Safety – C. Carter
Horticulture – B.S. Jatana, Z. Snipes, and B.K. Ward*
Plant Pathology – A. Keinath
Postharvest – K. Albornoz
Weed Science – M. Cutulle

FLORIDA A & M UNIVERSITY
Food Safety – K. Sarjeant

LOUISIANA STATE UNIVERSITY AGRICULTURAL CENTER
Horticulture – C.E. Coker, K. Fontenot*, and C. Motsenbocker
Plant Pathology – R. Singh
Sweet Potato Research Station – C. Gregorie
Weed Science – C. Blankenship

MISSISSIPPI STATE UNIVERSITY
Entomology – J. Perrier
Food Safety – S.B. White
Horticulture – T. Ayankojo and P. Orlinski
Plant Pathology – R. A. Melanson*
Weed Science – J. Byrd

NORTH CAROLINA STATE UNIVERSITY
Entomology – J.F. Walgenbach
Food Safety – E.T. Rogers
Horticulture – R.B. Batts, J.M. Davis*, R.C. Mauney, E.R. Eure, J.R. Schultheis, and E. Torres
Plant Pathology – A.M. Gorny, L.M. Quesada Ocampo, and I.M. Meadows
Weed Science – K. Jennings

OKLAHOMA STATE UNIVERSITY
Food Safety – R. Jadeja
Horticulture – T. Mason

TEXAS A & M UNIVERSITY
Food Safety – A. Castillo

TUSKEGEE UNIVERSITY
Food Safety – K.L. Woods

UNIVERSITY OF ARKANSAS
Entomology – A. Cato
Food Safety – A. Perez
Horticulture – T. Ernst, J. Lee, and A. McWhirt*
Weed Science – M.B. Bertucci

UNIVERSITY OF GEORGIA
Entomology – A. Sparks
Food Safety – L.L. Dunn
Horticulture – T.W. Coolong*, A. Deltsidis, T. P. McAvoy, and C. Tyson
Plant Pathology – B. Dutta, and I.A. Chowdhury
Postharvest – A. Deltsidis

UNIVERSITY OF KENTUCKY
Entomology – R.T. Bessin
Food Safety – P.V. Priyesh
Horticulture – R. Rudolph*
Plant Pathology – N. Gauthier

UNIVERSITY OF TENNESSEE
Entomology & Plant Pathology – J. Hayter
Horticulture & Food Safety – A.L. Wszelaki*
Weed Science – H. Wright-Smith

VIRGINA TECH
Entomology – T.P. Kuhar, and K. Sutton
Food Safety – L. Strawn
Nutrient Management/Soils – M.S. Reiter*
Plant Pathology – D.S. Higgins, and S.L. Rideout
Weed Science – V. Singh

The purpose of this book is to provide the best and most up-to-date information available for commercial vegetable growers in the southeastern US: Alabama, Arkansas, Florida, Georgia, Kentucky, Louisiana, Mississippi, North Carolina, Oklahoma, South Carolina, Tennessee, Texas, and Virginia. These recommendations are suggested guidelines for production in the above states. Factors such as markets, weather, and location may warrant modifications and/or different practices or planting dates not specifically mentioned in this book.

*State Coordinators

UPCOMING EVENTS FOR 2026

DATE/TIME	LOCATION	CONTACT INFO
ALABAMA		
Alabama Fruit & Vegetable Growers Association Annual Conference and Trade Show		
4 to 6 Feb	Lodge at Gulf State Park, Gulf Shores, AL	Blake Thaxton at bthaxton@alfafarmers.org http://www.afvga.org
Farmers Conference		
TBA	Montgomery, AL	https://www.cisc1881.org/our-events/
Professional Agricultural Workers Conference		
TBA (Nov)	Montgomery, AL	https://pawc.info
Alabama Black Belt Food Systems Alliance Annual Meeting		
TBA (Nov)	Selma, AL	https://pawc.info
ARKANSAS		
Northwest Arkansas High Tunnel Workshop		
Spring 2026	Fayettevillle, AR	Taunya Ernst (ternst@uada.edu)
High Tunnel Grower Skill-Building Workshop		
Fall 2026	Fayettevillle, AR	Taunya Ernst (ternst@uada.edu)
GEORGIA		
SE Regional Fruit & Vegetable Conference		
8 to 10 Jan	Savannah, GA	Georgia Fruit and Vegetable Growers Association, 877-994-3842 http://www.gfvga.org
Georgia Watermelon Association Conference		
22 to 24 Jan	St. Simons Island, GA	http://www.georgiawatermelonassociation.org
Sunbelt Ag. Expo		
20 to 22 Oct	Moultrie, GA	http://www.sunbeltexpo.com
KENTUCKY		
Kentucky Fruit & Vegetable Conference and Trade Show		
11 to 13 Jan	Bowling Green, KY	https://kyhortcouncil.org/2026-kentucky-fruit-and-vegetable-conference/
Organic Association of Kentucky Conference		
30 to 31 Jan	Kentucky State University Research Farm, Frankfort, KY	https://www.oak-ky.org/annual-conference
MISSISSIPPI		
Private Pesticide Applicator Trainings		
Anytime	Virtual	Gene Merkl at gm53@msstate.edu
Various (TBA)	Various	http://extension.msstate.edu/content/online-private-applicator-certification-program
Produce Safety Alliance Training		
Cleaning & Sanitation Workshop Hybrid	In person/Virtual (TBA)	Juan Silva at jsilva@foodscience.msstate.edu Angelica Abdallah-Ruiz at ama416@msstate.edu
Water Systems Workshop Hybrid		
Vegetable Short Course		
24 to 25 Feb	North Mississippi Research and Extension Center, Verona, MS	Timothy Ayankojo at ita10@msstate.edu
NORTH CAROLINA		
Business of Farming Conference		
28 Feb	Asheville, NC	https://asapconnections.org/farmer-resources/business-of-farming-conference/
Winter Vegetable Conference		
TBA (mid to late Feb)	Asheville, NC	https://www.ncagr.gov/divisions/marketing/tomatoes
Sweetpotato Regional Meeting		
TBA (late February)		https://fielddays.ces.ncsu.edu/
North Carolina Watermelon Association Meeting		
13 to 14 Feb	Wrightsville Beach, NC	https://www.ncmelons.com/index.php
Organic Growers School		
6 to 8 Mar	Mars Hill University, Mars Hill, NC	https://www.organicgrowersschool.org/spring-conference
Mountain Research Station Field Day		
TBA (mid-July)	Mountain Research Station, Waynesville, NC	https://www.ncagr.gov/divisions/research-stations

UPCOMING EVENTS FOR 2026 (cont'd)

DATE/TIME	LOCATION	CONTACT INFO
NORTH CAROLINA (cont'd)		
Mountain Horticultural Crops Field Day		
TBA (early to mid Aug)	Mountain Horticultural Crops Research and Extension Center, Mills River, NC	https://mountainhort.ces.ncsu.edu/
NC Sweetpotato Field Day		
TBA (Oct)	Horticultural Crops Research Station, Clinton, NC	https://fielddays.ces.ncsu.edu/
NC Greenhouse Vegetable Growers Annual Meeting		
TBA (late Oct/early Nov)	TBA	http://www.ncghvga.com/
Sustainable Agriculture Conference		
TBA (early Nov)	Durham Convention Center, Durham, NC	https://www.carolinafarmstewards.org/sustainable-agriculture-conference/
Southeast Vegetable & Fruit Expo		
TBA (late Nov to early Dec)	TBA	Allan Thornton at allan_thornton@ncsu.edu or Erin Eure at erin_eure@ncsu.edu Cathy Price at 919-413-9544 or cathy@seasag.com, https://ncvga.com/
SOUTH CAROLINA		
SC Watermelon Growers Association Meeting		
16 to 18 Jan		https://scgrower.com/upcoming-events/
Preplant Growers Meeting		
28 Jan		https://scgrower.com/upcoming-events/
Statewide Cucurbit Production Meeting		
5 Feb		https://scgrower.com/upcoming-events/
Midlands Spring Vegetable Meeting		
11 Feb		https://scgrower.com/upcoming-events/
Pee Dee Spring Vegetable Production Meeting		
19 Feb		https://scgrower.com/upcoming-events/
Statewide Pea and Butterbean Meeting		
5 Mar		https://scgrower.com/upcoming-events/
South Carolina Women in Agriculture Conference		
6 to 7 Mar		https://scgrower.com/upcoming-events/
Upstate Vegetable Meeting		
12 Mar		https://scgrower.com/upcoming-events/
Sweetpotato Meeting		
19 Mar		https://scgrower.com/upcoming-events/
TENNESSEE		
Pick TN Conference		
13 to 15 Jan	Lebanon, TN	https://www.picktnconference.com/
Grow and Tell Field Day		
18 June	East TN Research & Education Center	https://easttn.tennessee.edu
Hort, Hops and Crops Field Day		
13 Aug	East TN Research & Education Center	https://easttn.tennessee.edu
Steak and Potatoes Field Day		
20 Aug	Plateau Research & Education Center, Crossville, TN	https://plateau.tennessee.edu
VIRGINIA		
Eastern Shore Ag Conference and Trade Show		
14 to 15 Jan	Exmore Moose Lodge, Belle Haven, VA	https://northampton.ext.vt.edu/news/2025-eastern-shore-agricultural-conference-and-trade-show.html
Virginia Pumpkin Growers Association Annual Meeting		
24 Jan	Hillsville VFW Building, Hillsville, VA	

MOBILE APPS

CALIBRATE MY SPRAYER app for iOS and Android

Improperly calibrated pesticide spraying equipment may cause either too little or too much pesticide to be applied. This free mobile app was created to aid in the proper calibration of spraying equipment. Simply select the type of sprayer you want to calibrate (Broadcast or Banded), insert values in each input box, select what you want the app to calculate (Volume/Area or Catch/Nozzle), and tap 'Calculate'. Each input's units can be customized by tapping the units. Sprayers can be saved with user-defined names.

Download the free **Calibrate My Sprayer** app at the Apple or Google app stores.

MIX MY SPRAYER app for iOS and Android

Mix My Sprayer was created to aid with quick, accurate calculations of product mixes to be applied with spraying equipment. Users can create custom lists of favorite products by category. Simply add or select a product, insert values in each input box, and the app automatically calculates the amount of product to include in the user-defined mix size. Units for each input can be customized by tapping the unit buttons. Products are saved with the user settings last used.

Download the free **Mix My Sprayer** app at the Apple or Google app stores.

FARMING BASICS app for iOS and Android

The Farming Basics app from Alabama Extension is a gateway of information for small and beginning farmers. The user-friendly app includes:
- information about major insect pests and diseases
- general management tactics
- a fertilizer and irrigation calculator to assist in saving beginning farmers dollars on inputs

Clear pictures depict each crop, insect, and disease. Other features include crop and pest alerts, location services, with contact information for Extension regional agents, and an activity calendar linking to upcoming meetings and field days. This app is available for iOS and Android devices.

Download the free **Farming Basics** app at the Apple or Google app stores.

SoilWeb 2.0 app for iOS and Android and Web

Allows users to pinpoint soil types in their current location. The app is complete with data about the current soil type. With its embedded GPS system, SoilWeb provides convenient, instantaneous soil information from the exact soil the user is standing on. The app gives users greater accessibility to NRCS' soil survey information that has been collected across the US since the late 1890's. The app presents the data in mobile form, helping users make better-informed decisions about the location they are researching.
Web-based version -- *https://websoilsurvey.sc.egov.usda.gov/App/HomePage.htm*

Download the free **SoilWeb 2.0** app at the Apple or Google app stores.

TANK MIX CALCULATOR app for iOS, Android, and Web

Tank Mix Calculator is an easy-to-use farm app from TapLogic. This agriculture app is free for any farmer to use on their mobile device to quickly and easily generate a tank mix. Just enter your acreage, tank size, and carrier volume. Next, select your chemicals from our list or add your own. Tank Mix Calculator will then provides the number of loads required to spray your acreage, along with full and partial load mixes of the chemicals you selected.

Download the free **Tank Mix Calculator** app at the Apple or Google app stores.

MOBILE APPS

VEGDR app for iOS and Android

VegDr provides you with up-to-date information for vegetable diseases in Georgia. The app contains information on various diseases, images of the symptoms, and list what products are considered effective for control. At present time, the app includes cucurbits (cucumber, watermelon, squash, cantaloupe, pumpkin) and solanaceous crops (tomato and pepper). Recommendations are based on information on the manufacturer's label and performance data from trials the University of Georgia.

Download the free **VegDr** app at the Apple or Google app stores.

MyIPM for Vegetables app for iOS and Android

MyIPM for Vegetables is a pest management resource for commercial vegetable production. This free mobile app for iOS and Android devices provides integrated pest management (IPM) information for management of common diseases and insects in conventional and organic vegetable production systems. MyIPM for Vegetables currently contains information on diseases and insects of cucurbits (cantaloupe, cucumber, pumpkin, squash, and watermelon) and tomatoes. Disease and insect information for additional crops will be added in the future. Download the app today to access images of diseases and insects; descriptions of disease signs and symptoms, insects, and insect damage; descriptions of insect and pathogen life cycles; management recommendations; efficacies, rates, preharvest (PHI) and restricted-entry intervals (REI), and season limits for fungicides and insecticides. For more information, visit
https://myipm.app/vegetables/

Download the free **MyIPM** for Vegetables app at the Apple or Google app stores.

WEB-BASED APPS

CLEMSON CENTER PIVOT FERTIGATION CALCULATOR

Calculates the amount of liquid fertilizer to apply through center pivot irrigation systems.

https://precisionag.sites.clemson.edu/Calculators/Fertigation/Pivot/

CLEMSON DRIP FERTIGATION CALCULATOR

Calculates the amount of liquid fertilizer to apply through the drip system.

https://precisionag.sites.clemson.edu/Calculators/Fertigation/Drip/

Vegetable Production Information Web Sites

ALABAMA

Alabama SARE Program
http://www.southernsare.org/SARE-in-Your-State/Alabama

Alabama Cooperative Extension System
http://www.aces.edu

AU Plant Diagnostic Lab
http://offices.aces.edu/plantlabauburn/

AgWater Safety Program - Microbial Water Testing Lab
https://www.aces.edu/blog/topics/farming/agwater-safety-program/

Vegetable IPM Info
http://www.aces.edu/vegetableipm

Alabama Beginning Farms Program
http://www.aces.edu/beginningfarms

ARKANSAS

Arkansas Cooperative Extension Service
https://www.uaex.uada.edu/

Arkansas Fruit, Vegetable, and Nut Update, UA CES Blog
http://uaex.uada.edu/hortblog

UA CES Commercial Vegetable Production
https://www.uaex.uada.edu/farm-ranch/crops-commercial-horticulture/horticulture/vegetables/

UA Vegetable IPM - Insect Pest Monitoring
http://uaex.uada.edu/hort-IPM

FLORIDA

University of Florida Cooperative Extension Service
http://edis.ifas.ufl.edu

GEORGIA

University of Georgia Cooperative Extension Service
http://extension.uga.edu

UGA Vegetables
https://vegetables.caes.uga.edu/

University of Georgia Extension Publications
http://www.extension.uga.edu/publications.html

KENTUCKY

University of Kentucky Cooperative Extension Service
http://ces.ca.uky.edu/ces

Kentucky Vegetable Integrated Pest Management Program
http://ipm.ca.uky.edu/

UK Vegetable Crops Extension and Research
https://vegcrops.ca.uky.edu

UK Plant Pathology
https://plantpathology.ca.uky.edu

UK Horticulture
https://www.uky.edu/hort

UK Entomology
http://entomology.ca.uky.edu/

UK Center for Crop Diversification
https://ccd.uky.edu/

LOUISIANA

Louisiana SARE Program
http://www.southernsare.org/SARE-in-Your-State/Louisiana

LSU AgCenter
http://www.lsuagcenter.com

LSU AgCenter Plant Diagnostic
http://www.lsuagcenter.com/plantdiagnostics

LSU Soil Testing & Plant Analysis Lab
https://www.lsuagcenter.com/portals/our_offices/departments/spess/servicelabs/soil_testing_lab

MISSISSIPPI

Mississippi State University Extension Service
http://extension.msstate.edu

Mississippi State University Extension Plant Diagnostic Lab
http://extension.msstate.edu/lab

Mississippi Commercial Horticulture Information
http://extension.msstate.edu/agriculture/crops/commercial-horticulture

Farmers' Markets
http://extension.msstate.edu/agriculture/local-flavor/farmers-markets

Mississippi State University Extension Soil Testing Lab
http://extension.msstate.edu/agriculture/soils/soil-testing

NORTH CAROLINA

North Carolina Cooperative Extension Service
http://www.ces.ncsu.edu

Information on Herbs, Organics, & Specialty Crops
http://ncherb.org

NCSU Vegetable Pathology
http://go.ncsu.edu/veggiepathology

NCSU Extension Plant Pathology Portal
http://plantpathology.ces.ncsu.edu/

NCSU Plant Disease and Insect Clinic
https://pdic.ces.ncsu.edu/

NCSU Entomology Portal
http://entomology.ces.ncsu.edu/

NCSU IPM Portal
http://ipm.ces.ncsu.edu/

North Carolina Pest News
http://ipm.ncsu.edu/current_ipm/pest_news.html

Horticulture Information
https://www.ces.ncsu.edu/categories/agriculture-food/commercial-horticulture-nursery-turf/

NCSU Extension Growing Small Farms
https://growingsmallfarms.ces.ncsu.edu/

Fresh Produce Safety
http://ncfreshproducesafety.ces.ncsu.edu/

OKLAHOMA

Oklahoma Cooperative Extension Service
https://extension.okstate.edu/topics/plants-and-animals/crops/vegetables/

SOUTH CAROLINA

Clemson University Cooperative Extension Service
http://www.clemson.edu/extension

Clemson Coastal Research & Education Center
https://www.clemson.edu/cafls/research/coastal/

Clemson Edisto Research and Extension Center
https://www.clemson.edu/cafls/research/edisto/

SC Growers F&V News
https://scgrower.com/

TENNESSEE

University of Tennessee Extension
https://utextension.tennessee.edu/

UT Vegetable Production
https://utvegetable.com/

TEXAS

Texas Agricultural Extension Service
http://agrilifeextension.tamu.edu

VIRGINIA

Eastern Shore AREC, Virginia Tech
https://www.facebook.com/ESAREC

Virginia Cooperative Extension
http://www.ext.vt.edu

Virginia Tech Eastern Shore AREC
https://www.arec.vaes.vt.edu/arec/eastern-shore.html

Virginia Tech Soils and Nutrient Management
https://www.facebook.com/EasternShore.Soils

Virginia Tech Vegetable Entomology Facebook
https://www.facebook.com/VirginiaTechVIPRLab/

Virginia Tech Vegetable Pathology Facebook
http://www.facebook.com/vtvegpp
VirginiaTechVIPRlab

Table of Contents

UPCOMING EVENTS FOR 2026 .... iv - v

MOBILE AND WEB-BASED APPS .... vi - vii

VEGETABLE PRODUCTION INFORMATION WEB SITES viii

TABLES FOR GENERAL PRODUCTION RECOMMENDATIONS 1-47
    Table 1a. Vegetable Families 1
    Table 1. Soil Test Interpretations and Recommendations Based on Soil Test Results 3
    Table 2. General Fertilizer Suggestions for Vegetable Crops 6
    Table 3. Nutrient Values For Manure Applications and Crop Residues 10
    Table 3A. Nutrient Values for Various Plant, Animal, and Natural Products 10
    Table 4. Percentage Equivalents and Conversion Factors for Major, Secondary, and Micronutrient Fertilizer Sources 11
    Table 5. Optimum and Minimum Temperatures for Transplant Production After Germination ... 16
    Table 6. Vegetable Seed Sizes 19
    Table 7. Population of Plants Per Acre at Several Between-Row and In-Row Spacings 20
    Table 8. Critical Periods of Water Need for Vegetable Crop 20
    Table 9. Available Water-Holding Capacity Based on Soil Texture 20
    Table 10. Soil Infiltration Rates Based on Soil Texture 20
    Table 11. Hours Required to Apply 1" Water Based on Row Spacing 21
    Table 12. Maximum Application Time in Minutes for Drip Irrigated Vegetables 21
    Table 13. Predators and Parasites of Vegetable Pests 31
    Table 14. Recommended Storage Conditions And Cooling Methods for Maximum Postharvest Life of Commercially Grown Vegetables 40
    Table 15. Southeast Us Vegetable Crop Handbook Budget - Template 47

GENERAL PRODUCTION RECOMMENDATIONS 1-46
    Southeastern Vegetable Extension Workers li
    Vegetable Crop Handbook 2026 li
    Varieties 1
    Crop Rotation 1
    Soils and Soil Fertility 1
    Nutrient Management and Maximizing Production 3
    Minimum Tillage for Vegetable Production 12
    Cover Crops 12
    Helpful Resources 15
    Transplant Production 16
    Grafting in Vegetable Crops 18
    Disease Control in Plant Beds 19
    Seed Storage and Handling 19
    Plant Populations 19
    Irrigation 19
    Important Notes: 23
    Precision Agriculture for Vegetable Production 23
    Season Extension 24
    Pollination 26

Table of Contents (cont'd)

GENERAL PRODUCTION RECOMMENDATIONS (cont'd) 1-46

- Pesticides and The Endangered Species Act: What You Need to Know, ... 27
- How to Improve Pest Control, ... 27
- Beneficial Insects, ... 30
- Diagnosing Vegetable Crop Problems, ... 34
- Air Pollution Injury 35
- Produce Safety, ... 35
- What Are Good Agricultural Practices (Gaps)? 36
- Food Safety Modernization Act (Fsma) Produce Safety Rule (Psr), ... 36
- Agricultural Water Psr - Subpart E, ... 36
- Important Definitions: 37
- Produce Safety Alliance Training, ... 37
- On-Farm Readiness Review - National Association Of State Departments Of Agriculture, ... 38
- Food Safety Audits Versus PSR Inspections, ... 38
- Food Safety Certification for Specialty Crops Program, ... 38
- Produce Safety Resources in Your State, ... 38
- Additional Resources, ... 40
- Postharvest Handling, ... 40
- Optimizing Commercial Cooling, ... 42
- Cooling Methods, ... 43
- Genetic Selection, ... 44
- Methods to Maintain Produce Quality During Storage, ... 45
- Enterprise Budgets for Horticulture Crops, ... 45
- Additional Resource Can Be Found on the Following Websites, ... 46

CALIBRATING CHEMICAL APPLICATION EQUIPMENT 131-139

- Calibrating a Sprayer 134
- Calibrating a Granular Applicator 136
- Calibrating a Broadcast Spreader 138
- Calibration Variables 139

REGISTERED FUNGICIDES, INSECTICIDES, AND MITICIDES FOR VEGETABLES 140-148

- Resistance Management and the Insecticide Resistance Action Committee (IRAC) Codes for Modes of Action of Insecticides ... 140
- General Information 141
- Respiratory Protective Devices for Pesticides 144
- Protecting Our Groundwater 145
- Toxicity of Chemicals Used in Pest Control 147
- Conversion Information for Use of Pesticides on Small Areas 147
- Pesticide Dilution Tables 148

INSECT, DISEASE, AND WEED CONTROL TABLES 149-377

EMERGENCY NUMBERS BY STATE 378

Table of Contents (cont'd)

GENERAL INSECT, DISEASE, AND WEED CONTROL TABLES

ALL VEGETABLES

Table 2-24.	Fire Ant Management in Commercial Vegetable Fields	196
Table 2-25.	Relative Effectiveness of Insecticides and Miticides for Insect and Mite Control on Field-Grown Vegetables	196
Table 2-25 A.	Preharvest Intervals (In Days) for Pyrethroid Insecticides in Vegetable Crops	198
Table 2-26.	Virus-Based Insecticides and Lepidopteran Larvae Controlled	198
Table 2-27.	List of Generic Insecticides by Active Ingredient	199
Table 2-28.	Components of Insecticide Mixtures	200
Table 2-29.	Insect Control for Greenhouse Vegetables	201
Table 2-30.	Alternative IPM & Bioinsecticide Recommendations in Vegetable Crops	205
Table 3-40.	Nematode Control in Vegetable Crop	298
Table 3-41.	Efficacy of Fumigants or Fumigant Combinations for Managing Soilborne Nematodes, Diseases, and Weeds	299
Table 3-42.	Management of Soilborne Nematodes with Non-Fumigant Nematicides	299
Table 3-43.	Greenhouse Disease Control for Various Vegetable Crops	301
Table 3-44.	Efficacy of Products for Greenhouse Tomato Disease Control	317
Table 3-45.	Recommended Temperatures & Treatment Times for Hot Water Disinfestation of Vegetable Seed	319
Table 3-46.	Basic Cleaning and Sanitizing Procedures for Food Contact Surfaces	320
Table 3-47.	Water, Produce, and Equipment Sanitation	321
Table 3-48.	Various Fungicides for Use on Vegetable Crops	322
Table 3-49.	Biopesticides, Fungicide, and Nematicide Alternatives for Vegetables	324
Table 3-50.	Fungicide Modes of Action for Fungicide Resistance Management	329
Table 4-28.	Weed Response to Herbicides Used in Vegetable Crops	377

Table of Contents (cont'd)

SPECIFIC COMMODITY RECOMMENDATIONS

ASPARAGUS
- Varieties and Production Practices ... 48
- Insect Control (Table 2-1) ... 150
- Alternative Management Practices (Table 3-2) ... 208
- Disease Control (Table 3-1) ... 208
- Chemical Weed Control (Table 4-1) ... 332

BASIL
- Varieties and Production Practices ... 75
- Disease Control (Table 3-3) ... 209

BEANS: Lima/Butter and Snap
- Varieties and Production Practices ... 51
- Insect Control (Table 2-2) ... 150
- Disease Control (Table 3-4) ... 211
- Alternative Management Practices (Table 3-5) ... 214
- Chemical Weed Control (Table 4-2) ... 334

BEETS
- Varieties and Production Practices ... 54
- Insect Control (Table 2-3) ... 154
- Chemical Weed Control (Table 4-3) ... 337

BROCCOLI
- Varieties and Production Practices ... 55
- Insect Control (Table 2-4) ... 155
- Disease Control (Table 3-6) ... 214
- Efficacy Of Products (Table 3-7) ... 218
- Alternative Management Practices (Table 3-8) ... 219
- Chemical Weed Control (Table 4-8) ... 342

BRUSSELS SPROUTS
- Varieties and Production Practices ... 55
- Insect Control (Table 2-4) ... 155
- Disease Control (Table 3-6) ... 214
- Efficacy Of Products (Table 3-7) ... 218

BRUSSELS SPROUTS (cont'd)
- Alternative Management Practices (Table 3-8) ... 219
- Chemical Weed Control (Table 4-8) ... 342

CABBAGE
- Varieties and Production Practices ... 55
- Insect Control (Table 2-4) ... 155
- Disease Control (Table 3-6) ... 214
- Efficacy Of Products (Table 3-7) ... 218
- Alternative Management Practices (Table 3-8) ... 219
- Chemical Weed Control (Table 4-8) ... 342

CARROTS
- Varieties and Production Practices ... 61
- Insect Control (Table 2-5) ... 158
- Disease Control (Table 3-28) ... 269
- Alternative Management Practices (Table 3-29) ... 274
- Chemical Weed Control (Table 4-5) ... 340

CAULIFLOWER
- Varieties and Production Practices ... 55
- Insect Control (Table 2-4) ... 155
- Disease Control (Table 3-6) ... 214
- Efficacy Of Products (Table 3-7) ... 218
- Alternative Management Practices (Table 3-8) ... 219
- Chemical Weed Control (Table 4-8) ... 342

CELERY
- Varieties and Production Practices ... Coming in 2027!
- Insect Control (Table 2-6) ... 159
- Chemical Weed Control (Table 4-6) ... 341

CILANTRO
- Varieties and Production Practices ... 75
- Chemical Weed Control (Table 4-7) ... 342

COLLARD
- Varieties and Production Practices ... 55
- Insect Control (Table 2-4) ... 155
- Efficacy Of Products (Table 3-7) ... 218
- Alternative Management Practices (Table 3-8) ... 219
- Disease Control (Table 3-16) ... 236
- Chemical Weed Control (Table 4-8) ... 342

CORN, SWEET
- Varieties and Production Practices ... 116
- Insect Control (Table 2-8) ... 162
- Disease Control (Table 3-9) ... 220
- Chemical Weed Control (Table 4-9) ... 344

CUCUMBER
- Varieties and Production Practices ... 64
- Fertigation Schedule ... 65
- Insect Control (Table 2-9) ... 164
- Disease Control (Table 3-10) ... 223
- Efficacy Of Products (Table 3-11) ... 228
- Alternative Management Practices (Table 3-12) ... 230
- Chemical Weed Control (Table 4-10) ... 348

Table of Contents (cont'd)

SPECIFIC COMMODITY RECOMMENDATIONS (cont'd)

EGGPLANT
- Varieties and Production Practices 68
- Fertigation Schedule 69
- Insect Control (Table 2-10) 168
- Disease Control (Table 3-14) 232
- Chemical Weed Control (Table 4-11) 349

ENDIVE
- Varieties and Production Practices 81
- Disease Control (Table 3-17) 240

ESCAROLE
- Varieties and Production Practices 81

GARLIC AND ELEPHANT GARLIC
- Varieties and Production Practices 70
- Disease Control (Table 3-14) 234
- Chemical Weed Control (Table 4-11) 351

GREENS: Collard, Kale, Mustard and Turnip
- Varieties and Production Practices 72
- Insect Control (Table 2-7) 160
- Disease Control (Table 3-16) 236
- Chemical Weed Control (Table 4-13) 352

HERBS
- Varieties and Production Practices 75
- Basil Disease Control (Table 3-3) 209
- Cilantro and Parsley Disease Control (Table 3-21) 252
- Cilantro and Parsley Alternative Management Tools (Table 3-22) 253
- Cilantro Chemical Weed Control (Table 4-7) 342

HOPS
- Varieties and Production Practices 78
- Insect Control (Table 2-11) 172
- Chemical Weed Control (Table 4-14) 354

KOHLRABI
- Varieties and Production Practices 55
- Insect Control (Table 2-4) 155
- Disease Control (Table 3-6) 214
- Efficacy Of Products (Table 3-7) 218
- Alternative Management Practices (Table 3-8) 219
- Chemical Weed Control (Table 4-8) 342

LEEKS
- Varieties and Production Practices 80

LETTUCE
- Varieties and Production Practices 81
- Insect Control (Table 2-12) 173
- Disease Control (Table 3-17) 240
- Chemical Weed Control (Table 4-15) 355

MELONS, Specialty
- Varieties and Production Practices 84
- Fertigation Schedule 85
- Insect Control (Table 2-9) 164
- Disease Control (Table 3-10) 223
- Cantaloupe Chemical Weed Control (Table 4-4) 338

OKRA
- Varieties and Production Practices 87
- Fertigation Schedule 87
- Insect Control (Table 2-13) 175
- Disease Control (Table 3-18) 243
- Chemical Weed Control (Table 4-16) 356

ONIONS and GREEN ONIONS
- Varieties and Production Practices 89
- Insect Control (Table 2-14) 177
- Disease Control (Table 3-19) 245
- Efficacy of Products for Disease Control in Onion (Table 3-20) .. 251
- Chemical Weed Control (Table 4-17) 357

PARSNIP
- Varieties and Production Practices 91
- Disease Control (Table 3-28) 269

PEAS: English/Garden and Southern
- Varieties and Production Practices 93
- Insect Control (Table 2-15) 178
- Disease Control (Table 3-23) 254
- Chemical Weed Control (Table 4-18) 358

PEPPERS: Hot and Sweet
- Varieties and Production Practices 95
- Fertigation Schedule 97
- Insect Control (Table 2-17) 180
- Disease Control (Table 3-24) 258
- Efficacy Of Products (Table 3-25) 263
- Alternative Management Practices (Table 3-26) 263
- Chemical Weed Control (Table 4-19) 361

POTATOES, Irish
- Varieties and Production Practices 99
- Insect Control (Table 2-18) 182
- Disease Control (Table 3-27) 264
- Chemical Weed Control (Table 4-20) 364

Insect, Disease, & Weed Control Tables (cont'd)

SPECIFIC COMMODITY RECOMMENDATIONS (cont'd)

PUMPKINS
Varieties and Production Practices 84
Fertigation Schedule 85
Insect Control (Table 2-9) 164
Disease Control (Table 3-10) 223
Efficacy of Products for Disease Control (Table 3-11) 223
Chemical Weed Control (Table 4-21) 365

RADISHES
Varieties and Production Practices 108
Insect Control (Table 2-19) 187
Disease Control (Table 3-28) 269
Chemical Weed Control (Table 4-22) 367

RUTABAGAS
Varieties and Production Practices 108
Disease Control (Table 3-28) 269

SPINACH
Varieties and Production Practices 110
Insect Control (Table 2-20) 187
Disease Control (Table 3-30) 274
Chemical Weed Control (Table 4-23) 368

SQUASH: Summer and Winter
Summer Squash Varieties and Production Practices 112
Summer Squash Fertigation Schedule 113
Winter Squash Varieties and Production Practices 102
Insect Control (Table 2-9) 164
Disease Control (Table 3-10) 223
Efficacy of Products for Disease Control (Table 3-11) 223
Chemical Weed Control (Table 4-24) 369

SWEETPOTATO
Varieties and Production Practices 120
Insect Control (Table 2-21) 189
Disease Control (Table 3-31) 277
Sweetpotato Storage House Sanitation (Table 3-32) 280
Efficacy of Products for Disease Control (Table 3-33) 281
Alternative Management Tools (Table 3-34) 281
Chemical Weed Control (Table 4-25) 370

TOMATILLO
Disease Control (Table 3-35) 282

TOMATO
Varieties and Production Practices 124
Fertigation Schedule 125
Tomato Rootstocks and Disease Resistance 126
Insect Control (Table 2-22) 190
Disease Control (Table 3-36) 284
Alternative Management Tools (Table 3-37) 295
Efficacy of Products for Disease Control (Table 3-38) 296
Example Spray Program for Foliar Disease Control
   in Fresh-Market Tomato Production (Table 3-39) 298
Efficacy of Products for Greenhouse
   Tomato Disease Control (Table 3-44) 317
Chemical Weed Control (Table 4-26) 372

TURNIP
Varieties and Production Practices 124
Insect Control (Table 2-23) 194
Disease Control (Table 3-28) 269
Postemergence Chemical Weed Control (Table 4-13) 352

WATERMELON
Varieties and Production Practices 130
Fertigation Schedule 132
Insect Control (Table 2-9) 164
Disease Control (Table 3-10) 223
Efficacy of Products for Disease Control (Table 3-11) 228
Alternative Management Tools (Table 3-12) 230
Example Spray Program For Foliar Disease Control
   In Watermelon Production (Table 3-13) 231
Chemical Weed Control (Table 4-27) 375

General Production Recommendations

VARIETIES

New vegetable varieties are constantly being developed throughout the world. Since it is impossible to list and describe all of them, only some of the better performing commercial types are listed in the specific crop section. These varieties are believed to be suitable for commercial production under most conditions in the southeastern US.

The ultimate value of a variety for a particular purpose is determined by the grower, their management of the variety, and environmental conditions. Several years of trial plantings are suggested for any variety not previously grown. For a true comparison, always include a standard in the same field.

Disease Resistance or Tolerance. Natural variation within a crop species, particularly from wild types can be a source of disease resistance. Plant scientists have taken advantage of this natural variation to develop varieties that are resistant or tolerant. Superscripts appearing after the variety names refer to the disease resistance or tolerance which are spelled out in the footnotes.

Specialty Vegetables. Many producers are considering growing specialty or "gourmet" vegetables. A limited number of pesticides are registered for many specialty vegetables and herbs. Successful pest control in these crops is dependent on sanitation, seed treatment, crop rotation, planting site, mechanical cultivation, and the use of resistant varieties when available.

Promising specialty vegetable crops include asparagus, kale, Swiss chard, herbs, ethnic vegetables, red leaf lettuce, Romaine lettuce, scallions, snap peas, and snow peas.

Other promising types include bok choy, garlic, small melons (aside from cantaloupes), leeks, Pak choi, sweet onions, and sweetpotatoes (moist and dry types with unusual colors). Baby or miniature vegetables are vegetables that are harvested at an immature stage to ensure tenderness and often sweetness. These include beets, carrots (finger and round types), cucumbers, eggplant (little finger types), acorn squash, baby lettuce, pickling corn, snap beans (small sieve types), summer squash (immature with blossom attached), and winter squash.

Before planting a specialty crop, **growers must determine that specific retail, wholesale, restaurant, or processing markets exist.**

CROP ROTATION

Crop rotation is an effective and widely used cultural practice to prevent or reduce the buildup of soil-borne plant pathogens and weeds. An effective rotation sequence includes crops from different families that are poor or non-hosts of these pathogens. In addition, a diverse crop rotation puts varying selection pressure on weeds, preventing any one weed species from becoming problematic. In general, the longer the rotation, the better the results; a 3- to 5-year rotation is recommended. From a practical standpoint, this will depend upon the availability of land, the markets, the selection of alternate crops suited to grow in the area, the pathogen(s), and the purpose of the rotation (prevention versus reduction). When used to reduce pathogen populations, rotations of longer than 5 years may be required (see Table 1A).

SOILS AND SOIL FERTILITY

The best soils for growing vegetables are well-drained, deep soils that are high in organic matter. These soils should have good structure and have been limed and fertilized based on soil test results. Loamy sand and sandy loam soils are generally better suited for growing early market crops because they drain quickly and warm early in spring. Deep, well-drained organic soils are ideal for leafy vegetables, bulb and root crops that offer a high return per acre.

Soils that are not ideal for vegetable production may be made suitable for production by addressing the underlying problem(s). For example, poorly drained soils may require tiling to improve drainage. A large percentage of the vegetables grown in mineral soils of the Coastal Plain are grown in soils with

TABLE 1A. VEGETABLE FAMILIES

Allium Family	Composite Family	Gourd Family	Mustard Family	Parsley Family	Solanaceae Family
Chive	Artichoke	Cantaloupe	Broccoli	Carrot	Eggplant
Garlic	Chicory	Cucumber	Brussels sprout	Celery	Pepper
Leek	Dandelion	Muskmelon	Cabbage	Cilantro	Potato
Onion	Endive & Escarole	Pumpkin	Cauliflower	Parsley	Tomato
Shallot	Lettuce	Squash	Collard	**Pea Family**	
Asparagus Family	Jerusalem artichoke	Watermelon	Kale	Bean (lima, snap)	
Asparagus	**Goosefoot Family**	**Grass Family**	Kohlrabi	Cowpea or Southernpea	
Bindweed Family	Beet	Ornamental corn	Mustard		
Sweetpotato	Chard	Popcorn	Radish	Garden/English Pea	
	Spinach	Sweet Corn	Rutabaga	Soybean	
		Mallow Family	Turnip		
		Okra	Upland cress		

essentially no structure. These sandy soils also have little intrinsic fertility; however, they can be productive if managed properly. Adequate fertilization, adjusting pH as needed, and use of irrigation water are required.

Soil Management. In a good soil management program, proper liming and fertilization, good tillage practices, crop rotation, annual additions of organic matter with cover crops, and adequate irrigation are all necessary to maintain high levels of production. To prevent deterioration of the soil structure, it is essential to use winter cover crops, rotate with non-vegetable crops, utilize reduced tillage when possible, and periodically rest the land with summer cover crops between vegetable plantings.

Nutrient Management and the Environment. The sandy soils preferred for vegetable production in the southeastern US result in an aerated root zone and enable timely tillage, planting, and harvesting. The same drainage allows water and dissolved nutrients to move through the soil profile. Even with loams or clays, nutrients retained in surface soil may be carried with sediment or as dissolved run-off to surface water. Nitrogen (N) and phosphorus (P) remain the two agricultural nutrients of greatest environmental concern. Small losses of N & P can impact water quality, especially in eco-sensitive regions. Agronomically, other issues of potential concern include K fertilizer losses and accumulation of heavy metals such as copper, zinc, etc. supplied with amendments.

Ongoing research has documented increased costs and reduced profits, negative impacts on yield and quality, and human health risks, due to over-fertilization. It is critical that both nutrients and irrigation are managed to optimize vegetable production while minimizing their impacts on the environment. Careful nutrient management includes consideration of the following four issues: right rate, right timing, right placement, and right source, also called four R's of nutrient management. Land-grant university recommendations are based on calibrated crop response studies that can differ substantially across the region. Producers should consult guidelines prepared specifically for their state for the most appropriate nutrient management recommendations. A well-balanced nutrient management plan represents good stewardship and should satisfy any applicable environmental regulations.

Soil Acidity and Liming. Many soils in the southeast are naturally acidic, or become acidic with cropping, and need liming to attain optimum production levels. Soil acidity is the term used to express the quantity of hydrogen (H^+) and aluminum (Al^{3+}) cations (positively charged ions) in soils. Soil pH is determined by using a 1:1 soil-to-water solution. The pH of the solution is measured by a pH meter. Soil pH is an indicator of "soil acidity and soil alkalinity." Combined, the use of the soil pH and soil textural class determines the lime requirement. A pH of 7.0 is defined as neutral, with values below 7.0 being acidic and above 7.0 being basic or alkaline.

Root growth and plant development may be severely restricted if acidic cations, especially aluminum, occupy a large percentage of the negatively charged soil cation exchange capacity (CEC). This negative charge is due to the chemical make-up of the soil clay and organic matter and means that they can attract positively charged ions. Acidification occurs when H+ is added to soils by decomposition of plant residues and organic matter, during the nitrification of ammonium when added to soils as fertilizer (UAN solutions, urea, ammonium nitrate, ammonium sulfate, anhydrous ammonia), manures, or plant residues and from additions of acidic soil amendments. Declines of one pH unit can occur even in properly fertilized beds. Other soil processes, such as aluminum hydrolysis, will release H+ into the soil system, thereby increasing acidity. Lime is applied to neutralize soil acidity by releasing a base (HCO -, OH-) into the soil solution, which reacts with acid-forming ions (H+). Increasing soil pH reduces the concentration of dissolved aluminum, as well as influencing the concentrations of other ions.

Lime recommendations consider differences in acidity among soils as well as differences among various crops' tolerance to acidity. Both the soil pH and some measure of residual or exchangeable acidity are needed to calculate lime recommendations. Although portable soil test kits determine pH rapidly, it is not possible to make an accurate lime recommendation based solely on a pH measurement. Another issue to consider is that different soil laboratories may use different testing methods developed for their particular soil conditions. Due to these differences, producers should consult with their local Extension office or consulting laboratory about laboratory methods and target pH assumptions used in determining lime recommendations.

If soil pH is too high for the desired crop, elemental sulfur (S) is the most effective material to reduce soil pH. The amount of acidity generated by 640 pounds of elemental S is the same as that neutralized by 1 ton of lime. Soil pH can be lowered by applying aluminum sulfate or iron sulfate. Whether trying to increase or decrease the pH of your soil, always follow the manufacturer's instructions for appropriate rates. A slight pH reduction can be produced by using ammonium sulfate, ammonium nitrate, or urea as a fertilizer source of nitrogen. Liming materials containing only calcium carbonate ($CaCO_3$), calcium hydroxide [$Ca(OH)_2$], or calcium oxide (CaO) are called calcitic limes. Pure calcium carbonate is used as the standard for liming materials and is assigned a rating of 100 percent. This rating is also known as the "calcium carbonate equivalent" and is referred to as the CCE. All other liming materials are rated in relationship to pure calcium carbonate.

Liming materials with significant amounts of magnesium carbonate ($MgCO_3$) are called dolomitic limes. Dolomitic limes should be used on soils low in magnesium, as indicated by a soil test report. It is possible to use a magnesium fertilizer instead of dolomitic lime, but the costs of this source of magnesium may be higher. Because lime dissolves very slowly, it must be finely ground to effectively neutralize soil acidity. Lime laws in most states describe standards for composition and particles sizes.

The most commonly used liming materials are finely ground dolomitic or calcitic rock. Most agricultural lime is sold in bulk by the ton. Additional liming materials include burnt lime or hydrated lime, pelleted lime, liquid lime, wood ash, ground seashells, and industrial slags. Lime pellets and lime suspensions (liquid lime) can be convenient and fast-acting but are more expensive than ground limestones. Industrial by-product liming materials can be useful soil amendments capable of reducing soil acidity and supply a variety of nutrients including calcium, magnesium, potassium, phosphorus, and micronutrients. Each lot of such materials should be analyzed as considerable variation in CCE, fineness, and nutrient composition might be present.

Within a one-to-three-year time-period, lime moves little in the soil and neutralizes acidity only in the zone where it is applied. To be most effective, lime must be uniformly spread and thoroughly incorporated. In practice, rates are adjusted after checking the spreader pattern and making appropriate corrections. If the application is not correct, strips of under-limed soil could result, possibly reducing crop yields. The most commonly used lime incorporation tool is the disk. It will not incorporate lime as well as offset disks that throw the soil more vigorously. The best incorporation implement is a heavy-duty rotary tiller that mixes the soil throughout the root zone.

Overall, it is imperative that a soil is limed properly before soil is covered in vegetable systems; i.e., installing polyethylene mulch, as lime additions through drip irrigation or overhead are not practical or possible.

Lime and Fertilizer. Lime and fertilizer work together synergistically to produce high yields and better crops. Lime is not a substitute for fertilizer, and fertilizer is not a substitute for lime.

How to Use Plant Nutrient Recommendation Table 1 and 2.
Use Table 1 to determine the relative levels of phosphorus and potassium in the soil based on the soil test report. Use Table 2 as a guide in conjunction with specific soil test results. Plant nutrient recommendations for different vegetables, listed in Table 2, are expressed in terms of nitrogen (N), phosphate (P_2O_5), and potash (K_2O), rather than in specific grades and amounts of fertilizer. The phosphate (P_2O_5) and potash (K_2O) needs for each cropping situation can be determined by selecting the appropriate values under the relative soil test levels for phosphorus and potassium: very low, low, medium, high, or very high.

The cropping and manuring history of the field must be known before a fertilization program can be planned (see Table 3). This history is important in planning a nitrogen fertilization program because a reliable soil test for nitrogen is not available. Plant nutrient recommendations listed in Table 2 were developed for fields where no manure was applied and where no legume crop is being turned under prior to the planting of a new crop. If manure and/or legume crops are used, the plant nutrient recommendations listed in Table 2 should be reduced by the amounts of nitrogen (N), phosphate (P2O5), and potash (K2O) being contributed from these sources. See Tables 3 and 3A for nutrient values of various products.

Once the final fertilizer-plant nutrient needs are known, determine the grade and rate of fertilizer needed to fulfill these requirements. For example, if the final plant nutrient requirements that need to be added as a commercial fertilizer are 50 pounds of nitrogen (N), 100 pounds of phosphate (P_2O_5), and 150 pounds of potash (K_2O), a fertilizer with a 1-2-3 ratio, such as 5-10-15, 6-12-18, 7-14-21, is needed. Once the grade of fertilizer is selected, the quantity needed to fulfill the plant's nutrient requirements can be determined by dividing the percentage of N, P_2O_5, or K_2O contained in the fertilizer into the quantity of the respective plant nutrient needed per acre and multiplying the answer by 100.

For example, if a 5-10-15 fertilizer grade is chosen to supply the 50 pounds of N, 100 pounds of P_2O_5, and 150 pounds of K_2O needed, calculate the amount of 5-10-15 fertilizer needed as follows: Divide the amount of nitrogen (N) needed per acre (50 pounds) by the percentage of N in the 5-10-15 fertilizer (5 percent), and multiply the answer (10) by 100, which equals 1,000 pounds. This same system can be used for converting any plant nutrient recommendations into grades and amounts.

NUTRIENT MANAGEMENT AND MAXIMIZING PRODUCTION

Nitrogen Management. Nitrogen is one of the most difficult nutrients to manage in vegetable production systems. Nitrogen is readily leached in sandy texture soils. Nitrogen can be immobilized by soil microbes, volatilized if not quickly incorporated, or lost via denitrification under water-saturated soil conditions. Nitrogen recommendations are based on years of fertilizer trials and yield potential. Nitrogen application timings, application methods, and sources are also commonly tested in state university fertilizer trials and have resulted in recommendations for splitting nitrogen fertilizer application for increased fertilizer use efficiency.

Heavy rainfall, higher than normal yields, and production following non-legume cover crops are a few examples where nitrogen fertilizer may be immobilized or lost from the production system. When these nitrogen reduction scenarios arise, an additional application of nitrogen is warranted. Leaf tissue testing is the best option to determine if and how much additional nitrogen is needed to meet expected yields and is described below. Leaf tissue testing can help identify any *"hidden hunger"* that might

TABLE 1. SOIL TEST INTERPRETATIONS AND RECOMMENDATIONS BASED ON SOIL TEST RESULTS

Soil Test Rating	Relative Yield without Nutrient (%)	Recommendations
Low	50–75	Annual application to produce maximum response and increase soil fertility.
Medium	75–100	Normal annual application to produce maximum yields.
High	100	Small applications to maintain soil level to replace nutrients removed by crop. Amount suggested may be doubled and applied in alternate years.
Very high*	100	Depending on crop and state, none until level drops back into high range. This rating permits growers, without risk of loss in yields, to benefit economically from high levels added in previous years. Where no P or K is applied, soils should be resampled in 2 years. When phosphorus is extremely high, further additions may limit the availability of Fe and/or Zn.

* Some states recommend that no fertilizer P or K be added when the soil test rating is "Very High" in order to minimize losses in nutrient-sensitive watersheds. Some crops may still benefit by additions of starter fertilizer for P and K and state guidelines should be followed.

exist in the crop. A *"hidden hunger"* develops when a crop needs more of a given nutrient but has shown no visual deficiency symptoms. With most nutrients and crops, responses can be obtained even though no recognizable symptoms have appeared.

Evaluating the Effectiveness of Your Fertility Program— Using Plant Analysis/Leaf Tissue Testing.
Plant analysis is the chemical evaluation of essential element concentrations in plant tissue. Essential elements include those that are required to complete the life cycle of a plant. The elements carbon (C), oxygen (O), and hydrogen (H) are supplied by the atmosphere and water and generally are not considered limiting. Scientists place most emphases on essential elements supplied by soil or feeding solutions. Macronutrients — nitrogen (N), phosphorus (P), potassium (K), calcium (Ca), magnesium (Mg), and sulfur (S) — are required in the greatest quantities. Micronutrients — iron (Fe), manganese (Mn), zinc (Zn), copper (Cu), boron (B), molybdenum (Mo), and chlorine (Cl) — are required in minute quantities. Toxicities of micronutrients are equally important and yield limiting as deficiencies. Plant analysis is effective in diagnosing toxicities of micronutrients. The interpretation of plant analysis results is based on the scientific principle that healthy plants contain predictable concentrations of essential elements.

State and private soil testing laboratories can provide nitrogen concentrations as well as those of the other essential macro and micronutrients to aid in fertilizer application decisions. A program of periodic leaf tissue sampling and analysis will help you optimize your fertility program and often can allow you to correct deficiencies before symptoms become apparent. *Sufficiency Ranges* or *Critical Values* have been established for most economically important vegetable crops.

Critical Values have been defined as the concentration at which there is a 5 to 10% yield reduction. The use of critical values for practical interpretation has limited value. It is best suited to diagnose severe deficiencies and has little application in identifying hidden hunger. Symptoms are generally visibly evident when nutrient concentrations decrease below the critical value. Critical values establish lower limits of sufficiency ranges.

Sufficiency Range interpretation offers significant advantages over the use of critical values. First, hidden hunger in plants can be identified since the beginning of the sufficiency range is clearly above the critical value. Sufficiency ranges also have upper limits, which provide some indication of the concentration at which the element may be in excess.

Method for Collecting Leaf Tissue Samples for Analysis

It is best to send in two samples. One sample from the problem area and another from an area that is growing well. This will allow comparison for your particular situation.

Each vegetable crop has a specific corresponding plant part that is collected and used to determine foliar nutrient concentrations. Often this corresponds to sampling the most recently matured or fully expanded leaves. Careful sampling ensures the effectiveness of plant analysis as a diagnostic tool. For major crops, best indicator samples have been identified by stage of growth. For young seedlings, the entire plant is sampled 1 inch (2.5 cm) above the soil level. For larger plants, the most recent fully expanded or mature leaf is the best indicator of nutritional status. As some crops, including corn, approach flowering and fruiting, the best indicator of nutritional status is the leaf adjacent to the uppermost fruit (ear leaf). When unfamiliar with sampling protocol for a specific crop, it is generally acceptable to select the most recent mature leaf as the best indicator of nutritional status. Detailed information for sampling most vegetable crops can be found at

https://www.ncagr.gov/plant-tissue-analysis-guide/open

- Sample from 20 to 30 plants.
- Sample across the field, from different rows, and avoid problem areas (low spots, ridges, washed out areas, etc.) unless you are trying to diagnose a problem in one of those areas.
- Sample when the plants are actively growing (typically between 9 a.m. and 4 p.m.).
- Do not collect samples from water stressed plants.
- Send samples to a laboratory in a paper bag. DO NOT SEND SAMPLES IN A PLASTIC BAG. Plastic bags will cause your samples to spoil and will impact results. Contact your local Extension office for information on how to submit leaf tissue samples to your state's diagnostic labs.

Phosphorus Management.
Crops are very likely to respond to P fertilization when the soil test indicates that P is *deficient—very low* or *low*. A soil testing *deficient—medium* will sometimes respond to P fertilization and will sometimes not. Soils testing *optimum* or *exceeds crops needs* are unlikely to respond to P fertilizer, but P may be applied to maintain the fertility level in the *optimum* range. Crops are more likely to respond to P fertilizer when growing conditions are favorable for high yields. Some crops, such as tomato and Irish potato, benefit from crop removal amounts of starter P fertilizer even if the soil is testing optimum or very high. Consult your local Extension office for more information.

P is immobile in the soil meaning it does not move within the soil. It is often recommended that a band of P fertilizer be placed near the seed as a starter fertilizer regardless of the P fertility level. Banded P is especially helpful at low soil test levels. Even at P soil test levels that exceed crop needs, a small amount of banded P may benefit the crop's establishment. When the soil test level is *deficient*, P should generally be applied as a combination of broadcast and banded methods. When the level exceeds the crop's need, only a small amount of P should be applied as a band. Commonly, soils exceed the crop's needs for P due to previous fertilizer and manure applications. When applied in excess of crop removal, P accumulates in the soil. Phosphorus is strongly adsorbed to soil particles, and little is subject to loss via leaching. When the soil test level exceeds crop needs, growers can benefit economically by withholding P fertilizers.

Potassium Management. Crops are very likely to respond to K fertilizer when the soil test indicates that K is *deficient—very low* or *low*. A soil testing *deficient—medium* in K may or may not respond to K fertilizer. Crops are more responsive to K when growing under drought stress, with root injury, or with a soil hard pan than when growing under favorable conditions. Soils testing *optimum* or *exceeds crop needs* are unlikely to respond to K fertilizer, but K may be applied to maintain the soil fertility level in the *optimum* range.

In general, most of the K fertilizer should be broadcast. When the fertility level is *deficient*, it may be advantageous to apply a portion of the total K application as a band. There is usually no benefit to applying banded K when soil fertility levels are *optimum* or *exceeds crop needs*. Crops remove larger amounts of K than P from the soil during a growing season. In addition, sandy soils have low reserves of K, and potassium is susceptible to leaching. Therefore, frequent applications of K are needed to maintain K at an optimum level.

Secondary Nutrients. Calcium (Ca), magnesium (Mg), and sulfur (S) are included in the secondary element group. Calcium may be deficient in some soils that have not been properly limed, where excessive potash fertilizer has been used, and/or where crops are subjected to drought stress, with root injury, or with a soil hard pan. Magnesium is the most likely of these elements to be deficient in vegetable soils. Dolomitic or high-magnesium limestones should be used when liming soils that are low in magnesium. Magnesium should be applied as a fertilizer source on low-magnesium soils where lime is not needed (Table 4). Magnesium may be applied as a foliar spray to supply magnesium to the crop in emergency situations (2 TBSP of Epsom salts per gallon of water). Sulfur is known to be deficient in vegetable systems in coastal plain soils.

Sulfur deficiency is often mistaken for lack of nitrogen because the symptoms are similar. Onions and Brassicas generally require additional sulfur over what other vegetables normally require, particularly on coastal plain soils.

Micronutrients. Boron is the most widely deficient micronutrient in vegetable soils. Deficiencies of this element are most likely to occur in the following crops: asparagus, most bulb and root crops, Cole crops, and tomatoes. Excessive amounts of boron can be toxic to plant growth. This problem can occur when snap beans (a sensitive crop) follow sweetpotatoes (a crop where boron is applied late in the season). Do NOT exceed recommendations listed in Table 2.

Manganese deficiency often occurs in plants growing on soils that have been over limed. In this case, broadcast 20 to 30 pounds or band 4 to 8 pounds of manganese sulfate per acre. Alternatively, manganese can often be applied using a foliar application and mixed with crop protectants. Do not apply lime or poultry manure to such soils until the pH has dropped below 6.5 and be careful not to over lime again.

Molybdenum deficiency of cauliflower (which causes whiptail) may develop when this crop is grown on soils more acid than pH 5.5. An application of 0.5 to 1 pound of sodium or ammonium molybdate per acre will usually correct this. Liming acid soils to a pH of 6.0 to 6.5 will usually prevent the development of molybdenum deficiencies in vegetable crops.

Deficiencies of other micronutrients in vegetable crops in the Southeast are rare; and when present, are usually caused by over liming or other poor soil management practices. Contact Extension if a deficiency of zinc, iron, copper, or chlorine is suspected. Sources of fertilizers for the essential plant nutrients are found in in Tables 3A and 4.

Municipal Biosolids. *Biosolids should not be applied to land on which crops will be grown that will be entering the human food chain.* Municipal biosolids are the solid material removed from sewage in treatment processes. In some cases, application of biosolids may be allowed to land used for the production of non-food crops. Check with your local or state department of environmental management for latest regulations.

Foliar Fertilization. Foliar feeding of vegetables is usually not needed. Plants usually obtain their nutrients from the soil through their roots. Plants can also absorb a limited amount of some nutrients through aerial organs such as their leaves. Properly managed soils will supply the essential mineral nutrients the crop will need during its development. If, for some reason, one or more soil-supplied micronutrients becomes limiting or unavailable during the development of the crop, foliar nutrient applications may then be helpful.

TABLE 2. GENERAL FERTILIZER SUGGESTIONS FOR VEGETABLE CROPS*

CROP	Desirable pH	Nitrogen (N) lb/acre	Recommended Nutrients Based on Soil Tests								Nutrient Timing and Method
			Soil Phosphorus Level				Soil Potassium Level				
			Low	Med	High	Very High	Low	Med	High	Very High	
			P_2O_5 lb/acre				K_2O lb/acre				
ASPARAGUS											
	6.5 to 7.0	100	250	150	100	0	250	225	150	0	Total recommended.
		50	250	150	100	0	150	100	75	0	Broadcast before cutting season.
		50	0	0	0	0	100	125	75	0	Sidedress after cutting.
		Apply 2 lb boron (B) per acre every 3 years on most soils.										
BEAN, Lima											
...Single crop	6 to 6.5	75 to 110	120	80	40	20	160	120	80	20	Total recommended.
		25 to 50	80	40	20	0	120	80	60	0	Broadcast and disk-in.
		20	40	40	20	20	40	40	20	20	Band-place with planter.
		25 to 40	0	0	0	0	0	0	0	0	Sidedress 3 to 5 weeks after emergence.
BEAN, Snap											
	6 to 6.5	40 to 80	80	60	40	20	80	60	40	20	Total recommended.
		20 to 40	40	40	0	0	40	40	0	0	Broadcast and disk-in.
		20 to 40	40	20	40	20	40	20	40	20	Band-place with planter.
BEET											
	6 to 6.5	75 to 100	150	100	50	0	150	100	50	0	Total recommended.
		50	150	100	50	0	150	100	50	0	Broadcast and disk-in.
		25 to 50	0	0	0	0	0	0	0	0	Sidedress 4 to 6 weeks after planting.
		Apply 2 to 3 lb boron (B) per acre with broadcast fertilizer.										
BROCCOLI											
	6 to 6.5	125 to 175	200	100	50	0	200	100	50	0	Total recommended.
		50 to 100	150	100	50	0	150	100	50	0	Broadcast and disk-in.
		50	50	0	0	0	50	0	0	0	Sidedress 2 to 3 weeks after planting.
		25	0	0	0	0	0	0	0	0	Sidedress every 2 to 3 weeks after initial sidedressing.
		Apply 2 to 3 lb boron (B) per acre with broadcast fertilizer.										
BRUSSEL SPROUTS, CABBAGE, and CAULIFLOWER											
	6 to 6.5	100 to 175	200	100	50	0	200	100	50	0	Total recommended.
		50 to 75	200	100	50	0	200	100	50	0	Broadcast and disk-in.
		25 to 50	0	0	0	0	0	0	0	0	Sidedress 2 to 3 weeks after planting.
		25 to 50	0	0	0	0	0	0	0	0	Sidedress if needed, according to weather.
		Apply 2 to 3 lb boron (B) per acre and molybdenum per acre as 0.5 lb sodium molybdate per acre with broadcast fertilizer.										
CANTALOUPES and MELONS											
...Bareground	6 to 6.5	75 to 115	150	100	50	25	200	150	100	25	Total recommended.
		25 to 50	125	75	25	0	175	125	75	0	Broadcast and disk-in.
		25	25	25	25	25	25	25	25	25	Band-place with planter.
		25 to 40	0	0	0	0	0	0	0	0	Sidedress when vines start to run.
		Apply 1 to 2 lb boron (B) per acre with broadcast fertilizer.										
...Plasticulture		100 to 150	125	75	25	25	200	150	100	25	Total recommended.
		25	125	75	25	25	100	75	50	25	Broadcast and disk-in.
		75 to 125	0	0	0	0	100	75	50	0	Fertigate.
		Apply 1 to 2 lb boron (B) per acre with broadcast fertilizer. Drip fertilization: See "cantaloupe" in specific commodity recommendations later in this handbook.										
CARROT											
	6 to 6.5	90 to 120	150	100	50	0	150	100	50	0	Total recommended.
		50	150	100	50	0	150	100	50	0	Broadcast and disk-in.
		40 to 70	0	0	0	0	0	0	0	0	Sidedress if needed.
		Apply 1 to 2 lb boron (B) per acre with broadcast fertilzer.										

* Nitrogen and sulfur rates should be based on your local fertilizer recommendations.

TABLE 2. GENERAL FERTILIZER SUGGESTIONS FOR VEGETABLE CROPS* (cont'd)

CROP	Desirable pH	Nitrogen (N) lb/acre	Recommended Nutrients Based on Soil Tests								Nutrient Timing and Method
			Soil Phosphorus Level				Soil Potassium Level				
			Low	Med	High	Very High	Low	Med	High	Very High	
			P_2O_5 lb/acre				K_2O lb/acre				
CUCUMBER											
…Bareground	6 to 6.5	80 to 160	150	100	50	25	200	150	100	25	Total recommended.
		40 to 100	125	75	25	0	175	125	75	0	Broadcast and disk-in.
		20 to 30	25	25	25	25	25	25	25	25	Band-place with planter.
		20 to 30	0	0	0	0	0	0	0	0	Sidedress when vines begin to run, or apply in irrigation water.
…Plasticulture		120 to 150	150	100	50	25	150	100	50	25	Total recommended.
		25	125	75	25	0	150	100	50	0	Broadcast and disk-in.
		95 to 125	0	0	0	0	25	25	25	25	Fertigate.
		Drip fertilization: See "cucumber" in specific recommendations later in this handbook.										
EGGPLANT											
…Bareground	6 to 6.5	100 to 200	250	150	100	0	250	150	100	0	Total recommended.
		50 to 100	250	150	100	0	250	150	100	0	Broadcast and disk-in.
		25 to 50	0	0	0	0	0	0	0	0	Sidedress 3 to 4 weeks after planting.
		25 to 50	0	0	0	0	0	0	0	0	Sidedress 6 to 8 weeks after planting.
		Apply 1 to 2 lb boron (B) per acre with broadcast fertilizer.										
…Plasticulture		145	250	150	100	0	240	170	100	0	Total recommended.
		50	250	150	100	0	100	100	100	0	Broadcast and disk-in.
		95	0	0	0	0	140	70	0	0	Fertigate.
		Apply 1 to 2 lb boron (B) per acre with broadcast fertilizer. Drip fertilization: See "eggplant" in specific recommendations later in this handbook.										
ENDIVE, ESCAROLE, LEAF and ROMAINE LETTUCE											
	6 to 6.5	75 to 150	200	150	100	0	200	150	100	0	Total recommended.
		50 to 100	200	150	100	0	200	150	100	0	Broadcast and disk-in.
		25 to 50	0	0	0	0	0	0	0	0	Sidedress 3 to 5 weeks after planting.
HERBS (BASIL, PARSLEY, CILANTRO)											
Basil	6 to 6.5	100 to 175	200	150	100	0	200	150	100	0	Total recommended.
		50 to 75	200	150	100	0	200	150	100	0	Broadcast and disk-in.
Parsley and Cilantro	6.5 to 7.5	25 to 50	0	0	0	0	0	0	0	0	Sidedress after first cutting.
		25 to 50	0	0	0	0	0	0	0	0	Sidedress after each additional cutting.
HOPS											
	6.0	100 to 150	80 to 120	30 to 80	0 to 30	0	100 to 140	50 to 100	0 to 50	0	Total recommended.
		50 to 75	80 to 120	30 to 80	0 to 30	0	100 to 140	50 to 100	0 to 50	0	Broadcast
		50 to 75	0	0	0	0	0	0	0	0	Sidedress 4 weeks after bines emerge
		Apply 1 lb of boron (B) per acre with broadcast fertilizer. NOTE: First year planting N rate should be reduced to 75 lbs/acre. Nitrogen should not be applied after flowering.										
LEAFY GREENS, COLLARD, KALE, and MUSTARD											
	6 to 6.5	75 to 80	150	100	50	0	150	100	50	0	Total recommended.
		50	150	100	50	0	150	100	50	0	Broadcast and disk-in.
		25 to 30	0	0	0	0	0	0	0	0	Sidedress, if needed.
		Apply 1 to 2 lb boron (B) per acre with broadcast fertilizer.										
LEEK											
	6 to 6.5	75 to 125	200	150	100	0	200	150	100	0	Total recommended.
		50 to 75	200	150	100	0	200	150	100	0	Broadcast and disk-in.
		25 to 50	0	0	0	0	0	0	0	0	Sidedress 3 to 4 weeks after planting, if needed.
		Apply 1 to 2 lb boron (B) per acre with broadcast fertilizer.										

* Nitrogen and sulfur rates should be based on your local fertilizer recommendations.

TABLE 2. GENERAL FERTILIZER SUGGESTIONS FOR VEGETABLE CROPS* (cont'd)

CROP	Desirable pH	Nitrogen (N) lb/acre	Recommended Nutrients Based on Soil Tests								Nutrient Timing and Method
			Soil Phosphorus Level				Soil Potassium Level				
			Low	Med	High	Very High	Low	Med	High	Very High	
			P_2O_5 lb/acre				K_2O lb/acre				
OKRA											
	6 to 6.5	100 to 200	250	150	100	0	250	150	100	0	Total recommended.
		50 to 100	250	150	100	0	250	150	100	0	Broadcast and disk-in.
		25 to 50	0	0	0	0	0	0	0	0	Sidedress 3 to 4 weeks after planting.
		25 to 50	0	0	0	0	0	0	0	0	Sidedress 6 to 8 weeks after planting.
		Apply 1 to 2 lb boron (B) per acre with broadcast fertilizer.										
		NOTE: Where plastic mulches are being used, broadcast 50 to 100 lb nitrogen (N) per acre with recommended P_2O_5 and K_2O and disk incorporate prior to laying mulch. Drip fertilization: See "okra" in specific commodity recommendations later in this handbook.										
ONION											
...Bulb	6 to 6.5	125 to 175	200	100	50	0	200	100	50	0	Total recommended.
		50 to 75	200	100	50	0	200	100	50	0	Broadcast and disk-in.
		75 to 100	0	0	0	0	0	0	0	0	Sidedress twice 4 to 5 weeks apart.
		Apply 1 to 2 lb boron (B) and 20 lb sulfur (S) per acre with broadcast fertilizer.										
...Green		150 to 175	200	100	50	0	200	100	50	0	Total recommended.
		50 to 75	200	100	50	0	200	100	50	0	Broadcast and disk-in.
		50	0	0	0	0	0	0	0	0	Sidedress 4 to 5 weeks after planting.
		50	0	0	0	0	0	0	0	0	Sidedress 3 to 4 weeks before harvest.
		Apply 1 to 2 lb boron (B) and 20 lb sulfur (S) per acre with broadcast fertilizer.										
PARSNIP											
	6 to 6.5	50 to 100	150	100	50	0	150	100	50	0	Total recommended.
		25 to 50	150	100	50	0	150	100	50	0	Broadcast and disk-in.
		25 to 50	0	0	0	0	0	0	0	0	Sidedress 4 to 5 weeks after planting.
		Apply 1 to 2 lb boron (B) per acre with broadcast fertilizer.										
PEA, Garden/English											
	5.8 to 6.5	40 to 60	120	80	40	0	120	80	40	0	Total recommended. Broadcast and disk-in before seeding.
PEPPER											
...Bareground	6 to 6.5	100 to 130	200	150	100	0	200	150	100	0	Total recommended.
		50	200	150	100	0	200	150	100	0	Broadcast and disk-in.
		25 to 50	0	0	0	0	0	0	0	0	Sidedress after first fruit set.
		25 to 30	0	0	0	0	0	0	0	0	Sidedress later in season, if needed.
...Plasticulture		100 to 185	320	250	100	0	350	250	100	40	Total recommended.
		50	200	150	100	0	200	150	100	40	Broadcast and disk-in.
		50 to 135	0	0	0	0	150	100	0	0	Fertigate.
		Drip fertilization: See "pepper" in specific commodity recommendations later in this handbook.										
POTATO											
...Loams and silt loams	5.8 to 6.2	100 to 150	110	90	70	50	200	150	50	50	Total recommended.
		85 to 135	60	40	20	0	200	150	50	50	Broadcast and disk-in.
		15	50	50	50	50	0	0	0	0	Band-place with planter at planting.
...Sandy loams and loamy sands		150 to 180	200	150	100	50	300	200	100	50	Total recommended.
		50	200	150	100	50	300	200	100	50	Broadcast and disk-in.
		100	0	0	0	0	0	0	0	0	Sidedress 4 to 5 weeks after planting.
		0 to 30	0	0	0	0	0	0	0	0	Based on petiole nitrate testing at flowering
PUMPKIN and WINTER SQUASH											
...Bareground	6 to 6.5	80 to 100	150	100	50	0	200	150	100	0	Total recommended.
		40 to 50	150	100	50	0	200	150	100	0	Broadcast and disk-in.
		40 to 50	0	0	0	0	0	0	0	0	Sidedress when vines begin to run.
...Plasticulture		80 to 150	150	100	50	0	200	150	100	0	Total recommended.
		25 to 50	150	100	50	0	100	75	50	0	Broadcast and disk-in.
		55 to 100	0	0	0	0	100	75	50	0	Fertigate.

* Nitrogen and sulfur rates should be based on your local fertilizer recommendations.

TABLE 2. GENERAL FERTILIZER SUGGESTIONS FOR VEGETABLE CROPS* (cont'd)

CROP	Desirable pH	Nitrogen (N) lb/acre	Recommended Nutrients Based on Soil Tests								Nutrient Timing and Method
			Soil Phosphorus Level				Soil Potassium Level				
			Low	Med	High	Very High	Low	Med	High	Very High	
			P_2O_5 lb/acre				K_2O lb/acre				
RADISH											
	6 to 6.5	50	150	100	50	0	150	100	50	0	Total recommended. Broadcast and disk-in.
		Apply 1 to 2 lb boron (B) per acre with broadcast fertilizer.									
RUTABAGA and TURNIP											
	6 to 6.5	50 to 100	150	100	50	0	150	100	50	0	Total recommended.
		25 to 50	150	100	50	0	150	100	50	0	Broadcast and disk-in.
		25 to 50	0	0	0	0	0	0	0	0	Sidedress when plants are 4 to 6 in. tall.
		Apply 1 to 2 lb boron (B) per acre with broadcast fertilizer.									
SOUTHERNPEA											
	5.8 to 6.5	20	100	50	0	0	100	50	0	0	Total recommended. Broadcast and disk-in.
SPINACH											
...Fall	6 to 6.5	75 to 125	200	150	100	0	200	150	100	0	Total recommended.
		50 to 75	200	150	100	0	200	150	100	0	Broadcast and disk-in.
		25 to 50	0	0	0	0	0	0	0	0	Sidedress or topdress.
...Overwinter		80 to 120	0	0	0	0	0	0	0	0	Total recommended for spring application to an overwintered crop.
		50 to 80	0	0	0	0	0	0	0	0	Apply in late February.
		30 to 40	0	0	0	0	0	0	0	0	Apply in late March.
SQUASH, Summer											
	6 to 6.5	100 to 130	150	100	50	0	150	100	50	0	Total recommended.
		25 to 50	150	100	50	0	150	100	50	0	Broadcast and disk-in.
		50	0	0	0	0	0	0	0	0	Sidedress when vines start to run.
		25 to 30	0	0	0	0	0	0	0	0	Apply through irrigation system.
		Apply 1 to 2 lb boron (B) per acre with broadcast fertilizer.									
		Drip fertilization: See "summer squash" in specific commodity recommendations later in this handbook.									
SWEET CORN											
	6 to 6.5	110 to 155	160	120	80	20	160	120	80	20	Total recommended.
		40 to 60	120	100	60	0	120	100	60	0	Broadcast before planting.
		20	40	20	20	20	40	20	20	20	Band-place with planter.
		50 to 75	0	0	0	0	0	0	0	0	Sidedress when corn is 12 to 18 in. tall.
		Apply 1 to 2 lb boron (B) per acre with broadcast fertilizer. NOTE: On very light sandy soils, sidedress 40 lb N per acre when corn is 6 in. tall and another 40 lb N per acre when corn is 12 to 18 in. tall.									
SWEETPOTATO											
	5.8 to 6.2	50 to 80	200	100	50	0	300	200	150	120	Total recommended.
		0	150	60	30	0	150	50	30	0	Broadcast and disk-in.
		50 to 80	50	40	20	0	150	150	120	120	Sidedress 21 to 28 days after planting.
		Add 0.5 lb of actual boron (B) per acre 40 to 80 days after transplant.									
TOMATO											
...Bareground for Sandy loams and loamy sands	6 to 6.5	80 to 90	200	150	100	0	300	200	100	0	Total recommended.
		40 to 45	200	150	100	0	300	200	100	0	Broadcast and disk-in.
		40 to 45	0	0	0	0	0	0	0	0	Sidedress when first fruits are set as needed.
		Apply 1 to 2 lb boron (B) per acre with broadcast fertilizer.									
...Bareground for Loam and clay		75 to 80	200	150	100	0	250	150	100	0	Total recommended.
		50	200	150	100	0	250	150	100	0	Broadcast and disk-in.
		25 to 30	0	0	0	0	0	0	0	0	Sidedress when first fruits are set as needed.
		Apply 1 to 2 lb boron (B) per acre with broadcast fertilizer.									
...Plasticulture		130 to 210*	200	150	100	50	420	345	275	50	Total recommended.
		50	200	150	100	0	125	125	125	0	Incorporate into the plant bed before laying polyethylene mulch.
		80 to 160	0	0	0	0	295	220	150	0	Fertigate.
		Apply 1 to 2 lb boron (B) per acre with broadcast fertilizer.									
		Drip fertilization: See "tomato" in specific commodity recommendations later in this handbook.									
		*Local recommendations may vary. Contact your local Extension professionals for the recommendation in your state.									

* Nitrogen and sulfur rates should be based on your local fertilizer recommendations.

TABLE 2. GENERAL FERTILIZER SUGGESTIONS FOR VEGETABLE CROPS* (cont'd)

CROP	Desirable pH	Nitrogen (N) lb/acre	Soil Phosphorus Level — P_2O_5 lb/acre				Soil Potassium Level — K_2O lb/acre				Nutrient Timing and Method
			Low	Med	High	Very High	Low	Med	High	Very High	
WATERMELON											
...Nonirrigated	6 to 6.5	75 to 90	150	100	50	0	200	150	100	0	Total recommended.
		50	150	100	50	0	200	150	100	0	Broadcast and disk-in.
		25 to 40	0	0	0	0	0	0	0	0	Topdress when vines start to run.
...Irrigated		100 to 150	150	100	50	0	200	150	100	0	Total recommended.
		50	150	100	50	0	200	150	100	0	Broadcast and disk-in.
		25 to 50	0	0	0	0	0	0	0	0	Topdress when vines start to run.
		25 to 50	0	0	0	0	0	0	0	0	Topdress at first fruit set.
...Plasticulture		125 to 150	150	100	50	0	200	150	100	0	Total recommended.
		25 to 50	150	100	50	0	100	75	50	0	Disk in row.
		100	0	0	0	0	100	75	50	0	Fertigate.

NOTE: Excessive rates of N may increase the incidence of hollow heart in seedless watermelon.
Drip fertilization: See "watermelon" in specific commodity recommendations later in this handbook.

* Nitrogen and sulfur rates should be based on your local fertilizer recommendations.

TABLE 3. NUTRIENT VALUES FOR MANURE APPLICATIONS AND CROP RESIDUES

	N	P_2O_5	K_2O		N	P_2O_5	K_2O
	Pounds per Ton				Pounds per Ton		
Cattle manure	5-20[1]	8-10	14-19	Ladino clover sod	60	0	0
Poultry litter & manure	25-65[1]	20-60	10-55	Crimson clover sod	50	0	0
Swine manure	5-40[1]	2-17	2-15	Red clover sod	40	0	0
Horse manure	6-12[1]	3	6	Birdsfoot trefoil	40	0	0
Liquid poultry manure (5 - 15% solids)	7-15[1]	5-10	5-10	Lespedeza	20	0	0
				Soybeans			
Alfalfa sod	50-100[2]	0	0	Tops and roots	40	0	0
Hairy vetch	50-100	0	0	Grain harvest residue	15	0	0

[1] Manures are highly variable. Consult a state or private lab for a nutrient analysis prior to application.
[2] 75% stand = 100 - 0 - 0, 50% stand = 75 - 0 - 0, and 25% stand = 50 - 0 - 0.

TABLE 3A. NUTRIENT VALUES FOR VARIOUS PLANT, ANIMAL, AND NATURAL PRODUCTS

	Typical NPK Analysis			Release Time
	N	P_2O_5	K_2O	
Plant By-Products				
Alfalfa Meal or Pellets	2.0	1.0	2.0	1 to 4 months
Corn Gluten Meal	9.0	0.0	0.0	1 to 4 months
Cottonseed Meal	6.0	0.4 to 3.0	1.5	1 to 4 months
Soybean Meal	7.0	1.2 to 2.0	1.5 to 7.0	1 to 4 months
Kelp Powder	1.0	0.0	4.0	Immediate to 1 month
Animal By-Products				
Bat Guano (high N)	10.0	3.0	1.0	4 plus months
Bat Guano (high P)	3.0	10.0	1.0	4 plus months
Blood Meal	12.0 to 14.0	2.0	1.0	1 to 4 months
Bone Meal (raw)	3.0	22.0	0.0	1 to 4 months
Bone Meal (steamed)	1.0 to 2.0	11.0 to 15.0	0.0	1 to 4 months
Feather Meal	7.0 to 12.0	0.0	0.0	4 plus months
Fish Emulsion	5.0	2.0	2.0	1 to 4 months

TABLE 3A. NUTRIENT VALUES FOR VARIOUS PLANT, ANIMAL, AND NATURAL PRODUCTS (cont'd)

	Typical NPK Analysis			
	N	P_2O_5	K_2O	Release Time
Animal By-Products (cont'd)				
Fish Powder	12.0	0.3	1.0	Immediate to 1 month
Enzymatically Digested Hydrolyzed Liquid Fish	4.0	2.0	2.0	1 to 4 months
Fish Meal	10.0	6.0	2.0	1 to 4 months
Worm Castings	2.0	1.5	1.5	1 to 4 months
Natural Minerals				
"Soft" Rock Phosphate	0.0	14 to 16	0.0	Very slow (years)
Greensand	0.0	0.0	3.0	Very slow

TABLE 4. PERCENTAGE EQUIVALENTS AND CONVERSION FACTORS FOR MAJOR, SECONDARY, AND MICRONUTRIENT FERTILIZER SOURCES

Fertilizer Source Material	Required to Supply 1 Lb of Plant Food Contents,%	Lb of Material the Initially Listed Plant Nutrient	Fertilizer Source Material	Required to Supply 1 Lb of Plant Food Contents,%	Lb of Material the Initially Listed Plant Nutrient
Nitrogen Materials			**Sulphur Materials**		
Monoammonium phosphate*	11 (N) and 48 (P_2O_5)	9.1	Granulated sulfur	90 to 92 (S)	1.1
Nitrate of potash*	13 (N) and 44 (K_2O)	7.7	Ammonium sulfate*	23 (S) and 20.5 (N)	4.3
Nitrate of soda-potash*	15 (N) and 14 (K_2O)	6.7	Gypsum*	15-18 (S) and 19 to 23 (Ca)	6.1
Calcium nitrate*	15 (N) and 19 (Ca)	6.7			
Nitrate of soda	16 (N)	6.3	Epsom salts*	13 (S) and 10 (Mg)	7.7
Diammonium phosphate*	18 (N) and 46 (P_2O_5)	5.6	**Boron Materials**		
Nitrogen solution	20 (N)	5.0	Fertilizer Borate Granular*	14.30 (B)	7.0
Ammonium sulfate*	20.5 (N) and 23 (S)	4.9	Fertilizer Borate-48	14.91 (B)	6.7
Nitrogen solution	30 (N)	3.3	Solubor	20.50 (B)	4.9
Nitrogen solution	32 (N)	3.1	Fertilizer Borate-68	21.13 (B)	4.7
Ammonium nitrate	33.5 to 34.0 (N)	3.0	**Manganese Materials**		
Nitrogen solution	40 (N)	2.5	Manganese sulfate*	24.0 (Mn)	4.2
Urea	45 to 46 (N)	2.2	Manganese sulfate*	25.5 (Mn)	3.9
Anhydrous ammonia	82 (N)	1.2	Manganese sulfate*	29.1 (Mn)	3.4
Phosphorus Materials			Manganese oxide	48.0 (Mn)	2.1
Normal superphosphate*	20 (P_2O_5) and 11 (S)	5.0	Manganese oxide	55.0 (Mn)	1.8
Triple superphosphate*	44 to 46 (P_2O_5)	2.2	**Zinc Materials**		
Monoammmonium phosphate*	48 (P_2O_5) and 11 (N)	2.1	Zinc sulfate*	36 (Zn)	2.8
Diammonium phosphate*	46 (P_2O_5) and 18 (N)	2.2	Zinc oxide	73 (Zn)	1.4
Potassium Materials			**Molybdenum Materials**		
Nitrate of soda-potash*	14 (K2O) and 13 (N)	7.1	Sodium molybdate	39.5 (Mo)	2.5
Sulfate of potash-magnesia*	21.8 (K2O) and 11.1 (Mg)	4.6	Sodium molybdate	46.6 (Mo)	2.1
Nitrate of potash*	44 (K2O) and 13 (N)	2.3	Ammonium molybdate*	56.5 (Mo)	1.8
Sulfate of potash*	50 (K_2O) and 17 (S)	2.0			
Magnesium Materials					
Epsom salts*	10 (Mg) and 13 (S)	9.6			
Sulfate of potash-magnesia*	11.1 (Mg) and 21.8 (K_2O)	9.0			
Kieserite*	18.1 (Mg)	5.5			
Brucite	39 (Mg)	2.6			

* Supplies more than one essential nutrient.

MINIMUM TILLAGE FOR VEGETABLE PRODUCTION

The development of various types of tillage practices was an integral part of the evolution of modern farming practices. Tillage is helpful in crop production systems for purposes of weed management, incorporation of amendments such as lime and fertilizer, burial of crop residues to facilitate other field operations, disease management, and the preparation of a seedbed that is conducive to crop establishment. While the use of tillage practices provides a number of benefits to crop producers, researchers have also learned that the soil disturbance associated with tillage has some drawbacks. In a nutshell, tillage over time results in the degradation of several soil properties that are important to crop productivity.

One of these properties is organic matter content. Organic matter is important because it contributes appreciably to the water and nutrient holding capacity of soil and to the maintenance of a desirable soil structure. These soil properties, in turn, allow soil to better support the weight of equipment and workers. In warm southern climates the loss of organic matter due to tillage is even more pronounced than in cooler climates. Tilled soil is also less hospitable to a variety of soil organisms including microbes, insects, and other small animals. When present in adequate numbers these are beneficial for various reasons. When minimum tillage is used, soil structure is improved by the release of exudates of various organisms that glue soil particles together into larger, more desirable aggregates. Plant roots benefit from the increased presence of pore spaces in the soil such as earthworm channels, and plant diseases may also be reduced by the increased diversity of soil microorganisms.

Adoption of minimum tillage in vegetable production is possible but requires careful planning and preparation. Making a transition to minimum tillage will affect several vegetable production field operations. For example, one common objective of minimum tillage is to retain crop residues on the soil surface. These residues are beneficial for reducing soil erosion but also may interfere with the seeding of crops, particularly small-seeded vegetable crops. Similarly, cultivation, often an important measure for controlling weeds in vegetables, may require different equipment than what the farmer is able to use in conventionally tilled fields. In general, it may be best to start with those vegetables that are grown similarly to agronomic row crops or to use crops that can be established by transplanting through crop residues. Row crop examples include sweet corn and cowpeas. Examples of vegetables that are easily transplanted include tomato, pepper, squash, and watermelon.

Growers interested in adopting minimum tillage practices should begin by learning about the practices currently employed by agronomic crop producers and others who grow vegetables using reduced tillage. One such practice is to limit tillage and seedbed preparation to a narrow strip where the crop will be planted. This may be done in combination with the use of cover crops that are killed by rolling and crimping prior to tilling the strip. This method has been used successfully for vegetables such as tomatoes and cucurbits.

COVER CROPS

Many soils that are not productive due to poor physical properties can be restored and made more productive through the continued use of cover crops. Cover crops can provide many benefits to soils that include reducing the buildup of soilborne disease and arthropod pests, increasing soil organic matter, suppressing weeds, improving soil structure, promoting beneficial soil microorganisms, improving nutrient cycling, and reducing soil erosion. Each cover crop can offer different potential benefits to a production system and not every cover crop will work for each grower's intended purpose.

Many cover crops can reduce or limit the build-up of soilborne disease and insect pests that damage vegetable crops. Prevalent disease and insect pressure should be considered when selecting a cover crop as some cover crops could increase the severity of these issues. In some cases, specific cultivars of cover crops can differ in their host status to various plant-parasitic nematodes. For example, 'Cahaba White' common vetch is suppressive to southern root-knot nematodes while 'Vantage' common vetch is susceptible to southern root-knot nematode.

With intensive cropping, working the soil when it is too wet and excessive traffic from using heavy equipment will contribute to damaging soils. These practices cause soils to become hard and compact, resulting in poor seed germination, loss of transplants, and shallow root formation of surviving plants. Such soils can easily form crusts on the surface, become compacted, which make them difficult to irrigate properly. Combined, these practices will yield negative consequences for your soil; poor plant stands, poor crop growth, low yields, and loss of income. In some cases, sub-soiling in the row might help improve aeration and drainage, but its effect is limited and short term. Continued and dedicated use of cover crops will aid in preventing these conditions. It may take several years of continued use to observe some of the benefits that cover crops can provide to soils.

Cover crops can also be planted in strips for wind protection preceding the planting of the cash crop. Annual rye seeded before November can be a good choice for use in wind protection. Cover crops reduce nutrient loss during the winter and early spring. Cover crops may deplete the soil moisture. If this is a concern, be sure to disk or plow the cover crops before soil moisture is depleted.

Seeding dates suggested in the following section are for the central part of the Southeastern United States and will vary with elevation and northern or southern locations. For state specific recommendations for planting dates for cover crops, consult your local Extension office.

Summer Cover Crops

Summer cover crops can be useful in controlling weeds, soilborne diseases, and plant-parasitic nematodes. They also provide organic matter and improve soil tilth while reducing soil erosion. There are many potential summer cover crops available, but you will need to find one that will work well in your area and fit into your overall production scheme. Sudex (sorghum-sudan grass cross; do not allow to exceed 3 ft. before mowing), southernpeas (cowpeas), millet, and Lab Lab are summer cover crops that provide organic matter, control erosion, and will enhance the natural biota of your soil.

Summer cover crops, such as sudangrass or sudex, seeded at 20 to 40 pounds per acre are good green manure crops. Sunn hemp and pearl millet also provide a good green manure. They can be planted as early as field corn is planted requiring eight to 12 weeks of frostfree growth conditions. These crops should be clipped, mowed, or disked to prevent seed development that could lead to weed problems. Summer cover crops can be disked and planted to wheat or rye in September or allowed to winter-kill and tilled-in the following spring. Soil test to determine lime and fertilizer needs. For state specific recommendations for planting dates, seeding dates and management for cover crops, consult your local Extension office.

TYPES OF SUMMER COVER CROPS: SMALL GRAINS SEEDING RATES AND DEPTHS

SORGHUM-SUDANGRASS: broadcast 50 to 60 lbs/A; drill 45 to 50 lbs/A (seeding depth: ½ - 1½ in.)

SUDANGRASS: broadcast 40 to 50 lbs/A; drill 35 to 40 lbs/A (seeding depth: ½ - 1 in.)

JAPANESE MILLET: broadcast 25 to 35 lbs/A; drill 20 to 25 lbs/A (seeding depth: ½ - 1 in.)

GERMAN FOXTAIL MILLET: broadcast 30 to 40 lbs/A; drill 25 to 30 lbs/A (seeding depth: ½-1 in.)

PEARL MILLET: broadcast 10-25 lbs/acre; drill 5 to 15 lbs/A (seeding depth: ½ - 1 in.)

BUCKWHEAT: broadcast 50 to 100 lbs/A; drill 30 to 90 lbs/A (seeding depth: ½ in.)

TEFF: broadcast 30 to 40 lbs/A

TYPES OF SUMMER COVER CROPS: LEGUMES SEEDING RATES AND DEPTHS

COWPEAS: broadcast 70 to 120 lbs/A drill 40 to 50 lbs/A (seeding depth: 1- 1½ in.) (USDA Cowpea lines 1136, 1137 and 1138 are soft seeded [i.e., typically will not reseed as volunteers the following season] and nematode resistant; each plant covers 100 sq ft)

SESBANIA: broadcast 25 to 40 lbs/A; drill 20 to 25 lbs/A (seeding depth: ½ - 1 in.)

SOYBEAN: broadcast 80 to 100 lbs/A; drill 60 to 80 lbs/A (seeding depth: 1- 1½ in.)

SUNN HEMP: broadcast 30 to 40 lbs/A; drill 25 to 35 lbs/A (seeding depth: ½ - 1 in.) drill 60 lbs/A for no-till or for vegetable transplants

VELVETBEAN: broadcast 30 to 40 lbs/A; drill 25 to 35 lbs/A (seeding depth: ½ - 1½ in.) LAB LAB: broadcast 50 to 60 lbs/A; drill 40 to 45 lbs/A (seeding depth: ½ - 1½ in.)

LAB LAB: broadcast 50 to 60 lbs/A; drill 40 to 45 lbs/A (seeding depth: ½ - 1½ in.)

Winter Cover Crops

Just as like summer cover crops, there are many choices for winter cover crops. Rye, triticale, barley, wheat, oats, and ryegrass can be planted in the fall; expect to harvest or plow under anywhere from 1/2 ton to 4 tons of dry matter per acre. Soil test to determine lime and fertilizer needs.

TYPES OF WINTER COVER CROPS: SMALL GRAINS

RYE: Rye is probably used more as a winter cover crop than any other grain. Rye can be sown from late September through mid-November. Most ryes will grow well in the fall (even late fall) and in late winter/early spring. This makes rye a top choice for farmers who have late-season vegetable crops with little time left before winter to sow a cover. Spring growth provides excellent biomass to turn under for use in early potatoes, Cole crops, etc. Rye also provides a forage source for grazing animals and a straw source if harvested before mature seeds are formed or after rye seed harvest (typical seeding rate: 60-120 lbs/A).

BARLEY: Barley provides an excellent source of biomass in the spring. It grows shorter than rye, will tiller, and potentially produces as much straw/forage/plow-down as rye. Barley takes longer to catch up with equivalent rye biomass in the spring, and the possibility of winter kill will be greater with barley. Late fall planting of barley will often result in winter kill. Plant in September or early October for greatest survival (typical seeding rate: 80-120 lbs/A broadcast; 60-110 lbs/A drilled).

WHEAT: Using wheat as a cover crop works well and provides the additional option of a grain harvest. Wheat can be seeded late September through mid-November. Wheat produces biomass like barley but will be a week or two later. It can be grazed before turning under or harvested for grain and the straw removed. Problems may occur if the Hessian fly is abundant, so choose another small grain in areas where Hessian fly is present. (typical seeding rate: 60-120 lbs/A)

OATS: Oats can be managed to provide many options for the cover crop and good late spring biomass. Seeding spring oats during September or October provides a good cover crop that will winter-kill in the colder areas but may overwinter in warmer areas. It can be grazed, made into excellent hay, or the grain harvested, and oat straw produced. Planting spring oats in the early fall can provide good winter-killed mulch that could benefit perennial vegetables or small fruits. Spring oats survive through some milder winters; thus, herbicides may be necessary to kill spring oats in perennial plantings (typical seeding rate: 80-120 lbs/A).

RYEGRASS: This grass has great potential use as a green manure and as a forage/hay material, but ryegrass can potentially become a difficult weed in some farm operations. In the mountains, rye grass grows slowly in the fall and provides only moderate winter erosion protection, but in late spring it produces an abundant supply of biomass. Grazing and spring hay from ryegrass can be excellent, and a fine, extensive root system makes it a great source for plow-down (typical seeding rate: 5-10 lbs/A drilled; 15-30 lbs/A broadcast).

Italian ryegrass can be seeded and then bedder-disked for bare ground or plasticulture. The remaining ryegrass will form a mulch between the beds (typical seeding rate: 100 lbs/A).

TRITICALE: Triticale is a small grain resulting from a cross between wheat and rye. Triticale has similar characteristics to wheat, but the plant has the overall vigor and winter-hardiness of rye. Fall planting of triticale should follow similar recommendations as wheat, sowing 60 to 120 lbs/A. Triticale biomass can exceed wheat, thus plowing under or killing for no-till culture should occur at an earlier time in the spring.

NOTES on SMALL GRAINS: Determine small grain fertilizer and lime needs based on soil test results. Successful stand establishment generally can be obtained with planting dates later than those of legumes, even as late as early December in coastal plain regions. This permits establishment of the cover crop after late-fall-harvested crops such as sweetpotatoes. Remember, that some soil erosion protection may be sacrificed with late seeding dates. For sandier coastal plain soils, rye is the preferred small grain cover crop. As previously discussed, seeding depth varies from ½ to 1½ inches, depending on soil texture. Planting methods are the same as described for legumes.

TYPES OF WINTER COVER CROPS: LEGUMES

A wide range in planting dates exists for most legumes, though best results are obtained with early plantings. Early seeding dates are easy to meet with legume cover crops following spring vegetables. Because Cahaba White Vetch possesses little winter hardiness, it is not adapted to western NC and the northern regions of other southeastern states. Freeze damage has also occurred with Austrian Winter Pea in higher elevations (above 2,500 feet). Avoid planting late otherwise you increase the risk of winter kill. For state specific recommendations for planting dates, seeding dates and management for legume cover crops, consult your local Extension office.

Seeding Rates and Depths

REGAL GRAZE LADINO CLOVER: broadcast 15 to 20 lbs/A (seeding depth: 1/4 in.). Can be bedder-disked for bare ground or plasticulture forming a between row mulch useful for conventional or bare ground production. Inhibits yellow nutsedge and can survive year-round with moisture.

CRIMSON CLOVER: broadcast 20 to 25 lbs/A; drill 15 to 20 lbs/A (seeding depth: ¼ - ½ in.)

HAIRY VETCH: broadcast 20 to 30 lbs/A; drill 15 to 20 lbs/A (seeding depth: ½ - 1½ in.)

CAHABA WHITE VETCH: broadcast 20 to 30 lbs/A; drill 15 to 20 lbs/A (seeding depth: ½ - 1½ in.)

AUSTRIAN WINTER PEA: broadcast 25 to 35 lbs/A; drill 20 to 25 lbs/A (seeding depth: ¾ - 1½ in.)

When seeding, use shallow planting depths for finer-textured, clayey soils and deeper depths for coarse-textured, sandy soils. Drilling into a conventional seedbed is the most reliable way to obtain a uniform stand. A no-till grain drill can be used successfully, provided residue from the previous crop is not excessive and soil moisture is sufficient to allow the drill to penetrate to the desired planting depth. Seeds can be broadcast if the soil has been disked and partially smoothed. Cultipacking after broadcasting will encourage good soil/seed contact. Crimson clover responds favorably to cultipacking.

TYPES OF WINTER COVER CROPS: RADISH

Forage and oilseed radishes have become standard species in many cool and warm season cover crop mixes. Because radishes establish quickly even under moderate drought conditions, the plants provide good protection against wind and water erosion, which can be helpful for stabilizing wet clay or sandy soils. The forage radish has an aggressive, expanding rosette which spreads laterally filling available space. Radishes are excellent at breaking up shallow layers of compacted soils, earning them the nickname *tillage radishes*. They are excellent scavengers of nitrate, phosphorous, potassium and other nutrients from deep soil layers. These nutrients remain in place after radish decomposition. In addition, decomposed radish roots leave channels for deep water infiltration over the winter, allowing the soil surface to dry and warm more quickly in the spring. Radish, like most Brassica species, release chemical compounds called glucosinolates that are toxic to many soil-borne pathogens and pests, such as nematodes, fungi, and some weeds.

SEEDING RATES AND DEPTHS

A current seed test is recommended when planting radishes. Under good storage conditions, when relative humidity (%) and temperature (°F) added together are less than 100, radish seeds will last four years. Radish seeding rates of 8 *Pure Live Seed* (PLS) pounds per acre with a conventional, air, or no-till drill, or 12 PLS pounds per acre when broadcasting, typically result in good stands. When using a drill, radishes are best planted between 0.25 inches (when moisture conditions are adequate) and 1-inch deep (given dry conditions).

When planted in a mix with other large-seeded cover crop species, such as beans *Phaseolus* spp., and field peas *Pisum sativum*, a 1-inch planting depth is recommended as the larger-seeded seedlings will provide channels for the radish seedlings to reach the surface. Low planting rates (4 PLS pounds per acre) are generally recommended because of high seed cost and larger root size at lower plant densities. In some situations, high planting rates (12 PLS pounds per acre) may be more beneficial. These include cases where control of weeds, diseases, and nematode pests are the primary focus. In diverse cover crop mixes (generally 5 or more species), recommended rates are 1 to 2 PLS pounds per acre. To improve weed and pest management, avoid planting radishes in the same field more than two consecutive years.

Several trademark varieties in the U.S. include Tillage Radish, Nitro, and Groundhog. Common varieties are also sold under VNS labels (Variety Not Stated).

The above is excerpted from: Hybner, R.M. 2014. Plant Materials Technical Note No. MT-106 Radish *Raphanus sativus* L. An introduced cover crop for conservation use in Montana and Wyoming. USDA NRCS. Accessed at: *https://www.nrcs.usda.gov/plantmaterials/mtpmctn12456.pdf*

MIXING GRASS AND LEGUMES: Planting a single grass or legume may be necessary but combining a grass and legume together may prove better than either one alone. Grasses provide soil protection during the winter and produce great forage or plow-down organic matter. Legumes do not grow well during the winter, but late spring growth is abundant and produces high protein forage and nitrogen for the following crop. Crimson clover is a legume to grow in combination with a grass. Crimson clover's height matches well with barley, wheat, and oats, but may be shaded and outcompeted by rye. Hairy vetch has been sown with grass cover crops for many years, using the grass/vetch combination as a hay or plowdown.

BIOFUMIGATION AND COVER CROPS: Biofumigation refers to the suppression of soilborne pathogens and pests (such as plant-parasitic nematodes and weeds) using naturally occurring biocidal compounds or allelochemicals, particularly isothiocyanates. These compounds, which are chemically similar to the active ingredient of the chemical fumigant metam sodium, are released from bioactive cover crops. The biofumigant process is initiated by cellular disruption (chopping up crop tissue), incorporating the crop tissue into the soil, and irrigating the soil with the incorporated tissue. This activates a chemical reaction of naturally occurring plant compounds called glucosinolates. The chemical reaction releases gases into soil pores that are generally toxic to microbes.

Most biofumigant cover crops are in the Brassica family (also known as Crucifers or Cole crops) and mustards are the commonly utilized biofumigant crop. However, not all mustards are well-suited as biofumigants since concentrations of the bioactive compounds vary by species and cultivar.

Brassica crops have been used extensively as winter cover crops and as "break crops" where the residues are tilled into the soil for their biofumigation effect. They have also been used in rotations, where the Brassica crop is harvested for sale and then the remaining residue is tilled-in for the biofumigation effect. There are several commercially available cover crops that have been used for biofumigation. "Caliente 119" (a mixture of oilseed radish and mustard), oilseed radish, "Florida Broadleaf Mustard", garden cress, penny cress, "Dwarf Essex" rape, and several canola varieties have been reported to have biofumigation potential.

In much of the southeast region of the U.S., these crops can be seeded in fall and over-wintered, or direct seeded in early spring. In either case, the crop should be chopped and tilled-in when it is in the early flowering stage to achieve the maximum biofumigation potential. The early flowering stage is the point at which the allelochemical concentrations are their highest. Seeding rates range from 4 to 20 lbs/A and will vary with location and seed size (the smaller the seed size, the lower the rate).

These crops respond and produce more biomass and more biofumigation potential when provided with 30 to 90 lbs/A N fertilizer at planting. These crops grow rapidly and can normally be plowed down in 6 to 10 weeks. In areas where the average last spring frost is 1 May or later, only spring planting is recommended. Optimal results occur when the Brassica cover crop is thoroughly chopped and tilled completely into the top 6 to 8 inches of soil and then watered in thoroughly. Irrigating will help trap the volatile compounds into the soil. Brassica seed meals (specifically mustard seed meal) may also be utilized for biofumigation. Mustard seed meal is highly concentrated in volatile compounds and provides a partial source of organic fertilizer for the following crop. Mustard seed meal can be used as a biofumigant by spreading it like a fertilizer, tilling into the soil, and then irrigating to trap the volatiles.

PLOWDOWN: Plowing early defeats the purpose of growing cover crops as little biomass will have been produced by the cover crop. In the case of legume cover crops, they require sufficient time to develop biomass, which an early plowdown would prevent. If you need to plow early, use a grass cover crop (rye) that produces **sufficient** fall growth and will provide maximum biomass for incorporation. Allow 3-6 weeks between plowdown and planting.

HELPFUL RESOURCES

MANAGING COVER CROPS PROFITABLY:
https://www.sare.org/wp-content/uploads/Managing-Cover-Crops-Profitably.pdf

BUILDING BETTER SOILS FOR BETTER CROPS:
https://www.sare.org/news/updated-building-soils-for-better-crops-focuses-on-soil-health-fundamentals/

SOUTHERN COVER CROPS COUNCIL:
https://southerncovercrops.org/

COVERS UNDER COVER:
https://www.uky.edu/ccd/sites/ www.uky.edu.ccd/files/CoversUnderCover1.pdf

WARM-SEASON COVER CROPS FOR HIGH TUNNELS IN THE SOUTHEAST:
https://www.uky.edu/ccd/sites/www.uky.edu.ccd/files/warm-season_covercrops.pdf

COOL-SEASON COVER CROPS FOR HIGH TUNNELS IN THE SOUTHEAST:
https://ccd.uky.edu/sites/default/files/2024-11/ccd-sp-18_cool-season_covercrops.pdf

THE BASICS OF BIOFUMIGATION:
https://ccd.uky.edu/sites/default/files/2024-12/ccd-fs-20_biofumigation.pdf

Biofumigant Seed Sources:
HIGH PERFORMANCE SEEDS, INC.:
https://www.hpseeds.com/products

JOHNNY'S SELECTED SEED:
https://www.johnnyseeds.com/farm-seed/brassicas/

SEEDWAY:
https://www.seedway.com/product-category/vegetable-seed/cover-crop-seeds/

WELTER SEED & HONEY CO.:
https://welterseed.com/

Brassicaceous Seed Meal Source:
FARM FUEL INC.:
https://farm-fuel-inc.square.site/

TRANSPLANT PRODUCTION

The following recommendations apply to plants grown under controlled conditions such as in a greenhouse, high tunnel, or hotbed using soilless media. Information on how to grow plants once they are in the field is covered in the specific section for each crop. A transplant is affected by factors such as temperature, fertilization, water, and spacing. A good transplant is one that has been grown under the best possible conditions.

Table 5 presents optimum and minimum temperatures for seed germination and seedling growth, time and space (area) requirements, and number of plants per square foot for several economically important vegetable crops in the southeastern US.

Commercial Plant-growing Mixes. A number of commercial media formulations are available for growing transplants. Most of these mixes will produce high-quality transplants when used with good management practices. However, these mixes can vary greatly in composition, particle size, pH, aeration, nutrient content, and water-holding capacity. Avoid formulations that have fine particles, as these may hold excessive water and have poor aeration, which may cause transplants to stay too wet and put them at risk for disease. When trying a soilless media for the first time, it is a good idea to have the mix formulation tested by a reputable soil testing laboratory to determine the pH and the level of nutrients the mix contains. Many states have soil testing facilities at their land-grant universities.

Treatment of Flats. Flats and other containers used in the production of transplants should be new or as clean as new to avoid damping-off and other disease issues that could carryover from your previous crop. If flats are reused, thoroughly clean and sanitize them. See Table 3-49 – under the section on "*Equipment*" for methods and materials used to sanitize flats. There are several products available. Be sure to follow label instructions in order to successfully sanitize flats and to avoid possible crop injury. Some products must be rinsed from flats prior to use.

TABLE 5. OPTIMUM AND MINIMUM TEMPERATURES FOR TRANSPLANT PRODUCTION AFTER GERMINATION[2]

	°F Opt. Day	°F Min. Night	Weeks to Grow
Broccoli	65-70	60	5-7
Cabbage	65	60	5-7
Cantaloupe[1]	70-75	65	3-4
Cauliflower	65-70	60	3-4
Cucumber	70-75	65	2-3
Eggplants	70-85	65	5-8
Endive & Escarole	70-75	70	5-7
Lettuce	60-65	40	5-6
Onions	65-70	60	8-12
Peppers	70-75	60	5-9
Summer squash	70-75	65	2-3
Sweetpotato	75-85	ambient	4-6
Tomatoes	65-75	60	5-6
Watermelon, seeded	80-85	65	3-5
Watermelon, seedless	80-85	65	3-5

[1] Cantaloupe and other melons
[2] See individual crop recommendations for germination temperatures.

Plant Containers. There is a wide variety of containers available for starting seeds for transplants. Most growers start seeds either in flats or in cell packs. The main advantage of using flats is that more plants can fit into the same space compared to cell packs. However, if you start seeds in open flats, you will need to transplant them into larger cell packs or to individual pots as the seedlings grow.

Seeding directly into cell packs saves time because you avoid needing to transplant seedlings into a larger container later. Simply use a cell pack with your desired final cell size. Cell packs come in many different cell sizes, but the overall tray size is standardized.

For lettuce or smaller rooted vegetables, use 128-cell packs. For tomatoes and peppers, 72-cell packs work well. For larger-seeded vegetables, such as cucumbers, squash, and watermelons, use 50-cell packs.

Each vegetable crop has optimal cell sizes for containerized transplant production and requires a certain number of weeks before they are ready for transplanting (Table 5). For example, broccoli, Brussels sprouts, cabbage, cauliflower, and collards require a 0.8- to-1.0-inch cell and 5- to 7-weeks to reach an adequate size for transplanting; cantaloupe and watermelons require a 1.0-inch cell and 3- to 5-weeks; eggplant and tomato require a 1.0-inch cell and 4- to 6-weeks; and pepper requires a 0.5- to- 0.8-inch cell and 5- to 7-weeks. Other options are available depending on your crop and situation.

Seed Germination. Seed that is sown into open flats to be "pricked out" later should be germinated in vermiculite (horticultural grade, coarse sand size) or a plug-growing mix. It is recommended that no fertilizer be included when germinating seed. Do not fertilize the seedlings until their cotyledons, the first two leaf-like structures that emerge, are fully expanded and their true leaves are beginning to unfold. Fertilization should be in liquid form and at one-half the rate for any of the ratios listed in the following section on "Liquid Feeding." Seedlings can be held for a limited time if fertilization is withheld until 3 to 4 days before "pricking out." Seed that is sown into pots and cells that will not be "pricked out" can be germinated in a mix that contains fertilizer.

To get earlier, more uniform emergence, germinate and grow seedlings on benches or in a floor-heated greenhouse. Germination can be aided by using germination mats and chambers which provide heat directly to the trays. In the case of chambers, relative humidity can be controlled. These mats and chambers allow you to set the optimal temperature for germination. With supplemental heating, seedling emergence and uniformity can be enhanced, which decreases the amount of time required to produce a transplant. If heated floors or benches are not available, seed the trays, water, and stack them off the floor during germination. Additionally wrapping the trays in plastic to maintain a high relative humidity can be helpful for some crops like seedless watermelons but only wrap and stack them for up to 24 to 36 hr. Check for emergence and germination often. Unwrap and unstack as soon as germination is observed.

Heating and Venting. Exhaust from heaters must be vented to the outside. Be sure to have an outside fresh-air intake for the heaters. Be sure vents and fans are properly designed and positioned to avoid drawing exhaust gases into the greenhouse. Exhaust gases improperly adjusted heating systems can cause yellowing, stunting, distortion, and death of seedlings. Do not grow or hold seedlings in an area where pesticides are stored.

Liquid Feeding. The following materials dissolved in 5 gallons of water and used over an area of 20 square feet are recommended for use on transplants, if needed:

- 20-20-20   1.2 to 1.6 oz/5-gal water
- 15-15-15   2 oz/5-gal water
- 15-30-15   2 oz/5-gal water

Rinse leaves after liquid feeding. Fertilizers used for liquid feeding must be 100% water soluble. When transplanting to the field, use a "starter fertilizer" and follow the manufacturer's instructions for its use.

Watering. While growing seedlings, keep the soilless media moist but not continually wet. Water less in cloudy weather. Watering in the morning allows plant surfaces to dry before night and reduces the possibility of disease development. Watering twice per day (once in the morning and once in the afternoon) may be necessary on hot, sunny days.

Hardening and Transplant Height Control. Proper hardening of transplants stiffens stems and hardens the transplants, which decreases the chances of transplant shock and increases the chances for survival. This process allows for the seedlings to transition from a protected indoor environment to harsher outdoor conditions with wind and fluctuating temperatures.

There are several methods – chemical and cultural – used to harden transplants and the choice of which to use is often crop-dependent. At this time there is one chemical plant growth regulator available for use in producing vegetable transplants, but its use is limited to several Solanaceous crops. More details are in the section below *Chemical Hardening*.

Cultural Methods for Hardening. Hardening plants off can be done by gradually exposing plants to more sunlight. This can be done by placing transplants in a shady location and moving them each day to be exposed to more sunlight. You can also reduce the amount of water the transplants receive (but never allow them to wilt). A combination of these two methods is often used. By reducing the amount of water used and lowering ambient temperatures, one can cause a check or slow down in plant growth to prepare plants for the field setting. Never reduce or limit fertilizer as a means to harden transplants because it will often delay maturity. Pay attention to the weather forecast. If ambient temperatures are too low, transplants can develop chilling injury causing plant damage and delayed growth after transplanting.

Transplant Height Control. The goal of a producing transplants should be to produce a strong transplant with sturdy growth that can withstand transplanting into the field. Tall, spindly, or overgrown transplants can be difficult to remove from the transplant tray and might become entangled in the transplanting equipment. They may also be weak and not stand upright after being transplanted. There are a few methods available that can aid the producer with managing the top growth of developing transplants.

One method of height control is to use cold water for irrigation, 33-34°F, which has been shown to control top growth in some species.

Another method is called *Brushing*. Mechanically brushing the plants involves light mechanical contact to the tips of the growing plants. The easiest way to do this is to lightly drag a PVC or bamboo rod across the tops of the growing transplants. This needs to be done multiple times per day. Some species like tomatoes respond well to this with little to no physical damage to the brushed leaves whereas eggplant and pepper leaves tend to be more sensitive and easily damaged. The number of times transplants are treated to achieve height control depends on the crop. Brushing can also be accomplished by setting up fans so that the plants are moved (brushed by air). This results in a light mechanical stress to the plant's stem and can result in shorter plants. Again, the intensity and frequency of this treatment will have to be adjusted to avoid damage to developing foliage while still achieving height control.

DIF. Plant height can be held in check and hardening can be improved by using a process that reduces or increases ambient temperatures in the early morning over the course of several days. Plants elongate the most at daybreak. Raising the temperature before daybreak (2 hours before) or lowering the temperature just after daybreak (2 hours after) by 10°F will cause plants to be shorter and more hardened. This process is called DIF, because you are employing a difference in temperature. DIF can be positive or negative, but positive DIF is more commonly used for hardening transplants. Negative DIF can cause crop injury on cold-sensitive crops or bolting on cool-season crops.

Chemical Hardening. Uniconazole, under the trade name Sumagic, is a plant growth regulator (PGR) that is used to produce more compact growth and thickened stems in transplants. It is labeled for foliar sprays on the following vegetable transplants: tomato, pepper, ornamental pepper, eggplant, tomatillo, groundcherry, and pepino. But the new label is restrictive; the maximum total allowed application is 10 ppm at 2 quarts per 100 square feet. This means only one 10-ppm spray, two 5-ppm, or four 2.5-ppm sprays are allowed, and so on.

The following is taken from the *Supplemental Label for Sumagic Plant Growth Regulator for Use on Fruiting Vegetable Transplants:*

- Apply uniformly as a foliar spray at a spray volume of 2 qt/100 sq ft.
- Make initial foliar application when 2 to 4 true leaves are present.
- Sequential applications at lower recommended rates will generally provide more growth regulation than a single high-rate application.
- First time users should apply the lowest recommended rate in order to determine optimal rate for individual cultivars under local environmental conditions. If additional growth regulation is required a sequential spray application at the lowest recommended rate should be made 7- to 14-days after initial application.

- If multiple applications are made to the transplants, the total amount of uniconazole-P applied may not exceed that from a single application of a 10-ppm spray concentration at 2 qt/100 sq ft (equivalent to 0.000042 lb ai/100 sq ft or 0.018 lb ai/A). The final application may not occur later than 14 days after the 2 to 4 true leaf stage.

The Supplemental Label for Sumagic referred to above can be found at : *https://www.domyown.com/msds/sumagic_supp_label_for_fruiting_veg_transplant.pdf*

Uniconazole is a highly active PGR, it is critical to emphasize that caution is paramount while using uniconazole for vegetable transplant height control.

GRAFTING IN VEGETABLE CROPS

The phase-out of methyl bromide fumigation has driven the search for alternative methods to manage soilborne pathogens and plant-parasitic nematodes in vegetable crops. Although alternative pesticides and other physical treatments are being tested and developed, grafting with resistant rootstocks offers one of the best methods to protect against infection.

Grafting involves combining a desirable scion, which is the fruit-bearing portion of a grafted plant, with a rootstock, which provides resistance to various soilborne pathogens and plant-parasitic nematodes. The scion is generally from a plant that produces highly desirable fruit. However, grafting can also influence vegetative growth and flowering, affect fruit ripening and quality, enhance abiotic stress resistance, and enhance yield, especially under temperature extremes (high or low).

At present, most research is focused on grafting solanaceous and cucurbit crops. The primary motive for grafting tomato and watermelon is to manage soilborne pests and pathogens, when genetic or chemical approaches for management are not available. Grafting a susceptible scion onto a resistant rootstock creates a resistant cultivar without the need to breed a resistant cultivar. Further, grafting allows a rapid response to new races of pathogens and, in the short-term, provides a less expensive and more flexible solution for managing soilborne diseases than by breeding new, resistant cultivars.

Grafted transplants are more expensive than non-grafted transplants (as much as 4x) due to labor, material costs (grafting supplies, seed costs of rootstock and scion), and specialized facilities required to produce grafted plants. These specialized facilities include healing chambers, grafting rooms, and trained personnel to produce and care for these grafted transplants. Potential changes in fruit quality, which occur with some rootstocks, must also be considered.

Growers can produce their own grafted transplants. Before setting out to produce their own transplants, growers need to determine if the extra time and costs involved makes fiscal sense for their operation.

Commercial transplant producers offer grafting services and, with improved grafting techniques and mechanization, costs are decreasing. Grafted transplants are available across the US.

Current research and developments in grafting can be found at *http://www.graftingvegetables.org*. This site provides the latest findings on grafting solanaceous (tomatoes, peppers, eggplants, etc.) and cucurbit (watermelons, melons, etc.) crops. Details on how to graft and what equipment is needed is available as well as tables listing research on rootstocks and their specific attributes. New rootstocks are becoming commercially available every year while older rootstocks are taken off the market.

Grafting is gaining popularity in the US due to increasing limited land availability for rotation and the loss of soil fumigation as a means for managing soilborne diseases. Grafting presents an option to manage soilborne pathogens such as Fusarium wilt, Monosporascus vine decline, Phytophthora blight, and root-knot nematode.

Grafting can also enhance tolerance to abiotic stress, increase water and nutrient use efficiency, extend harvest periods, and improve fruit yield and quality in certain cucurbits. However, each rootstock cultivar provides advantages or disadvantages under certain environmental conditions. Consult with your local Extension agent or state vegetable specialist for rootstock recommendations for your area and specific issues.

Grafting Methods for Cucurbits. Cucurbit grafting is difficult, and continuous practice is needed to be successful. There are four commonly used methods to date: *tongue approach graft, hole insertion graft, one cotyledon graft, and side graft.*

Only the *one cotyledon graft method* is currently automated. The other three methods are labor intensive. Each method has its own advantages and disadvantages. Remember that grafting is an art that requires attention to detail. The major concern with cucurbit grafting is the constant threat of regrowth of the rootstock. This regrowth needs to be removed by hand. Further details and step-by-step instructions can be found at: *http://www.vegetable- grafting.org/resources/grafting-manual*.

Grafting Methods for Tomatoes. There are two primary techniques used for grafting tomatoes: splice grafting and cleft grafting. Cleft grafting and splice grafting (also called tube grafting) are similar in that the shoot of the fruit-producing scion is completely cut off from its own roots and attached to the severed stem of the rootstock. Silicone clips or tubes are used to join the scion and rootstock together while they heal together. Splice grafting is quicker and less complicated to do than cleft grafting because it only requires a single, diagonal cut on both the rootstock and the scion. In addition, because fewer intricate cuts are involved, this technique can be used on very small seedlings.

Grafting can be performed at various stages of seedling growth, but it is recommended at the 2- to 3-true leaf stage. With both cleft and splice grafting, the newly grafted plants must be protected from drying out until the graft union has healed. This usually involves covering the plants with a plastic cover or protecting them in some type of healing chamber where temperature and humidity can be regulated. Employ some method to reduce light intensity to the grafted plants for several days after the procedure. It is critical to maintain the relative humidity in the chamber to near 100% for the first two days. After two days, reduce humidity incrementally over the next five days. This will prohibit the formation of adventitious roots from the scion and provide time for the graft union to heal. Tomato grafts heal quickly, and the seedlings can be acclimated back into the greenhouse after 4- to 5-days.

With both cleft and splice grafting, it is important that the diameter of the cut ends (of the scion and the rootstock) match up perfectly. If the diameter does not match, the graft may not heal properly, if at all. Rootstock cultivars tend to have different growth habits than scion cultivars, so it is important to grow a small amount of rootstock and scion seed at first to determine their growth rates relative to each other. Rootstock cultivars tend to be more vigorous than scion cultivars, but this is not always the case. Cut rootstock seedlings below the cotyledons. If the cotyledons are left, they will generate suckers that will compete with the scion requiring removal. For step-by-step instructions, go to:

http://www.vegetablegrafting.org/resources/grafting-manual/.

DISEASE CONTROL IN PLANT BEDS

For the best control of all soilborne diseases, use a good commercial plant-growing mix. Do not re-use soilless growing media. If this is not possible, use one of the following procedures:

Preplant. The only practice that ensures complete sterilization of soil is the use of steam. When steam is used, a temperature of 180°F must be maintained throughout the entire mass of soil for 30 minutes. Further information on sanitizer use for equipment, storage houses, produce, and water can be found in Tables 3-48 and 3-49.

Seed Treatment. Seed treatment is important to manage seed-borne diseases. Use of untreated seed could lead to diseases in the plant bed, which could reduce plant stands or result in diseased transplants leading to potential crop failure. See section on Seed Treatments for detailed information on how to properly treat seeds and for materials labeled for use as seed treatments.

Postplant. Damping-off and foliar diseases can be a problem in plant beds. To prevent these diseases, it may be necessary to apply fungicide sprays especially as plants become crowded in plant beds. Refer to label clearance before use. The use of sphagnum moss as a top dressing will reduce damping-off because it keeps the surface dry. See the section on Greenhouse Vegetable Crop Disease Control for management option.

SEED STORAGE AND HANDLING

Both high temperature and relative humidity will reduce seed germination and vigor of stored seed. Do not store seed in areas that have a combined temperature and humidity value greater than 100 [e.g., 50°F + 50% relative humidity]. In addition, primed seed does not store well after shipment to the buyer. Therefore, if you do not use all the primed seed ordered in the same season, have the seed tested before planting in subsequent seasons.

Corn, pea, and bean seed are especially susceptible to mechanical damage due to rough handling. Bags of these seed should not be dropped or thrown because the seed coats can crack and seed embryos can be damaged, resulting in a nonviable seed. When treating seed with a fungicide, inoculum, or other chemical, use only gentle agitation to avoid seed damage.

PLANT POPULATIONS

For vegetable seed sizes and plant populations see Tables 6 and 7.

IRRIGATION

Basic Principles. The most common use of irrigation in vegetable production is to supply the crop water demand. However, irrigation events can also be used to modify the crop microclimate (i.e., frost protection), control weeds, leach salts, or deliver fertilizer and chemicals to the growing crops. To supply the crop water demand, maintaining proper soil moisture levels is important to maximize the productivity of vegetable crops. Plant stress caused by too much or too little soil moisture can lead to decreased size and weight of individual fruit and to defects such as: toughness; strong flavor; poor tip fill and pod fill; cracking; blossom-end rot; and misshapen fruit.

TABLE 6. VEGETABLE SEED SIZES

Crop	Seeds/Unit Weight	Crop	Seeds/Unit Weight	Crop	Seeds/Unit Weight
Asparagus	13,000-20,000/lb	Kale	7,500-8,900/oz	Radishes	40,000-50,000/lb
Beans:		Kohlrabi	9,000/oz	Rutabaga	150,000-192,000/lb
small seeded lima	1,150-1,450/lb	Leeks	170,000-180,000/lb	Spinach	40,000-50,000/lb
large seeded lima	440-550/lb	Lettuce:		Squash:	
snap	1,600-2,200/lb	head	20,000-25,000/oz	summer	3,500-4,800/lb
Beets	24,000-26,000/lb	leaf	25,000-31,000/oz	winter	1,600-4,000/lb
Broccoli	8,500-9,000/oz	Mustard	15,000-17,000/oz	Sweet corn:	
Brussels sprouts	8,500-9,000/oz	Okra	8,000/lb	normal and sugary enhanced	1,800-2,500/lb
Cabbage	8,500-9,000/oz	Onions:		supersweet (sh2)	3,000-5,000/lb
Cantaloupes	16,000-19,000/lb	bulb	105,000-144,000/lb	Tomatoes:	
Carrots	300,000-400,000/lb	bunching	180,000-200,000/lb	fresh	10,000-11,400/oz
Cauliflower	8,900-10,000/oz	Parsnips	192,000/oz	processing	160,000-190,000/oz
Collards	7,500-8,500/oz	Parsley	240,000-288,000/oz	Turnip	150,000-200,000/lb
Cucumbers	15,000-16,000/lb	Peas	1,440-2,580/lb	Watermelons:	
Eggplants	6,000-6,500/oz	Peppers	4,000-4,700/oz	small seed	8,000-10,400/lb
Endive, Escarole	22,000-26,000/oz	Pumpkins	1,900-3,200/lb	large seed	3,200-4,800/lb

Once irrigating according to the soil water status, it is imperative for growers to maintain the soil moisture level near field capacity at all times during the growing season. Field capacity is the maximum water the soil can hold. When the soil moisture *status/content/tension* is at field capacity it means soil water availability is ideal for crop development. However, to maintain field capacity always requires a strategic scheduling of irrigation events, so the irrigation system is capable of make uniform, *frequent*, and precisely timed applications. Commonly, more than one irrigation cycle per day may be needed to **maintain** field capacity. This is particularly true for fast growing crops grown in soils with little water holding capacity, such as sandy loams. Even relatively short periods of inadequate soil moisture can adversely affect many crops.

Another important aspect to take into consideration to determine irrigation events using the soil water status is the soil type. Different soil types have different moisture holding capacities and the use of soil moisture sensors is essential to ensure proper water availability. Soil moisture sensors are the most common technology used to monitoring the soil water availability for plants and recent research indicates that maintaining soil moisture levels in a narrow range, just slightly below field capacity (75% to 90% available soil moisture), maximizes crop growth. This may mean that more frequent irrigations of smaller amounts are better than delaying irrigations until the soil moisture reaches a lower level (40% to 50% available soil moisture) and then applying a heavy irrigation. It is important to remember that different soils also have different infiltration rates, which will

TABLE 7. POPULATION OF PLANTS PER ACRE AT SEVERAL BETWEEN-ROW AND IN-ROW SPACINGS

Between-row spacing (in.)	In-row spacing (in.)												
	2	4	6	8	10	12	14	16	18	24	30	36	48
7	448,046	224,023	149,349	112,011	89,609	74,674	64,006						
12	261,360	130,680	87,120	65,340	52,272	43,560	37,337	32,670	29,040	21,780	17,424	14,520	10,890
18	174,240	87,120	58,080	43,560	34,848	29,040	24,891	21,780	19,360	14,520	11,616	9,680	7,260
21	149,349	74,674	49,783	37,337	29,870	24,891	21,336	18,669	16,594	12,446	9,957	8,297	6,223
24	130,680	65,340	43,560	32,670	26,136	21,780	18,669	16,335	14,520	10,890	8,712	7,260	5,445
30	104,544	52,272	34,848	26,136	20,909	17,424	14,935	13,068	11,616	8,712	6,970	5,808	4,356
36 (3 ft)	87,120	43,560	29,040	21,780	17,424	14,520	12,446	10,890	9,680	7,260	5,808	4,840	3,630
42 (3.5 ft)	74,674	37,337	24,891	18,669	14,934	12,446	10,668	9,334	8,297	6,223	4,978	4,149	3,111
48 (4 ft)	65,340	32,670	21,780	16,335	13,068	10,890	9,334	8,167	7,260	5,445	4,356	3,630	2,722
60 (5 ft)			17,424	13,068	10,454	8,712	7,467	6,534	5,808	4,356	3,485	2,904	2,178
72 (6 ft)			14,520	10,890	8,712	7,260	6,223	5,445	4,840	3,630	2,904	2,420	1,815
84 (7 ft)			12,446	9,334	7,467	6,223	5,334	4,667	4,149	3,111	2,489	2,074	1,556
96 (8 ft)			10,890	8,167	6,534	5,445	4,667	4,084	3,630	2,722	2,178	1,815	1,361

TABLE 8. CRITICAL PERIODS OF WATER NEED FOR VEGETABLE CROP

Crop	Critical Period
Asparagus	Brush
Beans, Lima	Pollination and pod development
Beans, Snap	Pod enlargement
Broccoli	Head development
Cabbage	Head development
Carrots	Root enlargement
Cauliflower	Head development
Corn	Silking and tasseling, ear development
Cucumbers	Flowering and fruit development
Eggplants	Flowering and fruit development
Lettuce	Head development
Melons	Flowering and fruit development
Onions, Dry	Bulb enlargement
Peppers	Flowering and fruit development
Potatoes (Irish)	Tuber set and tuber enlargement
Radishes	Root enlargement
Southernpeas	Seed enlargement and flowering and English
Squash, Summer	Bud development and flowering
Sweetpotato	Root enlargement
Tomatoes	Early flowering, fruit set, and enlargement
Turnips	Root enlargement

TABLE 9. AVAILABLE WATER-HOLDING CAPACITY BASED ON SOIL TEXTURE

Soil Texture	Available Water Holding Capacity (water/inches of soil)
Coarse sand	0.02–0.06
Fine sand	0.04–0.09
Loamy sand	0.06–0.12
Sandy loam	0.11–0.15
Fine sandy loam	0.14–0.18
Loam and silt loam	0.17–0.23
Clay loam and silty clay loam	0.14–0.21
Silty clay and clay	0.13–0.18

TABLE 10. SOIL INFILTRATION RATES BASED ON SOIL TEXTURE

Soil Texture	Soil Infiltration Rate (inch/hour)
Coarse sand	0.75–1.00
Fine sand	0.50–0.75
Fine sandy loam	0.35–0.50
Silt loam	0.25–0.40
Clay loam	0.10–0.30

affect the frequency of irrigation events. For example, water is more readily held in clay soils; however, clay soils have a lower water infiltration rate as compared to sandy soils. Depending on the soil structure, high application rates will result in irrigation water running off the field, contributing to erosion and fertilizer runoff particularly on heavy clay soils. Application rates should follow values in Table 10.

Applying the proper amount of water at the correct time is critical for achieving the optimum benefits from irrigation. The crop water requirement, termed evapotranspiration, or ET, is equal to the quantity of water lost from the plant (transpiration) plus water that evaporated from the soil surface. The ET rate is also an important and effective strategy to schedule irrigation events. Numerous factors must be considered when estimating ET. The amount of solar radiation, which provides the energy to evaporate moisture from the soil and plant surfaces, is the major factor. Other factors include crop growth stage; day length; air temperature; wind speed; and humidity level. Plant factors that affect ET are crop species; canopy size and shape; leaf size, and shape. Soil factors must also be considered. Soils having high levels of silt, clay, and organic matter have greater water-holding capacities than sandy soils or compacted soils (Table 9). Soils with high water-holding capacities require less frequent irrigation than soils with low water-holding capacities. When such soils are irrigated less frequently, a greater amount of water must be applied per application.

Water loss from plants is much greater on clear, hot windy days than on cool, overcast days. During periods of hot, dry weather, ET rates may reach 0.25 inch per day or higher. ET can be estimated using a standard evaporation pan or using local weather networking systems. Check with your local Extension office for information.

In general, irrigation is beneficial in most years since rainfall is rarely uniformly distributed even in years with above-average precipitation. Moisture deficiencies occurring early in the crop cycle may delay maturity and reduce yields. Shortages later in the season often lower quality and yield. Over-irrigating, however, especially late in the season, can reduce quality and postharvest life of the crop. Table 8 shows the critical periods of crop growth when an adequate supply of water is essential to produce high-quality vegetables.

Drip Irrigation. Drip irrigation is used to maintain soil moisture whereas other types of irrigation are used more to replace depleted soil moisture. Drip irrigation is a method of slowly applying water directly to the plant's root zone. Water is applied frequently, often daily, to maintain favorable soil moisture conditions.

Even so, field operations can continue uninterrupted. Water is applied without wetting the foliage, thereby decreasing evaporative losses, and decreasing disease pressure due to damp foliage. Additionally, the use of drip irrigation can limit waste and potential contamination from overuse (or unnecessary use) of agricultural chemicals. In most cases, drip irrigation is considerably more uniform and efficient in its distribution of water to the crop than other irrigation methods. In addition, fertilizers applied through the drip irrigation system are conserved.

Drip irrigation is used on a wide range of fruit and vegetable crops. It is especially effective when used with mulches; on sandy soils; and on high value crops, such as cantaloupes, watermelons, squash, peppers, eggplants, and tomatoes. Drip irrigation systems have several other advantages over sprinkler and surface irrigation systems. Low flow rates and operating pressures are typical of drip systems. These characteristics lead to lower energy costs. Once in place, drip systems require little labor to operate, can be automatically controlled, and can be managed to apply the precise amount of water and nutrients needed by the crop. These factors also reduce operating costs. The areas between rows remain dry reducing weed growth between rows and reducing the amount of water lost to weeds.

The equipment used for drip irrigation systems must be routinely monitored and maintained to prevent any challenges. Drip irrigation tape and tubing can be damaged by insects, rodents, other wildlife, and laborers. Pressure regulation and filtration require equipment not commonly used with sprinkler or surface systems. The drip system, including a pump, headers, filters, and various connectors, must be checked and be ready to operate before planting. Failure to have the system operational could result in costly delays, poor plant survival, and irregular stands, reducing yield.

Calculating the length of time required to apply a specific depth of water with a drip irrigation system is more difficult than with sprinkler systems. Unlike sprinkler systems, drip systems apply water to only a small portion of the total crop acreage. Usually, a fair assumption to make is that the mulched width approximates the extent of the plant root zone. Although the root zone is confined, the plant canopy is vigorous and water use

TABLE 11. HOURS REQUIRED TO APPLY 1" WATER BASED ON ROW SPACING.

Drip Tube Flow Rate		Row Spacing (Ft.)				
gph/100 ft.	gpm/100 ft.	4	5	6	8	10
11.4	0.19	21.9	27.3	32.8	43.7	54.7
13.2	0.22	18.9	23.6	28.3	37.8	47.2
20.4	0.34	12.2	15.3	18.3	24.4	30.6
27.0	0.45	9.2	11.5	13.9	18.5	23.1
40.2	0.67	6.2	7.8	9.3	12.4	15.5
80.4	1.34	3.1	3.9	4.7	6.2	7.8

TABLE 12. MAXIMUM APPLICATION TIME IN MINUTES FOR DRIP IRRIGATED VEGETABLES.

Available Water Holding Capacity[1] Inch Of Water/Inch Soil Depth	Drip Tubing Flow Rate (gpm/100 ft.)				
	0.2	0.3	0.4	0.5	0.6
	Maximum Number of Minutes per Application[2]				
0.02	20	14	10	8	7
0.04	41	27	20	16	14
0.06	61	41	31	24	20
0.08	82	54	41	33	27
0.10	102	68	51	41	34
0.12	122	82	61	49	41
0.14	143	95	71	57	48
0.16	163	109	82	65	54
0.18	183	122	92	73	61

[1] Refer to Table 9 for Available Water Holding Capacity based on soil texture
[2] Assumes a 10-inch deep root zone and irrigation at 25% soil moisture depletion

and loss from evapotranspiration (ET) can far exceed the water applied if application is based on a banded or mulch width basis. Table 11 calculates the length of time required to apply 1-inch of water with drip irrigation based on the drip tape flow rate and crop row spacing. The use of this table requires that the drip system be operated at the pressure recommended by the manufacturer.

Table 12 has been prepared to calculate the maximum recommended irrigation period for drip irrigation systems. The irrigation periods listed assume that 25% of the available water in the plant root zone is depleted. Soil texture directly influences the water-holding capacity of soils and the consequential depth reached by irrigation water.

In drip tape systems, water is carried through plastic tubing (which expands when water flows through it) and distributed along the drip tape through built-in outlets or devices called emitters. The pressure-reducing flow path also allows the emitter to remain relatively large, allowing particles that could clog an emitter to be discharged.

Although modern emitter design reduces the potential for trapping small particles, emitter clogging can be a common occurrence with drip irrigation systems. Clogging can be attributed to physical, chemical, or biological contaminants. Proper filtration is a must and occasional water treatment might be necessary to keep drip systems from clogging. Further information on drip irrigation systems can be obtained from manufacturers, dealers, and your local Extension office.

Chlorination. Bacteria can grow inside drip irrigation tapes, forming a slime that can clog emitters. Algae present in surface waters can also clog emitters. Bacteria and algae can be effectively controlled by chlorination of the drip system. Periodic treatment before clogs develop can keep the system functioning efficiently. The frequency of treatments depends on the quality of the water source. Generally, two or three treatments per season are adequate. Irrigation water containing high concentrations of iron (greater than 1 ppm) can also cause clogging problems due to a type of bacteria that "feeds" on iron. In consuming the dissolved (ferrous) form of iron, the bacteria secrete a slime called ochre, which may combine with other solid particles in the drip tape and plug emitters. The precipitated (ferric) form of iron, known commonly as rust, can also physically clog emitters. In treating water containing iron, chlorine will oxidize the iron dissolved in water, causing the iron to precipitate so that it can be filtered and removed from the system.

Chlorine treatment should take place upstream of filters to remove the precipitated iron and microorganisms from the system. Chlorine is available as a gas, liquid, or solid. Chlorine gas is extremely dangerous, and caution should be exercised if this method of treatment is chosen. Solid chlorine is available as granules or tablets containing 65% to 70% calcium hypochlorite but might react with other elements in irrigation water to form precipitates which could clog emitters. Liquid chlorine is available in many forms, including household bleach (sodium hypochlorite), and is the easiest and often safest form to use for injection. Stock solutions can be bought that have concentrations of 5.25%, 10%, or 15% available chlorine. Use chlorine only if the product is labeled for use in irrigation systems.

Since chlorination is most effective at pH 6.5 to 7.5, some commercial chlorination equipment also injects buffers to maintain optimum pH for effective kill of microorganisms. This type of equipment is more expensive, but more effective than simply injecting sodium hypochlorite solution.

The required rate of chlorine injection is dependent on the concentration of microorganisms present in the water source, the amount of iron in the irrigation water, and the method of treatment being used. To remove iron from irrigation water, start by injecting 1 ppm of chlorine for each 1 ppm of iron present in the water. For iron removal, chlorine should be injected continuously. Adequate mixing of the water with chlorine is essential. For this reason, be certain to mount the chlorine injector a distance upstream from filters. An elbow between the injector and the filter will ensure adequate mixing.

For treatment of algae and bacteria, a chlorine injection rate that results in the presence of 1 to 2 ppm of "free" chlorine at the end of the furthest lateral will assure that the proper amount of chlorine is being injected. Free, or residual, chlorine can be tested using an inexpensive DPD (diethyl-phenylene-diamine) test kit. A swimming pool test kit can be used, but only if it measures free chlorine. Many pool test kits only measure total chlorine.

If a chlorine test kit is unavailable, one of the following schemes is suggested as a starting point:

For iron treatment:
- Inject liquid sodium hypochlorite continuously at a rate of 1 ppm for each 1 ppm of iron in irrigation water. In most cases, 3 to 5 ppm is sufficient.

For bacteria and algae treatment:
- Inject liquid sodium hypochlorite continuously at a rate of 5 to 10 ppm where the biological load is high.
- Inject 10 to 20 ppm during the last 30 minutes of each irrigation cycle where the biological load is medium.
- Inject 50 ppm during the last 30 minutes of irrigation cycles two times each month when biological load is low.
- Superchlorinate (inject at a rate of 200 to 500 ppm) once per month for the length of time required to fill the entire system with this solution and shut down the system. After 24 hours, open the laterals and flush the lines.

The injection rates for stock solutions that contain 5.25%, 10% and 15% can be calculated from the following equations:

FOR 5.25% STOCK SOLUTION:

Injection rate of chlorine, gph = [(Desired available chlorination level, ppm) x (Irrigation flow rate, gpm)] divided by 875.

FOR A 10% STOCK SOLUTION:

Injection rate of chlorine, gph = [(Desired available chlorination level, ppm) x (Irrigation flow rate, gpm)] divided by 1,667.

FOR A 15% STOCK SOLUTION:

Injection rate of chlorine, gph = [(Desired available chlorination level, ppm) x (Irrigation flow rate, gpm)] divided by 2,500.

It is important to note that chlorine will cause water pH to rise. This is critical because chlorine is most effective in acidic water. If your water pH is above 7.5 before injection, it must be acidified for chlorine injection to be effective.

- Approved backflow control valves, low pressure drains, and interlocks must be used in the injection system to prevent contamination of the water source.
- Chlorine concentrations above 30 ppm may kill plants.

IMPORTANT NOTES:

Fertilization. Before considering a fertilization program for mulched-drip irrigated crops, be sure to have the soil pH checked. If a liming material is needed to increase the soil pH, the material should be applied and incorporated into the soil as far ahead of mulching as practical. For most vegetables, adjust the soil pH to around 6.5. When using drip irrigation in combination with mulch, apply the recommended amount of preplant fertilizer and incorporate it 5 to 6 inches into the soil before laying the mulch. If equipment is available, apply the preplant fertilizer to the soil area that will be covered by the mulch. This is more efficient than a broadcast application to the entire field.

The most efficient method of fertilizing an established mulched crop is through a drip irrigation system, which is installed during the mulching operation. Due to the very small holes or orifices in the drip tape, high quality liquid fertilizers, or water-soluble fertilizers must be used. Since phosphorous is a stable non-mobile soil nutrient and can cause clogging of the drip tape emitters, it is best to apply 100% of the crop's phosphorous needs pre-plant. Additionally, apply 20 to 40% of the crop's nitrogen and potassium needs pre-plant. The remainder of the crop's nutrient needs can be applied through the drip system with a high-quality liquid fertilizer such as 8–0–8, 7–0–7, or 10–0–10. Generally, it is not necessary to add micronutrients through the drip system. Micronutrients can be best and most economically applied pre-plant or as foliar application if needed.

The amount of nutrients to apply through the drip system depends upon the plant's growth stage. In general, smaller amounts of nutrients are needed early in the plant's growth with peak demand occurring during fruit maturation. The frequency of nutrient application is most influenced by the soil's nutrient holding capabilities. Clay soils with a high nutrient holding capacity could receive weekly nutrient applications through the drip system while a sandy soil with low nutrient holding capacity will respond best with a daily fertigation program. Fertigation rates are provided under crop specific recommendations later in this handbook.

PRECISION AGRICULTURE FOR VEGETABLE PRODUCTION

Introduction

Precision agriculture (PA), often referred to as "smart farming" or "precision farming," represents a transformative paradigm shift in the field of horticultural sciences, particularly when applied to vegetable cultivation. The fundamentals of PA are site-specific management of inputs such as water, fertilizer, pesticide, etc. Nowadays, PA further leverages the advances in remote sensing, robotics, and data science, optimizing crop production, enhancing both yield and resource efficiency while minimizing environmental impacts. The following is a list of terms, sensors, and parameters commonly used in precision agriculture for vegetables.

SENSORS	DESCRIPTION
RGB camera	Most widely used camera with lots of ready-to-use application suitable for vegetable production such as crop counting, general crop scouting and crop growth monitoring, measuring plant size (cover and height), plant color, etc. The structure-from-motion algorithm allows fairly accurate terrain information to be extracted from RGB images.
Multispectral camera	Measures canopy reflectance at four to five bands including green, red, red edge, and near infrared (NIR) bands. NIR are invisible to human eyes and the vegetation indices calculated from NIR and other visible bands are good indications of plant health. It can be used to assess chlorophyll content, nutrient levels, and pest infestations.
Thermal camera	Measures temperature from plant canopy and soil. It can be used in assessing crop water stress caused by the changes in plant transpiration which acts like a cooling process.
Drone	Drones equipped with cameras enable high-resolution imaging of small to medium-sized vegetable fields. They offer flexibility in data collection frequency and can capture detailed, up-to-date images for assessing crop health, growth, and pest infestations.
Satellite	Captures large-scale information about vegetation health, soil moisture, and weather patterns. Commonly used satellites for agriculture include MODIS (Moderate Resolution Imaging Spectroradiometer), Landsat, and Sentinel, providing valuable data for monitoring crop growth.

Vegetation indices are mathematical formulas or ratios calculated using data collected by multispectral and thermal cameras to provide valuable insights into the health and condition of vegetation. They help identify stress factors, nutrient deficiencies, pest infestations, and diseases early in the growing season. Some of the most common VIs are listed in the following table.

Farmers use NDVI data to monitor the health of their vegetable crops throughout the growing season. Although NDVI doesn't diagnose stress, it allows problematic areas to be alerted and human investigation is required to identify potential issues caused by nutrient deficiencies, pest infestations, or water stress. Monitoring NDVI allows for timely interventions, such as adjusting irrigation, optimizing fertilization, or targeting specific areas for pest control.

Normalized Difference Vegetation Index (NDVI)	NDVI is one of the most widely used VIs. It measures the amount of healthy green vegetation in an area by comparing the reflectance of near-infrared (NIR) and red light. NDVI values range from -1 to 1, with higher values indicating healthier vegetation.
Enhanced Vegetation Index (EVI)	EVI is an improvement upon NDVI that corrects for atmospheric and canopy background influences. It provides a more accurate measure of vegetation cover and health.
Soil Adjusted Vegetation Index (SAVI)	SAVI is designed to reduce the sensitivity of vegetation indexes to soil brightness. It is useful in areas with exposed soil, where the presence of soil can affect the accuracy of NDVI.
Normalized Difference Water Index (NDWI)	NDWI is used to detect the presence and condition of water bodies or water stress in vegetation. It compares the reflectance of near-infrared and shortwave infrared (SWIR) bands.

Leaf Area Index (LAI)	LAI is an index that quantifies the amount of leaf material in a canopy. It helps in understanding canopy structure and light interception.
Crop Water Stress Index (CWSI)	CWSI is derived from thermal infrared data and measures plant water stress. It is used to optimize irrigation practices.
Chlorophyll Index (CI)	CI assesses the chlorophyll content in vegetation, providing information about plant health and nutrient status.
Green Normalized Difference Vegetation Index (GNDVI)	Similar to NDVI but specifically tailored to emphasize the greenness of vegetation. It is often used in vineyard and forestry applications.

SEASON EXTENSION

Mulches

Plastic mulches

Plastic mulches have been used in vegetable production for many years to control weeds, conserve moisture, improve yield and quality, and acquire several other benefits. The most widely used mulches for vegetable production are black, white-on-black, clear, and metalized polyethylene mulches. Black mulch is most widely used, especially for spring applications where both ele- vated soil temperatures and weed control are desired. Clear plas- tic mulch is used when maximum heat accumulation is desired and weed control is not as critical, as weeds will grow well un- der clear plastic. White-on-black plastic (with white-side of the plastic facing up) is used for late spring and summer plantings where the benefits of moisture retention and weed control are valued and heat accumulation may be detrimental. Growers of- ten apply diluted white latex paint (1 part paint to 5 parts water) to black mulch when double cropping. Metalized mulch, com- monly referred to as reflective or silver mulch, is used to combat aphids and thrips that vector viral diseases. Metalized mulch should reflect a recognizable image (that is, be mirror-like) to be most effective.

Degradable mulches

Biodegradable plastic mulches provide similar benefits of standard polyethylene mulches. They are becoming more common, but usually cost more than conventional polyethylene. This additional expense may be offset by reduced labor and disposal costs at the end of the season, as these do not need to be removed from the field for disposal. Biodegradable plastic mulches are designed to degrade in the soil by microorganisms when soil moisture and temperatures are favorable for biological activity. Biodegradable film will usually be retained on the surface of the soil rather than be blown away from the application site. In addition, all of the biodegradable film will eventually decompose, including the tucked edges buried in the soil, into carbon dioxide, water, and microbial biomass. It is recommended that biodegradable mulch be tilled into the soil at the end of the harvest or growing season. Cover crops can be planted the next day after biodegradable plastic mulch has been rototilled into the soil.

Photodegradable and oxodegradable mulches are made with conventional plastic, with additives that make the mulch break into pieces when exposed to the UV radiation in sunlight, heat, and/or oxygen, respectively. These fragments will not biodegrade in the soil, but rather will accumulate in the soil and in bodies of water, where they can enter the food chain.

For more information on biodegradable mulches, visit *www.biodegradablemulch.org*.

Organic mulches

Organic mulches such as straw, pine straw, compost, and coarse hay provide weed control and moisture retention, while keeping soils cooler than bare ground. One benefit of using organic mulches is that they add organic matter back to the soil when incorporated after the growing season. Using hay often introduces weed seeds into a field. Regardless of the organic mulch chosen, a thick layer is required to adequately cover the soil and prevent weed seed from germinating and breaking through the mulch layer, which can make weed management even more difficult than if no mulch were used. When using organic mulches, supplemental nitrogen may be needed to compensate for the nitrogen that is lost to soil microbes in the process of breaking down the organic mulch (i.e., nitrogen immobilization).

Bed Formation and Drip Irrigation

Bed formation and moisture are critical to the success of growing vegetables with plastic or biodegradable mulches. Beds should be smooth, free of clods and sticks, and of uniform height. Black mulches warm the soil by conduction, so mulches must fit tightly over the soil. Tight contact between the mulch and soil will allow better transmission of heat from the sun. Raised beds allow the soil to drain and warm more quickly.

Drip tape is commonly laid under the plastic mulch in the same field operation. The soil should be moist when the mulch is applied since it is difficult to add enough water to thoroughly wet the width of the bed when using drip irrigation. Steep slopes may limit row length when using drip tape; normally row lengths should not exceed 300 to 600 feet depending on the specifications of drip tape.

Fumigants and Herbicides

Follow label directions for fumigants and herbicides used with plastic mulches. Fumigants and herbicides often have a waiting period before seeds or transplants can be planted. Transplanters and seeders are available to plant through plastic or biodegradable mulch. In fields with a history of nutsedge, appropriate measures must be taken in order to reduce or eliminate infestations as plastic mulches, whether conventional or biodegradable cannot control nutsedge. Nutsedge will compromise plastic mulch by piercing it, though it cannot penetrate paper mulch.

Fertilization and Fertigation

Prior to laying plastic mulch, soil pH should be adjusted with lime as recommended by a soil test. Soil pH cannot practically be changed once plastic mulch is established within a field. Vegetables produced on plastic mulch, but without the ability to supply nutrients through the drip irrigation system, should have all of their required fertilizer incorporated into the beds prior to applying the mulch. Broadcasting the fertilizer before bedding has been shown to be an effective method of application since the bedding process moves most of the fertilizer into the bed.

Growers not able to fertilize with their irrigation system need to be careful not to overwater beds, as excessive irrigation can push soluble nutrients (i.e. nitrate, sulfate) below the plants' effective root zone. Growers using fertigation should follow the recommendations for each specific crop. Fertigation schedules are listed for cantaloupe, cucumber, eggplant, okra, pepper, summer squash, tomato, and watermelon later in this handbook. Also, refer to the previous section for further information on fertilization.

Double cropping

Growers frequently grow two crops on black plastic mulch to get the most out of their investment. The spring crop is killed and removed, and then a second crop is planted through the mulch. The new crop should be planted into new holes and fertilizer added based on soil test results and the new crop's nutrient requirements.

Plastic Mulch Removal and Disposal

Commercial mulch lifters are available. Plastic can be removed by hand by running a coulter down the center of the row and picking the mulch up from each side. There are many types of plastic mulch removal equipment that can be used to reduce the cost of removal. Following the list of best management practices below for plastic mulch retrieval is imperative if the plastic mulch will be recycled. Make sure a viable recycling market exists for mulch film in your region. If there is a viable market, the following BMP's will help you prepare your material in a way that is desirable to the recycler. Even if there is not a viable recycling market, the best management practices, if implemented, have shown as much as a $125 per acre savings in some areas. Transportation logistics and costs should be considered when deciding the process to be used. Sanitary landfills may accept plastic mulch in some areas. There are a few recycling projects which accept plastic mulch. Some states allow burning of mulch with a permit.

Plastic Mulch Retrieval Best Management Practices

- **Remove all top vegetation:** Mowing down the existing crop to eliminate debris on the plastic. Reel mowers seems to remove the debris better when compared to rotary types. Plant debris is difficult to remove in the recycling process and tends to remain with the plastic causing a "forest fire smell" and moisture retention, both of which are undesirable in the recycling process.
- **Remove as much soil as possible:** Soil adds weight making transport to the landfill or recycling plant more expensive. Proper soil conditions for the timing of removal is as dry a day as possible with a few days of soil drying underneath the plastic mulch before the retrieval process.
- **Keep it dry:** Pull film on dry days and keep it under cover. Water adds weight to the shipment and is the nemesis of the shredding and prime cleaning operation. Make sure when bundling that there are no pockets of water in the folds of the plastic.
- **Roll it tight:** The bundles should be tight, solid, and no longer than 24" long. The tight bundles will allow for more plastic per load and the bundles will fit better on the conveyors inbound to the shredder. Loading the conveyor is likely to be a manual activity so 25 to 50 pounds should be the target weight of each bundle.
- **Consider thicker films:** Increasing the film thickness aids retrieval plus we can get higher plastic recovery rates in the recycling process. Use 1.5 mil or thicker plastic. Using the appropriate width of plastic on the bed so that no more than 6" of plastic is buried on both sides of the bed reduces the soil and debris.
- **Proper retrieval equipment:** Adjustments to removal equipment might be required to optimize debris removal as well as tightness of bundles.

Row covers

Row covers are used to hasten the maturity of the crop, exclude certain insect pests, and provide a small degree of frost protection. There are two main types of row covers: vented clear or translucent polyethylene and floating row covers. Polyethylene types are supported by wire hoops placed at regular (5 to 6 ft) intervals. In addition, plastic can be placed loosely over the plants with or without wire supports. Floating row covers are porous, lightweight spunbonded materials placed loosely over the plants.

Floating covers are more applicable to the low-growing vine crops. Upright plants like tomatoes and peppers have been injured by abrasion when the floating row cover rubs against the plants or excess temperatures build-up. Erratic spring temperatures require intensive management of row covers to avoid blossom shed and other high-temperature injuries. In particular, clear plastic can greatly increase air temperatures under the cover on warm sunny days, resulting in a danger of heat injury to crop plants. Therefore, vented materials are recommended. Even with vents, clear plastic has produced heat injury, especially when the plants have filled a large portion of the air space in the tunnel. This has not been observed with the translucent materials. Usually, row covers are combined with plastic mulch.

High Tunnels

High tunnels are unheated, polyethylene-covered structures that provide a larger degree of frost protection than row covers. A properly built high tunnel with one or two layers of plastic, should afford 5-8°F of frost protection. As with row covers, high tunnels require intensive management to ensure that they are vented properly when warm spring temperatures can cause excessive heat to build up in tunnels, resulting in damage to the crop. Tomatoes are commonly produced in high tunnels as well as a variety of leafy greens crops, due to the premium prices obtained. Row covers are often combined with the use of high tunnels and plastic mulch.

High tunnels can provide 3-4 weeks or more of season extension for spring and fall crops such as tomatoes as well as year-round production of Cole crops and lettuce. High tunnels can reduce the incidence of certain diseases and insects due to protection from rain and changes in light interception, respectively, inside the tunnel; however, traditional greenhouse pests, such as leaf mold, aphids, spider mites, and whiteflies may be more prevalent in high tunnels. In many states, high tunnels are considered a greenhouse structure for the application of pesticides, which may reduce the number of chemicals available compared to field production. Be sure to determine if the pesticides you are applying are acceptable for use in high tunnels in your state.

Extensive information regarding construction, specifics of crop production, soil management, and economics of production for many fruits, vegetable, and cut flowers grown in high tunnels can be found at: *https://www.sare.org/resources/high-tunnels-and-other-season-extension-techniques/*.

Considerations for Using Mulch, Drip Irrigation, and Row Covers

Each grower considering mulches, drip irrigation, and/or row covers must weigh the economics involved. The long-term versus short-term opportunities must be considered.

- Does the potential increase in return justify the additional costs?
- Are the odds in favor of the grower getting the most benefit in terms of earliness and yield from the mulch, drip irrigation, and/or row covers?
- Does the market usually offer price incentives for early produce? Will harvesting early allow competition against produce from other regions?

Depending on the row spacing and bed width, polyethylene mulch can cost $350 to $700 per acre, including installation and removal. Biodegradable mulches can cost twice as much as polyethylene mulches. Additionally when using these mulches, drip irrigation must also be installed ($400 to $1,200/acre). Growers must determine costs for their situation and calculate their potential returns. For more help in determining exact costs for your operation, use the Mulch Calculator:

https://biodegradablemulch.tennessee.edu/wp-content/uploads/sites/214/2020/12/Chen-Mulch-calculcator-introduction.pdf

POLLINATION

Honey bees or other pollinating insects are essential for commercial production of cucurbit vine crops and may also improve the yield and quality of fruit in beans, eggplants, peas, and peppers. In other fruiting vegetables, such as okra and some legumes, pollination does not require insect visits. Lack of adequate pollination usually results in small or misshapen fruit in addition to low yields. The size and shape of the mature fruit is related to the number of seeds produced and each seed requires one or more pollen grains for normal development. Cucumbers, squash, pumpkins, and watermelons have separate male and female flowers, while cantaloupes and other specialty melons have male and hermaphroditic (perfect or bisexual) flowers. Adequate amount of the sticky pollen of the male flowers must be transferred to the female flowers to achieve fruit set. Cucurbit flowers are usually open and attractive to bees for no more than one day.

Flower opening, release of pollen, and commencement of nectar secretion normally precede bee activity. Pumpkin, squash, cantaloupe, and watermelon flowers normally open around daybreak and close by or before noon; whereas, cucumbers and melons generally remain open the entire day. Pollination must take place on the day the flowers open because pollen viability, stigmatic receptivity, and attractiveness to bees lasts only that day.

European honey bees and various native wild bees may visit the flowers of flowering vegetables. However, successful pollination requires that these insects visit male and female flowers frequently for pollen transfer to occur. Some insects present in flowers may not contribute greatly to pollination. Honey bees and bumblebees move frequently from one flower to another and placing hives of these bees into a crop at the correct time will greatly enhance pollination. Even though bumblebees and other species of wild bees are excellent pollinators, populations of these native pollinators usually are not adequate for large acreages grown for commercial production. Colonies of wild honey bees have been decimated by Tracheal and Varroa mites, and possibly other environmental factors, and therefore cannot be counted on to aid in pollination. One of the best ways to ensure adequate pollination is to keep or rent strong colonies of honey bee from a reliable beekeeper. Another option for pollination is the bumblebee. Bumblebees are becoming a popular grower's choice over the past decade and are being found to be effective as a pollinator alternative to honey bees. Commercial bee attractants are available but have not proven to be effective in enhancing pollination. Growers are advised to increase numbers of honey bee or bumblebee colonies and not to rely on such attractants. Suppliers of both honey bee and bumblebee colonies need ample notice prior to when the bees are needed for a given crop. Approximately 3 to 4 months advance notice is usually sufficient to meet crop pollinator needs. A written contract between the grower and beekeeper/ supplier can prevent misunderstandings and, thus, ensure better pollinator service. Pollinator contracts should specify the number and strength of colonies, the rental/purchase fee, time of delivery, and distribution of bees in the field.

Honey bee activity is determined to a great extent by weather and conditions within the hive. Bees rarely fly when the temperature is below 55°F. Flights seldom intensify until the temperature reaches 70°F. Wind speed beyond 15 miles per hour seriously slows bee activity. Cool, cloudy weather and threatening storms greatly reduce bee flights. In poor weather, bees foraging at more distant locations will remain in the hive, and only those that have been foraging nearby will be active. Ideally, colonies should be protected from wind and be exposed to the sun from early morning until evening. Colony entrances facing east or southeast encourage bee flight. The hives should be off the ground and the front entrances kept free of grass and weeds. For best results, hives should be grouped together. A clean water supply should be available within a quarter mile of the hive.

The number of colonies needed for adequate pollination varies with location, attractiveness of crop, density of flowers, length of blooming period, colony strength, and competing blossoms of other plants in the area. In vine crops, recommendations are one to two colonies per acre, with the higher number for higher density plantings. Each honey bee hive or colony should contain at least 40,000 - 50,000 bees. Eight or more bee visits per female flower are required to produce marketable fruit. Bumblebees have some advantages compared to honey bees in that the former fly in cool, rainy, and windy weather and often visit flowers earlier in the morning than honey bees. Bumblebees are also active later in the day when temperatures cool and they have a larger body size than honey bees, thus requiring fewer visits to achieve good pollination and fruit set. Bumblebee hives are sold as a quad or four hives per quad. A quad is the minimum order that can be purchased from a supplier. Generally one bumblebee hive contains 200 to 250 bees and is equivalent to one honey bee hive; however, research that can specifically document this is lacking. Thus, one quad of bumblebees would provide good pollination for four acres of a cucurbit crop if the recommendation is to use 1 bumblebee hive per acre. In some instances, two hives per acre or more may be recommended (i.e. triploid watermelon). In this case, one quad would provide good pollination for two acres. Bumblebee hives should not be placed in direct sunlight so that the bees work more efficiently. No more than two bumblebee quads should be placed in one location so that pollination is more uniform in the field. The quad locations should be at least 650 to 700 feet from each other.

Insecticides applied while a crop is in bloom pose a serious hazard to bees visiting flowers. If insecticides must be applied, select a product that will give effective control of the target pest but pose the least danger to bees. Apply these chemicals near evening when the bees are not actively foraging and avoid spraying adjacent crops while bees are active. If insecticide spraying is necessary, give the beekeeper 48 hours of notice of when you expect to spray so that precautions can be taken. Avoid leaving puddles of water around chemical mixing areas as bees may pick up and be harmed by insecticide contaminated water. As with honey bees, one must carefully plan when to spray insecticides so that the bumblebees are not injured or killed. Because bumblebees are most active from dawn until late morning, and again from about 4 PM until sunset, the hives need to be closed around 11 AM so that the bumblebees remain in the hive and are protected during a late evening spray application.

PESTICIDES AND THE ENDANGERED SPECIES ACT: WHAT YOU NEED TO KNOW

The following description has been endorsed by the Weed Science Society of America, Entomological Society of America, and American Phytopathological Society.

1. **What is the Endangered Species Act (ESA)?**
   The Endangered Species Act is a long-standing federal law, first passed in 1973, which requires government agencies to ensure any actions they take do not jeopardize a species that has been federally listed as endangered or threatened. When an agency has a proposed action that might affect a listed species or its habitat, they consult with one or both of the agencies that helps enforce the ESA, the U.S. Fish and Wildlife Services or the National Marine Fisheries Service (this is known as "*a consultation*" with "*the Services*"). The Services then may recommend changes to the project or action to protect listed species or habitats.

2. **How does the ESA affect pesticide use?**
   The Environmental Protection Agency (EPA) Office of Pesticide Programs (OPP) is the federal agency that regulates pesticide use. Because the use of pesticides can affect animals and plants (or their habitat), pesticide registrations are considered "actions" that would trigger an endangered species consultation.

3. **Why am I hearing about the ESA and pesticide use now?**
   Due to the complex nature of the process, the EPA has not fully completed the required endangered species consultations with the Services for pesticide registrations in the past, which has left many of those pesticides vulnerable to lawsuits. Courts have annulled pesticide registrations which has led to their removal from market. To make pesticide registrations more secure from litigation, ultimately all pesticide registrations will comply with the Endangered Species Act (*https://www.epa.gov/endangered-species*).

4. **How will this affect the pesticide I use today?**
   Many pesticide labels will likely have changes that could include:
   - Requirements to check the EPA's Bulletins Live! Two website and follow current ESA restrictions for the pesticide product in the bulletin (*https://www.epa.gov/endangered-species/bulletins-live-two-view-bulletins*)
   - Measures to reduce spray drift
   - Measures to reduce runoff/erosion
   - Other measures to reduce pesticide exposure to listed species and their habitat

   In short, farmers and applicators should expect to see some new application requirements on their pesticide labels. But there is no need to panic. To date, no pesticide has ever been fully removed from the market based solely on endangered species risks, and that remains an unlikely scenario in the future.

5. **Why does complying with the ESA matter?**
   By starting to fully comply with the ESA, **EPA anticipates that this will give farmers and applicators more stable, reliable access to the pesticides they need.** Furthermore, the ESA has been successful at bringing back some species Americans care about – such as the bald eagle or the Eggert sunflower – and restoring them to healthy populations, which has benefited the natural and cultivated ecosystems that agriculture (and society) rely on.

HOW TO IMPROVE PEST CONTROL

Failure to control an insect, mite, disease, or weed is often erroneously blamed on the pesticide when the cause frequently lies elsewhere. Several common reasons for failure to achieve control are the following:

1. Delaying applications until pest populations become too large or damage is too advanced.
2. Poor coverage caused by insufficient volume, inadequate pressure, or clogged or poorly arranged nozzles.
3. Selecting the wrong pesticide for the target pests.
4. Contaminated or high pH water source.

Suggested Steps For More Effective Pest Control:

1. *Scout fields regularly*. Know the pest situation and any build-ups in your fields. Frequent examinations (at least once or twice per week) help determine the proper timing of the next pesticide application. Do not apply a pesticide simply because a neighbor chooses to do so.
2. *Integrated Pest Management (IPM)*. Use an ongoing program of biological, physical, cultural, and chemical methods in an integrated approach to managing pests. IPM involves scouts visiting fields to collect pest population data. Use this updated information to decide whether insecticide applications or other management actions are needed to avoid economic loss from pest damage. Control decisions are based on factors such as:
   - The economic action threshold level (when the cost of control equals or exceeds potential crop losses attributed to real or potential damage)
   - Other factors are listed in the Recommended Control Guidelines section that follows

To employ an IPM program successfully, basic practices need to be followed. Whether participating in a university or grower-supported IPM program, hiring a private consultant, or doing the work personally, the grower still must practice:

- Frequent and regular examination of fields to assess pest populations
- Applying a control measure only when the economic threshold level has been reached
- Where possible, employing a cultural practice or a biological control or using a pesticide that is less harmful to natural enemies of the target pest

Resistance Management. The way pesticides are used affects the development of resistance. Resistance develops because intensive pesticide use kills the susceptible individuals in a population, leaving only resistant ones to breed. Usually, insecticides and miticides belonging to the same chemical group have a common target site in pests and therefore have a common Mode of Action (MoA). Resistance often develops based on a genetic modification of this target site. Then, the compound usually loses its pesticidal activity. Because all compounds within the chemical grouping share a common MoA, there is a high risk that this resistance will automatically confer cross-resistance to all the compounds in that group. When choosing insecticides and miticides for a resistance management strategy, the MoA classification can serve as a helpful guide. It was developed and endorsed by the Insecticide Resistance Action Committee (IRAC) to ensure growers can effectively alternate insecticides with different MoAs. More information can be found at: *https:// irac-online. org/mode-of-action/classification-online/*.

The Fungicide Resistance Action Committee (FRAC) developed and endorsed an equivalent mode of action classification for fungicidal compounds which can be found at: *https:// www.frac.info/*.

Adopting the following practices will also reduce the development of pest resistance:

1. Rotate crops to a non-host crop, thus reducing the need for pesticide treatment and, thereby, reducing the ratio of resistant to susceptible individuals in the breeding population.
2. Use control guidelines as an important tactic for reducing the pesticide resistance problem. For more information concerning control guidelines, refer to the following section and the crop-specific sections of this handbook.
3. Spot treat when possible. Early-season insects and mites are often concentrated in areas near their overwintering sites. Diseases often can be first detected in favorable microclimates, such as low or wet areas of the field. Perennial weeds and newly introduced or herbicide-resistant annual weeds often occur first in small numbers in a part of a field. Treating these areas, rather than the entire field, can prevent the development of resistant populations.
4. Control pests early because seedling weeds and immature insects are more susceptible to pesticides and less likely to develop resistance than older and more mature crop pests.
5. Do not overspray. Attempts to destroy every pest in the field by multiple applications or by using higher than labeled rates often eliminate the susceptible pests but not the resistant pests.
6. Rotate pesticides to reduce the development of resistance, particularly with pesticides that differ in their modes of action (MoA). Rotation among different chemical groups reduces resistance problems.
7. Use appropriate additives when recommended on the pesticide's label. For example, adding a crop oil concentrate or a surfactant to certain postemergence herbicides will increase the effectiveness of the herbicides.

Control Pests According to Recommended Control Guidelines or Schedule. Control guidelines provide a way to decide whether pesticide applications or other management actions are needed to avoid economic loss from pest damage. Guidelines for pests are generally expressed as a count of a given insect stage or as a crop damage level based on certain sampling techniques. They are intended to reflect the pest population that will cause economic damage and thus guarantee the cost of treatment. Guidelines are usually based on pest populations, field history, crop development stage, variety, weather conditions, life stage of the pest, parasite, and/or predator populations, resistance to chemicals, time of year, and other factors. Specific thresholds are given in this handbook for a number of pests of many crops.

Insect and mite population sampling techniques include:

- *Visual observation*. Direct counts of any insect and mite stages (eggs, larvae, nymphs, adults, etc.) are accomplished by examining plants or plant parts (leaves, stems, flowers, fruits). Counts can be taken on single plants or a prescribed length of row, which will vary with the crop. Usually, quick-moving insects are counted first, followed by those that are less mobile. This is the most common method to sample mite pests.

- *Shake cloth* (also known as a ground cloth). This sampling procedure uses a standard 3-foot by 3-foot shake cloth to assess insect populations. Randomly choose a site without disturbing the plants and carefully unroll the shake cloth between two rows. Bend the plants over the cloth one row at a time and shake the plants vigorously to dislodge insects on stems, leaves, and branches. Count only insects that have landed on the shake cloth. The number of sampling sites per field will vary with the crop.
- *Sweep net*. This sampling procedure uses a standard 15-inch diameter sweep net to assess insect populations. While walking along one row, swing the net from side to side with a pendulum-like motion to face the direction of movement. The net should be rotated 180 degrees after each sweep and swung through the foliage. Each pass of the net is counted as one sweep. The number of sweeps per field will vary with the crop.

Weed population sampling techniques include:

- *Weed identification*. This first step is frequently skipped. Perennial weeds and certain serious annual weeds should be controlled before they can spread. Common annual weeds need only be controlled if they represent a threat to yield, quality, or harvestability. It is critical to know the weed history of a field prior to planting as many herbicides need to be applied pre-plant. Your county Extension office is an excellent resource for reliable weed identification.
- *Growth stage determination*. The ability of weeds to compete with the crop is related to size of the weed and size of the crop. Control using herbicides or mechanical methods is also dependent on weed size. A decision to control or not to control a weed must be carried out before the crop is affected and before the weed is too large to be controlled.
- *Weed population*. Weeds interfere with crop production in many ways. They may compete for light, water, nutrients, and space. The extent of this interference is dependent on population and is usually expressed as *weeds per foot of row or weeds per square meter*. Control measures are needed when the weed population exceeds the maximum tolerable population of that species.

Disease monitoring involves determining the growth stage of the crop, observing disease symptoms on plants, and/or the daily weather conditions in the field.

Disease control is often obtained by applying crop protectants on a regular schedule. For many diseases, application must begin at a certain growth stage and must be repeated every 7 to 10 days. When environmental conditions are favorable for disease development, delaying a spray program will result in a lack of control if the disease has progressed too far. For certain diseases that do not spread rapidly, fields should be scouted regularly.

Predictive systems are available for a few diseases. Temperature, rainfall, relative humidity, and duration of leaf wetness period are monitored, and the timing of fungicide application is determined by predicting when disease development is most likely to occur. One such program for Downy Mildew is available at: *http://cdm.ipmpipe.org/*

Weather Conditions. These are important to consider before applying a pesticide. Spray only when wind velocity is less than 10 miles per hour. Do not spray when sensitive plants are wilted during the heat of the day. *If possible, make applications when ideal weather conditions prevail.*

Certain pesticides, including biological insecticides (e.g., BT's) and some herbicides, are ineffective in cool weather. Others do not perform well or may cause crop injury when hot or humid conditions are prevalent. Optimum results can frequently be achieved when the air temperature is in the 70°F range during application.

Strive for Adequate Coverage of Plants. Improved control of aphids can be achieved by adding and arranging nozzles so that the application is directed toward the plants from the sides as well as from the tops (also see *Alkaline Water and Pesticides*, which follows). In some cases, nozzles should be arranged so that the application is directed beneath the leaves. As the season progresses, plant size increases, as does the need for increased spray gallonage to ensure adequate coverage.

Applying insecticide and fungicide sprays with sufficient spray volume and pressure is critical. Spray volumes should increase as the crop's surface area increases. Sprays from high-volume-high-pressure rigs (airblast) should be applied at rates of 40 to 200 gallons per acre at 200 psi or greater. Sprays from low-volume-low-pressure rigs (boom type) should be applied at rates of 50 to 100 gallons per acre at 20 psi. The addition of a spreader-sticker improves coverage and control when wettable powders are applied to smooth-leaved plants, such as Cole crops and onions.

Use a dedicated sprayer for herbicides and a different sprayer for fungicides and insecticides. Herbicide sprays should be applied at between 15 and 50 gallons of spray solution per acre using low pressure (20 to 40 psi). Never apply herbicides with a high-pressure sprayer that was designed for insecticide or fungicide application because excessive drift can result in damage to non-target plants in adjacent fields and areas. Do not add oil concentrates, surfactants, spreader-stickers, or any other additive unless specified on the label, or crop injury is likely.

Select the Proper Pesticide. Know the pests to be controlled and choose the recommended pesticide and rate of application. If in doubt, consult your local Extension office. The herbicide choice should be based on weed species or cropping systems.

For insects or mites that are extremely difficult to control or resistant, it is essential to alternate labeled insecticides/miticides with different classes or modes of action (MOA). Be alert for a possible aphid or mite buildup following the application of certain insecticides, such as broad-spectrum chemicals (e.g., pyrethroids).

Caution: Proper application of soil systemic insecticides is extremely important. The insecticide should be applied according to the label instructions (which, in general, indicate application should be directed away from the seed) or crop injury may occur.

Be sure to properly identify the disease(s). Many fungicides control specific diseases but provide no control of others. For this reason, on several crops, fungicide combinations are recommended.

Pesticide Compatibility. To determine if two pesticides are compatible, use the following "jar test" before you tank-mix pesticides or tank-mix pesticides with fertilizers:

1. Add 1 pint of water or fertilizer solution to a clean quart jar, then add the pesticides to the water or fertilizer solution in the same proportion used in the field.

2. To a second clean quart jar, add 1 pint of water or fertilizer solution. Then add ½ teaspoon of an adjuvant to keep the mixture emulsified. Finally, add the pesticides to the water-adjuvant or fertilizer adjuvant in the same proportion as used in the field.

3. Close both jars tightly and mix thoroughly by inverting 10 times. Inspect the mixtures immediately and after standing for 30 minutes. If a uniform mix cannot be made, the mixture should not be used. If the mix in either jar remains uniform for 30 minutes, the combination can be used. If the mixture with adjuvant stays mixed and the mixture without adjuvant does not, use the adjuvant in the spray tank. If either mixture separates but readily remixes, constant agitation is required. If non-dispersible oil, sludge, or clumps of solids form, do not use the mixture.

Note: For compatibility testing, the pesticide can be added directly or premixed in water first. In actual tank-mixing for field application, unless label directions specify otherwise, add pesticides to the water in the tank in this order: first, wettable granules or powders, then flowables, emulsifiable concentrates, water solubles, and companion surfactants. If tank-mixed adjuvants are used, these should be added first to the fluid carrier in the tank. Thoroughly mix each product before adding the next product.

Select Correct Sprayer Tips. The choice of a sprayer tip for use with many pesticides is important. Flat fan-spray tips are designed for preemergence and postemergence application of herbicides. These nozzles produce a tapered-edge spray pattern that overlaps for uniform coverage when properly mounted on a boom. Standard flat fan-spray tips are designed to operate at low pressures (20-40 psi) to produce small-to-medium-sized droplets that do not have excessive drift. Flat fan-nozzle tips are available in brass, plastic, ceramic, stainless steel, and hardened stainless steel. Brass nozzles are inexpensive and are satisfactory for spraying liquid pesticide formulations. Brass nozzles are the least durable, and hardened stainless steel nozzles are most durable and are recommended for wettable powder formulations, which are more abrasive than liquid formulations. When using any wettable powder, it is essential to calibrate the sprayer frequently because, as a nozzle wears, the volume of spray material delivered through the nozzle increases.

Flood-type nozzle tips are generally used for complete fertilizers, liquid N, etc., and sometimes for spraying herbicides onto the soil surface prior to incorporation. They are less suitable for spraying postemergence herbicides or for applying fungicides or insecticides to plant foliage. Coverage of the target is often less uniform and complete when flood-type nozzles are used, compared with the coverage obtained with other types of nozzles. Results with postemergence herbicides applied with flood-type nozzles may be satisfactory if certain steps are taken to improve target coverage. Space flood-type nozzles a maximum of 20 inches apart, rather than the suggested 40-inch spacing. This will result in an overlapping spray pattern. Spray at the maximum pressure recommended for the nozzle. These techniques will improve target coverage with flood-type nozzles and result in more satisfactory weed control.

Full and hollow-cone nozzles deliver circular spray patterns and are used for application of insecticides and fungicides to crops where thorough coverage of the leaf surfaces is extremely important and where spray drift will not cause a problem. They are used when higher water volumes and spray pressures are recommended. With cone nozzles, the disk size, and the number of holes in the whirl plate affect the output rate. Various combinations of disks and whirl plates can be used to achieve the desired spray coverage.

Alkaline Water and Pesticides. At times, applicators have commented that a particular pesticide has given unsatisfactory results. Usually, these results can be attributed to poor application, a bad batch of chemicals, pest resistance, weather conditions, etc. However, another possible reason for unsatisfactory results from a pesticide may be the pH of the mixing water.

Some materials carry a label cautioning the user against mixing the pesticide with alkaline materials. The reason for this caution is that some materials (in particular the organophosphate insecticides) undergo a chemical reaction known as "alkaline hydrolysis." This reaction occurs when the pesticide is mixed with alkaline water, that is, water with a pH greater than 7. The more alkaline the water, the greater the breakdown (i.e., "hydrolysis").

In addition to lime sulfur, several other materials provide alkaline conditions: caustic soda, caustic potash, soda ash, magnesia or dolomitic limestone, and liquid ammonia. Water sources in agricultural areas can vary in pH from less than 3 to greater than 10.

To check the pH of your water, purchase a pH meter, or in most states, you can submit a water sample to your state's soil or water testing lab. If you have a problem with alkaline pH, there are several products available that are called nutrient buffers that will lower the pH of your water.

There are some instances when materials should not be acidified, namely, sprays containing fixed copper fungicides, including: Bordeaux mixture, copper oxide, basic copper sulfate, copper hydroxide, etc.

BENEFICIAL INSECTS

A number of environmental factors, such as weather, food availability, and natural enemies combine to keep insect populations under control naturally. In some human-altered landscapes, such as in agricultural crop fields, the levels of natural control are often not acceptable to us, and we have to intervene to lower pest populations. While some environmental factors, such as weather, cannot be altered to enhance control of pests, others, such as populations of natural enemies, can be affected. The practice of taking advantage of and manipulating natural enemies to suppress pest populations is called *biological control*.

TABLE 13. PREDATORS AND PARASITES OF VEGETABLE PESTS

Predators and Parasites	Approx. # North American Species	Pest(s) Controlled or Impacted
WASPS		
Aphelinid Wasps *	1,000	Aphids on some greenhouse and vegetable crops; mummies of parasitized aphids will turn black. Eretmocerus californicus has shown promise against sweetpotato whitefly.
Aphidiid Wasps *	114	Aphid parasites; mummies turn tan or golden.
Braconid Wasps *	1,000	Caterpillars on cole crops and potatoes, leafminers in greenhouse crops.
Chalcid Wasps *	1,500	Internal and external parasite of fly and moth larvae and pupae. A few species attack beetles.
Cotesia Wasps (Braconid Family) *	200	Parasites of caterpillars including armyworms, Alfalfa caterpillar, loopers, cabbageworms.
Encarsia formosa *	1	A commercially available Aphelinid wasp, whiteflies on greenhouse/some field crops. *Encarsia* can provide effective control of greenhouse whitefly, but not sweetpotato whitefly. Therefore, it is critical to identify which whitefly is present before releasing *Encarsia*.
Encyrtid Wasps	5 00	Aphids on some greenhouse crops, Cabbage looper, root maggot.
Eulophid Wasps	3,400	Colorado Potato Beetle, Mexican Bean beetle, Asparagus beetle, leafminers in field crops.
Ichneumonid Wasps *	3,100	Caterpillars, eggs, and beetle larvae in cole crops, corn, Asparagus whiteflies on cole crops & tomato.
Mymarid Egg Wasps	1,300	Lygus bug, Tarnished Plant bug, carrot weevil. Egg parasite of beetles, flies, grasshoppers, leafhoppers, and many true bugs (Stink bugs, lygaeids). Adults are 1/25 inch in size.
Pteromalid Wasps	3,000	Parasites of beetles, flies, and other wasps, cabbage worm, diamondback moth.
Scelionid Egg Wasps	300	Parasite of true bug and moth eggs.
Tiphiid Wasps	225	Parasites of Japanese beetles and Tiger beetles.
Trichogramma Wasps *	650	Moth eggs on cole crops, peppers, sweet corn, and tomatoes. Because only eggs are parasitized, releases must be timed to coincide with egg laying (use pheromone traps to determine timing).
Thripobius semiluteus (Eulophidae Family)*	t1	Controls thrips via parasitism.
Vespid Wasps (hornets, yellowjackets, etc.)	200	Caterpillars, flies, true bugs, beetles, and other wasps are fed to Vespid larvae.
FLIES		
Aphid Flies (Chamaemyiid Flies)	36	Feed on aphids, mealybugs, and soft scales.
Bombyliid Flies (Bee Flies)	250	Internal and external parasites of various caterpillars and wasp larvae, beetle larvae, some eggs.
Nemestrinid Flies (Tangle Veined Flies)	250	Internal parasites of locusts and beetle larvae and pupae.
Phorid Flies (Humpbacked Flies)	350	Internal parasites of ants, bees, caterpillars, moth pupae, and fly larvae.
Pipunculid Flies (Big-Headed Flies)	100	Internal parasites of leafhoppers and planthoppers.
Predatory Midges (Cecidomyiid Flies)*	10	Aphids and mites on some greenhouse crops.
Pyrgotidae (Pyrgotid Flies)	5	Internal parasites of June beetles and related Scarab beetles; nocturnal and rarely seen.
Syrphid Flies	1,000	Most larvae are predaceous upon aphids, whitefly pupae, and soft-bodied small insects.
Tachinid Flies	1,300	Internal parasite of beetle, butterfly, and moth larvae, earwigs, grasshoppers, and true bugs. Tachinids lay eggs directly on host or on a leaf that the host insect then eats. Some species parasitize Japanese
TRUE BUGS		
Assassin Bugs (Reduviidae)	160	Generalist predators against small and soft-bodied insects, eggs, and pupae.
Big-Eyed Bug, Geocoris spp.	25	Generalist predators feeding on a wide variety of insect eggs and small larvae. Both immature and adults are predaceous and feed on over 60 species of other insects.
Damsel Bugs (Nabidae)	34	Mites, aphids, caterpillars, leafhoppers, and other insects, especially soft-bodied insects.
Minute Pirate Bug, Orius insidiosis (aka Flower Bug)*	1	Thrips, spider mites, aphids, small caterpillars, small insects in sweet corn, potato, and some greenhouse crops.
Predatory Stink Bug (Perillus, Podisus spp.)	14	Look similar to plant feeding Stink bugs, but feed on caterpillar pests, small insects and insect eggs, and Colorado Potato beetle (larvae). Effective in solanaceous crops, beans, cole crops, and asparagus.
Spined Soldier Bug, *Podisus maculiventris**	1	Generalist predator on many vegetables (i.e., potato, tomato, sweet corn, cole crops, beans, eggplant, cucurbits, asparagus, onions). Attacks larvae of European Corn borer, Diamondback moth, Corn Earworm, Beet Armyworm, Fall Armyworm, Colorado Potato beetle, Cabbage Looper, Imported Cabbageworm, and Mexican Bean beetle. A pheromone to attract Spined Soldier Bug is also available.
BEETLES		
Ground Beetles (Carabid Beetles)	2,200	Both larvae and adults are predaceous, nocturnal. Feed on mites and snails, soil dwelling beetle and fly eggs and pupae, some caterpillars, and other soft bodied insects. Most beneficial in cole crops, root crops, and onions.
Lady Beetles (Coccenelidae)*	400	Aphids, mites, whitefly, small insects, and insect eggs in most vegetable crops (especially potatoes, tomatoes, sweet corn, and cole crops. Release purchased lady beetles in evening, in vicinity of pest, and cover with a light sheet or cloth overnight for best predator retention.

* Insects marked with an asterisk represent species available commercially for purchase
For a list of Biological Control (Beneficial Insects) Suppliers, see http://wiki.bugwood.org/Commercially_available_biological_controls

TABLE 13. PREDATORS AND PARASITES OF VEGETABLE PESTS (cont'd)

Predators and Parasites	Approx. # North American Species	Pest(s) Controlled or Impacted
Rove Beetles (Staphylinidae)	3,100	Distinguished by short outer wings and exposed abdomens, Rove beetles feed on a variety of eggs, pupae, larvae, and soft bodied insects (aphids, mites, whitefly).
Soft-Winged Flower Beetles (Melyridae)	450	Adults and larvae feed on aphids, leafhoppers, and other immature insects. Covered in fine hairs that give the insect a velvety appearance.
Soldier Beetles (Cantharids, aka Leather-Winged Beetles)	100	All larvae, and some adults, are predaceous. Other adults feed on nectar and pollen, so can be attracted by flower plantings. Predators of eggs and larvae of beetles, butterflies, moths, aphids, others. Most effective in cole crops, cucurbits, and sweet corn.
Tiger Beetles (Cicindelid Beetles)	40	Adults and larvae prey on a wide variety of insects.
OTHER BENEFICIAL ORGANISMS		
Praying Mantis *		Flies, crickets, bees, moths. All life stages are predatory. Commercially available mantis are usually Tenodera aridifolia, a Chinese species. Praying mantids ambush any prey small enough (usually mobile insects) to pass before them. They do not specifically attack targeted pests.
Lacewings*	27	Aphids, thrips, small caterpillars, leafhoppers, mealybugs, psyllids, whiteflies, and insect eggs. Release purchased lacewings as soon as target pest is noticed in field to achieve good results.
Parasitic nematodes*		Cutworms, beetle larvae, root maggots, some wireworms.
Predatory mites (Phytoseiidae & 3 other families)*	6	Releases most beneficial in strawberries and greenhouse vegetables; avoid carbamates and organophosphates to encourage natural populations in field. Primarily effective against spider mites and thrips.

* Insects marked with an asterisk represent species available commercially for purchase
For a list of Biological Control (Beneficial Insects) Suppliers, see http://wiki.bugwood.org/Commercially_available_biological_controls

Approaches To Biological Control. There are three general approaches to biological control: importation, augmentation, and conservation of natural enemies. These techniques can be used alone or in combination in a biological control program or as part of an IPM program.

Importation: Importation of natural enemies, also known as classical biological control, is used when a pest of exotic origin is the target of the biocontrol program. Pests are constantly being imported into countries where they are not native. Many of these introductions do not result in establishment, if they do, the organism may not become an invasive pest. However, some of these introduced organisms can become pests due to a lack of natural e nemies to suppress their populations. In these cases, importation of natural enemies can be highly effective.

Once the country of origin of the invasive pest is determined, exploration in the native region can be conducted to search for promising natural enemies. If such enemies are identified, they may be evaluated for potential impact on the pest organism in the native country or imported into the new country for further study. Natural enemies are imported into the U.S. only under permit from the U.S. Department of Agriculture. They must first be placed in quarantine for one or more generations to ensure no undesirable species are accidentally imported (diseases, hyperparasitoids, etc.). Additional permits are required for interstate shipment and field release.

Augmentation: Augmentation is the direct manipulation of natural enemies to increase pest suppression. This can be accomplished by mass production and/or periodic colonization of natural enemies. The most used of these approaches is the first, in which natural enemies are produced in insectaries, then released either inoculatively or inundatively. For example, in areas where a particular natural enemy cannot overwinter, an **inoculative** release each spring may allow the population to establish and adequately control a pest. **Inundative** releases involve the frequent release of large numbers of a natural enemy such that their population completely overwhelms the pest.

Augmentation is used where populations of a natural enemy are not present or cannot respond quickly enough to the pest population. Therefore, augmentation usually does not provide permanent suppression of pests, as may occur with importation or conservation methods.

Because most augmentation involves mass-production and periodic colonization of natural enemies, this type of biological control has lent itself to commercial development. There are hundreds of biological control products available commercially for dozens of pest invertebrates (insects, spider mites, etc.), vertebrates (deer, rodents, etc.), weeds, and plant pathogens.

A summary of these biocontrol products/agents and their target pests is presented in Table 13. The efficacy of these predators and parasites is dependent on many factors, including but not limited to:

- **Temperature:** Most biocontrol agents perform best under specific environmental conditions. Detailed knowledge about the optimal temperature range of the predator or parasite species is vital to ensure successful performance.

- **Relative humidity (RH):** Biocontrol agents adapted to temperate climates can survive at medium RH levels (40-60%), but subtropical and tropical species may require RH above 60% for their survival and development. Overall geographic RH and temperature, as well as conditions within the crop (under protected or semi-protected structures), are important for establishment.

- **Weather:** Heavy rain and strong winds can limit biocontrol agents' establishment in open field crops.

- **Food availability:** There are *generalist* predators and parasites that can feed on a wide range of food items, including prey (other insects or mites), flower products (pollen or nectar), or other (fungus or bacteria). As a preventive tactic, these agents can be established before the pests colonize the crop or large infestations are reached. Combining prey + alternative food resources (i.e., pollen) may increase their reproductive capacity and the likelihood of establishment. *Selective* predators and parasites feed only on one or a few prey and are limited in their establishment to the availability of that prey.

- **Crop development stage:** The stage of the crop (vegetative, reproductive, senescing) may determine food and shelter available for predators and parasites.
- **Pest infestation levels:** If pest populations are already high before introducing biocontrol agents, their performance may be limited, and pesticide applications may need to be scheduled before predator or parasite release.
- **Pest development stage:** Gathering information about the developmental stage of the pests' populations (proportion of eggs, immatures, and adults present) in the crop will allow you to schedule well-timed releases and/or pesticide applications. Natural enemies often feed on one or two life stages of the pests, and some are limited to feeding on eggs, while most pesticides target adults or immature stages, and few have ovicidal activity.
- **Plant/leaf anatomy:** This is particularly important when using predatory mites. Some predatory mite species, such as *Amblyseius swirskii* or *Neoseiulus cucumeris* prefer plant hosts with few leaf trichomes and target insects (thrips, spider mites other than twospotted) or mite pests that produce little webbing (tarsonemids). Other are selective predatory mites (e.g., *Phytoseiulus persimilis*) feeding only on mites that produce lots of webbing (twospotted spider mites and other *Tetranychus* species) and may be better adapted to plants with many leaf trichomes.

Management of the target pest to non-damaging levels is more likely than 100% control when using biocontrol agents. It is critical to familiarize yourself with the proper usage of these predators and parasites to achieve satisfactory results. Selection of products and suppliers should be done carefully, as with purchasing any product. See the following section (Incorporating Biological Control Into A Pest Management Program), contact the provider representative, or the closest Extension office for more information regarding optimal conditions for biocontrol agents.

Conservation: The most common form of biological control is conservation of natural enemies that already exist in a cropping situation. Conservation involves identifying the factor(s) which may limit the effectiveness of a particular natural enemy and modifying these factor(s) to increase the effectiveness of these natural enemies. In general, this involves either reducing factors that interfere with natural enemies or providing resources that natural enemies need in their environment. Naturally occurring and augmented predators and parasites are limited or promoted by similar factors and conditions, such as temperature and RH, food availability, and pest infestation levels. Additionally, the application of pesticides and some cultural practices often interferes with natural enemy effectiveness. For instance, tillage or burning of crop debris can also kill natural enemies or make the crop habitat unsuitable. In some crops, accumulation of dust deposits on leaves from repeated tillage or a location near roadways may kill small predators and parasites and cause increases in certain insect and mite pests (e.g., twospotted spider mites). In some cases, the chemical and physical defenses that plants use to protect themselves from pests may reduce the effectiveness of biological control.

Incorporating Biological Control Into A Pest Management Program: Biological control can be an effective, environmentally sound method of managing pests. When trying to make the best use of natural enemies in your crop, it may be helpful to consider the following suggestions.

- Pest problems must be anticipated and planned for by carefully monitoring pest population development.
- Effective trapping, monitoring, and field scouting should be used to determine when pests appear and the timing of natural enemy releases (if needed).
- Make sure you have your pest(s) accurately identified. Consulting your local Extension office is a good practice regardless of which pest control method you use.
- Determine if natural enemy releases are appropriate for your situation. Sometimes knowledge of crop and cultural practices that encourage naturally-occurring biological control agents can allow you to maximize the control they provide. By conserving these natural enemies, pesticide use (and therefore costs) might be minimized.
- If you decide to use commercially available biological control agents, choose your product and supplier carefully. Guidelines to find reliable providers can be found at:
  - Guide for Purchasing and Using Commercial Natural Enemies (*https://ipm.ifas.ufl.edu/*)
  - Commercially Available Biological Controls (*http:// wiki.bugwood.org/Commercially_available_biological_controls*)
- Details about some of the most common natural enemies found/released in cropping systems are provided in Table 13. Additional resources include:
  - Biological Control: A Guide to Natural Enemies in North America (*https://biocontrol.entomology.cornell.edu/index.php*)
  - How To Use Biocontrol (*https://cals.cornell.edu/new-york-state-integrated-pest-management/eco-resilience/biocontrol/how-use-biocontrol*)
- Usually, released natural enemies work best as a preventative pest management method. If they are introduced into your crop at the *beginning* of a pest infestation, they can prevent that population from developing to damaging levels.
- If pest infestations have become severe *before* releasing natural enemies, using natural enemies usually will not work.

- Once you have received your natural enemies, handle them carefully, following all instructions provided by your supplier.
- The number or rate of natural enemies to release can be determined through consultation with a reliable supplier, as can the application timing.
- Because natural enemies are living organisms, adverse conditions (e.g., stormy weather, pesticide residues) can kill or reduce their effectiveness.
- Because the actions of natural enemies are not as obvious as those of pesticides, it may be important to work with a supplier to evaluate the effectiveness of your releases.
- Remember, just because an organism is sold as a "natural" or "biological" control does not mean it will work as you expect. This does not mean that biological control will not work for your situation. There are various products and approaches that can provide satisfactory results.

Finally, if insecticides/miticides are necessary to control a pest, choose the least harmful product to the beneficial(s) you are trying to preserve or introduce. Pest control can be provided by natural enemies already present in the field when excessive pesticide applications are avoided, and selective pesticides are sprayed only when threshold levels are reached. Table 13A shows the relative safety of different pesticides to key groups of beneficial organisms in vegetable systems. For additional information about suppliers of organisms and related products, the purchase of natural enemies, and how to effectively use them, consult with Extension.

DIAGNOSING VEGETABLE CROP PROBLEMS

All vegetable problems, such as poor growth, leaf blemishes, wilts, rots, and other problems should be promptly diagnosed. It is critical that growers implement prompt and effective corrective measures, and to reduce the probability of these issues reoccurring in following crops, or spreading to susceptible neighboring crops. Follow the steps outlined below to diagnose problems.

1. Describe the problem.
2. Determine whether there is a pattern of symptomatic plants in the field.
    a. Does the pattern correlate with a certain area in the field, such as a low spot, poor-drainage area, or sheltered area?
    b. Does the pattern correlate with concurrent field operations, such as certain rows, time of planting, method of fertilization, or rate of fertilization?
3. Try to trace the history of the problem.
    a. On what date were the symptoms first noticed?
    b. Which fertilizer and liming practices were used?
    c. Which pest-management practices were used to manage diseases, undesirable insects, and weeds — which chemicals (if any), were applied, at what application rates, and what was the previous use of equipment that was used for application?
    d. What were the temperatures, soil moisture conditions, and level of sunlight? Were there any recent significant weather events?
    e. What was the source of seed or planting stock?
    f. Which crops were grown in the same area during the past 3 or 4 years?
4. Examine affected plants to determine whether the problem is related to insects, diseases, or cultural practices.
    a. Do the symptoms point to **insect** problems? Insect problems are usually restricted to the crop. (A hand lens is usually essential to determine this.)
        (1) Look for the presence of insects, webbing, and frass on foliage, stems, and roots.
        (2) Look for feeding signs such as chewing, sucking, or boring injuries.
    b. Do the symptoms suggest **disease** problems? These symptoms are usually not uniform; rather, they are specific for certain crops.
        (1) Look for necrotic (dead) areas on the roots, stems, leaves, flowers, and fruit.
        (2) Look for discoloration of the vascular tissue (plant veins).
        (3) Look for fungal growth.
        (4) Look for virus patterns; often these are similar to injury from 2,4-D or other hormones and nutritional problems.
        (5) Examine roots for twisting, galling, discoloration, or rot.
    c. Do the symptoms point to **cultural** problems? Look for the following:
        (1) Nutrient deficiencies. (A soil test from good areas and poor areas should be done as well as analysis of nutrient content of leaf tissue from the same areas.)
        - Nitrogen—light green or yellow foliage. Nitrogen deficiencies are more acute on lower leaves.
        - Phosphorus—purple coloration of leaves; plants are stunted.
        - Potassium— yellow or brown leaf margins and leaf curling.
        - Magnesium—interveinal chlorosis (yellowing between veins) of mid level or lower leaves.
        - Boron—development of lateral growth; hollow, brownish stems; cracked petioles.
        - Iron—light green or yellow foliage occurs first and is more acute on young leaves.
        - Molybdenum—"whiptail" leaf symptoms on cauliflower and other crops in the cabbage family.
        (2) Chemical toxicities.
        - Toxicity of minor elements—boron, zinc, manganese.
        - Soluble salt injury—wilting of the plant when wet, necrosis around the leaf margins, or death (usually from excessive fertilizer application or accumulation of salts from irrigation water).

(3) Soil problems. (Take soil tests of good and poor areas.)
- Poor drainage.
- Poor soil structure, compaction, etc.
- Hard pans or plow pans.

(4) Pesticide injury. (Usually uniform in the area or shows definite patterns, and more than one plant species, such as weeds, are often symptomatic.)
- Insecticide burning or stunting.
- Herbicide burning or abnormal growth.

(5) Climatic damage.
- High-temperature injury.
- Low-temperature (chilling) injury.
- Lack of water.
- Excessive moisture (lack of soil oxygen).
- Frost or freeze damage.

(6) Physiological damage.
- Air-pollution injury.
- Genetic mutations (scattered plants with stripes, variegation, or other physical irregularities). This is rare.

In summary, when trying to solve a vegetable crop problem, take notes of problem areas, look for a pattern to the symptoms, trace the history of the problem, and examine the plants and soil closely. These notes can be used to avoid the problem in the future or to assist others in helping solve their problem. Publications and bulletins designed to help the grower identify specific vegetable problems are available from Extension.

AIR POLLUTION INJURY

The extent of plant damage by particular pollutants in any given year depends on meteorological factors leading to air stagnation, the presence of a pollution source, and the susceptibility of the plants.

Some pollutants that affect vegetable crops are sulfur dioxide (SO_2), ozone (O_3), peroxyacetyl nitrate (PAN), chlorine (Cl), and ammonia (NH_3).

Sulfur dioxide. SO_2 causes acute and chronic plant injury. Acute injury is characterized by clearly marked dead tissue between the veins or on leaf margins. The dead tissue may be bleached, ivory, tan, orange, red, reddish brown, or brown, depending on plant species, time of year, and weather conditions. Chronic injury is marked by brownish red, turgid, or bleached white areas on the leaf blade. Young leaves rarely display damage, whereas fully expanded leaves are very sensitive.

Some crops sensitive to sulfur dioxide are: squash, pumpkin, mustard, spinach, lettuce, endive, Swiss chard, broccoli, bean, carrot, and tomato.

Ozone. A common symptom of O_3 injury is small stipplelike or flecklike lesions visible only on the upper leaf surface. These very small, irregularly shaped spots may be dark brown to black (stipplelike) or light tan to white (flecklike). Very young leaves are normally resistant to ozone. Recently matured leaves are most susceptible. Leaves become more susceptible as they mature, and the lesions spread over a greater portion of the leaf with successive ozone exposures. Injury is usually more pronounced at the leaf tip and along the margins. With severe damage, symptoms may extend to the lower leaf surface.

Pest feeding (red spider mite and certain leafhoppers) produces flecks on the upper surface of leaves much like ozone injury. Flecks from insect feeding, however, are usually spread uniformly over the leaf, whereas ozone flecks are concentrated in specific areas. Some older watermelon varieties and red varieties of potatoes and beans are particularly sensitive to ozone.

Peroxyacetyl nitrate. PAN affects the under surfaces of newly matured leaves, and it causes bronzing, glazing, or silvering on the lower surface of sensitive leaf areas.

The leaf apex of broadleaved plants becomes sensitive to PAN about 5 days after leaf emergence. Since PAN toxicity is specific for tissue of a particular stage of development, only about four leaves on a shoot are sensitive at any one time. With PAN only successive exposures will cause the entire leaf to develop injury. Injury may consist of bronzing or glazing with little or no tissue collapse on the upper leaf surface. Pale green to white stipplelike areas may appear on upper and lower leaf surfaces. Complete tissue collapse in a diffuse band across the leaf is helpful in identifying PAN injury.

Glazing of lower leaf surfaces may be produced by the feeding of thrips or other insects or by insecticides and herbicides, but differences should be detectable by careful examination.

Sensitive crops are: Swiss chard, lettuce, beet, escarole, mustard, dill, pepper, potato, spinach, tomato, and cantaloupe.

Chlorine. Injury from chlorine is usually of an acute type, and it is similar in pattern to sulfur dioxide injury. Foliar necrosis and bleaching are common. Necrosis is marginal in some species, but scattered in others either between or along veins. Lettuce plants exhibit necrotic injury on the margins of outer leaves, which often extends into solid areas toward the center and base of the leaf. Inner leaves remain unmarked. Crops sensitive to chlorine are: Chinese cabbage, lettuce, Swiss chard, beet, escarole, mustard, dill, pepper, potato, spinach, tomato, cantaloupe, corn, onion, and radish.

Ammonia. Field injury from NH_3 has been primarily due to accidental spillage or use of ammoniated fertilizers under plastic mulch on light sandy soils. Slight amounts of the gas produce color changes in the pigments of vegetable skin. The dry outer scales of red onions may become greenish or black, whereas scales of yellow or brown onions may turn dark brown. In addition, chicken litter may be high in ammonia (NH_3) and ammonium (NH_4), and if sufficient time is not allowed between litter application and planting, substantial damage from ammonia toxicity may occur to seeds or seedlings.

PRODUCE SAFETY

This section provides a simple overview of produce safety and provides some resources for your farm. Once contamination occurs, it cannot be removed from the produce, and, since many crops are eaten without cooking, prevention is the most important step in reducing outbreaks. Microorganisms that can cause foodborne illnesses are commonly found in human and animal feces, including *Salmonella*, pathogenic *E. coli* (e.g., O157:H7), and *Campylobacter* spp. Some viruses, such as norovirus and hepatitis A, can also survive for longer periods and can contaminate fresh produce. Human parasites are not commonly associated with produce contamination in the US; however, *Cyclospora cayetanensis* has become an emerging concern in the past few years, especially in irrigation water sources.

WHAT ARE *GOOD AGRICULTURAL PRACTICES* (GAPS)?

The term Good Agricultural Practices (GAPs) refers to basic practices that minimize food safety risks. The purpose of GAPs is to provide guidance for the safe production, packing, handling, and holding of fruits and vegetables resulting from a well-developed farm food safety plan. Worker hygiene and health, manure use, and water quality throughout the production and harvesting processes are areas that GAPs cover. Growers, packers, and shippers are urged to take a proactive role in minimizing food safety risks associated with fresh produce. Furthermore, operators should encourage the adoption of safe practices by their partners along the farm-to-table food chain. This includes distributors, exporters, importers, retailers, transporters, food service operators, and consumers. Being familiar with potential biological, chemical, and physical hazards that exist on the farm is the first step toward minimizing risk. This helps to protect the consumer, as well as the growers' families, communities, and livelihoods. Some of the areas of the farm that could be potential sources of contamination and should be considered are:

- **The growing site:** Take into consideration the surrounding area, topography, soil amendments, and irrigation water source used before harvest. Think about the effect of adjacent or surrounding land use on contamination of produce and how GAPs will be implemented to reduce risk long before planting. Recent research has shown that nearby animal production areas can contaminate produce through irrigation water and wind spread. Vegetative buffers, berms, diversion ditches, and woody areas may help reduce this risk.
- **Water use:** If water touches the harvestable portion of the crop or is used on surfaces the produce touches after harvest, then water quality becomes a key concern. Because every farm is different, the agricultural water system should be assessed to identify hazards and implement corrective actions or mitigation measures. In relation to FSMA's Produce Safety Rule, the FDA published the final provisions for pre-harvest agricultural water in 2024. These provisions replace the previous microbial quality criteria and testing requirements for pre-harvest agricultural water. If a farm is undergoing third-party food safety audits, there can be additional requirements in place.
- **Domesticated and wild animal intrusion:** Both domesticated and wild animals can carry and shed human pathogens. Sometimes, this is because they have carried pathogens from a nearby animal production area or human contamination source. It is important to minimize contamination from animals in the production and packing areas, including farm pets.
- **Worker hygiene and sanitation:** Provide workers with appropriate training to help them understand the importance of hygiene in handling produce. Providing convenient, clean, and well-stocked restrooms and handwashing facilities will enable workers to properly wash their hands after eating, smoking, handling non-produce items, using the restroom, and other times when hands may have become contaminated. Sick workers must be restricted from handling produce so that they do not pass along pathogens.
- **Equipment and transportation:** Equipment used in the production, harvest, postharvest, and transportation of produce can be a source of contamination. When possible, designate specific equipment for different areas of the farm. For example, designate a set of cleaning tools for bathroom use only and have a different set of cleaning tools for the packing line and harvest containers. The cleaning and sanitation of equipment should be conducted and documented on a regular schedule.

The above items are just a few that warrant attention in a farm food safety plan. Documentation of the practices and who performs them is essential for accountability and continuous improvement.

FOOD SAFETY MODERNIZATION ACT (FSMA) PRODUCE SAFETY RULE (PSR)

FSMA was signed into law on January 4, 2011, initiating a new public health mandate for the FDA to establish science-based standards for the prevention of foodborne illness. Per the FSMA, the FDA has set minimum standards for practices conducted by those who harvest, grow, process, transport, and store food. This law is also meant to ensure that foreign producers who import into the U.S. meet these same standards.

The most relevant rule to the produce industry under FSMA Rules is the "Standards for the Growing, Harvesting, Packing, and Holding of Produce for Human Consumption," commonly referred to as the Produce Safety Rule or PSR. The PSR outlines standards that minimize the possibility of consumers becoming ill from eating fresh fruits and vegetables contaminated with pathogens and is based on the GAPs described above. The PSR focuses on microbial hazards, such as foodborne illness pathogens. Key areas of the regulation include water quality, biological soil amendments, *sprout* production, domesticated and wild animals, worker health and hygiene, and *equipment,* tools, and buildings. The PSR went into effect on 26 Jan 2016 and allows some farms to qualify for an exemption from some of the requirements. More information can be found on the FDA website (in the link below).

AGRICULTURAL WATER PSR - SUBPART E

The FDA FSMA Produce Safety Rule (PSR) Subpart E establishes agricultural water requirements for 1) pre-harvest agricultural water provisions (effective May 2024) and 2) harvest and postharvest agricultural water requirements (effective July 2022) for non-sprout-covered produce operations to ensure that the water is safe and of adequate sanitary quality for its intended use (§ 112.41). As part of the revisions process for Subpart E, the FDA added two new definitions to § 112.3, Agricultural Water Assessment and Agricultural Water Systems.

IMPORTANT DEFINITIONS:

Agricultural Water: "Agricultural water" refers to water used in direct contact with produce or surfaces the produce will contact. This includes water used in growing activities (such as irrigation water applied via direct methods, water used for preparing crop sprays, and water used for growing sprouts) and in harvesting, packing, and holding activities (including water used for washing or cooling harvested produce and for preventing dehydration of covered produce).

Agricultural Water System: A source of agricultural water, the water distribution system, any building or structure that is part of the water distribution system (such as a well house, pump station, or shed), and any equipment used for application of agricultural water to covered produce during growing, harvesting, packing, or holding activities.

Agricultural Water Assessment: An evaluation of an agricultural water system, agricultural water practices, crop characteristics, environmental conditions, and other relevant factors (including test results, where appropriate) related to growing activities for covered produce (other than sprouts) to: (1) identify any condition(s) that are reasonably likely to introduce known or reasonably foreseeable hazards into or onto covered produce or food contact surfaces and (2) determine whether measures are reasonably necessary to reduce the potential for contamination of covered produce or food contact surfaces with such known or reasonably foreseeable hazards.

Pre-Harvest Agricultural Water Requirements

The final rule replaces the term production water (water used in contact with produce during the growing season) with the term pre-harvest agricultural water (agricultural water used while growing nonsprout-covered produce). The revision moves from reliance on testing pre-harvest water for decision-making to an agricultural water assessment of the whole water system. The expectation is that each grower understands the farm's water quality and how water management practices minimize risks to the crops they grow.

- Inspection and Maintenance of the Agricultural Water System
- **Agricultural Water Assessment:** Farms are required to evaluate the following factors using a risk-based tiered approach to identify conditions likely to introduce known or reasonably foreseeable hazards into or onto produce or food contact surfaces: 1) agricultural water use practices, 2) crop characteristics, 3) environmental conditions and 4) other relevant factors.

Covered farms must annually use the results of their pre-harvest agricultural water assessments to document whether corrective or mitigation measures are necessary to minimize the risk of contamination to covered produce (excluding sprouts) or food contact surfaces from hazards related to pre-harvest agricultural water. Farms are also required to maintain records of their pre-harvest agricultural water assessments each year or as needed for reassessment, which supervisors must review along with the determinations made based on these results.

Harvest and Postharvest Agricultural Water Requirements

The overall requirements for farms and packing houses for this section include:

- Ensure there is no detectable generic *E. coli* per 100 ml of agricultural water. Untreated surface water must never be used in harvest or post-harvest operations.
- At least once per season, inspect and maintain the agricultural water system to the extent it's under the farm's control to identify conditions that are reasonably likely to introduce known or reasonably foreseeable hazards.
- If water is treated, the process has to be monitored to ensure water is safe.
- Monitor water quality when washing produce.

For more information on the requirements for this section, go to Requirements for Harvest and Postharvest Agricultural Water in Subpart E for Covered Produce for Other than Sprouts or this resource from North Carolina State University's Produce Safety Program.

The revised rulemaking established the following compliance dates for agricultural water for covered farms:

FSMA PSR SUBPART E, AGRICULTURAL WATER COMPLIANCE DATES

Farm Size	Harvest/Postharvest	Pre-Harvest
All other businesses (>$500k)*	January 26, 2023	April 7, 2025
Small businesses (>$250k-$500k)*	January 26, 2024	April 6, 2026
Very small businesses (>$25k-$250k)*	January 26, 2025	April 5, 2027

* Based on the annual average monetary value of produce the farm sold during the previous 3-year period

PRODUCE SAFETY ALLIANCE TRAINING

One important aspect of the PSR is that it establishes training requirements. The PSR requires that at least one supervisor or other responsible party completes a PSR food safety training. The Produce Safety Alliance (PSA) Grower Training Course is one way to meet this requirement. PSA training provides an overview of GAPs and the standards set in the PSR that growers must follow. For a list of PSA training, visit:

Upcoming grower training:
https://producesafetyalliance.cornell.edu/training/grower-training-courses/upcoming-grower-trainings/

ON-FARM READINESS REVIEW - NATIONAL ASSOCIATION OF STATE DEPARTMENTS OF AGRICULTURE

Most states are also offering a free and voluntary educational visit called an On-Farm Readiness Review (OFRR) for farms to assess their readiness for a PSR inspection. State contacts to schedule an OFRR can be found on the National Association of State Departments of Agriculture website.

Find an OFRR contact in your state: *https://www.nasda.org/nasda-foundation/about-on-farm-readiness-review/*

FOOD SAFETY AUDITS VERSUS PSR INSPECTIONS

Food safety audits for produce growers and packers have been around since the 1990s as an industry response to several produce-associated foodborne illness outbreaks. There are several tiers of audits available now. USDA offers GAP/Good Handling Practice (GHP), Harmonized GAP (aligns with PSR), and Harmonized GAP Plus+ (aligns with both PSR and Global Food Safety Initiative [GFSI]). These require increasing amounts of grower investment. It is important to check with your buyers to see what they require. By comparison, PSR inspections are a requirement of the FSMA PSR and a relatively new concept, with inspections beginning in the summer of 2019. Although the practices required for audits and inspections are both based on GAPs, the specific practices required, and the scope can differ. There are also some notable differences between the two, namely:

- An audit is a way to voluntarily provide assurance to a buyer that produce is grown to minimize food safety risks, while a PSR inspection is required by law for farms covered by the PSR.
- Typically, an audit is conducted by an independent third party (e.g., USDA-AMS, Quality Certification Services (QCS), and PrimusLabs) or a produce buyer, while inspections in the southeast are conducted by state departments of agriculture.
- Audits tend to be expensive, while there is no charge for a PSR inspection.
- Audits require a food safety plan. While the PSR does not require a food safety plan, a plan can be helpful in developing a farm food safety program that will meet the PSR requirements.
- Audits cover specific crops chosen by the grower for market access purposes, while a PSR inspection is concerned with PSR-covered crops and covered activities.

FOOD SAFETY CERTIFICATION FOR SPECIALTY CROPS PROGRAM

The Farm Service Agency (FSA) currently has a program called *Food Safety Certification for Specialty Crops* to provide financial assistance for produce growers with on-farm food safety costs on obtaining and renewing a farm food safety certification. Since this is a reimbursement program, the expenses must have been paid or incurred before the application for reimbursement.

To be eligible for the program, applicants must meet the following criteria:

- Be a specialty crop operation;
- Have obtained or renewed a food safety certification in the year of application;
- Have paid eligible expenses;
- Meet the definition of a small- or medium-sized business;
- Be located in the United States or its territories.

Eligible expenses covered by the program include:

- Maintenance of food safety plan (75% up to $675).
- Food safety certification (75% up to $2,000) and upload fees (75% up to $375).
- Microbial testing for produce, water, and soil amendments (75% up to five tests/type).
- Training (100% up to $500).

The program does not cover infrastructure improvements, equipment, supplies, salaries and benefits, or late payment fees and penalties. Growers can find more information about the program through their local FSA office at *https://offices.usda.gov/*

PRODUCE SAFETY RESOURCES IN YOUR STATE

Alabama:

- **Food Safety & Quality - Agents and Specialists - Alabama Cooperative Extension System:** *https://ssl.acesag.auburn.edu/directory-new/programAgentSearch.php?program=8*
- **Produce Safety Training for Alabama Growers:** *https://www.aces.edu/blog/topics/food-safety/produce-safety-training-options-for-alabama-growers/*
- **ANR-2556 Food Safety Considerations During Flooding Gulf States Multistate Resource:** *https://www.aces.edu/wp-content/uploads/2019/08/ANR- 2556_FoodSafetyConsiderationsDuringFlooding_FOR_ WEB_082719L.pdf*
- **AgWater Safety Program - Auburn University - (agricultural water testing):** *https://www.aces.edu/blog/topics/farming/agwater-safety-program/*

Arkansas:

- **University of Arkansas Produce Safety Outreach, Education, and Technical Assistance Program:** *http://uaex.uada.edu/producesafety*
- **Arkansas Department of Agriculture Produce Safety Program (FSMA and GAP):** *https://www.agriculture.arkansas.gov/plant-industries/regulatory-section/produce-safety-program/*
- **Arkansas Water Resources Laboratory (agricultural water testing):** *https://awrc.uada.edu/*

Florida:
- **Southern Center for Food Safety Training, Outreach, and Technical Assistance:** *https://sc.ifas.ufl.edu/*
- **Florida Department of Agriculture and Consumer Services:** *https://www.fdacs.gov/*

Georgia:
- **Waters Agricultural Laboratories, Inc (Ag water testing):** *https://watersag.com/services/*
- **University of Georgia - Agricultural & Environmental Services Labs (agricultural water testing):** *https://aesl.ces.uga.edu/water.html*
- **Workshop and Training Calendar - Dept of Food Science & Technology – UGA:** *https://extension.uga.edu/programs-services/food-science/ workshops.html*

Kentucky:
- **Produce Safety Webpage for resources:** *https://www.uky.edu/ccd/foodsafety*
- **Extension Publications:** *https://afs.ca.uky.edu/foodscience/food-science- publications*

Louisiana:
- **Factsheet, PSA vs GAP:** *https://www.lsuagcenter.com/topics/food_health/food/safety*
- **Investigating Fresh Produce Cyclospora Outbreaks:** *https://www.afdo.org/wpcontent/uploads/2021/05/InvestigatingFresh-Produce-Cyclospora-Outbreaks.pdf*
- **Produce Safety Rule Inspections and Voluntary USDA Audits:** *https://www.afdo.org/wp-content/uploads/2022/07/AF-DO- Produce-Safety-Rule-Inspections-and-Voluntary-USDA- Audits.pdf*
- **Factsheet on Service Animals:** *https://www.lsuagcenter.com/topics/food_health/food/safety*
- **Louisiana Department of Agriculture and Forestry website:** *www.ldaf.state.la.us*

Mississippi:
- **Mississippi State University Extension:** *https://extension.msstate.edu/food-and-health/food-safety/produce-safety*
- **Mississippi State Chemical Laboratory:** *https://www.mscl.msstate.edu/*
- **Mississippi Department of Agriculture and Commerce:** *https://www.mdac.ms.gov/bureaus-departments/regulatory- services/produce-safety-division*
- **BonnerAnayltical Testing:** *https://batco.com*

North Carolina:
- **NC Fresh Produce Safety Resources:** *https://ncfreshproducesafety.ces.ncsu.edu*
- **Certified Laboratory Listing in NC:** *https://deq.nc.gov/about/divisions/water-resources/water resources-data/water-sciences-homepage/laboratory certification-branch/certified-laboratory-listings*

Oklahoma:
- **Robert M. Kerr Food & Ag Products Center:** *https://food.okstate.edu/expertise/index.html https:///food.okstate.edu/expertise/focus-areas.html*
- **Oklahoma Department of Agriculture, Food, and Forestry:** *https://ag.ok.gov/divisions/food-safety/*

South Carolina:
- **Clemson Cooperative Extension Produce Safety:** *https://www.clemson.edu/extension/food/psa_grower_training/index.html*
- **South Carolina Department of Agriculture Produce Safety:** *https://agriculture.sc.gov/divisions/consumer-protection/ produce-safety/*
- **Carolina Farm Stewardship Association:** *https://www.carolinafarmstewards.org/*

Tennessee:
- **UT Produce Safety:** *https://vegetables.tennessee.edu/food-safety/*
- **Tennessee Department of Agriculture Produce Safety:** *https://www.tn.gov/agriculture/consumers/food-safety/agconsumersproduce-safety.html*
- **Tennessee Water Quality Testing Labs:** *https://vegetables.tennessee.edu/wp-content/uploads/sites/167/2020/08/TN-water-quality-lab-map.pdf*

Texas:
- **Testing laboratories: Food Safety Net Services:** *https://fsns.com/services/*
- **Texas Department of Agriculture FSMA:** *https://www.texasagriculture.gov/RegulatoryPrograms/FoodSafetyModernizationAct(FSMA).aspx*
- **Texas International Produce Association:** *https://texipa.org/page/2/*

Virginia:
- **Produce Safety Science You Tube:** *https://www.youtube.com/channel/ UCY2J9s4mOfKwSian- lO4ELRg*
- **Virginia Fresh Produce Food Safety:** *https://ps.spes.vt.edu/*
- **Virginia Produce Safety Program:** *https://www.vdacs.virginia.gov/food-produce-safety.shtml*
- **Virginia Cooperative Extension (VCE) Resources (fact- sheets, info):** *https://www.pubs.ext.vt.edu/*
- **Virginia Water Testing Labs:** *https://www.vdacs.virginia.gov/pdf/producesfty-watertestinglabs.pdf*
- **VCE Is my farm covered or exempt from the Produce Safety Rule?:** *https://vce.az1.qualtrics.com/jfe/form/SV_emnhR0UpF- piVvlr*
- **VCE PSR Worker Training Curriculum:** *https://register.ext.vt.edu/search/publicCourseSearchDetails.do?method=load&courseId=1222895&selectedProgramAreaId=25579&selectedProgramStreamId=*

ADDITIONAL RESOURCES

- National GAPs Program: *https://gaps.cornell.edu/*
- Produce Safety Alliance: *https://producesafetyalliance.cornell.edu/*
- Food Safety Resource Clearinghouse: *https://foodsafetyclearinghouse.org/*
- AFDO Produce Safety: *https://www.afdo.org/resources_category/produce-safety/*
- USDA GAP Programs: *https://www.ams.usda.gov/services/auditing/gap-ghp*

POSTHARVEST HANDLING

Crop Shelf-Life

Fresh vegetables are living tissues, and their composition and physiology rapidly change after harvest. Once detached from the vine, vegetables will continue to undergo metabolic reactions, which lead to deterioration and crop death. Shelf-life is a critical component of postharvest physiology and is defined as the finite period of time a harvested crop can be consumed after harvest.

Each commodity has its own shelf-life, and once harvested, it is a race against the clock for growers, stakeholders and retailers to maintain shelf-life and crop quality. There are many factors that can impact crop shelf-life, including preharvest conditions like ambient environment (temperature, light and humidity), cultivation techniques, agronomic practices, harvest timing, cultivar selection, plant breeding/genetics, climactic events and integrated pest management. For instance, if produce is improperly handled during harvest and is damaged or bruised, shelf-life can be compromised due to increased water loss and respiration. Additionally, improper handling leads to a higher risk of crop contamination via bacteria, spoilage organisms and food safety pathogens. Thus, postharvest strategies are key to maintaining shelf-life and quality. Much of postharvest physiology focuses on temperature management, cold storage, modified atmosphere storage, optimal humidity and the use of packaging (e.g., clamshells and polyethylene bags), edible films and other postharvest storage treatments. *Any* deviation from the optimum care of a specific crop will lead to reduced quality and shorter shelf-life. The sections below go into greater detail on maintaining crop quality and extending shelf-life.

Storage Requirements

Horticultural crops may be grouped and stored into two broad categories based on sensitivity to storage temperatures. The broad categories are 1) chilling sensitive, indicating the crop cannot be stored below ~50-53°F (10-12°C), and 2) chill-hardy, indicating the commodity can handle storage temperatures ranging ~32- 39°F (0-4°C). Storing crops at the lowest acceptable temperature is essential in extending the shelf-life. If stored above the temperature, senescence and decay will intensify, leading to shortened storage periods. If the crop is stored below the temperature threshold, it may freeze or develop chilling-induced disorders.

Crops that are chilling sensitive include tomato, cucumber, eggplant, melons, okra, snap beans, basil, peppers, potatoes, summer squash, sweetpotatoes and many tropical and subtropical crops. Those that are chilling sensitive should be held at temperatures generally above 50°F (10°C). Storage below this threshold will give rise to chilling injury, which is characterized by the development of sunken lesions on the skin, browning or discoloration, increased susceptibility to decay, increased shriveling, and incomplete ripening (poor flavor, texture, aroma, and color). The extent of chilling symptoms is also dependent on the crop's maturity at harvest, cultivar, and length of exposure to low temperatures. Short exposure times will result in less injury than longer exposure to chilling temperatures.

Most leafy greens, Asian greens, Cole crops (e.g., broccoli, cauliflower, Brussels sprouts, mustards and radishes), beets, sweet corn, scallion, carrots and several other vegetables are chill-hardy. Crops not as sensitive to chilling may be stored at temperatures as low as 32°F (0°C). Cold storage is critical in reducing the rate of respiration and water loss as well as suppressing ethylene production and senescence.

In addition to maintaining storage rooms at proper temperatures, the relative humidity should also be controlled to reduce water loss from the crop. Low levels of relative humidity will drive moisture loss from the product, whereas excessive levels, along with poor temperature management, can lead to condensation and subsequent produce decay. In cold storage, the relative humidity for vegetable crops can range from 75 to 100%, with most preferring 85%. Optimal storage recommendations and precooling methods are included for a wide range of vegetable commodities in Table 14.

TABLE 14. RECOMMENDED STORAGE CONDITIONS AND COOLING METHODS FOR MAXIMUM POSTHARVEST LIFE OF COMMERCIALLY GROWN VEGETABLES

Crop	Temperature °F	Temperature °C	% Relative Humidity	Approximate Storage Life	Cooling Method [1]
Asparagus	32-35	0-2	95-100	2-3 weeks	HY
Bean, green or snap	40-45	4-7	95	7-10 days	HY, FA
Bean, lima (butterbean)	37-41	3-5	95	5-7 days	HY
Bean, lima, shelled	32	0	95-100	2-3 days	ROOM, FA
Beet, topped	32	0	98-100	4-6 months	ROOM
Broccoli	32	0	95-100	10-14 days	HY, ICE

[1] FA = Forced-air cooling; HY = Hydrocooling; ICE = Package ice, slush ice; ROOM = Room cooling; VAC = Vacuum cooling
[2] Curing required prior to long-term storage. 'Curing' of dry onions involves drying the outer bulb scales, reducing the fresh weight by 5-6%

TABLE 14. RECOMMENDED STORAGE CONDITIONS AND COOLING METHODS FOR MAXIMUM POSTHARVEST LIFE OF COMMERCIALLY GROWN VEGETABLES (cont'd)

Crop	Temperature °F	Temperature °C	% Relative Humidity	Approximate Storage Life	Cooling Method [1]
Cabbage, early	32	0	98-100	3-6 weeks	ROOM
Cabbage, Chinese	32	0	95-100	2-3 months	HY,VAC
Carrot, bunched	32	0	95-100	2 weeks	HY
Carrot, mature, topped	32	0	98-100	7-9 months	HY
Cauliflower	32	0	95-98	3-4 weeks	HY,VA
Collard	32	0	95-100	10-14 days	HY,ICE,VAC
Cucumber	50-55	10-13	95	10-14 days	HY
Eggplant	46-54	8-12	90-95	7-10 days	FA
Endive and escarole	32	0	95-100	2-3 weeks	HY,ICE,VAC
Garlic	32	0	65-70	6-7 months	ROOM
Greens, leafy	32	0	95-100	10-14 days	HY,ICE,VAC
Kale	32	0	95-100	2-3 weeks	HY,ICE,VAC
Kohlrabi	32	0	98-100	2-3 months	ROOM
Leek	32	0	95-100	2-3 months	HY,ICE,VAC
Lettuce	32	0	98-100	2-3 weeks	HY, VAC, ICE
Melon					
Cantaloupe, 3/4-slip	36-41	2-5	95	15 days	FA,HY
Mixed melons	45-50	6-10	90-95	2-3 weeks	FA,HY
Watermelon	50-60	10-15	90	2-3 weeks	ROOM, FA
Microgreens (common)					
Amaranth	37-39	3-5	90-95	14 days	ROOM
Basil	33-39	1-5	90-95	14 days	ROOM
Brassica (kale, mustard, arugula, broccoli, radish, kholrabi)	32-37	0-3	90-95	3-4 weeks	ROOM
Cilantro	37-39	3-5	90-95	2-3 weeks	ROOM
Lettuce	32-37	0-3	90-95	14 days	ROOM
Pea tendril	37-39	3-5	90-95	2-3 weeks	ROOM
Spinach	32-37	0-3	90-95	2-3 weeks	ROOM
Okra	45-50	7-10	90-95	7-10 days	FA
Onion, green	32	0	95-100	3-4 weeks	HY,ICE
Onion, dry [2]	32	0	65-70	1-8 months	ROOM
Parsley	32	0	95-100	2-2.5 months	HY,ICE
Parsnip	32	0	98-100	4-6 months	ROOM
Pea, green or English	32	0	95-98	1-2 weeks	HY,ICE
Southernpea	40-41	4-5	95	6-8 days	FA,HY
Pepper, sweet (bell)	45-55	7-13	90-95	2-3 weeks	FA, ROOM
Potato, White [2]	40	4	90-95	4-5 months	HY,ROOM,FA
Pumpkin	50-55	10-13	50-70	2-3 months	ROOM
Radish, spring	32	0	95-100	3-4 weeks	HY, FA
Radish, oriental	32	0	95-100	2-4 months	ROOM
Rutabaga	32	0	98-100	4-6 months	ROOM
Spinach	32	0	95-100	10-14 days	ICE,HY,VAC
Squash, summer	41-50	5-10	95	1-2 weeks	FA,HY
Sweet corn	32	0	95-98	5-8 days	HY,ICE,VAC
Squash, winter	50	10	50-70	Depending on type	ROOM
Sweetpotato [2]	55-60	13-16	85-90	4-7 months	ROOM
Tomato, mature-green	55-70	13-21	90-95	1-3 weeks	FA,ROOM
Tomato, firm-red	46-50	8-10	90-95	4-7 days	FA,ROOM
Turnip	32	0	95	4-5 months	FA,ROOM
Turnip greens	32	0	95-100	10-14 days	HY,ICE,VAC

[1] FA = Forced-air cooling; HY = Hydrocooling; ICE = Package ice, slush ice; ROOM = Room cooling; VAC = Vacuum cooling
[2] Curing required prior to long-term storage. 'Curing' of dry onions involves drying the outer bulb scales, reducing the fresh weight by 5-6%

Quality Changes After Harvest

Fresh fruits and vegetables are highly perishable once harvested. During storage, many compositional changes, both desired and undesired, will continue because of maturation and ripening. In fruit-type vegetables, like tomatoes, ripening leads to changes that make the product palatable, e.g., softer, juicier, sweeter, with a better aroma. However, it also results in a rapid decline in firmness, making it more susceptible to mechanical damage and spoilage. Softening of tissue has been well correlated to loss of shelf-life in many vegetables, including cucumber, squash and tomato. In harvested vegetables, water loss, which manifests as loss of texture via shriveling and wilting, is one of the greatest issues in postharvest. Some degree of water loss is acceptable, but at high levels, it makes produce unmarketable and has a significant economic impact on products sold by weight.

Quality changes are also related to crop color, such as yellowing and loss of green. These traits are undesirable in many vegetables such as cucumber, green zucchini, and leafy greens. Here, a change in color due to the breakdown of chlorophyll, carotenoids and anthocyanins is often associated with loss of freshness. Anthocyanins (e.g., red, purple and blue pigments) tend to increase during development in several vegetables, including amaranth, kale and cabbage. Conversely, carotenoids (e.g., orange and red pigments) and chlorophyll (e.g., green pigment) typically decrease. The harvest stage can also affect carotenoid content and color development, and once harvested, proper postharvest storage conditions are critical to decrease losses. If crops are harvested at an immature stage, cold storage can inhibit carotenoid accumulation and red color formation. For example, tomatoes harvested at the mature green stage and stored at 41°F (5°C) result in a decrease in carotenoids compared to those harvested at the pink or fully red stage stored at 50°F (10°C). Conversely, if a leafy vegetable such as lettuce or spinach is harvested when fully developed, storage temperatures of ~ 35-39°F (2-4°C) tend to decrease carotenoid losses.

Maintaining Quality Through Postharvest Management

Within hours of harvest, crops held at field or packing house temperatures can suffer irreversible losses in quality, reducing shelf-life. Additionally, many vegetables, such as greens and lettuce, are cut at harvest, and this wound further increases stress on the plant. Respiration is the process of life by which oxygen (O_2) is combined with stored carbohydrates and other components (reserves) to produce heat, chemical energy, water, carbon dioxide (CO_2) and other products. The relative perishability of a crop is reflected in its respiration rate. The respiration rate varies by commodity; those commodities with high respiration rates utilize the reserves faster and are more perishable than those with lower respiration rates. Therefore, vegetables with higher respiration rates, such as broccoli and sweet corn, must be rapidly cooled to the optimal storage temperature to slow metabolism, maintain quality and extend shelf-life during subsequent shipping and handling operations. Still, all perishable crops will benefit from cooling after harvest.

Rapid cooling (precooling) allows produce to be shipped to distant markets while maintaining high quality. Precooling is defined as the rapid removal of field heat to temperatures approaching optimal storage temperature, and it is the first line of defense in slowing down the biological processes that reduce vegetable quality. Precooling, in conjunction with refrigeration during subsequent handling operations, provides a "cold chain" from packinghouse to supermarket to maximize shelf-life and control diseases and pests. A fresh produce cold chain is defined as a temperature-controlled supply chain that involves all stages of production, from produce harvesting and precooling through to the transportation stages, storage, distribution processes, and delivery to the end customer. Timely and careful harvest, prompt transport to the packinghouse, rapid packing and cooling, and swift delivery to the market or buyer are also essential for maintaining quality. Everyone involved at each of the steps during product handling (e.g., shippers, truckers, receivers) must take care to ensure that the refrigerated cold chain is not broken. Cold chain breaks are common throughout the distribution and retail display, and can have detrimental effects on organoleptic, microbiological and nutritional quality of produce.

Many shippers are well equipped to rapidly cool their crops, and a growing number are incorporating precooling or improving their existing facilities. **Simple placement of packed vegetables in a refrigerated cooler is not sufficient to maintain quality for products destined for distant markets.** Neither should non-cooled vegetables be loaded directly into refrigerated trailers. In both situations, the product cools very slowly, at best. Refrigerated trailers are designed to maintain product temperature during transport, and they do not have the refrigeration capacity to quickly remove field heat. Therefore, only produce that has been properly precooled should be loaded, and only into trailers that have been cooled prior to loading.

OPTIMIZING COMMERCIAL COOLING
Cooling Concepts

Cooling is a vital postharvest practice for maintaining crop quality and extending shelf-life. It offers several key benefits, including slowing respiration and ripening, minimizing water loss, inhibiting microbial growth and decay, and reducing both

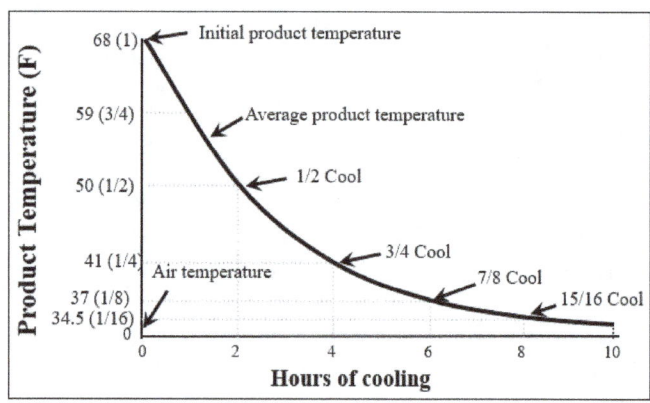

Figure 1. Cooling curve indicating product temperature reduction over time based on the proportion of field heat removed. The curve illustrates typical cooling milestones, e.g., ½ cool and ⅞ cool, starting from harvest temperature (68°F) and progressing to optimal storage temperature (37°F). Estimated times to reach each stage are indicated.

the production and impact of ethylene. To optimize postharvest cooling, it is essential to understand the primary sources of heat and how they influence cooling efficiency. Typically, about 50% of the heat load is attributed to sensible or field heat, while approximately 12% originates from crop respiration. These two sources represent the most significant forms of heat that must be effectively removed and managed during postharvest storage to maintain product quality.

Cooling is typically described in terms of crop temperature reduction, cooling rate, and the method used. As mentioned previously, most vegetable crops perform best when cooled and held at 37 °F (2.5 °C). Effective commercial cooling for perishable crops is defined as "the rapid removal of field heat using a compatible method, which significantly extends shipping life". For most commodities, cooling efficiency is expressed as "½ cool" or "⅞ cool", indicating the proportion of field heat removed and the time required to achieve it (**Fig. 1**). For example, a crop harvested at 68°F (20 °C) typically requires about 2 hours to reach ½ cool and 6 hours to reach ⅞ cool. Excluding chilling-sensitive commodities, achieving ⅞ cool is strongly recommended to ensure adequate shelf-life for long-distance shipping. The remaining ⅛ is typically eliminated during subsequent refrigerated storage and handling, with minimal impact on product quality. Even for growers shipping to local markets, cooling is the most important step in ensuring the quality and shelf-life of the product.

The rate of heat transfer, or the cooling rate, is critical for the efficient removal of field heat to achieve cooling. As a form of energy, heat naturally moves toward equilibrium; during cooling, this means transferring heat from the product to the cooling medium. The effectiveness of this process depends on three key factors: time, temperature, and contact. To achieve maximum cooling, the product must remain in the precooler for sufficient time to remove heat. The cooling medium (air, water, crushed ice) must be maintained at a constant temperature throughout the cooling period. The cooling medium also must have continuous contact with the surfaces of the individual vegetables. For reasonable cooling efficiency, the temperature of the cooling medium should be at least at the recommended storage temperature for the commodity (Table 14). Cooling efficiency is also affected by container design, as containers with inadequate venting or drainage, as well as poorly stacked pallets, can significantly restrict the flow of the cooling medium. This not only increases cooling time but may also compromise product quality.

COOLING METHODS

The cooling rate is not only dependent upon time, temperature, and contact with the commodity; it is also dependent upon the cooling method being used. The various cooling methods have different capacities to remove heat.

Room Cooling

The simplest but slowest cooling method is room cooling, in which the bulk or containerized commodity is placed in a refrigerated room for several hours or even days. Air is circulated by fans past the evaporator coil to the room, dropping the temperature of the produce items that are stored in it. Vented containers and proper stacking are critical to minimizing obstructions to airflow and ensuring maximum heat removal. Room cooling is generally used for the long storage of crops and is satisfactory only for commodities with low respiration rates, such as mature potatoes, dried onions, and cured sweetpotatoes. Even these crops may require precooling when harvested under high ambient temperatures. It is important to maintain high relative humidity in the room, as room cooling can also cause significant water loss, especially for prolonged storage periods. More information can be found on this in the forced-air cooling section below.

Forced-Air Cooling

The cooling efficiency of refrigerated rooms can be greatly improved by increasing the airflow through the product. This principle led to the development of forced-air cooling, in which refrigerated room air is drawn at a high flow rate through specially stacked containers or bins by means of a high-capacity fan. This method can cool as much as four times faster than conventional room cooling. Forced-air cooling is an efficient method for pre-cooling. In many cases, cold storage rooms can be retrofitted for forced-air cooling, which requires less capital investment and has fewer environmental implications (water/ice usage) compared to other available pre-cooling methods. However, to achieve such rapid heat removal, the refrigeration capacity of the room may need to be increased to be able to maintain the desired air temperature during cooling. Alternatively, portable systems can be taken to the field.

With either room cooling or forced-air cooling, precautions must be taken to minimize water loss from the product. The refrigeration system dehumidifies the cold-room air as water vapor in the air condenses on the evaporator coil. This condensation lowers the relative humidity in the room. As a result, the product loses moisture in the air, which is, in turn, removed from the cold room in the form of drained water. To minimize water loss during cooling and storage, the ambient relative humidity should be maintained at the recommended level for the specific crop (commercial humidification systems are available), and the product should be promptly removed from the forced-air precooler upon achieving *7/8 Cooling*. Forced-air cooling is recommended for most fruit-type vegetables and is especially appropriate for vegetables such as peppers and tomatoes but would not be appropriate for leafy crops.

Hydrocooling

Hydrocooling removes heat at a faster rate than forced-air cooling. The heat capacity of refrigerated water is greater than that of air, which means that a given volume of water can remove more heat than the same volume of air at the same temperature.

Hydrocooling is beneficial in that it does not remove water from the commodity. It is most efficient (and, therefore, most rapid) when individual vegetables are cooled by immersion in flumes or by overhead drench since the water completely covers the product surfaces. Cooling becomes less efficient when the commodity is hydrocooled in closed containers, and even less efficient when containers are palletized and hydrocooled. It is important to continuously monitor the hydrocooler water and product temperatures (incoming and outgoing) and adjust the amount of time the product is in the hydrocooler accordingly to achieve thorough cooling.

Sanitation of the hydrocooling water is critical since it is re-circulated. Decay organisms and foodborne illness pathogens present on the vegetables can accumulate in the water and contaminate products being hydrocooled. Cooling water should be changed frequently. Commodities that are hydrocooled must be sufficiently resistant to withstand the force of the water drench. The container must also have sufficient strength to resist the application of water. Crops recommended for hydrocooling include sweet corn, snap beans, cucumbers, and summer squash.

Contact Icing

Contact icing has been used for both cooling and temperature maintenance during shipping. The heat from the product is absorbed by the ice, causing it to melt. Provided the contact between the ice and produce is maintained, cooling is rapid, and the melted ice serves to maintain a high humidity level in the package, which keeps the produce fresh and crisp. However, the non-uniform distribution of ice reduces the cooling efficiency. There are two types of contact icing: *top icing and package icing*.

Top icing involves the placement of crushed ice over the top layer of the product in a container prior to closure. Although relatively inexpensive, the cooling rate can be slow since the ice only directly contacts the product on the top layer. For this reason, it is recommended that top icing be applied after precooling to crops with lower respiration rates, such as celery, but is not appropriate for chilling sensitive crops. Prior to shipping, ice is blown on top of containers loaded in truck trailers to aid in cooling and maintenance of higher relative humidity. However, care should be taken to avoid blockage of vent spaces in the load; this restricts airflow, which results in the warming of the product in the center of the load during shipment. Ice should also be "tempered" with water to bring the temperature to 32°F (0°C) to avoid freezing of the product.

Package Icing. Crushed ice distributed within the container is known as package icing. Cooling is faster and more uniform than for top icing, but it can be more labor-intensive to apply.

A modified version of package icing utilizes a slurry of refrigerated water and finely chopped ice drenched over either bulk or containerized produce or injected into side hand holds. This "slush ice" method has been widely adopted for commodities tolerant to direct contact with water and requiring storage at 32°F (0°C). The water acts as a carrier for the ice so that the resulting slush, or slurry, can be pumped into a packed container. The rapidly flowing slush causes the product in the container to float momentarily until the water drains out the bottom. As the product settles in the container, the ice encases the individual vegetables by filling air voids, thus providing good contact for heat removal. Slush icing is somewhat slower than forced-air cooling, but it does reduce pulp temperatures to 32°F (0°C) within a reasonable amount of time and maintains an environment of high relative humidity. Container selection is critical. The container must be oversized to accommodate sufficient ice to provide cooling. Corrugated fiberboard cartons must be resistant to contact with water (usually impregnated with paraffin wax) and must be of sufficient strength so as not to deform when filled with the slurry. Shipping operations must also plan for water dripping from the melting ice during handling and storage. These pallets cannot be double stacked because the melting ice could carry contamination. Package icing is successfully used for leafy crops, broccoli, sweet corn, green onions, and asparagus.

Vacuum Cooling

Vacuum cooling is a very rapid method of cooling and is most efficient for commodities with a high surface-to-volume ratio, such as leafy crops. This method is based on the principle that, as the atmospheric pressure is reduced, the boiling point of water decreases. Containerized or bulk product is thoroughly wetted, placed in a vacuum chamber (tube), and sealed. The pressure in the chamber is reduced until the water on the product surface evaporates at the desired precooling temperature. As water on the product surface evaporates, it removes field heat; the resultant vapor is condensed on evaporator coils within the vacuum tube to increase cooling efficiency. Any water that evaporates from the vegetable tissue is removed uniformly throughout the product. Therefore, it does not tend to result in visible wilting in most cases.

Precautions must be taken so as not to cool the products below their chilling temperature threshold. Vacuum coolers are costly to purchase and operate, and are normally used only in high-volume operations or are shared among several growers. Commodities that can be cooled readily by vacuum cooling include leafy crops, such as spinach, lettuce, and collards.

Cooling Summary

When selecting an appropriate cooling method, several factors must be considered, including the maximum volume of product requiring precooling on a given day, the compatibility of the method with the commodities to be cooled, subsequent storage and shipping conditions, and fixed/variable costs of the system.

GENETIC SELECTION

Just like variety selection for preharvest and harvest attributes, postharvest genetic selection can offer a long-term strategy to extend shelf-life in vegetables while reducing the need for intensive cooling treatments. Through traditional breeding and modern molecular tools, traits such as delayed senescence, enhanced cuticle thickness, and improved water retention can be introduced or selected for. These genetic improvements help vegetables maintain firmness, reduce susceptibility to chilling injury, and preserve nutritional quality under milder storage conditions. As a result, varieties with enhanced shelf-life not only reduce reliance on energy-demanding cooling systems but also minimize postharvest losses across the supply chain.

METHODS TO MAINTAIN PRODUCE QUALITY DURING STORAGE

Controlled Atmosphere (CA) Storage And Modified Atmosphere Packaging (MAP)

Vegetables, once harvested, continue to respire, consuming oxygen (O_2) from the atmosphere, utilizing stored carbohydrates, and producing carbon dioxide (CO_2). While cooling is the primary method of slowing down respiration rate and extending shelflife, additional techniques like Controlled Atmosphere (CA) Storage and Modified Atmosphere Packaging (MAP) are often used in conjunction with cooling to further enhance the preservation of fresh vegetables. By reducing the availability of O_2 and/or increasing the concentration of CO_2 around the produce, these methods can significantly extend shelf-life beyond what cooling alone can achieve.

CA storage is particularly useful for the long-term storage of bulk quantities of vegetables. However, it can be cost-prohibitive due to the need for specialized storage facilities and equipment. Despite this, CA storage can dramatically extend the shelf-life of vegetables, ranging from an additional 1-2 weeks for leafy greens (compared to cooling alone) to several months for root vegetables, onions, or cabbage. Typically, CA conditions involve reducing O_2 levels from 21% (the normal atmospheric level) to 2-5% and increasing CO_2 levels from 0.04% to 2-10%, with the remainder of the atmosphere being nitrogen, an inert gas. These gas concentrations are lethal to humans, necessitating careful handling, proper personal protective equipment (PPE), and strict safety procedures.

MAP is a more accessible and practical alternative to CA storage, especially for individual packages rather than bulk storage. There are two primary methods within MAP: Active MAP and Passive MAP. Active MAP involves packaging vegetables in materials that are not permeable to gases. Before the package is sealed, a special mixture of gases (similar to those used in CA storage) is introduced into the package, or the entire packaging process is conducted under controlled atmospheric conditions. Passive MAP, on the other hand, typically uses semi-permeable plastic containers or films that allow gas exchange between the inside of the package and the surrounding air. The vegetables inside these packages naturally respire, consuming O_2 and releasing CO_2. Over time, the atmosphere inside the package adjusts to a modified state, which slows the respiration rate and extends the shelf-life of the produce.

ENTERPRISE BUDGETS FOR HORTICULTURE CROPS

In the pages provided, producers can calculate budgets for individual vegetable crops they plan to grow. It is important to note these calculations are not exact (as they are calculated before the season happens) but it can be important for growers to figure out what is financially feasible. If it is a challenge to show profitability on paper, a grower may decide to grow something different or may want to strategize ways to improve their outcomes. Knowing how much is invested in a crop can help a grower decide yield or prices needed to be profitable. A budget determines what the break-even amounts are for each crop. Calculations made from a budget are the first step to a successful marketing plan.

In the budget template, categories are listed for common expenses on a farm. Each farm will be different as to which categories are used. At times it can be a challenge to figure out the right amounts, prices, or expected costs. Guidelines and practical tips are provided in this section to help growers calculate their numbers. Your state Cooperative Extension Service may have enterprise budgets available for use. Extension Agents, lenders, and other trusted advisors can help. However, with a non-traditional crop or dated budget files, it may be the farmer has to start from scratch. Either way, the purpose of this section is to help farms estimate potential costs and returns for their enterprise.

Gross Revenue:

It is likely that a farm will market their crop several different ways. A local farmers market, restaurants, and wholesale markets are examples. Likely, the price received at each location will differ. Determining how much of a crop will go to a particular market and the expected price helps determine potential revenue earned. It is important the farm sells above their break-even amounts to earn a profit from their activities. Historical sales prices are good to reference as are conversations with chefs, business owners, and customers. It is important not to think of this as a guarantee but an estimate of expected revenue.

Another important item is how much the crop is expected to produce. If the crop has been grown in the past that is the best indicator, but crop production guides, Extension Agents, seed companies or co-ops can give estimates on expected yield.

Weather, pests, and other factors can impact actual yield, so it is best to be conservative when calculating this.

Direct Costs:

These are all the expenses that are needed to produce and market an item. The include inputs like seeds, plants, fertilizers, sprays, labor, utilities, fuel, oil, and marketing costs. As you produce more you will incur more of these costs, though they tend to get cheaper *per-unit* the larger the scale of production. Some of these may be fairly accurate, like knowing the number of seeds or plants needed and the current price, while others are a rough calculation. Resources like the Southeast Vegetable Handbook can help give guidelines on fertilizer and pesticide rates.

As an example, if roughly half a gallon of *X* is expected to be used, what is the current price of *X,* and how many (labor) hours will it take to apply it?

It is important to calculate all labor hours associated with the crop. This includes an owners individual time spent working as well. Account for any difference in wages earned by machine operator(s) or skilled labor versus general farm labor. Often labor can be one of the highest expenses, so it is important not to underestimate this. It may also help an owner determine how to use labor most efficiently.

Other direct expenses may be marketing costs like containers that will be needed. As an example, if 10,000 lbs. is the expected yield and the crop is loaded into 20 lb. boxes, at least 500 boxes will be needed to be purchased. Marketing materials could include boxes, crates, bags, trays, labels, shipping containers, or other items needed to haul the product for sale.

Often machinery and equipment will be used in the operation. For the template, it is included in indirect costs though there is a direct and indirect nature to these costs. Machinery and equipment will depreciate over time but will depreciate faster with more use. It can be difficult to calculate just how much an item depreciates while it is being used.

   a. One method is taking the price paid (or replacement cost) for an item, $10,000 for a small tractor, and estimating its useful life, 500 hours. $10,000/500 hours = $20. Each time the tractor is used for an hour, $20 is the estimated cost. Then it will be estimated how many hours the tractor will be used for each crop.

   b. Another method, if the price was $10,000 and it is expected to be used for 5 years, $2,000 per year would be an estimate. It can then be applied per crop. If 50% of the time the equipment is used for tomatoes, 50% of $2,000 would be applied to the tomato budget that year.

   c. The simplest method is taking a percentage each year as a general depreciation amount. For example, with the value of all equipment, estimating that 15% will be depreciated each year (which equates to replacing equipment every 6 – 7 years). Then a percentage of that 15% annual depreciation would be applied to individual crops based on the farm. For instance, if all items of equipment cost $100,000 for the farm, each year it is estimated 15% of that, or $15,000 is the depreciation expense. Taking a smaller percentage of that to apply to a crop gives you a general estimate. If tomatoes were 50% of production, then $7,500 ($15,000 x 50%), would be considered depreciation expense for tomatoes.

It is important to note with the depreciation methods mentioned above it is for enterprise profitability purposes and is not the same as what may show up on a tax return for depreciation.

Indirect Costs:

Each business will have certain costs that are incurred by the business that are not attributable to one activity. Often, they will be the same types of expenses and may not vary much from year to year. For instance, insurance for a farm building is important but is not solely attributable to one crop as the building may be used by all enterprises on the farm. Still the insurance will need to be paid so it is important that it is included.

There are several methods for being able to estimate and assign overhead. If one crop is expected to be 50% of the production of the farm, then indirect costs could be attributed 50% to that crop budget. If the producer knows items are more attributable to one activity or another, they may decide to assign it that way. The percentages can vary crop-by-crop, but it is important to ensure 100% of these costs are included.

As a very rough calculation, if direct costs are known, calculating an additional 10-15% as "overhead" is one method to include these expenses. For instance, if all direct costs are $50,000, multiplying that by 10% will give you an estimate of overhead cost, $5,000. This is an area where a farmer may be able to pencil in a more accurate percentage based on their historical production. Typically, indirect costs are not as large as direct costs but still need to be paid for to calculate profitability.

Net Margin:

The goal is that expected revenues exceed potential costs. In some cases, there may be a comfortable margin, and others may be close or even negative. If the expected margins are tighter than hoped for, examine activities you can modify to improve potential profitability. This may include other marketing avenues, raising prices, exploring alternative chemicals or nutrients, machinery or equipment replacing farm labor, or cooperative working agreements with other businesses or producers. Anything that can raise revenues and decrease expenses will have a positive impact on profitability. When budgets for each crop on the farm are calculated, this makes up the total farm budget and can give a full picture of profitability for the year.

ADDITIONAL RESOURCE CAN BE FOUND ON THE FOLLOWING WEB SITES:

Alabama Cooperative Extension System:
https://www.aces.edu/blog/topics/farm-management/enterprise-budgets-for-horticulture-crops/

Clemson University:
https://www.clemson.edu/extension/agribusiness/resources/request-budget.html

Louisiana State University:
https://www.lsuagcenter.com/portals/our_offices/research_stations/deanlee/features/enterprise-budgets

Mississippi State:
https://www.agecon.msstate.edu/whatwedo/budgets.php

North Carolina State University:
https://cals.ncsu.edu/agricultural-and-resource-economics/extension/business-planning-and-operations/enterprise-budgets/

Oklahoma State University:
https://extension.okstate.edu/programs/farm-management-and-finance/budgets/

Texas A&M:
https://agecoext.tamu.edu/resources/crop-livestock-budgets/

University of Arkansas:
https://www.uaex.uada.edu/farm-ranch/economics-marketing/farm-planning/budgets/

University of Georgia:
https://agecon.uga.edu/extension/budgets.html

University of Kentucky:
https://ccd.uky.edu/resources/budgets

University of Tennessee:
https://arec.tennessee.edu/extension/budgets/

Virginia Tech:
https://www.pubs.ext.vt.edu/tags.resource.html/pubs_ext_vt_edu:enterprise-budgets

TABLE 15. SOUTHEAST US VEGETABLE CROP HANDBOOK BUDGET - TEMPLATE

	UNIT	AMOUNT	PRICE	TOTAL
GROSS REVENUE				
Market 1				
Market 2				
Market 3				
			TOTAL REVENUE	$
DIRECT COSTS				
Seeds / Plants				
Fertilizer				
Crop Protection				
Herbicides				
Fungicides				
Insecticides				
Plastic Mulch				
Irrigation				
Machinery				
Fuel				
Repairs				
Labor				
Planting / General Labor				
Operator Labor				
Harvest Cost				
Harvest Labor				
Harvesting & Hauling				
Packaging				
Other				
Irrigation Supplies				
Tools & Accessories				
Interest On Op. Cap.				
			TOTAL DIRECT COSTS	$
INDIRECT COSTS				
Depreciation				
General Overhead				
			TOTAL INDIRECT COSTS	$
			TOTAL COSTS (INDIRECT + INDIRECT):	$
			NET MARGIN (TOTAL REVENUE - TOTAL COST):	$

Specific Crop Recommendations

For further information about Insect, Disease and Weed Control, see the appropriate control section after these specific crop recommendations.

ASPARAGUS (*Asparagus officinalis*)

VARIETIES[1]	AL	GA	KY	LA	MS	NC	OK	SC	TN	VA
ASPARAGUS										
Jersey Giant [2]	A	G	K	L	M	N		S	T	V
Jersey King [2]						N				V
Jersey Knight [2]	A	G	K	L	M		O	S	T	V
Jersey Supreme [2]	A	G	K	L		N		S	T	V
Purple Passion	A		K	L	M	N			T	V
Millennium [2]			K			N				V
UC157 F[1]	A	G	K	L	M		O			

[1] Abbreviations for state where recommended   [2] Male hybrid

Soil Preparation. Be sure to soil test in order to determine liming and fertilizer requirements. The ideal pH for asparagus is between 6.7 and 7.0. Asparagus does not tolerate acidic soils and will not grow well at or below a pH of 6.0. Fungal diseases that contribute to asparagus decline (Fusarium Crown and Root rot) survive better at lower pH. Liming the soil 7.0 – 7.5 will reduce the survivability of Fusarium. Apply 100 lbs/acre of nitrogen. If no soil test is performed, supply sufficient phosphorus and potassium so that the soil contains 250 lbs/acre of available phosphorus and 300 lbs/acre of available potassium. Phosphorus does not move readily in the soil and cannot be incorporated into the soil after the asparagus is planted, so it must be incorporated prior to planting.

Asparagus grows and yields best in a deep, well-drained sandy loam soil, but will tolerate heavier soils as long as the soil has good internal drainage and the water table does not come within four feet of the soil surface as this would interfere with the extensive and deep root system.

Broadcast the fertilizer and plow it under when preparing the land for the planting furrows. Then, each year after harvest is complete, broadcast 100 lbs/acre nitrogen and other nutrients (if needed). Lime can also be added at this time. For the first four years, soil test yearly to determine if fertility and pH adjustments are necessary. Fertilizing in the spring before spears emerge will not benefit the developing crop since the buds on the crown were formed utilizing nutrients from the previous year. After four years, soil test every two years.

Planting. An optimal soil temperature of 50°F is critical for rapid growth by crowns. See "Asparagus Planting Dates" table for suggested dates. Avoid planting crowns into cold soil. Prolonged exposure to cool, wet soils will make crowns more susceptible to Fusarium Crown and Root Rot. **If crowns are received before the field is ready to plant, crowns must be stored between 33 - 38°F. Otherwise, the buds on the crowns will sprout, causing the fleshy crown roots to shrivel and die.**

Asparagus crowns and transplants are placed into furrows. Make furrows 6" deep. On a heavy soil, plant crowns no deeper than 5" and on a light textured soil, no more than 6". Apply fertilizer in the bottom of the furrow before planting crowns. Place crowns in the bottom of the furrow and cover with 1 to 2" of soil. The fertilizer will not burn the crowns. Although crown orientation is not important, crowns placed with their buds oriented upward will emerge faster. Research shows that pre-plant applications of phosphorus below the crown are an important factor in long-term asparagus production. Omitting the phosphorus placed in the bottom of the furrow will reduce yields in subsequent years as compared to not applying the additional phosphorus.

NOTE: Asparagus crowns are received in bulk or in bundles of 25 crowns per bundle. After receiving, separate the crowns by size into small, medium, or large. When ready to plant, plant all the smalls together in the same row, all the mediums together, and all the large crowns together. Do not plant a small crown next to a medium or large sized crown. This will cause the larger crown to shade the smaller one, which will then never attain its full growth potential.

Spacing. Crowns can be spaced 12" to 18" within the row. Research shows that there is no advantage of planting 9" between crowns in the row. Although a larger yield is obtained earlier with 9" spacing, after 4 or 5 years, the yield will be the same as with 18" in row spacing. Also, the closer the crowns are spaced in the row, the more crowns needed, increasing cost (for example, 18" in row x 5 feet between row = 5,808 crowns per acre; 12" crowns in row x 5 feet between row = 8,712 crowns per acre).

Asparagus crowns should not be planted in a solid block; rather, plant the field with drive rows spaced between a block of five rows. In order to obtain optimal spray coverage, an air-blast sprayer is needed to evenly apply insecticides and fungicides into the dense fern canopy from both sides of the five-row block. Boom sprayers usually cannot be set high enough to prevent the knocking over of ferns causing damage.

The furrows can be filled-in completely to soil level after planting without damaging the crowns. Do not drive on or compact the soil over the newly planted furrows, however; or emergence of the spears will be severely delayed or reduced. With good soil moisture, the new spears will break through the soil in 1-2 weeks.

ASPARAGUS PLANTING DATES*

	Planting Dates
AL North	2/15–4/15
AL South	1/15–3/15
AR North	3/1-5/1
AR South	2/15-4/1
GA North	2/15–4/15
GA South	1/15–3/15
LA	3/15–4/15
KY East	3/20–4/1
KY Central	3/15–3/25
KY West	3/10–3/20
MS	3/15-4/15
NC East	2/15–3/31
NC West	4/1–5/31
OK	4/1–5/1
SC East	2/1–3/15
SC West	3/1–4/15
TN East	3/1–3/31
TN West	2/25–3/15
VA East (coastal)	3/1–4/15
VA West (mountains)	4/1-5/31

* After the last frost date for your area of the state.

SPECIAL NOTES FOR PEST MANAGEMENT
WEED MANAGEMENT

Weed control is critical in asparagus. If young plants compete with weeds, these young plants will become stressed preventing them from developing good fern growth. Cultivation is not recommended as there are effective herbicides labeled for use. Research shows that even the shallowest of cultivations between asparagus rows cuts and injures roots, predisposing them to Fusarium root rot fungus that eventually will kill the asparagus. Apply a pre-emergence and post-emergence herbicide over the entire field after planting crowns and again after the old fern growth is mowed each spring. Apply an herbicide three weeks prior to the emergence of new spears and ferns, so that these newly emerging spears and fern growth will not compete with weeds.

Although asparagus is highly salt tolerant and salt can be used to control weeds, salt will cause severe soil crusting; impeding water infiltration and percolation. Additionally, salts can leach horizontally through the soil killing other vegetables adjacent to the asparagus which are not as salt tolerant.

INSECT MANAGEMENT
Cutworms feed on the spear tips at night before emerging from the soil. They feed on one side of the spear, causing the tip to bend over. Cutworms can easily be managed with approved insecticides.

Asparagus Beetle adults chew on ferns reducing photosynthesis. Any reduction in leaf area causes a loss of stored food reserves in the crown which is needed for next year's crop. Asparagus beetles also lay eggs on the spears during harvest and will result in further damage. During this period, the best way to manage the beetle is to pick on a timely basis preventing any spear getting tall and spindly, or allowing them to fern out.

DISEASE MANAGEMENT
Accurate diagnosis is necessary for effective disease management. Incorporate appropriate and effective cultural practices and pesticides, as detailed in the Disease Management section, in a season-long integrated disease management program.

Cercospora Needle Blight: This fungal disease produces spores that are wind-blown during the summer when hot and humid. Needles of the fern first turn yellow, then brown, and then defoliate. This severely reduces the photosynthetic capability of the ferns to manufacture carbohydrates which are critical for next year's spears. Spray an approved fungicide to manage Cercospora when reddish-brown, football-shaped lesions on the fern stalks are first noticed. Research in NC has demonstrated that spraying once every 7 days from early July through September, alternating with chlorothalonil and mancozeb weekly, successfully manages Cercospora. Spray once every 7 to 10 days through September. Neglecting to spray might reduce spear yields up to 40% the following year. Burning the old ferns off instead of mowing them off and letting the residue remain on the ground will not prevent Cercospora. Be prepared to spray, regardless if the old ferns are burned or not.

Fusarium Crown and Root Rot: This disease can be a destructive disease of asparagus and can take fields out of production. There are no effective management options once plants succumb to this disease. Prevent infection by limiting plant stress caused by overharvesting, low soil pH and fertility, frost damage to spears, waterlogged soil, and pests. Consider establishing beds with hybrid varieties that have exhibited tolerance to the disease.

Purple Spot: This fungal disease commonly affects ferns and spears. Disease should be managed on ferns to reduce inoculum available for disease development on spears. Additionally, fungicides are only labeled for use on ferns and have lengthy pre-harvest intervals. Fungicide applications should be planned accordingly.

HARVESTING AND STORAGE
During the second year, about 3 weeks before the spears emerge, mow off the dead ferns and spray a tank-mix of an approved preemergence and post-emergence herbicide. Mow the dead ferns off as close to the ground as possible. Do not cut ferns down in the fall because the dead ferns will capture moisture in the winter and will keep the soil temperature about 5 degrees colder than the temperature of bare soil. This colder soil temperature will delay spear emergence in the early spring when warm day temperatures force the growth of new spears in bare soil, causing frost injury on spears.

With air temperatures (<70°F; <21°C), harvesting might be done once every 2 to 3 days, harvesting a 7 to 9" tall spear with tight tips. Air temperatures over 70°F (21°C) will cause the tips of the spears to open up or "fern out" at a shorter height, causing fiber development in the spears and making them tough. Fiber development is determined by the tightness of the spear tip but has no bearing on spear toughness. Harvesting under warm temperatures forces the grower to pick shorter, 5 to 7" tall spears before the spear tips fern out. Otherwise, the spears will lack tenderness and quality. This involves harvesting in the morning and evening of the same day as spears elongate rapidly under high temperatures.

Asparagus can be harvested safely for 2 weeks during the second year (the year following the initial establishment of crowns), 4 weeks during the third year, 6 weeks during the fourth year, and 8 weeks during the fifth year. It is best to determine when to stop harvesting by looking at the spear diameter. When ¾ of the spears are pencil-sized in diameter, stop harvesting. This will take some experience to determine.

Asparagus can be harvested with a knife, below the soil, but snapping is preferred. Using a knife, results in a tough and fibrous butt being produced that has to be trimmed off. Cutting below the soil with a knife increases the chances of cutting into other buds on the crown that would normally produce more spears. Snapping involves using the thumb and index finger together to gently break the spear just above the soil line. Snapped asparagus contains no fibrous butt that needs to be removed. The harvested spear is all usable.

Do not allow any small spindly spears to grow into ferns while harvesting. These ferns can provide a site for asparagus beetles to lay their eggs, develop into larvae, and then into adult beetles. The field *should* be free of any tall, spindly spears or fern growth during harvest, except for new spears coming up or ones ready to be harvested.

Harvest asparagus in the morning when the temperatures are cool. Asparagus has an extremely high respiration rate, just like a fresh cut flower. Place harvested spears into plastic containers that have holes in them to let water pass through. Plunge them into ice-cold water for about 5 minutes. This will remove the field heat out of the spears. Next, allow containers to drain and put them into plastic bags. Place into refrigerated storage set at 36°F (~2°C) with 95-100% relative humidity.

Storage life at 36°F (~2°C) is about 2 weeks, but growers should try to sell the asparagus soon after it is picked, as it may develop chilling injury symptoms after 10 days, allowing the consumer to hold it for 2 weeks if needed.

Asparagus should be stored upright to prevent the bending of tips, as spears will continue developing during postharvest storage. See Table 14 for further postharvest information.

BEANS: LIMA/BUTTER (*Phaseolus lunatus*) AND SNAP (*Phaseolus vulgaris*)

VARIETIES[1]	AL	GA	KY	LA	MS	NC	OK	SC	TN	VA
BEANS- Lima										
Bush (small seeded)										
Bridgeton	A	G		L	M	N		S	T	
Dixie Butterpea	A	G			M			S		
Early Thorogreen	A	G		L	M	N		S		
Henderson Bush	A	G		L	M	N		S		
Jackson Wonder	A			L	M			S	T	
Bush (large seeded)										
Fordhook 242	A	G	K			N		S	T	
Dixie Speckled Butterpea	A	G			M	N		S	T	
Pole (large seeded)										
Carolina Sieva	A	G		L	M	N		S		
Florida Speckled Butter		G		L	M	N		S		
King of the Garden	A	G		L	M	N		S	T	
Willow Leaf	A			L						
BEANS- Snap										
Bush (Fresh Market)										
Bartram		G				N				
Bronco	A	G	K	L	M	N	O	S	T	
Bush Blue Lake 274	A	G	K	L		N				
Caprice	A	G	K			N		S	T	V
Colter		G	K							
Hialeah	A		K	L		N		S		
Jackson		G								
Jade II	A		K			N				
LaSalle		G								
Lewis	A		K							
Lynx				L		N		S		
Magnum	A	G	K		M	N			T	
Momentum		G				N				
Prevail	A	G								
Provider						N				
PV 857		G						S		
Roma II (flat pod)	A	G	K		M	N		S		
Strike	A	G		L		N		S		
Sybaris		G								
Valentino		G		L			O		T	
Pole										
Cornfield Greasy/Cornfield Beans [4]						N			T	
Greasy Beans/Greasy Cutshorts [4]						N			T	
Half Runners [4]						N				
Kentucky Blue			K	L	M	N			T	
Louisiana Purple Pole				L	M				T	
McCaslan	A	G		L		N			T	
Mountain Half Runner						N				
Rattlesnake				L	M				T	
Red Noodle [2]	A				M	N			T	
State Half Runner		G	K			N			T	
Stringless Blue Lake	A	G		L		N		S		
Volunteer/Tennessee Half Runner [3]		G	K			N			T	
White Seeded Kentucky Wonder 191	A	G	K	L	M	N		S	T	

[1] Abbreviations for state where recommended
[2] Yard long/Asparagus bean
[3] Not for Coastal Plain areas
[4] Heirlooms

Seed Treatment. To protect against root rots and damping off, use treated seed or treat with an appropriate protectant at manufacturer's recommendation. Further information on seed treatments can be found under Seed Treatments in the disease management section of the handbook. Rough handling of seed greatly reduces germination.

MARKET SNAPS PLANTING DATES

	Spring	Fall
AL North	4/1–7/15	NR
AL South	2/10–4/30	8/15–9/20
GA North	5/1–7/15	NR
GA South	2/15–4/30	7/15–9/15
KY East	5/1–7/15	NR
KY Central	4/25–7/25	NR
KY West	4/10–8/1	NR
LA North	4/1–5/15	8/15–9/15
LA South	3/1–5/31	8/15–9/15
MS North	3/30–5/10	8/15–9/1
MS South	2/10–5/1	8/15–9/20
NC East	3/20–6/15	8/1–9/15
NC West	5/1–8/15	NR
OK	3/31–4/30l	8/15–9/15
SC East	4/1–6/1	8/1–9/1
SC West	4/15–7/1	7/20–8/1
TN East	4/20–6/20	7/15–8/20
TN West	4/1–6/1	NR
VA (coastal)	3/30–7/20	NR
VA (mountains)	4/10–7/30	NR

NR = Not recommended

PROCESSING SNAPS PLANTING DATES

	Spring	Fall
AL North	4/1–7/15	NR
AL South	2/10–4/30	8/15–9/20
GA North	5/1–7/15	NR
GA South	2/15–4/30	7/15–9/15
KY East	5/1–7/15	NR
KY Central	4/25–7/25	NR
KY West	4/10–8/1	NR
MS North	4/1–5/15	9/5–9/20
MS South	2/10–4/30	8/15–9/20
NC East	4/1–6/15	NR
NC West	5/15–7/31	NR
SC East	4/1–6/1	8/1–9/1
SC West	4/15–7/1	7/20–8/1
TN East	4/20–8/20	7/15–8/20
TN West	4/1–7/15	NR

NR = Not recommended

LARGE & SMALL LIMAS PLANTING DATES

	Spring	Fall
AL North	4/1–7/1	NR
AL South	2/10–5/1	8/15–9/20
GA North	5/1–7/1	NR
GA South	3/1–5/1	7/15–9/1
KY East	5/10–7/10	NR
KY Central	5/1–7/20	NR
KY West	4/15–7/1	NR

LARGE & SMALL LIMAS PLANTING DATES (cont'd)

	Spring	Fall
LA	4/1–7/25	NR
MS North	4/1–7/25	NR
MS South	3/1–8/15	NR
NC East	4/10–6/15	7/15–8/1
NC West	6/1–7/15	NR
SC East	4/15–6/1	7/15–8/1
SC West	5/1–6/15	7/1–7/15
TN East	5/1–6/30	7/15–8/20
TN West	4/15–7/15	NR

NR = Not recommended

SOIL AND FERTILITY

Snap beans grow best on medium-textured, well drained soils. Commercial producers generally apply 65 lbs N/A by banding at planting 2 inches to each side and 3 inches below the seed. Direct contact with seed can cause injury or kill germinating seed.

SPACING

Snap Beans: With rows 30 to 36 inches apart, plant 5 to 7 seeds per foot. To increase yield plant in rows 18 to 24 inches apart with 4 to 6 seeds per foot. Calibrate planter according to seed size. Sow 1 to 1.5 inches deep in light sandy soil; shallower in heavier soil.

Lima Beans, Large Seeded: Plant in rows 30 to 36 inches apart, 2 seeds per foot, 1 to 1.5 inches deep.

Lima Beans, Small Seeded: Space rows 30 to 36 inches apart, 2 seeds per foot, 0.75 to 1.25 inches deep (deeper if soil is dry). For mechanically harvested irrigated fields: Rows 18 to 30 inches apart, 4 to 5 inches between plants.

Edamame: Edamame are green, immature soybeans sold as fresh vegetables with the seeds in the pods. Grown like bush beans, the pods are harvested when the seeds have reached full size but before the pods begin to yellow. Some commonly grown varieties include Midori Giant, Tohya, Lanco, and Envy.

SPECIAL NOTES FOR PEST MANAGEMENT
INSECT MANAGEMENT

Seed Maggot: See the "Seed Treatment" section, or use approved soil systemic insecticides at planting time if probability of pest outbreak is high.

Experience has shown that effective insect control with systemics usually lasts from 4 to 6 weeks after application. Frequent field inspections are necessary after this period to determine pest incidence and the need for additional spray controls.

Thrips: Treatments should be applied if thrips are present from cotyledon stage to when the first true leaves are established and/or when first blossoms form.

Mites: Spot treat areas along edges of fields when white stippling along veins on undersides of leaves is first noticed and 10 mites per trifoliate are present.

Aphids: Treat only if aphids are distributed throughout the field (50% or more of terminals with five or more aphids), and if beneficial species are lacking.

Leafhoppers: Treat only if the number of adults plus nymphs exceeds 1 to 2 adults per sweep.

Tarnished Plant Bug (Lygus): Treat only if the number of adults and/or nymphs exceeds 15 per 50 sweeps from the pin pod stage until harvest.

Beet Armyworm (BAW), Cabbage Looper (CL): Treat if the number of worms (BAW and CL) averages 15 per 3 feet of row.

European Corn Borer (ECB)–Snap Beans Only: Treat when moth catches in local blacklight traps average five or more per night. The first application should be applied during the bud–early bloom stage and the second application during the late bloom–early pin stage. Additional sprays may be needed between the pin spray and harvest. Consult a pest management specialist for local blacklight trap information and recommended spray intervals.

Corn Earworm (CEW), Fall Armyworm (FAW): In snap beans, treat every 5 to 7 days if CEW catches in local blacklight traps average 20 or more per night and most corn in the area is mature. The use of pheromone (insect sex attractants) and blacklight traps is very helpful in detecting population build-up.

For limas, treat when CEW populations exceed one per 6 feet of row from the late flat pod stage to harvest.

For both lima bean types, treatment should be timed when 50% or more of the CEW and/or FAW populations reach a length of 1/2 inch or longer. Treating too early for young CEW/FAW populations will eliminate natural control and may result in the need for additional sprays for reinfestations. Consult a pest management specialist for more refined decision-making.

Whiteflies: Treat when whiteflies exceed five adults per fully expanded leaflet.

DISEASE MANAGEMENT

Accurate diagnosis is necessary for effective disease management. Incorporate appropriate and effective cultural practices and pesticides, as detailed in the Disease Management section, in a season-long integrated disease management program.

Root and Crown Rots (Pythium, Rhizoctonia, Fusarium, Southern Blight): Several fungi and oomycetes can cause root and crown rots in beans. Some may also cause damping-off resulting in poor stand development. Management of these diseases relies on crop rotation with nonhost crops, deep-tilling (southern blight), minimizing soil compaction, and use of fungicides applied as a seed treatment or in-furrow, at-planting application.

Pythium (Cottony) Leak: Pythium leak is a common disease of snapbean pods, primarily caused by warm season *Pythium* spp. Effective disease management involves selecting cultivars with an upright architecture and high bean set and using narrow row spacing to keep plants erect and susceptible pod tissue from touching the soil. Fungicides can be useful in managing the disease, but the number of available fungicides is limited.

Anthracnose: Anthracnose can cause up to 100% yield loss when contaminated seeds are planted and prolonged cool, wet weather conditions persist during the growing season. The most effective way to reduce disease severity is with resistant varieties. Additional management measures include cultural practices, such as burying infected crop residue and employing a two-year rotation with a non-host crop to reduce disease inoculum, and application of fungicides.

White Mold: White mold can affect all aboveground parts of lima and snapbeans, both in the field and after harvest. Management includes avoiding fields with a history of severe infestation, rotating with non-host crops like cereals or corn for at least three years, and minimizing late-season irrigation. Harvesting infested fields last and using timely harvest, rapid cooling, and refrigeration can reduce postharvest losses. Fungicides can also be effective if applied during flowering.

NEMATODE MANAGEMENT

Soybean Cyst Nematodes: Soybean cyst nematode, races I and III, are present in soybeans in some areas. Snap beans are susceptible, but small seeded lima beans are resistant to this nematode. Growers who rotate snap beans with soybeans should be alert to the possibility of problems in infested fields.

Root-knot Nematodes (RKN): RKN can be a problem in beans. Yield suppression (up to 90%) can occur on beans when high population densities of RKN are combined with stressful environmental conditions. Damaging levels of nematodes usually occur after several years of continuous planting of susceptible crops. Growing non-hosts crops can lower nematode populations below damaging levels for a sustained period. Only a few lima bean varieties are resistant to RKN. Clean and disinfest equipment between fields. Use appropriate crop rotations and cover crops with non-host crops to reduce nematode populations. Avoid fields with high populations of nematodes or use nematode control tables (Tables 3-42 to 3-44) listed in the Disease Control section.

WEED MANAGEMENT

Since beans are a quickly maturing crop, early weed control is essential. Weeds can reduce yield, quality and the efficiency of the mechanical harvester. Preparing a weed-free seedbed is the first step of a weed control program, which allows the bean plants to get off to a quick start without competition from weed growth. Carefully read herbicide labels ensure that beans can tolerate the material. Be sure to avoid planting beans after other crops for which herbicides with a long residual have been used.

MINIMUM TILLAGE

When planning to use minimum tillage, give consideration to bean variety, date of planting, soil fertility practices, planting equipment, cover crop, and weed species in the field. Minimum tillage might not be suited to all growing areas around the SE US. Soil type and other environmental conditions might limit its success. Consult with your local Extension for the latest recommendations.

HARVESTING AND STORAGE

See Table 14 for postharvest information.

BEETS (*Beta vulgaris*)

VARIETIES[1]	AL	GA	KY	LA	MS	NC	SC	TN	VA
BEETS									
Avalanche (white interior)			K	L					
Bohan							S		
Boldor (yellow interior)	A				M		S		
Boro			K	L					
Bresko							S		
Bull's Blood (for greens)	A	G	K	L	M	N	S	T	V
Chioggia (red and white interior)	A	G	K	L		N		T	V
Detroit Dark Red	A	G	K	L	M		S	T	V
Early Wonder Tall Top	A			L			S		
Red Ace	A	G	K	L		N	S	T	V
Red Cloud			K	L					
Ruby Queen	A	G	K	L		N	S	T	V
Touchstone Gold (yellow interior)			K			N			

[1] Abbreviations for state where recommended

Note: For consistent root size, beets should be thinned. Most beet seeds are naturally "multigerm" varieties. Multigerm seed contain more than one embryo. When these germinate, they will produce 2 to 5 seedlings at once.

Light, well-drained soils are best for root development in beets. Beets are frost tolerant and produce the best commercial quality when grown during cool temperatures (50° to 65°F).

Lighter color and wider zoning within the roots occur during periods of rapid growth in warm temperatures. If plants are exposed to 2 or 3 weeks of temperatures below 50°F after several true leaves have formed, seedstalks (undesirable because they will reduce root quality) will form. Cultivars vary in their sensitivity to this problem with newer cultivars generally being less sensitive to it.

Beets are susceptible to boron deficiency and will develop internal black spot if soil boron is not adequate. If boron is deficient, apply 2 to 3 lb. of boron per acre with broadcast fertilizer, or for smaller plantings, apply ½ oz. Borax per 100 square feet of row with initial fertilizer application.

Seeding and Spacing. Optimum germination temperature for beets ranges from 50° to 85°F, but early plantings can be made 4 to 6 weeks before the average last spring frost. Germination takes between 10-14 days, but can be hastened by soaking seed in warm water prior to planting. Sow seed ½ to 3/4 in deep at the rate of 15 to 18 seeds per foot of row. Space rows 15 to 20 inches apart; thin plants to 3 inches apart. Seeds remain viable for 2-3 years when stored properly.

BEET PLANTING DATES

	Spring	Fall
AL North	3/15–5/30	8/1–9/15
AL South	2/1–3/31	8/1–9/30
GA North	4/15–5/30	7/15–8/15
GA South	2/1–3/31	8/1–9/30
KY East	3/20–4/15	NR
KY Central	3/15–4/10	NR
KY West	3/10–4/1	NR
LA	2/1–3/31	9/15–11/15
MS	NR	8/1–9/1

BEET PLANTING DATES (cont'd)

	Spring	Fall
NC East	3/1–4/15	8/1–9/15
NC West	4/1–5/31	7/15–8/15
SC East	2/15–3/31	8/15–9/30
SC West	3/15–5/31	7/15–8/31
TN East	3/15–4/15	9/1–9/30
TN West	3/1–4/1	9/15–10/1
VA East (coastal)	3/15–4/15	8/1–8/31
VA West (mountains)	4/1–5/31	7/15–8/15

NR = Not recommended

SPECIAL NOTES FOR PEST MANAGMENT

DISEASE MANAGEMENT

Seed Rot and Damping-off: Seed rot and damping-off may be a problem, especially in early spring plantings when soils are cool. Seeds should be treated with an appropriate fungicide to protect the seed.

Cercospora: Cercospora leaf spot is the most common disease that occurs on beets. Circular spots with reddish brown or purplish margins are the first signs. Spray every 2 to 3 weeks with an appropriate crop protectant.

INSECT MANAGEMENT

The most common insect pests of beets are aphids, leafminers, flea beetles, and webworms. Sanitation and crop rotation should be practiced to avoid pest build up.

HARVESTING AND STORAGE

Market beets are hand-harvested when 1-3/4 to 2 inches in diameter, usually about 50-75 days after planting. Expected yield per 100-row feet is 100 lbs. See Table 14 for further postharvest information.

BROCCOLI, BRUSSELS SPROUTS, CABBAGE, CAULIFLOWER, COLLARDS, KALE, AND KOHLRABI (*Brassica* spp.)

Varieties[1]	AL	AR	GA	KY	LA	MS	NC	OK	SC	TN	VA
BROCCOLI											
Abrams (69 DTH)							N		S		
Arcadia [6, 10] (69 DTH)	A	R	G	K	L	M	N		S	T	
Belstar (65 DTH)			G			M	N		S	T	V
Blue Wind (49 DTH)	A			K							V
Burgundy [15, 18] (37 DTH)	A									T	
Burney (60 DTH)	A						N		S		
Castle Dome (50 DTH)	A				L	M	N	O	S	T	V
Destiny [2, 5] (70-75 DTH)						M			S		
Diplomat [10] (68 DTH)	A		G		L		N		S	T	V
DuraPak [15] (90-100 DTH)							N		S		
Eastern Crown (94 DTH)	A		G	K			N		S		
Emerald Crown (59 DTH)	A		G	K		M	N		S		V
Emerald Pride (97 DTH)	A		G		L		N		S		
Emerald Star (105 DTH)			G				N				
Emperor (70 DTH)	A				L	M					
Green Magic [2, 11] (57 DTH)	A	R	G	K	L	M	N		S	T	V
Gypsy [10] (60 DTH)	A	R	G	K	L	M	N		S	T	V
Imperial [13] (66 DTH)	A			K		M		O	S		
Ironman [5] (75 DTH)	A						N		S		V
Jacaranda [15] (50 DTH)	A									T	
Lieutenant [5] (55-65 DTH)	A					M	N	O	S	T	V
Marathon [5, 6, 7, 10] (68 DTH)	A			K	L		N	O			V
Packman (50 DTH)	A	R	G	K	L	M	N			T	V
Patron [2, 10] (94 DTH)	A		G				N		S		
Premium Crop [10, 12] (65 DTH)	A			K	L	M	N				
Roxanne (53 spring/71 fall DTH)							N		S		
BRUSSELS SPROUTS											
Attis [6] (140 DTH)	A										
Capitola (130 DTH)					L		N		S		
Churchill (90 DTH)							N				
Dagan (100 DTH)			G		L		N		S	T	
Dimitri (105 DTH)				K							V
Price Marvel (85 DTH)	A					M			S	T	
Franklin (125 DTH)					L				S	T	V
Gustus (100 DTH)					L				S		
Hestia (93 DTH)			G						S	T	V
Jade Cross E (85 DTH)	A			K		M	N		S	T	V
Marte (130 DTH)											
Nautic (150 DTH)					L				S		
CABBAGE: green											
Blue Dynasty [4, 6, 9]	A		G	K	L		N			T	V
Blue Vantage [4, 6, 8, 9]	A		G	K	L	M	N		S	T	V

*DTH = Days to harvest

[1] Abbreviations for state where recommended
[2] Bolting tolerant
[3] Bolting susceptible
[4] Tip burn tolerant
[5] Hollow stem tolerance/resistance
[6] Black rot tolerance/resistance
[7] Bacterial leaf spot tolerance/resistance
[8] Bacterial speck tolerance/resistance
[9] Fusarium yellows tolerance/resistance
[10] Downy mildew tolerance/resistance
[11] Powdery mildew tolerance/resistance
[12] Suitable for side shoot production
[13] Warm weather tolerance
[14] Orange
[15] Purple
[16] Miniature
[17] Fall only
[18] Sprouting broccoli

Varieties[1]	AL	AR	GA	KY	LA	MS	NC	OK	SC	TN	VA
CABBAGE: green (cont'd)											
Bravo [6,9]	A		G	K	L	M	N		S	T	
Bronco [4,9]	A		G	K			N			T	
Cheers [6,8,9]	A		G	K	L	M			S	T	
Varieties[1]	AL	AR	GA	KY	LA	MS	NC		SC	TN	VA
Emblem [4,6,9]					L						
Platinum Dynasty [4,6,9]	A		G			M	N		S	T	V
Royal Vantage [4,6,8,9]	A		G		L						V
Savoy Ace Improved	A		G	K		M				T	V
Solid Blue 780 [6,9]					L	M	N		S		
Thunderhead [4,6,9]	A		G				N				
Vantage Point [4,6,8,9]					L						V
CABBAGE: red											
Red Dynasty [4,6]	A		G	K	L	M	N		S	T	V
Ruby Perfection [6]	A		G				N				V
CHINESE CABBAGE											
Napa Cabbage											
Allred (red) [17]	A										
China Express	A					M					
Minuet [16]	A					M	N				
Rubicon			G		L	M	N		S	T	
Michihli											
Green Rocket	A				M		T				
Monument	A				M						
Choi											
Joi Choi	A		G	K	L	M	N		S	T	
Mei Qing Choi	A		G		L	M					
Win Win Choi	A		G			M					
CAULIFLOWER											
Candid Charm	A		G	K	L						V
Cheddar [14]	A		G		L		N		S		
Early Snowball	A		G				N		S		
Flamenco [17]			G	K							
Freedom	A		G	K							V
Fremont			G							T	V
Fujiyama [17]			G								
Graffiti [15]	A		G	K	L	M			S	T	
Minuteman			G								V
Symphony			G		L						V
Snow Crown	A	R	G	K	L	M	N		S	T	
White Magic	A		G		L				S		
KOHLRABI											
Early Purple Vienna [15]	A		G		L	M	N		S	T	V
Grand Duke	A				L	M	N		S	T	V
Kolibri [15]	A			K							
Kossak			G	K		M	N			T	
Quickstar	A		G			M	N		S		

*DTH = Days to harvest

[1] Abbreviations for state where recommended
[2] Bolting tolerant
[3] Bolting susceptible
[4] Tip burn tolerant
[5] Hollow stem tolerance/resistance
[6] Black rot tolerance/resistance
[7] Bacterial leaf spot tolerance/resistance
[8] Bacterial speck tolerance/resistance
[9] Fusarium yellows tolerance/resistance
[10] Downy mildew tolerance/resistance
[11] Powdery mildew tolerance/resistance
[12] Suitable for side shoot production
[13] Warm weather tolerance
[14] Orange
[15] Purple
[16] Miniature
[17] Fall only
[18] Sprouting broccoli

Seed Treatment. Check with seed supplier to determine if seed is hot water treated for black rot control. If not, soak seed at 122°F. Use a 20-minute soak for broccoli, cauliflower, collards, kale, and Chinese cabbage. Soak cabbage for 25 minutes.

Note. Hot water seed treatment may reduce seed germination.

Following either treatment above, dry the seed, then treat with a labeled fungicide to prevent damping-off. Further information on seed treatments can be found under Seed Treatments in the disease management section of the handbook.

BROCCOLI PLANTING DATES

	Spring	Fall
AL North	3/1–7/1	8/1-9/1
AL South	2/1–3/31	8/1-9/30
GA North	3/15–7/1	7/25-8/15
GA South	2/1–3/31	8/1-9/30
KY East	4/10–4/30	7/1-7/15
KY Central	4/5–4/20	7/15-8/1
KY West	3/30–4/10	8/1-8/15
LA	1/15–3/15	8/1-10/31
MS North	2/15–3/15	7/25-8/15
MS South	1/15–3/10	8/5-9/15
NC East	2/15–4/15	8/1-9/15
NC West	4/1–8/15	NR
OK	3/7-3/28	7/31-8/20
SC East	3/1–4/10	9/1-9/30
SC West	3/20–4/30	8/15-9/15
TN East	3/25–4/25	8/1-8/31
TN West	3/15–4/5	8/10-8/31
VA East (coastal)	3/15–4/15	7/1-7/31
VA West (mountain)	4/1–4/20	6/20-7/20

Note: Planting dates in the spring could be earlier if low tunnels or other season extension measures are used in some locations

NR = Not recommended

BRUSSELS SPROUTS PLANTING DATES

	Spring	Fall
AL North	NR	6/30-8/1
AL South	NR	8/1-9/15
GA North	NR	6/30-8/1
GA South	NR	8/1-9/15
KY East	NR	6/15-7/15
KY Central	NR	7/1-8/1
KY West	NR	7/15-8/15
LA North	NR	8/1-10/1
LA South	NR	8/15-10/15
MS North	NR	7/15-8/15
MS South	NR	8/1-9/1
NC East	NR	7/15-8/15
NC West	4/15-5/15	NR
SC East	NR	9/15-10/15
SC West	NR	8/15-9/15
TN East	5/15-6/30	7/15-8/15
TN West	4/20-6/20	8/1-9/1
VA East (coastal)	NR	8/1-9/15
VA West (mountain)	NR	8/1-9/1*

Note: Planting dates in the spring could be earlier if low tunnels or other season extension measures are used in some locations. Use transplants.

NR = Not recommended

CABBAGE PLANTING DATES

	Spring	Fall
AL North	3/15–7/1	8/1-9/1
AL South	2/1–3/31	8/1-10/31
GA North	3/15–7/1	7/25-8/15
GA South	2/1–3/31	8/1-10/31
KY East	4/1–4/15	6/20-7/1
KY Central	3/15–3/25	7/1-7/15
KY West	3/01–3/15	7/15-8/01
LA	1/15–3/15	8/1-11/30
MS North	2/5–4/1	7/25-8/15
MS South	1/15–3/15	8/5-9/15
NC East	2/15–4/15	8/1-9/15
NC West	4/1–8/15	NR
SC East	2/1–3/31	8/15-9/30
SC West	3/15–4/30	7/15-8/30
TN East	3/25–4/25	7/25-8/15
TN West	3/15–4/15	8/25-9/15
VA East (coastal)	3/15– 4/15	7/1-8/15
VA West (mountain)	4/1– 4/30	6/20-7/20

Note: Planting dates in the spring could be earlier if low tunnels or other season extension measures are used in some locations

NR = Not recommended

CAULIFLOWER PLANTING DATES

	Spring	Fall
AL North	3/15–7/1	NR
AL South	2/1–3/31	8/1-9/30
GA North	3/15–7/1	7/25-8/15
GA South	2/1–3/31	8/1-9/30
KY East	4/10–4/30	7/1-7/15
KY Central	4/5–4/20	7/15-8/1
KY West	3/30–4/10	8/1-8/15
LA	2/1–3/15	7/15-10/31
MS North	2/15–3/15	7/25-8/15
MS South	1/15–3/10	8/5-9/15
NC East	2/15–4/15	8/1-9/30
NC West	4/1–8/15	NR
SC East	3/1–4/10	8/15-8/30
SC West	3/20–4/30	7/15-8/30
TN East	3/25–4/25	7/15-8/15
TN West	3/15–4/15	8/1-8/20
VA East (coastal)	NR	7/1-7/30
VA West (mountain)	4/1–4/20	6/20-7/20

Note: Planting dates in the spring could be earlier if low tunnels or other season extension measures are used in some locations

NR = Not recommended

COLLARDS PLANTING DATES

	Spring	Fall
AL North	2/15–6/30	7/15-10/15
AL South	1/15–5/31	7/15-10/31
GA North	3/15–7/31	7/25-8/15
GA South	2/1–3/31	8/1-10/31
KY East	3/15–4/30	7/1-7/15
KY Central	3/10–4/25	7/15-8/1
KY West	3/1–4/15	8/1-8/15
LA	1/15–3/15	7/15-10/31
MS North	1/20–4/1	7/25-8/20
MS South	1/15–3/1	8/10-9/15

COLLARDS PLANTING DATES (cont'd)

	Spring	Fall
NC East	2/15–6/30	8/1–9/15
NC West	4/1–8/15	NR
SC East	2/1–6/15	8/1–10/30
SC West	3/15–6/30	8/1–9/30
TN East	3/15–5/1	7/15–8/15
TN West	2/15–4/15	8/1–8/20
VA East (coastal)	4/1– 4/30	7/1–8/31
VA West (mountain)	4/20– 5/20	7/15–8/20

Note: Planting dates in the spring could be earlier if low tunnels or other season extension measures are used in some locations

NR = Not recommended

KALE PLANTING DATES

	Spring	Fall
AL North	3/15–4/30	8/1–9/15
AL South	2/1–3/31	8/1–10/31
GA North	3/15–4/30	NR
GA South	2/1–3/31	8/1–10/31
KY East	4/1–4/30	7/1–7/15
KY Central	3/20–4/15	7/15–8/1
KY West	3/10–4/10	8/1–8/15
LA	2/1–3/15	7/15–10/31
MS North	1/20–4/1	7/25–8/20
MS South	1/15–3/1	8/10–9/15
NC East	2/15–6/30	8/1–9/15
NC West	4/1–8/15	NR
SC East	2/1–6/15	8/1–10/30
SC West	3/15–6/30	8/1–9/30
TN East	3/15–5/1	8/1–9/1
TN West	2/15–4/15	8/15–9/15
VA East (coastal)	4/1–4/30	7/1–8/31
VA West (mountain)	4/20–5/20	7/15–8/20

Note: Planting dates in the spring could be earlier if low tunnels or other season extension measures are used in some locations

NR = Not recommended

KOHLRABI PLANTING DATES

	Spring	Fall
AL North	3/15–7/1	NR
AL South	2/1–3/31	8/1–9/30
GA North	3/15–7/1	NR
GA South	2/1–3/31	8/1–9/30
KY East	4/10–4/30	NR
KY Central	4/5–4/20	NR
KY West	3/30–4/10	NR
LA	2/1–3/15	7/15–10/31
MS	2/1–3/31	8/1–9/30
NC East	2/15–6/30	8/1–9/15
NC West	4/1–8/15	NR
SC East	2/1–6/15	8/1–9/30
SC West	3/15–6/30	8/1–9/15
TN East	3/25–4/25	8/1–8/15
TN West	3/15–4/15	8/15–8/30
VA East (coastal)	3/15–4/15	7/1–7/31
VA West (mountain)	4/1– 4/20	6/25–7/15

Note: Planting dates in the spring could be earlier if low tunnels or other season extension measures are used in some locations

NR = Not recommended

CROP INFORMATION

Broccoli. Space raised beds 36 to 40 inches apart; plant twin rows per bed 6 to12 inches apart within row and 12 inches between row; for bunch broccoli use the shorter within row spacing which will also require a more aggressive pest management program; for organic production single rows per bed can be used and within row spacing increased to as much as 24-inch spacing to aid in pest prevention and maximize fertilizer management.

Brussels Sprouts. Use transplants for best results. Sow seed 4 to 6 weeks prior to setting seedlings out. Space transplants 18 to 24 inches between plants into rows 30 to 36 inches apart. Using raised mulch covered beds will allow tighter plant spacings and enhance earliness. It is critical to maintain soil moisture and fertility levels for successful production.

Cabbage. The early cabbage crop is grown from transplants seeded at the rate of 1 ounce for 3,000 plants. Transplants are ready for field planting 4 to 6 weeks after seeding. Storage of pulled, field-grown cabbage transplants should not exceed 9 days at 32°F or 5 days at 66°F prior to planting in the field. Precision seeders can be used for direct seeding. However, seed should be sown 15 to 20 days in advance of the normal transplant date for the same maturity date. Early varieties require 85 to 90 days from seeding to harvest, and main-season crops require 110 to 115 days. Set transplants in rows 2 to 3 feet apart and 9 to 15 inches apart in the row for early plantings and 9 to 18 inches apart for late plantings, depending on variety, fertility, and market use. With cabbage for slaw or kraut, wider in-row spacings are used to ensure larger heads.

Chinese Cabbage. Chinese cabbage is a diverse group of growth types of *Brassica rapa*. Napa and Michihli resemble romaine lettuce with densely overlapping broad leaves with flat mid-veins. Napa cabbage is barrel shaped while Michihli is tall and slender. Choi types have a vase shaped growth with broad petioles of white or green with green leaves at the top of the vase. Production inputs are the same as ball head cabbage. Both types grow more quickly than ball head cabbage with most cultivars maturing in 50 days from seeding.

Cauliflower. Cauliflower can be more challenging to grow than other Brassicas in this section. Consider a small test plot to determine the best methods (planting dates, varieties, etc.) to use in your area. Start seed in greenhouse or protected frames 4 to 6 weeks before planting. Use 1 ounce of seed for 3,000 plants. Set transplants in rows 3 to 4 feet apart, and plants are set 18 to 24 inches apart in the row. Make successive plantings in the field at dates indicated in preceding table.

Kohlrabi. Transplants may be used for a spring crop. Seed 6 weeks before expected transplant date. Use precision seeder for hybrid varieties. Space rows 18 to 24 inches apart and 6 to 8 between plants.

PLASTIC MULCH

Early spring cabbage, cauliflower, and broccoli are frequently grown using plastic mulch, with black mulch used in the spring and white on black or black mulch painted white used in the fall.

BOLTING

Bolting in cabbage and buttoning in cauliflower, can occur if the early-planted crop is subjected to 10 or more continuous days of temperatures between 35 to 50°F. However sensitivity to bolting depends upon the variety. Note: A low level of calcium in plant tissue can also induce bolting.

SPECIAL NOTES FOR PEST MANAGEMENT

Note: The use of a **spreader-sticker** is recommended for cole crops in any case; the heavy wax coating on the leaves reduces deposition of spray materials. These adjuvants allow the spray to spread out and stick to the leaves. Multiple nozzles per row or bed will provide the underleaf coverage and high coverage rates necessary to manage caterpillar pests of Cole crops. See labels for exceptions to this recommendation, such as for products with translaminar activity. In these cases, **penetrating adjuvants** are recommended instead, and spreader-sticker products may inhibit their efficacy.

INSECT MANAGEMENT

Aphids: The cabbage aphid can be a serious problem on these crops and should be treated immediately if noticed. Other aphid species are found on these crops and should be treated if the crop is near harvest, or if their level of infestation is increasing. Often parasitic wasps take out these species if broad-spectrum insecticide use is avoided.

Cabbage Root Maggot: Root maggots and other similar insects such as the seed corn maggot can be a problem in heavier soils in the Southeast especially during cool, damp times of the year. Avoid planting into soils with freshly plowed down crop residue or high levels of organic matter.

Caterpillars: A number of moth and butterfly larvae feed on cole crops. The major ones are the cabbage looper (CL), the imported cabbageworm (ICW), and the diamondback moth (DBM) referred to as the cabbageworm complex. Other caterpillars found on cole crops are the cross-striped cabbageworm, corn earworm, armyworms, and webworms. Webworms often damage the bud of the young plants and should be treated immediately; very young larvae are much more easily managed than older ones.

Scouting and using a threshold for spray applications is a cost-effective method of managing these pests. Contact your Extension personnel for information on thresholds developed in specific crops and regions. The use of a threshold to determine the need for treatment usually reduces the number of sprays per crop without loss of crop quality and improves the profit margin. Broad-spectrum insecticides that reduce the natural enemies in the field should be avoided, if at all possible.

Note: *Bacillus thuringiensis* (BT) preparations are effective against most of these pests but must be eaten by the larvae. Thorough coverage of the plant, particularly the under-surface of the leaf is essential, and the use of a **spreader-sticker** is strongly recommended.

Note: Several of these insects are prone to develop resistance to insecticides. Growers must rotate among classes of insecticides for each pest generation. See the section on resistance management.

DISEASE MANAGEMENT

Accurate diagnosis is necessary for effective disease management. Incorporate appropriate and effective cultural practices and pesticides, as detailed in the Disease Management section, in a season-long integrated disease management program. Brassicas grown in the Southeast are affected by several diseases that may infect foliage, stems, heads, and/or roots. These diseases are primarily caused by bacteria, oomycetes, and fungi. However, root-knot nematodes may also impact production if populations are high.

Soil-borne Diseases: Club root and white mold can significantly impact Brassica production in some areas of the Southeast. Both diseases are difficult to manage once the pathogens become established in a field. The use of disease-free transplants and sanitation of tools between fields should be used to prevent introducing the clubroot pathogen. Fields known to be infested with the pathogens should be avoided, if possible. Crop rotation for a minimum of two years can reduce white mold pathogen populations. Resistant varieties are available against club root in some hosts. Soil management to increase pH and calcium levels may help reduce club root. Delayed planting may help to suppress or limit white mold development in cooler climates. Fungicide applications must be combined with cultural practices for effective management.

Foliar and Head Diseases: The main foliar diseases are Alternaria leaf spot (black spot), downy mildew, and black rot. Alternaria leaf spot can affect heads of broccoli and cauliflower pre- and post-harvest, as well as bagged collard and kale greens. Both Alternaria leaf spot and downy mildew start on older leaves, so scouts must check the older leaves which may be hidden by younger leaves. Crop rotation for two years can reduce pathogen populations. Black rot can be introduced on seed or transplants, so seed should be purchased from companies that advertise seed health testing. Fungicide applications for Alternaria leaf spot and downy mildew should begin when the first symptoms are seen. Fungicides/bactericides have limited efficacy against black rot.

NEMATODE MANAGEMENT

Nematode populations are highest in the late summer and fall. Collect soil samples for nematode analysis during this period from fields to be planted with Brassicas the following year to determine if root-knot nematodes or other plant parasitic nematodes are present. Avoid fields with high populations of nematodes or use appropriate fumigants and other nematicides for management (see fumigant and nematicide tables). Clean and disinfest equipment between fields. Use appropriate crop rotations and cover crops with non-host crops to reduce nematode populations.

HARVESTING AND STORAGE

Cabbage: Fresh market cabbage should be harvested when heads are firm and weigh 2.5 to 3.0 pounds. Most markets require one to three wrapper leaves to remain. The heads should be dense and free of insect damage. Cabbage for slaw or kraut usually has much larger heads and weighs 3 to 12 pounds.

Broccoli: Broccoli is harvested in several ways. Bunched broccoli are stalks bound together to form a single unit. Broccoli crowns are the heads of stalks that have cut off or shortened stems. Broccoli florets are bud clusters or pieces of bud clusters closely trimmed from the head, with the remaining stalk usually being one inch or less. Broccoli is ready to *harvest* when the beads (flower buds) are still tight, but a few outer beads have begun to loosen. Bunched broccoli is the most typical method. Stalks should be 7 inches long from top of the crown to the butt. Broccoli is usually bunched in 1.5 pound bunches with 2 to 3 heads per bunch. Secure bunches with a rubber band or twist tie.

Brussels sprouts: Brussels sprouts are ready to harvest when they are firm, green, and 1 to 2 inches in diameter. Remove sprouts by twisting them until they break away from the plant.

Kohlrabi: Kohlrabi should be harvested when the bulbs are 2 to 3 inches in diameter and before internal fibers begin to harden.

Cauliflower: Cauliflower is harvested while the heads are pure white and before the curds become loose and ricey. Heads are blanched by tying outer leaves over the heads when heads are 3 to 4 inches in diameter. Blanching takes about 1 week in hot weather and 2 weeks in cooler weather.

See Table 14 for further postharvest information on these crops.

CARROTS (*Daucus carota*)

VARIETIES[1]	AL	GA	KY	LA	MS	NC	SC	TN	VA
CARROTS									
Apache [3]	A	G		L	M	N	S	T	V
Baltimore [4]	A	G							
Belgrado [4,8]	A	G							
Bolero [4]		G	K	L	M		S	T	V
Danvers 126 [5]	A			L		N	S		V
Deep Purple [2,3]	A			L		N			
Maverick [6]	A	G		L					V
Mellow Yellow [4,7]	A								
Mokum [4]			K					T	
Napoli [4]			K					T	V
Purple Haze [2,3]	A	G	K	L	M	N	S	T	
Sugarsnax 54	A	G	K	L		N	S	T	V
Yellow Bunch [3,7]	A			L		N			

[1] Abbreviations for state where recommended
[2] Purple
[3] Imperator type: 7-8 inches long w/ long conical shape and narrow shoulders
[4] Nantes type: Smooth, cylindrical over entire length w/ a blunt end
[5] Danvers type: Tapering root w/ a semi-blunt tip
[6] Nantes x Imperator type
[7] Yellow
[8] Processing type

Seeding Dates. Small carrot seedlings up to six leaves cannot withstand hard freezes but are somewhat frost tolerant. Optimum temperatures are in the range of 60-70°F, with daytime highs of 75°F and nighttime lows of 55°F ideal. Although the crop can be grown outside this range with little or no effect on tops, temperatures differing drastically from the above can adversely affect root color, texture, flavor, and shape. Lower temperatures in this range may induce slow growth and make roots longer, more slender and lighter in color. Carrots with a root less than one inch in diameter are more susceptible to cold injury than larger roots. Soil temperatures should be above 40°F and below 85°F for best stand establishment.

CARROT PLANTING DATES (cont'd)

	Spring	Fall
AL North	3/1–4/15	NR
AL South	NR	8/1–11/30
GA North	3/1–4/15	NR
GA South	NR	8/1–11/30
KY East	4/1-4/30	NR
KY Central	3/20-4/15	NR
KY West	3/10-4/10	NR
LA	1/15–2/28	9/15–10/15
MS North	2/15–4/1	NR
MS South	1/15–3/15	NR
NC East	2/15–3/31	6/15–8/15
NC West	4/1–6/15	7/21-8/15
SC East	2/1–3/15	9/1–9/15
SC West	2/15–3/31	8/1–9/15
TN East	3/15-5/1	NR
TN West	3/1-4/30	NR
VA East (coastal)	3/15–4/30	6/1–7/31
VA West (mountain)	4/1– 6/15	7/1–7/31
NR = Not recommended		

SPACING

Spatial arrangements for planting can differ markedly. Carrots can be planted with vacuum, belt, or plate seeders. Often a special attachment called a scatter plate or spreader shoe is added to the plate planters to scatter the seed in a narrow band. Carrots should be spaced 1½ to 2 inches apart within the row. Carrot seed should be planted no deeper than ¼-½ inch. A final stand of 14 to 18 plants per foot of twin row is ideal. Ideal patterns are twin rows that are 2½ -3½ inches apart. Three or four of these twin rows are situated on one bed, depending on the width of the bed. One arrangement is to plant three twin rows on beds that are on 72-inch centers. Another arrangement is to plant four twin rows on a 92-inch bed (center to center). The sets of twin rows are 14 to 18 inches apart. Beds on 72-inch centers will have approximately 48 inches of formed bed. Row spacing wider than 18 inches will reduce total plant stand per acre and thus, will reduce total yield. Ideal plant populations should be in the range of 400,000 for fresh market carrots and 250,000 for processing carrots.

PLANTING AND LAND PREPARATION

Beds that are **slightly** raised are advantageous because they allow for good drainage. Beds should be firmed and not freshly tilled before planting and soil should be firmed over the seed at planting. A basket or roller attachment is often used to firm the soil over the seed as they are planted. Light irrigation will be required frequently during warm, dry periods for adequate germination.

Windbreaks are almost essential in areas with primarily sandy soils. Sand particles moved by wind can severely damage young carrot plants, reducing stands. Small grain strips planted between beds or at least planted between every few beds can help reduce sandblasting injury.

Begin by deep turning soils to bury any litter and debris and breaking soils to a depth of 12-14 inches. Compacted soils or those with tillage pans should be subsoiled to break the compacted areas. If uncorrected, compact soil or tillage pans can result in restriction of root expansion. It is best to apply lime after deep turning to prevent turning up acid soil after lime application.

Prepare a good seedbed using bed-shaping equipment. Do not use disks or rototiller to avoid soil compaction. Carrots should be planted on a slightly raised bed (2-3 inches) to improve drain- age. After beds are tilled and prepared for seeding, it is best to allow the beds to settle slightly before planting. Avoid other tillage practices that can increase soil compaction. Following in the same tracks for all field operations will help reduce compaction in planting areas.

SPECIAL NOTES FOR PEST MANAGEMENT
DISEASE MANAGEMENT

Accurate diagnosis is necessary for effective disease management. Incorporate appropriate and effective cultural practices and pesticides, as detailed in the Disease Management section, in a season-long integrated disease management program.

Soil-Borne Root Diseases

Depending on the cropping history of the field, Pythium, southern blight, and Sclerotinia blight may cause problems. It is advisable to avoid fields where these diseases have been identified in the previous crop. Deep turning is also necessary to help prevent root diseases.

Pythium Blight: This disease is usually characterized by flagging of the foliage indicating some root damage is occurring. Under wet conditions, Pythium may cause serious problems to the root; a white mycelial mat may develop on the infected root, which rapidly develops a watery soft rot. Forking of the root system is also a common symptom associated with Pythium infection. Rotation is a major factor for reducing Pythium along with the use of fungicides.

Southern Blight: Southern blight causes severe damage to carrots. This disease is usually associated with carrots remaining in the field after the soil begins to warm in the spring. This disease causes a yellow top to develop with a cottony white fungal growth associated with the upper part of the carrot root. The top of the root and the surrounding soil may be covered with a white mycelium with tan sclerotia developing as the disease progresses. Southern blight is best managed by using rotation and deep turning.

Sclerotinia Blight: Sclerotinia blight causes severe damage to the roots of carrots. This disease is usually worse under wet soil conditions. White mycelium forms around the infected area, and later, dark sclerotia develop on the mycelium; this fungal growth is a good indicator of Sclerotinia rot. This disease causes a progressive watery soft rot of the carrot root tissue and is considered a potential problem in the production of carrots. Rotation and deep turning of the soil are recommended to reduce losses to this disease.

Rhizoctonia Rot: This fungal disease causes brown to black lesions to develop on the sides of the carrot root. The disease is much worse under cool, wet conditions. Saturated soil conditions often enhance all soil-borne diseases, which are potential problems in carrot production. Rhizoctonia damage can be minimized by using rotation and good cultural practices. Soil fumigation will prevent damage from any of the soil-inhabiting fungi.

Foliar Diseases

Bacterial Blight: Bacterial blight causes irregular brown spots on the leaves and dark brown streaks on the petioles and stems. The lesions on the foliage begin as small yellow areas with the centers becoming dry and brittle and an irregular halo. Leaflets, stems, and petioles may become infected as disease progresses. Some lesions may crack open and ooze the bacteria. These bacteria may be washed down to the crown of the plant causing brown lesions on the top of the root. The earlier the infection occurs, the greater the damage to the root. The bacterium is spread by splashing water, and it takes about 10-12 days after inoculation before symptoms appear. Disease development progresses rapidly between 77° F and 86°F. Crop rotation is a important factor in bacterial blight management.

Alternaria Blight: Alternaria blight causes small dark brown to black lesions with yellow edges on leaf margins. Lesions increase in size as disease progresses and, in some cases, entire leaflets may be killed. In moist weather, the disease can spread so rapidly it resembles frost injury. Such conditions can reduce the efficiency of mechanical harvesters, which require strong healthy tops to effectively remove carrots from the soil. Alternaria may also cause damping-off of seedlings and a black decay of roots. Fungal spores and mycelium are spread by splashing rain, through movement of infested soil, and use of infested tools. Alternaria blight can manifest itself about 10 days after infection. The optimum temperature for disease development is 82°F.

Cercospora Leaf Blight: This fungal disease causes lesions to form on the leaves, petioles, and stems of the carrot plant. The symptoms appear to mimic that of Alternaria blight but these diseases can be distinguished from one another if samples are examined with a compound microscope. Cercospora blight progresses in warm, wet weather and lesions appear about 10 days after infection. The youngest leaves are usually more susceptible to Cercospora infection.

NEMATODE MANAGEMENT

Root-knot Nematodes (RKN): By far, the most destructive problem in carrots are RKN. Root-knot nematodes are small eel-like worms that live in the soil and feed on plant roots. Since the root of the carrot is the harvested portion of the plant, root-knot damage is not acceptable. Root-knot nematodes cause poor growth and distorted or deformed root systems, which results in a non-marketable root. Root-knot nematode damage also allows entry of other pathogen, such as Fusarium, Pythium, and Erwinia. If any RKN are found in a soil assay, treatment is recommended. Good success has been obtained using field soil fumigation to eradicate RKN in the root zone of carrots. Use nematicides or fumigants listed in the "Nematode Control in Vegetable Crops" tables in the Disease Management section.

INSECT MANAGEMENT

Soil Insects: Wireworms, white grubs, and the granulate cutworm *may* be partially controlled with good cultural practices. Soil should be deep turned in sufficient time prior to planting to allow destruction of previous crop residue that may harbor soil insects. When possible, avoid planting just after crops that are slow to decompose such as tobacco and corn. Avoid planting behind peanuts and root crops such as sweetpotatoes and turnips. If a field has a history of soil insect problems, either avoid these or, broadcast incorporate a soil insecticide prior to planting. Plantings in fields that were recently in permanent pasture should be avoided as should fields recently planted to sod/turf, although these are not as critical. Fields with a history of whitefringed beetle larvae should not be planted to carrots because there are no currently registered insecticides effective on this pest.

Flea beetle larvae can damage *roots* by feeding from the surface into the cortex. The damage will take on the appearance of narrow "s" shaped canals on the surface. Flea beetle larvae can be prevented easily with soil insecticides.

The seedcorn maggot is an opportunistic pest that takes advantage of crops that are under stress or where there is decaying organic matter. At planting, soil insecticides will prevent the development of maggot infestations for several weeks after planting. Seedcorn maggots cannot be effectively controlled after the infestation begins. If plants become stressed during the period of high root maggot potential, preventive applications of insecticides should be made every 7 days until the stress is minimized.

Foliar Insects: Foliar insect pests may be monitored and insecticides applied as needed. Carrots should be scouted at least once per week for developing populations of foliage pests.

Aphids: Several species of aphids may develop on carrots. The most common aphids to inhabit carrots are the green peach aphid and the cotton or melon aphid. Often parasitic wasps and fungal diseases will control these aphids. If populations persist and colonize plants rapidly over several weeks and honeydew or sooty mold is observed readily, then foliar insecticides are justified.

Flea Beetles: Fleas beetle adults may cause severe damage to the foliage on occasion. If carrots are attacked during the seedling stage and infestations persist over time, an insecticide application may be necessary. If plants are in the cotyledon to first true leaf stage, treatments should be made if damage or flea beetles are observed on more than 5% of the plants. After plants are well established, flea beetles should be controlled only if foliage losses are projected to be moderate to high, e.g., 15% or more.

Vegetable Weevil: The adult and larvae of the vegetable weevil may attack carrots. The adult and larvae feed on the foliage. Vegetable weevil larvae often will feed near the crown of plants and, if shoulders are exposed at the soil surface, larvae will feed on tender carrots. Treatments are justified if adults or larvae and damage are easily found in several locations.

Armyworms: The armyworm can cause damage in carrots. Armyworms may move from grain crops or weeds into carrots or adults may lay eggs directly on carrot plants. Armyworms are easily managed with foliar insecticides.

Beet Armyworm: The beet armyworm infests carrots in the late spring. Usually natural predators and especially parasites regulate beet armyworm populations below economically damaging levels.

Whiteflies: The silverleaf whitefly can be a problem during the early seedling stage of fall plantings. Silverleaf whitefly migrates from agronomic crops and other vegetables during the late summer. Infestation may become severe on carrots grown in these production areas. Often whiteflies may be controlled by several natural enemies and diseases by early fall so, treatments may not be justified. However, if whiteflies develop generally heavy populations, treatment of young plantings is justified.

HARVESTING AND STORAGE

Topped Carrots: will last 4 to 5 months at 32°F (0°C) and 90 to 95% relative humidity. See Table 14 for further postharvest information.

CUCUMBERS (*Cucumis sativus*)

VARIETIES[1]	AL	GA	KY	LA	MS	NC	SC	TN	VA
CUCUMBERS									
Slicer / Fresh Market									
Brickyard [2, 4, 5, 6, 7, 8, 9, 10, 14]						N			
Bristol [2, 3, 4, 5, 6, 7, 8, 10, 14]	A	G	K	L		N	S		V
Cobra [2, 3, 5, 6, 7, 8, 9, 10, 14]				L		N			
Cortez [2, 3, 5, 6, 7, 8, 9, 10, 14]	A	G	K	L			S	T	V
Dasher II [2, 3, 4, 5, 6, 10, 14]	A	G	K	L	M	N	S	T	V
Diamondback [2, 3, 4, 5, 6, 7, 8, 9, 10, 11, 14]		G							
Dominator [2, 3, 4, 5, 6, 10, 14]				L					V
General Lee [4, 5, 6, 10, 14]	A		K	L	M		S		V
Mamba [2, 3, 4, 5, 6, 7, 8, 9, 10, 11, 14]		G							
Poinsett 76 [2, 3, 5, 10]	A						S		
Raceway [2, 3, 5, 4, 6, 7, 8, 9, 10, 11, 14]						N			
Speedway [2, 3, 5, 6, 10, 14]	A	G	K	L			S	T	V
Stonewall [2, 3, 4, 5, 6, 10, 14]		G	K	L		N	S		V
SV 4142CL [2, 3, 4, 6, 8, 10]	A	G				N			
SV 4719CS [2, 3, 4, 5, 8, 10, 14]	A	G							
Thunder [3, 4, 5, 6, 8, 10, 14]	A		K	L	M		S	T	V
Pickling Types - Multiple Harvest									
Calypso [2, 3, 4, 5, 6, 10, 14]	A	G	K	L	M				
Eureka [2, 3, 4, 5, 6, 7, 8, 9, 10]			K						V
Fancipak [2, 3, 4, 5, 6, 10, 14]	A	G					S	T	V
Vlasstar [2, 3, 5, 6, 10]						N			
Pickling Types - Multiple or Once-over Harvest									
Arabian [2, 3, 4, 5, 6, 10, 14]	A					N			
Peacemaker [2, 3, 4, 5, 6, 10, 14]						N			
Parthenocarpic Types - Seedless Pickling – Multiple Harvest									
Bernstein [5, 10]						N			
Liszt [5, 6, 10]						N			
Parthenocarpic Types – Seedless Pickling – Multiple or Once-over Harvest									
Gershwin [5, 10]	A					N	S		
Rubinstein [5, 6, 10]						N			
Parthenocarpic Types – Seedless Pickling – Once-over Harvest									
Bowie [5, 10]						N			
GREENHOUSE/HIGH TUNNEL CUCUMBERS									
Long Dutch/English Types									
Bologna [5, 10, 11]	A		K	L	M		S	T	
Camaro [5, 13, 14]	A		K	L	M		S	T	
Cumlaude [3, 5, 10, 14]	A		K	L	M		S	T	V
Kasja [5, 10, 11]	A		K						
Verdon [5, 6, 10, 11, 12]	A		K	L	M			T	
Beit Alpha/Mini Types									
Deltastar [5, 6, 10, 11, 12]	A		K	L	M		S	T	
Jawell [5, 6, 14]	A		K	L	M		S	T	
Katrina [5, 6, 12]	A		K	L	M		S	T	
Manar [5, 6, 14]	A		K	L	M		S	T	
Picowell [5]	A		K	L	M		S	T	
Sarig [5]	A		K						

[1] Abbreviations for state where recommended
[2] Anthracnose tolerance/resistance
[3] Angular leaf spot tolerance/resistance
[4] Downy mildew tolerance/resistance
[5] Powdery mildew tolerance/resistance
[6] Cucumber mosaic virus tolerance/resistance
[7] Papaya ringspot virus tolerance/resistance
[8] Zucchini yellows mosaic virus tolerance/resistance
[9] Watermelon mosaic virus tolerance/resistance
[10] Scab and gummosis tolerance/resistance
[11] Target spot tolerance/resistance
[12] Cucumber vein yellowing virus tolerance/resistance
[13] Low light tolerant
[14] All female (gynoecious)

Field Production. For earlier cucumber production and higher, more concentrated yields, use gynoecious varieties. A gynoecious plant produces only female flowers. Upon pollination female flowers will develop into fruit. To produce pollen, 10% to 15% pollenizer plants must be planted; seed suppliers add this seed to the gynoecious variety. Both pickling and slicing gynoecious varieties are available. For machine harvest of pickling cucumbers, high plant populations (55,000 per acre or more) concentrate fruit maturity for increased yields.

Planting Dates. For earliness container-grown transplants are planted when daily mean soil temperatures have reached 60°F but most cucumbers are direct seeded. Consult the following table for planting dates for transplants in your area. Early plantings should be protected from winds with hot caps or row covers. Growing on plastic mulch can also enhance earliness.

CUCUMBER *SLICERS* PLANTING DATES

	Spring	Fall
AL North	4/1–7/15	8/1–8/30
AL South	3/1–4/30	8/1–9/15
GA North	4/15–7/15	8/1–8/30
GA South	3/1–4/30	8/1–9/15
KY East	5/10-6/1	6/1-6/15
KY Central	5/5-6/1	6/1-7/1
KY West	4/25-5/15	5/15-7/15
LA North	3/15–5/15	7/15–8/31
LA South	3/1–5/15	8/1–9/15
MS North	4/1–5/15	7/25–8/21
MS South	3/15–5/1	8/14–9/14
NC East	4/15–5/15	7/15–8/15
NC West	5/15–7/31	NR
SC East	3/15–5/15	8/1–8/30
SC West	4/15–6/5	8/1–8/30
TN East	5/5-6/15	7/1-8/10
TN West	5/1-6/1	7/25-8/25
VA East (coastal)	4/15-6/15	7/1-7/31
VA West (mountains)	5/15-6/30	6/15-7/31

NR = Not recommended

CUCUMBER *PICKLING* PLANTING DATES

	Spring	Fall
AL North	4/15–7/15	8/1–8/30
AL South	3/1–4/30	8/1–9/15
GA North	4/15–7/15	NR
GA South	3/1–4/30	8/1–9/15
KY East	5/10-6/1	6/1-6/15
KY Central	5/5-6/1	6/1-7/1
KY West	4/25-5/15	5/15-7/15
LA North	4/1–5/15	7/15–8/31
LA South	3/15–5/15	8/1–9/15
MS South	4/1–4/15	NR
NC East	4/20–5/20	7/15–8/15
NC West	5/25–7/31	NR
SC East	3/15-5/15	8/1–8/30
SC West	4/15–6/15	8/1–8/30
TN East	5/5-6/15	7/1-8/10
TN West	5/1-6/1	7/25-8/25
VA East (coastal)	4/15-6/15	7/1-7/31
VA West (mountains)	5/15-6/30	6/15-7/31

NR = Not recommended

Spacing. *Slicers:* Space rows 3 to 4 feet apart with plants 9 to 12 inches apart. *Pickles*: For hand harvest, space 3 to 4 feet apart; for machine harvest, space three rows 24–28 inches apart on a bed. Plants for hand harvest should be 6 to 8 inches apart in the row; 2 to 4 inches apart for machine harvest. Close spacing increases yields, provides more uniform maturity and reduces weed problems, but require slightly higher fertilizer rates. *Seeding for slicers*: 1.5 pounds per acre. *Seeding for pickles*: 2 to 5 pounds per acre.

Bitterness. Bitterness can be a common problem in cucumber. Cucumbers (and all other cucurbits) produce a group of chemicals called cucurbitacins which can cause bitterness to develop. As cucurbitacin concentrations increase, the more bitter the cucumber will taste. Generally, the amount of cucurbitacin in a cucurbit fruit is low and cannot be tasted. Mild bitterness results from higher levels of cucurbitacin often triggered by environmental stresses, including high temperatures, wide temperature swings, or too little water. Uneven watering practices (too wet followed by too dry), low soil fertility, and low soil pH are also possible factors. Over-mature or improperly stored cucumbers may also develop a mild bitterness, although it is not usually severe.

Mulching. Fumigated soil aids in the control of weeds and soil-borne diseases. Black plastic mulch laid before field planting conserves moisture, increases soil temperature, and increases early and total yield. Plastic and fumigant should be applied on well-prepared planting beds 2 to 4 weeks before field planting. Plastic should be placed immediately over the fumigated soil. The soil must be moist when laying the plastic. Fumigation alone may not provide satisfactory weed control under clear plastic. Herbicides labeled and recommended for use on cucumbers may not provide satisfactory weed control when used under clear plastic mulch on nonfumigated soil. Black plastic can be used without a herbicide. Fertilizer must be applied during bed preparation. At least 50% of the nitrogen (N) should be in the nitrate (NO_3) form.

Foil and other reflective mulches can be used to repel aphids that transmit viruses in fall-planted (after July 1) cucumbers. Direct seeding through the mulch is recommended for maximum virus protection. Fumigation will be necessary when there is a history of soilborne diseases in the field. Growers should consider drip irrigation with plastic mulch. For more in- formation, see the section on "Irrigation".

SUGGESTED FERTIGATION SCHEDULE FOR CUCUMBER* (N:K,1:2)

Days after planting	Daily nitrogen	Daily potash	Cumulative	
			Nitrogen	Potash
	(lb / A)			
Preplant			25.0	45.0
0–14	0.9	1.8	37.6	75.2
15–63	1.5	3.0	110.3	196.6
64-77	0.7	1.4	120.1	216.6

ALTERNATIVE FERTIGATION SCHEDULE FOR CUCUMBER* (N:K,1:1)

Days after planting	Daily nitrogen	Daily potash	Cumulative Nitrogen	Potash
			(lb / A)	
Preplant			24.0	24.0
0-7	1.0	1.0	31.0	31.0
8-21	1.5	1.5	52.0	52.0
22-63	2.0	2.0	136.0	136.0
64-70	1.5	1.5	150.0	150.0

*Adjust based on tissue analysis

Greenhouse Cucumber Production. If you plan on growing cucumbers to maturity in the greenhouse, you need to select a greenhouse variety. This is because these varieties have been bred specifically for greenhouse conditions – lower light, higher humidity and temperature, etc., and they have better disease resistance than field types.

Nearly all greenhouse cucumber varieties are gynecious, parthenocarpic hybrids. This means that these varieties produce only female flowers and the fruit are seedless. Since they are all female, no pollination is needed. The seedless characteristic makes the fruit very tender to eat. Greenhouse cucumbers are also thin skinned which makes them more desirable than field varieties. While non-greenhouse types would grow in the greenhouse, the yield and quality would be reduced, and therefore they may not be profitable.

Variety selection is based on yield, fruit size, uniformity, disease resistance, and lack of physiological disorders, as well as the market demand for the type grown. In some markets the long, European types sells better, while in others, the small beit alpha types, also referred to as "minis", are preferred. For suggestions on varieties, see the variety table above. Insect and disease control methods for greenhouse vegetables can be found in Tables 2-29 (in the Insect section) and 3-45 (in Disease section), respectively.

SPECIAL NOTES FOR PEST MANAGEMENT
INSECT MANAGEMENT

Cucumber Beetle: Cucumber beetles can transmit bacterial wilt; however, losses from this disease vary greatly from field to field and among different varieties. Pickling cucumbers grown in high-density rows for once-over harvesting can compensate for at least 10% stand losses. On farms with a history of bacterial wilt infections and where susceptible cultivars are used, foliar insecticides should be used to control adult beetles before they feed extensively on the cotyledons and first true leaves. Begin spraying shortly after plant emergence and repeat applications at weekly intervals if new beetles continue to invade fields. Treatments may be required until stems begin vining (usually about 3 weeks after plant emergence), at which time plants are less susceptible to wilt infections.

Pickleworm, Melonworm: Make one treatment prior to fruit set, and then treat weekly.

Aphids: Aphids transmit several viruses (CMV, WMV, PRSV, etc.) and can delay plant maturity. Thorough spray coverage beneath leaves is important. For further information on aphid controls, see the preceding "Mulching" section. Treat seedlings every 5 to 7 days or as needed.

Mites: Mite infestations generally begin around field margins and grassy areas. CAUTION: DO NOT mow or maintain these areas after midsummer because this forces mites into the crop. Localized infestations can be spot-treated. Begin treatment when 50% of the terminal leaves show infestation. Note: Continuous use of pyrethroids may result in mite outbreaks.

Whiteflies: In addition to transmitting viral diseases (see Disease Management for more details), high populations of whiteflies can cause significant direct damage to cucumber plants, even in the absence of viruses. Effective control can be achieved using neonicotinoids (IRAC group 4A) and butanolides (4D), but it is critical to rotate these with other modes of action to prevent resistance. Diamides (28) and pyriproxyfen (7C) can be a management strategy for rotation and sustained control.

DISEASE MANAGEMENT

Accurate diagnosis is necessary for effective disease management. Incorporate appropriate and effective cultural practices and pesticides, as detailed in the Disease Management section, in a season-long integrated disease management program.

Cucurbit Downy Mildew (CDM): CDM can be a devastating disease of cucurbits in the southeastern and eastern U.S. Successful CDM management often requires the use of an appropriate fungicide spray program. Fungicide efficacy is influenced by crop, regional, and seasonal differences, as well as resistance management practices. A useful tool for downy mildew management (maps, text alerts, identification resources, etc.) is the CDM reporting system (http://cdm.ipmpipe.org). If you think you have CDM or have questions about disease management, please contact your local Extension office.

Phytophthora Fruit Rot: Phytophthora fruit rot can be a serious issue in cucumbers in some locations, especially in production systems that use surface water and overhead irrigation. To minimize the occurrence of this disease, fields should be adequately drained to ensure that soil water does not accumulate around the base of the plants. Just before plants begin vining, subsoil between rows to allow for faster drainage following rainfall.

Avoid moving infested soil between fields. Crop rotation is not effective. Resistance to some fungicides (e.g. mefenoxam) has been reported and should be considered when selecting fungicides. Apply appropriate fungicides when fruit are small.

Belly Rot: Belly rot is a soil-borne disease. Deep disking or moldboard plowing to turn over the top few inches of soil will reduce disease pressure. Application of appropriate crop protectant at last cultivation may be helpful.

Whitefly-transmitted Viruses: Several whitefly-transmitted viruses, including cucurbit chlorotic yellows virus (CCYV), cucurbit leaf crumple virus (CuLCrV), and cucurbit yellow stunting disorder virus (CYSDV), are emerging in various areas throughout the Southeast when unusually high populations of whiteflies are present. These viruses may produce similar symptoms; molecular testing is necessary for virus identification.

These viruses often also occur in mixed infections. Management of whitefly-transmitted viruses focuses on prevention and management of the whitefly vector. Additional information on whitefly-transmitted viruses in cucurbits is available at https://eCucurbitviruses.org.

NEMATODE MANAGEMENT

Various root-knot nematodes, such as the south- ern root-knot and guava root-knot, as well as other nematodes can infect cucumber. Nematode populations are highest in the late summer and fall. Collect soil samples for nematode analysis during this period from fields to be planted with cucumbers the following year. Clean and disinfest equipment between fields.

Use appropriate crop rotations and cover crops with non-host crops to reduce nematode populations. Avoid fields with high populations of nematodes or use appropriate fumigants and other nematicides for management (see fumigant and nematicide tables).

POLLINATION

Bees are critical for insuring that pollination and cucumber fruit set occurs. Supplementing a field with bee hives can be especially helpful when native bee populations are low or lacking. Having sufficient bees provides the opportunity to maximize cucumber yields and quality. Lack of sufficient pollination can result in a variety of misshapen fruits; dogbone, crooks, nubs, etc.

Rented honey bee hives are often placed in cucumber fields as plants begins to flower. The timing of hive placement is im- portant because cucumber flowers are not that attractive to hon- eybees. If the honey bee hives are placed by cucumber fields pre- maturely before the crop flowers, the honey bees may forage to wild flowers nearby which are more attractive due to their higher nectar and pollen supply. If this occurs, the honey bees may be predisposed to visit these wild flowers even though cucumber flowers are in full bloom a few days later. Assuming that the honey bee hive is a healthy hive, one hive per acre is recommended for hand-harvested pickling and slicing cucumbers with recommended plant populations of approximately 25,000 to 30,000 plants per acre. For mechanical or once-over harvested pickling cucumbers, the recommended plant populations are generally 55,000 to 60,000 plants per acre. Therefore, two honey bee hives should be placed per acre to account for the increased number of flowers from the increased plant population used for mechanically harvested cucumbers. When hybrid cucumbers are grown at high plant populations for machine harvest, flowers require 15 to 20 visits for maximum fruit set. Generally, as the number of vis- its increase, there will be an increase in the numbers of fruit set and an increase in number of seed per fruit, as well as improved fruit shape and fruit weight.

Bumblebees are an effective pollinator alternative to honeybees in cucumber production. Bumblebees have some advantages compared to honey bees; flying under more adverse weather conditions in which it is cool, rainy or windy. They will also visit flowers earlier in the morning than honey bees, and fly later in the afternoon and early evening when the temperatures cool. Because bumblebees have a larger body size than honey bees, fewer flower visits are required by bumblebees in order to achieve good pollination and fruit set.

As with honey bees, bumblebees should be placed in the cucumber field shortly after the crop begins to flower. Bumblebees will typically last for 6 to 12 weeks and will meet the pollination needs of 2 to 3 sequentially planted cucumber crops.

Bumblebee hives are sold as a quad or four hives per quad. A quad is the minimum order that can be purchased from a supplier. Generally one bumblebee hive contains 200 to 250 bees and is equivalent to one honey bee hive. Thus, one quad of bumblebees (minimum order, contains 4 bumblebee hives) would provide good pollination for four acres of hand-harvested cucumbers. For machine-harvest pickling cucumbers, one quad would provide good pollination for every two acres. Bumblebee hives should not be placed in direct sunlight so that the bees work more efficiently. No more than two bumblebee quads should be placed in one location so that pollination is more uni-*form* in the field. As with honey bees, one must carefully plan when to spray insecticides so that the bumblebees are not killed. Because bumblebees are most active from dawn until late morning and from about 4 PM to sunset, the hives need to be closed around 11 AM so that the bees in the hive remain protected during a late evening spray application. Bumblebee quads should be located a minimum of 650 to 700 feet away from the other quads in order to maximize pollinator efficiency.

See the section on "Pollination" in the General Production Recommendations for additional information.

HARVESTING AND STORAGE

See Table 14 for postharvest information.

EGGPLANT (*Solanum melongena*)

VARIETIES [1]	AL	GA	KY	LA	MS	NC	SC	TN	VA
EGGPLANT									
Asian									
Calliope [3, O]	A	G		L	M	N	S	T	V
Cambodian Green Giant [R]	A	G		L					
Ichiban [2, E]	A	G	K	L	M	N	S	T	V
Kermit [6, R]	A	G	K	L		N	S	T	V
Pingtung Long [7, E]	A	G			M	N			
Italian									
Black Bell [2, GL]				L	M			T	
Classic [2, GL]	A	G	K	L	M	N	S	T	V
Dusky [2, GL]	A		K	L	M	N	S	T	V
Epic [2, GL]	A	G	K	L	M	N	S	T	V
Night Shadow [2, GL]	A	G	K				S	T	V
Santana [2, GL]	A	G	K	L		N	S	T	V
Miniature/Specialty									
Casper [4, E]	A	G			M			T	V
Fairy Tale [5, E, M]	A	G	K	L	M			T	
Ghostbuster [4, OL]	A	G	K				S	T	
Gretel [4, E, M]	A	G	K		M	N		T	V
Hansel [2, E, M]	A	G	K		M	N	S	T	V
Little Fingers [2, E, M]	A		K		M	N		T	
Rosita [7, T]	A	G	K		M	N			

[1] Abbreviations for state where recommended
[2] Purple/black exterior
[3] White exterior with purple streaks
[4] White exterior
[5] Purple exterior with white stripes
[6] Green and white exterior
[7] Lavender exterior
[R] Round fruit
[E] Elongated fruit
[M] Miniature fruit
[T] Tear drop fruit
[O] Oval fruit
[OL] Long, oval fruit
[GL] Large, oval fruit

Eggplant is a warm-season crop that grows best at temperatures between 70 to 85°F. Temperatures below 65°F result in poor growth and fruit set.

Seed Treatment. Information on seed treatments can be found under Seed Treatments in the disease management section of the handbook.

EGGPLANT PLANTING DATES

	Spring	Fall
AL North	4/1–7/15	NR
AL South	3/1–4/30	7/15–8/31
GA North	4/15–7/15	NR
GA South	3/1–4/30	7/15–8/31
KY East	5/15-6/1	NR
KY Central	5/10-6/15	NR
KY West	5/1-7/1	NR
LA North	4/15–5/15	7/1–8/15
LA South	3/15–5/15	7/1–8/30
MS North	4/15-6/15	NR
MS South	3/1-4/30	8/1-8/31
NC East	4/15–5/10	8/1–8/15
NC West	5/15–7/15	NR
SC East	4/1–4/30	8/1–8/31
SC West	5/1–6/30	NR

EGGPLANT PLANTING DATES (cont'd)

	Spring	Fall
TN East	4/25-7/15	NR
TN West	4/15-6/15	NR
VA East (coastal)	5/1-6/30	NR
VA West (mountains)	5/15-6/30	NR

NR = Not recommended

Spacing. *Rows:* 4 to 5 feet apart; plants 2 to 3 feet apart in the row.

Staking. Staking eggplant improves quality and yield, while reducing decay. Use a 5-foot tomato stake between every other plant and place string along each side of the plants as they grow. This is described in detail in the tomato section of this guide. Side branches of eggplant should be pruned up to the first fruit and 2 main stems should be used. If additional stems grow too large, remove them. The first fruit should be pruned off until the flower is at least 8 inches above the ground, this will allow for straight fruit to form.

Transplant Production. Sow seed in the greenhouse 8 to 10 weeks before field planting. Three to 4 ounces of seed are necessary to produce plants for 1 acre. Optimum temperatures for germination and growth are 70 to 75°F. Seedlings should be transplanted to 2-inch or larger pots or containers any time after the first true leaves appear, or seed can be sown directly into the pots and thinned to a single plant per pot. Control aphids on seedlings in greenhouse before transplanting to field.

Transplanting Dates. Harden plants for a few days at 60 to 65°F and set in field after danger of frost and when average daily temperatures have reached 65 to 70°F.

Drip Irrigation and Fertilization. After mulching and installing the drip irrigation system, the soluble fertilizer program should be initiated using the following table. On low to low-medium boron soils, also include 0.5 pound per acre of actual boron.

The first soluble fertilizer application should be applied through the drip irrigation system within a week after field-transplanting the eggplant. Continue fertigating until the last harvest.

SUGGESTED FERTIGATION SCHEDULE FOR EGGPLANT*
(high soil potassium)

Days after planting	Daily nitrogen	Daily potash	Cumulative Nitrogen	Cumulative Potash
		(lb / A)		
Preplant			50.0	100.0
0-22	0.5	0.5	60.5	110.5
22-49	0.7	0.7	80.1	130.1
50-70	1.0	1.0	101.1	151.1
71-91	1.1	1.1	124.2	174.2
92-112	1.0	2.0	145.2	195.2

ALTERNATIVE FERTIGATION SCHEDULE FOR EGGPLANT*
(low soil potassium)

Days after planting	Daily nitrogen	Daily potash	Cumulative Nitrogen	Cumulative Potash
		(lb / A)		
0-22	0.5	0.5	60.5	111.0
22-49	0.7	1.4	80.1	150.2
50-70	1.0	2.0	101.1	192.5
71-91	1.1	2.2	124.2	238.7
92-112	1.0	2.0	145.2	280.7

*Adjust based on tissue analysis

SPECIAL NOTES FOR PEST MANAGEMENT
INSECT MANAGEMENT
Colorado Potato Beetle (CPB), Flea Beetles (FB): CPB has the ability to rapidly develop resistance to insecticides. Refer to "Eggplant" insecticide section for management options. The use of row covers can be highly effective for flea beetle management early in the season while plants become established and before flowering at which time row covers will need to be removed for optimum pollination.

Silverleaf Whitefly: Treat when an average of 5 or more adults are found per leaf.

DISEASE MANAGEMENT
Accurate diagnosis is necessary for effective disease management. Incorporate appropriate and effective cultural practices and pesticides, as detailed in the Disease Management section, in a season-long integrated disease management program.

Soil-borne Diseases (Bacterial Wilt, Southern Blight, Verticillium Wilt): Several soil-borne bacterial and fungal pathogens can affect eggplant and other solanaceous crops. Many are difficult to manage when fields become infested. Limited fumigants and fungicides are available for some diseases. Pathogens can be spread via movement of contaminated soil, water, and equipment/tools. Clean and disinfest tools when moving between fields, when possible. Direct water flow away from other fields and sources of surface water for irrigation. Practice appropriate crop rotations with nonhost crops.

Foliar and Fruit Diseases: Many foliar and fruit diseases are caused by fungal and oomycete pathogens. Use cultural practices that minimize water splash, leaf wetness, and canopy density, which contribute to pathogen infection and spread. Preventative, season-long spray programs using products from multiple FRAC groups may be necessary and should be used in conjunction with appropriate cultural practices.

WEED MANAGEMENT
See "Mulching" section for further infor- mation on weed control under plastic mulch.

RATOONING EGGPLANT: PRODUCING A FALL CROP FROM A SPRING PLANTED CROP
Ratooning eggplants can be done after the first crop is complete to allow a second crop to develop. Depending on the location, the first crop may be completed by June or July. Plants at this point will appear "topped out," not producing any more flowers and any subsequent fruits. Mow plants 6 to 8 inches above the soil line, being sure to leave two to three leaf axils. Next, fertilize with 50 to 60 pounds of nitrogen per acre and 80 to 100 pounds of potash per acre (K_2O). This combination will produce vigorous re-growth and stimulate flowering. Plants will begin producing fruit 4 to 6 weeks after ratooning and should produce eggplants until frost.

HARVESTING AND STORAGE
Eggplant may be harvested once the fruit has reached one-half to full size for a given variety. However, harvesting prior to full size may reduce potential yields.

Harvest-ready fruit have a glossy appearance and are firm, without wrinkles. Harvest eggplant fruit before they become overmature. When overmature, the fruit is dull in color, seeds are hard and dark, and the flesh is characteristically spongy. Although the fruit can often be "snapped" from the plant, they should be clipped with a sharp knife or scissors to prevent damage. When harvesting, cut the stem approximately ¼ inch from the fruit. Eggplant skin is tender and easily bruised, so handle with care. The recommended storage temperature for eggplant is 45-54°F and 90-95% relative humidity for a shelf-life of 7-10 days. Eggplant is susceptible to chilling injury, which will manifest as accelerated softening, surface pitting, discoloration, development of sunken areas, browning of seeds, flesh, and calyx, and increased decay susceptibility. See Table 14 for further postharvest information.

GARLIC (*Allium sativum*) AND ELEPHANT GARLIC (*A. ampeloprasum*)

VARIETIES[1]	AL	GA	KY	LA	MS	NC	OK	SC	TN	VA
GARLIC										
Chesnok Red [4,6]	A					N	O		T	
Creole [5]				L						
Duganski [4,6]	A									
Elephant (also called Tahiti) [3,5]	A	G	K	L	M	N		S	T	V
German Extra Hardy [4,7]	A		K			N	O	S	T	V
Italian [2,5]				L						V
Korean Red [4,6]	A									
Music [4,5]		G				N			T	
New York White Neck [2,5]	A					N		S	T	V
Spanish Roja [4,5]						N				

[1] Abbreviations for state where recommended
[2] Softneck
[3] *Allium ampeloprasum* (Broadleaf Wild Leek)
[4] Hardneck
[5] White cloves
[6] Purple striped cloves
[7] White outer skin, red cloves

Most garlic that is available from retail markets tends to be softneck types. When selecting softneck garlic for planting be sure to secure a strain of softneck garlic from a local grower who has had success with fall-planted garlic. Unlike many strains sold commercially, such a strain should be well adapted to your area to overwinter. Avoid planting the Creole types of softneck garlic in the northern range (also called Early, Louisiana, White Mexican, etc.), because they are not very winter-hardy and do not store well. Both the Italian and Creole types have a white outer skin covering the bulb, but the Italian type has a pink skin around each clove, whereas the skint around each Creole clove is white. Elephant-type garlic (milder than regular garlic and up to four times larger) may not yield very well when fall-planted in areas with severe cold or extensive freezing and thawing cycles, which cause heaving. Elephant garlic has performed well, however, in western North Carolina when it is well-hilled with soil or mulched with straw. The Italian and Elephant types take about 220 days to mature.

Many of the most productive Italian garlic cultivars produce seed heads prior to harvest. Whether removed as they form or left intact, they have produced satisfactory yields.

Research in Kentucky and North Carolina has shown that hardneck types of garlic produce superior yields and are more winter-hardy than softneck types. Unlike softneck types, which will produce large numbers of small cloves per bulb; hardneck garlic will produce bulbs with 7-10 large cloves. Hardneck types have a hard "seedstalk" (called a "scape") that is typically removed prior to harvest. Scapes are sometimes sold at farmers markets as a specialty item.

Seed pieces for hardneck garlic are often more expensive and harder to find than softneck types, but improved winter hardiness and bulb quality in the spring in Kentucky suggests that these are preferred for production at more northern latitudes. Results from these states might not translate to all areas of the southeastern US. Consult with your local Extension office to find appropriate cultural information for your area.

Soil Fertility. Maintain a soil pH of 6.2 to 6.8. Fertilize according to soil test recommendations for garlic. In moderately fertile soils, apply about 75 pounds nitrogen (N) per acre, 150 pounds phosphate (P_2O_5) per acre and 150 pounds potash (K_2O) per acre and disk about 6 inches deep before planting. When plants are about 6 inches tall (about March 15), topdress with 25 pounds per acre nitrogen and repeat the top dressing about May 1. Apply all top dressings to dry plants at midday to reduce chance of fertilizer burn.

Because sulfur may be partially associated with the extent of pungency, you may wish to use ammonium sulfate for the last top dressing (May 1). If ammonium sulfate is used, make sure pH is 6.5 to 6.8.

Garlic is commonly grown on muck, sandy, or fine textured soils as long as they are loose and friable. Use of organic matter or cover cropping is important.

Planting. Garlic cloves should be planted during the fall because a chilling requirement must be met for good bulb development. Plant according to the times listed in the following table to ensure that good root systems are established prior to winter.

Final bulb size is directly related to the size of the cloves that are planted. Avoid planting the long, slender cloves from the center of the bulb and cloves weighing less than 1 gram.

GARLIC PLANTING DATES

AL North	9/15–11/10
AL South	10/1–11/30
GA North	9/15–11/10
GA South	10/1–11/30
KY East	9/1–10/1
KY Central	9/10–10/15
KY West	9/15–11/1
LA	9/1–11/30
MS	9/15–10/30
NC East	9/15–11/10
NC West	8/15–10/15

GARLIC PLANTING DATES (cont'd)

OK	10/1-11/15
SC East	10/1–11/30
SC West	8/15–10/15
TN East	9/1-11/1
TN West	9/15-11/1
VA East (coastal)	9/15-11/15
VA West (mountains)	9/1-9/30

Spacing. Garlic should be planted 4 by 4 inches apart in triple rows or multiple beds 16 to 18 inches apart. Between-row spacing depends on the equipment available. Clove tops should be covered with 1 to 1.5 inches of soil. The cloves must not be so deep that the soil will interfere with the swelling of the bulbs, nor so shallow that rain, heaving from alternate freezing and thawing, and birds will dislodge them. Vertical placement of cloves by hand gives optimal results. Cloves dropped into furrows are likely to lie in all positions and may produce plants with crooked necks. Garlic has also been grown successfully in Kentucky using plastic mulch as this helps reduce weed pressure during the long growing season.

SPECIAL NOTES FOR PEST MANAGEMENT
INSECT MANAGEMENT

Thrips: During hot, dry weather, the population of thrips increases following harvest of adjacent alfalfa or grain. Thrips could therefore present the most serious insect problem on garlic. (See "Onions" in the Insect Control section of this publication). Read and follow specific label directions for use on garlic; if not listed, do not use. Treat if thrips counts exceed an average of 5 thrips per plant.

PREHARVEST PRACTICES

Garlic naturally goes dormant after maturing. This dormancy allows the plant to survive harsh environmental conditions during the summer and is essential for long-term storage. Several factors can prevent garlic from maturing properly, thereby reducing its storage potential. The main factors are irrigation and nitrogen fertilization. Excessive irrigation before harvest can cause too much moisture accumulation, leading to poor curing and an in- creased risk of mold. Overly heavy or late nitrogen fertilization promotes excessive vegetative growth at the expense of bulb development, resulting in immature bulbs with thinner skins that do not store well.

HARVESTING AND STORAGE

Garlic is ready for harvest in mid-May to mid-June—it must be harvested when around 30% of foliage is starting to yellow, or the bulbs will split and be more susceptible to disease. When a few tops fall over, push all of them down and pull a sample. There are only about 10 days to 2 weeks for optimal garlic harvest. Before then, the garlic is unsegmented; much after that period, the cloves can separate so widely that the outer sheath often splits and exposes part of the naked clove. Picked at the proper time, each clove should be fully segmented and yet fully covered by a tight outer skin.

Run a cutter bar under the bulbs to cut the extensive root system and partially lift them. The bulbs are usually pulled and gathered into windrows. Tops are placed uppermost in the windrow to protect bulbs from the sun, and the garlic is left in the field for a week or more to dry or cure thoroughly. Curing can also be accomplished in a well-ventilated shed or barn. The bulbs must be thoroughly dried before being shipped or stored. Outdoor curing is not recommended where morning dew can keep it too damp. Bring in for drying immediately from the field. Emphasize gentle handling. Cure for about 6 weeks.

After curing garlic, discard diseased and damaged bulbs. Clean the remaining bulbs to remove the outer loose portions of the sheath and trim the roots close to the bulb. Do not tap or bang bulbs together to remove soil. Braid or bunch together by the tops of the bulbs or cut off the tops and roots and bag the bulbs like dry onions.

When properly cured, garlic keeps well under a wide range of temperatures. Storage in open-mesh sacks in a dry, well-ventilated storage room at 60-90°F is satisfactory. However, garlic is best stored under the temperature and humidity conditions required for onions—32-35°F and 65% relative humidity.

Garlic cloves sprout quickly after bulbs have been stored at temperatures near 40°F, so avoid prolonged storage at this temperature. Garlic stored at above 70% relative humidity at any temperature will mold and begin to develop roots.

Marketing. New growers should develop a local retail market (roadside stands, night markets, gourmet restaurants), wholesale shipper, or processing market before planting. The demand for garlic is increasing due to recent reports about the health and medical benefits of garlic.

The markets of the northern and eastern United States will take the bulbs trimmed like dry onions and known as "loose garlic." Frequently, 30 to 50 bulbs are tied in bunches. Bulbs should be graded into three sizes—large, medium, and small. Each string or bunch should contain bulbs of uniform size and of the same variety.

First-class garlic bulbs must be clean and have unbroken outer sheaths. Many of the larger vegetable markets, such as the large chain stores, could retail garlic in the form of clean, uniform cloves, two dozen to a mesh bag. Processors are not particular about having the cloves enclosed in a neat sheath and occasionally accept sprouted bulbs.

Garlic growing can be very profitable when freshness is stressed and if the tops are braided, tied together, or placed into long, narrow, plastic mesh bags so they can be effectively displayed at roadside or night-market stands

GREENS: COLLARDS, KALE, MUSTARD, AND TURNIP (*Brassica* spp.)

VARIETIES[1]	AL	AR	GA	KY	LA	MS	NC	SC	TN	VA
COLLARDS										
Champion				K	L	M	N	S	T	
Flash	A		G	K	L	M	N	S	T	
Georgia Southern [4]	A		G	K	L		N	S	T	
Morris Heading							N	S		
Top Bunch 2.0 [5]	A	R	G	K	L	M	N	S	T	V
Vates	A	R	G	K	L	M	N	S	T	V
KALE										
Black Magic [6]	A			K		M	N			
Blue Ridge [5]	A		G							
Darkibor	A		G	K				S		
Lacinato [6]	A		G	K			N	S	T	
Premier							N	S		
Red Russian				K		M	N		T	
Redbor	A									
Siberian	A		G	K	L	M	N	S	T	
Starbor								S		
Vates	A		G	K	L	M	N	S	T	V
Winterbor	A		G	K		M	N	S	T	V
MUSTARD										
Carolina Broadleaf [2]	A						N			
Florida Broadleaf	A		G	K	L	M	N	S		
Garnet Red/Garnet Giant	A					M				
Green Wave	A			K	L				T	
Southern Giant/Southern Giant Curled	A		G	K	L	M	N	S	T	
Tendergreen [3]	A		G	K	L	M	N	S	T	
TURNIP GREENS										
Alamo	A		G	K	L	M		S		
All Top	A		G	K	L		N	S	T	
Just Right	A					M	N			
Purple Top White Globe	A		G	K	L	M	N	S	T	
Seven Top	A		G	K	L		N	S		
Shogoin	A				L		N			
Southern Green	A		G	K	L	M	N	S		
Topper				K	L	M		S	T	
Tokyo Cross			G		L			S	T	
Vates	A		G	K	L	M	N	S	T	V
Winterbor	A		G	K		M	N	S	T	V

[1] Abbreviations for state where recommended
[2] Bacterial leaf blight resistance
[3] Spinach-mustard
[4] Bolting susceptible
[5] Bolting tolerant
[6] Dinosaur or Tuscan Kale

Seeding. Greens can be seeded in succession throughout the times listed in the table below. The next seeding date should be made when the previous crop is 50% emerged. Seeds emerge in 3 to 12 days: emergence is temperature dependent, with rapid emergence in warm weather (fall planting) and slower in cool temps (spring planting).

Seed mustard and turnips at the rate of 3 to 4 pounds per acre. Plant seeds at a depth of ¼- to ½-inch with row spacing of 1 to 3 feet apart. When transplanting, use plant spacing of 2 to 6 inches within the row.

Seed collards and kale at the rate of 2 pounds per acre and thin to desired spacing. For precision, air-assist planters use ⅓ to ½ pound per acre for twin rows on 3-foot centers, or use half of this rate for single rows on 3-foot centers. When using transplants, set plants in rows 16 to 24 inches apart and 6 to 18 inches apart within the row.

Soils. Loamy soils will produce greatest yields, but many soil types are suitable. Sandy soils are preferred for cool season and overwintering production. Greens grown in sandy soils are easier to pull from the soil, and easier to clean off soil residue, than those grown in clay soils. Soil pH of 6.0 to 6.5 is desirable.

Fertilizers. Quality greens require quick, continuous growth. A continual supply of nitrogen is essential for good color and tenderness. Applications of nitrogen at planting followed by additional sidedress applications during the growing season, are essential to produce consistent, high-quality greens.

Cultivation. In addition to adequate nutrition, consistent irrigation is necessary for good leaf formation. Overhead irrigation should be avoided as it causes favorable conditions for the development of several diseases.

MUSTARD AND TURNIP PLANTING DATES

	Spring	Fall
AL North	2/1–4/30	8/1–9/15
AL South	2/1–5/15	8/1–10/31
GA North	3/15–4/30	8/1–9/15
GA South	2/1–5/15	8/1–10/31
KY East	3/15-4/30	7/1-7/15
KY Central	3/10-4/25	7/15-8/1
KY West	3/1-4/15	8/1-8/15
LA	2/1–3/15	7/15–10/31
MS North	1/20–4/1	7/25–8/20
MS South	1/15–3/1	8/10–9/15
NC East	2/15–6/30	8/1–9/15
NC West	4/1–8/15	NR
SC East	2/1–6/15	8/1–10/15
SC West	3/15–9/15	NR
TN East	4/1-5/30	7/1-7/30
TN West	2/15-4/15	8/1-8/31

Note: Planting dates in the spring could be earlier if low tunnels or other season extension measures are used in some locations
NR = Not recommended

COLLARD PLANTING DATES

	Spring	Fall
AL North	2/15–6/30	7/15–10/15
AL South	1/15–5/31	7/15–10/31
GA North	3/15–4/30	NR
GA South	2/1–3/31	8/1–10/31
KY East	3/15–4/30	7/1-7/15
KY Central	3/10-4/25	7/15-8/1
KY West	3/1-4/15	8/1-8/15
LA	1/15–3/15	7/15–10/31
MS North	1/20–4/1	7/25–8/20
MS South	1/15–3/1	8/10–9/15
NC East	2/15–6/30	8/1–9/15
NC West	4/1–8/15	NR
SC East	2/1–6/15	8/1–10/30
SC West	3/15–6/30	8/1-9/30
TN East	3/15–5/1	8/1–9/1
TN West	2/15–4/15	8/15–9/15
VA East (coastal)	4/1– 4/30	7/1–8/31
VA West (mountain)	4/20– 5/20	7/15–8/20

Note: Planting dates in the spring could be earlier if low tunnels or other season extension measures are used in some locations
NR = Not recommended

KALE PLANTING DATES

	Spring	Fall
AL North	3/15–4/30	8/1–9/15
AL South	2/1–3/31	8/1–10/31
GA North	3/15–4/30	NR
GA South	2/1–3/31	8/1–10/31
KY East	4/1–4/30	7/1–7/15
KY Central	3/20–4/15	7/15–8/1
KY West	3/10–4/10	8/1–8/15

KALE PLANTING DATES (cont'd)

	Spring	Fall
LA North	2/1–3/15	7/15–10/31
MS North	1/20–4/1	7/25–8/20
MS South	1/15–3/1	8/10–9/15
NC East	2/15–6/30	8/1–9/15
NC West	4/1–8/15	NR
SC East	2/1–6/15	8/1–10/30
SC West	3/15–6/30	8/1–9/30
TN East	3/15–5/1	8/1–9/1
TN West	2/15–4/15	8/15–9/15
VA East (coastal)	4/1–4/30	7/1–8/31
VA West (mountain)	4/20–5/20	7/15–8/20

Note: Planting dates in the spring could be earlier if low tunnels or other season extension measures are used in some locations
NR = Not recommended

SPECIAL NOTES FOR PEST MANAGEMENT
INSECT MANAGEMENT

Aphids: These insects can be serious pests of greens crops. Frequent examinations of the crops are necessary to avoid undetected infestations. Broad-spectrum insecticides used for caterpillar management can lead to aphid infestations. Numerous effective and selective aphid insecticides are available (e.g. IRAC 9, 29). See Table 2-25.

Caterpillars: Many of the same caterpillars that feed on the large cole crops (cabbage, collard, etc.) will feed on greens. Action thresholds for greens crops are currently lacking, but low levels of caterpillars can be tolerated during the early stages of growth. The use of BTs and other soft materials are encouraged in order to maintain natural enemy populations in the crops.

Flea Beetles: These small insects can be serious pests of greens crops. They are often associated with heavier soils and weedy areas. BTs are ineffective against beetle pests, but neonicoti- noids insecticides work well with reduced effects on natural enemies compared to pyrethroids. Treatment should begin when the infestation is first noticed. Frequent use of broad-spectrum insecticides for flea beetle management often leads to resurgence of other pests. Reflective mulches have been found to be effective in repelling flea beetles.

HARVESTING AND STORAGE

Collards: They may be harvested at any stage of growth. Recommended storage temperature is 32°F (0°C), and 95-100% relative humidity for a shelf-life of up to 14 days. Collards have a high sensitivity to ethylene, and storage with products that release ethylene should be avoided.

Kale: It is harvested by cutting off the entire plant near ground level, or lower leaves may be stripped from the plant. Kale can be stored at 32°F (0°C) for up to 21 days. For short-term storage or marketing within 14 days, storage at 41°F (5°C) will provide acceptable quality. Relative humidity of 95-100% is recommended. Condensation should be avoided, particularly in curly varieties, like 'Winterbor', as the accumulation of water on leaves can lead to decay during storage. Kale is moderately sensitive to ethylene gas, and storage with ethylene-releasing products should be avoided. Ethylene exposure induces leaf yellowing.

Turnip greens: They can be harvested by cutting off the entire plant when the leaves reach the desired length. Leaves are highly susceptible to bruising and must be handled carefully to avoid discoloration and decay during storage. Wilting and water loss are common problems during postharvest storage. Storage at 32 and 95-100% relative humidity is recommended for a shelf-life of 10-14 days. See Table 14 for further postharvest information on these crops.

Fresh-cut leafy greens: Fresh-cut leafy greens, also known as value-added, minimally processed, cut, or pre-cut, have been physically altered from their initial state at harvest. They have been subjected to processing operations like chopping, slicing, or shredding, and they do not undergo additional processing like blanching or cooking. They can be sold as *ready-to-use* when they can be consumed after additional processing, which may include washing or cooking, or as *ready-to-eat* greens when they can be consumed directly without washing or additional preparation steps. Fresh-cut leafy greens are often packaged in plastic clamshell containers or bags to protect the product. Cutting operations induce physical damage and expose leafy green tissues to stress, which elevates their rates of respiration and transpiration. This increased metabolic activity promotes dehydration, yellowing, browning, and spoilage, significantly reducing shelf-life compared to the whole, intact product. Cutting also favors the release of compounds from inside the cells, creating a nutrient-rich environment that promotes the growth of microorganisms that can increase the risk of foodborne illness. To slow and minimize these undesirable quality outcomes, fresh-cut leafy greens must be stored at temperatures close to 32 °F.

Cold chain breaks and severe temperature fluctuations are notoriously harmful for fresh-cut leafy greens. For instance, Brassica vegetables, e.g., cabbage, kale, collards and others, are rich in sulfur-containing compounds that are responsible for their unique taste and aroma. Temperature increases that result from cold chain breaks can limit gas exchange in bagged vegetables and promote fermentation. This may result in the development of off-odors and off-flavors that contribute to consumer rejection.

Maintaining strict sanitation practices as well as the cold chain is crucial throughout the entire process—from fresh-cut leafy greens processing, storage and transportation to distribution and sale.

HERBS
BASIL (*Ocimum basilicum*), CILANTRO (*Coriandrum sativum*), AND PARSLEY (*Petroselinum crispum*)

VARIETIES[1]	AL	GA	KY	LA	MS	NC	SC	TN	VA
BASIL									
Sweet									
Aroma II [2,3]	A	G	K	L	M	N	S	T	
Genovese [2]	A	G	K	L	M	N	S	T	
Italian Large Leaf [3]	A	G	K	L	M	N	S	T	
Mammoth						N			
Nufar [2,3]	A	G	K	L	M	N	S	T	
Purple Ruffles	A	G	K	L	M	N	S	T	
Rutgers Devotion DMR [4]	A						S		
Rutgers Obsession DMR [4]	A								
Rutgers Passion DMR [4]	A								
Rutgers Thunderstruck DMR [4,5,6]	A						S		
Specialty									
Mrs. Burns' Lemon	A	G		L	M	N	S	T	
Sweet Thai (Horapha, Hun Que)	A	G	K	L	M	N	S	T	
Cinnamon	A	G		L	M	N	S		
CILANTRO									
Calypso [2,3]	A		K			N	S		
Cruiser [7]							S		
Marino			K						
Turbo II [4]									V
Santo/Santo Long Standing [3]	A		K	L		N	S	T	V
PARSLEY									
Curly Leaf									
Banquet	A	G	K	L	M	N	S	T	V
Forest Green	A			L		N	S	T	V
Moss Curled	A					N	S		V
Flat Leaf									
Dark Green Italian	A			L			S		
Giant of Italy	A			L		N	S		V
Plain Italian Green	A	G	K	L	M	N	S	T	V

[1] Abbreviations for state where recommended
[2] Fusarium tolerance/resistance
[3] Suitable for high tunnel production
[4] Downy mildew tolerance
[5] Suitable for microgreen production
[6] Suitable for processing
[7] Bolting resistant

CULTIVATION

Basil. Basil is an easy to grow tender annual. Plant basil in *late spring after all danger of frost is past*. Grow in full sun in warm, well-drained soil, preferably in raised beds. A light sand to silt loam with a pH of 6.4 is best. Basil may be grown in the field from seed or transplants. Sow seed 1/8 inch deep. Trim transplants to encourage branching and plant in the field when about six inches tall (4 to 6 weeks old).

Double-row plantings on 2- to 4-foot wide beds increase yields per acre and helps to shade out weeds. Planting dates may be staggered to provide a continuous supply of fresh leaves throughout the growing season. For fresh-cut basil production, the use of black plastic mulch is highly recommended. Basil will not tolerate moisture stress; provide a regular supply of water through drip or overhead irrigation.

Do not over fertilize basil. It is generally suggested that 100 pounds each of N, P_2O_5, and K_2O per acre be broadcast and incorporated at time of planting or follow guidelines for fertilization of salad greens. If more than one harvest is made, sidedress with 15 to 30 pounds N per acre shortly after the first or second cutting.

BASIL PLANTING DATES

	Spring	Fall
AL North	4/1-7/31	NR
AL South	3/15-7/31	NR
GA North	4/1-7/31	8/1-8/15
GA South	3/15-5/1	8/1-9/15
KY East	5/15-6/30*	NR
KY Central	5/1-6/30*	NR
KY West	4/20-6/20*	NR

BASIL PLANTING DATES (cont'd)

	Spring	Fall
LA North	4/1-7/31	NR
LA South	3/15-7/31	NR
MS North	4/1-7/31	NR
MS South	3/15-7/31	NR
NC East	5/1-7/15*	NR
NC West	6/1-6/30*	NR
TN East	5/15-6/30*	8/1-8/15*
TN West	4/20-6/20*	8/1-8/15*
VA East (coastal)	5/1-6/30	NR
VA West (mountain)	5/15-6/30	NR

* Using 6 to 8 wk-old transplants
NR = Not recommended

Cilantro. Cilantro is a fast growing annual that is cultivated for its fresh leaves. The seeds of the cilantro plant are referred to as the spice coriander. Cilantro is best cultivated as a cool season crop in the Southeast.

Cilantro cultivars are divided into "temperature sensitive" and "slow-bolt" groups. When high temperatures and daylight greater than twelve hours occur, temperature sensitive cultivars tend to set flowers in as little as three weeks following germination. Cilantro responds well to growth stimulators (gibberellic acid, folcyteine, extracts of marine algae) to maximize leaf production. Premature bloom can be delayed through the use of these foliar sprays.

Parsley. Parsley is a biennial grown as an annual. There are two varietal types of parsley: flat leaf and curled leaf. Flat leaf parsley tends to be more aromatic than the curled leaf and is used for flavoring in cooking. Curled leaf parsley is more attractive and is primarily used as a garnish. Parsley is also best cultivated as a cool season crop in the Southeast.

CILANTRO/PARSLEY PLANTING DATES

	Spring	Fall
AL North	3/15–5/30	NR
AL South	2/1–3/31	8/1–9/30
GA North	3/15–5/30	NR
GA South	2/1–3/31	8/1–9/30
KY East	5/10-7/10	NR
KY Central	5/1-7/20	NR
KY West	4/15-7/1	NR
LA North	2/15–4/15	9/15–10/31
LA South	2/1–4/15	9/15–10/31
MS	NR	8/1–9/30
NC East	2/15–4/15	8/1–9/30
NC West	4/1–8/15	NR
SC East	2/15-4/15	9/1–11/15
SC West	4/1-8/15	8/15–9/30
TN East	5/15-6/30*	8/1-8/15*
TN West	4/20-6/20*	8/1-8/15*
VA East (coastal)	3/15-5/15	8/1-9/15
VA West (mountain)	4/15-8/15	NR

NR = Not recommended

Parsley and cilantro grow best in a well-drained, organic loam soil with soil pH between 6.5 and 7.5. Overhead irrigation is essential for stand establishment. Irrigation during the germination period and the 2-3 weeks following emergence are critical. Too little water at any point will result in diminished leaf yield. Long, warm periods with too little water results in bolting which is undesirable since the plants are grown for their leaves. In addition, bolting reduces the amount, quality, and fla- vor of the leaves.

Seeding and Spacing. Neither parsley nor cilantro transplant well due to their taproots which are typical of plants in the Apiaceae. Direct seeding is recommended and is best achieved when using a precision seeder. Multiple plantings every 1-3 weeks are necessary for a season-long supply. Parsley seed is slow to germinate (12-25 days, temperature dependent). Seed is viable for 3-5 years but its percentage germination reduces quickly after 1 year.

Seed is sown 1/3 to ½ inches deep in a well-prepared seed bed. Seeding rates are from 16 to 24 pounds per acre (1/4 oz. per 100 row feet) for parsley and 15 to 50 pounds per acre (1-2 oz. per 100 row feet) for cilantro. Spacing between single rows is 15 to 18 inches. Parsley and cilantro can be precision seeded into raised beds with 3 to 4 rows per bed. Final in-row spacing should be 6 to 8 inches for parsley and 2 to 5 inches for cilantro. Research has shown that maximum yields can be achieved with more closely spaced plants.

SPECIAL NOTES FOR PEST MANAGEMENT

There are few agricultural chemicals registered for use on basil, cilantro, and parsley.

DISEASE MANAGEMENT

Keep foliage as dry as possible by watering early in the day, or by using drip irrigation to reduce fungal disease. Rotate herbs to different parts of the field each year and remove and destroy all plant debris to reduce soilborne disease.

Fusarium Wilt. Plants infected with this disease usually grow normally until they are 6 to 12 inches tall, then they become stunted and suddenly wilt. Fusarium wilt may persist in the soil for 8 to 12 years. Growers should use Fusarium wilt tested seed or resistant or tolerant varieties.

Basil Downy Mildew. Use clean seed and less susceptible varieties as they become available. Minimize leaf wetness as much as possible. Become familiar with disease symptoms. Consider applying fungicides proactively when your local Extension agent or agricultural advisor indicates that cucurbit downy mildew is active in your area. Although these are distinctly different diseases, active cucurbit downy mildew indicates that conditions are favorable for basil downy mildew development. You can also monitor for basil downy mildew on Ag Pest Monitor: Basil at *https://basil.agpestmonitor.org/*.

Parsley and cilantro are prone to leaf blights, leaf spots, and mildews. Any approved fungicides should be sprayed as soon as symptoms appear. Cultural controls include the use of drip irrigation, crop rotation, and limited movement through the fields during wet conditions.

Root and crown rot of parsley is best controlled by a two-year crop rotation with non-susceptible plants.

INSECT MANAGEMENT

BT products can be used to control various worms and caterpillars. Swallowtail caterpillars feed on parsley and are present in large numbers in late summer months. Row covers while swallowtail butterflies are present may reduce damage by blocking butterfly access to plants for egg laying.

Genovese, Italian Large Leaf, and lettuce leaf basil varieties are susceptible to Japanese beetles. Japanese beetle traps set about 20 feet away from the basil will help prevent damage. Reflective mulches, beneficial insects, insecticidal soaps, traps, and handpicking may give some level of control of other insect pests.

WEED MANAGEMENT

In basil, to keep weed pressure down, use high plant populations, shallow cultivation, and/or mulch.

Both parsley and cilantro are weak competitors with other plants. Weed control is critical throughout the season and will also make harvest more efficient. Weeds are best managed by using black plastic mulch and cultivation.

HARVESTING AND STORAGE

Basil: Leaf yields range from 1 to 3 tons per acre dried or 6 to 10 tons per acre fresh. Foliage may be harvested whenever four sets of true leaves can be left after cutting to initiate growth, but when harvesting for fresh or dried leaves, always cut prior to bloom. The presence of blossoms in the harvested foliage reduces quality. Frequent trimming helps keep plants bushy.

For small-scale production of fresh-market basil, the terminal 2- to 3-inch-long whorls of leaves may be cut or pinched off once or twice a week. This provides a high-quality product with little stem tissue present. Basil can also be cut and bunched like fresh parsley. A sickle bar type mower with adjustable cutting height is commonly used for harvesting large plantings for fresh and dried production.

The optimum storage temperature for fresh basil is 50 to 55°F with a high (90-95%) relative humidity to reduce water loss. Basil is highly susceptible to chilling injury when stored at chilling temperatures. Chilling injury symptoms include the development of dark lesions, leaf and stem discoloration, and loss of glossiness and aroma. Basil has a high sensitivity to ethylene gas, and storage with ethylene-releasing products should be avoided. Exposure to ethylene will promote stem curvature, leaf abscission and yellowing.

Parsley and Cilantro: Both herbs are usually harvested by hand and bunched with rubber bands or twist ties in the field. Cutting entire plants 1.25 to 3 inches above the crown may result in secondary growth sufficient to allow for another harvest. Average yield for both parsley and cilantro is 30-40 pounds per 100 row feet of row. Maximum biomass usually occurs at 40-45 days after germination for cilantro and at 75-90 days for parsley. Multiple harvests are more likely with parsley than cilantro. Store parsley and cilantro at 32°F with high humidity. See Table 14 for further postharvest information.

Microgreens: Microgreens should be harvested at the peak of their cotyledon stage, just as the first true leaves begin to emerge, typically 7–14 days after sowing, depending on the species. Harvesting is best done in the early morning using sanitized scissors or blades to minimize wilting and microbial contamination. After harvest, microgreens should be handled gently and packaged immediately into a plastic vented clamshell or storage bag, then cooled. It is not recommended to rinse microgreens before storage, as excess moisture and water buildup increase the rate of decay and pathogen growth. For optimal shelf-life and quality, microgreens should be stored at 34–40°F with 90–95% relative humidity to prevent dehydration, and preserve texture and nutritional value. Shelf-life of microgreens depends on the species and cultivar being grown and can range between 14 (*Amaranthus*) and 28 days (*Brassica*).

HOPS (*Humulus lupulus*)

VARIETIES[1]	AL	AR	GA	KY	MS	NC	SC	TN
HOPS								
Cascade	A	R	G	K	M	N	S	T
Chinook	A					N		T
Galena						N		
Nugget						N		T
Zeus		R				N		T

[1] Abbreviations for state where recommended

Hops (*Humulus lupulus*) are a new crop for the Southeast. Most hops in the United States are grown in the Pacific Northwest Washington, Oregon, and Idaho. Hops are herbaceous perennials with long-lived underground crowns. Each year the plants send up multiple shoots (called bines) which bear papery cones (flowers) that are the plant parts that are harvested for making beer and herbal products. When mature, the cones contain bright yellow, sticky lupulin glands that contain the fragrance and bittering compounds that hops are valued for. Each year the bines can grow to be up to 25 feet long, so they need to be trellised. Hops have male and female plants, but only female plants are grown commercially so the cones do not contain seeds.

Hops are an expensive crop to establish because of the need for a permanent, tall trellis system. Short hop varieties are being bred, but at this time the common varieties need to be grown on tall trellises. There are several trellis designs available that are suitable for production in the Southeast. Most trellises are 16 to 18 feet tall and composed of locust or cedar posts and wire. The bines are trained to strings (often coir twine that is replaced annually) suspended from the top wires of the trellis.

Hops require a well-drained, fertile soil with a pH of 6.0 to 6.5. Hops are heavy feeders and soil tests should be taken annually to determine how to provide adequate nutrition. Hops also require irrigation which is usually supplied as drip irrigation.

Hops are photoperiod sensitive plants (short-day plants). Commercial hops varieties produce the highest yields in the most northern states. In short-day areas (below 35° latitude), flowering occurs too soon when the required number of nodes for a particular variety are produced. As a result, yields are not maximized. In longer day areas (above 35° latitude), vegetative growth is maximized prior to the point where day length begins to shorten in mid- to late summer.

Yields noticeably decrease the further south the plants are grown, particularly below the 35° latitude. Breeding efforts are underway to produce varieties specifically for the Southern U.S. In the meantime, there are cultural practices that can be used to increase yields in Southern hop yards. Most of the information currently available for hops production pertains to large-scale production in the Pacific Northwest or from the emerging industries in the Great Lakes and the Northeast. That information can be helpful, but the differences in photoperiod, disease incidence, lack of infrastructure for processing, and scale of production require adapting it to suit conditions in the Southeast.

Varieties. Cascade is the variety that is the most reliable throughout the Southeast. It is an aroma hop and is used by most brewers. Growers are encouraged to talk to their buyers to identify other varieties to grow. Growers in the Southeast have not been very successful growing the Noble varieties. New varieties bred specifically for the Southeast have been developed and should be available in the near future.

Pruning. To encourage flowering at the proper time for increased yields, emerging hop shoots are often cut to the ground in the spring until late April. Then several shoots are selected to be trained to each string. The remaining emerging shoots are kept pruned away. The foliage from the lower four feet or so of the plant is mechanically or chemically removed to encourage good air movement around the plants.

Harvesting. Cones are harvested in mid to late summer. Small-scale growers often hand-pick cones multiple times during the season. As the hop yard expands, however, this becomes impractical and most growers move to a one-time harvest which involves cutting the bines, removing them from the yard, and running them through a mechanical harvester which separates the cones from the foliage, bines, and strings.

Cones can be sold as fresh (wet), whole cones to brewers for making seasonal ales. More commonly, hops are dried in a dryer called an oast. A few brewers use whole dried cones, but most brewers require dried hop pellets. Hops quality is determined by chemical analysis which includes alpha and beta acids and essential oils. Most brewers will want these numbers before purchasing hops. How the hops are grown, when they are harvested, and how they are handled after harvesting and stored will greatly affect these values.

SPECIAL NOTES FOR PEST MANAGEMENT
Disease, insect, and weed control strategies for the Southeast are still being developed.

DISEASE MANAGEMENT
Downy mildew is the primary disease that growers need to be prepared to manage and should be a major consideration when choosing varieties. There are many other diseases, including viruses and viroids, that affect hops and that a grower should be scouting for.

INSECT MANAGEMENT

Spider Mites: Spider mites are the most impactful arthropod pest of hops in the Southeast. Infestations generally arise during hot and dry periods of the summer on the basal portions of plants. Feeding leads to stippling on leaves and cones, which ultimately effects whole plant production. Avoid overuse of nitrogen fertilizer and underwatering of plants to help prevent early outbreaks of mites. Many broad-spectrum insecticides can kill natural enemies such as predatory mites, ultimately leading to outbreaks of spider mites. Utilize scouting to make any insecticide applications in hops to avoid spider mite outbreaks. Many miticides have long pre-harvest intervals and should be used to suppress mites early (5-10 mites per leaf).

Japanese Beetles: Japanese Beetles (*Popillia japonica*) can be a serious defoliator of hops. Control should be considered based on the infestation density of Japanese beetle and the amount of feeding that is observed. Healthy and mature hops plants can withstand substantial amounts of defoliation before insecticides are warranted. All insecticides labeled for hops that are effective in controlling Japanese beetle are likely to flare mites, including organophosphates (IRAC 1B), pyrethroids (IRAC 3A), and neonicotinoids (IRAC 4A). Smaller plantings are more likely to see a larger impact from Japanese beetle.

Caterpillar Pests: Several caterpillar pests commonly feed on hops grown in the Southeast. Yellow-striped armyworm (*Spodoptera ornithogalli*), question mark caterpillars (*Polygonia interrogationis*), and comma caterpillars (*Polygonia comma*) are commonly observed feeding on hops leaves and cones. Hops leaves should be scouted regularly after shoots are pruned back in the late spring, and insecticides are warranted when eggs or young caterpillars are observed. Prioritize insecticides that are less likely to flare mites such as those in the diamide insecticide class (IRAC 28) or spinosads (IRAC Group 5).

WEED MANAGEMENT

Weed control should be planned for in advance and may include use of herbicides, landscape fabric, and other mulches. A good air-blast sprayer will be needed to provide good spray coverage up to the top of the trellis.

LEEKS (*Allium porrum*)

VARIETIES[1]	AL	GA	KY	NC	SC	TN
LEEKS						
Bandit			K	N		T
Chinook	A					
King Richard			K			T
Lancelot	A	G		N	S	T
Rally	A			N	S	
Tadorna	A			N	S	

[1] Abbreviations for state where recommended

Transplants. Transplants are used for early spring plantings. For summer planting, sow in seed beds as indicated in following table. About 2 pounds of seed are required to provide enough plants to set an acre. Seed should be planted $1/3$- to $1/2$-inch deep 8 to 12 weeks before field setting. Plants will be ready to set in early August. Plug cells have worked well.

LEEK PLANTING DATES

	Spring	Fall
AL North	3/15–4/30	9/15–10/31
AL South	2/1–3/31	NR
GA North	3/15–4/30	9/15–10/31
GA South	2/1–3/31	NR
KY East	4/1–6/15	NR
KY Central	3/25–7/1	NR
KY West	3/15–7/15	NR
MS	NR	NR
NC East	2/15–6/30	NR
NC West	4/1–8/15	NR
SC East	2/1–6/15	NR
SC West	3/15–6/30	NR
TN East	4/1–6/30	NR
TN West	3/15–8/1	NR

NR = Not recommended

Field Spacing. Set plants in trenches 3 to 4 inches deep with rows 20 to 30 inches apart and plants 4 inches apart in the row.

Culture. Leeks grow slowly for the first 2 or 3 months. To develop a long white stem, start to gradually fill in trenches and then hill soil around stems to 3 or 4 inches.

SPECIAL NOTES FOR PEST MANAGEMENT
INSECT MANAGEMENT

Thrips: During hot, dry weather, thrips often move into crops following harvest of adjacent alfalfa or grain and can present the most serious insect problem on onions and leeks. (See "Onions" in the Insect Control section of this publication). Treat if thrips counts exceed an average of 5 thrips per plant.

Allium leafminer is an invasive pest that has recently become a problem in Virginia and North Carolina. This fly pest deposits eggs in stems where larvae mine and tunnel plants causing wilting and plant death and contamination of the product. (See "Onions" in the Insect Control section of this publication).

HARVESTING AND STORAGE

Spring-transplanted leeks are ready for harvest in July. Fall-transplanted leeks are ready to harvest by July. Fall-planted leeks are ready by November and can be overwintered. See Table 14 for postharvest information.

LETTUCE (*Lactuca sativa*), ENDIVE (*Cichorium endivia*), AND ESCAROLE (*C. endivia*)

VARIETIES[1]	AL	AR	GA	KY	LA	MS	NC	SC	TN	VA
LETTUCE										
Green Leaf										
Grand Rapids [6]	A		G	K	L		N	S		
Green Star [3, 6, 7]	A					M			T	V
Nevada [3, 6]	A				L					V
Salad Bowl [3]	A		G	K	L	M	N		T	
Salanova Green Batavia [7, 11]		R	G		L	M	N	S	T	
Sierra Batavia [3] (green w/ red edges)	A				L					
Slobolt [3]	A						N	S	T	
Tango	A			K	L		N		T	
Tehama [3, 6]					L		N	S		V
Tropicana [6]						M	N		T	
Two Star [3, 6]	A				L	M	N	S		
Red Leaf										
Cherokee [3, 4]					L		N		T	
New Red Fire [3, 4, 6]	A		G	K	L	M	N	S	T	
Pomegranate Crunch [7, 8]	A			K		M			T	
Red Express [4]							N	S		V
Red Sails [4]	A			K	L		N		T	
Ruby [4]					L		N		T	
Salanova Hydroponic Red [7]		R	G		L	M		S	T	
Cos / Romaine										
Bluerock [3, 7, 9]	A									
Breen [7, 10]	A									
Coastal Star	A							S	T	V
Green Forest [3, 6]	A			K	L		N	S	T	V
Green Towers	A		G	K	L	M	N		T	V
Parris Island Cos	A			K	L				T	
Salvius [3, 7]								S		
Sparx [3, 7]						M		S		
Truchas [4, 7, 9, 11]	A									
Valley Heart [2]	A						N	S		
Winter Density [3, 5]			G	K						
Butterhead										
Adriana [3, 6, 7]	A				L	M	N		T	V
Buttercrunch [3]	A		G	K	L	M	N	S	T	V
Ermosa	A			K	L		N	S		
Harmony [3, 6, 7]				K	L		N			V
Nancy	A			K		M	N		T	V
Salanova Red Butter [3, 7]	A					M	N		T	
Salanova Green Butter [7]	A					M	N		T	V
Salanova Green Oakleaf [7]		R	G		L	M	N	S	T	
Salanova Red Oakleaf [7]		R	G		L	M	N	S	T	
Sangria [3, 6]	A			K					T	
ENDIVE										
Galia Frisse	A		G				N	S		
Salad King	A		G	K	L		N	S	T	

[1] Abbreviations for state where recommended
[2] Recommended for fall production only (bolting susceptible)
[3] Bolting tolerance/resistance
[4] Red leaves
[5] Bibb-Romaine type
[6] Tipburn tolerance/resistance
[7] Downy mildew tolerance/resistance
[8] Cross between Cos and Butterhead types
[9] Tomato bushy stunt virus (TBSV) tolerance/resistance
[10] Bronze-red leaves
[11] Lettuce mosaic virus (LMV) tolerance/resistance

VARIETIES[1]	AL	AR	GA	KY	LA	MS	NC	SC	TN	VA
ESCAROLE										
Full Heart Batavian	A			K	L		N	S	T	
Full Heart NR65 [3,6]	A						N	S		

[1] Abbreviations for state where recommended
[2] Recommended for fall production only (bolting susceptible)
[3] Bolting tolerance/resistance
[4] Red leaves
[5] Bibb-Romaine type
[6] Tipburn tolerance/resistance
[7] Downy mildew tolerance/resistance
[8] Cross between Cos and Butterhead types
[9] Tomato bushy stunt virus (TBSV) tolerance/resistance
[10] Bronze-red leaves
[11] Lettuce mosaic virus (LMV) tolerance/resistance

Lettuce and endive are cool season crops. Properly hardened lettuce transplants can tolerate temperatures as low as 20 to 25°F. Temperatures above 85°F for several days will cause seed stalk formation and bolting in lettuce. Temperatures below 70°F during the seedling stage promote premature stalk formation in endive and escarole.

Due to a number of factors such as length of time to harvest, the production of head lettuce is not recommended in the regions covered by this handbook.

Seeding and Transplanting. *Spring crop.* Lettuce transplants are started in frames or greenhouses. Seed for the lettuce crop is sown in heated greenhouses at the rate of 4 to 6 ounces of seed for 1 acre of plants.

Direct-seeded lettuce is sown in prepared beds as early in the spring as the ground can be worked. Seed should be sown shallow—some of the seed will actually be uncovered and visible. Pelleted seed can be sown slightly deeper and should be watered at night during high-temperature periods (soil temperatures above 80°F) until germination occurs.

LETTUCE COS / ROMAINE PLANTING DATES

	Spring	Fall
AL North	4/15–5/30	8/1–9/15
AL South	2/1–3/31	8/1–9/30
GA North	4/15–5/30	NR
GA South	2/1–3/31	8/1–9/30
KY East	4/1–4/30	8/1–9/1
KY Central	3/25–4/15	8/15–9/15
KY West	3/15–4/1	8/30–9/30
LA	1/15–3/15	9/15–10/30
MS	1/15–3/15	9/15–10/30
NC East	2/1–4/10	8/28–9/15
NC West	3/15–8/1	NR
SC East	2/1–4/15	9/15–11/1
SC West	3/1–5/15	NR
TN East	3/15–4/30	8/1–9/1
TN West	3/1–4/15	8/15–9/15
VA East	4/15–8/1	NR

NR = Not recommended

LETTUCE LEAF AND BUTTERHEAD PLANTING

	Spring	Fall
AL North	4/15–5/30	8/1–9/30
AL South	2/1–4/15	8/1–10/15
GA North	4/15–5/30	8/1–8/30
GA South	2/1–4/15	8/1–10/15
KY East	4/1–4/30	8/1–9/1
KY Central	3/25–4/15	8/15–9/15
KY West	3/15–4/1	8/30–9/30
LA	1/15–3/15	9/15–10/30
MS North	3/15–4/30	8/1–9/30
MS South	2/1–4/15	8/1–10/15
NC East	2/1–4/20	8/25–10/1
NC West	3/1–8/25	NR
SC East	2/1–4/15	9/15–11/1
SC West	3/1–5/15	NR
TN East	3/15–4/30	8/1–9/1
TN West	3/1–4/15	8/15–9/15
VA East	4/15–8/1	NR

NR = Not recommended

ENDIVE/ESCAROLE PLANTING DATES

	Spring	Fall
AL North	4/15–5/30	8/1–9/15
AL South	2/1–3/31	8/1–9/30
GA North	4/15–5/30	NR
GA South	2/1–3/31	8/1–9/30
KY East	4/1–4/30	NR
KY Central	3/25–4/15	NR
KY West	3/15–4/1	NR
LA	1/15–3/15	9/15–10/30
MS	NR	NR
NC East	3/20–6/15	8/1–9/15
NC West	5/1–8/15	NR
SC East	2/1–4/15	9/15–11/1
SC West	3/1–5/15	NR
TN East	3/15–4/30	8/1–9/1
TN West	3/1–4/15	8/15–9/15
VA East	4/15–8/1	NR

NR = Not recommended

Mulching. Using polyethylene mulch can be very beneficial for all types of lettuce and endive, in that the plastic reduces the amount of soil that gets inside the leaves. Most leaf lettuce varieties can be planted in 3 or 4 rows to the 30-inch bed top. In-row spacing should be 9 to 12 inches and between row spacing should be 9 to 12 inches. Romaine types do best with 2 or 3 rows per bed and 12 to 15 inches in-row spacing.

SPACING

Lettuce: Leaf and Butterhead type lettuce are planted 3 to 4 rows per bed with beds spaced 66 to 72 inches on centers. Space plants 9 to 12 inches apart in the row.

Endive/Escarole: Plant three to four rows per bed and space beds 66 to 72 inches center-to-center. Space plants 9 to 15 inches apart in-row.

SPECIAL NOTES FOR PEST MANAGEMENT
DISEASE MANAGEMENT

Accurate diagnosis is necessary for effective disease management. Incorporate appropriate and effective cultural practices and pesticides, as detailed in the Disease Management section, in a season-long integrated disease management program.

Lettuce Drop (Sclerotinia Rot): Lettuce drop is a common disease of lettuce in spring-planted lettuce and can be problematic. Disease can be difficult to manage when fields become infested. When possible, use resistant varieties. The pathogens can be spread via movement of contaminated soil, water, and equipment/tools. Clean and disinfest equipment/tools when moving between fields. Practice appropriate crop rotations with nonhost crops. In sites with a history of disease, successful management often requires the use of an appropriate fungicide spray program.

Powdery Mildew: Powdery mildew can be a problem in enclosed structures and in field plantings in some areas. Plant resistant varieties, when possible. Successful management often requires the use of a preventative fungicide spray program.

Downy Mildew: Downy mildew can be a devastating disease of lettuce in some areas. Plant resistant varieties, when possible. *Note:* Multiple races of the downy mildew pathogen exist; races vary by location, and resistance may not be effective against all races. Cultural practices, such as drip irrigation, crop rotation, and sanitation, can slow disease spread. Successful management often requires the use of an appropriate fungicide spray program, particularly in sites with a history of disease.

INSECT MANAGEMENT

Keep lettuce fields isolated from endive and escarole for spray purposes.

Thrips: Scout for thrips and begin treatments when observed. To prevent thrips infestation, do not produce vegetable transplants with bedding plants in the same greenhouse.

Leafhopper: Control of leafhoppers will prevent the spread of lettuce yellows. In the spring, spray when plants are one-half inch tall; repeat as needed. In the fall, spray seedlings 4-5 times at 5-day intervals.

Corn Earworm (CEW): Note: Head lettuce seedlings, in the 7 to 18 leaf stage, are vulnerable to CEW attack in August to September. Control must be achieved before center leaves start to form a head (15- to 18-leaf stage).

Tarnished Plant Bug: This insect can cause serious damage to the fall crop; it is usually numerous where weeds abound.

HARVESTING AND STORAGE

Lettuce can be harvested at both an immature and mature stage. Harvest should be done in the morning to decrease postharvest losses. Prior to harvest, clean and sanitize the clippers or knife with a 10% bleach solution. The immature stage is when the lettuce is 4 to 6 inches in height. Cut the lettuce 1 inch from the soil or substrate surface. After harvest, rinse the lettuce in cold water and thoroughly dry. It is best to dry immature lettuce using a commercial lettuce spinner. Once dry, bag the lettuce in a large plastic bag and store at 32°F for 10 to 14 days.

For mature lettuce, cut the head just above the soil or substrate surface with a sanitized knife. Remove any discolored or damaged leaves. Rinse the heads in cold water and shake off excess moisture. Mature lettuce is best stored in a plastic clamshell with vented holes or polyethylene wrap. Head lettuce can be stored at 32°F for 2 to 3 weeks (see Table 14 for postharvest information).

MELONS (*Cucumis melo*)

VARIETIES[1]	AL	GA	KY	LA	MS	NC	SC	TN	VA
CANTALOUPES and MIXED MELONS									
Eastern									
Accolade [4, 5, 7, 8, 9, 11]		G				N			
Avatar [2, 4, 5, 7, 8, 9]	A		K					T	
Ambrosia [2, 3, 6]	A			L	M	N	S	T	
Aphrodite [4, 5, 7, 8, 9, 11]		G	K	L	M	N	S	T	
Astound [4, 5, 7, 8, 9, 11]		G				N			
Atlantis [2, 4, 5, 7, 8, 9]	A		K			N		T	
Athena [4, 5, 7, 8, 9, 11]	A	G	K	L	M	N		T	V
Da Vinci [7, 9, 14]	A	G					S		
Tirreno [4, 5, 7, 8, 9, 14]	A		K						
Long Shelf-life									
Anna's Charentais [7, 8, 9, 17, *]	A								
Caribbean Gold [5, 7, 8, 9, 12]	A	G				N	S	T	
Caribbean King [5, 6, 7, 8, 9, 12]		G				N			
D'Artagnan [7, 8, 9, *]	A								
Infinite Gold [4, 5, 7, 8, 9, 12]	A	G				N	S		
Savor [7, 8, 9, *, 17]	A								
Honeydew									
252 HQ [4, 5, 7, 9]	A	G				N			
Dream Dew [4, 5, 8, 9]	A						S		
Full Moon [5, 9]	A								
Honey Orange [13]	A	G	K			N	S	T	
Santa Fe						N			
Summer Dew [4, 5, 7, 9]	A	G	K			N	S	T	
Temptation [7, 9, 13]			K					T	
Galia									
Galia [*, &]	A	G				N	S		
Juan/Juane Canary									
Brilliant	A								
Camino Europa [7, 8, 9]						N			
Gladial [4, 5, 7, 9]						N	S		
Golden Beauty [6]	A	G	K			N		T	
Halo [*, &]						N	S		
Natal [4, 5, 7, 9]	A					N	S		
Sunbeam							S		
Tikal [4, 5, 7, 8, 9]	A					N	S	T	
Tweety [8, 9, 15, 16]						N			
Piel De Sapo/Santa Claus									
Lambkin [7, 8]	A					N			

[1] Abbreviations for state where recommended.
[2] Local markets only.
[3] Downy Mildew tolerance/resistance.
[4, 5, 6] Powdery mildew race 1, 2, or 3 tolerance/resistance.
[7, 8, 9, 10] Fusarium Wilt race 0, 1, 2, or 5 tolerance/resistance.
[11] Tolerant to sulphur.
[12] Extended shelf-life type/LSL.
[13] Orange-fleshed honeydew.
[14] Tuscan/Italian netted type.
[15] Gummy stem blight tolerance/resistance
[16] Anthracnose tolerance/resistance.
[17] Charentais type.
[*] Powdery Mildew tolerance/resistance (non-race specific).
[&] Fusarium Wilt tolerance/resistance (non-race specific).

Melon Types. There are many categories of specialty melons, all members of the cucurbit (Cucurbitaceae) family. The scientific name for specialty melons, which includes cantaloupes is *Cucumis melo*. *Cucumis melo* is subdivided into several botanical variants but only two variants, reticulatous and indorous have commercial importance in the Southeast.

The most important reticulatous melon grown in the South-east is what is commonly called cantaloupe or muskmelon. Cantaloupes are typically separated into two categories; Eastern and Western. The Eastern-type cantaloupe varieties, with limited netting and deep sutures, have traditionally been grown for the local market as many have a short shelf-life and bruise easily during shipping. Improved Eastern varieties such as 'Athena' have a longer shelf-life and can be shipped to distant markets. The Western cantaloupe has a uniform netted rind but lacks sutures. Although the Western-type cantaloupe has traditionally been grown in the Western states, there are varieties suitable for commercial production in the Southeast. While Eastern-type cantaloupes are typically harvested when the fruit stem pulls easily away from the fruit (full-slip), Western-type cantaloupes are harvested when the fruit stem partially pulls free (half-slip).

The Galia melon is another reticulatous variant grown in the Southeast. When mature the Galia melon has a yellow to orange skin that is covered in light golden-tan netting. The aromatic flesh is pale green and has a distinct sweet flavor. Galia melons are unfortunately prone to rapid softening and consequently most varieties are more suited for local markets.

Within the indorous variant there are two notable melons grown in the Southeast, honeydew and canary melons. Of these two the honeydew melon is the most widely grown. Although the rind color of honeydew can vary among varieties, typically the rind turns from green to whitish yellow as it matures. Unlike cantaloupe, the honeydew fruit does not slip from the vine when mature. The very sweet flesh of a mature honeydew is pale green with the deepest shade occurring immediately below the skin. Honeydew fruit are susceptible to cracking or splitting due to the uneven, high moisture conditions found in the Southeast. Several new varieties have helped lessen this issue.

The canary melon, also marketed as the Juan Canary, is an indorous variant. The canary melon is oval-shaped, with a smooth bright yellow skin. The canary like the honeydew does not slip from the vine. When mature the canary will turn dark yellow and develop a slightly corrugated skin appearance. The flesh is a pale ivory in color and slightly firmer than honeydew. A distinct advantage of the canary melon is that they have a very long shelf-life and can be shipped to distant markets.

Transplant Production. Plug flats that provide a cell depth of 1.5 inches are optimal for melon transplant production. Although plug flats with larger cells can be used a 200-square plug flat provides ample room for root growth. Although varieties may vary in transplant production time, under ideal conditions 2 to 3 weeks is enough time for transplant production.

Planting and Spacing. Consult the following Melon Planting Dates table for suggested planting dates in your area. Early season annual temperatures can vary tremendously so be aware that temperatures below 45°F will slow or stunt plant growth.

Melon plant recommended between row and in-row spacing can vary depending on the production method and equipment. The use of plastic mulch and drip irrigation allows for a closer row and in-row spacing compared to bare ground production. A typical planting scheme for plasticulture production will have rows spaced 5 to 6 feet with in-row spacing of 1.5 to 2 feet. For bare ground production, in-row spacing will need to be wider, 3 to 4 feet. An average of 10 to 12 ft^2 per plant for plasticulture production is typical while on bare ground 15 to 24 ft^2 per plant should be adequate.

MELON PLANTING DATES* (cont'd)

	Spring	Fall*
AL North	4/15–6/15	8/1–8/30
AL South	3/1–6/30	8/1–9/15
GA North	4/15–6/15	NR
GA South	3/1–4/30	8/1–9/15
KY East	5/15-6/15	NR
KY Central	5/10-7/1	NR
KY West	4/25-7/15	NR

MELON PLANTING DATES* (cont'd)

	Spring	Fall*
LA North	4/1–6/30	7/1–7/31
LA South	3/15–6/30	7/1–8/15
MS North	4/1–4/10	NR
MS South	3/1–3/15	NR
NC East	4/15–5/15	7/1–7/15
NC West	5/15–7/31	NR
SC East	3/15–5/15	7/1–7/30
SC West	4/15–6/5	NR
TN East	5/5-6/15	NR
TN West	4/15–6/1	NR
VA East (coastal)	3/30-7/20	NR
VA (mountain)	4/10-7/10	NR

*Use transplants for later season plantings
NR = Not recommended

Nutrient Application. A soil test to determine soil pH and nutrient availability is critical for commercial production of melons. Soil pH should be adjusted to 6.5. Pre-plant nutrient application should provide at least 25 pounds per acre of N and K$_2$O and all the P$_2$O$_5$ recommended by the soil test results. Applied nutrients should be thoroughly incorporated into the soil to reduce root damage to the melon transplants. Follow soil test results for minor nutrient application such as boron and sulphur.

Drip Fertigation. Liquid nutrient application through the drip tape (fertigation) should begin at or shortly after transplanting or direct seeding melons. Suggested fertigation schedules, based on plant growth stage, are provided in the below tables. Continue fertigating until the last harvest.

SUGGESTED FERTIGATION SCHEDULE FOR MELON*
(low potassium soil)

Plant growth stage	Days after planting	Daily nitrogen	Daily potash	Cumulative Nitrogen	Cumulative Potash
		(lb/A)			
Preplant				25.0	50.0
Planting to Vining	0-14	0.9	1.0	32.0	64.0
Vining to Flowering	14-28	1.0	2.0	46.0	92.0
Flowering to Fruit Set	29-49	1.5	3.0	76.0	152.0
Fruit Set to Ripening	50-77	2.0	4.0	130.0	260.0
Harvest	78-91	1.0	2.0	143.0	286.0

SUGGESTED FERTIGATION SCHEDULE FOR MELON*
(high potassium soil)

Plant growth stage	Days after planting	Daily nitrogen	Daily potash	Cumulative Nitrogen	Cumulative Potash
		(lb/A)			
Preplant				25.0	50.0
Planting to Vining	0-14	0.5	0.5	32.0	57.0
Vining to Flowering	14-28	1.0	1.0	46.0	91.0
Flowering to Fruit Set	29-49	1.5	1.5	76.0	101.0
Fruit Set to Ripening	50-77	2.0	2.0	130.0	155.0
Harvest	78-91	1.0	1.0	143.0	168.0

*Adjust based on tissue analysis

Plastic Mulch. The use of plastic mulch is particularly beneficial for melon production. Black plastic mulch used in the spring increases soil temperature, reduces soil moisture loss and reduces weed competition except for nutsedge. The use of black plastic mulch in the fall can result in detrimentally high soil temperatures, consequently white plastic mulch is recommended for fall melon production. Both colors of mulch help reduce fruit rots and provide significant yield increases compared to bare ground crop production.

SPECIAL NOTES FOR PEST MANAGEMENT
DISEASE MANGEMENT
Accurate diagnosis is necessary for effective disease management. Incorporate appropriate and effective cultural practices and pesticides, as detailed in the Disease Management section, in a season-long integrated disease management program.

Cucurbit Downy Mildew (CDM): CDM can be a devastating disease of cucurbits in the southeastern and eastern U.S. Successful CDM management often requires the use of an appropriate fun- gicide spray program. Fungicide efficacy is influenced by crop, regional, and seasonal differences, as well as resistance management practices. A useful tool for downy mildew management (maps, text alerts, identification resources, etc.) is the CDM reporting system (http://cdm.ipmpipe.org). If you think you have CDM or have questions about disease management, please contact your local Extension office.

Gummy Stem Blight: Gummy stem blight is a common foliar disease on melons in the Southeast. Crop rotation away from cucurbits for two full years will improve the efficacy of most fungicides. Bury crop debris as soon as possible after harvest; do not leave crowns of plants on plastic-mulched raised beds, which allows the fungus to survive. See Example Spray Program for Foliar Disease Control in Watermelon Production.

Whitefly-transmitted Viruses: Several whitefly-transmitted viruses, including cucurbit chlorotic yellows virus (CCYV), cucurbit leaf crumple virus (CuLCrV), and cucurbit yellow stunting disorder virus (CYSDV), are emerging in various areas throughout the Southeast when unusually high populations of whiteflies are present. These viruses may produce similar symptoms; molecular testing is necessary for virus identification. These viruses often also occur in mixed infections. Resistance to these viruses is either limited or not available in commercial varieties. Management of whitefly-transmitted viruses focuses on prevention and management of the whitefly vector. Additional information on whitefly-transmitted viruses in cucurbits is available at *https://eCucurbitviruses.org*.

NEMATODE MANAGEMENT
Various root-knot nematodes, such as the southern root-knot and guava root-knot, as well as other nematodes can infect melon. Nematode populations are highest in the late summer and fall. Collect soil samples for nematode analysis during this period from fields to be planted with melons the following year. Clean and disinfest equipment between fields. Use appropriate crop rotations and cover crops with non-host crops to reduce nematode populations. Avoid fields with high populations of nematodes or use appropriate fumigants and other nematicides for management (see fumigant and nematicide tables).

INSECT MANAGEMENT
Seed Corn Maggot (SCM): Use insecticide treated seed or at-planting soil-insecticide treatments to avoid SCM in the early season. SCM problems subside with later plantings.

Cucumber Beetle: Cucumber beetles transmit bacterial wilt, and most cultivars of muskmelons are highly susceptible to this disease. Also, adult beetles can cause direct feeding injury to young plants. Foliar insecticides should be used to control adult beetles before they feed extensively on the cotyledons and first true leaves. Begin spraying shortly after plant emergence and repeat applications at weekly intervals if new beetles continue to invade fields. Treatments may be required until vining, at which time plants are less susceptible to wilt infections.

Pickleworm, Melonworm: Make one treatment prior to fruit set, and then treat weekly.

Aphids: Aphids can delay plant maturity. Thorough spray coverage beneath leaves is important. For further information on aphid controls, see the preceding section on "Mulches and Row Covers." Treat seedlings every 5 to 7 days or as needed.

Squash Bug: Begin treatments shortly after vining. Treat every 7 to 10 days or as needed.

Leafhoppers: High numbers of leafhoppers cause leaf yellowing (chlorosis) known as hopper burn, resulting in yield loss.

POLLINATION
Honey bees are important for pollination, high yields, and quality fruit. Populations of pollinating insects may be adversely affected by insecticides applied to flowers or weeds in bloom. Apply insecticides only in the evening hours or wait until blooms have closed before application. See section on "Pollination" in the General Production Recommendations.

HARVESTING AND STORAGE
Cantaloupes should be harvested at quarter-to half-slip for shipping. Healthy vines and leaves must be maintained until melons are mature to obtain high-quality. Harvest daily or twice daily in hot weather. Many other types of melons do not slip and judging maturity can be difficult. Many will change their water not color. It is critical to be familiar with the unique character-istics of each melon. See Table 14 for further postharvest information.

OKRA (*Abelmoschus esculentus*)

VARIETIES[1]	AL	AR	GA	KY	LA	MS	NC	SC	TN
OKRA									
Burgundy/Red Burgundy [2]	A		G	K	L	M	N		T
Carmine Splendor [2, 3]	A				L	M	N		T
Clemson Spineless 80	A		G	K	L	M	N	S	T
Emerald	A	R	G		L		N	S	
Jambalaya		R		K		M			
Lee	A	R					N	S	

[1] Abbreviations for state where recommended [2] Red pods [3] Edible flowers

Okra is a tropical annual that is widely adapted, however, it is very sensitive to frost and cold temperatures and should not be planted until soil has warmed in the spring.

Seeding and Spacing. Generally, only one planting is made. For cooler areas, seed in the greenhouse in cells and transplant to the field through black plastic mulch. When direct seeding, okra seed require warm a soil temperature (70°F) for optimal germination and emergence.

For medium and tall varieties, space the rows 4 to 4.5 feet apart. Drill seeds 1- to 1.5- inch deep, with 3 or 4 seed per foot of row (5 to 7 pounds per acre). Thin plants when they are 5 inches high. Plants of medium and tall varieties should be spaced 18 to 24 inches apart in-row.

OKRA PLANTING DATES

	Spring	Fall
AL North	4/15–6/15	7/15–8/15
AL South	3/1–4/30	8/1–8/30
GA North	5/1–7/15	7/15–8/15
GA South	3/15–4/30	8/1–8/30
KY East	5/15–7/1	NR
KY Central	5/10–7/15	NR
KY West	4/20–8/1	NR
LA North	4/15–5/31	7/1–7/31
LA South	3/15–5/31	8/1–7/31
MS	4/15–6/1	8/1–9/1
NC East	5/1–5/30	8/1–8/30
NC West	5/25–7/31	NR
SC East	5/1–6/30	NR
SC West	5/15–7/15	NR
TN East	5/15–6/15	7/1–7/31
TN West	4/15–6/15	7/25–8/25

NR = Not recommended

Ratooning Okra: Producing a fall crop from a spring planting
Market price for okra typically declines sharply as the summer progresses. After the market price drops, consider ratooning or cutting back your okra. Ratooning okra will allow the plants to rejuvenate and produce a crop in the fall, when okra prices are generally higher. Cut plants back using a mower, leaving 6 to 12 inches of each plant above the ground. Re-fertilize with 15-0-14, 8-0-24, or 13-0-44 to encourage re-growth and the development of side branches. Fall yields of cutback okra will often exceed that of spring crops or the yields of a crop that is not cut back.

Drip Fertilization. Before mulching, adjust soil pH to 6.5 and in the absence of a soil test apply fertilizer to supply 25 pounds per acre of N, P_2O_5 and K_2O, (some soils will require 50 pounds per acre of K_2O), then thoroughly incorporate into the soil. Apply 1 to 2 pounds per acre of actual boron. After mulching and installing the drip irrigation system, the soluble fertilizer program should be followed as described in the tables below. The first soluble fertilizer application should be applied through the drip irrigation system within a week after field transplanting or direct seeding the okra. Continue fertigating until the last harvest.

SUGGESTED FERTIGATION SCHEDULE FOR OKRA*
(low potassium soil)

Days after planting	Daily nitrogen	Daily potash	Cumulative Nitrogen	Cumulative Potash
		(lb / A)		
Preplant			25.0	50.0
0-14	0.9	1.8	50.2	100.4
15-28	1.3	2.6	77.5	155.0
29-84	1.5	3.0	119.5	239.0
85-91	0.7	1.4	129.3	258.6

SUGGESTED FERTIGATION SCHEDULE FOR OKRA*
(high potassium soil)

Days after planting	Daily nitrogen	Daily potash	Cumulative Nitrogen	Cumulative Potash
		(lb / A)		
Preplant			25.0	50.0
0-14	0.9	0.9	50.2	75.2
15-28	1.3	1.3	77.5	102.5
29-84	1.5	1.5	119.5	144.5
85-91	0.7	0.7	129.3	154.5

*Adjust based on tissue analysis

Plastic Mulching. Polyethylene (black plastic) mulch can offer growers several advantages, such as increased yield and earlier harvest. Drip irrigation systems must be used with plastic mulch. On plastic mulch, transplant at the three-to four-leaf stage into staggered double rows spaced 15 to 18 inches apart between the double rows. Place plants 12 inches apart.

SPECIAL NOTES FOR PEST MANAGEMENT
DISEASE MANAGEMENT
Accurate diagnosis is necessary for effective disease management. Incorporate appropriate and effective cultural practices and pesticides, as detailed in the Disease Management section, in a season-long integrated disease management program.

Foliar Diseases: Various fungal leaf spots can affect okra and can sometimes warrant management. If leaf spots become problematic, apply fungicides.

Powdery Mildew: Powdery mildew can be a problem in okra and can sometimes warrant management. If powdery mildew is a persistent problem, apply fungicides preventatively.

INSECT MANAGEMENT
Aphids: These insects can be an occasional pest of okra that suck sap from the plant and produce honeydew. Mostly found on undersides of leaves. Broad-spectrum insecticides used for caterpillar management can reduce natural enemies leading to aphid infestations. Use a foliar spray when needed. See Table 2-25.

Japanese Beetle: Can be serious defoliators feeding between leaf veins of upper leaves. Use foliar treatments as needed and repeat applications if new beetles continue to invade fields in high numbers.

Stink bugs and leaffooted bugs: These pests are sap feeders that can cause misshapen (curled, bumpy) pods.

NEMATODE MANAGEMENT
Root-knot Nematodes (RKN): Root-knot nematodes can be a problem in okra. Nematode populations are highest in the late summer and fall. Collect soil samples for nematode analysis during this period from fields to be planted with okra the following year. Clean and disinfest equipment between fields. Use appropriate crop rotations and cover crops with non-host crops to reduce nematode populations. Avoid fields with high populations of nematodes or use appropriate fumigants and other nematicides for management (see fumigant and nematicide tables).

HARVESTING AND STORAGE
An okra pod usually reaches harvesting maturity 4 to 6 days after the flower opens. The pods are 3 to 3.5 inches long at this stage and are tender and free of fiber. Pick pods at least every second day to avoid the development of large, undesirable pods. Okra is susceptible to mechanical damage and should be gently handled during harvest. Bruising will promote browning and discoloration. Okra should be kept at temperatures between 50 to 55°F and of 85 to 90% relative humidity, since it is highly suscept-ible to water loss and shriveling. Okra pods are subject to chilling injury below 50°F, manifested as discoloration, surface pitting and increased decay susceptibility.

ONIONS (*Allium cepa*) AND GREEN ONIONS (*A. cepa*)

VARIETIES[1]	AL	GA	KY	LA	MS	NC	SC	TN	VA
GREEN ONIONS									
Evergreen Bunching [2]	A		K	L		N	S	T	
Ishikura Improved	A		K		M	N	S	T	
Parade			K				S		
Toyota Long White Bunching [2]						N	S		
White Spear				L			S		
ONIONS (Short Day)									
Bejo 369		G**							
Cabernet	A		K						
Candy Ann	A	G**							
Candy Joy	A	G**							
DP Sweet F₁ (1407)	A	G**							
Expression			K						
Georgia Boy		G**		L					
Granex Yellow PRR	A	G**		L		N	S	T	
Macon		G**							
Maragogi		G**							
Miss Megan		G**		L					
Mr. Buck		G**		L					
Red Burgundy [4]	A			L					
Red Hunter [4]	A						S		
Sapelo Sweet		G**					S		
Sierra Blanca	A		K						
Sweet Caroline	A	G**		L					
Tania		G**							
Texas Early Grano 502				L		N	S	T	
Texas Grano 1015Y	A			L	M	N	S		
Vidora		G**							
Yellow Granex [3]				L			S		
ONIONS (Intermediate Day)									
Candy			K	L	M			T	V
Monastrell [4]									
Ovation									
Redwing [4]			K						
Super Star (white)			K	L		N			V

[1] Abbreviations for state where recommended
[2] Bulbing type
[3] Also designates a "type" of onion and performance may vary
[4] Red

** Georgia Growers note: To be marketed as "Vidalia," varieties must be on the Georgia Department of Agriculture's "Recommended Vidalia Onion List" and grown in the Vidalia area. All of these varieties can be used for green onions.

Planting and Seeding Dates. Sets and seed for dry bulb onion production should be planted as soon as soil conditions are favorable. Typically, planting occurs late fall in the southern range of the Southeast, but can continue through the early spring in the northern range. Suggested direct seed and transplanting dates are indicated in the following tables.

Seed for bunching onions can be planted as soon as soil conditions are favorable in the spring and successive plantings can be made throughout the summer in the cooler parts of the Southeast.

On-farm transplant production can be performed in most conditions for dry bulb onion production. In the northern range of the Southeast it may be preferable to purchase transplants. Transplant production should begin by seeding plantbeds from late August to the end of September. A common method of producing transplants is to seed in high density plantings with 30 to 70 seed per linear foot. Four to five rows are planted 12 to 14 inches apart on beds prepared on 6-foot centers.

For dry bulb onion production from transplants follow planting dates recommended in the following table. Onion production from sets has not worked as well because it is difficult to mechanically orient the sets with the growing point up. Hand planting sets, however, works well for smaller operations.

Direct seeding dry bulb onions can save money on labor and materials. See seeding dates in table below. It is recom- mended that coated or encrusted seed be used with a vacuum planter to insure good seed singulation. It is critical that the beds be properly prepared without any previous plant debris. Preplant fertilizer application of 1/5 to 1/4 of required amount with proper bed moisture is recommended. Care should be taken so that the seed is singulating properly, soil is not clogging the seeder, and planting depth is correct (~ 0.25 in.). Watering is required to insure germination and emergence. It may be necessary to apply water more than once a day during periods of hot, dry weather.

Seeding dates for green onions are listed in the table below. Green onions during winter production will require 12-14 weeks. Spring production may be shorter. Green onions can also be produced from transplants.

Spacing. A typical planting arrangement for dry bulb onions is to plant four rows, 12-14 in. apart on raised beds (6 in high) prepared on six-foot centers. In-row spacing should be 4-6 inches. For direct seeded onions, set the planter to sow seed with a 3-4 in. in-row spacing.

For green onions, space rows 12 to 16 inches apart and space seed 0.75 to 1.5-inches apart (2-6 pounds per acre). A vacuum planter with a double row planter or a scatter shoe will work well. Seed depth should be 0.25-0.5 inches. Place transplants or sets 1.5 to 2.5 inches deep.

ONION *DIRECT SEED* PLANTING DATES

	Green Onions	Onions (dry)
AL North	NR	NR
AL South	8/15-10/15	10/5-10/25
GA North	NR	NR
GA South	8/15-10/15	10/5-10/25
LA North	9/15-10/31	9/15-10/31
LA South	10/1-10/31	10/1-10/31
MS North	2/15-3/30	9/15-10/15
MS South	10/15-2/15	9/15-10/30
NC East	8/1-6/15	9/15-10/31
NC West	4/1-8/15	9/1-9/30
SC East	2/15-10/15	9/15-11/15
SC West	3/15-7/30	NR
TN East	9/1-9/30	NR
TN West	NR	NR

NR = Not recommended

ONION *DIRECT SEED* PLANTING DATES

	Onions (dry)		Onions (dry)
AL North	11/1-12/31	MS North	12/15-3/1
AL South	11/1-1/31	MS South	10/1-2/15
GA North	11/1-12/31	NC East	10/1-3/1
GA South	11/1-1/31	NC West	9/15-10/15
KY East	4/1-6/15	SC East	10/1-11/15
KY Central	3/25-7/1	SC West	9/15-10/15
KY West	3/15-7/15	TN East	9/15/10/15
LA	12/15-1/31	TN West	3/1-3/30

Cultivation. For bunching onions, hill with 1 to 2 inches of soil to ensure white base.

SPECIAL NOTES FOR PEST MANAGEMENT
INSECT MANAGEMENT

Seedcorn Maggot: An early season problem in high organic matter soils and is common following winter injury to plants or in fields where planting occurs soon after a cover crop has been plowed under.

Thrips: During hot, dry weather, thrips often move into crops following harvest of adjacent alfalfa or grain and can present the most serious insect problem on onions and leeks. (See "Onions" in the Insect Control section of this publication). Treat if thrips counts exceed an average of 5 thrips per plant.

Allium leafminer is an invasive pest that has recently become a problem in Virginia and North Carolina. This fly pest deposits eggs in stems where larvae mine and tunnel plants causing wilting and plant death and contamination of the product. (See "Onions" in the Insect Control section of this publication).

PREHARVEST PRACTICES

Onions naturally go dormant after maturing, allowing the plant to withstand harsh summer conditions, which is also crucial for extended storage. Several factors can prevent onion from maturing properly, thereby reducing its storage potential. The main factors are irrigation and nitrogen fertilization. Excessive watering before harvest can cause onions to accumulate too much moisture, leading to poor curing and a higher risk of mold. Applying nitrogen too heavily or too late in the season can delay bulb formation, resulting in thicker necks that do not dry properly, making the onions more susceptible to rot and reducing their storage longevity.

HARVESTING AND STORAGE
See Table 14 for postharvest information

PARSNIP (*Pastinaca sativa*)

VARIETIES[1]	AL	GA	KY	LA	MS	NC	SC	TN
PARSNIP								
All American	A	G				N	S	T
Harris Model	A		K			N	S	
Javelin						N	S	T

[1] Abbreviations for state where recommended

Seeding and Spacing. Seed 3 to 5 pounds per acre at a depth of ¼ to 3/8 inch in rows 18 to 30 inches apart. Adjust seeder to sow 8 to 10 seeds per foot of row. Thin seedlings to 2-4 inches apart in the row. This will result in parsnips of similar shape and size to a plump carrot. To produce the huge roots popular in some areas, provide a much greater spacing of up to 12" between plants. Do not transplant parsnips.

Seeds germinate very slowly (taking up to 18 days). Seed more than one-year-old will not germinate. Parsnips need 120 to 180 days to mature and need to mature during cool weather.

Cultivation. Cultivate parsnips in a similar manner as to carrots. Do not let the roots dry out too much, as this will lead to cracked, unmarketable roots and bitter flavor.

Yield. Expected yield is 50-75 pounds per 100 row feet or 4 to 4.5 tons per acre.

HARVESTING AND STORAGE

Roots are ready for harvest when tops start to die back in autumn. Parsnips may be dug, topped, and then stored at 32°F at 90 to 95% relative humidity. Roots can be stored for up to 6 months. Parsnips left in the ground over winter should be removed before growth starts in the spring. See Table 14 for fur- ther postharvest information.

Note: Many people develop a rash after contact with the juice that parsnip leaves exude when crushed or torn, especially when handling leaves in the sun. Consider wearing gloves during harvest and handling; do not display parsnips with leaves still attached as is common for fresh market carrots.

PARSNIP PLANTING DATES

	Spring	Fall
AL North	3/15–4/30	8/1–9/15
AL South	2/1–5/15	8/1–9/30
GA North	3/15–4/30	8/1–9/15
GA South	2/1–5/15	8/1–9/30
KY East	4/1–6/1	NR
KY Central	3/20–6/15	NR
KY West	3/10–7/1	NR
NC East	2/15-4/15	8/1–9/30
NC West	4/1–8/15	NR
SC East	2/1–3/31	8/15–10/15
SC West	3/15–4/30	7/15–9/30
TN East	4/1–8/15	NR
TN West	2/15–4/15	8/1–9/30

NR = Not recommended

PEAS (ENGLISH/GARDEN) (*Pisum sativum*)

VARIETIES[1]	AL	GA	KY	LA	MS	NC	SC	TN
ENGLISH/GARDEN PEAS								
Green Arrow	A	G	K	L	M	N	S	T
Knight		G		L			S	T
Oregon Sugar Pod II [2,3]	A	G	K	L	M	N	S	T
Sugar Ann [3]		G	K	L	M	N	S	T
Sugar Bon [3]	A			L	M	N		
Sugar Snap [3]	A	G	K	L	M	N	S	T
Tall Telephone/Alderman				L	M	N		

[1] Abbreviations for state where recommended   [2] Flat podded - snow pea   [3] Edible pod type

Garden peas thrive in cool weather and are frost tolerate. Early plantings can be made as soon as soil can be tilled in the spring. Inoculation of seed can enhance early nodule formation and improve plant development.

Seed Treatment. Use seed already treated with an approved seed treatment, or treat seed with a slurry or dust that contains an approved fungicide. For a list of approved treatments, see Seed Treatments in the disease management section of the handbook.

Seeding and Spacing. For garden peas and processing peas, plant 3 to 4 seeds per foot in rows 6 to 8 inches apart, requiring seed 80-120 pounds per acre in 30-inch rows. Seed at a depth of no more than one inch unless soil is dry. Use press wheel drill or seeder to firm seed into soil.

Seedlings will emerge in 6 to 14 days, weather dependent. Harvesting usually begins 50-75 days after emergence. Average yield of garden peas is approximately 20 pounds per 100 row feet.

Cultivation. Avoid overfertilization. Too much nitrogen will reduce yields. Garden peas need some type of support structure for best performance and speedier picking. Garden peas should not follow beans or another Legume crop.

Harvesting and Storage. Harvest often. Picking is labor intensive and may need to happen almost daily during peak production periods. Allowing garden peas to get too large on the vines will greatly reduce production. Larger acreages of garden peas require mechanical harvesting to be profitable. Leafless type garden peas, with more tendrils than true leaves, are easier to harvest. Cool garden peas as soon as possible after picking as their sugars convert to starch at higher temperatures. See Table 14 for further postharvest information.

ENGLISH/GARDEN PEAS PLANTING DATES

	Spring	Fall
AL North	3/15–4/30	8/1–8/31
AL South	2/1–3/31	8/1–9/30
GA North	3/15–4/30	8/1–8/31
GA South	2/1–3/31	8/1–9/30
KY East	3/15-4/15	NR
KY Central	3/1-4/1	NR
KY West	2/20-3/20	NR
LA	11/15–2/1	NR
MS North	2/10-4/25	NR
MS South	1/25-4/5	NR
NC East	2/15–4/15	8/1–9/30
NC West	4/1–6/15	NR
SC East	2/1–3/15	8/15–11/30
SC West	3/1–4/15	8/15–10/30
TN East	3/15-4/30	NR
TN West	2/15-3/30	NR

NR = Not recommended

PEAS, SOUTHERN (*Vigna unguiculata*)

VARIETIES[1]	AL	AR	GA	KY	LA	MS	NC	SC	TN
PEAS, SOUTHERN									
Blackeyes									
California Blackeye #5 [2,5]	A	R	G				N	S	
Queen Anne [2,5]	A		G	K	L	M	N		T
Pinkeyes									
Coronet [2,5]	A	R						S	
Pinkeye Purple Hull [4]			G		L	M	N	S	T
Pinkeye Purple Hull - BVR [4]	A	R	G		L	M	N		T
QuickPick Pinkeye [2,5]	A		G		L	M	N	S	T
Texas Pinkeye	A	R			L				
Top Pick Pinkeye [2]	A	R	G		L	M			
Creams									
Elite [2,5]		R			L				T
Mississippi Cream [2,5]					L	M			T
Texas Cream 8			G			M		S	T
Texas Cream 12	A		G						
Top Pick Cream	A	R	G		L				T
White Acre-BVR	A		G						
Crowders									
Clemson Purple							N	S	
Colossus 80 [2,5]							N	S	
Dixie Lee					L		N		T
Hercules	A		G				N	S	T
Knuckle Purple Hull			G				N	S	
Mississippi Purple [3]	A		G	K	L	M	N	S	T
Mississippi Silver [3]	A	R	G	K	L	M	N	S	T
Top Pick Crowder	A	R	G		L		N		
Zipper Cream [4]	A	R	G		L	M	N	S	T

[1] Abbreviations for state where recommended
[2] Suitable for mechanical harvest
[3] Semi-vining
[4] Vining
[5] Bush

Southern peas originated in India in prehistoric times and moved to Africa, then to America. In India, southern peas are known by 50 common names and in the United States are called field peas, crowder peas, cowpeas, and blackeyes, but southern peas is the preferred name. Southern peas require relatively warm soils for good germination.

Seeding and Spacing. Sow when soil temperature reach a minimum of 65°F (70°F optimal) and continue sowing until 80 days before fall frost. Seeding too early causes poor stands and you may need to replant. Bush types should be seeded 4 to 6 per foot or 30 to 50 pounds of seed per acre. Vining types should be seeded 1 to 2 per foot or 20 to 30 pounds of seed per acre. Plant seeds 3/4 to 1 1/4 inch deep in rows spaced 20 to 42 inches apart, depending on cultivation requirements.

Fertility. Most soils will produce a good crop, but medium fertility with pH of 5.8 to 6.5 is desirable. High fertility produces excessive vine growth and poor yields. Inoculants of specific N fixing bacteria may increase yield especially in soils where southern peas have not been grown. Crop rotation or fumigation is important for nematode control.

SOUTHERN PEA PLANTING DATES

	Spring	Fall
AL North	4/15–7/31	NR
AL South	3/15–6/15	7/15–8/30
GA North	5/15–7/15	NR
GA South	3/15–5/15	7/15–8/30
KY East	5/10–6/15	NR
KY Central	5/5–7/1	NR
KY West	4/20–7/15	NR
LA North	4/15–7/31	7/1–7/31
LA South	4/1–5/31	7/15–8/15
MS North	4/15–7/15	NR
MS South	3/15–6/15	8/1–8/30
NC East	3/25–6/15	8/1–8/30
NC West	4/15–7/15	NR
SC East	4/1–6/15	7/15–8/1
SC West	4/15–7/15	NR
TN East	5/10–7/15	NR
TN West	4/15–7/31	NR

NR = Not recommended

INSECT MANAGEMENT
Cowpea Curculio: At first bloom, make three insecticides applications at five-day intervals for curculio control.

HARVESTING AND STORAGE
Depending on variety and weather, harvest will begin 65 to 80 days after seeding and continue for 3 to 5 weeks. Begin harvest when a few pods are beginning to change color and harvest only pods with well-formed peas. This is the best stage for shelling and eating. Southern peas are sold in bushel hampers or mesh bags. Do not use burlap sacks because they are not properly ventilated.

Southern peas weigh 22 to 30 pounds per bushel. One person can harvest 12 to 20 bushels per day if yields are average. Average production is 60 to 200 bushels per acre. See Table 14 for further postharvest information.

PEPPERS (*Capsicum annuum* and related species)

VARIETIES [1]	AL	GA	KY	LA	MS	NC	SC	TN	VA
PEPPER (open pollinated)									
Bell									
Capistrano	A			L	M	N	S		
Jupiter	A			L	M	N	S	T	
Purple Beauty [9]	A				M	N			
Frying type									
Cubanelle	A	G	K			N	S	T	
Sweet Banana	A	G	K	L		N	S	T	
HOT/PUNGENT TYPES (open pollinated)									
New Mexican/Anaheim type									
Anaheim	A	G	K	L		N		T	V
Cayenne type									
Carolina Cayenne [10]						N	S		
Charleston Hot [10]				L	M	N	S		
Large Red Thick				L					
Long Slim Cayenne	A	G			M	N	S		
Habanero/Scotch Bonnet type									
Habanero	A	G	K	L	M	N	S	T	V
Bonda Ma Jacque		G							
Wax type									
Long Hungarian Wax	A	G	K	L		N	S	T	V
Jalapeño type									
Jalapeño M	A	G		L	M	N	S	T	
Tula [4]	A	G		L		N	S	T	
PEPPER (Hybrid)									
Bell									
Alliance [4, 8 b-f, 11, 13, 14, 15]	A		K			N	S	T	V
Antebellum [3, 8 a-k, 16]		G							
Aristotle [4, 8 b-d]	A	G	K	L	M	N	S	T	V
Autry [3, 5, 8 a-k]		G		L			S		
Currier [2, 4, 5, 8 b-d, 11, 15]			K						
Camelot X3R [8 b-d]	A		K	L	M	N	S	T	
Daciana (ivory color maturing to red)							S		
Declaration [2, 3, 8 b-f, 11]	A	K		L		N		T	V
Excursion II [4, 8 b-d, 13]				L		N		T	
Flamingo [12, 13]	A	K				N			
Flavorburst [7]	A			L		N		T	
Galileo [2, 8 b, c, d]		G							
Green Machine [3, 8 a-k, 16]	A	G		L					
Karisma [4, 8 b-d, 11, 13, 15]	A	G		L					
King Arthur [4, 6, 8 c, 13]	A		K	L	M	N	S	T	V
Mecate [4, 7, 8 b-d, 13, 15]	A					N			
Mercer [2, 5, 8 b, c, d]		G							
Milena [3, 4, 6, 7, 16]	A		K						
Nitro [2, 3, 5, 8 a-k]		G							
Paladin [2, 13]	A	G	K	L		N		T	V
Playmaker [2, 8 a-k]		G							

[1] Abbreviations for state where recommended
[2] Phytophthora root rot tolerance/resistance
[3] Tomato spotted wilt virus tolerance/resistance (TSWV)
[4] Potato virus Y tolerance/resistance (PVY)
[5] Tomato mosaic virus tolerance/resistance (ToMV)
[6] Tobacco etch virus tolerance/resistance (TEV)
[7] Mature yellow fruit or mature orange fruit
[8a, b, c, d, e, f, g, h, i, j, k] Bacterial leaf spot resistance for races 0, 1, 2, 3, 4, 5, 6, 7, 8, 9, 10, respectively
[9] Mature purple fruit
[10] Southern root-knot nematode resistance (N)
[11] Cucumber mosaic virus tolerance/resistance (CMV)
[12] Fruit mature from white to red
[13] Tobacco mosaic virus (TMV) tolerance/resistance
[14] Pepper yellow mosaic virus tolerance/resistance (PYMV)
[15] Pepper mottle virus tolerance/resistance (PMV)
[16] Tobamovirus tolerance/resistance (TM)

VARIETIES [1]	AL	GA	KY	LA	MS	NC	SC	TN	VA
PEPPER (Hybrid) (cont'd)									
Bell (cont'd)									
PS 09979325 [8 a-k, 16]		G					S		
Red Knight [4, 8 b-d]			K			N	S		V
Revolution [2, 8 b-f, 11]	A	G	K	L	M	N		T	V
Standout [3, 5, 8a-k]		G							
SV 3255PB [8a-k, 16]		G							
Tarpon [2, 8 a-k, 16]		G							
Tequila [9, 13]	A			L	M	N		T	
Turnpike [2, 8 a-e,g,]			K	L				T	V
Vanguard [2, 8 b-f, 11]	A					N	S		V
Frying type									
Aruba		G							V
Banana Supreme	A	G	K	L		N	S	T	
Biscayne	A	G			M	N	S		
Gypsy	A	G	K	L	M	N			
Key Largo	A				M	N	S		V
Ancho/Poblano									
Tiburon	A	G	K		M			T	V
HOT/PUNGENT TYPES (Hybrid)									
Cayenne type									
Mesilla [4, 6]								T	V
Jalapeño type									
Compadre [5, 8 c, f]	A	G							
Inferno	A	G		L		N			
Jedi [8 b, c, d]									
Mixteco [8 b-d, 16]		G							
Mitla [4]	A	G		L	M	N	S	T	
Orizaba [8 b-d]		G							
Tormenta [4, 6, 8 b-d]	A		K	L				T	

[1] Abbreviations for state where recommended
[2] Phytophthora root rot tolerance/resistance
[3] Tomato spotted wilt virus tolerance/resistance (TSWV)
[4] Potato virus Y tolerance/resistance (PVY)
[5] Tomato mosaic virus tolerance/resistance (ToMV)
[6] Tobacco etch virus tolerance/resistance (TEV)
[7] Mature yellow fruit or mature orange fruit
[8a, b, c, d, e, f, g, h, i, j, k] Bacterial leaf spot resistance for races 0, 1, 2, 3, 4, 5, 6, 7, 8, 9, 10, respectively
[9] Mature purple fruit
[10] Southern root-knot nematode resistance (N)
[11] Cucumber mosaic virus tolerance/resistance (CMV)
[12] Fruit mature from white to red
[13] Tobacco mosaic virus (TMV) tolerance/resistance
[14] Pepper yellow mosaic virus tolerance/resistance (PYMV)
[15] Pepper mottle virus tolerance/resistance (PMV)
[16] Tobamovirus tolerance/resistance (TM)

Peppers are a warm-season crop that grow best at temperatures of 70° to 75°F. This crop is sensitive to temperature extremes. Poor fruit set and blossom drop can be expected when night temperatures drop below 60° or day temperatures rise above 85°F.

Seed Treatment. If seed is not treated in order to minimize the occurrence of bacterial leaf spot, dip seed in a solution containing 1 quart of household bleach and 4 quarts of water plus 1 teaspoon of surfactant for 15 minutes. Provide constant agitation. Use at the rate of 1 gallon of solution per pound of seed. Prepare a fresh solution for each batch of seed. Wash seed in running water for 5 minutes and dry seed thoroughly. Plant seed soon after treatment. Further information on seed treatments can be found under Seed Treatments in the disease management section of the handbook.

Planting and Spacing. Space rows 4 to 5 feet apart. Set plants 12 to 18 inches apart in double rows. Select fields with good drainage. Plant on raised, dome-shaped beds to aid in disease control.

To minimize sunscald when growing pepper on sandy soils and on plastic mulch without drip irrigation, plant varieties that have excellent foliage. Furthermore, staking can help reduce sunscald, as well as support a heavy fruit load, increase airrflow, reduces disease, and eases harvest.

PEPPER PLANTING DATES*

	Spring	Fall
AL North	5/15–6/30	7/1–8/1
AL South	3/1–5/15	7/15–8/30
GA North	5/15–6/30	7/1–8/1
GA South	3/1–4/30	7/15–8/30
KY East	5/20-6/15	NR
KY Central	5/10–7/1	NR
KY West	5/1–7/15	NR
LA North	4/1–5/15	6/15–7/31
LA South	3/1–5/15	6/15–7/31

PEPPER PLANTING DATES* (cont'd)

	Spring	Fall
MS North	4/20–6/30	NR
MS South	3/1–4/30	8/1–8/15
NC East	4/15–5/10	8/1–8/15
NC West	5/15–7/15	NR
SC East	4/1–5/15	7/10–8/1
SC West	5/1–6/30	NR
TN East	5/15-7/1	NR
TN West	4/20-6/30	NR
VA East (coastal)	4/1-4/30	7/1-8/1
VA West (mountain)	5/1-6/15	NR

* Using transplants. Note: Planting dates in the spring could be earlier if low tunnels or other season extension measures are used in some locations.
NR = Not recommended

Drip Fertilization. Before mulching, adjust soil pH to 6.5, and in the absence of a soil test, apply enough fertilizer to supply 50 pounds per acre of N, P_2O_5 and K_2O, (some soils will require 100 pounds per acre of K_2O) and incorporate into the soil. After transplanting the soluble fertilizer program described in the following table should then be initiated. On soils testing low-medium for boron, also include 0.5 pound per acre of actual boron. The first soluble fertilizer application should be applied through the drip irrigation system within a week after transplanting. Continue fertigating until the last harvest.

SUGGESTED FERTIGATION SCHEDULE FOR PEPPER*
(low soil potassium)

Days after planting	Daily nitrogen	Daily potash	Cumulative Nitrogen	Cumulative Potash
	(lb / A)			
Preplant			50.0	100.0
0–14	0.5	0.5	57.0	107.0
15–28	0.7	1.4	66.8	126.6
29–42	1.0	2.0	80.8	154.6
43–56	1.5	3.0	101.8	196.6
57–98	1.8	3.6	177.4	347.8

SUGGESTED FERTIGATION SCHEDULE FOR PEPPER*
(high soil potassium)

Days after planting	Daily nitrogen	Daily potash	Cumulative Nitrogen	Cumulative Potash
	(lb / A)			
Preplant			50.0	100.0
0–14	0.5	0.5	57.0	107.0
15–28	0.7	0.7	66.8	116.8
29–42	1.0	1.0	80.8	130.8
43–56	1.5	1.5	101.8	151.8
57–98	1.8	1.8	177.4	227.4

*Adjust based on tissue analysis

SPECIAL NOTES FOR PEST MANAGEMENT

INSECT MANAGEMENT

Green Peach and Melon Aphid: For best green peach aphid control during periods of drought, apply insecticide 2 to 3 days after irrigation. Thorough spray coverage beneath leaves is critical.

Pepper Maggot: Pepper maggot flies are active from June 1 to mid-August.

Pepper Weevil (PW): PW is a pest occasionally imported on older transplants or transplants with flowers or fruit.

European Corn Borer (ECB): The use of pheromone insect traps is recommended, treat when more than ten moths per trap per week are found. Follow the table in Insect Control section.

DISEASE MANAGEMENT

Accurate diagnosis is necessary for effective disease management. Incorporate appropriate and effective cultural practices and pesticides, as detailed in the Disease Management section, in a season-long integrated disease management program.

Phytophthora Blight: Phytophthora may affect roots, crowns, stems, and fruits and can be a serious issue in peppers, especially in production systems that use surface water and overhead irrigation. When available, plant resistant varieties. To minimize the occurrence of this disease, fields should be adequately drained to ensure that soil water does not accumulate around the base of the plants. Subsoil between rows to allow for faster drainage following rainfall. Avoid moving infested soil between fields. Crop rotation is not effective. Resistance to some fungicides (e.g. mefenoxam) has been reported and should be considered when selecting fungicides to protect both crowns and fruit.

Bacterial Spot: This disease can be severe in peppers. The pathogen is seed-borne and easily spread by overhead irrigation and rain events. Purchase seeds from companies that test seed lots for the pathogen and transplants from reputable sources. When available, plant varieties with resistance to all races of the pathogen. A preventative spray program is recommended.

Anthracnose Fruit Rot: Anthracnose can be a problem on all types of pepper fruit when weather conditions favor disease development. Avoid overhead irrigation. Practice a 2- to 3-year crop rotation. An effective fungicide program should be followed from bloom through harvest.

Viruses: Several viruses, including mechanically-transmitted (tobacco mosaic virus (TMV)), aphid-transmitted (cucumber mosaic virus (CMV), potato virus Y (PVY), etc.), or thrips-transmitted viruses (tomato spotted wilt virus, TSWV), can impact pepper production. When available, use tolerant or resistant varieties to these viruses in areas with a history of the virus. Generally, aphid-transmitted viruses cannot be adequately managed with insecticide applications, but symptom expression can be delayed through their use combined with the use of reflective mulches. However, effective insecticides and appropriate cultural practices should be used to manage thrips when TSWV-susceptible varieties are grown. Production areas should be scouted regularly. Do not grow ornamental bedding plants in the same greenhouse as pepper transplants.

NEMATODE MANAGEMENT

Various root-knot and other nematodes can infect peppers. Nematode populations are highest in the late summer and fall. Collect soil samples for nematode analysis during this period from fields to be planted with peppers the following year. Clean and disinfest equipment between fields. Use appropriate crop rotations and cover crops with non-host crops to reduce nematode populations. Avoid fields with high populations of nematodes or use appropriate fumigants and other nematicides for management (see fumigant and nematicide tables).

HARVESTING AND STORAGE

Bell peppers are harvested when they exhibit a firm texture and glossy exterior. Most commonly, they are harvested at the green stage; however, depending on market demands, the harvest window can be extended until the peppers transition to red, yellow, or other mature hues. Harvesting is done by hand to avoid bruising or damaging the fruit, using gentle clipping or twisting to remove the stem from the plant. Bell peppers are chilling-sensitive and are best stored at 45–55°F with 90% relative humidity. Under proper storage conditions, they maintain market quality for up to 2–3 weeks. Consistent temperature control, adequate ventilation, and careful handling are essential to extending shelf-life and minimizing postharvest losses. See Table 14 for postharvest information.

POTATOES (IRISH) (*Solanum tuberosum*)

VARIETIES [1]	AL	GA	KY	LA	MS	NC	SC	TN	VA
POTATOES									
Adirondack Blue [2, 5]	A					N			
Atlantic [4, 5, 9, 11]	A	G		L	M	N	S	T	V
Chieftain [3, 4, 6, 12, 13]									V
Colomba [4, 10]						N			V
Dark Red Norland/Norland/Red Norland 4, 7 (all from same clone)	A		K	L		N	S	T	V
Envol [5]						N			V
Harley Blackwell [4, 8]		G				N			
Huron Chipper [4, 6, 8, 11]						N			
Katahdin [5]							S		V
Kennebec [6, 8]		G	K	L	M	N	S	T	V
La Rouge [4]	A			L				T	V
Mackinaw [4, 6, 11]						N			
Mountain Rose [3]						N	S		
Natascha [10]						N			
Purple Majesty [2]	A	G		L		N	S		
Red LaSoda [5]	A	G	K	L	M	N	S	T	
Red Pontiac [5]		G	K		M	N			
Snowden [5]						N			
Soraya [4, 6, 10]						N			
Superior [4, 8]		G	K		M	N			V
Yukon Gold [5, 7, 9, 10]	A	G	K	L	M	N	S	T	V
Fingerling Types									
French Fingerling	A	G	K			N	S	T	V
Russian Banana [4]	A	G	K			N	S	T	V

[1] Abbreviations for state where recommended
[2] Purple flesh when mature
[3] Red flesh when mature
[4] Scab tolerance/resistance
[5] Susceptible to scab
[6] Late blight tolerance/resistance
[7] Ozone sensitive
[8] Tolerant to heat necrosis
[9] Susceptible to heat necrosis
[10] Yellow flesh when mature
[11] Chipping
[12] Rhizoctonia tolerance/resistance
[13] Verticillium tolerance/resistance

Planting and Spacing. The recommended planting dates for potatoes are in the following table.

POTATO PLANTING DATES

	Spring	Fall
AL North	2/15–4/30	NR
AL South	1/15–3/31	NR
GA North	3/15–4/30	NR
GA South	2/1–3/31	NR
KY East	3/20–6/15	NR
KY Central	3/15–7/1	NR
KY West	3/15–7/15	NR
LA	1/15–2/28	NR
MS North	1/20–3/15	NR
MS South	1/20–3/1	NR
NC East	2/15–3/31	NR
NC West	4/1–6/15	NR
SC East	2/1–3/31	NR
SC West	3/15–4/30	NR
TN East	3/20–4/30	NR
TN West	2/15–3/31	NR
VA East (coastal)	3/10–4/5	NR
VA West (mountains)	4/1–6/15	NR
NR = Not recommended		

Space seed pieces 9 to 12 inches apart in 34- or 36- inch rows. Use closer spacing for large, cut seed pieces and wider spacing for whole (B-size) seed. Use close spacing for potatoes being marketed in 5- and 10-pound consumer packs and for Katahdin and Kennebec, which tend to set few tubers and produce oversize tubers.

Seed Pieces Treatment. Use certified seed. Warm potato seed 65 to 70°F for a period of 2 to 3 weeks before planting to encourage rapid emergence. Do not use seed pieces that weigh less than 1.5 oz each. Plant seed pieces immediately after cutting or store under conditions suitable for rapid healing of the cut surfaces (60 to 70°F plus high humidity). Dust seed pieces immediately after cutting with fungicide. Some fungicide seed-piece treatments are formulated with fir or alder bark. Bark formulations have been effective treatments to reduce seed pieces decay. Further information on seed treatments can be found under Seed Treatments in the disease management section of the handbook.

SPECIAL NOTES FOR PEST MANAGEMENT
INSECT MANAGEMENT
Colorado Potato Beetle (CPB): Rotation to non-solanaceous crops (crops other than potato, tomato, eggplant, and pepper) is extremely important in reducing CPB problems.

The further fields can be planted from last year's solanaceous crop, the more beneficial it will be in reducing CPB problems. Avoid the application of late-season sprays to prevent the buildup of insecticide-resistant beetles.

Beginning at plant emergence, sample fields weekly for CPB to determine the need to spray. Select at least 10 sites per field along a V- or W-shaped path throughout the field. At each site, select one stem from each of five adjacent plants and count and record all adults, large larvae (more than half-grown), and small larvae (less than half-grown). As a general guideline, if more than 25 adults or 75 large larvae or 200 small larvae are counted per 50 stems, a treatment is recommended. The amount of yield loss as a result of CPB feeding depends on the age of the potato plant. The Superior variety (short season) cannot compensate for early season defoliation by overwintered beetles, but, during the last 30 days of the season, Superior can withstand up to 50% defoliation without yield loss.

Note: Several insecticides may no longer be effective in certain areas due to CPB resistance. Alternate insecticide classes from one year to the next to avoid resistance. Check with the county Extension agent in your area for the most effective control.

Flea Beetles and Leafhoppers: Treatment is suggested if leafhopper counts exceed three adults per sweep or one nymph per 10 leaves.

European Corn Borer (ECB): Continued treatment for ECB may significantly increase CPB insecticide resistance. However, for proper timing of ECB sprays, consult your local county Extension office for further information.

Potato Aphid and Green Peach Aphid: Insecticide treatments are recommended when aphid counts exceed two per leaf prior to bloom, four aphids per leaf during bloom, and 10 aphids per leaf within two weeks of vine kill.

Potato Tuberworm: Treat when foliage injury is first noted. Potato tuberworms are primarily a problem with late potatoes, in cull piles, or potatoes in storage. Sanitation is very important.

Cutworms: Cutworms are especially troublesome to tubers where soil cracking occurs. Variegated cutworms feed on lower leaves and petioles.

Wireworms: Wireworms are a generic term used for the larvae of several species of click beetles, which burrow into potato tubers. In the Southeast U.S., wireworms attacking potato are typically the corn wireworm, or one of five species in the genus Conoderus. Wireworm problems are most prevalent when potatoes follow corn, any cereal crop, sod, or pasture. Wireworms do not move quickly from one field to the next, but can remain as larvae in a field for 2-5 years, depending on the species. Since it is difficult and laborious to monitor for a pest that lives in the ground, field history is an important tool for determining the need to treat for wireworms. Treatment must begin at or before planting to be effective. Options include preplant broadcast or at-planting furrow application of an appropriate insecticide. There is no control for wireworms once they have infested potato tubers.

DISEASE MANAGEMENT
Early Blight: Fungicide applications can slow the spread of early blight, but cannot eliminate it, and will be effective only when application begins when airborne spores are first present. Spore formation is most prevalent during repeated wet/ dry cycles, such as caused by overhead irrigation or frequent dew. Minimizing plant stress can reduce damage caused by early blight, especially in younger plants. Early blight spreads more rapidly in young plants than mature ones. If a field is infested with early blight, potatoes may still be harvested if adequate time between vine kill and harvest allows the skins to set and great care is taken not to bruise the tubers in the field. Tubers can be infected during the harvest process, and will decay more rapidly in storage than noninfected tubers. There are some potato cultivars resistant to early blight.

White Mold: High fertility and frequent rain or overhead irrigation are conditions conducive to white mold, caused by the fungus *Sclerotinia sclerotiorum*. This organism can survive in the soil for three or more years, and impacts other produce crops such as lettuce, beans, broccoli, peppers, and others. Fungicides should be applied as a protectant in fields with a history of white mold. In addition, eliminate canopy wetness by reducing overhead irrigation or aligning rows with prevailing winds to promote rapid evaporation of water in the canopy. Avoid fields with poor air movement or poor drainage. Rotate with a non-susceptible crop for three yields after a heavy infestation of white mold.

Common Scab: Common scab is characterized by brown lesions on tuber skin that may be slightly raised or sunken in relation to surrounding surface. The causal agent is a bacteria (*Streptomyces* spp.), which can be seedborne or soilborne. Scab occurs most commonly in warm dry soils with pH 5.5 to 7.5, and does not affect yield. If lesions are significant, marketability and quality are affected. Scab is difficult to manage. In addition to seed treatments at planting, management strategies include rotation out of the field for 3-4 years, maintaining low soil pH, using resistant varieties, and maintaining high soil moisture during tuber formation.

Late Blight: Caused by the fungal pathogen *Phytophthora infestans*, late blight is the disease responsible for the potato famine and continues to plague potato crops worldwide. Environmental conditions that favor late blight include frequent rainfall (or overhead irrigation), cool weather (50-75°F), and high humidity. Spread is usually quite rapid and complete defoliation can occur within 2-3 weeks of initial infestation. Infection can occur at any growth stage, and in any plant part including the tuber.

Fungicides should be applied protectively; after infection, only systemic fungicides (those that penetrate plant tissue) can inhibit the spread of the disease. Additional management strategies include reducing periods of leaf wetness by decreasing overhead irrigation or increasing air movement. Use resistant cultivars when they are available. A critical strategy is reducing the initial amount of inoculum available in the field. Use certified dis- ease-free tubers, and dispose of infected tubers and volunteer vines in a pit with at least 2 ft soil coverage to avoid sprouting. Consider fungicide treatment of seed pieces.

In recent years, forecasting and reporting for this disease have become available, particularly valuable since Phytophthora spores can travel long distances on air currents. One such model is *http://www.usablight.org*. Use late blight modeling to forecast when disease pressure is most likely to be present and time fungicide sprays accordingly for most efficient use of chemicals.

HARVESTING AND STORAGE

Harvest indicators: Tuber formation of potatoes ends when soil temperatures are consistently over 80°F, regardless of whether or not vine tops have died back. For most "new" potatoes, tubers will be well developed between 65-75 days. Flowering is not a reliable indicator of tuber formation, as some varieties may flower little or not at all. Check for readiness by hand harvesting a few tubers for evaluation.

Vine Kill: Also known as desiccation, many growers chemically (labeled herbicide application) or physically (roto beating or chopping) defoliate potato vines once the optimum marketable size of tubers has been achieved. This allows the tuber skin to set and mature and helps minimize skinning prior to digging. This technique provides benefits including efficiency in harvest, better control over harvest timing, skin set to reduce harvest injury, and reduced impact of diseases like late blight. Vine killing halts the translocation of nutrients and sugar accumulation from the leaves, triggering the conversion of tuber sugars to starch for storage. Vine killing also weakens the juncture of the tuber and stolon, making tubers fall from the plant more easily. If vine killing is used, harvest of tubers should occur at 2-3 weeks after vines are completely dead. Harvesting before this and tuber skin may not have had adequate time to set, while harvesting later increases the chance for rotting organisms to attack the crop while in the ground. Care should be taken to monitor this period and harvest at the optimum time to minimize mechanical damage and breakdown. See Table 14 for further postharvest information.

PUMPKINS AND WINTER SQUASH (*Cucurbita* spp.)

VARIETIES[1]	AL	AR	GA	KY	LA	MS	NC	SC	TN	VA
PUMPKIN										
Miniature <2 lbs										
Apprentice [B, H, R]	A		G				N		T	V
Baby Boo [H, V, W, FL]	A		G	K		M	N		T	V
Bumpkin [H, S, FL, PM]	A		G			M	N	S	T	V
Casperita [B, W, FL, PM, WMV]	A					M	N		T	
Crunchkin [B, H, FL]	A						N		T	V
Gold Dust [H, S, FL, PM]	A						N		T	
Gooligan [H, V, W, FL]	A		G	K					T	
Jack-Be-Little [V, H, FL]	A		G	K	L	M				V
Jill-Be-Little [V, H, FL, PM]	A					M	N			
Lil Pump-ke-mon [H, B, FL, W w/ orange stripes]	A		G	K			N		T	V
Munchkin [V, H, FL]	A	R	G		L	M	N	S	T	V
WeeeeOne [B, R, PM] (carvable)	A					M			T	V
Small 2-6 lbs										
Baby Moon [S, W, R, PM]	A						N			
Blanco [S, W, R, PM]							N			
Cannon Ball [V, P, PM]	A		G	K			N		T	V
Darling [V, H, O, PM]	A				L				T	
Early Abundance [S, H, R, PM]	A				L					
Field Trip [S, FL-R, PM]	A	R		K			N		T	V
Iron Man [V, H, R, PM]	A	R	G	K			N	S	T	V
Little Giant [S, H, R]	A						N			V
Prankster [S, H, R, PM]	A		G							
Small Sugar [V, O]	A		G		L		N			V
Speckled Hound [S, VR, FL, ZYMV]		R					N		T	
Sunlight [V, Y, R, PM]	A				L					
Medium 6-12 lbs										
Autumn Gold [S, R]	A	R				M	N			
Cotton Candy [V, W, R]	A		G	K			N		T	V
Goosebumps II [V, WA, R]	A			K			N	S	T	V
Grey Ghost [V, BL]							N		T	
Grizzly Bear [V, BU, WA, FL-R, PM]	A							S		
Gumdrop [S, H, R, PM]	A						N		T	
Honeymoon [S, W, R]	N									
Hybrid Pam [B, H, R]			G	K			N		T	V
Jamboree [V, BL, FL-O, CMV, PRSV]	A								T	
Jarrahdale [V, BL, FL]	A	R	G	K			N		T	V
Long Island Cheese [V, BU, FL]	A			K			N		T	V
Lumina [V, W, FL-R]	A		G	K			N		T	
Mystic Plus [V, FL-R, PM]	A		G				N		T	V
Neon [S, R]	A		G			M	N			
Orange Bulldog [V, H, O-R, PM]	A		G							

[1] Abbreviations for state where recommended

Growth habit:
- [B] Bush growth habit
- [SB] Semi-bush growth habit
- [V] Vining growth habit
- [S] Semi-vining growth habit

Skin features:
- [BL] Blue skin
- [BU] Buff skin
- [G] Green skin
- [GR] Gray skin
- [H] Hardshell
- [P] Pink skin
- [W] White skin
- [WA] Warts
- [RS] Red Skin
- [VR] Variegated
- [Y] Yellow skin

Shape:
- [FL] Flat (Cinderella, pancake)
- [P] Pie pumpkin - suitable for cooking
- [O] Oblong
- [R] Round

Disease Tolerance/Resistance:
- [DM] Downy mildew tolerance/resistance
- [CMV] Cucumber Mosaic virus tolerance/resistance
- [F] Fusarium tolerance/resistance
- [PH] Phytophthora tolerance/resistance
- [PM] Powdery mildew tolerance/resistance
- [PRSV] Papaya Ringspot Virus tolerance/resistance
- [PY] Precocious yellow
- [VT] Virus tolerance (non-specific)
- [WMV] Watermelon Mosaic Virus (Strain 2) tolerance/resistance
- [ZYMV] Zucchini Yellows Mosaic Virus tolerance/resistance

VARIETIES[1]	AL	AR	GA	KY	LA	MS	NC	SC	TN	VA
Medium 6-12 lbs (cont'd)										
Rouge Vif D' Etampes V, RS, FL	A			K			N		T	V
Skidoo Gold S, H, R, PM	A						N		T	
Large 12-20 lbs										
Aspen S, R-O	A		G		L				T	
Bayhorse Gold V, R, PM							N		T	
Blue Bayou V, BL, DM, PM	A						N		T	
Blue Doll V, BL, DM, PM	A						N		T	
Cinderella V, RS, FL	A	R	G		L		N		T	
Fairytale V, BU, FL	A	R	G				N		T	
Flat White Boer Ford V, W, FL		R							T	
Gold Medal S, W, R	A		G		L				T	V
Knuckle Head S, WA, R-O				K			N		T	
Magic Lantern S, R, PM	A	R	G	K			N	S	T	V
Magic Wand S, R-FL, PM	A	R	G	K			N		T	V
Magician S, R, PM, ZYMV	A	R	G	K			N	S	T	V
Marina Di Chioggia V, BL-WA, FL	A	R							T	
Mint Prince V, BL, FL	A	R							T	
Porcelain Doll V, P, FL	A	R					N		T	
Orange Rave S, H, O, PM							N		T	
Orange Sunrise S, PM, PY			G				N		T	
Racer Plus S, H, R, PM			G				N			
Royal Blue V, BL, FL		R							T	
Specter S, W/WA, O-R, PM							N		T	
Extra Large 20-50 lbs										
Aladdin SV, R-O, PM	A	R	G	K	L		N	S	T	V
Apollo S, O, PM	A			K			N		T	V
Big Max V, R-O	A		G		L	M	N		T	
Camaro V, R, PM	A	R		K		M	N	S	T	
Cronus V, R, PM (poor performer under high temps)	A	R					N	S	T	V
Gladiator S, R, PM (poor performer under high temps)	A			K			N		T	V
Gold Medallion V, R-O	A		G			M			T	
Gold Rush V, R	A		G			M			T	
Howden Biggie V, O	A			K		M			T	V
Kratos S, H, FL-R, PM							N		T	V
Mustang S, R, PM	A						N	S	T	
Rhea V, H, FL-R, PM	A	R					N		T	
Super Herc V, R-O, PM	A		G	K					T	
Warlock V, H, R-O, PM				K			N		T	
Giant >50 lbs +										
Atlantic Giant V, R-O	A			K	L		N		T	V
Big Moose V, H, R							N		T	
First Prize V, O-R			G				N			V
Full Moon V, W, R	A		G				N		T	
New Moon V, O-R	A		G				N		T	V
Prizewinner V, RS (red-orange skin), FL-R	A		G	K	L		N		T	V

[1] Abbreviations for state where recommended

Growth habit:
- B Bush growth habit
- SB Semi-bush growth habit
- V Vining growth habit
- S Semi-vining growth habit

Skin features:
- BL Blue skin
- BU Buff skin
- G Green skin
- GR Gray skin
- H Hardshell
- P Pink skin
- W White skin
- WA Warts
- RS Red Skin
- VR Variegated
- Y Yellow skin

Shape:
- FL Flat (Cinderella, pancake)
- P Pie pumpkin - suitable for cooking
- O Oblong
- R Round

Disease Tolerance/Resistance:
- DM Downy mildew tolerance/resistance
- CMV Cucumber Mosaic virus tolerance/resistance
- F Fusarium tolerance/resistance
- PH Phytophthora tolerance/resistance
- PM Powdery mildew tolerance/resistance
- PRSV Papaya Ringspot Virus tolerance/resistance
- PY Precocious yellow
- VT Virus tolerance (non-specific)
- WMV Watermelon Mosaic Virus (Strain 2) tolerance/resistance
- ZYMV Zucchini Yellows Mosaic Virus tolerance/resistance

VARIETIES[1]	AL	AR	GA	KY	LA	MS	NC	SC	TN	VA
WINTER SQUASH										
Acorn										
Autumn Delight [S, PM]	A						N	S	T	V
Celebration [B, PM]	A				L	M	N	S	T	V
Mashed Potato [B, W, O]				K	L	M	N		T	
Table Ace [S]			G	K	L		N	S	T	V
Table Queen [V]	A		G	K	L	M	N	S	T	V
Taybelle [PM S, PM]	A		G	K	L	M	N	S	T	V
Buttercup										
Buttercup [V]	A		G				N	S	T	V
Butternut										
Atlas [S]			G							
Avalon [V]	A								T	V
Butterfly [S, PMt]							N			
Betternut 900 (local markets only) [S, PM]							N			
Early Butternut [S]	A				L			S		
Polaris [V, BU]				K			N			
Quantum [V, BU]							N			
Waltham Butternut [V]	A	R	G	K	L	M	N	S	T	V
Delicata										
Bush Delicata [B, PM]	A		G	K			N		T	V
Hubbard										
Golden Hubbard [V]	A		G				N		T	V
True Green Improved Hubbard [V]	A		G				N			V
Ultra [V]	A						N			V
Spaghetti										
Angel Hair (personal size) [V]									T	
Primavera [S]	A		G				N	S	T	V
Pinnacle [S]	A						N	S	T	V
Stripetti [V, VR]	A		G				N	S	T	V
Small Wonder (personal size) [R, V]				K			N		T	
Vegetable Spaghetti [V]	A	R	G	K		M	N	S	T	V
Miscellaneous Types										
Cushaw Green Striped [V]	A			K	L		N		T	V
Cushaw Orange Striped [V]							N		T	V
Kabocha										
Golden Butta Bowl [B, H, FL-O, PM, ZYMV]	A		G	K	L		N		T	
Shokichi Green Mini [B, GR, O-R]			G	K	L				T	
Speckled Pup S [VR, R, PM, ZYMV]				K			N		T	
Sunshine [S, FL-R]	A		G		L		N			
Sweet Mama [S]	A	R	G	K	L	M	N	S	T	V
Calabaza										
La Estrella [V]	A	R	G	K		M	N	S	T	V

[1] *Abbreviations for state where recommended*

Growth habit:
[B] Bush growth habit [SB] Semi-bush growth habit
[V] Vining growth habit [S] Semi-vining growth habit

Skin features:
[BL] Blue skin [H] Hardshell [RS] Red Skin
[BU] Buff skin [P] Pink skin [VR] Variegated
[G] Green skin [W] White skin [Y] Yellow skin
[GR] Gray skin [WA] Warts

Shape:
[FL] Flat (Cinderella, pancake) [O] Oblong
[P] Pie pumpkin - suitable for cooking [R] Round

Disease Tolerance/Resistance:
[DM] Downy mildew tolerance/resistance
[CMV] Cucumber Mosaic virus tolerance/resistance
[F] Fusarium tolerance/resistance
[PH] Phytophthora tolerance/resistance
[PM] Powdery mildew tolerance/resistance
[PRSV] Papaya Ringspot Virus tolerance/resistance
[PY] Precocious yellow
[VT] Virus tolerance (non-specific)
[WMV] Watermelon Mosaic Virus (Strain 2) tolerance/resistance
[ZYMV] Zucchini Yellows Mosaic Virus tolerance/resistance

Seeding and Spacing. Research around the SE U.S. has demonstrated that fruit size can vary for each variety among locations due to a number of environmental factors. To best determine how well a variety performs in your area, trial it before planting out a large acreage. Seed in the field as indicated below:

Bush types: Rows–5 to 6 feet apart; plants–2 to 3 feet apart in row; seed–4 to 6 pounds per acre.

Semi-vine types: Rows- 6 to 8 feet apart; plants–2 to 4 feet apart in row; seed–2 to 4 pounds per acre.

Vine types: Rows–8 to 10 feet apart; plants–4 to 5 feet apart in row; seed–2 to 4 pounds per acre.

PUMPKIN/WINTER SQUASH PLANTING DATES

	Pumpkin*	Winter Squash
AL North	6/15–7/15	4/15–6/15
AL South	6/15–7/15	3/15–5/15
AR North	6/15–7/5	5/15–6/30
AR South	6/15–7/5	4/15–6/30
GA North	6/15–7/15	4/15–6/15
GA South	6/15–7/15	3/15–5/15
KY East	5/10-6/1	5/15–6/15
KY Central	5/5-6/15	5/10–7/10
LA North	6/15–7/15	4/15–5/15
LA South	6/15–7/15	3/15–6/15
MS North	6/20–7/5	4/15–5/15
MS South	6/20–7/5	3/15–5/15
NC East	6/15–7/10	4/15–5/20
NC West	5/25–6/30	5/25–6/30
SC East	6/19–7/10	4/15–5/20
SC West	5/25–6/30	4/15–6/15
TN East	6/1-7/15	5/15–6/30
TN West	6/15–7/15	4/25–6/30
VA East (coastal)	6/15–7/15	5/1-5/20
VA West (mountain)	6/15–7/5	5/15–6/15

* Pumpkin dates target Halloween market

For Soil Strips between Rows of Plastic Mulch. Use the following land preparation, treatment, planting sequences, and herbicides labeled for pumpkins or squash or crop injury may result.

1. Complete soil preparation and lay plastic and drip irrigation (optional) before herbicide application. In some cases, overhead irrigation can be used if small holes are punched into the plastic.

2. Spray preemergence herbicides on the soil and the shoulders of the plastic strips in bands before weeds germinate. DO NOT APPLY HERBICIDE TO THE SURFACE OF THE PLASTIC. Herbicides may wash from a large area of plastic into the plant hole and result in crop injury.

3. Incorporate preemergence herbicide into the soil with 0.5 to 1 inch of rainfall or overhead irrigation within 48 hours of application and BEFORE PLANTING OR TRANSPLANTING.

4. Apply selective postemergence herbicides broadcast or in bands to the soil strips between mulch to control susceptible weeds.

Minimum Tillage. No-tillage is the most commonly used minimum tillage practice with pumpkins. No-till planters currently in use with row crop production will plant pumpkin seed but seed plates or feed cups need to match up with seed size. Improper seed plates or cups will break pumpkin seed. Type of winter cover crop residue can affect pumpkin seed depth. Inspect seed placement and adjust for correct depth. Early spring planting with no-tillage in pumpkin may delay growth and days to harvest. Planting after soils warm in the spring will improve vigor (pumpkins are normally planted after soil warms so this may not be a management problem). Use of small grain cover residue may require additional nitrogen fertilizer (20 to 30 lbs N/acre in addition to the normal recommendation) if cover crop is fairly mature when killed. Normal pumpkin nitrogen fertilizer recommendations can be used if a legume cover crop (hairy vetch, winter peas, or crimson clover) is used as residue.

SPECIAL NOTES FOR PEST MANAGEMENT
DISEASE MANAGEMENT

Accurate diagnosis is necessary for effective disease management. Incorporate appropriate and effective cultural practices and pesticides, as detailed in the Disease Management section, in a season-long integrated disease management program.

Powdery Mildew: Powdery mildew is an annual problem on pumpkins and winter squash. Management of powdery mildew relies on the use of resistant varieties in conjunction with an effective season-long fungicide program. Fungicides should be applied with adequate coverage prior to the development of disease.

Phytophthora Fruit and Crown Rot: Phytophthora fruit and crown rot can be a serious issue in squash in some locations, especially in production systems that use surface water and overhead irrigation. To minimize the occurrence of this disease, fields should be adequately drained to ensure that soil water does not accumulate around the base of the plants. Just before plants begin vining, subsoil between rows to allow for faster drainage following rainfall. Avoid moving infested soil between fields.

Crop rotation is not effective. Resistance to some fungicides (e.g. mefenoxam) has been reported and should be considered when selecting fungicides to protect both crowns and fruit.

Cucurbit Downy Mildew (CDM): CDM can be a devastating disease of cucurbits in the southeastern and eastern U.S. Successful CDM management often requires the use of an appropriate fungicide spray program. Fungicide efficacy is influenced by crop, regional, and seasonal differences, as well as resistance management practices. A useful tool for downy mildew management (maps, text alerts, identification resources, etc.) is the CDM reporting system (http://cdm.ipmpipe.org). If you think you have CDM or have questions about disease management, please contact your local Extension office.

Mosaic Viruses: Once plants become infected with a virus, they cannot be cured. Chemicals are not an effective management option for virus infections. Many mosaic viruses in cucurbits (cucumber mosaic virus [CMV], papaya ringspot mosaic virus [PRSV]), watermelon mosaic virus [WMV], zucchini yellow mosaic virus [ZYMV]) are transmitted by aphids. Managing aphids is not an effective way to manage virus diseases. Use resistant varieties when possible. Even resistant varieties may not escape virus diseases. Most varieties do not have resistance to all four viruses. Virus diseases cannot be accurately diagnosed based on symptoms. It is important to send samples of plants with virus symptoms to a diagnostic lab when symptoms are observed so that in the following year, varieties that have resistance to those viruses that may be more prone to occur in your production region can be selected for production.

Whitefly-transmitted Viruses: Several whitefly-transmitted viruses, including cucurbit chlorotic yellows virus (CCYV), cucurbit leaf crumple virus (CuLCrV), and cucurbit yellow stunting disorder virus (CYSDV), are emerging in various areas throughout the Southeast when unusually high populations of whiteflies are present. These viruses may produce similar symptoms; molecular testing is necessary for virus identification. These viruses often also occur in mixed infections. Management of whitefly-transmitted viruses focuses on prevention and management of the whitefly vector. Additional information on whitefly-transmitted viruses in cucurbits is available at https://eCucurbitviruses.org.

NEMATODE MANAGEMENT

Various root-knot nematodes, such as the southern root-knot and guava root-knot, as well as other nematodes can infect pumpkin. Nematode populations are highest in the late summer and fall. Collect soil samples for nematode analysis during this period from fields to be planted with pumpkins or squash the following year. Clean and disinfest equipment between fields. Use appropriate crop rotations and cover crops with non-host crops to reduce nematode populations. Avoid fields with high populations of nematodes or use appropriate fumigants and other nematicides for management (see fumigant and nematicide tables).

INSECT MANAGEMENT

Cucumber Beetle: Cucumber beetles cause direct feeding damage to the foliage. Young plants need to be protected with insecticide as soon as they emerge or are transplanted to prevent bacterial wilt disease. Consider soil applications of neonicotinoid insecticides at planting or transplanting in addition to foliar insecticides of other modes-of-action to prevent early infection. Cucumber beetles also feed on pumpkin and winter squash rinds. Fall treatments with foliar insecticides to prevent rind feeding may also reduce incidence of bacterial wilt. While Hubbard squash, butternut squash and processing pumpkins are susceptible to bacterial wilt, Jack-o-lantern pumpkins and most other varieties of squash are rarely susceptible to bacterial wilt. Some pumpkin varieties, such as *Cinderella*, are more susceptible to rind feeding than others.

Squash Vine Borer: Squash vine borer is a serious pest in small plantings. Pheromone baited sticky traps can be used soon after planting to monitor the activity of the adult moths. Start inspecting plants closely for squash vine borer eggs (1mm [1/25 inch] diameter oval, flattened, dull-red to brownish) as soon as plants emerge in the field, or after moths are caught in the traps. Insecticide applications should occur after finding eggs or when moths are observed in the field. Effective insecticide residue is necessary 7-10 days after the first observation of eggs, which could warrant a second application of insecticide depending on the residual activity of the product used. Applications should be made in afternoons or evenings after flowers close to reduce the spraying of valuable pollinators, especially bees. If pheromone traps are not used, a preventive treatment should be applied when vines begin to run. Re-apply insecticide every seven days for four weeks, depending on residual activity. Continue monitoring the pheromone traps into August to detect the emergence of the new moths. When moths are caught, inspect plants for secondgeneration eggs, and begin the insecticide applications when eggs first begin to hatch or just prior to hatching. Many insecticides used to manage cucumber beetles and squash bug may also suppress squash vine borer.

Melonworm and Pickleworm: Melonworm, *Diaphania hyalinata*, and pickleworm, *Diaphania nitidalis*, are tropical moth pest species that impact pumpkins and winter squash. Melonworm feed on foliage and the rind of fruit, while pickleworm feed on flowers and burrow into fruit. Although melonworm are known to feed on foliage, they will quickly move from mature plants to feed on pumpkin rinds. Impact from these pests will vary by location and generally by latitude, with some areas not observing melonworm or pickleworm until they immigrate North in August-September. Insecticides are necessary when larvae or moths are observed, although severity of infestations will vary depending on location.

Aphids: Aphids are often an induced pest of pumpkins and winter squash, occurring after broad-spectrum insecticides kill natural enemies. Aphid feeding can delay plant maturity, may lead to the transmission of viruses, and can lead to a build-up of sooty mold. Thorough spray coverage, especially on the underside of the leaves is important. Unfortunately, insecticide use for aphids does not reduce the spread of virus. Repeated use of stylet oil or utilization of reflective mulches may help reduce the impact of viruses. Refer to the "Mulches" section for more information.

Squash Bug: Begin scouting shortly after plant emergence. Treat every 7 to 10 days when eggs, adults or nymphs are observed every few plants. The control of squash bugs is particularly important where yellow vine disease occurs since squash bugs vector the pathogen responsible for this disease. Adults and older nymphs are harder to control, therefore insecticides should prioritize targeting nymphs as they hatch from eggs.

Spider Mites: Mite infestations generally begin around field margins and grassy areas. CAUTION: DO NOT mow these areas after midsummer because this forces mites into the crop. Localized infestations can be spot-treated. Note: Continuous use of pyrethroid sprays may result in mite outbreaks.

POLLINATION

Honey bees are important for pollination, high fruit yields, fruit size, and quality. Populations of pollinating insects may be adversely affected by insecticides applied to flowers or weeds in bloom. Use one hive per acre to get good pollination. Apply insecticides only in the evening hours or wait until blooms have closed before application. See section on "Pollination" in the General Production Recommendations.

HARVESTING AND STORAGE

Use clean storage bins and sanitize with a 5 to 10% bleach solution. Be sure to thoroughly clean and sanitize bins prior to usage to decrease postharvest diseases.

Harvest as soon as fruits are mature and prior to frost. Use care in handling fruit to prevent wounds. Winter squash and some pumpkin varieties need to be cured after harvest at temperatures between 80 to 85°F with a relative humidity of 75 to 80% for 10 days. All winter squash and pumpkin are chilling sensitive and should not be stored below 50°F. When storing, avoid direct contact with concrete.

For non-edible pumpkins, such as Jack-o-lanterns, stem rot can be an issue, especially in the southeastern USA. A 10% bleach solution diluted in water can be sprayed on the stems to decrease rot and extend shelf-life. Pumpkins are best stored in a cool, dry and well-vented area. Ideal storage temperature ranges from 50-60°F with 50-60% humidity for up to 2 to 3 months. High humidity can cause condensation and disease.

The hard-shelled squash varieties, such as 'Butternut', 'Delicious', 'Cinderella', 'Blue Jarrahdale', and the Hubbard strains, can be stored for up to 6 months. Store in a well-vented space held at 50-60°F and 50-60% relative humidity. For home gardeners, storage is best in a dimly lit room such as a pantry or cupboard with temperatures ranging from 68 to 75°F. See Table 14 for further postharvest information.

RADISHES (*Raphanus sativus*), RUTABAGAS (*Brassica napus*), AND TURNIPS (*Brassica rapa* var. *rapa*)

VARIETIES[1]	AL	AR	GA	KY	LA	MS	NC	SC	TN
RADISH: Salad, Daikon, and Icicle Types									
Amethyst [2]		R	G		L		N	S	T
Bacchus [2]	A				L		N	S	T
Cherriette [2]	A		G				N	S	
Cherry Belle [2]	A		G	K	L	M	N	S	T
Champion [2]	A		G			M	N		
Cook's Custom Blend [2] (mixture of 4 root colors & shapes)	A			K	L				
Crunchy Royale [2]				K		t	N	S	T
Easter Egg II [2] (mixture of 5 - 6 root colors)	A	R			L		N	S	T
Early Scarlet Globe [2]	A					M	N		
Fireball [2]								S	
Nero Tondo [2] (black exterior w/ white interior)	A	R					N		
Ostergruss Rosa [2,3]								S	
Ping Pong (white exterior & interior)	A				L		N		
Pink Beauty [2]	A	R		K	L		N		
Red Head [2]		R							T
Sora [2]							N	S	T
Stargazer [2,8]		R							
Sparkler [2] (half red, half white)	A			K		M	N	S	
Valentine [8]								S	
White Icicle [4]	A		G	K		M	N		T
Watermelon [2,7]		R	G					S	
RADISH: Storage Types									
April Cross [3]	A		G	K			N		
Everest [3]	A			K			N		
Long Black Spanish [5]	A					M	N		T
Round Black Spanish [5]	A					M	N		T
RUTABAGAS									
American Purple Top	A		G				N	S	
Laurentian	A		G				N	S	
TURNIPS									
Hakurei [6]			G	K	L				T
Purple Top White Globe	A		G		L	M	N	S	T
Royal Crown			G	K	L			S	T
Scarlet Queen Red Stems [6]	A				L				T
Shogoin	A					M		S	
Tokyo Cross	A			K	L	M	N	S	T
White Egg							N	S	
White Lady					L	M		S	T

[1] Abbreviations for state where recommended
[2] Garden radish
[3] Daikon radish
[4] Icicle radish
[5] Spanish radish
[6] Small root type; best when harvested at 2" to 3" diameter
[7] White exterior with red interior
[8] Green and white exterior with red or pink interior

Seed Treatment. Soak seed in hot water at 122°F. Soak rutabagas for 20 minutes and turnips for 25 minutes. Dry the seed, then dust with a labeled fungicide to prevent damping-off. Further in- formation on seed treatments can be found under Seed Treatments in the disease management section of the handbook.

SPACING AND SEEDING

Radishes: *Salad or garden radish* roots are normally red skinned, round, less than two inches in diameter and grow rapidly, generally taking less than one month from seeding to harvest. *Icicle* types are elongated root forms of garden radishes. *Daikon radishes* are Asian storage radishes that produce large, white cylindrical roots which can exceed twelve inches in length and can weigh over one pound. *Spanish radishes* have round or elongated large storage roots with black skin. *Storage radishes* can take up to ninety days from seeding to harvest.

Radishes are a quick-growing, cool-season crop producing its best quality when grown at temperatures of 50 to 65°F. Many radish types are ready for harvest 23 to 28 days after sowing. Radishes must be grown with an adequate moisture supply; otherwise, when growth is checked radishes become hot, tough, and pithy. Warm temperature and longer day-lengths induce seed-stalk formation.

Seed radish as early in the spring as soil can be worked, then in order to maintain a continual supply make additional plantings at 8to 10-day intervals. Space rows 8 to 15 inches apart and sow 12 to 15 seed per foot within a row. This will require 10 to 15 pounds of seed per acre.

RADISH PLANTING DATES

	Spring	Fall
AL North	2/15–5/15	8/1–10/15
AL South	1/15–3/31	8/1–10/31
GA North	3/15–5/15	8/1–9/15
GA South	2/1–3/31	8/1–10/15
KY East	3/15-5/15	8/1-9/1
KY Central	3/10-5/10	8/15-9/15
KY West	3/10-4/1	9/15-10/1
LA North	2/1–3/15	8/1–10/30
LA South	1/15–3/15	8/1–10/30
MS North	3/5-4/30	8/1-9/15
MS South	2/1-3/31	8/1-9/30
NC East	2/15-6/30	8/1-9/15
NC West	4/1-8/15	NR
SC East	2/1-6/15	8/1-9/30
SC West	3/15-6/30	8/1-9/15
TN East	4/1-5/30	8/1-9/15
TN West	3/1-5/1	8/1-9/30
NR = Not recommended		

Rutabagas: A cool-season crop that develops best at temperatures of 60 to 65°F. Usually considered a fall crop, it can be grown in the spring. Seed at least 90 days before the early freeze date in the fall. Sow 1.5 to 2 pounds of seed per acre at a depth of 1 inch in rows 30 to 36 inches apart. Thin to 4 to 8 inches in the row when plants are 2 to 3 inches tall.

This is a long-term root crop normally requiring up to 120 days or more to mature.

RUTABAGA PLANTING DATES

	Spring	Fall
AL North	2/15–5/15	8/1–9/15
AL South	1/15–3/31	8/1–10/15
GA North	3/15–5/15	8/1–9/15
GA South	2/1–3/31	8/1–10/15
KY East	3/15-5/15	NR
KY Central	3/10-5/10	NR
KY West	3/10-4/1	NR

RUTABAGA PLANTING DATES (cont'd)

	Spring	Fall
LA North	2/1–3/15	7/15–10/30
LA South	1/15–3/15	7/15–10/30
MS North	2/15–5/15	8/1–9/15
MS South	1/15–3/31	8/1–10/15
NC East	2/15–4/15	8/1–9/30
NC West	4/1–8/15	NR
SC East	2/1–3/31	8/15–10/15
SC West	3/15–4/30	7/15–9/30
TN East	3/15-5/15	NR
TN West	3/10-4/1	NR
NR = Not recommended		

Turnips: Seed as early in the spring as soil can be worked or at least 70 days before the early freeze date in the fall. Seed in rows 1 to 2 pounds per acre, 0.25 to 0.5 inch deep, in rows 14 to 18 inches apart. Plants should be 2 to 3 inches apart in the row. Seed can also be broadcast at the rate of 2.5 pounds per acre.

TURNIP (ROOTS) PLANTING DATES

	Spring	Fall
AL North	2/15–5/15	8/1–10/15
AL South	1/15–3/31	8/1–10/30
GA North	3/15–5/15	8/1–9/15
GA South	2/1–3/31	8/1–10/15
KY East	3/15-4/15	7/1-7/15
KY Central	3/10-4/10	7/15-8/1
KY West	3/1-4/1	8/1-8/15
LA North	2/1–3/15	7/15–10/31
LA South	1/15–3/15	7/15–10/31
MS North	1/20-4/1	7/25-8/20
MS South	1/15-3/1	8/10-9/15
NC East	2/15-6/30	8/1-9/15
NC West	4/1-8/15	NR
SC East	2/1-4/1	8/1-9/30
SC West	3/15-4/30	8/1-9/15
TN East	3/15-5/30	7/15-8/10
TN West	3/1-5/1	8/1-8/25
NR = Not recommended		

HARVESTING AND STORAGE

Rutabagas: Pull and trim tops in the field. Bruised, damaged, or diseased rutabagas will not store well. Wash rutabagas in clean water, spray-rinse with clean water, then dry as rapidly as possible before waxing and shipping. Rutabagas can be stored for 2 to 4 months at 32°F and at 90 to 95% relative humidity.

Turnips: The crop is dug mechanically and either bunched or topped. Turnips can be stored at 32 to 35°F and at 90 to 95% relative humidity. For further postharvest information on radish, rutabaga, and turnip, see Table 14.

SPINACH (*Spinacia oleracea*)

VARIETIES[1]	AL	GA	KY	LA	MS	NC	SC	TN
SPINACH								
Bloomsdale Long Standing [2,5]	A	G	K			N	S	T
Bolero [5]						N		
Early Hybrid #7 [2,4,6]	A	G				N	S	
Space [3,4]			K			N		
Tyee [3,4]	A	G	K	L	M	N	S	T
Whale [2,4,5]							S	

[1] Abbreviations for state where recommended
[2] Savoy leaf type
[3] Semi-savoy leaf type
[4] Downy mildew tolerance/resistance
[5] Bolting tolerant
[6] Cucumber mosaic virus tolerance/resistance

Spinach may be divided by use into fresh and processed types. Spinach cultivars are either upright or spreading in habit and can be further subdivided by leaf type into savoy (wrinkled), crushed, crumpled, long-season, semi-savoy, and smooth varieties. The processing types are usually smooth leaved; the semi-savoy types can be used for both purposes; while the fresh market prefers the savoy types. In addition, spinach cultivars can be classified as being fast bolters or slow bolters. Spinach usually matures in 30 to 50 days.

Geographic/Climate Requirements. Since spinach is a hardy, cool season plant growing best where temperatures are moderate. Spinach is frost tolerant and cold hardy to 20°F. Germination is maximized at 41°F with emergence taking 23 days.

Higher temperatures reduce germination. Spinach may be an early spring, late fall, or winter crop, where the conditions permit surviving winter killing temperatures. Longer day length triggers bolting.

SPINACH PLANTING DATES

	Spring	Fall
AL North	3/15–4/30	8/1–9/15
AL South	2/1–3/31	8/15–9/30
GA North	3/15–4/30	8/1–9/15
GA South	2/1–3/31	8/1–9/30
KY East	3/10-4/10	8/1–8/15
KY Central	3/1–4/1	8/15-9/1
KY West	2/15-3/15	9/1-9/15
LA North	2/1–3/15	9/1–11/15
LA South	2/1–3/15	9/15–11/15
MS	NR	8/15-9/30
NC East	2/15–6/30	8/1–9/15
NC West	4/1–8/15	NR
SC East	2/1–4/1	8/15–10/15
SC West	3/15–4/15	8/1–9/30
TN East	3/1-3/31	8/15-9/15
TN West	2/15-3/15	9/1-9/15

NR = Not recommended

Soil and Fertilizer Requirements. Spinach is sensitive to acidic soils, preferring a pH range from 6.0 to 7.5. Warmer sandy soils are preferred for overwintering spinach. Fertilizer is more important during the slower growing winter period. Before any fertilizer is worked into a spinach field, careful soil sampling and analysis should be obtained to determine P and K levels.

Cultural Practices. Spinach is usually direct seeded in rows using either precision methods and coated seed or regular drilled uncoated seed. In some areas spinach is simply broadcast on beds. The rate of germination fluctuates widely depending on methods of seeding but also due to the risk of damping-off. Despite the size of spinach seed, it is sown fairly shallowly (0.8 to 1-inch), in soil moisture conditions ranging from slightly above permanent wilting to field capacity.

Where spinach is pulled by hand for harvest, it is possible to select the larger, more vigorous plants, leaving space for the slower, crowded plants to grow. Seeding rates for non-clipped: 10 to 14 pounds per acre and for clipped: 18 to 25 pounds per acre. Spacing within rows is generally 12 inches. For smaller stands, sow 1 oz. of seed per 100 row feet; average yields are 40 lb per100 row feet, 152 cwt per acre.

Irrigation and Drainage. Spinach is shallow rooted and, therefore, may become water-stressed if irrigation is not avail- able between rain showers. More often than not the first irrigation is necessary to germinate the seed. Since spinach is sensitive to overwatering or waterlogging, provision for drainage either through seedbed preparation or tiling is essential. The large transpiring leaf surface of maturing spinach plants coupled with warm temperatures can readily deplete the available moisture reserves.

SPECIAL NOTES FOR PEST MANAGEMENT
INSECT MANAGEMENT

Insect pests of spinach include: Aphids, Leaf miners, Cabbage loopers, mites, and Sweet corn maggots. Control methods include crop rotation, destruction of crop residues, and the use of pesticides.

Seed Corn Maggot: To prevent maggot damage to spring-seeded plants, treat seed with an approved commercially available insecticide or use a broadcast application of a soil-incorporated insecticide.

DISEASE MANAGEMENT

Spinach is vulnerable to attacks by blight (CMV), downy mildew, leaf spot, damping-off, seed rot, nematodes, and root rot. Resistant or tolerant cultivars help to ward off diseases. Where possible the use of the right seed dressing will protect the germinating seed and developing roots. Fumigation may sometimes be necessary, coupled with good rotational practices.

DISEASE MANAGEMENT

Weed management is especially important in young spinach stands to reduce competition. Spinach competes poorly with many weeds, and the presence of weeds can significantly reduce yield. Growers should include both cultural and chemical controls for weed management, as no one practice consistently provides control.

Cultural controls include choosing a planting site that has low weed pressure, sanitation, and crop rotation. If field cultivation is necessary prior to planting, it should be shallow so as not to expose buried weed seeds to the sun. Consumer tolerance for weeds in bagged spinach is low; cultivation and hand-weeding are sometimes employed in addition to other controls. Hand-weeding is costly but is often the only option as the plants mature.

As with other leafy greens, herbicide options after seedlings have emerged are limited. In addition, spinach is sensitive to damage from certain herbicides, and should not be planted if residues may be present in the soil. Preplant application of an appropriate herbicide is recommended. Proper identification of the weeds present in the field is crucial to selection of the best herbicide for the location, as there are few options, and none that address all the weed types that may be present.

Some growers have been able to reduce weed pressure with pre-irrigation, saturation of the field before the spinach seed is planted. This causes a flush of weeds to emerge, which can be eliminated by burning or herbicide application. The spinach seed is planted after the weedy plants have been removed with the expectation of less competition.

HARVESTING AND STORAGE

See Table 14 For further postharvest information.

SUMMER SQUASH (*Cucurbita pepo*)

VARIETIES[1]	AL	GA	KY	LA	MS	NC	SC	TN	VA
SUMMER SQUASH									
Yellow Crook Neck									
Destiny III [3, 4, 5, 6]	A	G	K	L		N	S	T	
Dixie	A			L	M	N	S	T	
Gentry	A	G	K	L	M	N		T	
Gold Star [6, 8]	A	G	K		M	N	S		
Prelude II [3, 4, 5, 8]	A	G	K	L		N	S	T	
Yellow Straight Neck									
Conqueror III [3, 4, 5, 6, 7]	A	G	K			N		T	
Cougar [2, 4, 7]	A		K	L	M	N	S	T	
Enterprise	A	G				N	S	T	
Grandprize [4, 5, 8]	A	G			M	N	S		
Goldprize [4, 5]	A	G				N	S		V
Fortune [2]			K			N		T	
Lioness [4, 5, 6, 7]		G	K	L		N			
Lazor [4]							S		
Multipik [2]	A		K	L		N		T	
Solstice [4, 5]	A	G		L		N	S	T	
Superpik [2]	A			L		N		T	
Zucchini									
Cashflow [4]						N	S	T	
Cue Ball [4, 5, 8, 12]			K			N		T	
Dunja [4, 5, 8]			K	L	M		S		
Eight Ball [4, 5, 8, 12]			K			N		T	
Everglade [4, 5, 6, 7, 10]		G							
Fortress [4, 5, 6, 7, 10]		G							
Golden Delight [4, 5, 9]			K		M				
Goldy [8, 9]	A	G					S		
Justice III [3, 4, 5, 6]	A	G	K			N			
Ladoga [4, 5, 7]	A					N			
Leopard [4, 7]						N	S	T	
Mexicana [4, 5, 6, 7, 8, 11]			K						
One Ball [4, 5, 9, 12]			K			N		T	
Paycheck [4, 5, 6, 7]	A	G	K			N		T	V
Payroll [4, 5, 8]	A	G	K		M	N	S	T	
Payload [4, 6, 8]	A	G		L		N	S	T	
Sebring [8, 9]	A			L					
Senator	A			L	M		S	T	
Spineless Beauty	A	G	K		M	N	S	T	V
Spineless Perfection [4, 5, 8]	A	G	K			N	S	T	V
Tigress [4, 5, 7]	A		K	L	M	N	S	T	
Zephyr (bi-color)	A	G	K	L	M	N	S	T	
Zucchini Elite	A		K	L		N		T	
Grey Zucchini									
Ishtar [4, 5]	A	G						T	

[1] Abbreviations for state where recommended
[2] *Py* - Precocious yellow gene; has a prominent yellow stem. The *Py* gene helps mask infection from CMV and WMV II.
[3] Transgenic
[4] Zucchini yellows mosaic virus tolerance/resistance
[5] Watermelon mosaic virus tolerance/resistance
[6] Cucumber mosaic virus tolerance/resistance
[7] Papaya ringspot virus tolerance/resistance
[8] Powdery mildew tolerance/resistance
[9] Yellow zucchini
[10] Downy mildew tolerance/resistance
[11] Speckled light green
[12] Round

VARIETIES[1]	AL	GA	KY	LA	MS	NC	SC	TN	VA
Scalloped									
Benning's Green Tint	A			L		N	S		
Peter Pan	A	G	K			N	S	T	
Sunburst	A	G	K	L		N	S	T	
Total Eclipse	A							T	

[1] Abbreviations for state where recommended
[2] *Py* - Precocious yellow gene; has a prominent yellow stem. The *Py* gene helps mask infection from CMV and WMV II.
[3] Transgenic
[4] Zucchini yellows mosaic virus tolerance/resistance
[5] Watermelon mosaic virus tolerance/resistance
[6] Cucumber mosaic virus tolerance/resistance
[7] Papaya ringspot virus tolerance/resistance
[8] Powdery mildew tolerance/resistance
[9] Yellow zucchini
[10] Downy mildew tolerance/resistance
[11] Speckled light green
[12] Round

Seed Treatment. Check with seed supplier to determine if seed has been treated with an insecticide and/or fungicide. Information on seed treatments can be found under Seed Treatments in the disease management section of the handbook.

Seeding, Transplanting, and Spacing. Use 4 to 6 pounds of seed per acre. Seed or container-grown transplants are planted when daily mean temperatures have reached 60°F. Containerized squash transplants are susceptible to stunting if they are held in the container for too long. Sowing seed will eliminate this concern. Seed as indicated in following table. Early plantings should be protected from winds with row covers, rye strips, or wind breaks. Space rows 3 to 6 feet apart with plants 1.5 to 2.5 feet apart in the row.

SUMMER SQUASH PLANTING DATES

	Spring	Fall
AL North	4/15–8/15	8/1–8/30
AL South	3/1–4/30	7/15–9/15
GA North	4/15–8/15	NR
GA South	3/1–4/30	7/15–9/15
KY East	5/15–7/15	NR
KY Central	5/10–8/1	NR
KY West	4/20–8/15	NR
LA North	3/15–5/15	7/15–8/31
LA South	3/1–5/15	8/1–9/15
MS North	4/15–6/15	7/25–8/14
MS South	2/15–5/1	8/14–9/14
NC East	4/1–5/30	7/15–8/15
NC West	5/15–7/31	NR
SC East	3/15–7/30	8/1–8/30
SC West	4/15–7/30	7/30–8/15
TN East	5/10–8/1	NR
TN West	4/15–7/15	NR
VA East (coastal)	3/30–9/10	NR
VA West (mountain)	4/10–8/30	NR
NR = Not recommended		

Mulching. Plastic mulch laid before field planting conserves moisture, increases soil temperature, reduces mechanical damage to fruit, and increases early and total yield. Plastic should be applied on well-prepared planting beds. The soil must be moist when laying the plastic. Black plastic mulch can be used without a herbicide. In most situations, 50 percent of the nitrogen(N) should be in the nitrate (NO_3) form.

Reflective, plastic mulches can be used to repel aphids that transmit viruses in fall-planted (after July 1) squash. Direct seeding through the mulch is recommended for maximum virus protection.

Growers should consider drip irrigation. See the section on "Irrigation" in this handbook.

SUGGESTED FERTIGATION SCHEDULE FOR SUMMER SQUASH*
(N:K;1:2)

Days after planting	Daily nitrogen	Daily potash	Cumulative	
			Nitrogen	Potash
	(lb / A)			
Preplant			24.0	24.0
0–14	0.9	1.8	36.6	49.2
8–28	1.3	2.6	54.8	85.6
29–63	1.5	3.0	107.3	190.6

* Adjust based on tissue analysis

Harvest. Summer squash is a fast growing crop and when temperatures are optimal (85 to 90°F during day and 65 to 70°F during at night), and water is readily available, fruit will be ready to harvest 35 to 40 days after seeding. When growing conditions are optimal, yellow squash should be harvested every two to three days while zucchini should be harvested every one to two days.

SPECIAL NOTES FOR PEST MANAGEMENT

INSECT MANAGEMENT

Cucumber Beetle: Cucumber beetles cause direct feeding damage to the foliage. Young plants need to be protected with insecticide as soon as they emerge or are transplanted.

Squash Vine Borer: Pheromone baited sticky traps can be used soon after planting to monitor the activity of the adult moths. Start inspecting plants closely for squash vine borer eggs (1mm [1/25 inch] diameter oval, flattened, dull-red to brownish) as soon as moths are caught in the traps. The first application of insecticide should occur when eggs begin to hatch or just prior to hatching. Applications should be made in afternoons or evenings after flowers close to reduce the spraying of valuable pollinators, especially bees. If pheromone traps are not used, a preventive treatment should be applied when vines begin to run. Re-apply insecticide every seven days for four weeks. Continue monitoring the pheromone traps into August to detect the emergence of the new moths. When moths are caught, inspect plants for second-generation eggs, and begin the insecticide applications when eggs first begin to hatch or just prior to hatching.

Aphids: Aphid feeding can delay plant maturity. Thorough spray coverage, especially on the underside of the leaves is important. Treat seedlings every five to seven days, or as needed. The transmission of plant viruses by aphids has the potential to be the most damaging to the crop. Unfortunately, insecticide use for aphids does not reduce the spread of virus. A better approach is the application of Stylet Oil to fill tiny grooves between the leaf cells. When the aphid probes the leaf surface, its stylet must pass through a layer of oil. This reduces the infectivity of the virus resulting in less disease in the squash plant. The application of Stylet Oil can delay virus infection, but requires application every other day, thorough coverage and high-pressure sprays. Squash varieties with virus resistance are recommended for all season production when aphids and virus transmission are prevalent. Also, refer to the preceding "Mulches" section for information on metallized reflective mulch used to repel or disorient aphids that can spread viruses.

Squash Bug: Begin scouting shortly after plant emergence. Treat every 7 to 10 days when adults or nymphs appear. The control of squash bugs is particularly important where yellow vine disease occurs since squash bugs vector the pathogen responsible for this disease.

Spider Mites: Mite infestations generally begin around field margins and grassy areas. CAUTION: DO NOT mow these areas after midsummer because this forces mites into the crop. Localized infestations can be spot-treated. **Note:** Continuous use of pyrethroid sprays may result in mite outbreaks.

Whiteflies: Even at low populations, whiteflies can transmit viruses such as Cucurbit leaf crumple virus (CuLCrV), Cucurbit yellow stunting disorder virus (CYSDV), and Cucumber chlorotic yellows virus (CCYV), which can cause severe damage to summer squash. Additionally, silverleaf disorder can further reduce fruit production and quality. A good chemical control solution is the rotation of neonicotinoids or butenolides with insect growth regulators to manage whitefly populations effectively.

DISEASE MANGEMENT

Accurate diagnosis is necessary for effective disease management. Incorporate appropriate and effective cultural practices and pesticides, as detailed in the Disease Management section, in a season-long integrated disease management program.

Mosaic Viruses: Once plants become infected with a virus, they cannot be cured. Chemicals are not an effective management option for virus infections. Many mosaic viruses in cucurbits (cucumber mosaic virus [CMV], papaya ringspot mosaic virus [PRSV], watermelon mosaic virus [WMV], zucchini yellow mosaic virus [ZYMV]) are transmitted by aphids. Managing aphids is not an effective way to manage virus diseases. Use resistant varieties when possible; choose conventional or GMO resistance, depending on your market. Even resistant varieties may not escape virus diseases. Most varieties do not have resistance to all four viruses. Virus diseases cannot be accurately diagnosed based on symptoms. It is important to send samples of plants with virus symptoms to a diagnostic lab when symptoms are observed so that in the following year, varieties that have resistance to those viruses that may be more prone to occur in your production region can be selected for production.

Phytophthora Fruit and Crown Rot: Phytophthora fruit and crown rot can be a serious issue in squash in some locations, especially in production systems that use surface water and overhead irrigation. To minimize the occurrence of this disease, fields should be adequately drained to ensure that soil water does not accumulate around the base of the plants. Just before plants begin vining, subsoil between rows to allow for faster drainage following rainfall. Avoid moving infested soil between fields. Crop rotation is not effective. Resistance to some fungicides (e.g. mefenoxam) has been reported and should be considered when selecting fungicides to protect both crowns and fruit.

Cucurbit Downy Mildew (CDM): CDM can be a devastating disease of cucurbits in the southeastern and eastern U.S. Successful CDM management often requires the use of an appropriate fungicide spray program. Fungicide efficacy is influenced by crop, regional, and seasonal differences, as well as resistance management practices. A useful tool for downy mildew management (maps, text alerts, identification resources, etc.) is the CDM reporting system (http://cdm.ipmpipe.org). If you think you have CDM or have questions about disease management, please contact your local Extension office.

Whitefly-transmitted Viruses: Several whitefly-transmitted viruses, including cucurbit chlorotic yellows virus (CCYV), cucurbit leaf crumple virus (CuLCrV), and cucurbit yellow stunting disorder virus (CYSDV), are emerging in various areas throughout the Southeast when unusually high populations of whiteflies are present. These viruses may produce similar symptoms; molecular testing is necessary for virus identification. These viruses often also occur in mixed infections. Management of whitefly-transmitted viruses focuses on prevention and management of the whitefly vector. Additional information on whitefly-transmitted viruses in cucurbits is available at *https://eCucurbitviruses.org*.

NEMATODE MANAGEMENT

Various root-knot nematodes, such as the southern root-knot and guava root-knot, as well as other nematodes can infect squash. Nematode populations are highest in the late summer and fall. Collect soil samples for nematode analysis during this period from fields to be planted with squash the following year. Clean and disinfest equipment between fields. Use appropriate crop rotations and cover crops with non-host crops to reduce nematode populations. Avoid fields with high populations of nematodes or use appropriate fumigants and other nematicides for management (see fumigant and nematicide tables).

WEED MANAGEMENT

See the previous "Mulching" section for further information on weed control under clear plastic mulch.

For Seeding into Soil without Plastic Mulch. Stale bed technique: Prepare beds 3 to 5 weeks before seeding. Allow weed seedlings to emerge and spray with paraquat a week prior to seeding. Then seed beds without further tillage.

For Soil Strips between Rows of Plastic Mulch. Use the following land preparation, treatment, planting sequences, and herbicides labeled for squash, or crop injury may result.

1. Complete soil preparation and lay plastic and drip irrigation before herbicide application.

2. Spray preemergence herbicides on the soil and the shoulders of the plastic strips in bands before weeds germinate. DO NOT APPLY HERBICIDE TO THE BED SURFACE OF THE PLASTIC. Herbicides may wash from a large area of plastic into the plant hole and result in crop injury.

3. Incorporate herbicide into the soil with 1/2 to 1 inch of rainfall or overhead irrigation within 48 hours of application and BEFORE PLANTING OR TRANSPLANTING.

4. Apply selective postemergence herbicides broadcast or in bands to the soil strips between mulch to control susceptible weeds.

POLLINATION

Bees are important for producing high yields and quality fruit. Move honey bees or bumble bees into the field as flowers on squash plant open. Populations of pollinating insects may be adversely affected by insecticides applied to flowers or weeds in bloom. Apply insecticides only in the evening hours or wait until bloom is completed before application. See section on "Pollination" in the General Production Recommendations.

HARVESTING AND STORAGE

See Table 14 for postharvest information.

SWEET CORN (*Zea mays*)

VARIETIES[1]	AL	AR	GA	KY	LA	MS	NC	SC	TN	VA
CORN, SWEET										
White - Early										
Sweet Ice (se)	A		G				N	S		
White - Mid-Season										
Argent (se)			G	K	L		N		T	
Avalon (se)	A								T	
Devotion (sh_2)	A		G				N		T	
Devotion II (sh_2)[3]	A						N			
Glacial (sh_2)			G				N			
Milky Way (se)[3]							N			
XTH 3674 (sh_2)			G							
Xtra-Tender Brand 378A (sh_2)			G				N	S	T	
White - Late season										
Silver King (se)	A			K			N	S	T	
Silver Queen (su)	A	R	G	K	L	M	N	S	T	
Yellow - Early										
Bodacious (se)						M	N	S		
Yellow - Mid-Season										
Aspire (se)[3]		R		K			N			
GSS 1170	A		G				N	S		
Honey Select (se)	A	R	G						T	
Incredible (se)			G	K	L	M	N	S	T	
Jubilation (sh_2)							N			
Merit (su)	A	R	G	K	L	M	N	S	T	
Passion (sh_2)	A		G				N			
Passion II (sh_2)[2,3]	A			K			N			
Protector[2,3]							N			V
Seminole Gold (sh_2)			G							
SC1336			G							
Vision MXR	A					M		S	T	V
Bicolor - Early										
Anthem XR II (sh_2)[3]			G				N			
Precious Gem (se)	A				L		N	S	T	
Temptation (se)	A						N	S	T	
Temptation II (se)[2,3]	A						N			
Bicolor - Mid-Season										
Affection (sh_2)	A		G				N		T	
American Dream (sh_2)			G							
Awesome XR (sh_2)	A		G				N		T	V
BSS8021 (sh_2)	A		G				N		T	
Cameo (sh_2)[2]				K	L				T	
Coastal (sh_2)	A		G				N		T	
Everglades (sh_2)	A		G				N		T	
Flagler (sh_2)	A		G				N	S	T	
Mirai 301BC (sh_2)	A	R			L	M		S	T	
Montauk (sh_2)				K					T	
Obsession (sh_2)	A		G	K			N	S	T	
Obsession II (sh_2)[2,3]	A	R		K	L	M	N	S	T	
Patriarch (sh_2)[3]							N			
Providence (sh_2)				K					T	V

[1] Abbreviations for state where recommended   [2] BT sweet corn (transgenic)   [3] RoundUp Ready sweet corn (transgenic)

VARIETIES[1]	AL	AR	GA	KY	LA	MS	NC	SC	TN	VA
CORN, SWEET										
Remedy (se)[3]	A	R					N			
Seminole Sweet XR (sh_2)	A		G				N		T	
Superb MXR (sh_2)	A		G				N		T	
Bicolor - Mid-Season (cont'd)										
Sweet Chorus (se)	A		G				N		T	
Sweet G90 (su)		R			L	M		S	T	
Sweet Rhythm (se)	A		G					S		

[1] Abbreviations for state where recommended   [2] BT sweet corn (transgenic)   [3] RoundUp Ready sweet corn (transgenic)

There are three primary genes contributing to sweetness in sweet corn. They are; normal sugary (*su*), sugary enhanced (*se*), and supersweet or shrunken-2 (sh_2).

Normal sugary sweet corn (*su*) has been enjoyed for many years. *Su* sweet corn is known for its creamy texture and mild sugars; however, sugars in these cultivars are rapidly converted into starch if not cooked the day of harvest. These cultivars are commonly sold in farmer's markets and roadside stands.

The sugary enhanced (*se*) sweet corn gene, known under trade names such as Everlasting Heritage have varying degrees of increased sugar content with a creamier kernel texture as compared to *su* sweet corn types. This translates into increased sweetness with a smoother kernel texture. Another advantage is that *se* sweet corn types maintain their quality for a longer period of time than normal sugary sweet corn types (*su*).

Cultivars of "Supersweet" or "shrunken" sweet corn (sh_2) derive their name from the appearance of the dried kernel, which is much smaller than kernels of *su* or *se* sweet corn types. Recently, germination of sh_2 sweet corn cultivars has been improved and is now comparable with the su and se types. Seed of supersweet (sh_2) sweet corn cultivars should be handled very gently and the use of plateless planter is recommended to prevent damage to seed. Many older supersweet cultivars require warm soil (70°F or higher) to germinate since they are less vigorous than the *se* or *su* genotypes. Supersweet sweet corn (sh_2) cultivars have a crunchier kernel, are sweeter than su and se cultivars, and will delay the conversion of sugar to starch extending their shelf life.

Xtra-tender, *Ultrasweet*, and *Triplesweet* are names for the latest development in sweet corn cultivars. These new types of sweet corn combine the genetics of sh_2, *se*, and *su* genotypes. These cultivars are high in sugar levels, hold well in storage, and have a pericarp that is tender (this improves the eating quality of the sweet corn). Plant these cultivars using the same recommendations as those of the sh_2 types of sweet corn.

Another important development in sweet corn cultivar development is the incorporation of the *BT* gene (called *BT* sweet corn). *BT* sweet corn has been genetically modified by incorporating a small amount of genetic material from another organism through modern molecular techniques. In sweet corn, the incorporated *BT* genes is particularly effective in providing protection against European corn borer and corn earworm. The protein produced by the *BT* gene is very selective, generally not harming insects in other orders (such as beetles, flies, bees, or wasps) but more importantly this protein is safe for consumption by humans, other mammals, fish, and birds. Syngenta Seeds has incorporated the *BT* gene into several sweet corn cultivars that are sold commercially under the trade name of *Attribute* followed by a series of numerals to identify the cultivar. Certain restrictions such as isolation, minimum acreage requirements, and destruction of the crop are part of the terms of contract when purchasing *BT* sweet corn seed.

In general, when selecting a cultivar, be sure to evaluate its acceptance in the market. Plant small acreages of new cultivars to test market their acceptance.

Isolation requirements for the sweet corn genotypes are important in order to obtain the highest quality sweet corn. Supersweet (sh_2) sweet corn must be isolated by a minimum distance of 300 feet or 12 days difference in silking date to avoid cross pollination from field corn, pop corn, normal sugary (*su*), and/or sugar enhanced (*se*) types. Failure to properly isolate the sh_2 genotype will result in it producing starchy, tough kernels. Isolation of sugary enhanced from normal sugary sweet corn types is recommended to maximize quality; however, quality is usually very minimally affected should cross pollination occur. It is recommended that augmented sweet corn types be isolated from all other sweet corn types for best quality.

Seed Treatment. Check with seed supplier to ensure seed was treated with an insecticide and fungicide. Information on seed treatments can be found under Seed Treatments in the disease management section of the handbook.

Seeding and Spacing. Seed is sown as early as February in more southern regions on light, sandy soils. Use a high vigor seed variety for early plantings. Seed is drilled in the field about 1 inch deep. Varieties are spaced 30 to 42 inches apart between rows depending on cultural practices, equipment, and seed size. In-row spacings range from 6 to 12 inches apart, with small-eared, early seasons varieties planted closest.

SWEET CORN PLANTING DATES

	Spring	Fall
AL North	4/15–5/30	NR
AL South	2/1–4/30	7/15-8/15
GA North	4/1–4/30	NR
GA South	2/1–3/31	7/15–8/15
KY East	5/1-6/15	NR
KY Central	4/20-7/10	NR
KY West	4/10-7/20	NR

SWEET CORN PLANTING DATES (cont'd)

	Spring	Fall
LA North	3/1–5/15	NR
LA South	2/15–5/1	NR
MS North	3/20–4/9	NR
MS South	2/21–3/14	NR
NC East	3/15–4/30	NR
NC West	4/15–6/15	NR
SC East	3/1–4/15	7/15-8/15
SC West	3/30–5/30	6/15-7/15
TN East	4/15-6/30	NR
TN West	4/15-6/15	NR
VA East (coastal)	3/20-8/10	NR
VA West (mountain)	3/30-7/30	NR

NR = Not recommended

Mulching. The use of clear plastic mulch will improve stands, conserve moisture, and produce earlier maturity. Corn is seeded in the usual manner, except 10 to 20 days earlier in double rows 14 inches apart and on 5- to 6-foot centers. Apply herbicide and then cover with clear, 4-foot-wide plastic. Allow plastic to remain over plants for 30 days after emergence, then cut and remove plastic from field. Plants can then be cultured in the usual manner. A nematode assay is recommended before using this system. If nematodes are present in the soil, control measures are necessary before planting. Use a high vigor seed variety to avoid uneven and reduced stand.

Minimum Tillage. No-tillage is the most commonly used minimum tillage practice with sweet corn. No-till planters currently in use with row crop production will plant sweet corn seed with minimal modifications. Type of winter cover crop residue can affect sweet corn seed depth. Inspect seed placement and adjust for correct depth. Early spring planting with no-tillage in sweet corn may delay growth and days to harvest. Planting after soils warm in the spring will improve vigor. Use of small grain cover residue may require additional nitrogen (20 to 30 lbs N/acre in addition to the normal recommendation) if cover crop is fairly mature when killed. No additional nitrogen above recommendations is required if a legume cover crop (hairy vetch, winter peas, or crimson clover) is used as residue.

SPECIAL NOTES FOR PEST MANAGEMENT
(listed as "Corn, Sweet" in the Pest Management section)

INSECT MANAGEMENT

Corn Earworm (CEW): CEW initiates egg laying when the plants begin to silk and ends when the silks wilt. Eggs are laid singly on the fresh silks. Begin to control CEW when 10% of the ears are silked. Repeat sprays at 3-5 day intervals until 90% of the silks have wilted. Control is more difficult late in the season. Direct sprays toward the middle third of the plant. Corn hybrids having a long, tight-fitting shuck appear to suffer less damage than those with loose shucks.

Another management tactic for CEW and European corn borer (ECB) control is the use of BT sweet corn. These hybrids produce their own natural insecticide for control of these pests. However, under high pressure, supplemental sprays may be needed to achieve damage-free ears. Minimum acreage and resistance management practices are required with BTs sweet corn. Some markets may not accept these hybrids.

Corn Flea Beetle: Flea beetles transmit a bacterial wilt disease, known as Stewart's Wilt, and these beetles are numerous after mild winters. Treat susceptible varieties at spike stage when 6 or more beetles per 100 plants can be found. Repeat every 3 to 5 days as needed. **Note:** Soil-applied insecticides may be ineffective during the first week of plant growth if soil temperatures are cool. Foliar applications of an insecticide may be necessary during this period.

European Corn Borer (ECB): Thorough spray coverage in whorls and on plants is essential. Many insecticides are highly toxic to bees. Granular formulations, if applied over the whorl, are generally more effective than liquid formulations for ECB control.

Sap Beetle (SB): Loose-husked varieties tend to be more susceptible to sap beetle attack. Ears damaged by other insects attract SB. Begin sampling at pollen shed and treat when 5% of the ears have adults and/or eggs. **Note:** Insecticides used for worm control at silk may not control SB infestations.

Fall Armyworm (FAW): Direct granules over the plants so that they fall into leaf whorls when FAW first appear and repeat application, if necessary. For foliar spray applications, high-spray gallonage (50 to 75 gallons per acre) is necessary for effective FAW control.

INSECT MANAGEMENT DECISION-MAKING

Whorl/Tassel Infestation: In general, insect larval feeding (ECB and FAW) during the whorl stage of sweet corn development has a greater impact on early planted, short-season varieties. For ECB on early plantings, apply first spray or granular application when 15% of the plants show fresh feeding signs. Additional applications may be necessary if infestation remains above 15%. An early tassel treatment is usually more effective than a whorl treatment because larvae are more exposed to the chemicals.

The impact of infestation on mid-and late-season plantings depends on the stage of the plants when the infestation occurs. Treat for FAW during the early whorl stage when more than 15% of the plants are infested. During mid to late-whorl stages, treatment for both FAW and ECB may be necessary if more than 30% of the plants are infested. Treat fields in early tassel stage if more than 15% of the emerging tassels are infested with ECB, FAW, or young corn earworm (CEW) larvae.

Ear Infestation: Direct sampling for CEW, FAW, and ECB during silking is not practical because of the low thresholds for ear damage. Begin treatment when 10% of the ears show silk. If CEW populations are heavy, it may be necessary to begin treatments when the very first silks appear. Silk sprays should continue on a schedule based on area blacklight and pheromone trap counts, geographical location, and time of year. Early in the season, silk sprays may be required on a 3 to 6-day schedule. When CEW populations are heavy, it may be necessary to treat on a 1-to 3-day schedule. Applications during low populations can end up to 5 days before last harvest. During heavy populations and high temperatures, treatments will need to be made according to the legal "days to harvest" of the chemical.

For best control during heavy populations, maximize the gallonage of water per acre, use a wetting agent, and make applications with a high-pressure sprayer (200+ psi) with drop nozzles directed at the silks.

HARVESTING AND STORAGE

Sweet corn respiration rate is classified as 'extremely high'; therefore, its shelf-life is very short. The conversion of sugars to starch in kernels is accelerated at high temperatures, promoting the loss of sweetness and sensory quality. When stored at high temperatures, sweet corn experiences a loss of tenderness, dehydration, discoloration, development of off-flavors and off-odors, and increased microbial growth. To prevent this, sweet corn must be cooled immediately after harvest—ideally within one hour. The optimal storage conditions are temperatures of 32-34.7°F and 95-98% relative humidity. Sweet corn is not chilling sensitive, but temperatures below 31°F may promote freezing injury, manifested as water-soaked areas. See Table 14 for postharvest information.

SWEETPOTATO (*Ipomoea batatas*)

VARIETIES [1]	AL	AR	GA	KY	LA	MS	NC	SC	TN	VA
SWEETPOTATO										
Beauregard [2, 5, 6, 8, 9, 11]	A	R	G	K	L	M	N	S	T	V
Bayou Belle [2, 6, 7, 8, 9, 11]	A		G		L	M	N	S		
Bellevue [2, 6, 7, 8]						M		S	T	
Bonita [3, 6, 7, 8, 9, 12]	A				L	M	N	S		V
Burgundy [2, 6, 7, 8]								S		V
Covington [2, 6, 7, 11]	A	R	G	K		M	N	S		V
Evangeline [2, 5, 6, 7, 8, 11]			G		L	M	N			V
Hernandez [2, 6, 7, 8, 11]	A		G	K			N	S	T	
Jewel [2, 7, 8, 9, 10, 11]	A	R	G	K		M		S		V
Murasaki [4, 6, 7, 8, 9, 11]	A		G	K		M	N			
O' Henry [3, 5, 6, 8, 9, 11]	A		G	K		M		S		V
Orleans [2, 6, 8, 9, 11]	A				L	M	N	S	T	V
Purple Splendor [6, 7, 8, 13]							N			
Vermilion [2, 6, 7, 8, 9]					L					

[1] Abbreviations for state where recommended
[2] Red or copper skin, orange flesh
[3] Tan or cream, white flesh
[4] Purple skin, white flesh
[5] Sclerotial blight resistance
[6] Soil rot resistance
[7] Root knot resistance
[8] Fusarium wilt resistance
[9] Rhizopus resistance
[10] Bacterial soft rot resistance
[11] Fusarium root rot resistance
[12] Suitable choice for organic production
[13] Purple skin, purple flesh

Variety Selection. Selection of a variety depends on the intended market. Most varieties require 90 to 150 days to produce maximum yields. Sweetpotatoes are cold sensitive and should not be planted until all danger of frost is past. The optimum temperature to achieve the best growth of sweetpotatoes is between 70°F and 88°F, although they can tolerate temperatures as low as 55°F and as high as 105°F.

Soil. Well-drained sandy and sandy loam soils will produce the best-shaped sweetpotatoes. Heavy soils or compacted soils will induce misshapen roots, while soils with high levels of organic matter can promote scurf. Use long rotations to decrease the incidence of scurf, Fusarium wilt, black rot, certain nematode species and other diseases. Avoid fields that have produced a crop of sweetpotatoes in the past two years and fields that have high nematode populations and are seriously eroded or grassy. Select a soil that is well drained but not prone to drought. Waterlogged, poorly drained soils prevent roots from obtaining sufficient oxygen, which can cause "souring" of roots.

Fertilizer and Lime. Get a soil test. The amount of fertilizer needed will vary by variety and location and on available nutrients as indicated by a soil test. The optimum pH range is 5.8 to 6.2. If your soil needs lime, incorporate the appropriate amount several months before planting. This will allow sufficient time for the lime to increase your soil's pH.

Broadcast or band half of the required nitrogen (N) before planting and then sidedress the remainder at layby when the vines begin to run. N is very mobile in the soil and special attention should be given to sandy soils and regions with frequent, heavy rainfall events during the season. An appropriate total seasonal rate of N range for sweetpotato is between 40 and 80 pounds per acre. Some varieties require less N fertilizer than others. For the variety Beauregard, 40 to 50 pounds per acre per growing season is sufficient while for the variety Covington, 80 pounds per acre in preferred. Follow the recommended rate of fertilizer because high fertilizer concentrations may result in salt burn and plant damage. Additionally, applying surplus fertilizer can cause excessive vine growth and be a waste of resources due to added costs that will not result in higher yields.

CROP ESTABLISHMENT

Propagating. Sweetpotatoes are propagated from sprouts or slips (vine cuttings). Only purchase certified, disease-free seed stock or slips. Select seed (the word *seed* refers to the roots used for slip production), that is free from insect and disease damage, that has a uniform flesh with variety appropriate skin color, and that is free from veins. Your profitability depends on starting with the highest quality seed stock available. Using quality roots for seed is essential for producing quality sweetpotatoes. Quality sweetpotatoes are not produced from poor-quality seed.

In most years, it is not possible to purchase sufficient certified seed stock to produce slips for your entire planned production. Thus, seed must be saved from each year's crop. When possible, isolate your seed planting from that of your commercial planting to minimize viruses. Save seed from the highest quality roots that you produce. Carefully inspect roots for defects (no off-types), disease, insect damage, etc., as listed above. Each year purchase a portion of your required seed stock for slip production as certified seed, supplementing your total need with seed from the previous year's crop. Certified slips are available from several growers in around the Southeastern US. Consult your local Extension office for information.

Additionally, sweetpotato is part of the National Clean Plant Network and several Clean Plant Centers are located in the Southeastern U.S. that support the production of virus-tested, certified planting material. The following Universities are participating in the National Clean Plant Network for Sweet Potato: NC State University, LSU AgCenter, Mississippi State University, University of Arkansas at Pine Bluff, UC Davis, University of Hawaii. Please contact your local Extension office for more information.

Presprouting. Presprouting is a technique that produces two to three times more slips than seed stock that is not presprouted. Some refer to presprouting as "waking up" the sweetpotatoes after they have been in storage. Presprouting encourages more prolific sprouting in roots. This can decrease production costs by decreasing the amount of seed stock required. In addition to increasing the number of slips produced, presprouting produces slips faster. Conditions required for presprouting are similar to those required for curing sweetpotatoes. Presprouting involves placing seed stock in a controlled storage area, such as a curing room. You must be able to control temperature and relative humidity and be able to provide ventilation. Be sure that you are able to replace the air one to two times per day because the roots require a significant amount of oxygen to facilitate presprouting. A rule of thumb: if there is not enough oxygen for a match to stay lit, there is likely not enough oxygen for the sweetpotatoes. To presprout, place seed stock in a presprouting room for 21 to 35 days at 70 to 80°F with 90% relative humidity. Spraying the walls and floors with water two times per day can help maintain relative humidity. Mechanical humidifiers (automatic humidifiers, misting systems) can help establish and maintain the required relative humidity. Avoid humidity near 100 percent or wetting of the surface of the roots as this can lead to the development of rots.

Bedding. Provide 4 to 5 pounds of 8-8-8 or 10-10-10 type fertilizer per 100 square feet of bed area. Treat seed with appropriate fungicides to prevent bedding root decay. After presprouting, place roots into beds, being careful not to damage them. Be sure to cover roots completely with 2 to 3 inches of soil. Do not be concerned if a few sprouts are above the soil line. Keep beds moist but not wet. After planting roots, cover beds immediately with black or clear plastic to warm the soil. Punch holes in plastic for ventilation as needed. Slips are ready to harvest when they have 6 to 10 leaves (8 to 12 inches long) and adventitious roots are initiated (roots projected from the nodes or joints of the stem). Slips from presprouted roots are generally ready one week earlier.

Preparing Slips for Transplanting. To harvest, cut the slips about 1 inch above the bed surface. Cutting is preferred to pulling slips. Always pull the knife up and away from the soil to prevent contamination from the seedbeds from moving into the production field. Clean knives frequently by dipping them into a 1:1 (v/v) solution of bleach and water. This will also prevent the spread of diseases from the seedbed into the field. Set the slips in the field within three days after harvesting them from the plant beds. About 500 slips can be produced from one bushel of seed. One bushel of seed requires 20 to 30 square feet of bed area.

Transplanting. Avoid planting slips until all danger of frost is past because they are very frost sensitive. Beds should be 4 to 8 inches high and as wide as equipment will allow. Narrow beds tend to dry quickly and may reduce overall yields. High beds will aid in promoting drainage, thus preventing water damage to roots. The most economical method to set a large number of plants is with a mechanical transplanter. Space slips 6 to 16 inches apart within rows spaced 3½ to 4 feet apart on row centers. The number of slips needed per acre will depend on your desired spacings. Be sure to manage water carefully to avoid transplant shock. Slips set more widely apart in-row will facilitate root enlargement, while closer in-row spacing can result in increased competition and delay root sizing.

SWEETPOTATO PLANTING DATES

AL North	5/1–6/30
AL South	3/15–6/15
GA North	5/15–6/15
GA South	4/1–6/15
KY East	5/20-6/1
KY Central	5/10-6/10
KY West	5/1-7/15
LA North	5/1–6/30
LA South	4/15–6/30
MS	4/15-6/20
NC East	5/1–7/15
NC West	5/25–6/30
SC East	4/15–6/15
SC West	5/1–6/15
TN East	5/15-6/30
TN West	5/1-6/30
VA East (coastal)	5/15-6/30
VA West (mountains)	6/1-6/20

SPECIAL NOTES FOR PEST MANAGEMENT
DISEASE MANAGEMENT

Black Rot: Symptoms are generally seen at harvest, after curing, or after storage and consist of a dry, firm, dark-colored rot that does not extend into the cortex of the sweetpotato root. Black rot is typically transmitted from infected parent material (roots or slips), infested soil, infested equipment, and some insects. The causal agent, *Ceratocystis fimbriata*, can also survive in the soil for several years or on the roots of wild morning glories. Infections can also take place on wounded sweetpotatoes during washing and packing if any equipment or the wash tank is contaminated with the pathogen. Prevention is the best method of management. Use disease free fields for bedding and cut slips at least 2 inches above the soil line. Cleaning and sterilization of harvest, storage, and packing equipment should be completed on a regular basis. Limited fungicides are available for chemical management.

Fusarium Root Rot: Root rot of sweetpotato generally appears as circular lesions with light and dark brown concentric rings. Unlike surface rots of sweetpotato, root rot extends past the periderm and into the cortex of the root often forming open cavities in the tissue. Root rot may be initiated at the proximal or distal end of the sweetpotato. The causal agent, *Fusarium solani*, is a common soil inhabitant and can survive in the absence of a host plant for up to five years. Infections can occur in wounds during harvest. *F. solani* may also survive on contaminated harvesting and packing equipment, allowing for subsequent infections after packing. *F. solani* can be transmitted from infected slips to sweetpotato roots. Prevention is the best method of management. Use disease free fields for bedding and cut slips at least 2 inches above the soil line. Cleaning and sterilization of harvest, storage, and packing equipment should be completed on a regular basis. Limited fungicides are available for chemical management.

Geotrichum Sour Rot: Geotrichum sour rot is reliant on wet or anoxic environmental conditions. Chilling injury may also promote disease. The symptoms produced are highly varied and thought to be dependent on the duration of favorable conditions. The most obvious symptom of an active infection is a wet, soft rot of the storage root combined with a distinct sour odor. The best method of management is avoiding high moisture, high temperatures, and low oxygen environments in all stages of sweetpotato production.

Rhizopus Soft Rot: Rhizopus soft rot is typically observed in post-harvest handling and shipping. Symptoms originate at a wounded area of the sweetpotato and consist of a soft, watery rot that progresses quickly under favorable conditions and can result in full decay of an infected root. Rhizopus stolonifer requires wounds in the sweetpotato root, particularly bruising wounds, for the disease to initiate. Disease progression and sporulation are then significantly influenced by storage conditions. Reduction of wounds as well as proper sanitization of equipment during harvest, storage, and packing can greatly reduce disease outbreaks. Limited fungicides are available for chemical management.

Southern Blight: Southern blight is a foliar disease affecting plants in greenhouse and field beds. Symptoms are usually observed around isolated circular hot spots which can spread over time. Symptoms manifest with a soft rot affecting the seed root, followed by the sudden wilt and subsequent death of a plant after sprouts have breached the soil line. Thick, white mycelial mats will begin to grow on the soil surface soon producing white to tan colored sclerotia that serve as survival structures. The causal agent, *Athelia rolfsii*, is a prominent soil microbe as it can infect many host plants and can survive in the soil without a host for many years. Use disease free fields for bedding and cut slips at least 2 inches above the soil line. Limited fungicides are available for chemical management.

Scurf: Scurf, caused by *Monilochaetes infuscans*, grows only in the skin of the sweetpotato and produces spores on the surface under high humidity conditions, typically in storage. The diseased areas are purple or grayish-brown to black, are only skin-deep, and enlarge slowly during storage. In the field, scurf is only found under the soil line. *M. infuscans* can survive in the soil for 1 to 3 years. Use disease free fields and seed roots for beds or treat seed roots with fungicide if soil is infested. Sanitize contaminated equipment.

NEMATODE MANAGEMENT
Guava Root-knot Nematode: Guava root-knot nematodes, *Meloidogyne enterolobii*, as well as other root-knot nematodes can infect sweetpotato. Nematode populations are highest in the late summer and fall. Collect soil samples for nematode analysis during this period from fields to be planted with sweetpotato the following year. Clean and sanitize equipment between fields.

Use crop rotations and non-host cover crops, when available, to reduce nematode populations. *M. enterolobii* has been able to overcome some of the resistance genes previously effective for other root-knot nematodes. Avoid fields with high populations of nematodes or use appropriate fumigants and other nematicides for management (see fumigant and nematicide tables).

INSECT MANAGEMENT
Lepidoptera Larvae: Sweetpotato hornworm, corn earworm, southern armyworm, yellowstriped armyworm, beet armyworm, fall armyworm, and soybean looper all feed on foliage leaving small to large holes. In plant beds and newly set fields, damage may be serious. Mid to late season foliar feeding may reduce yields or delay sizing of roots when coupled with plant stress. After harvest, larvae may continue feeding on sweetpotatoes left in the field and in storage. Apply insecticide to plant beds and in fields as needed. Cuttings should be free of insects before planting. Where worms are abundant at harvest, spray fields 2 to 3 days before digging. Remove harvested sweetpotatoes from the field immediately.

Cucumber Beetles (rootworms): Adults and larvae of the banded cucumber beetle, *Diabrotica balteata*, and the spotted cucumber beetle, *Diabrotica undecimpunctata* feed on sweetpotato. Both species are highly mobile and will also feed on several other host plants including, various vegetable plant species, soybeans, and corn. Adult beetles feed on sweet potato foliage, creating irregular holes in the leaves. Adult beetles lay eggs in the soil and larvae developing in the soil feed on developing sweet potato roots. Feeding on the roots can occur throughout the production season, but damage from these insects increases late season. Feeding injury results in unsightly blemishes on the roots at harvest. The larval stage lasts from 8-30 days depending on the temperature and food supply. Pupae are found just below the soil surface. Adults will emerge in approximately one week. Numerous generations of these insects can develop and injure sweet potatoes throughout the production season. Soil applied insecticides can reduce damage from these insects if applied close to planting. Adults should be scouted weekly during the production season and labeled insecticides should be applied when the number of beetles sampled reaches or exceeds the treatment threshold of 2 beetles/100 sweeps.

Tortoise Beetle: Generally, damage by tortoise beetles threatens newly set plants or plants under stress. Leaves of infested plants are riddled with large, round holes. Adults and larvae which feed on sweetpotato foliage include: mottled tortoise beetle, striped tortoise beetle, and argus tortoise beetle, blacklegged tortoise, and golden tortoise beetle. Isolate plant beds and control morningglory. Monitor movement of ornamental sweetpotatoes which often contain tortoise beetles and other insects. Apply insecticides to young plants if needed. Control beetles in plant beds and fields.

Sweetpotato Weevil: This is the most serious worldwide pest of sweetpotatoes, however, this pest is effectively management in commercial production areas where it occurs in the US, due to comprehensive IPM programs that include (monitoring with pheromone traps, field sanitation, cultural practices, clean seed programs and the effective use of labeled insecticides). Adults and larvae feed on foliage, but prefer stems and roots. Infested sweetpotatoes are riddled with small holes and galleries especially in the stem end. They turn bitter and are unfit for consumption by either humans or livestock. Use only "seed" and plants produced in approved and trapped weevil-free areas. All purchased roots/plants, including those produced out-of-state, must be certified. Use pheromone traps in plant beds, greenhouses, and in fields to detect sweetpotato weevil. Some varietal tolerance exists. Chemical control with weekly or biweekly sprays is difficult. Most states with commercial production areas of sweetpotato have strict quarantine rules in place for sweetpotato weevil. Chemical control with biweekly or weekly sprays can be difficult to attain depending on other factors and should be used in conjunction with other management tools as indicated above, in areas where the weevil does occur.

Sweetpotato Flea Beetle: Adult beetles overwinter in debris, along fence rows, and at the edges of wooded areas. In the spring, eggs are laid in the soil near host plants. There are several generations per year. Adults feed on foliage leaving channels on the upper leaf surfaces. Larvae feed on roots etching shallow, winding, sunken trails on the surface, which enlarge, darken and split. Monitor adults with yellow sticky cups. Control morningglories and weeds along field margins and plow under crop debris. Use resistant or tolerant varieties. In fields with a history of infestation use a preplant or a side-dressed soil insecticide over the foliage up to the last cultivation. Control adults with insecticides.

Whitefringed Beetle: Larvae feed on roots causing damage similar to that of wireworms and white grubs. Only flightless, female adults occur and feed at the base of plants leaving scars on the stem. They also feed and notch leaves. They are most active in July and August and produce eggs in groups without mating. Avoid infested fields and rotate crops. Only grasses are not suitable as hosts. Monitor for adults or leaf notching. Limited control may be achieved by using tolerant varieties, foliar insecticides applied every two weeks and soil insecticides. Record whitefringed beetle sites and do not plant sweetpotatoes in these locations.

Wireworms: Tobacco wireworm, southern potato wireworm, corn wireworm leave small, irregular, shallow or deep holes in the surface of sweetpotato roots. Larvae are identified by differences in their last abdominal segment. Wireworm adults (click beetles) lay their eggs in grassy, undisturbed soil. Adults feed on weed seeds (pigweed) and corn pollen. Avoid land previously in sod or fallow. Wireworms may be detected prior to planting using corn, wheat, or oatmeal bait stations. If necessary, broadcast and incorporate a preplant insecticide, or use a granular material at root swell. Timed foliar sprays are of limited value, as adults do not feed on sweetpotato and are only controlled when sprays contact adults or larvae move into a treated area. Control weeds and do not allow them to mature to seed. Avoid planting in fields with corn wireworm. Avoid planting behind corn, grain, and grain sorghum. Tobacco wireworm adults can be monitored with yellow sticky cups. Wireworm adults are attracted to black-light insect traps.

White Grubs: These can cause large, shallow, irregular damage on the surface of sweetpotatoes. Species include Japanese beetle, spring rose beetle, and green June beetle. Adults lay eggs in grassy areas (also see section on wireworms). Pheromone traps are under evaluation. Japanese beetles are attracted to traps. White bucket traps attract spring rose beetles. Use a preplant insecticide and foliar sprays when adults are active.

Fruit Fly: Fruit flies may be a nuisance in storage houses when sweetpotatoes decay due to other causes such as souring, chilling, and Rhizopus soft rot. Fruit flies feed on decaying vegetables. Maggots may be seen in decaying roots. Fruit flies may become established in cull piles and spread to the storage house. They do not cause rots. Harvest, cure and store only sound sweetpotatoes. Dispose of culls, inspect the storage house and use traps. If necessary, spray with an appropriate insecticide.

HARVESTING AND STORAGE

A 3 to 4-month growing season is required for root development. After the roots are dug, they should be cured in the storage house at 80 to 85°F and 90% relative humidity for 6 to 8 days. Curing time can be adjusted for varieties that have a propensity to sprout quickly under curing conditions following harvest. After curing, the temperature should be lowered to 55°F, but relative humidity should be maintained at 85%. Temperature should never go below 55°F, or chilling injury may result, depending on the length of exposure. Above 60°F, sprouting will occur and root weight decrease. See Table 14 for further postharvest information.

TOMATOES (*Solanum lycopersicum*)

VARIETIES [1]	AL	AR	GA	KY	LA	MS	NC	SC	TN	VA
TOMATOES										
Fresh Market										
Amelia VR [2, 6, 7, 8, 10, 11, 13]	A	R	G	K	L	M	N	S	T	V
Bella Rosa [2, 3, 5, 7, 8, 11, 13]	A		G		L	M				V
BHN 589 [6, 7, 13, 15, 18]	A			K		M			T	V
BHN 602 [2, 6, 7, 8, 13]	A	R	G	K	L	M	N	S	T	V
Big Beef [5, 7, 8, 10, 11, 13, 15]		R		K	L	M				
Carolina Gold (yellow fruit) [7, 8, 13]	A	R	G		L		N		T	V
Celebrity [7, 10, 13, 18]	A	R		K	L		N		T	
Camaro [3, 5, 7, 8, 8a, 11, 13, 16]			G							
Defiant PhR [7, 8, 13, 14, 17]	A	R					N	S		
Florida 47R [5, 7, 8, 11, 13]	A	R	G	K	L	M	N	S	T	V
Florida 91 [3, 5, 7, 8, 11, 13]					L	M			T	V
Grand Marshal [5, 7, 8, 13, 16]	A		G					S		V
Jolene [7, 8, 9, 13, 16]	A		G				N	S	T	
Mountain Gem [2, 7, 8, 13, 16, 17, 18]	A					M	N		T	
Mountain Magic (Campari) [6, 7, 8, 13, 14, 17]	A	R	G	K		M			T	
Mountain Majesty [2, 7, 8, 13, 18]	A		G				N	S	T	
Mountain Merit [2, 6, 7, 8, 10, 13, 17]	A	R								V
Mountain Rouge [10, 17]	A				L	M	N	S		
Mountain Spring [7, 8, 13, 18]	A	R	G	K		M	N			V
Patsy [2, 7, 8, 9, 13, 16]	A									
Phoenix [3, 5, 7, 8, 11, 13]	A	R	G	K	L		N		T	V
Primo Red [2, 7, 8, 13, 15]	A			K			N		T	V
Red Bounty [2, 7, 8, 10, 11, 13]	A		G					S		
Red Defender [2, 5, 7, 8, 11, 13, 18]	A	R	G	K	L				T	
Red Deuce [5, 7, 8, 13, 15, 18]				K		M				
Red Morning [2, 7, 8, 13, 15]	A			K					T	
Red Mountain [2, 7, 8, 10, 13]	A		G							
Red Snapper [2, 5, 7, 8, 13, 16]	A		G				N	S		V
Rocky Top [6, 7, 8, 11, 13, 18]	A			K		M	N	S		V
Roadster [2, 5, 6, 7, 8, 9, 11, 13, 22]	A							S		V
Cherry Types										
Matt's Wild Cherry [14, 17]	A							S	T	
Sun Gold (yellow fruit) [7, 8, 12]	A	R	G	K	L	M	N	S	T	V
Sun Sugar (yellow fruit) [7, 8, 12]	A	R	G	K	L	M	N	S	T	
Grape Types										
Brad's Atomic Grape	A					M		S	T	
Cupid [5, 7, 11, 21]	A		G			M			T	
Elfin [4]	A		G				N	S		
Gold Spark [11, 15, 19a-e] (yellow fruit)	A									
Mountain Honey [2, 4, 6, 7, 8, 13, 17]	A			K			N	S		
Mountain Vineyard [2, 4, 6, 7, 8, 11, 13]	A						N			
Smarty [7, 8, 13]	A	R	G	K			N	S		
Roma Types										
Daytona [5, 7, 8, 10, 13, 16]	A		G							
Granadero [2, 6, 7, 10, 11, 13, 15, 20]	A		G					S		
Mariana [5, 7, 8, 10, 11, 13]	A						N		T	

[1] Abbreviations for state where recommended
[2] Tomato spotted wilt virus resistance (TSWV)
[3] Heat set (heat tolerance)
[4] Determinant or short internode grape tomato
[5] Alternaria stem canker tolerance/resistance (ASC)
[6, 7, 8, 8a] Fusarium Wilt race 0, 1, 2, 3 tolerance/resistance (F)
[9] Fusarium crown root rot tolerance/resistance (FCRR)
[10] Nematode resistance (N)
[11] Gray leaf spot resistance (SL, SBL, SS)
[12] Tobacco mosaic virus resistance (TMV)
[13] Verticillium wilt resistance (V)
[14] Early blight tolerance/resistance
[15] Tomato mosaic virus tolerance/resistance (ToMV)
[16] Tomato yellow leaf curl virus tolerance/resistance (TYLCV)
[17] Late blight tolerance/resistance
[18] Suitable for high tunnel production
[19a-e] Tomato leaf mold race a, b, c, d, e tolerance/resistance
[20] Powdery mildew tolerance/resistance
[21] Bacterial speck tolerance/resistance (BSK-0)
[22] Tomato brown rugose fruit virus (ToBRFV) tolerance/resistance

VARIETIES [1]	AL	AR	GA	KY	LA	MS	NC	SC	TN	VA
TOMATOES										
Roma Types (cont'd)										
Picus [2, 5, 7, 11, 13]	A	R	G					S		
Plum Regal [2, 7, 8, 13, 14, 17]	A	R		K	L		N	S	T	
Pony Express [6, 7, 8, 10, 13, 15, 21]				K					T	
Greenhouse Types – Beefsteak										
Bigdena [7, 8, 9, 12, 13, 15, 19a-e]	A		G	K	L	M			T	
Geronimo [7, 8, 12, 13, 19a-e, 20]	A		G	K	L	M	N	S	T	
Torero [7, 8, 9, 13, 15, 19a-e, 20]	A				L	M				
Trust [7, 8, 9, 12, 13, 19a-e]	A	R	G	K		M	N	S	T	

[1] Abbreviations for state where recommended
[2] Tomato spotted wilt virus resistance (TSWV)
[3] Heat set (heat tolerance)
[4] Determinant or short internode grape tomato
[5] Alternaria stem canker tolerance/resistance (ASC)
[6, 7, 8, 8a] Fusarium Wilt race 0, 1, 2, 3 tolerance/resistance (F)
[9] Fusarium crown root rot tolerance/resistance (FCRR)
[10] Nematode resistance (N)
[11] Gray leaf spot resistance (SL, SBL, SS)
[12] Tobacco mosaic virus resistance (TMV)
[13] Verticillium wilt resistance (V)
[14] Early blight tolerance/resistance
[15] Tomato mosaic virus tolerance/resistance (ToMV)
[16] Tomato yellow leaf curl virus tolerance/resistance (TYLCV)
[17] Late blight tolerance/resistance
[18] Suitable for high tunnel production
[19a-e] Tomato leaf mold race a, b, c, d, e tolerance/resistance
[20] Powdery mildew tolerance/resistance
[21] Bacterial speck tolerance/resistance (BSK-0)
[22] Tomato brown rugose fruit virus (ToBRFV) tolerance/resistance

Seed Treatment. Seed should be treated to minimize the occurrence of bacterial and viral pathogens. If the seed you purchase is not treated, you can do it yourself. See Seed Treatments in the disease management section of this handbook for details on how to treat your seed.

TOMATO PLANTING DATES*

	Spring	Fall
AL North	4/15–6/15	7/1-8/1
AL South	3/1–4/30	7/15–8/15
AR North	5/15-6/20	7/1-7/15
AR South	4/1–6/15	7/15-8/1
GA North	4/15–6/15	7/1–8/1
GA South	3/1–4/30	7/15–8/30
KY East	5/15–6/1	NR
KY Central	5/5–6/15	NR
KY West	4/20-7/1	NR
LA North	3/15–6/30	7/1–8/10
LA South	3/1–6/30	7/15–8/15
MS North	4/20–6/30	NR
MS South	3/1–3/15	NR
NC East	4/15–5/10	8/1–8/15
NC West	5/15–7/15	NR
SC Coastal Island	3/1-4/30	7/1-7/15
SC East	3/15–4/30	7/1–7/15
SC West	5/1–6/30	NR
TN East	4/15-6/15#	7/1-7/15
TN West	4/1–6/15#	7/1-7/15
VA (coastal)	4/15–5/31	6/1–7/15
VA (mountains)	4/10–7/30	NR

* Using transplants. Note: Planting dates in the spring could be earlier if low tunnels or other season extension measures are used in some locations.
# Early planting date based on average frost-free date. Adjust for your location.
NR = Not recommended

Hardening Transplants. Harden tender tomato seedlings prior to planting them in the field. The ideal method to harden tomato transplants is by intermittently withholding water over a period of several days prior to transplanting or exposing seedlings to direct sunlight for several days before planting. Allow seedlings to wilt slightly between waterings. Additionally, research has shown that hardening tomato seedlings by exposure to cool temperatures (60 to 65°F/ day and 50 to 60°F/night) for one week reduces the incidence of catfacing. Never harden tomato transplants by withholding fertilizer from them.

Fertilization and Fertigation. Pre-plant fertilization, liming, and fertigation should be based on a soil test. In the absence of a soil test, adjust soil pH to 6.0 to 6.5 and apply enough pre-plant fertilizer to supply 50 lbs per acre of N and K_2O and 125 lbs per acre of P_2O_5. Be sure to thoroughly incorporate fertilizer and lime into the soil. On soils testing low to low-medium in boron, also include 0.5 pound per acre of actual boron. Be sure to fully incorporate the pre-plant fertilizer to the soil, without leaving any on the surface of the bed.

The following table provides two suggested fertigation schedules for tomatoes depending on the level of potassium in the soil. Begin the soluble fertilizer program within a week after transplanting. Timing of applications in this schedule is based on the relative growth of the crop. You might need to advance or retreat timing based on the crop's development. Tissue analysis/leaf tissue sampling will help you fine tune your application times to match your crop's need. Continue fertigating until one week before the last harvest.

SUGGESTED FERTIGATION SCHEDULE FOR TOMATO* (low soil potassium)

Days after planting	Daily nitrogen	Daily potash	Cumulative	
			Nitrogen	Potash
	(lb / A)			
Preplant			50.0	125.0
0–14	0.5	0.5	57.0	132.0
15–28	0.7	1.4	66.8	151.6
29–42	1.0	2.0	80.8	179.6
43–56	1.5	3.0	101.8	221.5
57–77	2.2	4.4	148.0	313.9
78–98	2.5	5.0	200.5	418.9

SUGGESTED FERTIGATION SCHEDULE FOR TOMATO*
(high soil potassium)

Days after planting	Daily nitrogen	Daily potash	Cumulative Nitrogen	Cumulative Potash
	(lb / A)			
Preplant			50.0	125.0
0–14	0.5	0.5	57.0	132.0
15–28	0.7	0.7	66.8	141.8
29–42	1.0	1.0	80.8	155.8
43–56	1.5	1.5	101.8	176.5
57–77	2.2	2.2	148.0	223.0
78–98	2.5	2.5	200.5	275.5

*Adjust based on tissue analysis.

Plasticulture. Yield, fruit size, and quality of fresh market tomatoes can be increased and disease and weed issues reduced by using raised beds (16 to 24 inches at the base by 8 to 12 inches in height) covered with plastic mulch in combination with drip irrigation, on-center, under the plastic. Early planted crops usually benefit from soil warming provided by black plastic mulch. By mid-June, or when air temperatures exceed 85°F during the growing season, use white on black, fully white, or reflective plastic mulch. Paper mulch is also available for summer planting, but the length of the roll is limited due to the increased weight compared to plastic mulch.

Grafted Tomatoes. The use of grafted tomato plants is a strategy to combat soil-borne diseases such as bacterial wilt, fusarium wilt, verticillium wilt as well as other diseases and abiotic stresses, especially on heirloom tomatoes. Grafted plants are more expensive per transplant. The following table provide a current list of commonly used rootstocks and their diseases resistance. For more information, see the *Grafting in Vegetable Crops* section in this handbook.

Greenhouse Tomatoes. If you plan on growing tomatoes to maturity in the greenhouse, select proven greenhouse varieties. These varieties have been bred specifically for greenhouse conditions and have better disease resistance than field types grown in the greenhouse environment. Nearly all greenhouse varieties are indeterminate hybrids so that they will yield over a long harvest season. While non-greenhouse types (determinate) will grow in the greenhouse, the yield and quality will usually be reduced or compressed into a shorter production cycle, and therefore, their profitability might be strongly influenced by short-term prices.

Variety selection is based on yield, fruit size, uniformity, disease resistance, and lack of physiological disorders, as well as the market demand for the type of tomato grown. For suggestions on varieties, see the variety table in this section above. Insect and disease control methods for greenhouse vegetables can be found in Tables 2-29 and 3-45 (Insect and Diseases Control sections, respectively). For further information on greenhouse tomato production, see *https://extension.msstate.edu/sites/de-fault/files/publications/publications/p1828_web.pdf*

Ground Culture. Growing plants directly on the ground in the wet, humid Southeast is not recommended. If you choose to do so, space *determinate* varieties in rows 4 to 5 feet apart with plants 18 to 24 inches apart in the row. For *indeterminate* varieties, space rows 5 to 6 feet apart with plants 24 to 36 inches apart in the row.

Staked Culture. Staking tomatoes is a specialized production system highly recommended for the growing conditions in the Southeast. The recommendations below are for the *short-stake cultural system* using determinate cultivars that grow 3 to 4 feet in height. Use between-row spacings of 5 to 6 feet and in-row spacings of 18 to 24 inches. By employing longer stakes, the short-stake cultural system can also be used for indeterminate varieties.

Pruning/Suckering: Pruning is practiced to establish a desired balance between vine growth and fruit growth. The extent of pruning needed is dictated by the growth habit of the variety being grown. For most commercial varieties, little to no pruning results in a plant with a heavy load of smaller fruit, whereas moderate pruning results in fewer fruits that are larger and easier

COMMERCIALLY AVAILABLE TOMATO ROOTSTOCKS AND THEIR DISEASE RESISTANCE

Rootstock	Southern RKN [1]	Bacterial wilt	Corky root rot	Fusarium crown & root rot	Fusarium wilt [2]	Verticillium wilt [3]	Tomato mosaic virus	Organic seed available	Untreated seed available
Armada		X							X
Arnold	I		X	X	1,2	0,1,2	X		
Balance	I		X	X	1,2,3	1,2	X		X
Booster									
Bowman	I	X	X	X	0,1,2,3	1,2	X		
DRO141TX	I		X	X	2	0	X		X
Estamino	I		X	X	0,1,2,3	0	X	X	X
Fortamino	I		X	X	0,1,2	0	X	X	X
Maxifort	T		X	X	0,1,2	1	X		X
Multifort	T		X		0,1,2	1			X
Shin Cheong Gang	H	X	X	X	1,2,3	0			
Submarine	I			X	1	0	X		
Synergy	I			X	1,2,3	1			

[1] Southern root-knot nematode (RKN) is the most damaging plant-parasitic nematode for tomato. High resistance (H): indicates low to zero infection rate of RKN; intermediate resistance (I): indicates moderate infection rate of RKN; tolerant (T): indicates high infection rate while still maintaining high fruit yield.
[2] There are currently four races of Fusarium wilt (Race 0, 1, 2, 3, 4). If known, the resistance to each race is noted on each rootstock.
[3] There are three races of Verticillium wilt (Race 0, 1, 2). If known, the resistance to each race is noted on each rootstock.

to harvest. Pruning can cause earlier maturity of the crown fruit and improved spray coverage and pest control.

Removing all suckers (side-shoots) up to the one immediately below the first flower cluster is adequate for most determinate cultivars. Removing the sucker immediately below the first flower cluster or pruning above the first flower cluster can result in severe leaf curling and stunting of the plant and should be avoided for most varieties.

Prune when the suckers are no more than 2 to 4 inches long. A second pruning may be required to remove suckers that are too small to be easily removed during the first pruning and to remove ground suckers that may develop. Pruning when suckers are too large requires more time and can damage the plants, delay maturity, and increase disease incidence. Do not prune plants when they are wet to avoid spread of diseases. Pruning should be done before the first stringing because the string can slow the pruning process. Pruning is variety-and fertility-dependent.

Less-vigorous determinate cultivars generally require less pruning. Growers should experiment with several degrees of pruning on a small scale to determine pruning requirements for specific cultivars and cultural practices.

Staking: Staking improves fruit quality by keeping plants and fruit off the ground and providing better spray coverage. Staked tomatoes are also easier to harvest than ground tomatoes. Staking tomatoes consists of inserting a series of stakes within the plant row and weaving twine around the stakes to train the plants to grow vertically off the ground. This method is commonly referred to as the Florida-Weave.

In the traditional Florida-Weave system, wooden stakes (4 to 4.5-feet long by 1-inch square) are driven into the soil between plants in the row. A stake placed between every other plant is adequate to support most determinate varieties. Vigorous cultivars may require larger and longer stakes and indeterminate varieties may require a stake for every plant. Placing an additional stake at an angle and tied to the end stake of each row section will strengthen the trellis system. Stakes can be driven by hand with a homemade driving tool or with a commercially available, power-driven stake driving tool. Drive stakes to a consistent depth so that spray booms can be operated in the field without damaging the trellis system.

The use of fiberglass for stakes is growing in popularity because they are durable, do not rot, are somewhat flexible (can withstand strong winds), lightweight, can be easily sanitized, and are easier to store than wood. Fiberglass stakes are made of fiberglass and resin and are ¼ inch to an inch in diameter. There are fiberglass stake-pullers available that are not as damaging as the wood stake pullers.

Select "tomato twine" that is resistant to weathering and stretching and that binds well to the wooden or fiberglass stakes. Tomato twine is available in 3- to 4-pound boxes. Approximately 30 pounds of twine is required per acre. To make tying convenient, use a homemade stringing tool. This tool can be made from a length of metal conduit, schedule 40 PVC pipe, broom handle, or wooden dowel. With conduit and PVC pipe, the string is fed through the pipe. With a broom handle or wooden dowel, two small parallel holes, each about 1 inch from the end, must be drilled to feed the string through one hole along the length of the tool and through the other hole. The tool serves as an extension of the worker's arm (the length cut to the worker's preference) and helps to keep the string tight.

Proper stringing consists of tying the twine to an end stake passing the string along one side of the plants, and then looping the twine around each stake until the end of a row or section (100-foot sections with alleys may be helpful for harvesting) is reached. The same process continues on the other side of the row. The string tension must be tight enough to hold the plants upright. Note: If strings are too tight, they can make harvesting fruit difficult and can scar fruit.

The first stringing should be 8 to 10 inches above the ground when plants are 12 to 15 inches tall and before they fall over. Run the next string 6 to 8 inches above the preceding string before plants start to fall over. Three to four stringings are required for most determinate varieties. Stringing should be done when the foliage is dry to prevent the spread of diseases.

Heirloom Tomatoes. Heirloom tomatoes are varieties that have been available for 50 years or more, are open-pollinated, and grow "true to type" from seed saved from fruit each year. They are generally indeterminate, requiring trellising and constant pruning. Most varieties have little disease resistance. The fruit are usually thin-skinned, soft, and tend to crack. Consumers are attracted to heirloom tomatoes because many varieties are very flavorful, colorful, come in many sizes and shapes, and have interesting names. For growers, heirloom tomatoes are challenging to produce and difficult to ship but can bring high prices on the local market.

There are hundreds of varieties of heirloom tomatoes available. Some of the most popular varieties include Brandywine, German Johnson, Mr. Stripey, Cherokee Carbon, Cherokee Purple, and Green Zebra. Tomato breeders have developed several improved versions of several heirloom types by incorporating traits from more modern varieties into heirlooms. These have the benefit of heirloom characteristics with improved disease resistance and fruit quality. Most of these are hybrids, referred to as heirloom-type hybrids, and will not come back true to type from saved seed.

Because most heirloom tomatoes are indeterminate, they must be grown in a very sturdy tall-stake system or on a tall, strong trellis. A trellis can be constructed of 3-inch diameter, or larger, posts set 10-15 feet apart within the row. Use 7-8 ft. long posts, leaving 6-7 ft. above ground. Run a stout wire (12 gauge) across the tops of the posts and secure it with staples. Pieces of twine, long enough to reach the ground, should be tied to the top wire above each plant. The twine can be anchored with a loop to each plant or to a bottom line of twine that is strung about 6 in. off the ground and secured to the posts. To use the standard string and weave-staked culture system for heirloom tomatoes, as described for the determinate tomatoes, use 6-ft. long stakes instead of the normal 4-ft. long stakes, and double stake the ends.

In a trellis system, plants are usually spaced 8-10 in. apart within the row and pruned to a single stem. A two-stem system may also be used, in which the plants should be spaced 18-30 in. apart within the row. If using a standard staking system, plants should be spaced 18-24 in. apart. Once the plants are established, suckers must be removed several times a week. If the main growing point is broken off, a sucker can be trained to take its place.

Because most heirloom tomatoes have little disease resistance, it is important to maintain a good fungicide spray schedule. For organic production, it might be necessary to grow heirloom tomatoes under high tunnels, especially in areas with high disease pressure. Grafting heirloom varieties onto diseases resistant rootstocks might also increase your success at growing organically. Disease resistant heirloom-type hybrids are often successful in organic systems when paired with a timely spray program.

TOMATO DISORDERS

Your local county Extension office has bulletins that describe fruit disorders in detail. Here are several common disorders of tomato and their causes:

- **catfacing** – cool day and/or night temperatures or very hot dry days
- **internal browning, graywall and blotchy ripening** – tobacco mosaic virus, overcast cloudy environment, high N, low K or soil compaction
- **yellow shoulder** – direct sun exposure, worse on green shouldered varieties
- **sunburn and sunscald** – direct rapid exposure to the sun
- **weathercheck** – fruit exposed to dew and/or rain
- **blossom end rot** – localized calcium deficiency in the fruit often initiated by moisture stress
- **cracking** – variety, irregular water, growth, and/or nutrition

Check out the following link for images and discussions for several common problems in tomato production -- http://extension.msstate.edu/publications/publications/tomato-troubles-common-problems-tomatoes

SPECIAL NOTES ON PEST MANAGEMENT
INSECT MANAGEMENT

Colorado Potato Beetle (CPB), Flea Beetles (FB): While flea beetles are a common pest of tomato throughout the southeastern US, Colorado potato beetles are most common in areas where significant acreage of potatoes is also grown. Flea beetles are primarily a problem early in the season shortly after planting and are usually controlled by insecticides applied to control other insects. Adults feed on foliage, resulting in small round holes on leaves. In most situations this damage does not affect early season growth or subsequent yields, but control may be necessary when populations are high (20-30% defoliation).

Colorado potato beetle adults and larvae feed on tomato foliage and can cause extensive defoliation if not controlled. CPB feed only on solanaceous plants, and populations tend to be concentrated in areas where potato, eggplant, and tomato have previously been grown. Consequently, rotation to non-solanaceous crops is very effective in helping to avoid infestations. Thoroughly scout fields and spray only when necessary. Treatment should be made if populations exceed 15 adults per 10 plants or a combination of 20 CPB larvae and/or adults per 10 plants.

Insecticide sprays should be made after most egg masses have hatched, but before larvae become large. CPB have developed resistance to many different insecticides, so knowledge of the resistance status of populations in your area is essential in choosing which insecticides to use.

Tomato Fruitworm: The tomato fruitworm, also known as the corn earworm and cotton bollworm, is potentially the most damaging pest of tomato. However, there are many insecticides that provide excellent control. The key to controlling this insect is to ensure that there is a toxic pesticide residue on the plant during egg laying periods so that larvae are killed shortly after hatching, because larvae feed on leaf tissue for only a short time before boring into fruit. Tomato fruitworm moth activity can be monitored with pheromone traps and serves as a measure of the adult population within an area. Corn that is in the silking stage is a preferred host of fruitworm, but when corn silks begin to dry, moths will switch egg laying to other hosts, including tomato.

Armyworms: At least three species of armyworms are potential pests of tomato, including the beet armyworm, southern armyworm, and yellowstriped armyworm. Infestations are usually sporadic in the northern regions of the southeastern US but are an annual problem in more southern areas. In contrast to tomato fruitworm, armyworms will feed extensively on foliage as well as fruit, and the presence of feeding damage on leaves can help differentiate between fruitworm and armyworm damage. Beet armyworm is notorious for exhibiting resistance to a wide range of insecticides, but the recent registration of newer insecticides has greatly aided the management of this pest.

Tomato Pinworm: The tomato pinworm is more common in the south compared with northern regions of the Southeast, but late-season infestations are common in northern areas. Moths lay eggs on foliage, and larvae feed within leaves, creating blotchy mines. As larvae age, they bore into stems and/or fruit. The use of pheromone-based mating disruption is an effective control method. Initiate mating disruption at the first sign of mines on foliage. Numerous insecticides also control pinworm.

Stink Bugs: The green and brown stink bug can be important direct pests of tomato, but they are sporadic in occurrence. Stink bugs are most common in smaller fields (i.e., 5 acres or less) that are surrounded by weedy borders, or fields that are adjacent to soybeans. In fact, chemical control of stink bugs is often not necessary in fields that do not fit the previous description. Unfortunately, there is not a good sampling method to assess population densities before damage occurs, and preventive strategies are used. Depending on the surrounding habitat and abundance of stink bugs within an area, one to three applications of an insecticide are necessary to prevent damage.

Thrips: Thrips can cause direct damage to tomato fruit by their feeding or oviposition scars on small fruits and are also indirect pests of tomato due to their ability to transmit tomato spotted wilt virus (TSWV). The tobacco thrips and western flower thrips are vectors of tomato spotted wilt virus. The majority of virus infections are the result of primary spread (thrips transmitting the virus from surrounding weeds directly to tomatoes), and insecticides do not kill thrips quickly enough to prevent inoculation. However, an aggressive insecticide control program early in the season (3 to 4 weeks after transplanting) and the use of reflective mulches can help to reduce the incidence of TSWV in tomatoes. Thrips can also cause direct damage to tomato fruit. This is the result of thrips feeding and/or laying eggs in small fruits before stamens are shed from flowers. This damage appears as small dimples in fruit. Sample thrips in tomato flowers by placing a white index card below flowers and tapping the flowers with a finger. An average of 1 thrips per flower has worked well as a treatment threshold level.

Whiteflies: The greenhouse whitefly and silverleaf whitefly can both infest tomatoes in the Southeast. Generally, the silverleaf whitefly is more common in the southern region and the greenhouse whitefly is more common in the northern region of the Southeast. Once whitefly populations of either species become established on a crop, they are very difficult to control. Therefore, preventive control is usually necessary for effective, season-long management. Preventive control can be achieved with soil-applied systemic insecticides or the application of other insecticides when populations are low.

Mites: Mites have become an increasingly important problem on tomatoes and other vegetables grown in the Southeast. Two-spotted spider mite is the most common mite pest, but the broad mite and carmine spider mite can also infest tomatoes.

Mites overwinter on weeds and move into tomatoes in the spring as weeds die. Mites can also move from other crops (including other tomato fields) into tomatoes throughout the season. Localized infestations can be spot treated, but thorough coverage of foliage is important. Mites can be sampled by using a sample of 10 leaflets (terminal leaflet on a leaf from the upper one-third of the plant), from a minimum of 5 sample sites per field. When mites reach an average of 2 mites/leaflet, a miticide should be applied. Note that certain pesticides, such as pyrethroids and some neonicotinoids, aggravate mite populations and can lead to high mite densities.

DISEASE MANAGEMENT

Accurate diagnosis is necessary for effective disease management. Incorporate appropriate and effective cultural practices and pesticides, as detailed in the Disease Management section, in a season-long integrated disease management program.

Soil-borne Diseases (bacterial wilt, Fusarium wilt, southern blight, etc.): Several soil-borne bacterial, fungal, and oomycete pathogens can significantly affect tomatoes. Many are difficult to manage when fields become infested. When possible, use resistant varieties or rootstocks. Limited fumigants and fungicides are available for some diseases. Pathogens can be spread via movement of contaminated soil, water, and equipment/tools. Clean and disinfest tools when moving between fields, when possible. Direct water flow away from other fields and sources of surface water for irrigation. Practice appropriate crop rotations with non-host crops.

Foliar and Fruit Diseases: Many foliar and fruit diseases are caused by bacterial, fungal, and oomycete pathogens. Use cultural practices that minimize water splash, leaf wetness, and canopy density, which contribute to pathogen infection and spread. Resistant varieties are available for some diseases, including early blight, gray leaf spot, late blight, and leaf mold. Preventative, season-long spray programs using products from multiple FRAC groups should be implemented prior to the development of symptoms and in conjunction with appropriate cultural practices.

Viruses: Once plants become infected with a virus, they cannot be cured. Chemicals are not an effective management option for virus infections. Many viruses are transmitted by insects, such as aphids (tobacco mosaic virus (TMV), cucumber mosaic virus (CMV), tobacco etch virus (TEV)), thrips (tomato spotted wilt virus (TSWV)), and whiteflies (tomato yellow leaf curl virus (TYLC)), and can cause significant losses in tomatoes. Other viruses may be mechanicallyor seed-transmitted. Use resistant varieties, when available, to manage viruses in areas where they are prevalent or when high insect vector pressure is expected. Generally, aphid-transmitted viruses cannot be adequately managed with insecticide applications; however, appropriately timed insecticide applications can be an effective in the management of thripsand whitefly-transmitted viruses. Use of reflective mulches may help with aphid-transmitted viruses. Avoid using tobacco products in operations producing TMV- or tomato mosaic virus (ToMV)-susceptible varieties. Purchase seeds from reputable sources, and do not save seeds from diseased plants.

NEMATODE MANAGEMENT

Various root-knot nematodes (RKN), such as the southern RKN and guava RKN, as well as other nematodes can infect tomatoes. Nematode infection may predispose plants to other soil-borne diseases, such as Fusarium crown and root rot, Fusarium wilt, and Phytophthora root rot. Sample fields the previous year when nematodes are active (late-summer or fall).

Some varieties and rootstocks have resistance to the southern RKN and should be used in infested fields. Clean and disinfest equipment between fields. Use appropriate crop rotations and cover crops with non-host crops to reduce nematode populations. Use appropriate fumigants and other nematicides for management (see fumigant and nematicide tables).

HARVESTING AND STORAGE

Tomatoes can be harvested at different stages of maturity, depending on the market. 'Mature green' stage is the earliest stage at which harvested fruit will continue ripening. Mature green fruit should have completely developed seeds that will not be cut when the fruit is sliced, and present gel formation in the locular area. Fruit harvested earlier will not fully ripen and have poor quality. Vine-ripened tomatoes will have better sensory quality (flavor and aroma) than those picked earlier but will have lower firmness, thus requiring careful handling to avoid mechanical damage. Tomato is susceptible to chilling injury when stored at temperatures below 55°F. Fruit harvested earlier will be more susceptible to chilling injury than at advanced stages of ripening. If held at 50°F, storage should be less than 14 days and less than 5 days if kept at 40°F. Chilling injury symptoms include failure to ripen and develop surface color, surface pitting, seed browning, flavor reduction, and increased susceptibility to decay.

Postharvest Diseases: Many postharvest diseases are the result of field infections. Harvest and cool fruits in a timely manner. Avoid harvesting overripe or damaged fruit. Handle fruits carefully. Follow appropriate storage conditions. See Table 14 for further postharvest information.

WATERMELON (*Citrullus lanatus*)

VARIETIES[1]	AL	AR	GA	KY	LA	MS	NC	SC	TN	VA
WATERMELONS										
Diploid, Open-pollinated										
AU Producer [A, IR]	A		G							
Crimson Sweet [MS]	A	R	G	K	L	M	N	S	T	
Jubilee II [IR]	A	R	G		L	M	N	S	T	
Diploid, Hybrid										
Escarlett [A, IR]							N	S		
Estrella [A, R]	A		G				N	S	T	
Jamboree [A, IR]			G		L		N	S	T	
Lemon Krush [A, 2, IR, PM]	A		G			M				
Nunhems 800 [A, R]	A						N	S	T	
Nunhems 860 [A, R]	A						N	S		
Sangria [A, IR]	A	R	G	K	L		N	S	T	
Sentinel [S]							N			
Starbrite [A, S]	A	R	G	K	L	M	N	S	T	
Top Gun [A, IR]	A		G				N	S		
Icebox										
Mickey Lee [R]	A		G		L	M	N	S		
Sugar Baby [S]		R		K	L	M			T	
Triploid/Seedless										
Affirmed	A			K			N	S		
Bottle Rocket [R]	A	R	G				N			
Buttercup [2, S]	A						N	S	T	V
Captivation	A	R	G				N	S	T	
Crackerjack [IR]	A	R	G				N	S		
Crunchy Red [A, S]	A	R	G	K	L		N		T	
Dark Knight [A, IR, PM]			G				N	S		
Exclamation [A, IR]	A	R	G	K			N	S	T	
Excursion [A, IR]		R	G					S		
Fascination [A, IR]	A	R	G	K			N	S	T	
Jet Ski [IR]		R	G					S		
Joy Ride [R]	A	R	G		L		N	S		
LaJoya [A, IR]							N	S		
Miramonte [R]		R	G				N	S		
Orange Crisp [3]	A			K		M	N		T	
Road Trip [A, R]	A		G		L		N	S		V
Superseedless 7167 [A]	A		G	K			N	S		
Superseedless 7197HQ [A]	A		G	K			N	S		
Sweet Dawn	A	R	G		L		N	S		
Sweet Gem	A		G	K			N	S	T	
Treasure Chest [2, 4]	A			K	L					
Tri-X 313 [A, S]	A	R		K	L	M	N	S	T	
Troubadour [A]	A	R	G	K			N	S	T	
Warrior [IR]			G					S		
Valor [A, IR]							N			
Firm Flesh Triploid/Seedless *										
Cato			G						T	
Crisp Delight [IR]								S	T	
Golden Crisp [IR, 2]			G					S	T	

[1] Abbreviations for state where recommended
[2] Yellow flesh fruit
[3] Orange flesh fruit
[4] Local markets only
[A] Anthracnose tolerance/resistance
[MS] Moderately susceptible to Fusarium wilt race 1
[IR] Intermediate resistance to Fusarium wilt race 1
[S] Susceptible to Fusarium wilt race 1
[R] Resistant to Fusarium wilt race 1
[PM] Powdery mildew tolerance/resistance

* Firm fleshed varieties withstand rigors of packaging and shipping. The flesh retains its water better than traditional varieties, resulting in less moisture loss. Reduced juice leakage improves the shelf-life of fresh-cut products, making these ideal for processors

VARIETIES[1]	AL	AR	GA	KY	LA	MS	NC	SC	TN	VA
WATERMELONS										
Firm Flesh Triploid/Seedless * (cont'd)										
Scarlet Crisp [IR]								S	T	
Shoreline [IR]								S	T	
Triploid Mini/Seedless Mini										
Extazy [S]		A	G				N	S	T	V
Leopard [S]				K			N		T	
Mielheart [S]		A	G				N			
Ocelot [S]			G	K			N	S		
Proxima [S]			G					S		
Sirius [IR]		A	G				N	S		

[1] Abbreviations for state where recommended
[2] Yellow flesh fruit
[3] Orange flesh fruit
[4] Local markets only
[A] Anthracnose tolerance/resistance
[MS] Moderately susceptible to Fusarium wilt race 1
[IR] Intermediate resistance to Fusarium wilt race 1
[S] Susceptible to Fusarium wilt race 1
[R] Resistant to Fusarium wilt race 1
[PM] Powdery mildew tolerance/resistance

* Firm fleshed varieties withstand rigors of packaging and shipping. The flesh retains its water better than traditional varieties, resulting in less moisture loss. Reduced juice leakage improves the shelf-life of fresh-cut products, making these ideal for processors

Seed Treatment. Check with seed supplier to determine if seed has been treated with an insecticide or fungicide. Be sure that seeds have been assayed for bacterial fruit blotch. Further information on seed treatments can be found under Seed Treatments in the disease management section of the handbook.

Direct Seeding & Transplant Production.
Direct seeding for diploids only: Seed when the soil temperature at the 4-inch depth is 60-65 °F. The recommended spacing for watermelons, both direct seeded and transplant, is 6-12 feet between rows providing 24-30 square feet per plant. Typically, a wider row spacing will produce a larger-sized watermelon.

Transplant production for triploids: For seedless watermelons, transplants should be grown in containers that provide a space of at least 1.5 inches by 1.5 inches for each plant. Smaller pots or cells will restrict root growth and provide less protection to the newly set transplants. If the seed is of good quality with high germination, one seed per pot is sufficient. The seed coat of seedless watermelons tends to adhere to the seedling as it emerges, at times slowing growth, or reducing stand.

For maximum germination, seedless watermelon seed should be planted with the point of the seed facing up (root end). However, this is not possible for any large-scale operation. Growing media should be pre-watered and allowed to drain before seeding and no additional moisture applied for at least 48 hours. Place trays in a germination chamber at 85-90°F for 48 hours. Next, place the trays in a greenhouse with day temperatures 70-80°F and night temperatures 65-70°F.

The required amount of seed for direct seeding and transplant production can be estimated using Table 6.

Planting. *Transplants:* Ideally, transplant container-grown plants into plastic mulch after the last killing frost. This is difficult to predict and planting dates vary, so consult the following table for your area. Plantings should be protected from winds with row covers, rye strips, or windbreaks.

Grafting and Diseases. In recent years, grafted watermelon plants have emerged as a strategy to combat soilborne diseases. Grafting is used against Fusarium wilt. Grafted plants are more expensive per transplant. For more information, see the section Grafting in Vegetable Crops in this handbook.

Pollination and Planting Arrangement with Triploids.
Fruit set and enlargement in watermelon is dependent upon growth regulators from pollen grains and from embryos in the developing seeds within the fruit. Inadequate pollination results in triploid fruit that are triangular in shape and of inferior quality. Additionally, inadequate pollination increases the incidence of hollow heart. Triploid flowers do not produce viable pollen, which is required to induce fruit set and development. Therefore, fields should be interplanted with pollenizer plants or diploid/seeded plants in order to provide sufficient, viable pollen for fruit set and development.

Several seed companies have developed new varieties for use solely as a pollenizer, non-edible special pollenizers. These pollenizers can be interplanted into a field totally devoted towards the production of triploid watermelons. Unique, compact growth habits prevent these pollenizer from competing for space with triploid plants.

The arrangement for incorporating pollenizer plants in a field of seedless watermelons has changed over the years. Currently most growers plant a pollenizer plant between every third or fourth plant within each row without changing the plant spacing of the seedless/triploid watermelon. Due to the less competitive nature of the special pollenizer plants mentioned above, they often work best with this seedless/pollenizer arrangement. However, growers can use a normal diploid/seeded watermelon as a pollenizer. The diploid/seeded watermelon must be marketable and have a fruit rind pattern and/or shape easily distinguished from the seedless fruit. Not all diploid/seeded watermelons are suitable for use as a pollenizer due to excessive plant vigor, reduced pollen production, and lack of disease resistance. (See recommendations below for Edible Diploids). It is also advisable that the grower reach an agreement with his harvesting crew as to how both types of watermelons will be handled at harvest.

Many growers prefer special pollenizers because they do not have markets for seeded watermelons. In addition, using a special pollenizer makes harvesting easier for crews who can more easily distinguish between fruit produced from the special pollenizer and fruit from the seedless watermelon varieties. Most fruit of the special pollenizer is much smaller (< 10 lbs) than the fruit produced by the seedless variety. If mini-seedless watermelons are planted, the rind pattern of both mini-seedless fruit and special pollenizer fruit will need to be easy to distinguish. Pollenizers found to work well in the southeast include:

	AL	GA	KY	LA	MS	NC	SC	TN	VA
Pollenizers for Triploid Watermelon									
Edible Diploids									
Crimson Sweet	A		K		M	N		T	V
Estrella	A				M	N			V
Escarlett [A, IR]						N	S		
Mickey Lee [MS]	A	G	K		M	N	S	T	V
Nunhems 800	A		K	L	M	N	S		V
Regency	A		K		M	N			
Sangria [A,IR]	A		K	L				T	V
Stargazer	A	G					S	T	V
Special Pollenizer (Non-palatable)									
Accomplice	A	G	K		M	N	S		
Ace Plus [A, IR]	A	G			M	N	S		V
Jenny	A	G			M	N	S		
Minipool	A	G						T	
SideKick	A	G	K		M	N	S		
SP-7 [A, IR, PM]	A	G				N	S	T	
Wild Card Plus [A, IR]	A	G				N	S		
Wingman [IR]						N	S		

[A] Anthracnose tolerance/resistance
[MS] Moderately susceptible to Fusarium wilt race 1
[IR] Intermediate resistance to Fusarium wilt race 1
[S] Susceptible to Fusarium wilt race 1
[PM] Powdery mildew tolerance/resistance

Be sure to follow the seed suppliers' instructions when using a special pollenizer. New, improved specialized pollenizer varieties are continually being developed with better germination, flowering habit, and/or disease resistances/tolerances. **Do not plant your pollenizer variety and seedless (triploid) varieties in separate or adjacent blocks. Plant your pollenizer variety within 10 to 15 feet of triploid varieties to assure good pollination. Specialized pollenizer varieties should be placed within 10 feet of triploids as these varieties tend to have less aggressive vining than normal seeded pollenizers.**

It is important that pollen from the pollenizer variety be available when the female blossoms on the triploid plants are open and ready for pollination. As a rule, a diploid pollenizer variety should be seeded on the same day that the triploid is seeded in the greenhouse. Some of the special pollenizers are slow to germinate; consequently, they might need to be planted earlier than the seedless. Check with your seed supplier for the grow out vigor of your special pollenizer.

Honey bees are important for high fruit yields and quality. Populations of pollinating insects may be adversely affected by insecticides app lied to flowers or weeds in bloom. Apply insecticides only when bees are not foraging in the evening hours or wait until bloom is completed before application. See section on Pollination in the General Production Recommendations for further information.

WATERMELON PLANTING DATES

AL North	5/15–6/15	MS North	4/15–5/15
AL South	3/1–6/30	MS South	2/15–5/1
AR North	5/15-6/20	NC East	4/15–6/30
AR South	4/15-6/15	NC West	5/25–6/30
GA North	5/15–6/15	SC Eastt	4/1–7/31
GA South	3/1–6/30	SC West	4/15–6/15
KY East	5/15-6/15	TN East	5/10-6/30
KY Central	5/5-7/1	TN West	4/20-7/30
KY West	4/20-7/15	VA East (coastal)	4/10-7/30
LA North	3/10–6/30	VA West (mountain)	4/20-7/20
LA South	3/1–7/5		

Drip Fertilization and Mulching. Before mulching, adjust soil pH to 6.5, and in the absence of a soil test, apply enough fertilizer to supply 50 pounds per acre of N, P_2O_5 and K_2O, (some soils will require 100 pounds per acre of K_2O) then thoroughly incorporate into the soil. On soils testing low to low-medium boron, also include 0.5 pound per acre of actual boron.

After mulching and installing the drip irrigation system, the soluble fertilizer program should then be initiated according to the following table. The first soluble fertilizer application should be applied through the drip irrigation system within a week after field transplanting or direct-seeding the melons. Continue fertigating until the last harvest.

SUGGESTED FERTIGATION SCHEDULE FOR WATERMELONS*

Growth Stage [1]	Days after planting	Daily nitrogen	Daily potash	Cumulative nitrogen	Cumulative potash
		(lb/A)			
Preplant				35.0	35.0
Planting to Vining	0 - 14	0.5	0.5	42.0	42.0
Vining to Flowering	15 - 28	1.0	1.0	56.0	56.0
Flowering to Fruit Set	29 - 49	1.5	1.5	86.0	86.0
Fruit Set to Initial Ripening	50 - 77	2.0	2.0	140.0	140.0
Harvest	78 - 91	1.0	1.0	153.0	153.0

* Adjust based on tissue analysis.
[1] Growth Stage can vary from season to season. For optimal results, fertigate watermelons based on their growth stage as opposed to days after planting.

MINI-SEEDLESS WATERMELON

The mini-seedless watermelon was introduced in 2003 and the demand for this product is year-round. These fruit generally range from 3 to 7 pounds and offer an attractive alternative for the consumer that has limited refrigerator space or a small family. Mini-seedless fruit must be handled carefully to minimize bruising. Some varieties of mini-seedless watermelons are grown under specific labels such as "PureHeart" but are only available under a contract basis. The varieties Extazy, Leopard, Sweet Bite, and Mielheart are readily available and are recommended for production in certain states. These varieties have performed well commercially or in University trials.

Plant spacing requirements for mini-seedless watermelon are considerably less than spacing requirements for normal seedless watermelon. Some mini seedless varieties will need to be planted close in order to maintain an average size of 7 pounds or less. As little as 10 square feet per plant might be needed. For example, if rows are spaced on 6 ft. centers, mini-watermelon plants should be spaced 1 ½ to 2 ft. apart in-row.

SPECIAL NOTES FOR PEST MANAGEMENT
DISEASE MANGEMENT

Accurate diagnosis is necessary for effective disease management. Incorporate appropriate and effective cultural practices and pesticides, as detailed in the Disease Management section, in a season-long integrated disease management program.

Fusarium Wilt: Fusarium wilt of watermelon is widespread throughout the southeastern US. Most varieties of watermelon, other than many heirloom varieties, are resistant to race 0. Many seeded (hybrid diploid) and seedless (triploid) varieties are resistant to race 1. Race 2 is present throughout Florida, Georgia, South Carolina and likely in other states. No watermelon varieties are resistant to race 2. Currently, grafting is the only cultural practice available to manage Fusarium wilt race 2. The superscripts "S" for susceptible, "MS" for moderately susceptible, "IR" for intermediate resistance, and "R" for resistant are listed next to each recommended variety. These superscripts indicate the reaction of commonly grown diploid and triploid varieties to Fusarium wilt race 1. Growers should choose resistant varieties whenever possible, including pollenizers for seedless watermelon production. Delaying transplanting until soil warms to 81°F at 4-inch depth will reduce the overall number of diseased plants.

Gummy Stem Blight: Gummy stem blight is the most common foliar disease on watermelon in the Southeast. Crop rotation away from cucurbits for two full years will improve the efficacy of most fungicides. Remove plant residue and/or bury crop debris with tillage as soon as possible after harvest. See Example Spray Program for Foliar Disease Control in Watermelon Production.

Cucurbit Downy Mildew (CDM): CDM can be a devastating disease of cucurbits in the southeastern and eastern U.S. Successful CDM management often requires the use of an appropriate fungicide spray program. Fungicide efficacy is influenced by crop, regional, and seasonal differences, as well as resistance management practices. A useful tool for downy mildew management (maps, text alerts, identification resources, etc.) is the CDM reporting system (http://cdm.ipmpipe.org). If you think you have CDM or have questions about disease management, please contact your local Extension office.

Phytophthora Fruit Rot: Phytophthora fruit rot can be a serious issue in watermelon in some locations, especially in production systems that use surface water and overhead irrigation. To minimize the occurrence of this disease, fields should be adequately drained to ensure that soil water does not accumulate around the base of the plants. Just before plants begin vining, subsoil between rows to allow for faster drainage following rainfall. Avoid moving infested soil between fields. Crop rotation is not effective. Resistance to some fungicides (e.g., mefenoxam) has been reported and should be considered when selecting fungicides. Apply appropriate fungicides when fruit are small.

Whitefly-transmitted Viruses: Several whitefly-transmitted viruses, including cucurbit chlorotic yellows virus (CCYV), cucurbit leaf crumple virus (CuLCrV), and cucurbit yellow stunting disorder virus (CYSDV), are emerging in various areas throughout the Southeast when unusually high populations of whiteflies are present. These viruses may produce similar symptoms; molecular testing is necessary for virus identification. These viruses often also occur in mixed infections. Management of whitefly-transmitted viruses focuses on prevention and management of the whitefly vector. Squash vein yellowing virus (SqVYV), which causes vine decline in watermelon, can also be problematic and can lead to near complete crop loss. Additional information on whitefly-transmitted viruses in cucurbits is available at *https://eCucurbitviruses.org*.

INSECT MANAGEMENT

Cucumber Beetle: Watermelons are resistant to bacterial wilt; however, control may be needed to prevent feeding damage to seedlings. Treat when an average of two beetles per plant is found. Insecticide seed treatments have systemic properties and will protect new seedlings for approximately 2 weeks.

Aphids: Aphid feeding can stunt young transplants and delay fruit maturation. Thorough spray coverage beneath leaves is important. Treat seedlings every 5 to 7 days or as needed. Insecticide seed treatments will protect seedlings from aphids.

Mites: Mite infestations generally begin around field margins and grassy areas. CAUTION: DO NOT mow or maintain these areas after midsummer because this forces mites into the crop. Localized infestations can be spot-treated. Note: Continuous use of pyrethroids may result in mite outbreaks.

Rindworm complex: Several insects can feed on the rinds, reducing fruit marketability. Insect feeding should not be confused with physical damage to rinds or disease symptoms. Rindworms primarily include various caterpillars and cucumber beetles (larvae and adults). Correct insect identification is critical in selecting the appropriate insecticide (Table 2-25). Cucumber beetle larvae feed at the fruit-soil interface. Caterpillar and adult beetle feeding can be visually differentiated. Contact your local Extension personnel for assistance with proper identification.

HARVESTING AND STORAGE

Watermelons must be harvested at their fully ripened stage. Avoid rigorous harvesting as fruit can be bruised or damaged. After harvest, watermelon is best stored in ridged bulk containers. Storage temperature can range from 45 to 59°F (7-15°C) and 85% relative humidity. The typical shelf-life of watermelon ranges from 2 to 3 weeks with some of the crisp varieties storing upwards of 4 weeks. Watermelons are sensitive to ethylene gas and should never be stored with crops that release ethylene. Exposure to ethylene can cause watermelons to lose quality and firmness. During storage, bulk bins should be checked for rotten or blemished fruit. See Table 14 for postharvest information.

Calibrating Chemical Application Equipment

PURPOSE
To determine if the proper amount of chemical is being applied, the operator must measure the output of the application equip- ment. This technique is known as *calibration*. Calibration not only ensures accuracy, a critical factor with regard to many chemicals, but it can also save time and money and benefit the environment.

GETTING STARTED
Careful and accurate control of ground speed is important for any type of chemical application procedure. From large self-propelled sprayers and spreaders to small walk-behind or backpack units, precise ground speed is a key for success. Ground speed can be determined by one of two methods. The first method requires a test course and stopwatch. For this procedure, measure a suitable test course in the field and record the time it takes to cover the course with the equipment. The course should be between 100 and 300 feet long. Drive or walk the course at least twice, once in each direction and average the times for greater accuracy. Calculate the speed with Equation 1 below.

Equation 1. Ground Speed (MPH) =
$$\frac{\text{Distance} \times 60}{\text{Seconds} \times 88}$$

The second method is to use a true ground speed indicator such as a tractor-mounted radar or similar system. Do not rely on transmission speed charts and engine tachometers. They are not accurate enough for calibration.

CALIBRATING A SPRAYER:
PREPARING TO CALIBRATE
For calibration to be successful, several items need to be taken care of before going to the field. Calibration will not be worthwhile if the equipment is not properly prepared. Whenever possible, calibration should be performed using water only. If you must calibrate using spray mixture, calibrate the equipment on a site listed on the chemical label and with wind speeds less than 5 MPH. Follow the steps outlined below to prepare spraying equipment for calibration.

1. Inspect the sprayer. Be sure all components are in good working order and undamaged. On backpack sprayers, pay particular attention to the pump, control wand, strainers, and hoses. On boom sprayers, pay attention to the pump, control valves, strainers, and hoses. On airblast sprayers, be sure to inspect the fan and air tubes or deflectors as well. Be sure there are no obstructions or leaks in the sprayer.

2. Check the label of the product or products to be applied and record the following:
   - *Application Rate*, Gallons per Acre (GPA)
   - *Nozzle Type*, droplet size and shape of pattern
   - *Nozzle Pressure*, Pounds per Square Inch (PSI)
   - *Type of Application*, broadcast, band, or directed

3. Next, determine some information about the sprayer and how it is to be operated. This includes:
   - *Type of Sprayer:* backpack, boom, or airblast. The type of sprayer may suggest the type of calibration procedure to use.
   - *Nozzle Spacing (inches):* for broadcast applications, nozzle spacing is the distance between nozzles.
   - *Nozzle Spray Width (inches):* For broadcast applications, nozzle spray width is the same as nozzle spacing—the distance between nozzles. For band applications, use the width of the sprayed band if the treated area in the band is specified on the chemical label; use nozzle spacing if the total area is specified. For directed spray applications, use the row spacing divided by the number of nozzles per row. Some directed spray applications use more than one type or size of nozzle per row. In this case, the nozzles on each row are added together and treated as one. Spray width would be the row spacing.

   In most cases, a backpack sprayer uses a single nozzle. Some sprayers use mini-booms or multiple nozzles. The spray width is the effective width of the area sprayed, being sure to account for overlap. If you are using a sweeping motion from side to side, be sure to use the full width sprayed as you walk forward. If you are spraying on foliage in a row, use the row spacing. Dyes are available to blend with the spray to show what has been covered.
   - *Spray Swath (feet):* The width covered by all the nozzles on the boom of a sprayer. For airblast or other boomless sprayers, it is the effective width covered in one pass through the field.
   - *Ground Speed, miles per hour (MPH).* When using a backpack sprayer, walk a comfortable pace that is easy to maintain. Slow walking speeds will take longer to complete the task while high speeds may be tiresome. Choose a safe, comfortable speed that will enable you to finish the job in a timely manner. On tractor-mounted sprayers, select a ground speed appropriate for the crop and type of sprayer used. Slow speeds will take longer to complete the task, while high speeds may be difficult to control and unsafe. Choose a safe, controllable speed that will enable you to finish the job in a timely manner. Ground speed can be determined from Equation 1.

4. The *discharge rate*, gallons per minute (GPM), required for the nozzles must be calculated in order to choose the right nozzle size. Discharge rate depends on the application rate; ground speed; and nozzle spacing, spray width, or spray swath.

For applications using nozzle spacing or nozzle spray width (inches), use Equation 2.

Equation 2. Discharge Rate =
$$\frac{\text{Application Rate} \times \text{Ground Speed} \times \text{Nozzle Spray Width}}{5{,}940}$$

For applications using the spray swath (feet):

Equation 3. Discharge Rate =
$$\frac{\text{Application Rate} \times \text{Ground Speed} \times \text{Spray Swath}}{495}$$

5. Choose an appropriate nozzle or nozzles from the manufacturer's charts and install them on the sprayer. Check each nozzle to be sure it is clean and that the proper strainer is installed with it.

6. Fill the tank half full of water and adjust the nozzle pressure to the recommended setting. Measure the discharge rate for the nozzle. This can be done by using a flow meter or by using a collection cup and stopwatch. The flow meter should read in gallons per minute (GPM). If you are using the collection cup and stopwatch method, the following equation is helpful to convert ounces collected and collection time, in seconds, into gallons per minute.

Equation 4. Discharge Rate =
$$\frac{\text{Ounces Collected} \times 60}{\text{Collection Time} \times 128}$$

7. Whenever possible, calibrate with water instead of spray solution. Do not calibrate with spray solution unless required by the chemical label. Follow all recommendations on the label. If the spray solution has a density different than water, the rate can be corrected using the procedure shown in Calibration Variables.

8. On boom sprayers or sprayers with multiple nozzles, average the discharge rates of all the nozzles on the sprayer. Reject any nozzle that has a bad pattern or that has a discharge rate 10 percent more or less than the overall average. Install a new nozzle to replace the rejected one and measure its output. Calculate a new average and recheck the nozzles compared to the new average. Again, reject any nozzle that is 10 percent more or less than the average or has a bad pattern. When finished, select a nozzle that is closest to the average to use later as your "quick check" nozzle.

On backpack sprayers or sprayers with a single nozzle, compare the discharge rate of the nozzle on the sprayer to the manufacturer's tables for that nozzle size. Reject any nozzle that has a bad pattern or that has a discharge rate 10 percent more or less than the advertised rate. Install a new nozzle to replace the rejected one and measure its output.

Once the sprayer has been properly prepared for calibration, select a calibration method. When calibrating a sprayer, changes are often necessary to achieve the application rates needed.

The sprayer operator needs to understand the changes that can be made to the adjust rate and the limits of each adjustment. The adjustments and the recommended approach are:

- *Pressure:* if the error in application rate is less than 10 percent, adjust the pressure.
- *Ground speed:* if the error is greater than 10 percent but less than 25 percent, change the ground speed of the sprayer.
- *Nozzle size:* if the error is greater than 25 percent, change nozzle size. The goal is to have application rate errors less than 5 percent.

Calibration Methods

There are four methods commonly used to calibrate a sprayer:

The *basic*, *nozzle*, and *128th acre* methods are "time-based methods" which require using a stopwatch or watch with a second hand to ensure accuracy. The area method is based on spraying a test course measured in the field. Each method offers certain advantages. Some are easier to use with certain types of sprayers. For example, the basic and area methods can be used with any type of sprayer. The 128th acre and nozzle methods work well for boom and backpack sprayers. Choose a method you are comfortable with and use it whenever calibration is required.

BASIC METHOD

1. Accurate ground speed is very important to good calibration with the basic method. For tractor-mounted sprayers, set the tractor for the desired ground speed and run the course at least twice. For backpack sprayers, walk the course and measure the time required. Walk across the course at least twice. Average the times required for the course distance and determine ground speed from Equation 1.

2. Calculate the application rate based on the average discharge rate measured for the nozzles, the ground speed over the test course, and the nozzle spacing, nozzle spray width, or spray swath on the sprayer.

When using nozzle spacing or nozzle spray width measured in inches, use the following equation:

Equation 5. Application Rate =
$$\frac{5{,}940 \times \text{Discharge Rate}}{\text{Ground Speed} \times \text{Nozzle Spray Width}}$$

For spray swath applications measured in feet:

Equation 6. Application Rate =
$$\frac{495 \times \text{Discharge Rate}}{\text{Ground Speed} \times \text{Spray Swath}}$$

3. Compare the application rate calculated to the rate required. If the rates are not the same, choose the appropriate adjustment and reset the sprayer.
4. Recheck the system if necessary. Once you have the accuracy you want, calibration is complete.

NOZZLE METHOD

1. Accurate ground speed is very important to good calibration with the nozzle method. For tractor-mounted sprayers, set the tractor for the desired ground speed and run the course at least twice. For backpack sprayers, walk the course and measure the time required. Walk across the course at least twice. Average the times required for the course distance and determine ground speed from Equation 1.
2. Calculate the nozzle discharge rate based on the application rate required the ground speed over the test course, and the nozzle spacing, spray width, or spray swath of the sprayer. For nozzle spacing or spray width measured in inches.

Equation 7. Discharge Rate =
$$\frac{\text{Application Rate} \times \text{Speed} \times \text{Spray Width}}{5{,}940}$$

For spray swath measured in feet:

Equation 8. Discharge Rate =
$$\frac{\text{Application Rate} \times \text{Speed} \times \text{Spray Swath}}{495}$$

Set the sprayer and determine the average nozzle rate.

3. Compare the rate calculated to the average rate from the nozzles. If the two don't match, choose the appropriate adjustment and reset the system.
4. Recheck the system if necessary. Once you have the accuracy you want, calibration is complete.

128TH ACRE METHOD

1. The distance for one nozzle to cover 128th of an acre must be calculated. The nozzle spacing or spray width in inches is used to determine the spray distance. Spray distance is measured in feet. On backpack sprayers, be sure to measure the full width sprayed as you walk forward. Use Equation 9.

Equation 9. Spray Distance =
$$\frac{4{,}084}{\text{Spray Width}}$$

2. Measure the spray distance on a test course in the field. Check the ground speed as you travel across the course. Be sure to maintain an accurate and consistent speed. Travel the course at least twice and average the time to cover the course.
3. For backpack sprayers, collect the output from the nozzle for the time measured in step 2. For tractor-mounted sprayers, park the sprayer, select the nozzle closest to the average, and collect the output for the time determined in step 4. Ounces collected will equal application rate in GPA.

4. Compare the application rate measured to the rate required. If the rates are not the same, choose the appropriate adjustment method and reset the sprayer.
5. Recheck the system. Once you have the accuracy you want, calibration is complete.

AREA METHOD

1. Determine the distance that can be sprayed by one tank using the full spray swath measured in feet.

Equation 10. Tank Spray Distance (ft) =
$$\frac{\text{Tank Volume (gal)} \times 43{,}560}{\text{Application Rate (GPA)} \times \text{Swath (ft)}}$$

2. Lay out a test course that is at least 10 percent of the tank spray distance from Step 1. Fill the sprayer tank with water only, mark the level in the tank, set the sprayer as recommended, and spray the water out on the course. Be sure to maintain an accurate and consistent speed.
3. After spraying the test course, carefully measure the volume of water required to refill the tank to the original level. Calculate the application rate as shown:

Equation 11. Application Rate (GPA) =
$$\frac{\text{Volume Sprayed (gal)} \times 43{,}560}{\text{Test Course Distance (ft)} \times \text{Swath (ft)}}$$

4. Compare the application rate measured to the rate required. If the rates are not the same, choose the appropriate adjustment method and reset the sprayer.
5. Recheck the system. Once you have the accuracy you want, calibration is complete.

CALIBRATING A GRANULAR APPLICATOR:
PREPARING TO CALIBRATE

Granular application calibration is usually done with the chemical to be applied. It is difficult to find a blank material that matches the granular product. Extra care should be taken in handling this product. Minimize worker exposure and take precautions against spills during calibration.

To prepare for calibration, follow these steps:

1. Before calibrating, carefully inspected the equipment to ensure that all components are in proper working order. Check the hopper, the metering rotor, the orifice, and the drop tubes. Be sure there are no leaks or obstructions.
2. Determine the type of application required for the product: ·
   - Broadcast: treats the entire area (includes band applications based on broadcast rates).
   - Band: treats only the area under the band.
   - Row: treats along the length of the row.
3. Determine the application rate needed: ·
   - Broadcast: pounds per acre. ·
   - Band: pounds per acre of treated band width. ·
   - Row: pounds per acre or pounds per 1,000 feet of row length.

4. What type of drive system does the applicator use?
   - Independent: uses PTO, hydraulic, or electric motor drive.
   - Ground Drive: uses ground driven wheel.
5. Regardless of how the application rate is expressed or type of application, calibration is easier if the rate is expressed in terms of pounds per foot of row length. Use one of the following steps to determine the correct row rate in pounds per foot.

For broadcast and row applications
(Application Rate = lb/ac):

Equation 12. Row Rate, lb/ft =
$$\frac{\text{Application Rate} \times \text{Row Width (ft)}}{43{,}560}$$

For banded applications
(Application Rate = lb/ac of Band Width):

Equation 13. Row Rate, lb/ft =
$$\frac{\text{Application Rate} \times \text{Band Width (ft)}}{43{,}560}$$

For directed (row) applications
(Application Rate = lb per 1,000 ft):

Equation 14. Row Rate, lb/ft =
$$\frac{\text{Application Rate}}{1{,}000}$$

6. Choose a calibration distance to work with and measure a test course of this distance in the field you will be working in. Choose an area that is representative of field conditions. The calibration distance should be at least 50 feet but not more than 500 feet. Longer distances are generally more accurate.
7. Calculate the weight of material that should be collected for the calibration distance chosen.

Equation 15. Weight Collected =
Row Rate × Calibration Distance

8. Select a ground speed appropriate for the crop and type of equipment used. Slow speeds take longer to finish the task, while high speeds may be inefficient and unsafe. Consult your equipment manual for a recommended speed. Even ground-driven application equipment can be sensitive to changes in speed. Maintaining an accurate and consistent speed is very important. Choose a safe, controllable speed that will enable you to complete the job in a timely and efficient manner.
9. Set your equipment according to recommendations from the equipment or chemical manufacturer. Most equipment manufacturers and chemical manufacturers provide rate charts to determine the correct orifice setting or rotor speed for each applicator. Fill the hopper at least half full to represent average capacity for calibration.
10. Attach a suitable collection container to each outlet on the applicator. You should be able to collect all material discharged from the applicator. Locate a scale capable of weighing the samples collected in calibration. Some samples may be very small, so a low-capacity scale may be needed. An accurate scale is very important.

Calibration Methods

Two methods for calibrating granular applicators are commonly used. The first is the *distance method*. This method is preferred by many operators because it applies to any type of granular machine and is easy to perform. The second method is the *time method*. This method is similar to sprayer calibration and can be used for applicators driven by PTO, hydraulic, or electric motors.

DISTANCE METHOD

1. On the test course selected in the field, collect the output from the applicator in a container as you travel the course and weigh the material collected. Record the time required to travel the course also. Run the course twice, once in each direction, and average the results for both weight and time.
2. Determine the weight of the product that should be collected for the calibration distance.

Equation 16. Weight Collected (lb) =
Row Rate (lb/ft) × Calibration Distance (ft)

3. Compare the weight of the product actually collected to the weight expected for the calibration distance. If the rates differ by more than 10 percent, adjust the orifice, rotor speed, or ground speed and repeat. Bear in mind, speed adjustments are not effective for ground-driven equipment.
4. Repeat the procedure until the error is less than 10 percent.

TIME METHOD

1. On the test course selected in the field, record the time required to travel the course. Run the course twice, once in each direction, and average the results. Accurate ground speed is very important to good calibration with the time method.
2. With the equipment parked, set the orifice control as recommended and run the applicator for the time measured to run the calibration distance. Collect and weigh the output of the applicator for this time measurement.
3. Determine the weight of the product that should be collected for the calibration distance.

Equation 17. Weight Collected (lb) =
Row Rate (lb/ft) × Calibration Distance (ft)

4. Compare the weight of the product actually collected during the time it took to cover the calibration distance to the weight expected for the calibration distance. If the rates differ by more than 10 percent, adjust the orifice, rotor speed, or ground speed and repeat. Bear in mind, speed adjustments are not effective for ground-driven equipment.

5. Repeat the procedure until the error is less than 10 percent.

CALIBRATING A BROADCAST SPREADER:
PREPARING TO CALIBRATE

Broadcast spreaders include machines designed to apply materials broadcast across the surface of the field. They include *drop*, *spinner*, and *pendulum* spreading devices. Calibration of a broadcast spreader is usually done using the product to be applied. Blank material is available and can be used, but may be hard to find. Use extra care and preparation when calibrating with the chemical. To begin, follow these steps:

1. Carefully inspect all machine components. Repair or replace any elements that are not in good working order.

2. Determine the type of drive system that is being used: ground drive or independent PTO. This may help determine the method of calibration.

3. Determine the application rate and the bulk density of the product to be applied.

4. Determine the spreader pattern and swath of the spreader. Check the pattern to ensure uniformity. To check the pattern, place collection pans across the path of the spreader. For drop spreaders, be sure to place a pan under each outlet. For centrifugal and pendulum spreaders, space the pans uniformly with one in the center and an equal number on each side. The pattern should be the same on each side of the center and should taper smoothly as you go to the outer edge. The swath would be set as the width from side to side where a pan holds 50 percent of the maximum amount collected in the center pan.

5. Fill the hopper half full to simulate average conditions.

6. Set the ground speed of the spreader.

7. Set the spreader according to the manufacturer's recommendations and begin calibration.

Calibration Methods

There are two common methods used to calibrate broadcast spreaders. The first method is the *discharge* method. To use this procedure, collect and measure the total discharge from the spreader as it runs across a test course. The second method, the *pan* method, is used on centrifugal and pendulum spreaders. The pattern test pans used to determine pattern shape and swath are used to determine the application rate.

DISCHARGE METHOD

1. Determine the test distance to use. Longer distances may give better accuracy but may be difficult to manage. A distance of 300 to 400 feet is usually adequate. Use shorter distances if necessary to avoid collecting more material than you can reasonably handle or weigh.

2. Set the ground speed. Be sure to maintain a constant ground speed at all times.

3. If using a ground drive spreader, attach a collection bin to the discharge chute or under the outlets and collect all the material discharged from the spreader as it runs across the test distance. If using an independent drive spreader, record the time required to run the test course. Park the spreader at a convenient location and measure the discharge from the spreader for the time measured on the test distance. The course should be run twice and the times averaged for better accuracy.

4. Calculate the application rate (pounds per acre):

Equation 18. Application Rate, lb/ac =
$$\frac{\text{Weight Collected (lb)} \times 43{,}560}{\text{Distance (ft)} \times \text{Swath (ft)}}$$

5. Compare the application rate measured to the rate required. Adjust and repeat as necessary.

PAN METHOD

1. Place pans in the field across the swath to be spread. Pans should be uniformly spaced to cover the full swath. One pan should be at the center of the swath with equal numbers of pans on each side. Use enough pans, 11 or more, to get a good measurement.

2. Make three passes with the spreader using the driving pattern to be used in the field. One pass should be directly over the center pan and the other passes at the recommended distance, lane spacing, to the left and right of the center pass.

3. Combine the material collected in the pans and determine the weight or volume collected. Divide by the number of pans used to determine the average weight or volume per pan.

4. Calculate the application rate.

If you are measuring the weight in the pans in grams:

Equation 19. Application Rate, lb/ac =
$$\frac{13{,}829 \times \text{Weight (grams)}}{\text{Pan Area (inches)}^2}$$

If you are measuring the volume in the pans in cubic centimeters (cc):

Equation 20. Application Rate, lb/ac =
$$\frac{13{,}829 \times \text{Bulk Density (lb/ft}^3\text{)} \times \text{Volume (cc)}}{\text{Pan Area (inches)}^2 \times 62.4}$$

5. Compare the rate measured to the rate required.

CALIBRATION VARIABLES

Several factors can affect proper calibration. The ground speed of any type of PTO-powered machine can make a difference. On the other hand, ground-driven machines are usually only slightly affected by changes in ground speed. If using dry or granular material, product density will affect the discharge rate and may change the pattern for broadcast spreaders. For liquids, calibration can be affected by pressure, nozzle size, density and viscosity of the liquid, and application type—band or broadcast. The following adjustments may help in adjusting these variables.

SPEED

For PTO-powered equipment or other equipment in which the discharge rate is independent of ground speed, Equation 10 is useful.

Equation 21. New Application Rate =
Old Application Rate x (Old Speed/New Speed)

For ground-driven equipment, there should be little or no change in application rate when speed is changed.

PRESSURE

For liquids in sprayers, the discharge rate changes in proportion to the square root of the ratio of the pressures.

Equation 22. New Discharge Rate =

$$\text{Old Discharge Rate} \times \sqrt{\frac{\text{New Pressure}}{\text{Old Pressure}}}$$

DENSITY

For liquids in sprayers, the discharge rate changes if the specific gravity (S.G.) of the liquid changes. Use water for calibration and adjust as shown below. Calibrate with spray solution only if recommended by the supplier.

Equation 23. Water Discharge Rate =

$$\text{Spray Discharge Rate} \times \sqrt{\text{S.G. of Spray Solution}}$$

BAND APPLICATION VERSUS BROADCAST APPLICATION

Some pesticide application recommendations are based on area of cropland covered. Other recommendations are based on area of land treated in the band covered. Check the label for the product you are using to see how it is listed.

Broadcast application is based on area of cropland covered. Nozzle spacing is the distance between nozzles. Band applications in which the area of covered cropland is used for calibration and those applications in which multiple nozzles per row are used are both treated like broadcast applications. Divide the row spacing by the number of nozzles used per row to get a nozzle spacing for calibration.

For band applications in which area of treated land—not cropland covered—is specified, use the width of the band at the ground as the spacing for calibration.

DETERMINING UPPER AND LOWER LIMITS

Upper and lower limits provide a range of acceptable error. To set these limits for a given sample size, use the equations below. First, however, you must decide upon the degree of accuracy you wish to achieve. Select a percent error: 2 percent, 5 percent, 10 percent, or any other level of accuracy.

Equation 24. Upper Limit =
Target Rate x (1 + Percent Error/100%)

Equation 25. Lower Limit =
Target Rate x (1 − Percent Error/100%)

Registered Fungicides, Insecticides, And Miticides For Vegetables

Recommendations of specific chemicals are based upon information on the manufacturer's label and performance in a limited number of trials. Because environmental conditions and methods of application by growers may vary widely, perfor- mance of the chemical will not always conform to the safety and pest control standards indicated by experimental data.

Recommendations for the use of agricultural chemicals are included in this publication as a convenience to the reader. The use of brand names and any mention or listing of commercial products or services in this publication does not imply endorsement by Auburn University, Clemson University, Louisiana State University, Mississippi State University, North Carolina State University, Oklahoma State University, Texas A&M, Univer- sity of Florida, University of Georgia, University of Kentucky, University of Tennessee, and Virginia Tech nor discrimination against similar products or services not mentioned. Individuals who use agricultural chemicals are responsible for ensuring that the intended use complies with current regulations and conforms to the product label. Be sure to obtain current information about usage regulations and examine a current product label before applying any chemical. For assistance, contact your local county Extension office.

BE SURE TO CHECK THE PRODUCT LABEL BEFORE USING ANY PESTICIDE.

RESISTANCE MANAGEMENT AND THE INSECTICIDE RESISTANCE ACTION COMMITTEE (IRAC) CODES FOR MODES OF ACTION OF INSECTICIDES

Many insecticides affect a particular chemical involved in the function of an insect's nervous, digestive, respiratory, or other system. Some broad-spectrum insecticides affect chemicals that occur in many places within the insect and have a wide ranging effect on the insect. Usually, these are older insecticides that have been in use for many years. The chemicals that these insecticides affect are often found in other animals as well. This can result in the insecticide having undesirable effects on these other animals (non-target effects). Also, non-target effects and persistence in nature have contributed to concerns about these older insecticides.

Many new insecticides have been developed over the last decade, specifically to minimize non-target effects and reduce persistence in the environment compared to older insecticides. This limited persistence in the environment also reduces the potential for non-target effects. However, the primary means of reducing non-target effects has been to make these newer insecticides very specific for a particular chemical (usually an enzyme produced by a single gene) found only in certain insects or groups of insects; thus making the insecticide selective for a particular type of insect. Unfortunately, there is a negative aspect to this specificity. Because only one enzyme is affected, the natural process of mutation can result in genetic modifications that alter the enzyme so that it is unaffected by the insecticide. Insects possessing the modified gene will not be affected by the particular insecticide. These insects will reproduce and, in time with continued exposure to the insecticide, will produce a population of insects that is resistant to the insecticide. Since most of the new insecticides have been developed to be very specific, resistance will develop much more quickly than with previous insecticides.

Different insecticides affect different enzymes, and insecticides are placed into classes based on which enzymes are affected. These classes are called **Modes of Action (MOA)**. Although insecticides may have different names, they can have the same mode of action and affect the same enzyme or system. It is the mode of action to which the insect will become resistant. Because of this, an insect management program **MUST** rotate the modes of action of the insecticides used during the cropping cycle. To prevent the development of resistance, it is important not to apply insecticides with the same mode of action to successive generations of the same insect. Insect development time can vary by species and environmental conditions, and generations often overlap in the field; proper scouting is necessary to determine when modes of action should be rotated. To make it easier to determine an insecticide's mode of action, the **IRAC** has developed a numerical code with a different number corresponding to each mode of action. New packaging has been developed with a colored banner on the top of the package and label giving the **IRAC code.** For example, the insecticide, Movento®, has a new mode of action and the package says:

GROUP 23 INSECTICIDES

Growers can now easily identify the mode of action of a specific insecticide. This will help them to plan their rotation of materials to avoid rapid development of insecticide resistance and help prolong the life of these important new crop protection materials while providing adequate management of their pest problems. More information about insecticide resistance and a concise chart of all of the **IRAC codes** can be found at the website: *www.irac-online.org*.

GENERAL INFORMATION

LAWS AND REGULATIONS

Be sure to check current state and federal laws and regulations regarding the proper use, storage, and disposal of pesticides before applying any chemicals. For restricted-use pesticides, an applicator is required to be certified or to work under the direct supervision of a certified individual. Additional information on Worker Protection Standards (WPS) can be found at *https://www.epa.gov/pesticide-worker-safety/agricultural-work- er-protection-standard-wps*.

CERTIFICATION–PESTICIDE APPLICATORS

The Federal Insecticide, Fungicide, and Rodenticide Act of 1972 (FIFRA) requires each state to certify individuals who use restricted use pesticides. Pesticide applicators are required to demonstrate competency by completing training and examinations certified by their state. Certified applicators are classified as private or commercial. The certification process is somewhat different for each group. The definitions of private and commercial applicators are as follows:

Private Applicator: Any person who uses, or supervises the use of a restricted-use pesticide for the purpose of raising some type of agricultural commodity. The application can be done on land owned or rented by the applicator or the applicator's employer. However, any applications done on a "for-hire" basis are considered commercial applications. Examples of private applicators are dairy farmers, vegetable or fruit growers, greenhouse growers, and ranchers.

Commercial Applicator: Any person who uses, or supervises the use of, pesticides on a "for-hire" basis; any person who applies pesticides as a part of his or her job with any governmental agency (public operator). Examples of commercial applicators are: pest control operators, ornamental and turf operators, and anyone that has a "fee" based service.

For detailed information on certification of pesticide applicators, call your state's Department of Agriculture or your local Extension office for information.

HANDLING PESTICIDES

Before opening a pesticide container, all applicators should carefully read the label, and accurately follow all directions and precautions specified on the labeling. In order to handle and apply pesticides safely, it is essential to use proper personal protective equipment (PPE). Always follow the label for the minimum PPE required.

Your physician should be advised of the types of pesticides used in your work. Before the start of the spray season, each applicator should have a baseline blood cholinesterase level determined if you will be applying any organophosphate or carbamate insecticides.

When applying pesticides, be sure to have a decontamination site as required by the EPA's Worker Protection Standards (WPS) and a supply of clean water and liquid detergent available for drenching and washing in case of an accident. A single drop of certain pesticides in the eye is extremely hazardous. Be prepared to wash a contaminated eye with clean water for 15 minutes.

Only an experienced applicator wearing the protective clothing and safety equipment required by the manufacturer should handle highly toxic pesticides, such as Guthion, Lannate, and Temik.

APPLYING PESTICIDES

Before using a pesticide, read and obey all labeling instructions. Always have the most up-to-date label available when applying a pesticide.

Do not handle or apply pesticides if you have a headache or do not feel well. Never smoke, eat or drink while using pesticides. Avoid inhaling pesticide sprays, dusts, and vapors.

If hands, skin, or other body parts become contaminated or exposed, wash the area immediately with clean water and a liquid detergent. If clothing becomes contaminated, remove it immediately. Wash contaminated clothing separately. After each spraying or dusting, bathe and change clothing; always begin the day with clean clothing.

Always have someone present or in close contact when using highly toxic pesticides -those with the signal word **DANGER** plus the skull and crossbones symbol.

APPLY THE CORRECT DOSAGE

- To avoid excessive residues on crops for feed and food
- To achieve optimum pest control and minimum danger to desirable organisms
- To avoid chemical damage to the crops
- To obtain the most economical control of pests.

Use pesticides for only those crops specified on the label, and use only those that have state and federal registration. Avoid drift to nontarget areas. Dusts drift more than sprays; airblast sprays drift more than boom sprays. When cleaning or filling application equipment, **do not contaminate** streams, ponds, or other water supplies. Keep a record of all pesticides used.

TREATED AREAS

Be sure all treated areas are posted so as to keep out unauthorized personnel. This should be a regular procedure for greenhouse operators.

REENTRY PERIOD

Persons must not be allowed to enter the treated area until after sprays have dried or dusts have settled and until sufficient time has passed to ensure that there is no danger of excessive exposure. This time period is listed on the pesticide label as the Restricted Entry Interval (REI). In no case during the reentry period are farm workers allowed to enter the treated area to engage in activity requiring substantial contact with the treated crop.

PPE is required for any early entry into the treated area and is only allowed for trained applicators.

FARM WORKER SAFETY

Federal pesticide legislation sets an interval during which unprotected persons may not reenter areas treated with certain pesticides to ensure that there is no danger to excessive expo- sure. These intervals (days to reentry) are listed on each pesti- cide's label. Points for special attention are:

1. No pesticide shall be applied while any person not involved in the application is in the field being treated.

2. No owner shall permit any worker not wearing protec- tive clothing (that is, PPE) to enter a field treated with pesticides until sprays have dried or dusts have settled, unless they are exempted from such. **Protective cloth- ing:** hat or head covering; woven, long-sleeved shirt and long-legged pants; and shoes and socks. Additional safety equipment may be needed.

3. Pesticides classified in EPA Category 1 have a reentry time of at least 24 hours.

4. If the label states a longer reentry time or has more strin- gent requirements than indicated here, the label restric- tions must be followed. Existing safety standards speci- fied on the label remain in force.

5. When workers are expected to be working in the vicinity of a field treated or to be treated with a pesticide, a timely (written or oral) warning to such workers shall be given.

    a. For all pesticides, workers must be warned by post- ing a bulletin board at all point(s) where workers might as- semble. This bulletin board should include a map of the farm which designates the different areas of the farm that might be treated and listing of the following information:

        i. Location and name of crop treated

        ii. Brand and common chemical name of pesticide ap- plied.

    b. Date of application

    c. Date of safe reentry into treated area

    d. When a pesticide having a reentry time greater than 7 days is applied, warning signs must be posted for the duration of the reentry time. The signs must be clearly readable at a distance of 25 feet and printed in English and the language of the worker, if other than English.

    e. The sign must contain the words:

        Danger

        Name of the pesticide

        Treatment date

        Do not enter until _____

6. The sign must not be removed during the reentry time, but must be removed before workers are allowed to have contact with the treated plants.

For additional information on these and other state farm worker safety regulations, contact the Pesticide Control Program office or the Cooperative Extension pesticide office in your state.

STORAGE

Pesticides should always be stored in their original con- tainers and kept tightly closed. For the protection of others, espe- cially firefighters, the storage area should be posted as *Pesticide Storage* and kept securely locked.

Herbicides, especially hormone-like weedkillers such as 2,4-D, should not be stored with other pesticides—primarily in- secticides and fungicides—to prevent the accidental substitution of the herbicide for these chemicals.

Store the pesticides in a cool, dry, well-ventilated area that is not accessible to children and others who do not know and un- derstand the safe and proper use of pesticides. Pesticides should be stored under lock and key. Special precautions may be needed in case of a fire in these storage areas.

Any restricted use pesticide (RUP) or container contami- nated by restricted pesticides **must** be stored in a secure, locked enclosure while unattended. This enclosure **must** bear a warning that pesticides are stored there. In many states, it is illegal to store any pesticide in any container other than its original con- tainer.

Keep an inventory of all pesticides held in storage and lo- cate the inventory list in an accessible place away from the stor- age site so that it may be referred to in case of an emergency at the storage site.

Keep your local fire department informed of the location of all pesticide storages. Fighting a fire that includes smoke from burning pesticides can be extremely hazardous to firefighters. Firefighters should be cautioned to avoid breathing any smoke from such a fire. A fire with smoke from burning pesticides may endanger the people of the immediate area or community. The people of an area or community may have to be evacuated if the smoke from a pesticide fire-drifts in their direction. To ob- tain Prefire Planning Guides, contact the US National Response Team (NRT) at at *http://www.nrt.org* or at *http://ipm.ncsu.edu* (under "Information for Pesticide Applicators/Dealers").

Pesticide Formulation	General Signs of Deterioration
EC	Evidence of separation is such as a sludge or sediment Milky appearance does not occur when water is added.
Oils	Milky appearance does not occur when water is added.
WP, SP, WGD	Excessive lumping; powder does not suspend in water.
D, G, WDG	Excessive lumping or caking.

After freezing, place pesticides in warm storage [50°-80°F] and shake or roll container every few hours to mix product or eliminate layering. If layering persists or if all crystals do not completely dissolve, do not use the product. If in doubt, call the manufacturer.

PESTICIDE TRANSPORT

Containers must be well-secured to prevent breakage or spillage. An adequate supply of absorbent material, a shovel, and a fire extinguisher must be available. While under transport, pes- ticides must be stored in a separate compartment from the driver. All pesticide containers and equipment must be secured to the vehicle so as to prevent removal by unauthorized person(s) when the vehicle is unattended. The door or hatch of any service vehicle tank containing a pesticide must be equipped with a cover that will prevent spillage when the vehicle is in motion.

The above requirements do not apply if the pesticide is being transported within the application equipment tank.

For additional information on pesticide transport, contact your state Pesticide Control Program office, Department of Transportation, or your States Pesticide Safety Education Program.

DISPOSAL

Pesticides should not be disposed of in sanitary landfills or by incineration, unless these locations and equipment are especially designed and licensed for this purpose by the state.

The best method to dispose of a pesticide is to use it in accordance with current label registrations. The **triple rinse-and-drain** (see below) procedure or the **pressure-rinse procedure** (see below) is the recommended method to prepare pesticide containers for safe disposal. This method can save money as well as protect the environment.

Crush or puncture the container for disposal in a sanitary landfill or deposit in landfills that accept industrial waste, or deliver the intact container to a drum reconditioner or recycling plant. Check with the landfill operator prior to taking empty containers for disposal. For additional information on the disposal of pesticides themselves or unrinsed containers or rinsate, call the state agency responsible for hazardous wastes. See back cover for telephone numbers.

Triple Rinse–and–Drain Method. To empty a pesticide container for disposal, drain the container into the spray tank by holding the container in a vertical position for 30 seconds. Add water to the pesticide container. Agitate the container thoroughly, then drain the liquid (rinsate) into the spray tank by holding in a vertical position for 30 seconds. Repeat two more times. Puncture or otherwise create a hole in the bottom of the pesticide container to prevent its reuse.

Pressure Rinse Method. An optional method to rinse small pesticide containers is to use a special rinsing device on the end of a standard water hose. The rinsing device has a sharp probe to puncture the container and several orifices to provide multiple spray jets of water. After the container has been drained into the sprayer tank (container is upside down), jab the pointed pressure rinser through the bottom of the inverted container. Rinse for at least 30 seconds. The spray jets of water rinse the inside of the container and the pesticide residue is washed down into the sprayer tank for proper use. Thirty seconds of rinse time is equivalent to triple rinsing. An added benefit is that the container is rendered unusable.

PROTECT OUR ENVIRONMENT

- Do not burn pesticides. The smoke from burning pesticides is dangerous and can pollute air.
- Do not dump pesticides in sewage disposal or storm sewers because this will contaminate water.
- Avoid using excess quantities of pesticides. Calibrate sprayers to make sure of the output.
- Adjust equipment to keep spray on target. Chemicals off-target pollute and can do harm to fish, wildlife, honey bees, and other desirable organisms.

Keep pesticides out of ponds, streams, and water supplies, except those intended for such use. A small amount of drift can be hazardous to food crops and to wildlife. Empty and clean sprayers away from water areas (such as ponds, lakes, streams, etc.)

MINIMIZE SPRAY DRIFT

- Avoid spraying when there is strong wind.
- Use large orifice nozzles at relatively low pressure.
- Use nozzles that do not produce small droplets.
- Adjust boom height as low as practical.
- Do not spray at high travel speeds.
- Spray when soil is coolest and relative humidity is highest.
- Use nonvolatile pesticides.
- Use drift control additives when permitted by the pesticide label.
- Always take a copy of your pesticide label and SDS (Safety Data Sheet) when seeking medical assistance.

PESTICIDE POISONING

If any of the following symptoms are experienced during or shortly after using pesticides: headache, blurred vision, pinpoint pupils, weakness, nausea, cramps, diarrhea, and discomfort in the chest, seek medical assistance immediately. Be sure to take a copy of the pesticide label. For minor symptoms, call the appropriate Poison Control Center in your state. See back cover for emergency telephone numbers. Prompt action and treatment may save a life.

IN CASE OF AN ACCIDENT

Remove the person from exposure:

- Get away from the treated or contaminated area immediately
- Remove contaminated clothing.
- Wash with soap and clean water.
- Call a physician and the state Poison Control Center or Agency. See back cover for emergency telephone numbers.
- Be prepared to give the active ingredient name (common name)

PESTICIDE SPILLS

Keep a supply of absorbent on hand to scatter over liquid spills in the storage room. Sawdust or janitorial sweeping compound works well in absorbing the liquids in a cleanup. Use a respirator and rubber gloves to clean up spills; cover the contaminated surface with household lye, trisodium phosphate, or liquid detergent. Let it soak a couple of hours and reabsorb the solution from the floor. This procedure is recommended for cleaning truck beds that are contaminated. Specific information concerning pesticide cleanup can be obtained by calling the manufacturer directly. The phone numbers for emergencies are listed on every product label. Information can also be obtained by calling CHEMTREC at 800/424-9300. Report pesticide spills to the proper state agency. See back cover for telephone numbers.

RESPIRATORY PROTECTIVE DEVICES FOR PESTICIDES

For many toxic chemicals, inhalation is one of the quickest routes of entry into the circulatory system. From the blood capillaries of the lungs, these toxic substances are rapidly transported throughout the body.

Respiratory protective devices vary in design, use, and protective capability. In selecting a respirator, the user must first consider the degree of hazard associated with breathing the toxic substance, and then understand the specific uses and limitations of the available equipment. Select a respirator that is designed for the intended use, and always follow the manufacturer's instructions concerning the use and maintenance of that particular respirator. Different respirators may be needed for application of different chemicals or groups of chemicals. Select only equipment approved by the National Institute of Occupational Safety and Health (NIOSH). The NIOSH approval numbers begin with the letters TC. *NOTE*: The label will specify which respirator is needed for that particular pesticide.

TYPES OF RESPIRATORS

Respiratory protective devices can be categorized into three classes: air-purifying, supplied-air, and self-contained. Because most pesticide contaminants can be removed from the atmosphere by air-purifying devices, we will look at these in greatest detail.

Air-purifying devices include chemical cartridge respirators, mechanical filters, gas masks (also referred to as canister filter respirators), and battery powered respirators. They can be used only in atmospheres containing sufficient oxygen to sustain life.

- Chemical cartridge respirators provide respiratory protection against certain gases and vapors in concentrations not greater than 0.1% by volume, provided that this concentration does not exceed an amount that is immediately dangerous to life and health. They are for use only when exposure to high continual concentrations of pesticide is unlikely, such as when mixing pesticides outdoors. They are available either as halfmasks, covering only the nose and mouth, or as full-facepiece respirators for both respiratory and eye protection.

- Mechanical filter respirators (dust masks) provide respiratory protection against particulate matter such as mists, metal fumes, and nonvolatile dusts. They are available either as disposable or reusable halfmasks that cover the nose and mouth, or as reusable full-facepieces. Dust masks should never be used when mixing or applying liquids because splashed or spilled liquids, or pesticide vapors can be absorbed by the mask.

- Many respiratory protective devices are combinations of chemical cartridge and mechanical filter (prefilter) respirators. These can provide respiratory protection against both gases and particulate matter.

- Full-face piece respirators provide respiratory protection against particulate matter, and/or against certain specific gases and vapors, provided that their concentration does not exceed an amount that is immediately dangerous to life and health. Gas masks, like full-facepieces, cover the eyes, nose, and mouth, but will last longer than cartridges when continuously exposed to some pesticides. A gas mask will not, however, provide protection when the air supply is low. A special respirator with a self-contained air supply should be worn in these situations.

- Battery powered air-purifying respirators equipped with pesticide filters/cartridges are also effective in filtering out pesticide particles and vapors. They are available as halfmasks, full-face masks, hoods, and protective helmets, and are connected by a breathing hose to a battery powered filtration system. This type of filtration system has the additional advantage of cooling the person wearing it. But, like other air purifying devices, this system does not supply oxygen and must be worn only when the oxygen supply is not limited.

Chemical cartridge respirators protect against light concentrations of certain organic vapors. However, no single type of cartridge is able to remove all kinds of chemical vapors. A different type of chemical cartridge (or canister) must be used for different contaminants. For example, cartridges and canisters that protect against certain organic vapors differ chemically from those that protect against ammonia gases. Be sure that the cartridge or canister is approved for the pesticide you intend to use. Cartridge respirators are not recommended for use against chemicals that possess poor warning properties. Thus, the user's senses (smell, taste, irritation) must be able to detect the substance at a safe level if cartridge respirators are to be used correctly.

The effective life of a respirator cartridge or canister depends on the conditions associated with its use—such as the type and concentration of the contaminants, the user's breathing rate, and the humidity. Cartridge longevity is dependent on its gas and vapor adsorption capacity. When the chemical cartridge becomes saturated, a contaminant can pass through the cartridge, usually allowing the user to smell it. At this point, the cartridge must be changed immediately. There are times when the mechanical prefilter also needs to be changed. A prefilter should be replaced whenever the respirator user feels that breathing is becoming difficult. Dispose of all spent cartridges to avoid their being used inadvertently by another applicator who is unaware of their contaminated condition.

Chemical cartridge respirators cannot provide protection against extremely toxic gases such as hydrogen cyanide, methyl bromide, or other fumigants. Masks with a self-contained air supply are necessary for these purposes.

USE AND CARE OF RESPIRATORS

Respirators are worn as needed for protection when handling certain pesticides. The use of respirators is now regulated requiring a health screening prior to their use by a health professional. This is due in part to the Fumigant re-registration decisions by EPA. These prerequisites are outlined in the OSHA Respiratory Protection Act. Prior to using a respirator, read and understand the instructions on the cartridge or canister and all supplemental information about its proper use and care. Be sure the filter is approved for protection against the pesticide intended to be used. Respirators labeled only for protection against particulates must not be used for gases and vapors. Similarly, respirators labeled only for protection against gases and vapors should not be used for particulates. Remember, cartridges and filters do not supply oxygen. Do not use them where oxygen may be limited. All respirators must be inspected for wear and deterioration of their components before and after each use. Special attention should be given to rubber or plastic parts which can deteriorate. The facepiece, valves, connecting tubes or hoses, fittings, and filters must be maintained in good condition.

All valves, mechanical filters, and chemical filters (cartridges or canisters) should be properly positioned and sealed. Fit the respirator on the face to ensure a tight but comfortable seal. Facial hair will prevent a tight seal and consequently OSHA regulations prohibit the use of a respirator when the user has a beard or facial hair. Two tests can be done to check the fit of most chemical cartridge respirators. The first test requires that you place your hand tightly over the outside exhaust valve. If there is a good seal, exhalation should cause slight pressure inside the facepiece. If air escapes between the face and facepiece, readjust the headbands until a tight seal is obtained. Readjusting the headbands may at times not be sufficient to obtain a good seal. It may be necessary to reposition the facepiece to prevent air from escaping between the face and facepiece. The second test involves covering the inhalation valve(s) by placing a hand over the cartridge(s). If there is a good seal, inhalation should cause the facepiece to collapse. If air enters, adjust the headbands or reposition the facepiece until a good seal is obtained.

Get to fresh air immediately if any of the following danger signals are sensed:

- Contaminants are smelled or tasted
- Eyes, nose, or throat become irritated
- Breathing becomes difficult
- The air being breathed becomes uncomfortably warm
- Nauseous or dizzy sensations are experienced

Cartridges or filters may be used up or abnormal conditions may be creating contaminant concentrations which exceed the capacity of the respirator to remove the contamination.

After each use of the respirator, remove all mechanical and chemical filters. Wash the facepiece with soap and warm water, and then immerse it in a sanitizing solution such as household bleach (two tablespoons per gallon of water) for two minutes, followed by a thorough rinsing with clean water to remove all traces of soap and bleach. Wipe the facepiece with a clean cloth and allow to air dry.

Store the respirator facepiece, cartridges, canisters, and mechanical filters in a clean, dry place, preferably in a tightly sealed plastic bag. ***Do not store respirators with pesticides or other agricultural chemicals.***

Handle respirators with the same care given to other protective equipment and clothing.

PROTECTING OUR GROUNDWATER

Groundwater is the water contained below the topsoil. This water is used by 90% of the rural population in the United States as their sole source of drinking water. Contamination of the water supply by pesticides and other pollutants is becoming a serious problem. One source of contamination is agricultural practices. Protection of our groundwater by the agricultural community is essential.

Groundwater collects under our soils in aquifers that are comprised of layers of sand, gravel, or fractured bedrock which, by their nature, hold water. This water comes from rainfall, snowfall, etc., that moves down through the soil layers to the aquifer. The depth of the aquifer below the surface depends on many factors. Where it is shallow, we see lakes, ponds and wetlands. In areas where it is deep, we find arid regions.

FACTORS THAT AFFECT MOVEMENT OF WATER AND CONTAMINANTS

The depth of aquifers, in conjunction with soil types, influences how much surface water reaches the aquifer. Their depth also affects how quickly water and contaminants reach an aquifer. Thus, shallow water tables tend to be more vulnerable to contamination than deeper ones.

This tendency, however, depends on the soil type. Soils with high clay or organic matter content may hold water longer and retard its movement to the aquifer.Conversely, sandy soils allow water to move downward at a fast rate. High levels of clay and/or organic content in soils also provide a large surface area for binding contaminants that can slow their movement into groundwater. Soil texture also influences downward water movement. Finer textured soils have fewer spaces between particles than coarser ones, thus decreasing movement of water and contaminants.

CHEMISTRY PLAYS A ROLE

The characteristics of an individual pesticide affect its ability to reach groundwater. The most important characteristics are solubility in water, adsorption to soils, and persistence in the environment.

Pesticides that are highly soluble in water have a higher potential for contaminating groundwater than those that are less soluble. The water solubility of a chemical indicates how much chemical will dissolve in water and is measured in parts per million (ppm). Those chemicals with a water solubility greater than 30 ppm may create problems. Be sure to read the Environmental Precautions on each pesticide label.

A chemical's ability to adhere to soil particles plays an important role. Chemicals with a high affinity for soil adsorption are less likely to reach the aquifer. Adsorption is also affected by the amount of organic matter in the soil. Soils with high organic matter content are less vulnerable than those with low organic matter content.

Finally, how persistent a chemical is in the environment may affect its ability to reach groundwater. Those that persist for a long time may be more likely to cause contamination than materials that breakdown quickly. Persistence is measured by the time it takes half of a given pesticide to degrade. This is called the chemical's half-life. Chemicals with an overall estimated half-life longer than 3 weeks pose a threat to groundwater.

HOW TO PREVENT CONTAMINATION OF GROUND WATER

Examine the chemical properties of the pesticides used. If using materials that persist for long periods of time, are very water soluble, or are not tightly held by the soil, then your groundwater may become contaminated. Another material may be selected that has a shorter persistence, lower water solubility, or higher potential for soil adsorption. The following chart assists with these decisions.

1. Determine the local soil and geologic circumstances. If in an area with a shallow water table or the soil is low in organic matter or sandy in nature, there is a greater risk of contaminating your groundwater. In these cases, choose a pesticide that has a low water solubility and is not persistent.

2. Evaluate management practices. These practices maybe the most important factors in determining the risk of contaminating groundwater. If the same materials are used year after year, or many times a season, the potential for contamination can be increased due to the amount of pesticide in the soil. The timing of pesticide applications has an effect on groundwater contamination. If applications during periods of high rainfall or heavy irrigation are made, it is more likely that contamination may occur. Also, the water table in the spring may be higher than at other times. Early season applications, therefore, may pose a greater chance for groundwater contamination.

3. The method of application may have an effect. Direct injection, incorporation, and chemigation all increase the chance of contamination. If using these techniques, be sure to follow the procedures listed on the material's label.

4. The location of wells can be important. If the sprayer loading area or pesticide storage building is too close to a well, the risk of contamination may be greater. Wells should be located a minimum distance from all pesticide storage and loading areas. This distance differs between states but is generally between 50 and 100 ft. In the event of an accident, this distance should prevent contamination. This minimum distance should also be followed for field irrigation wells. If they are too close to application areas, contamination might occur.

5. Check the condition of any wells in the vicinity of sprayer loading areas, pesticide storage areas, or field applications. If they have cracked casings trouble is being invited. Cracks in a well casing provide a direct point of entry for pesticide-contaminated water that is in the soil.

6. Use some type of anti back-flow device in any system used for chemigation or to fill the sprayer with water. In the event of a pump shutoff or other failure, if any back-flow into the water system occurs, these devices will prevent pesticides from entering the well. Many state laws require that anti back-flow devices be placed on all sprayer water intake systems prior to the water entering the tank. The use of an air gap only is no longer acceptable in some states.

7. Care and maintenance of equipment is also an important consideration. If the equipment does not function properly, over-delivery may occur, which increases the chance of groundwater contamination. Prior to the beginning of the season, inspect all of the working parts of the sprayer or chemigation system. Check the pump to ensure that it is working properly. For both sprayers and chemigation systems, check the water lines for clogs and leaks. For sprayers, check the nozzles for wear and clogs. Clogged, leaking, or worn lines and nozzles can cause pesticides to be delivered in too high an amount or into unwanted areas. Be sure to calibrate equipment. Uncalibrated equipment can cause over-delivery as well. Equipment should be calibrated at the beginning of the season, periodically during the remainder of the season, and any time changes or adjustments are made to the equipment.

8. Apply materials only when needed. The use of pesticides, when not needed, can increase the threat of contamination. Check irrigation practices as well. Do not irrigate immediately after a pesticide application, unless required by a pesticide's label. The increased water content in the soil might speed up the downward movement of a pesticide.

9. One of the best ways to reduce indirect or direct runoff would be to only purchase and use what is needed to completed a treatment. Runoff and poisoning can stem from not properly disposing of extra pesticides.

REMEMBER, GROUNDWATER MUST BE PROTECTED.

TOXICITY OF CHEMICALS USED IN PEST CONTROL

The danger in handling pesticides does not depend exclusively on toxicity values. Hazard is a function of both toxicity and the amount and type of exposure. Some chemicals are very hazardous from dermal (skin) exposure as well as oral (ingestion). Although inhalation values are not given, this type of exposure is similar to ingestion. A compound may be highly toxic but present little hazard to the applicator if the precautions are followed carefully.

Toxicity values are expressed as acute oral LD_{50} in terms of milligrams of the substance per kilogram (mg/kg) of test animal body weight required to kill 50 percent of the population. The acute dermal LD_{50} is also expressed in mg/kg. These acute values are for a single exposure and not for repeated exposures such as may occur in the field. Rats are used to obtain the oral LD and the test animals used to obtain the dermal values are usually rabbits.

CATEGORIES OF TOXICITY[1]

Categories	Signal Word	LD_{50} Value (mg/kg)	
		Oral	Dermal
I	Danger-Poison	0 – 50	0 – 200
II	Warning	50-500	200-2,000
III	Caution	500-5,000	2,000-20,000
IV	None[2]	5,000	5,000 20,000

[1] EPA accepted categories
[2] No signal word required based on acute toxicity; however, products in this category usually display "Caution"

Read all labels and become familiar with the symptoms of pesticide poisoning. For help in a pesticide emergency, seek immediate medical attention and call the appropriate poison information number on the back cover of this book.

TOXICITY AND LD_{50} CALCULATIONS WEIGHT CONVERSIONS

1 ounce (oz)	= 28 grams (gr)
1 pound (lb)	= 454 grams (gr)
1 gram (gr)	= 1,000 milligrams (mg)
1,000 mg	= 0.035 oz
1 mg	= 0.000035 oz

CONVERSIONS: BODY WEIGHT IN POUNDS (LB) TO BODY WEIGHT IN KILOGRAMS (KG)

(lb)		(kg)
25	=	11.25
50	=	22.5
75	=	33.75
100	=	45

To determine an exact weight, multiply known body weight in pounds by 0.45. *Example*: 100 lb x 0.45 = 45 kg

Note: All the following calculations use a body weight of 100 pounds. To determine the LD_{50}, first convert body weight to kilograms; to do this multiply weight in lb by 0.45.
Example: 100 x 0.45 = 45 kg

Next, multiply given LD_{50} by body weight in kg.
Note: LD_{50} numbers are given by the manufacturer.
Example: LD_{50} of 11 x 45 kg = 495 mg

Next, to convert milligrams (mg) to ounces (oz), multiply mg by 0.000035. *Example*: 495 mg x 0.000035 = 0.017 oz.

The following is a chart of LD_{50} figures converted to ounces for three commonly used products in the agricultural industry.

		Body Weight in Pounds				
	LD_{50}	30	60	100	150	200
		Ounces				
Insecticide Furadan	11	0.005	0.010	0.017	0.026	0.035
Herbicide Micro-Tech/Partner	1,800	0.9	1.7	2.8	4.3	5.7
Fungicide Chlorothalonil	10,000	4.9	9.5	15.7	23.8	31.5

CONVERSION INFORMATION FOR USE OF PESTICIDES ON SMALL AREAS

LIQUID MATERIALS

Recommended Rate per acre	Approximate Rate per 1,000 sq. ft.	Approximate Rate per 100 sq. ft.
1 pint	¾ tablespoons	¼ teaspoon
1 quart	1½ tablespoons	½ teaspoon
2 quarts	3 tablespoons	1 teaspoon
1 gallon	6 tablespoons	2 teaspoons
25 gallons	4½ pints	1 cup
50 gallons	8 pints	1 pint
75 gallons	7 quarts	1½ pints
100 gallons	9 quarts	1 quart

For dry materials, such universal conversions are not possible because these materials vary widely in density. You can use volume measurements such as teaspoons, tablespoons, and cups, but you must first weigh a tablespoon of each product so that you will know what volume measurement to use to obtain the desired weight. Remember that there are 43,560 square feet in an acre. To convert a per-acre rate to 1,000 square feet, divide the peracre rate by 43.56. To convert a per-acre rate to 100 square feet, divide the per-acre rate by 435.6.

Example: A rate of 2 pounds of Dithane DF per acre is desired for a planting of 1,000 square feet. Divide the per-acre rate of 907 grams (453.6 grams per pound) by 43.56 to get 20.8 grams. Since Dithane DF weighs about 10 grams per tablespoon, you would need two tablespoons. Knowing the weight per tablespoon for each product you work with, you can use a tablespoon for measuring, rather than weighing.

PESTICIDE DILUTION TABLES

The following tables provide quantity of either liquid or wettable powder concentrates to use per acre to give desired dosage of an active ingredient per acre.

HOW TO USE THESE TABLES

Example: Reading the product label, you determine that you need to apply 0.50 lbs of actual Guthion per acre to treat a specific problem. You have Guthion 2L liquid that contains 2 lb. of Active Ingredient per gallon of product. Referring to the "Liquid Concentrate" table find "2 lb" in the first column. Next locate the "0.50" column in the heading across the top of the table. These two columns intersect at "2.0 pints". Thus, you need to add 2 pints of Guthion 2L in enough water to treat one acre. The other two tables work the same way.

TABLE OF MEASURES

3 teaspoons (tsp) = 1 tablespoon	2 cups (c) = 1 pint
2 tablespoons (tbl) = 1 fluid once	2 pints (pt) = 1 quart
16 tablespoons (tbl) = 1 cup	4 quarts (qt) = 1 gallon
8 fluid ounces (fl oz) = 1 cup	

LIQUID CONCENTRATE – AMOUNT TO USE IN PINTS PER ACRE

Pounds A.I. /gallon	Pounds per acre of A.I. (Active Ingredient) Recommended							
	0.125	0.25	0.50	0.75	1.0	2.0	3.0	4.0
1 lb.	1.0	2.0	4.0	6.0	8.0	16.0	24.0	32.0
1½ lb.	6.7	1.3	2.6	4.0	5.3	10.6	16.0	21.3
2 lb.	0.5	1.0	2.0	3.0	4.0	8.0	12.0	16.3
3 lb.	0.34	0.67	1.3	2.0	2.7	5.3	8.0	10.7
4 lb.	0.25	0.50	1.0	1.5	2.0	4.0	6.0	8.0
5 lb.	0.20	0.40	0.80	1.2	1.6	3.2	4.8	6.4
6 lb.	0.17	0.34	0.67	1.0	1.3	2.6	4.0	5.3
7 lb.	0.14	0.30	0.60	0.90	1.1	2.3	3.4	4.6
8 lb.	0.125	0.25	0.50	0.75	1.0	2.0	3.0	4.0
9 lb.	0.11	0.22	0.45	0.67	0.90	1.8	2.7	3.6
10 lb.	0.10	0.20	0.40	0.60	0.80	1.6	2.4	3.2

WETTABLE POWDER – AMOUNT TO USE IN POUNDS PER ACRE

%A.I.	Pounds per acre of A.I. (Active Ingredient) Recommended							
	0.125	0.25	0.50	0.75	1.0	2.0	3.0	4.0
15%	13/16	1¾	3⅓	5	6½	13	20	26½
25%	½	1	2	3	4	8	12	16
40%	5/16	⅝	1¼	1¾	2½	5	7½	10
50%	¼	½	1	1½	2	4	6	8
75%	3/16	⅜	11/16	1	1⅓	2⅔	4	5½

DUST OR GRANULES– AMOUNT TO USE IN POUNDS PER ACRE

%A.I.	Pounds per acre of A.I. (Active Ingredient) Recommended							
	0.125	0.25	0.50	0.75	1.0	2.0	3.0	4.0
2½%	5	10	20	30	40	80	120	160
5%	2½	5	10	15	20	40	60	100
10%	1¼	2½	5	7½	10	20	30	40
20%	5/8	1¼	2½	3¾	5	10	15	20
25%	½	1	2	3	4	8	12	16

Insect Control for Commercial Vegetables

Read the pesticide label before application. High pressure (200 psi) and high volume (minimum 50 gallons per acre) aid in vegetable insect control. Ground sprays with airblast sprayers or sprayers with hollow cone drop nozzles are suggested. Incorporate several methods of control for best results. In recent years, the number of generic products has increased significantly. For brevity, these generic products typically are not listed within each section. The trade names listed are intended to aid in identification of products and are intended neither to promote use of specific trade names nor to discourage use of generic products. A list of active ingredients and generic brand names appears in Table 2-27 at the end of this section.

Insecticides are placed into IRAC MOA classes based on their mode of action (insecticides in the same MOA class have the same mode of action). Effective insecticide resistance management involves the use of alternations, rotations, or sequences of different insecticide MOA classes. To prevent the development of resistance, it is important not to apply insecticides with the same MOA to successive generations of the same insect.

THE FOLLOWING ONLINE DATABASES PROVIDE CURRENT PRODUCT LABELS AND OTHER RELEVANT INFORMATION:

Database	Web Address
Agrian Label Database	https://home.agrian.com/
Crop Data Management Systems	http://www.cdms.net/Label-Database
EPA Pesticide Product and Label System	https://iaspub.epa.gov/apex/pesticides/f?p=PPLS:1
Greenbook Data Solutions	https://www.greenbook.net/
Kelly Registration Systems [2]	http://www.kellysolutions.com

[1] Additional databases not included in this list may also be available. Please read the database terms of use when obtaining information from a particular website
[2] Available for AK, AR, AZ, CA, CO, CT, DE, FL, GA, HI, IA, ID, LA, MA, MD, MN, MO, MS, MT, NC, ND, NE, NJ, NV, OK, OR, PA, SC, SD, VA, VT, WI, and WY
    Kelly Registration Systems works with State Departments of Agriculture to provide registration and license information

PESTICIDES AND THE ENDANGERED SPECIES ACT: WHAT YOU NEED TO KNOW

The following description has been endorsed by the Weed Science Society of America, Entomological Society of America, and American Phytopathological Society.

1. What is the Endangered Species Act (ESA)?

The Endangered Species Act is a long-standing federal law, first passed in 1973, which requires government agencies to ensure any actions they take do not jeopardize a species that has been federally listed as endangered or threatened. When an agency has a proposed action that might affect a listed species or its habitat, they consult with one or both of the agencies that helps enforce the ESA, the U.S. Fish and Wildlife Services or the National Marine Fisheries Service (this is known as "a consultation" with "the Services"). The Services then may recommend changes to the project or action to protect listed species or habitats.

2. How does the ESA affect pesticide use?

The Environmental Protection Agency (EPA) Office of Pesticide Programs (OPP) is the federal agency that regulates pesticide use. Because the use of pesticides can affect animals and plants (or their habitat), pesticide registrations are considered "actions" that would trigger an endangered species consultation.

3. Why am I hearing about the ESA and pesticide use now?

Due to the complex nature of the process, the EPA has not fully completed the required endangered species consultations with the Services for pesticide registrations in the past, which has left many of those pesticides vulnerable to lawsuits. Courts have annulled pesticide registrations which has led to their removal from market. To make pesticide registrations more secure from litigation, ultimately all pesticide registrations will comply with the Endangered Species Act (https://www.epa.gov/endangered-species).

4. How will this affect the pesticide I use today?

Many pesticide labels will likely have changes that could include:
- Requirement to check the EPA's Bulletins Live! Two website and follow current ESA restrictions for the pesticide product in the bulletin (https://www.epa.gov/endangered-species/bulletins-live-two-view-bulletins)
- Measures to reduce spray drift
- Measures to reduce runoff/erosion
- Other measures to reduce pesticide exposure to listed species and their habitat.

In short, farmers and applicators should expect to see some new application requirements on their pesticide labels. But there is no need to panic. To date, no pesticide has ever been fully removed from the market based solely on endangered species risks, and that remains an unlikely scenario in the future.

5. Why does complying with the ESA matter?

By starting to fully comply with the ESA, **EPA anticipates that this will give farmers and applicators more stable, reliable access to the pesticides they need.** Furthermore, the ESA has been successful at bringing back some species Ameri- cans care about – such as the bald eagle or the Eggert sun- flower – and restoring them to healthy populations, which has benefited the natural and cultivated ecosystems that agriculture (and society) rely on.

TABLE 2-1. INSECT CONTROL FOR ASPARAGUS

Insecticide, Mode of Action Code, and Formulation	Amount of Formulation Per Acre	Restricted Entry Interval (REI)	Pre harvest Interval (PHI) (Days)	Precautions and Remarks
APHID				
dimethoate 400, MOA 1B	1 pt	48 hrs	180	Do not exceed 2 pints per acre per year.
malathion, MOA 1B (various) 57 EC	1.5 to 2 pt	12 hrs	1	Aphid colonies appear by early September.
pymetrozine, MOA 9B (Fulfill) 50 WDG	2.75 oz	12 hrs	180	For aphid control on ferns after harvest.
acetamiprid, MOA 4A (Assail) 30 SC	2.5 to 5.3 oz	12	1	Do not make more than two applications per calendar year.
ASPARAGUS BEETLE, JAPANESE BEETLE, GRASSHOPPER				
carbaryl, MOA 1A (Sevin) XLR Plus	1 to 2 qt	12 hrs	1	Low rate to be used on seedlings or spears. Do not apply more often than once every 3 days. With established beetle populations, three consecutive weekly sprays are required for beetles and grasshoppers in the fall.
acetamiprid, MOA 4A (Assail) 30 SC	2.5 to 5.3 oz	12 hrs	1	Use higher rate for Japanese beetle. Do not make more than two applications per calendar year.
dimethoate 400, MOA 1B	1 pt	48 hrs	180	Do not exceed 5 pints per acre per year.
malathion, MOA 1B (various) 57 EC	2 pt	12 hrs	1	
methomyl, MOA 1A (Lannate) 2.4 LV	1.5 pt	48 hrs	1	Let a row on edge of field near overwintering sites of asparagus beetles fern out. This will attract and hold beetles for that directed insecticide spray (trap and destroy).
pyrethroid, MOA 3A		12 hrs		See **Table 2-25 A** for a list of registered pyrethroids and pre-harvest intervals.
spinetoram, MOA 5 (Radiant) 1 SC	4 to 8 fl oz	4 hrs	60	For asparagus beetle only. This use is only for asparagus ferns; do not apply within 60 days of spear harvest.
BEET ARMYWORM, CUTWORM, YELLOWSTRIPED ARMYWORM				
Bacillus thuringiensis, MOA 11A (Dipel) DF	0.5 to 1 lb	4 hrs	0	
chlorantraniliprole, MOA 28 (Coragen eVo) 5 SC	1.2 to 2.5 fl oz	4 hrs	1	
cyantraniliprole, MOA 28 (Exirel) 0.83 EC	7 to 13.5 fl oz	12 hr	1	Do not make applications within 25 ft of water sources.
methomyl, MOA 1A (Lannate) 2.4 LV (Lannate) 90 SP	1.5 to 3 pt 0.5 to 1 lb	48 hrs	1	
spinetoram, MOA 5 (Radiant) 1 SC	4 to 8 fl oz	4 hrs	60	This use is only for asparagus ferns; do not apply within 60 days of spear harvest.
spinosad, MOA 5 (Entrust) 2 SC	4 to 6 fl oz	4 hrs	60	This use is only for asparagus ferns; do not apply within 60 days of spear harvest. **OMRI listed**.

TABLE 2-2. INSECT CONTROL FOR BEANS (SNAP, LIMA, POLE, EDAMAME)

Insecticide, Mode of Action Code, and Formulation	Amount of Formulation Per Acre	Restricted Entry Interval (REI)	Pre harvest Interval (PHI) (Days)	Precautions and Remarks
APHID				
acetamiprid MOA 4A (Assail) 30 SC	2.1 to 4.5 fl oz	12 hrs	7	
dimethoate 4 EC, MOA 1B	0.5 to 1 pt	48 hrs	0	On foliage as needed. Re-entry interval of 48 hours
imidacloprid, MOA 4A **Soil treatment** (Admire Pro) 4.6 F (various) 2F	7 to 10.5 fl oz 16 to 24 fl oz	12 hrs	21	See label for soil application instructions. Also controls leafhoppers and thrips.
Foliar treatment (Admire Pro) 4.6 F (various) 1.6 F	1.2 fl oz 3.5 fl oz	12 hrs	7	

TABLE 2-2. INSECT CONTROL FOR BEANS (SNAP, LIMA, POLE, EDAMAME) (cont'd)

Insecticide, Mode of Action Code, and Formulation	Amount of Formulation Per Acre	Restricted Entry Interval (REI)	Pre harvest Interval (PHI) (Days)	Precautions and Remarks
APHID (cont'd)				
sulfoxaflor, MOA 4C (Transform) 50 WG	0.75 to 1.0 oz	24 hrs	7	
flonicamid MOA 29 (Beleaf) 50 SG	2.8 oz	12 hr	7	Do not exceed 3 applications per season.
flupyradifurone (Sivanto Prime)	7 to 14 fl	4 hrs	7	
spirotetramat, MOA 23 (Movento) 2 SC	4 to 5 fl oz	24 hrs	1 (succulent) 7 (dried)	
THRIPS				
acephate, MOA 1B (Orthene) 97 PE	0.5 to 1 lb	24 hrs	14	Lima beans may be treated and harvested the same day. Do not apply more than 2 pounds a.i. per acre per season.
acetamiprid MOA 4A (Assail) 30 SC	3.8 to 4.5 fl oz	12 hrs	7	
pyrethroid, MOA 3A		12 hrs		Not effective against western flower thrips. See **Table 2-25 A** for a list of registered pyrethroids and pre-harvest intervals.
methomyl, MOA 1A (Lannate) 90 SP (Lannate) 2.4 LV	0.5 lb 1.5 pt	48 hrs	1	
novaluron MOA 15 (Rimon) 0.83 EC	12 fl oz	12 hrs	1	Effective against immature thrips only.
spinetoram, MOA 5 (Radiant) 1 SC	5 to 6 fl oz	4 hrs	3 (succulent) 28 (dried)	Do not apply more than 28 fluid ounces per acre per season on succulent beans or more than 12 fluid ounces on dried beans.
spinosad, MOA 5 (Blackhawk)	2.5 to 3.3 oz	4 hrs	3 (succulent) 28 (dried)	Do not apply more than 20 ounces per acre per season on succulent beans or more than 8.3 ounces on dried beans.
CORN EARWORM, EUROPEAN CORN BORER, LESSER CORNSTALK BORER, ARMYWORM				
chlorantraniliprole, MOA 28 (Coragen eVo) 5 SC (Vantacor) 5 SC	1.2 to 2.5 fl oz 1.2 to 2.5 fl oz	4 hrs	1	
cyantraniliprole, MOA 28 (Exirel) 0.83 SE	10 to 20.5 fl oz	12 hrs	1 (succulent) 7 (dried)	
methoxyfenozide, MOA 18 (Intrepid) 2F	4 to 16 fl oz	4 hrs	7	Use lower rates for early season applications to young crops, and higher rates for mid- to late-season applications and heavier infestations. Do not apply more than 16 fl oz per acre per season.
spinetoram, MOA 5 (Radiant) 1 SC	4.5 to 6 fl oz	4 hrs	3 (succulent) 28 (dried)	Do not apply more than 28 fluid ounces per acre per season on succulent beans or more than 12 fluid ounces on dried beans.
spinosad, MOA 5 (Blackhawk)	2.2 to 3.3 oz	4 hrs	3 (succulent) 28 (dried)	Do not apply more than 20 ounces per acre per season on succulent beans or more than 8.3 ounces on dried beans.
pyrethroid, MOA 3A		12 hrs		See **Table 2-25 A** for a list of registered pyrethroids and pre-harvest intervals. Not effective against armyworm.
viruses		4 hrs		See **Table 2-26** for virus-based insecticides effective against specific insects. **OMRI listed.**
COWPEA CURCULIO				
pyrethroid, MOA 3A		12 hrs		See **Table 2-25 A** for a list of registered pyrethroids and pre-harvest intervals. Control may be poor in areas where resistant populations occur, primarily in the Gulf Coast areas. Addition of piperonyl butoxide synergist (Exponent) may improve control of pyrethroids.
lambda-cyhalothrin, MOA 3A + chlorantraniliprole, MOA 28 (Besiege) ZC	6 to 10 fl oz	24 hrs	7 (succulent) 21 (dried)	
bifenthrin, MOA 3A + chlorantraniliprole, MOA 28 (Elevest) SC	5.6 to 9.6 fl oz	12 hrs	3 (succulent) 14 (dried)	
CUCUMBER BEETLE, BEAN LEAF BEETLE, JAPANESE BEETLE, CUTWORM				
carbaryl, MOA 1A (Sevin) XLR Plus	0.5 to 1 qt	12 hrs	3 (succulent) 21 (dried)	
pyrethroid, MOA 3A		12 hrs		See **Table 2-25 A** for a list of registered pyrethroids and pre-harvest intervals.

TABLE 2-2. INSECT CONTROL FOR BEANS (SNAP, LIMA, POLE, EDAMAME) (cont'd)

Insecticide, Mode of Action Code, and Formulation	Amount of Formulation Per Acre	Restricted Entry Interval (REI)	Pre harvest Interval (PHI) (Days)	Precautions and Remarks
GRASSHOPPER				
pyrethroid, MOA 3A		12 hrs		See **Table 2-25 A** for a list of registered pyrethroids and pre-harvest intervals.
acephate, MOA 1B (Orthene) 97 PE	0.5 to 1 lb	24 hrs	14	Lima beans may be treated and harvested the same day. Do not apply more than 2 pounds a.i. per acre per season.
chlorantraniliprole, MOA 28 (Vantacor) 5 SC	0.7 to 1.7 fl oz	4 hrs	1	
LEAFMINER				
cyromazine, MOA 17 (Trigard) 75 WP	2.66 oz	12 hrs	7	
spinetoram, MOA 5 (Radiant) 1 SC	4 to 8 fl oz	4 hrs	3 (succulent) 28 (dried)	Do not apply more than 28 fluid ounces per acre per season on succulent beans or more than 12 fluid ounces on dried beans.
spinosad, MOA 5 (Blackhawk)	2.5 to 3.3 oz	4 hrs	3 (succulent) 28 (dried)	Do not apply more than 20 ounces per acre per season on succulent beans or more than 8.3 ounces on dried beans.
PLANT BUG (Lygus, Alydidae)				
pyrethroid, MOA 3A		12 hrs		See **Table 2-25 A** for a list of registered pyrethroids and pre-harvest intevals.
acetamiprid, MOA 4A (Assail) 30 SC	2.1 to 4.5 fl oz		7	
dimethoate, MOA 1B (Dimethoate) 4 EC	1 pt	48 hrs	7	Do not apply if bees are visiting area to be treated when crops or weeds are in bloom.
flonicamid MOA 29 (Beleaf) 50 SG	2.8 oz	12 hr	7	Do not exceed 3 applications per season.
sulfoxaflor, MOA 4C (Transform) CA	1.5 to 2.25 oz	24 hrs	7	
MEXICAN BEAN BEETLE				
acetamiprid MOA 4A (Assail) 30 SC	2.1 to 4.5 fl oz	12 hrs	7	
pyrethroid, MOA 3A		12 hrs		See **Table 2-25 A** for a list of registered pyrethroids and pre-harvest intevals.
novaluron, MOA 15 (Rimon) 0.83 EC	9 to 12 fl oz	12 hrs	1	Apply when population is at egg hatch to small larval stage.
POTATO LEAFHOPPER				
acetamiprid MOA 4A (Assail) 30 SC	2.1 to 4.5 fl oz	12 hrs	7	
carbaryl, MOA 1A (Sevin) XLR Plus	1 qt	12 hrs	3 (succulent) 21 (dry)	On foliage as needed.
dimethoate 4 EC, MOA 1B	0.5 to 1 pt	48 hrs	7	
methomyl, MOA 1A (Lannate) 90 SP (Lannate) 2.4 L	0.5 lb 1.5 to 3 pt	48 hrs	1 1 to 3	Do not graze before 3 days or use for hay before 7 days.
pyrethroid, MOA 3A		12 hrs		See **Table 2-25 A** for a list of registered pyrethroids and their re-entry and pre-harvest intervals.
SEEDCORN MAGGOT, WIREWORM				
Use seed pretreated with insecticide for seedcorn maggot control.				Seed can be purchased pretreated. Pretreated seed will not control wireworms.
bifenthrin MOA 3A (Empower) 1.15G	3.5 to 8.7 lb	9 days	9	Apply preplant broadcast incorporated in the top 1 to 3 inches of soil.
phorate, MOA 1B (Thimet) 20 G	4.5 to 7.0 oz/ 1,000 ft row	12 hrs	60	Drill granules to the side of seed at planting. Avoid contact with seed.
SPIDER MITE				
abamectin, MOA 6 (Agri-Mek) 0.7SC	1.75 to 3.5 fl oz	12 hrs	7	Do not allow leaves to be used as livestock feed. Active against immatures and adults.
bifenazate MOA 20D (Acramite) 4 SC	16 to 24 fl oz	12 hrs	3	Active against immatures and adults. Some egg activity.

TABLE 2-2. INSECT CONTROL FOR BEANS (SNAP, LIMA, POLE, EDAMAME) (cont'd)

Insecticide, Mode of Action Code, and Formulation	Amount of Formulation Per Acre	Restricted Entry Interval (REI)	Pre harvest Interval (PHI) (Days)	Precautions and Remarks
SPIDER MITE (cont'd)				
acequinocyl, MOA 20B (Kanemite) 15 SC	31 fl oz	12 hrs	7	Active against all life stages.
fenpyroximate, MOA 21 (Portal) 0.4 EC	2 pt	12 hrs	1	For use on snap bean only. Effective against early instars. Active against immatures and adults.
fenazaquin, MOA 21A (Magister) 1.7	24 to 36 fl oz	12 hrs	3	Do not make more than one application per year. Active against all life stages.
STINK BUG, KUDZU BUG				
pyrethroid, MOA 3A		12 hrs		See **Table 2-25 A** for a list of registered pyrethroids and pre-harvest intervals.
imidacloprid (Admire Pro)	1.2 fl oz	12 hrs	1	
naled, MOA 1B (Dibrom) 8 EC	1.5 pt/ 100-gal water	48 hrs	1	
WHITEFLIES				
acetamiprid MOA 4A (Assail) 30 SC	3.4 to 4.5 fl oz	12 hrs	7	
buprofezin, MOA 16 (Courier) 40 SC	9 to 13.6 fl oz	12 hrs	14	For use on snap beans only.
chlorantraniliprole, MOA 28 (Coragen eVo) 5 SC (Vantacor) 5 SC	2.5 fl oz 2.5 fl oz	4 hrs	1	
cyantraniliprole, MOA 28 **Soil treatment** (Verimark) 1.67 SC **Foliar treatment** (Exirel) 0.83SE	6.75 to 13.5 fl oz 13.5 to 20.5 fl oz	4 hrs 12 hrs	1 (succulent) 7 (dried)	
flupyradifurone, MOA 4D (Sivanto Prime)	10.5 to 14 fl	4 hrs	7	
imidacloprid, MOA 4A **Soil treatment** (Admire Pro) 4.6 F (various) 2 F	7 to 10.5 fl oz 16 to 24 fl oz	12 hrs	21	See label for soil application instructions.
novaluron, MOA 7C (Knack IGR)	**8-10 fl oz**	12 hrs	7	
spirotetramat, MOA 23 (Movento)	4 to 5 fl oz	24 hrs	1(succulent) 7(dry)	PHI is 1 day for succulent beans and 7 days for dry beans.

TABLE 2-3. INSECT CONTROL FOR BEET

Insecticide, Mode of Action Code, and Formulation	Amount of Formulation Per Acre	Restricted Entry Interval (REI)	Pre harvest Interval (PHI) (Days)	Precautions and Remarks
APHID				
flonicamid, MOA 29 (Beleaf) 50 SG	2 to 2.8 oz	12 hrs	3	Begin applications before populations begin to build and before damage is evident. Use higher rates for high populations or dense foliage.
imidacloprid, MOA 4A **Soil treatment** (Admire Pro) 4.6 F (various) 2 F	4.4 to 10.5 fl oz 10 to 24 fl oz	12 hrs	21	See label for soil application instructions. Will also control flea beetle. Do not make exceed one application per season.
Foliar treatment (Admire Pro) 4.6 F (various) 2 F	1.2 fl oz 3.5 fl oz	12 hrs	7	
flupyradifurone, MOA 4D (Sivanto Prime) 1.67 SL	7.0 to 14 fl oz	4 hrs	7	Do not exceed 28 fl oz per acre per season.
thiamethoxam, MOA 4A **Soil treatment** (Platinum) 75 SG	1.7 to 4 oz	12 hrs	30	Soil application only. Platinum may be applied to direct-seeded crops in-furrow at seed or transplant depth, post seeding or transplant as a drench, or through drip irrigation. Do not exceed 3.67 fl oz per acre per season of Platinum. Check label for plant back restrictions for several crops. Will also control flea beetle.
Foliar treatment (Actara) 25 WDG	1.5 to 3 oz	12 hrs	7	Foliar application only. Do not exceed 4 oz per acre per season. Will also control flea beetle.
ARMYWORMS, BEET WEBWORM				
chlorantraniliprole MOA 28 (Coragen eVo) 5 SC (Vantacor) 5 SC	1.2 to 2.5 fl oz 1.2 to 2.5 fl oz	4 hrs	1	
indoxacarb MOA 22A (Avaunt eVo) 30 DG	3.5 to 6 oz	12 hrs	7	Do not use adjuvants with Avaunt eVo.
methoxyfenozide MOA 18 (Intrepid) 2 F	6 to 16 fl oz	4 hrs	1	
spinetoram, MOA 5 (Radiant) 1 SC	6 to 8 fl oz	4 hrs	7	Do not apply more than 32 fluid ounces per acre per season.
spinosad, MOA 5 (Blackhawk)	1.7 to 3.3 oz	4 hrs	3	
viruses		4 hrs		See **Table 2-26** for virus-based insecticides effective against specific insects. **OMRI listed.**
BLISTER BEETLE, FLEA BEETLE				
carbaryl, MOA 1A (Sevin) XLR	1 to 1.5 qt	12 hrs	7	
pyrethroid, MOA 3A		12 hrs		See **Table 2-25 A** for a list of registered pyrethroids and pre-harvest intervals.
LEAFMINER				
spinetoram, MOA 5 (Radiant) 1 SC	6 to 10 fl oz	4 hrs	7	Control will be improved with addition of a spray adjuvant.

TABLE 2-4. INSECT CONTROL FOR BROCCOLI, BRUSSELS SPROUT, CABBAGE, AND CAULIFLOWER

Insecticide, Mode of Action Code, and Formulation	Amount of Formulation Per Acre	Restricted Entry Interval (REI)	Pre harvest Interval (PHI) (Days)	Precautions and Remarks
APHID				
Where whitefly resistance is an issue (or any other insect with a high potential for resistance to Group 4A MOA insecticides), a foliar-applied Group 4A insecticide program and a soil-applied Group 4A program should not be used in the same season. Also, if using a foliar-applied program, avoid using a block of more than three consecutive applications of any products belonging to Group 4A insecticides.				
acetamiprid, MOA 4A (Assail) 30 SC	1.7 to 4.5 fl oz	12 hrs	7	
afidopyropen, MOA 9D (Versys) 0.83 DC	1.5 fl oz	12 hrs	0	Do not make more than 2 sequential applications before using a different mode of action.
clothianidin, MOA 4A (Belay) 2.13 EC **Soil treatment** **Foliar treatment**	9 to 12 fl oz 3 to 4 fl oz	12 hrs	NA 7	Soil application at planting only.
cyantraniliprole, MOA 28 (Exirel) SE	13.5 to 20.5 fl oz	12 hrs	1	Will suppress aphids when applied for lepidopteran larvae.
dimethoate 4 EC, MOA 1B	0.5 to 1 pt	48 hrs	7	Not for use on cabbage
flonicamid, MOA 29 (Beleaf) 50 SG	2 to 2.8 oz	12 hrs	0	
flupyradifurone, MOA 4D (Sivanto Prime) 1.67	7.0 to 14.0 fl oz	4 hrs	1	
imidacloprid, MOA 4A **Soil treatment** (Admire Pro) 4.6 F (various) 2 F	4.4 to 10.5 fl oz 10 to 24 fl oz	12 hrs	21	Do not follow soil applications of Admire with foliar applications of any neonicotinoid insecticide. Use only one application method. See label for soil application instructions. Imidacloprid also controls whiteflies.
Foliar treatment (Admire Pro) 4.6 F (various) 1.6 F	1.3 fl oz 3.75 fl oz	12 hrs	7	Imidacloprid also controls whiteflies. Not effective against fleabeetle.
pymetrozine, MOA 9B (Fulfill) 50 WDG	2.75 oz	12 hrs	7	
pyrifluquinazon, MOA 9B (PQZ) 1.87EC	2.4 to 3.2 fl oz	12 hrs	1	See label for rotational crop restrictions. Do not exceed 4.8 fl oz per acre per crop cycle.
spirotetramat, MOA 23 (Movento) 2 SC	4 to 5 fl oz	24 hrs	1	Do not exceed 10 fluid ounces per season. Requires surfactant.
thiamethoxam, MOA 4A **Soil treatment** (Platinum) 75 SG	1.66 to 3.67 oz	12 hrs	30	Platinum may be applied to direct-seeded crops in-furrow at seed or transplant depth, post seeding or transplant as a drench, or through drip irrigation. Do not exceed 3.67 oz per acre per season.
Foliar treatment (Actara) 25 WDG	1.5 to 3.0 oz	12 hrs	0	Thiamethoxam also controls whiteflies and certain thrips species.
tolfenpyrad, MOA 21A (Torac) 1.29EC	17 to 21 fl oz	12 hrs	1	
DIAMONDBACK MOTH, CABBAGE LOOPER, IMPORTED CABBAGEWORM, CORN EARWORM, CROSS-STRIPED CABBAGEWORM, CABBAGE WEBWORM, ARMYWORMS				
Insecticide-resistant populations of diamondback may not be controlled with some registered insecticides. To manage resistance, avoid transplants from Georgia and Florida, and avoid applying more than 2 sequential applications of insecticides with the same mode of action (MOA) before switching to another MOA. After two applications, rotate to an insecticide with a different mode of action. Do not allow populations to reach high densities before treatments are initiated. Thorough spray coverage is important for achieving effective control and can be improved by the use of a wetting agent. Use of pyrethroid insecticides destroys natural enemies and aggravates diamondback moth infestations.				
Bacillus thuringiensis, MOA 11A (Dipel) DF (Javelin) WG (Xentari) WDG	0.5 to 2 lb 0.12 to 1.5 lb 0.5 to 1.5 lb	4 hrs	0	On foliage every 7 days. On summer or fall plantings, during periods when eggs and larvae are present. This usually occurs when true leaves appear; on other plantings, it may occur later. A spreader/sticker will be helpful. Not effective against Cabbage webworm. **OMRI listed.**
chlorantraniliprole, MOA 28 (Coragen eVo) 5 SC	1.2 to 2.5 fl oz	4 hrs	3	Foliar, drip or soil application. See label for soil application instructions.
cyclaniliprole, MOA 28 (Harvanta) 50 SL	10.9 to 16.4 fl oz	4 hrs	1	Harvanta is for foliar application only.

TABLE 2-4. INSECT CONTROL FOR BROCCOLI, BRUSSELS SPROUT, CABBAGE, AND CAULIFLOWER (cont'd)

Insecticide, Mode of Action Code, and Formulation	Amount of Formulation Per Acre	Restricted Entry Interval (REI)	Pre harvest Interval (PHI) (Days)	Precautions and Remarks
DIAMONDBACK MOTH, CABBAGE LOOPER, IMPORTED CABBAGEWORM, CORN EARWORM, CROSS-STRIPED CABBAGEWORM, CABBAGE WEBWORM, ARMYWORMS (cont'd)				
cyantraniliprole, MOA 28 **Soil treatment** (Verimark) 1.67 SC	5 to 13.5 fl oz	4 hrs	NA	Verimark is for soil application only and may be applied as transplant drench no earlier than 72 hr before planting, or as an in-furrow spray at planting. Higher rates will control aphids.
Foliar treatment (Exirel) 0.83SE	7 to 17 fl oz	12 hrs	1	Exirel is for foliar application only. Use higher rates for cabbage looper.
emamectin benzoate, MOA 6 (Proclaim) 5 WDG	3.2 to 4.8 oz	12 hrs	7	
indoxacarb, MOA 22A (Avaunt eVo) 30 DG	2.5 to 3.5 oz	12 hrs	3	Add a wetting agent to improve spray coverage. Do not apply more than 14 ounces (0.26-lb a.i.) per acre per crop. The minimum interval between sprays is 3 days.
methoxyfenozide, MOA 18 (Intrepid) 2F	4 to 16 fl oz	4 hrs	7	Use lower rates for early season applications to young crops, and higher rates for mid- to late-season applications and heavier infestations. If used for diamondback moth expect only suppression of the infestation. Do not apply more than 16 fl oz per acre per season.
novaluron, MOA 15 (Rimon) 0.83 EC	6 to 12 fl oz	12 hrs	7	Use lower rates when targeting eggs or small larvae, and use higher rates when larvae are large. Make no more than three applications or 24 fluid ounces per acre per season.
spinetoram, MOA 5 (Radiant) 1 SC	5 to 10 fl oz	4 hrs	1	
methomyl MOA 1A (Lannate) 2.4 LV	1.5 to 3 pts	48 hrs	See remarks	PHI varies with crop – check label: Cabbage 1 day, broccoli, brussels sprouts, and cauliflower 3 days. Recommended in a rotational scheme for diamondback moth when multiple modes of action are required over the course of a season.
naled, MOA 1B (Dibrom) 8 EC	1 to 2 pts	48 hrs	1	Do not apply within 25 ft of bodies of water (lakes, rivers, streams, ponds, marshes, etc.) where wind is blowing or gusting towards these areas. Rec- ommended in a rotational scheme for diamondback moth when multiple modes of action are required over the course of a season.
tolfenpyrad, MOA 21A (Torac) 1.29 EC	17 to 21 fl oz	12 hrs	1	Do not make more than 2 applications per crop, or 4 applications per year. Recommended in a rotational scheme for diamondback moth when multiple modes of action are required over the course of a season.
DBM pheromone (CheckMate DBM-F)	2 to 3 fl oz	0	0	This is a pheromone product for mating disruption, not an insecticide. It works by reducing the ability of male moths to locate females and is spe- cific to diamondback moth. Preliminary information suggests application intervals of 1 to 2 wk intervals, but research is underway to assess this frequency.
viruses		4 hrs		See **Table 2-26** for virus-based insecticides effective against specific insects. **OMRI listed.**
FLEA BEETLE, HARLEQUIN BUG, VEGETABLE WEEVIL, YELLOWMARGINED LEAF BEETLE				
clothianidin, MOA 4A (Belay) 2.13 EC **Soil treatment**	9 to 12 fl oz	12 hrs	NA	Soil applications may only be made at planting. See label for application methods.
Foliar treatment	3 to 4 oz	12 hrs	7	Do not exceed 6.4 oz per acre per season. Do not combine soil and foliar applications.
cyantraniliprole, MOA 28 **Soil treatment** (Verimark) 1.67 SC	6.75 to 13.5 fl oz	4 hrs	NA	Verimark is for at planting soil application only. See label for application options. Use higher rates for flea beetle. Not for harlequin or stink bug.
Foliar treatment (Exirel) 0.83 SE	13.5 to 20.5 fl oz	12 hrs	1	Exirel is for foliar application only. Not for harlequin or stink bug.
dinotefuran, MOA 4A **Foliar treatment** (Venom) 70 SG (Scorpion) 35 SL	1 to 4 oz 2 to 7 oz	12 hrs	1	See label for application options. Do not combine soil and foliar applica- tions.
Soil treatment (Venom) 70 SG (Scorpion) 35 SL	5 to 6 oz 9 to 10.5 fl oz	12 hrs	21	
dimethoate 4 EC, MOA 1B	0.5 to 1 pt	48 hrs	7	
pyrethroid, MOA 3A		12 hrs		See **Table 2-25 A** for a list of registered pyrethroids and pre-harvest intervals
spinosad, MOA 5 (Entrust) SC	5 to 10 fl oz	4 hrs	1	Spinosad is one of the few effective OMRI-listed products against yellowmargined leaf beetle larvae and adults. YMLB is primarily a pest in low-input systems. See label restrictions for Georgia.

TABLE 2-4. INSECT CONTROL FOR BROCCOLI, BRUSSELS SPROUT, CABBAGE, AND CAULIFLOWER (cont'd)

Insecticide, Mode of Action Code, and Formulation	Amount of Formulation Per Acre	Restricted Entry Interval (REI)	Pre harvest Interval (PHI) (Days)	Precautions and Remarks
ROOT MAGGOT				
diazinon, MOA 1B (Diazinon 50 W) 50 WP	0.25 to 0.5 lb/ 50 gal	4 days	—	**Transplant water:** Apply in transplant water or drench water at 4- to 6-oz per plant at transplanting.
cyantraniliprole, MOA 28 (Verimark) 1.67 SC	10 to 13.5 fl oz	4 hrs	—	Apply to soil at planting as an in-furrow spray, transplant tray drench, transplant water, hill drench, surface band, or soil shank.
bifenthrin, MOA 3A (Capture) LFR	3.4 to 6.8 fl oz	12 hrs	—	Apply as a 5-to 7-inch band over the open seed or transplant furrow, or in furrow with the transplant. May be applied through transplant water.
tolfenpyrad, MOA 21A (Torac) 1.29EC	21 fl oz	12 hrs	1	Apply to soil at planting as in-furrow spray or surface band.
THRIPS				
dimethoate 4 EC, MOA 1B	0.5 to 1 pt	48 hrs	7	
acetamiprid, MOA 4A (Assail) 30 SC	3.4 to 4.5 fl oz	12 hrs	7	Efficacy will vary depending on thrips species.
imidacloprid, MOA 4A (Admire Pro) 4.6F (various) 2F	1.3 fl oz 3.0 fl oz	12 hrs	7	Check label for rates for other formulations. Foliar applications only. Efficacy will vary depending on thrips species.
methomyl, MOA 1A (Lannate) 2.4 LV	1.5 pt	48 hrs	3 (1 for cabbage)	
novaluron, MOA 15 (Rimon) 0.83 EC	6 to 12 fl oz	12 hrs	7	Make no more than two applications, or 24 fl oz per acre per season.
spinetoram, MOA 5 (Radiant) 1 SC	6 to 10 fl oz	4 hrs	1	
WHITEFLY				

Where whitefly resistance is an issue (or any other insect with a high potential for resistance to Group 4A MOA insecticides), a foliar-applied Group 4A insecti- cide program and a soil-applied Group 4A program should not be used in the same season. Also, if using a foliar-applied program, avoid using a block of more than 3 consecutive applications of any product belonging to Group 4A insecticides.

Insecticide, Mode of Action Code, and Formulation	Amount of Formulation Per Acre	Restricted Entry Interval (REI)	Pre harvest Interval (PHI) (Days)	Precautions and Remarks
acetamiprid, MOA 4A (Assail) 30 SC	2.5 to 4.0 oz	12 hrs	7	Use spreader stick to improve control.
afidopyropen, MOA 9D (Versys) 0.83 DC	2.1 to 4.5 fl oz	12	0	Do not make more than 2 sequential applications before using a different mode of action.
buprofezin (Courier SC)	9.0 to 13.6 fl oz	12	1	Do not make more than two applications per crop cycle.
cyantraniliprole, MOA 28 **Soil treatment** (Verimark) 1.67 SC	6.75 to 10 fl oz	4 hrs	NA	Verimark is for at planting soil application or drip chemigation. See label for application options.
Foliar treatment (Exirel) 0.83 SE	13.5 to 20.5 fl oz	12 hrs	1	Exirel is for foliar application only.
dinotefuran, MOA 4A **Foliar treatment** (Venom) 70 SG (Scorpion) 35 SL	1 to 4 oz 2 to 7 oz	12 hrs	1	Do not follow soil applications with foliar applications of any neonicotinoid insec- ticide. Use only one application method. Do not apply more than 6 oz per acre per season using foliar applications, or 12 oz per acre per season using soil applications. Soil applications may be applied by a narrow band below or above the seed line at planting; a post-seeding or transplant drench with sufficient water to ensure incorporation to the root zone; or through drip irrigation
Soil treatment (Venom) 70 SG (Scorpion) 35 SL	5 to 6 oz 9 to 10.5 fl oz		21	
flupyradifurone, MOA 4D (Sivanto Prime) **Foliar treatment**	10.5 to 14 fl oz	4 hrs	1	Do not exceed 28 fl oz per acre per season. Do not combine foliar and soil applica- tions. See label for application options.
Soil treatment	21 to 28 fl oz	4 hrs	21	
spiromesifen, MOA 23 (Oberon) 2 SC	8.5 fl oz	12 hrs	14	Do not exceed 22.4 fl oz per acre per season.
spirotetramat, MOA 23 (Movento) MPC	7 to 8 fl oz	24 hrs	1	Do not exceed 16 fl oz per season. Requires surfactant. Use a spread- er-penetrator adjuvant.
pyrifluquinazon, MOA 9B (PQZ) 1.87 EC	2.4 to 3.2 fl oz	12 hrs	1	See label for rotational crop restrictions. Do not exceed 4.8 fl oz per acre per crop cycle.
pyriproxyfen, MOA 7C (Knack) 0.86 EC	8 to 10 fl oz	12 hrs	7	Only treat whole fields, and do not apply to any crop other than those for which Knack is registered within 30 days after the last application. Will not control adults.

TABLE 2-5. INSECT CONTROL FOR CARROT

Insecticide, Mode of Action Code, and Formulation	Amount of Formulation Per Acre	Restricted Entry Interval (REI)	Pre harvest Interval (PHI) (Days)	Precautions and Remarks
APHID, LEAFHOPPER				
imidacloprid, MOA 4A **Soil treatment** (Admire Pro) 4.6 F (various) 1.6 F	4.4 to 10.5 fl oz 10 to 24 fl oz	12 hrs	21	Must be applied to the soil. May be applied via chemigation into the root zone through low-pressure drip, trickle, micro sprinkler, or equivalent equipment; in-furrow spray or shanked-in 1 to 2 inches below seed depth during planting; or in a narrow band (2 inches or less) 1 to 2 inches directly below the eventual seed row in a bedding operation 14 or fewer days before planting. Higher rates provide longer-lasting control. See label for information on approved application methods and rate per 100 row feet for different row spacing.
thiamethoxam, MOA 4A **Soil treatment** (Platinum) 75 SG	1.66 to 3.67 oz	12 hrs	30	Platinum may be applied to direct-seeded crops in-furrow at seeding, immediately after seeding with sufficient water to ensure incorporation into the root zone, or through trickle irrigation
Foliar treatment (Actara) 25 WDG	1.5 to 3 oz	12 hrs	7	Actara is applied to foliage. Do not exceed 4 ounces Actara per acre per season.
flonicamid, MOA 29 (Beleaf) 50 SG	2 to 2.8 fl oz	12 hrs	3	
flupyradifurone, MOA 4D (Sivanto Prime) 1.67 SL	7.0 to 10.5 fl oz	4 hrs	7	
cyantraniliprole, MOA 28 (Exirel) 0.83 SE	13.5 to 20.5 fl oz	12 hrs	1	
ARMYWORMS, PARSLEYWORM				
pyrethroid, MOA 3A		12 hrs		See **Table 2-25 A** for a list of registered pyrethroids and pre-har- vest intervals.
carbaryl, MOA 1A (Sevin) XLR Plus	1 to 2 qt	12 hrs	7	On foliage as needed
chlorantraniliprole, MOA 28 (Coragen eVo) 5 SC	1.2 to 2.5 fl oz	4 hrs	1	Coragen may be used for foliar or drip chemigation.
cyantraniliprole, MOA 28 (Exirel) 0.83 SE	13.5 to 20.5 fl oz	12 hrs	1	
emamectin benzoate, MOA 6 (Proclaim) 5 SG	3.2 to 4.8 fl oz	12 hrs	7	
methomyl, MOA 1A (Lannate) 2.4 LV (Lannate) 90 SP	0.75 to 1.5 pt 0.25 to 0.5 lb	48 hrs	1	
methoxyfenozide, MOA 18 (Intrepid) 2 F	4 to 10 fl oz	4 hrs	1	Use higher rates against large larvae.
spinetoram, MOA 5 (Radiant) 1 SC	6 to 8 fl oz	4 hrs	3	Radiant will not control leafhoppers. Do not make more than 4 applications per year.
viruses		4 hrs		See **Table 2-26** for virus-based insecticides effective against specific insects. **OMRI listed.**
LEAFMINER				
spinetoram, MOA 5 (Radiant) 1 SC	6 to 8 fl oz	4 hrs	3	
chlorantraniliprole, MOA 28 (Coragen eVo) 5 SC	1.2 to 2.5 fl oz	4 hrs	1	Coragen may be used for foliar or drip chemigation.
cyantraniliprole, MOA 28 (Exirel) 0.83 SE	13.5 to 20.5 fl oz	12 hrs	1	
WIREWORM				
diazinon, MOA 1B (Diazinon) (AG 500)	4 qt	3 days	—	Broadcast and incorporate preplant.

TABLE 2-6. INSECT CONTROL FOR CELERY

Insecticide, Mode of Action Code, and Formulation	Amount of Formulation Per Acre	Restricted Entry Interval (REI)	Pre harvest Interval (PHI) (Days)	Precautions and Remarks
APHID, LEAFHOPPER, FLEA BEETLE				
afidopyropen, MOA 9D (Versys) DC	1.5 fl oz	12 hrs	0	Do not make more than 2 sequential applications before using a different mode of action. Will not control flea beetle
imidacloprid, MOA 4A (Admire Pro) 4.6 F (various) 2 F	7 to 10.5 fl oz 16 to 24 fl oz	12 hrs	21	Apply via chemigation into the root zone, as an in-furrow spray at planting on/or below the seed, or as a post-seeding or transplant drench.
thiamethoxam, MOA 4A (Actara) 25 WDG	1.5 to 3 oz	12 hrs	7	
acetamiprid, MOA 4A (Assail) 30 SC	1.7 to 3.4 fl oz	12 hrs	7	
clothianidin, MOA 4A (Belay) 2.13 SC		4 hrs	21	
Soil treatment	9 to 12 fl oz			
Foliar treatment	3 to 4 fl oz	4 hrs	1	
sulfoxaflor, MOA 4C (Transform) WG	.75 to 2.75 fl oz	12 hrs	3	
flonicamid, MOA 29 (Beleaf) 50 SG	2 to 2.8 oz	12 hrs	0	Will not control flea beetle.
flupyradifurone, MOA 4D (Sivanto Prime) 1.67 SL	10.5 to 12.0 fl oz	4 hrs	1	Will not control flea beetle
pyrifluquinazon, MOA 29 (PQZ) 1.87EC	2.4 to 3.2 fl oz	12 hrs	1	See label for rotational crop restrictions. Do not exceed 4.8 fl oz per acre per crop cycle. Will not control flea beetle
spirotetramat, MOA 23 (Movento) 2 SC	4 to 5 fl oz	24 hrs	3	Do not exceed 10 fluid ounces per season. Not for flea beetle. Requires surfactant. Will not control flea beetle
cyantraniliprole, MOA 28 (Exirel) 0.83 SE	13.5 to 20.5 fl oz	12 hrs	1	
cyclaniliprole, MOA 28 (Harvanta) 50 SL	10.9 to 16.4 fl oz	4 hrs	1	
tolfenpyrad, MOA 21A (Torac) 1.29 EC	17 to 21 fl oz	12 hrs	1	
ARMYWORMS, CORN EARWORM, LOOPER				
chlorantraniliprole, MOA 28 (Coragen eVo) 5 SC	1.2 to 2.5 fl oz	4 hrs	1	Foliar or drip chemigation. Drip chemigation must be applied uniformly to the root zone. See label for instructions.
cyantraniliprole, MOA 28 (Exirel) 0.83 SE	7 to 13.5 fl oz	12 hrs	1	
cyclaniliprole, MOA 28 (Harvanta) 50 SL	10.9 to 16.4 fl oz	4 hrs	1	
emamectin benzoate, MOA 6 (Proclaim) 5 WDG	2.4 to 4.8 oz	12 hrs	7	Do not make more than two sequential applications without rotating to another product with a different mode of action.
indoxacarb, MOA 22A (Avaunt) eVo	3.5 oz	12 hrs	3	
methomyl, MOA 1A (Lannate) 2.4 LV	3 pt	48 hrs	7	Methomyl may induce leafminer infestations.
methoxyfenozide, MOA 18 (Intrepid) 2 F	4 to 10 fl oz	4 hrs	7	For early season applications only to young crop and small plants. For mid- to late-season applications and heavier infestations and under conditions in which thorough coverage is more difficult. Do not apply more than 16 fl oz per application and do not exceed 64 fl oz per season. See Rotational Crop Restrictions on label.
pyrethroid, MOA 3A		12 hrs		See **Table 2-25 A** for registered pyrethroids and pre-harvest intervals.
spinetoram, MOA 5 (Radiant) 1 SC	5 to 10 fl oz	4 hrs	1	Use higher rates for armyworms.
viruses		4 hrs		See **Table 2-26** for virus-based insecticides effective against specific insects. **OMRI** listed.
LEAFMINER				
abamectin, MOA 6 (Agri-Mek) 0.15EC	1.75 to 3.5 fl oz	12 hrs	7	
chlorantraniliprole, MOA 28 (Coragen eVo) 5 SC	5 to 7.5 fl oz	4 hrs	1	Foliar or drip chemigation. Drip chemigation must be applied uniformly to the root zone. See label for instructions.
cyromazine, MOA 17 (Trigard) 75 WP	2.66 oz	12 hrs	7	
spinetoram, MOA 5 (Radiant) 1 SC	6 to 10 fl oz	4 hrs	1	

TABLE 2-7. INSECT CONTROL FOR COLLARD, KALE, AND MUSTARD GREENS

Insecticide, Mode of Action Code, and Formulation	Amount of Formulation Per Acre	Restricted Entry Interval (REI)	Pre harvest Interval (PHI) (Days)	Precautions and Remarks
APHID				
acetamiprid, MOA 4A (Assail) 30 SC	1.7 to 4.5 fl oz	12 hrs	7	
afidopyropen, MOA 9D (Versys) DC	1.5 fl oz	12 hrs	0	Do not make more than 2 sequential applications before using a different mode of action.
clothianidin, MOA 4A (Belay) 2.13EC			21	
Soil treatment	9 to 12 fl oz	9 to 12 fl oz		
Foliar treatment	3 to 4 fl oz	3 to 4 fl oz	7	
flonicamid, MOA 29 (Beleaf) 50 SG	2 to 2.8 fl oz	12 hrs	0	
flupyradifurone, MOA 4D (Sivanto Prime) 1.67 SL	10.5 to 14.0 fl oz	4 hrs	1	
imidacloprid, MOA 4A				See label for soil application instructions. Admire Pro will also control flea beetle.
Soil treatment		12 hrs	21	
(Admire Pro) 4.6 F	4.4 to 10.5 fl oz			
(various) 2 F	10 to 24 fl oz			
Foliar treatment		12 hrs	7	
(Admire Pro) 4.6 F	1.3 fl oz			
(various) 2 F	3.8 fl oz			
pymetrozine, MOA 9B (Fulfill) 50 WDG	2.75 oz	12 hrs	7	
pyrifluquinazon, MOA 29 (PQZ) 1.87 EC	2.4 to 3.2 fl oz	12 hrs	1	See label for rotational crop restrictions. Do not exceed 4.8 fl oz per acre per crop cycle.
sulfoxaflor, MOA 4C (Closer) SC	4.25 to 5.75 fl oz	12 hrs	3	
spirotetramat, MOA 23 (Movento) 2 SC	4 to 5 fl oz	24 hrs	1	Do not exceed 10 fluid ounces per season. Requires surfactant.
DIAMONDBACK MOTH, CABBAGE LOOPER, IMPORTED CABBAGEWORM, CROSS-STRIPED CABBAGEWORM, CABBAGE WEBWORM, ARMYWORMS				

Insecticide-resistant populations of diamondback may not be controlled with some registered insecticides. To manage resistance, avoid transplants from Georgia and Florida, and avoid applying more than 2 sequential applications of insecticides with the same mode of action

(MOA) before switching to another MOA. After two applications, rotate to an insecticide with a different mode of action. Do not allow populations to reach high densities before treatments are initiated. Thorough spray coverage is import- ant for achieving effective control and can be improved by the use of a wetting agent. Use of pyrethroid insecticides destroys natural enemies and aggravates diamondback moth infestations.

Insecticide, Mode of Action Code, and Formulation	Amount of Formulation Per Acre	Restricted Entry Interval (REI)	Pre harvest Interval (PHI) (Days)	Precautions and Remarks
Bacillus thuringiensis, MOA 11A		4 hrs	0	Use a spreader/sticker. **OMRI listed (excluding Crymax).**
(Crymax) WDG	0.5 to 1.5 lb			
(Dipel) 2X, DF	8 oz			
(Dipel)	1 pt			
(Xentari) DF	0.5 to 2 lb			
chlorantraniliprole, MOA 28 (Coragen eVo) 5 SC	1.2 to 2.5 fl oz	4 hrs	1	Foliar or drip chemigation. Drip chemigation must be applied uniformly to the root zone. See label for instructions.
cyantraniliprole, MOA 28		12 hrs	1	Verimark is for soil application only and may be applied as transplant drench no earlier than 72 hr before planting, or as an in-furrow spray at planting. Higher rates will control aphids.
Soil treatment (Verimark) 1.67 SC	5 to 13.5 fl oz			
Foliar treatment (Exirel) 0.83SE	7 to 17 fl oz	4 hrs	1	Exirel is for foliar application only. Use higher rates for cabbage looper
cyclaniliprole, MOA 28 (Harvanta) 50 SL	10.9 to 16.4 fl oz	4 hrs	1	Harvanta is for foliar application only.
emamectin benzoate, MOA 6 (Proclaim) 5 WDG	2.4 to 4.8 oz	12 hrs	14	
indoxacarb, MOA 22A (Avaunt eVo) 30 WDG	3.5 oz	12 hrs	3	Do not apply Avaunt more than twice to any generation of diamondback moth. After two applications, rotate to an insecticide with a different mode of action. Do not make more than 6 applications (4 in GA) or exceed 14 ounces per season per crop.
spinetoram, MOA 5 (Radiant) 1 SC	5 to 10 fl oz	4 hrs	1	Do not use on Collard, Kale, and Mustard Greens in Georgia.

TABLE 2-7. INSECT CONTROL FOR COLLARD, KALE, AND MUSTARD GREENS (cont'd)

Insecticide, Mode of Action Code, and Formulation	Amount of Formulation Per Acre	Restricted Entry Interval (REI)	Pre harvest Interval (PHI) (Days)	Precautions and Remarks
DIAMONDBACK MOTH, CABBAGE LOOPER, IMPORTED CABBAGEWORM, CROSS-STRIPED CABBAGEWORM, CABBAGE WEBWORM, ARMYWORMS (cont'd)				
methoxyfenozide, MOA 18 (Intrepid) 2 F	4 to 10 fl oz	4 hrs	1	Use lower rates for early season applications to young crops and higher rates for mid- to late-season applications and heavier infestations. If used for diamondback moth expect only suppression of the infestation. Do not apply more than 16 fl oz per acre per season.
methomyl MOA 1A (Lannate 2.4 LV)	1.5 to 3 pt	48 hrs	10	PHI varies with crop – check label: Cabbage 1 day, broccoli, brussels sprouts, and cauliflower 3 days. Recommended in a rotational scheme for diamondback moth when multiple modes of action are required over the course of a season.
naled, MOA 1B (Dibrom) 8 EC	1 to 2 pts	48 hrs	1	Do not apply within 25 ft of bodies of water(lakes, rivers, streams, ponds, marshes, etc.) where wind is blowing or gusting towards these areas. Recommended in a rotational scheme for diamondback moth when multiple modes of action are required over the course of a season.
tolfenpyrad, MOA 21A (Torac) 1.29 EC	17 to 21 fl oz	12 hrs	1	Do not make more than 2 applications per crop, or 4 applications per year. Recommended in a rotational scheme for diamondback moth when multiple modes of action are required over the course of a season.
DBM pheromone (CheckMate DBM-F)	2 to 3 fl oz	0	0	This is a pheromone product for mating disruption, not an insecticide. It works by reducing the ability of male moths to locate females and is specific to diamondback moth. Preliminary information suggests application intervals of 1 to 2 wk intervals, but research is underway to assess this frequency.
viruses		4 hrs		See **Table 2-26** for virus-based insecticides effective against specific insects. **OMRI listed**.
FLEA BEETLE				
carbaryl, MOA 1A (Sevin) SL (Sevin) XLR Plus	0.5 to 2 qt 0.5 to 1 qt	12 hrs	3 to 14	See label for crop specific restrictions on PHI.
dinotefuran, MOA 4A **Foliar treatment** (Venom) 70 SG (Scorpion) 35 SL	1 to 4 oz 2 to 7 fl oz	12 hrs	7	Do not follow soil applications with foliar applications. Use only one application method. Do not apply more than 6 ounces per acre per season using foliar applications, or 12 ounces per acre per season using soil applications.
Soil treatment (Venom) 70 SG (Scorpion) 35 SL	2 to 3 oz 9 to 10.5 fl oz		21	Soil applications may be applied by a narrow band below or above the seed line at planting; a post-seeding or transplant drench with sufficient water to ensure incorporation to the root zone; or through drip irrigation.
pyrethroid, MOA 3A		12 hrs		See **Table 2-25 A** for a list of registered pyrethroids and pre- harvest inter- vals. May flare diamond back moth populations.
GRASSHOPPER				
pyrethroid, MOA 3A		12 hrs		See **Table 2-25 A** for a list of registered pyrethroids and pre-harvest inter- vals. May flare diamond back moth populations.
HARLEQUIN BUG, STINK BUG, YELLOWMARGINED LEAF BEETLE				
clothianidin, MOA 4A (Belay) 2.13EC **Soil treatment** **Foliar treatment**	9 to 12 fl oz 3 to 4 fl oz	12 hrs 12 hrs	21 7	
dinotefuran, MOA 4A (Venom) 70 SG (Scorpion) 35 SL	2 to 3 oz 2 to 7 fl oz	12 hrs	7	Dinotefuran recommendations are for foliar applications.
pyrethroid, MOA 3A		12 hrs		See **Table 2-25 A** for a list of registered pyrethroids and pre-harvest inter- vals. May flare diamond back moth populations.
thiamethoxam, MOA 4A (Actara) 25 WDG	3 to 5.5 oz	12 hrs	7	
spinosad, MOA 5 (Entrust) SC	5 to 10 fl oz	4 hrs	1	Effective against yellowmargined leaf beetle only and is one of the only OMRI labelled insecticides that can suppress this pest. See label restrictions for Georgia.
ROOT MAGGOT				
tolfenpyrad, MOA 21A (Torac)	21 fl oz	12 hrs	1	Read soil application guidelines on label.
cyantraniliprole, MOA 28 (Verimark)	10 to 13.5 fl oz	4 hrs	at-planting only	

TABLE 2-7. INSECT CONTROL FOR COLLARD, KALE, AND MUSTARD GREENS (cont'd)

Insecticide, Mode of Action Code, and Formulation	Amount of Formulation Per Acre	Restricted Entry Interval (REI)	Pre harvest Interval (PHI) (Days)	Precautions and Remarks
WHITEFLY				
acetamiprid, MOA 4A (Assail) 30 SC	2.1 to 4.5 fl oz	12 hrs	7	Apply against adults, before nymphs are present. Use a spreader/sticker to improve control.
afidopyropen, MOA 9D (Versys) DC	5 to 7 oz	12 hrs	0	Do not make more than 2 sequential applications before using a different mode of action.
buprofezin, MOA 16 (Courier) 40 SC	9 to 13.6 fl oz	12 hrs	1	
chlorantraniliprole, MOA 28 (Coragen eVo) 5 SC (Vantacor) 5 SC	2.5 fl oz 2.5 fl oz	4 hrs	3	
cyantraniliprole, MOA 28 **Soil treatment** (Verimark) 1.67 SC	6.75 to 13.5 fl oz	4 hrs	1	
Foliar treatment (Exirel) 0.83 SE	13.5 to 20.5 fl oz	12 hrs	1	
dinotefuran, MOA 4A (Venom) 70 SG (Scorpion) 35 SL	2 to 3 oz 2 to 7 fl oz	12 hrs	7	
flupyradifurone, MOA 4D (Sivanto Prime)	10.5 to 14.0 fl oz	4 hrs	1	Do not make more than 3 applications or apply more than 28 fluid oz per season.
pyriproxyfen, MOA 7C (Knack) 0.86 EC	8 to 10 fl oz	12 hrs	7	Do not apply Knack more than twice per season or exceed 0.134 pound per acre per season.
pyrifluquinazon, MOA 29 (PQZ) 1.87 EC	2.4 to 3.2 fl oz	12 hrs	1	See label for rotational crop restrictions. Do not exceed 4.8 fl oz per acre per crop cycle.
spiromesifen, MOA 23 (Oberon) 2 SC	7 to 8.5 fl oz	12 hrs	7	Do not make more than 3 applications or apply more than 25.5 fl oz per season.
spirotetramat, MOA 23 (Movento) MPC	7 to 8 fl oz	24 hrs	1	Do not exceed 16 fluid ounces per season. Requires surfactant.

TABLE 2-8. INSECT CONTROL FOR CORN, SWEET

Insecticide, Mode of Action Code, and Formulation	Amount of Formulation Per Acre	Restricted Entry Interval (REI)	Pre harvest Interval (PHI) (Days)	Precautions and Remarks
CORN EARWORM, FALL ARMYWORMS, EUROPEAN CORN BORER				
The consistency of pyrethroid insecticides in controlling corn earworm populations has declined in recent years. If reduced efficacy is observed, switch to insecticides with different modes of action. Likewise, similar loss of efficacy has been reported to group 28 diamides in some areas.				
transgenic sweet corn varieties expressing *Bt* protein	Highly effective against European corn borer. Effectiveness against corn earworm will vary among BT traits and there is evidence that resistance in corn earworm to commonly used traits is becoming common. Varieties containing the *Vip3A* gene (Attribute II and Attribute Plus Series) are still effective at controlling corn earworm. Additional insecticide applications may be required to prevent damage to the ear tips of varieties without the *Vip3A* gene.			
pyrethroid, MOA 3A		12 hrs		Check label for variety limitations and grazing restrictions. In addition, instances of corn earworm resistance to pyrethroids are becoming more prevalent in recent years. To protect ears, begin sprays when tassel shoots first appear. The frequency of sprays will vary depending on location and intensity of earworm populations, ranging from daily to twice weekly in higher elevations. Corn earworms & fall armyworms present in the late whorl stage must be controlled before tassel emergence to prevent migration to ears.
chlorantraniliprole MOA 28 (Coragen eVo) 5 SC (Vantacor) 5 SC	1.2 to 2.5 fl oz 1.7 to 2.5 fl oz	4 hrs	1	Do not apply more than 15.4 fl oz of Coragen or 0.2 lb a.i. of chlorantraniliprole per acre per year.
bifenthrin, MOA 3A + chlorantraniliprole, MOA 28 (Elevest) SC	5.6 to 9.6 fl oz	12 hrs	1	Do not make more than two applications per acre or exceed 0.2 lb [AI] per calendar year.

TABLE 2-8. INSECT CONTROL FOR CORN, SWEET (cont'd)

Insecticide, Mode of Action Code, and Formulation	Amount of Formulation Per Acre	Restricted Entry Interval (REI)	Pre harvest Interval (PHI) (Days)	Precautions and Remarks
CORN EARWORM, FALL ARMYWORMS, EUROPEAN CORN BORER (cont'd)				
lambda-cyhalothrin, MOA 3A + chlorantraniliprole, MOA 28 (Besiege) ZC	6 to 10 fl oz	24 hrs	1	Do not allow livestock to graze in treated areas or harvest treated corn foliage as feed for meat or dairy animals within 1 day after last treatment. Do not feed treated corn fodder or silage to livestock within 21 days of last application.
methoxyfenozide, MOA 18 + spinetoram, MOA 5 (Intrepid Edge)	8 to 12 fl oz	4 hrs	3	
methomyl, MOA 1A (Lannate) 90 SP (Lannate) 2.4 LV	4 to 8 oz 0.75 to 1.5 pt	48 hrs	0	Do not use methomyl for European corn borer control.
indoxacarb, MOA 22AA (Avaunt eVo) 30 WDG	2.5 to 3.5 oz	12 hrs 14 days for hand harvesting	3	For control of fall armyworm and European corn borer in whorl stage only. Do not apply more than 14 ounces Avaunt (0.26 lb a.i.) per acre per crop. Minimum interval between sprays is 3 days. Make no more than 4 applications per season.
spinetoram, MOA 5 (Radiant) 1 SC	3 to 6 fl oz	4 hrs	1	Do not apply more than 36 ounces per acre per year.
spinosad, MOA 5 (Blackhawk)	2.2 to 3.3 oz	4 hrs	1	
viruses		4 hrs		See **Table 2-26** for virus-based insecticides effective against specific insects. **OMRI listed.**
CUTWORM				
pyrethroid, MOA 3A		12 hrs		See **Table 2-25 A** for a list of registered pyrethroids and pre-harvest intervals.
SAP BEETLE, FLEA BEETLE, GRASSHOPPER, JAPANESE BEETLE, ROOTWORM BEETLE				
pyrethroid, MOA 3A		12 hrs		See **Table 2-25 A** for a list of registered pyrethroids and pre-harvest intervals.
acetamiprid 4A (Assail 30 SC)	3.4 to 4.5 fl oz	12 hrs	7	Do not exceed 9.4 oz (0.21 lb AI) per acre per season.
carbaryl, MOA 1A (Sevin) XLR Plus	1 to 2 qt	24 hrs	2	Sap beetle infestations are usually associated with prior ear damage. Popu- lations build on over mature and damaged fruit and vegetables. Sanitation is critical. No hand harvest.
CORN SILK FLY				
The corn silk fly is a tropical and subtropical pest and most common in Florida. However, in recent years it been reported AL, KY, GA, LA and SC.				
acetamiprid, MOA 4A (Assail 30 SC)	4.5 fl oz	12 hrs	7	Maximum of 11.2 fl oz per growing season.
malathion, MOA 1B (Fayfanon ULV AG)	6 to 8 fl oz	12 hrs	5	Maximum of 2 applications per year.
acetamiprid, MOA 4A, + novaluron, MOA 3A (Cormoran)	12 fl oz	12 hrs	7	Maximum of 40 fl oz/acre season, or 0.21 lb AI acetamiprid and 0.39 lb AI novaluron per year.
pyrethroids, MOA 3A		12 hrs		See **Table 2-25 A** for a list of registered pyrethroids and pre- harvest intervals.
SOUTHERN CORN BILLBUG, ROOTWORM, WIREWORM, CHINCH BUGS				
Seed treatments: clothianidin, MOA 4A (Poncho 600)	1.13 to 2.26 fl oz per 80,000 seeds		—	Seed treatments are applied by commercial seed treaters only. Not for use in hopper bins, slurry mixes, or any other type of on farm treatment.
imidacloprid, MOA 4A (Gaucho 600)	3.6 to 6 oz per cwt seed			
pyrethroid, MOA 3A		12 hrs		See **Table 2-25 A** for a list of registered pyrethroids and pre-harvest intervals.
terbufos, MOA 1B (Counter 20 G)	Banded or In-Furrow: 4.5 to 6.0 oz per 1,000 ft row for any row spacing		60	Place granules in a 7-inch band over the row directly behind the planter shoe in front of press wheel. Place granules directly in the seed furrow behind the planter shoe. Rotation is advised.
broflanilide, MOA 30 (Nurizma)	In furrow: 0.05 to 0.07 fl oz per 1000 ft row	12 hrs	NA	For in-furrow use only. Spray into open seed furrow between the planter furrow openers and press wheels. Do not apply more than 0.0445 lb active ingredient per application or per year.

TABLE 2-8. INSECT CONTROL FOR CORN, SWEET (cont'd)

Insecticide, Mode of Action Code, and Formulation	Amount of Formulation Per Acre	Restricted Entry Interval (REI)	Pre harvest Interval (PHI) (Days)	Precautions and Remarks
SPIDER MITES				
abamectin, MOA 6 (Agri-Mek SC)	1.75 to 3.5 fl oz	12 hrs	7	Thorough cover is important for good control. Do not make than 2 applications or exceed 7 fl oz per acre per season,
etoxazole, 10B (Zeal SC)	2 to 6 fl oz	12 hrs	21	Do not make for than 1 application per season.
spiromesifen, MOA 23 (Oberon 2 SC)	5.7 to 16.0 fl oz	12 hrs	5	Do not exceed 17 fl oz per acre per season or make more than 2 applications.
STINK BUG				
pyrethroids, MOA 3A				See **Table 2-25 A** for a list of registered pyrethroids and pre-harvest intervals.
methomyl, MOA 1A (Lannate) 90 SP	0.5 lb	48 hrs	0	Certain hybrid varieties of sweet corn are susceptible to methomyl injury.

TABLE 2-9. INSECT CONTROL FOR CUCURBITS (CUCUMBER, CANTALOUPE, PUMPKIN, SQUASH, & WATERMELON)

Insecticide, Mode of Action Code, and Formulation	Amount of Formulation Per Acre	Restricted Entry Interval (REI)	Pre harvest Interval (PHI) (Days)	Precautions and Remarks
NOTE: Insecticide applications in cucurbits should be made in late evening to protect pollinating insects. Refer to the pollination section of the general production recommendations in this publication for more information about protecting pollinators.				
"Rindworms" are a pest complex that feed on the rinds of watermelons and other cucurbits. Damage is primarily caused by various caterpillars or cucumber beetles (both larvae and adults). It is possible to identify the responsible insect from damage feeding patterns. Identifying which insect(s) are responsible for damage is critical as insecti- cide recommendations differ.				
APHID				
Where whitefly resistance is an issue (or any other insect with a high potential for resistance to Group 4A MOA insecticides), a foliar applied Group 4A insecti- cide program and a soil-applied Group 4A program should not be used in the same season. Also, if using a foliar-applied program, avoid using a block of more than three consecutive applications of any products belonging to Group 4A insecticides.				
acetamiprid MOA 4A (Assail) 30 SC	2.1 to 3.4 fl oz	12 hrs	0	Do not exceed 0.5 pound per acre per season.
afidopyropen, MOA 9D (Sefina) DC	3 oz	12 hrs	0	Do not make more than 2 sequential applications before using a different mode of action.
cyantraniliprole MOA 28 (Verimark) 1.67 SC	6.75 to 13.5 fl oz	4 hrs	1	Applied to the soil at planting or later via drip irrigation system. See label for appli- cation options.
flonicamid, MOA 29 (Beleaf) 50 SG	2 to 2.8 oz	12 hrs	0	Begin applications before populations begin to increase and before damage is evident.
flupyradifurone, MOA 4D (Sivanto Prime) 1.67 **Soil treatment**	21 to 28 fl oz	4 hrs	21	Soil applications through drip irrigation, injected below the seed level at plant- ing, or drench at transplanting.
Foliar treatment	7 to 14 fl oz	4 hrs	1	DO NOT make foliar applications of Sivanto to muskmelon, cantaloupe, or hon- eydew melon. See label for additional melons to which it should not be applied.
imidacloprid, MOA 4A (Admire Pro) 4.6 F	7 to 10.5 fl oz	12 hrs	21	Must be applied to the soil. May be applied preplant; at planting as a post-seed- ing drench, transplant water drench, or hill drench; subsurface side- dress or by chemigation using low-pressure drip irrigation methods. Will also control cucumber beetles, thrips, and whiteflies.
pymetrozine, MOA 9B (Fulfill) 50 WDG	2.75 oz	12 hrs	0	Apply before aphids reach damaging levels. Do not exceed 5.5 ounces per acre per season.
pyrifluquinazon, MOA 9B (PQZ) 1.87 EC	2.4 to 3.2 fl oz	12 hrs	1	See label for rotational crop restrictions. Do not exceed 4.8 fl oz of product per acre per crop cycle. See supplemental label for aerial application.
sulfoxaflor, MOA 4C (Transform) 50WG	0.75 oz	12 hrs	3	Limit application to times when managed and native pollinators are least active, e.g., 2 hrs before sunset or when temperature is below 50°F
thiamethoxam, MOA 4A (Platinum) 75 SG	1.66 to 3.67 oz	12 hrs	30	Platinum is for soil application and may be applied to direct-seeded crops in-furrow at seed or transplant depth, post seeding. See plant-back restrictions for several crops.

TABLE 2-9. INSECT CONTROL FOR CUCURBITS (CUCUMBER, CANTALOUPE, PUMPKIN, SQUASH, & WATERMELON) (cont'd)

Insecticide, Mode of Action Code, and Formulation	Amount of Formulation Per Acre	Restricted Entry Interval (REI)	Pre harvest Interval (PHI) (Days)	Precautions and Remarks
ARMYWORMS, CABBAGE LOOPER				
Bacillus thuringiensis, MOA 11A (Crymax) WDG (Dipel) 2X (Xentari) DF	0.5 to 1.5 lb 8 oz 0.5 to 2 lb	4 hrs	0	On foliage as needed. **OMRI listed (excluding Crymax).**
chlorantraniliprole, MOA 28 (Coragen eVo) 5 SC	1.2 to 2.5 fl oz	4 hrs	1	Coragen may be used for foliar or drip chemigation.
cyantraniliprole, MOA 28 **Soil treatment** (Verimark) 1.67 SC	5 to 13.5 fl oz	4 hrs	1	Verimark is for soil application only. It may be applied to the soil at planting at 6.75 to 13.5 ounces, or via drip chemigation at 5 to 10 fluid ounces. Do not make more than two soil or chemigation. See label for application options.
Foliar treatment (Exirel) 0.83 SE	7 to 17 fl oz	12 hrs	1	
indoxacarb, MOA 22A (Avaunt eVo) 30 WDG	2.5 to 6.0 oz	12 hrs	3	Use higher rates for beet armyworm or large larvae.
methoxyfenozide, MOA 18 (Intrepid) 2 F	4 to 10 fl oz	4 hrs	3	Apply at first sign of infestation, targeting eggs and small larvae.
novaluron, MOA 15 (Rimon) 0.83 EC	9 to 12 fl oz	12 hrs	1	Apply when peak population is at egg hatch through second instar.
spinetoram, MOA 5 (Radiant) 1 SC	5 to 10 fl oz	4 hrs	3	Use the higher rate for heavy infestations or large larvae.
viruses		4 hrs		See **Table 2-26** for virus-based insecticides effective against specific insects. **OMRI listed.**
CUCUMBER BEETLE				
acetamiprid MOA 4A (Assail) 30 SC	2.1 to 4.5 fl oz	12 hrs	0	Do not exceed 0.5 pound per acre per season.
carbaryl MOA 1A (Sevin) XLR Plus	1 qt	12 hrs	3	
clothianidin, MOA 4A (Belay) 2.13 **Soil treatment**	9 to 12 fl oz	12 hrs	21	Soil application at planting or through chemigation. See label for application options.
Foliar treatment	3 to 4 fl oz	12 hrs		Do not use an adjuvant with foliar applications. Do not spray after the fourth true leaf.
dinotefuran MOA 4A **Foliar treatment** (Venom) 70 SG (Scorpion) 35 SL	1 to 4 oz 2 to 7 fl oz	12 hrs	1	
Soil treatment (Scorpion) 35 SL	9 to 10.5 fl oz	12 hrs	21	
imidacloprid, MOA 4A (Admire Pro) 4.6 F	7 to 10.5 fl oz	12 hrs	21	Must be applied to the soil. May be applied preplant; at planting as a post-seeding drench, transplant water drench or hill drench, subsurface side-dress or by chemigation using low pressure drip irrigation methods. Will also control aphids and whiteflies.
thiamethoxam, MOA 4A (Platinum) 75 SG	1.66 to 3.67 oz	12 hrs	30	Apply as a soil treatment either at planting as an in-furrow spray at seeding or transplant depth, or post planting as a hill drench or through drip irrigation. Will also control aphids and whiteflies.
pyrethroid, MOA 3A		12 hrs		See **Table 2-25 A** for a list of registered pyrethroids and pre-harvest intervals. In some areas of the Mid-Atlantic, there has been a decline in efficacy of pyre- throids against cucumber beetles.
cyclaniliprole MOA 28 (Harvanta) 50SL	10.9 to 16.4 fl oz	4 hrs	1	Make no more than 3 applications per year. Do not exceed the minimum inter- val between treatment of 7 days [14 days in California].
LEAFMINER				
abamectin, MOA 6 (Agri-Mek) 0.7 SC	1.75 to 3.5 fl oz	12 hrs	7	To avoid illegal residues, Agri-Mek must be mixed with a nonionic activator-type wetting, spreading, and/or penetrating spray adjuvant. For resistance management do not make more than two sequential applications.
cyromazine, MOA 17 (Trigard) 75 WS	2.7 oz	12 hrs	0	
chlorantraniliprole, MOA 28 (Coragen eVo) 5 SC	5 to 7.5 fl oz	4 hrs	1	For foliar or soil application or drip chemigation. Drip chemigation must be applied uniformly to the root zone. See label for instructions.

TABLE 2-9. INSECT CONTROL FOR CUCURBITS (CUCUMBER, CANTALOUPE, PUMPKIN, SQUASH, & WATERMELON) (cont'd)

Insecticide, Mode of Action Code, and Formulation	Amount of Formulation Per Acre	Restricted Entry Interval (REI)	Pre harvest Interval (PHI) (Days)	Precautions and Remarks
LEAFMINER (cont'd)				
cyclaniliprole, MOA 28 (Harvanta) 50 SL	10.9 to 16.4 fl oz	4 hrs	1	
spinetoram, MOA 5 (Radiant) 1 SC	6 to 10 fl oz	4 hrs	3	
PICKLEWORM, MELONWORM, CUTWORM				
Bacillus thuringiensis, MOA 11A (Crymax) WDG (Dipel) 2X (Xentari) DF	0.5 to 1.5 lb 8 oz 0.5 to 2 lb	4 hrs	0	PW/MW pressure can be severe and repeated applications every 5-7 days is necessary to protect fruit . **OMRI listed (excluding Crymax).**
pyrethroid, MOA 3A		12 hrs		See **Table 2-25 A** for a list of registered pyrethroids and pre-harvest intervals.
carbaryl MOA 1A (Sevin) XLR Plus	0.5 to 2 qt	12 hrs	3	Apply to foliage when worms appear in blossoms. Repeat as needed. Protect pollinators by applying early morning or late evening when pollinators are not active. Rarely a problem before July.
chlorantraniliprole, MOA 28 (Coragen eVo) 5 SC	2.5 to 7.5 fl oz	4 hrs	1	For foliar application or drip chemigation. Drip chemigation must be applied uniformly to the root zone. See label for instructions. Use high rate for pickleworm.
cyclaniliprole, MOA 28 (Harvanta) 50 SL	10.9 to 16.4 fl oz	4 hrs	1	
cyantraniliprole, MOA 28 (Verimark) 1.67 SC (Exirel) 0.83 SE	5 to 10 fl oz 7 to 13.5 fl oz	4 hrs 12 hrs	1 1	Verimark is for drip chemigation only for these pests. Not effective at preventing fruit injury. Do not make more than two chemigation applications. See label for application options. Exirel is for foliar application only.
methoxyfenozide, MOA 18 (Intrepid) 2 F	4 to 10 fl oz	4 hrs	3	
spinetoram, MOA 5 (Radiant) 1 SC	5 to 10 fl oz	4 hrs	3	
SPIDER MITES				
Spider mites are primarily a problem on cucumber and watermelon, and less of an issue on squash and pumpkin.				
abamectin, MOA 6 (Agri-Mek) 0.7 SC	1.75 to 3.4 fl oz	12 hrs	7	To avoid illegal residues, Agri-Mek must be mixed with a nonionic activator type wetting, spreading and/or penetrating spray adjuvant. For resistance management, do not make more than two sequential applications. Active against immatures and adults.
acequinocyl, MOA 20B (Kanemite 15 SC)	31 fl oz	12 hrs	1	Do not use less than 30 gallons of water volume per acre. Do not make more than 2 applications or apply more than 62 fl oz per acre per year.
bifenazate, MOA 20D (Acramite) 50 WS	0.75 to 1.0 lb	12 hrs	3	Do not make more than one application per season. Active against immatures and adults. Some egg activity.
etoxazole, MOA 10B (Zeal) 72 WSP	2 to 3 oz	12 hrs	7	Does not kill adults. Do not make more than one application per season.
fenazaquin, MOA 21A (Magister) 1.7	24 to 36 fl oz	12 hrs	3	Do not make more than one application per year. Active against all life stages.
fenpyroximate, MOA 21A (Portal 0.4 EC)	2.0 pts	12 hrs	1	Use a minimum of 40 gallons of water per acre. Do not make more than two applications per year.
spiromesifen, MOA 23 (Oberon) 2 SC	7 to 8.5 fl oz	12 hrs	7	Active against immatures. Some adult activity.
SQUASH BUG				
Squash bug is a common pest of cantaloupe, pumpkin, and squash. Although cucumber and watermelon are occasionally reported as hosts of squash bug, rarely do infestations occur.				
acetamiprid MOA 4A (Assail) 30 SC	4.5 fl oz	12 hrs	0	Assail is most effective against newly laid eggs and nymphs.
clothianidin, MOA 4A (Belay) 2.13	3 to 4 fl oz	12 hrs	21	Do not spray after the 4" true leaf. See label for restrictions for protection of pollinators.
dinotefuran, MOA 4A **Foliar treatment** (Venom) 70 SG (Scorpion) 35 SL	1 to 4 oz 2 to 7 fl oz	12 hrs	1	Do not make an oil and foliar application – use one or the other. Do not exceed 6 ounces (foliar) or 12 oz (soil) of Venom per acre per season. Do not exceed 10.5 fl oz (foliar) or 21 fl oz (soil) of Scorpion per acre per season. See label for application restrictions for protection of pollinators.
Soil treatment (Venom) 70 SG (Scorpion) 35 SL	5 to 7.5 oz 9 to 10.5 fl oz	12 hrs	21	

TABLE 2-9. INSECT CONTROL FOR CUCURBITS (CUCUMBER, CANTALOUPE, PUMPKIN, SQUASH, & WATERMELON) (cont'd)

Insecticide, Mode of Action Code, and Formulation	Amount of Formulation Per Acre	Restricted Entry Interval (REI)	Pre harvest Interval (PHI) (Days)	Precautions and Remarks
SQUASH BUG (cont'd)				
flupyradifurone, MOA 4D (Sivanto Prime)	10.5 to 14.0 fl oz	12 hrs	1	Do not apply Sivanto Prime to cantaloupe, honeydew melon, or muskmelon. See label for other additional melons to which it should not be applied.
pyrethroid, MOA 3A		12 hrs		See **Table 2-25 A** for a list of registered pyrethroids and pre-harvest intervals.
SQUASH VINE BORER				
Squash vine borer only attacks squash and pumpkin and is more common in-home gardens opposed to commercial plantings.				
acetamiprid MOA 4A (Assail) 30 SC	4.5 fl oz	12 hrs	0	
pyrethroid, MOA 3A		12 hrs		See **Table 2-25 A** for a list of registered pyrethroids and pre-harvest intervals.
THRIPS				
methomyl, MOA 1A (Lannate) 2.4 LV (Lannate) 90 SP	0.75 to 1.5 pt 0.25 to 0.5 lb	48 hrs	0	
spinetoram, MOA 5 (Radiant) 1 SC	6 to 10 fl oz	4 hrs	3	
tolfenpyrad, MOA 21A (Torac EC)	21 fl oz	12 hrs	1	Do not make more than 2 applications per crop cycle. See label for restrictions for protection of pollinators.
WHITEFLIES				
acetamiprid MOA 4A (Assail) 30 SC	2.1 to 4.5 fl oz	12 hrs	0	
Afidopyropen, MOA 9D (Sefina)	14 fl oz	12	0	
buprofezin, MOA 16 (Courier) 40 SC	9 to 13.6 fl oz	12 hrs	1	Use sufficient water to ensure good coverage. Do not apply more than twice per crop cycle.
chlorantraniliprole, MOA 28 (Coragen eVo) 5 SC	5 to 7.5 fl oz	4 hrs	1	For foliar or soil application or drip irrigation. Drip chemigation must be applied uniformly to the root zone. See label for instructions.
cyantraniliprole, MOA 28 **Soil treatment** (Verimark) 1.67 SC	6.8 to 13.5 fl oz	4 hrs	1	Verimark is for soil application only. It may be applied to the soil at planting at 6.75 to 13.5 fl oz, or via drip chemigation at 10 fl oz. See label for application options.
Foliar treatment (Exirel) 0.83SE	13.5 to 20.5 fl oz	12 hrs	1	Exirel is for foliar application only. Use an adjuvant for best results.
dinotefuran, MOA 4A (Venom) 70SG **Soil treatment**	5 to 7.5 oz	12 hrs	21	Soil applications may be made with irrigation systems, including drip, or over-head irrigation. Do not apply while bees are foraging. Residues may remain toxic to bees up to 38 hrs following treatment.
Foliar treatment	1 to 4 oz	12 hrs	1	
flupyradifurone, MOA 4D (Sivanto Prime) 1.67 **Soil treatment**	21 to 28 fl oz	4 hrs	21	Soil applications by injection below the seed level at planting, drench at transplanting, or drip chemigation.
Foliar treatment	10.5 to 14 fl oz	4 hrs	7	Do not make foliar applications of Sivanto to muskmelon, cantaloupe, or honeydew melon. See label for additional melons to which it should not be applied.
imidacloprid, MOA 4A (Admire Pro) 4.6 F	7 to 10.5 oz	12 hrs	21	Must be applied to the soil. May be applied preplant; at planting; as a post-seeding drench or hill drench; subsurface sidedress; or by chemigation using low-pressure drip or trickle irrigation. Will also control aphids and cucumber beetles. Will also control wireworms.
pyriproxyfen, MOA 7C (Knack) 0.86 EC	8 to 10 fl oz	12 hrs	7	Do not make more than two applications per season, and do not make applications closer than 14 days apart.
pyrifluquinazon, MOA 9B (PQZ) 1.87 EC	2.4 to 3.2 fl oz	12 hrs	1	See label for rotational crop restrictions. Do not exceed 4.8 fl oz per acre per crop cycle.
spiromesifen, MOA 23 (Oberon) 2 SG	7 to 8.5 fl oz	12 hrs	7	Does not control adults. Apply when colonies first appear and before leaf damage or discoloration. Do not exceed 3 applications per season.
thiamethoxam, MOA 4A **Soil treatment** (Platinum)	5 to 11 fl oz	12 hrs	30	Platinum is for soil application and may be applied to direct-seeded crops in-furrow at seed or transplant depth, post seeding or transplant as a drench, or through drip irrigation. Do not exceed 11 fl oz per acre per season of Platinum. Check label for plant-back restrictions for several crops.

TABLE 2-9. INSECT CONTROL FOR CUCURBITS (CUCUMBER, CANTALOUPE, PUMPKIN, SQUASH, & WATERMELON) (cont'd)

Insecticide, Mode of Action Code, and Formulation	Amount of Formulation Per Acre	Restricted Entry Interval (REI)	Pre harvest Interval (PHI) (Days)	Precautions and Remarks
WHITEFLIES (cont'd)				
Foliar treatment (Actara) 25 WDG	3 to 5.5 oz	12 hrs	0	Actara is for foliar application. See label for application restrictions for protection of pollinators.
WIREWORM				
diazinon MOA 1B (Diazinon) AG 500	3 to 4 qt	3 days		Broadcast on soil just before planting and thoroughly work into upper 4 to 8 inches of soil.
imidacloprid (MOA 4A) (Admire Pro) 4.6F	7 to 10.5 fl oz	12 hrs	21	Soil application only on cucurbits. May be applied preplant; at planting as a post-seeding drench, transplant water drench, or hill drench; subsurface sidedress or by chemigation using low pressure drip irrigation methods. Will also control cucumber beetles, thrips, and whiteflies.

TABLE 2-10. INSECT CONTROL FOR EGGPLANT

Insecticide, Mode of Action Code, and Formulation	Amount of Formulation Per Acre	Restricted Entry Interval (REI)	Pre harvest Interval (PHI) (Days)	Precautions and Remarks
APHID				
Where whitefly resistance is an issue (or any other insect with a high potential for resistance to Group 4A MOA insecticides), avoid making foliar applications of Group 4A insecticides when a soil-applied Group 4A program is used, i.e., do not make both foliar and soil applications of Group 4A insecticides. Also, if using a foliar-applied program, avoid using a block of more than three consecutive applications of any products belonging to Group 4A insecticides.				
acetamiprid, MOA 4A (Assail) 30 SC	1.7 to 3.4 fl oz	12 hrs	7	Thoroughly cover foliage to control aphids effectively. Do not apply more than once every 7 days, and do not exceed a total of 7 oz per season.
flonicamid, MOA 29 (Beleaf) 50 SG	2 to 4.8 oz	12 hrs	0	
afidopyropen, MOA 9D (Sefina) DC	3 oz	12 hrs	0	Do not make more than 2 sequential applications before using a different mode of action.
Cyantraniliprole, MOA 28A (Exirel) 0.83SE	13.5 to 20.5 fl oz	12	1	
flupyradifurone, MOA 4D (Sivanto Prime) 1.67 SL	7.0 to 12.0 fl oz	4 hrs	1	
sulfoxaflor, MOA 4C (Closer) 2 SC	1.5 to 2.0 fl oz	12 hrs	1	
imidacloprid, MOA 4A **Soil treatment** (Admire Pro) 4.6 F (various) 1.6 F	7 to 10.5 oz 16 to 24 fl oz	12 hrs	21	See label for soil application instructions. For short-term protection of transplants at planting, apply Admire Pro (0.44 ounces per 10,000 plants) not more than 7 days before transplanting by (1) uniformly spraying transplants, followed immediately by sufficient overhead irrigation to wash product into potting media, or (2) injection into overhead irrigation system with adequate volume to thoroughly saturate soil media.
Foliar treatment (Admire Pro) 4.6 F (various) 1.6 F	1.3 to 2.2 fl oz 3.75 fl oz	12 hrs	0	
pymetrozine, MOA 9B (Fulfill) 50 WDG	2.75 oz	12 hrs	14	Apply before aphids reach damaging levels. Do not exceed 5.5 ounces per acre per season.
pyrifluquinazon, MOA 9B (PQZ) 1.87 EC	2.4 to 3.2 fl oz	12 hrs	1	See label for rotational crop restrictions. Do not exceed 4.8 fl oz per acre per crop cycle.
spirotetramat, MOA 23 (Movento) 2 SC	4 to 5 fl oz	24 hrs	1	Do not exceed 10 fluid ounces per season. Requires surfactant.
thiamethoxam, MOA 4A **Soil treatment** (Platinum) 75 SG	1.66 to 3.67 oz	12 hrs	30	Platinum may be applied to direct-seeded crops in-furrow at seed or transplant depth, post seeding or transplant as a drench, or through drip irrigation. Do not exceed 8 oz per acre per season. Check label for plant-back restrictions.
Foliar treatment (Actara) 25 WDG	2 to 3 oz	12 hrs	0	Actara is for foliar application.

TABLE 2-10. INSECT CONTROL FOR EGGPLANT (cont'd)

Insecticide, Mode of Action Code, and Formulation	Amount of Formulation Per Acre	Restricted Entry Interval (REI)	Pre harvest Interval (PHI) (Days)	Precautions and Remarks
BLISTER BEETLE				
pyrethroid, MOA 3A		12 hrs		See **Table 2-25 A** for a list of registered pyrethroids and pre-harvest intervals.
COLORADO POTATO BEETLE				
Resistance to many insecticides is widespread in Colorado potato beetle. To reduce risk of resistance, scout fields and apply insecticides only when needed to prevent damage to the crop. Crop rotation will help prevent damaging Colorado potato beetle infestations. If control failures or reduced levels of control occur with a particular insecticide, do NOT make a second application of the same insecticide at the same or higher rate. If an additional insecticide application is necessary, a different insecticide representing a different MOA class should be used. Do NOT use insecticides belonging to the same class 2 years in a row for Colorado potato beetle control.				
abamectin, MOA 6 (Agri-Mek) 0.7 SC	1.75 to 3.5 fl oz	12 hrs	7	Apply when adults and small larvae are present but before large larvae appear. For resistance management, use the higher rate.
acetamiprid, MOA 4A (Assail) 30 SC	1.3 to 2.1 fl oz	12 hrs	7	Do not apply more than once every 7 days, and do not exceed 7 ounces of formulation per season.
pyrethroid, MOA 3A		12 hrs		See **Table 2-25 A** for a list of registered pyrethroids and pre-harvest intervals.
chlorantraniliprole, MOA 28 (Coragen eVo) 5 SC	1.2 to 2.5 fl oz	4 hrs	1	Foliar or drip chemigation. Drip chemigation must be applied uniformly to the root zone. See label for instructions.
cyantraniliprole, MOA 28	7 to 13.5 fl oz	12 hrs	1	
Soil treatment (Verimark) 1.67 SC	5 to 10 fl oz	4 hrs	1	
Foliar treatment (Exirel) 0.83SE				
cyclaniliprole, MOA 28 (Harvanta) 50 SL	10.9 to 16.4 fl oz	4 hrs	1	
dinotefuran, MOA 4A **Foliar treatment** (Venom) 70 SG (Scorpion) 35SL	1 to 4 oz 2 to 7 oz	12 hrs	1	Do not follow soil applications with foliar applications on any neonicotinoid insecticide. Use only one application method. Do not apply more than 6 ounces per acre per season using foliar applications, or 12 ounces per acre per season using soil applications.
Soil treatment (Venom) 70 SG (Scorpion) 35 SL	5 to 6 fl oz 9 to 10.5 fl oz	12 hrs	21	Soil applications may be applied by (1) a narrow band below or above the seed line at planting, (2) a post-seeding or transplanting drench with sufficient water to ensure movement to the root zone, or (3) drip irrigation.
imidacloprid, MOA 4A **Soil treatment** (Admire Pro) 4.6 F (various) 2 F	7 to 10.5 fl oz 16 to 24 fl oz	12 hrs	21	See application methods under Aphids, Thrips.
Foliar treatment (Admire Pro) 4.6 F (various) 1.6 F	1.3 fl oz 3.75 fl oz	12 hrs	0	
novaluron, MOA 15 (Rimon) 0.83 EC	9 to 12 fl oz	12 hrs	1	
spinetoram, MOA 5 (Radiant) 1 SC	5 to 10 fl oz	4 hrs	1	
thiamethoxam, MOA 4A (Platinum) 75 SG	1.66 to 3.67 oz	12 hrs	30	
(Actara) 25 WDG	2 to 3 oz	12 hrs	0	See application methods under Aphids.
EGGPLANT LACE BUG				
imidacloprid, MOA 4A (Admire Pro) 4.6 F (various) 1.6 F	1.3 to 2.2 fl oz 3.8 to 6.2 fl oz	12 hrs	0	
FLEA BEETLE				
malathion, MOA 1B (various brands) 57 EC	3 pt	12 hrs	3	
pyrethroid, MOA 3A		12 hrs		See **Table 2-25 A** for a list of registered pyrethroids and pre-harvest intervals.
carbaryl, MOA 1A (Sevin) SL (Sevin) XLR Plus	0.5 to 1 qt 0.5 to 1 qt	12 hrs	3	

TABLE 2-10. INSECT CONTROL FOR EGGPLANT (cont'd)

Insecticide, Mode of Action Code, and Formulation	Amount of Formulation Per Acre	Restricted Entry Interval (REI)	Pre harvest Interval (PHI) (Days)	Precautions and Remarks
FLEA BEETLE (cont'd)				
cyantraniliprole, MOA 28 (Verimark) 1.67 SC	6.75 to 13.5 fl oz	4 hrs	1	Verimark for soil application only. Apply at planting or via drip chemigation. See label for application options.
dinotefuran, MOA 4A **Foliar treatment** (Venom) 70 SG (Scorpion) 35 SL	1 to 4 oz 2 to 7 fl oz	12 hrs	1	Do not follow soil applications with foliar applications on any neonicotinoid insecticide. Use only one application method. Do not apply more than 6 ounces per acre per season using foliar applications, or 12 ounces per acre per season using soil applications.
Soil treatment (Venom) 70 SG (Scorpion) 35 SL	5 to 6 fl oz 9 to 10.5 fl oz	12 hrs	21	Soil applications may be applied by (1) a narrow band below or above the seed line at planting, (2) a post-seeding or transplanting drench with sufficient water to ensure movement to the root zone, or (3) drip irrigation.
thiamethoxam, MOA 4A (Platinum) 75 SG	1.66 to 3.67 oz	12 hrs	30	
(Actara) 25 WDG	2 to 3 oz	12 hrs	0	See application methods under Aphids
HORNWORM, EUROPEAN CORN BORER, BEET ARMY WORM, CORN EARWORM				
chlorantraniliprole, MOA 28 (Coragen eVo) 5 SC	1.2 to 2.5 fl oz	4 hrs	1	Foliar or drip chemigation. Drip chemigation must be applied uniformly to the root zone. See label for instructions.
cyantraniliprole, MOA 28 **Soil treatment** (Verimark) 1.67 SC	5 to 10 fl oz	4 hrs	1	Verimark is for soil application only. Drip chemigation must be applied uniform- ly to the root zone. See label for application options.
Foliar treatment (Exirel) 0.83SE	7 to 13.5 fl oz	12 hrs	1	Exirel is for foliar application only.
cyclaniliprole, MOA 28 (Harvanta) 50 SL	10.9 to 16.4 fl oz	4 hrs	1	
indoxacarb, MOA 22A (Avaunt eVo) 30 WDG	2.5 to 3.5 oz	12 hrs	3	Do not apply more than 14 ounces per acre per season.
methomyl, MOA 1A (Lannate) 2.4 LV	1.5 to 3 pt	48 hrs	5	
methoxyfenozide, MOA 18 (Intrepid) 2 F	4 to 16 fl oz	4 hrs	1	Apply at rates of 4 to 8 fluid ounces early in season when plants are small. Apply at rates of 8 to 16 ounces to large plants or when infestations are heavy. During periods of continuous moth flights, re-treatments at 7 to 14 days may be required. Do not apply more than 16 fluid ounces per application or 64 fluid ounces of Intrepid 2F per acre per season.
spinetoram, MOA 5 (Radiant) 1 SC	5 to 10 fl oz	4 hrs	1	
pyrethroid, MOA 3A		12 hrs		See **Table 2-25 A** for a list of registered pyrethroids and pre-harvest intervals.
viruses		4 hrs		See **Table 2-26** for virus-based insecticides effective against specific insects. **OMRI listed.**
LEAFMINER				
abamectin, MOA 6 (Agri-Mek) 0.15 EC	8 to 16 fl oz	12 hrs	7	Use low rates for low to moderate infestations, and high rates for severe infestations
chlorantraniliprole, MOA 28 (Coragen eVo) 5 SC	5 to 7.5 fl oz	4 hrs	1	Foliar, soil, or drip chemigation. Drip chemigation must be applied uniformly to the root zone. See label for application instructions.
oxamyl, MOA 1A (Vydate) 2 L	1 to 2 qt	48 hrs	7	
spinetoram, MOA 5 (Radiant) 1 SC	5 to 10 fl oz	4 hrs	1	
STINK BUG, LEAFFOOTED BUG				
pyrethroid MOA 3A		12 hrs		See **Table 2-25 A** for a list of registered pyrethroids and pre-harvest intervals.
dinotefuran, MOA 4A **Foliar treatment** (Venom) 70 SG (Scorpion) 35 SL	1 to 4 oz 2 to 7 fl oz	12 hrs	1	
Soil treatment (Venom) 70 SG (Scorpion) 35 SL	5 to 6 fl oz 9 to 10.5 fl oz	12 hrs	21	
thiamethoxam, MOA 4A (Actara) 25 WDG	3 to 5.5 oz	12 hrs	0	Do not exceed 11 ounces Actara per acre per season.

TABLE 2-10. INSECT CONTROL FOR EGGPLANT (cont'd)

Insecticide, Mode of Action Code, and Formulation	Amount of Formulation Per Acre	Restricted Entry Interval (REI)	Pre harvest Interval (PHI) (Days)	Precautions and Remarks
SPIDER MITE				
abamectin, MOA 6 (Agri-Mek) 0.7 SC	1.75 to 3.5 fl oz	12 hrs	7	Use low rates for low to moderate infestations, and high rates for severe infestations. Active against immatures and adults.
acequinocyl, MOA 20B (Kanemite) 15SC	31 fl oz	12 hrs	1	Active against all life stages.
bifenazate, MOA 20D (Acramite) 50 WS	0.75 to 1.0 lb	12 hrs	3	Do not make more than one application per season. Active against immatures and adults. Some egg activity.
etoxazole, MOA 10B (Zeal)	2 to 3 oz	12 hrs	7	Do not make more than one Zeal application per season. Active against immatures and eggs.
fenpyroximate MOA 21A (Portal) 0.4EC	2 pts	12 hrs	3	Do not make more than two applications per season. Active against immatures and adults.
fenazaquin, MOA 21A (Magister) 1.7	24 to 36 fl oz	12 hrs	3	Do not make more than one application per year. Active against all life stages.
hexakis, MOA 12B (Vendex) 50 WP	2 to 3 lb	48 hrs	3	Active against immatures and adults.
spiromesifen, MOA 23 (Oberon) 2 SG	7 to 8.5 fl oz	12 hrs	7	Active against immatures. Some adult activity.
THRIPS				
dinotefuran, MOA 4A **Foliar treatment** (Venom) 70 SG (Scorpion) 35 SL	1 to 4 oz 2 to 7 fl oz	12 hrs	1	Will not control western flower thrips, only tobacco thrips, which are common early in the season. See Whitefly for application instructions. Soil applications are more effective against thrips than foliar applications are.
Soil treatment (Venom) 70 SG (Scorpion) 35 SL	5 to 6 oz 9 to 10.5 fl oz		21	
cyantraniliprole, MOA 28 (Verimark) 1.67 SC	5 to 10 fl oz	4 hrs	1	Soil applications of Verimark will suppress western flower thrips. Foliar applications of Exirel are less effective.
cyclaniliprole, MOA 28 (Harvanta) 50 SL	10.9 to 16.4 fl oz	4 hrs	1	Foliar applications will help suppress western flower thrips when used in a rotational program.
imidacloprid, MOA 4A (Admire Pro) 4.6 F (various) 2 F	7 to 10.5 fl oz 16 to 24 fl oz	12 hrs	21	Will not control western flower thrips, only tobacco thrips, which are common early in the season. See Aphids for application instructions.
methomyl, MOA 1A (Lannate) 2.4 LV	1.5 to 3 pt	48 hrs	3	
spinetoram, MOA 5 (Radiant) 1 SC	6 to 10 fl oz	4 hrs	1	
tolfenpyrad, MOA 21A (Torac) 1.29 EC	21 fl oz	12 hrs	1	
WHITEFLY				
acetamiprid, MOA 4A (Assail) 30 SC	2.1 to 3.4 fl oz	12 hrs	7	Begin applications when significant populations of adults appear. Do not wait until heavy populations have become established. Do not apply more than once every 7 days, and do not exceed 4 applications per season. Do not apply more than 7 ounces per season.
afidopyropen, MOA 9D (Sefina) DC	14 oz	12 hrs	0	Do not make more than 2 sequential applications before using a different mode of action.
buprofezin (Courier SC)	9.0 to 13.6 fl oz	12 hrs	1	Applications must be made at least 7 days apart. 4 Applications per year max (2 per crop cycle)
chlorantraniliprole, MOA 28 (Coragen eVo) 5 SC	5 to 7.5 fl oz	12 hrs	1	For foliar or drip chemigation. Drip chemigation must be applied uniformly to the root zone. See label for instructions.
cyantraniliprole, MOA 28 **Soil treatment** (Verimark) 1.67 SC	6.75 to 13.5 fl oz	4 hrs	1	Verimark for soil application only. Apply at planting or via drip chemigation. See label for application options.
Foliar treatment (Exirel) 0.83SE	13.5 to 20.5 fl oz	12 hrs	1	Exirel for foliar application only.

TABLE 2-10. INSECT CONTROL FOR EGGPLANT (cont'd)

Insecticide, Mode of Action Code, and Formulation	Amount of Formulation Per Acre	Restricted Entry Interval (REI)	Pre harvest Interval (PHI) (Days)	Precautions and Remarks
WHITEFLY (cont'd)				
dinotefuran, MOA 4A **Foliar treatment** (Venom) 70 SG (Scorpion) 35 SL	1 to 4 oz 2 to 7 fl oz	12 hrs	1	Use only one application method (foliar or soil) of Group 4A insecticides. Soil applications may be applied in a narrow band on the plant row in bedding op- erations, as a post-seeding or transplant drench, as a sidedress after planting and incorporated 1 or more inches, or through the drip irrigation system.
Soil treatment (Venom) 70 SG (Scorpion) 35 SL	5 to 6 fl oz 9 to 10.5 fl oz	12 hrs	21	
flupyradifurone, MOA 4D (Sivanto Prime) 1.67 SL	10.5 to 14.0 fl oz	4 hrs	1	
imidacloprid, MOA 4A (Admire Pro) 4.6 F (various) 2 F	7 to 10.5 fl oz 16 to 24 fl oz	12 hrs	21	Do not follow soil applications with applications of other neonicotinoid insecticides (Assail or Venom). See Aphids for application methods and restrictions.
pyriproxyfen, MOA 7C (Knack) 0.86 EC	8 to 10 fl oz	12 hrs	14	Knack prevents eggs from hatching. It does not kill whitefly adults. Applica- tions should begin when 3 to 5 adults per leaf are present. Do not make more than 2 applications per season, and do not apply a second application within 14 days of the first application. Do not exceed 20 fluid ounces of Knack per acre per season. Check label for plant-back restrictions.
pyrifluquinazon, MOA 9B (PQZ) 1.87 EC	2.4 to 3.2 oz	12 hrs	1	See label for rotational crop restrictions. Do not exceed 4.8 fl oz per acre per crop cycle.
spirotetramat, MOA 23 (Movento) 2 SC	4 to 5 fl oz	24 hrs	1	Do not exceed 10 fl oz per season. Requires surfactant.
spiromesifen, MOA 23 (Oberon) 2 SC	7 to 8.5 fl oz	12 hrs	7	Do not exceed 3 applications or 25.5 fluid ounces per season.
thiamethoxam, MOA 4A **Soil treatment** (Platinum) 75 SG	1.66 to 3.67 oz	12 hrs	30	Platinum is for soil applications and may be applied to direct-seeded crops in-furrow at seed or transplant depth, at post seeding or transplant as a drench, or through drip irrigation. Do not exceed 11 ounces per acre per season. Check label for plant-back restrictions for several plants.
Foliar treatment (Actara) 25 WDG	3 to 5.5 oz	12 hrs	0	Actara is for foliar application.

TABLE 2-11. INSECT CONTROL FOR HOPS

Insecticide, Mode of Action Code, and Formulation	Amount of Formulation Per Acre	Restricted Entry Interval (REI)	Pre harvest Interval (PHI) (Days)	Precautions and Remarks
APHIDS and LEAFHOPPERS				
imidacloprid, MOA 4A (Admire) 4.6 F **Soil treatment**	2.8 to 8.4 fl oz	12 hrs	60	Soil applications can be made by drip chemigation, Subsurface sidedress shanked into root-zone, or a hill drench in sufficient water to ensure incorporation into the root zoon by irrigation.
Foliar treatment	2.8 fl oz	12 hrs	28	
pymetrozine, MOA 9B (Fulfill) 50 WDG	4 to 6 oz	12 hrs	14	For aphids only. Will not control leafhoppers.
spirotetramat, MOA 23 (Movento) 2 F	5 to 6 fl oz	24 hrs	7	Do not exceed 12.5 fl oz per acre per season. Will also control twospotted spider mite.
malathion, MOA 1B (5 EC) (8 EC)	1 pt 0.63 pt	12 hrs	7	May suppress twospotted spider mite.
pyrethrins, MOA 3A (Pyganic EC) 1.4 II (Pyganic EC) 5 II	16 to 64 fl oz 4.5 to 17 fl oz	12 hrs	0	**OMRI listed.** Pyrethrins degrade very quickly in sunlight. Do not expect residual control.
JAPANESE BEETLE				
bifenthrin, MOA 3A (Brigade) 2 EC (Brigade) WSB	3.8 to 6.4 fl oz 9.6 to 16 fl oz	12 hrs	14	See label.

TABLE 2-11. INSECT CONTROL FOR HOPS (cont'd)

Insecticide, Mode of Action Code, and Formulation	Amount of Formulation Per Acre	Restricted Entry Interval (REI)	Pre harvest Interval (PHI) (Days)	Precautions and Remarks
JAPANESE BEETLE (cont'd)				
imidacloprid, MOA 4A (Admire) 4.6 F (generics) 2	2.8 fl oz 6.4 fl oz	12 hrs	28	
ARMYWORMS, CUTWORMS, LOOPERS, LEAFROLLER, QUESTION MARK BUTTERFLY				
Bacillus thuringiensis, MOA 11A (Xentari) DF (Crymax) WDG	0.5 to 2 lb 0.5 to 2 lb	4 hrs	0	**OMRI listed (excluding Crymax).**
chlorantraniliprole, MOA 28 (Coragen eVo) 5 SC	1.2 to 2.5 fl oz	4 hrs	0	Foliar or drip chemigation. Drip chemigation must be applied uniformly to the root zone. See label for instructions.
spinosad, MOA 5 (Entrust) SC	4 to 6 fl oz	4 hrs	1	**OMRI listed.**
spinetoram, MOA 5 (Delegate) 25 WG	2.5 to 4 oz	4 hrs	1	For use on dry cones only.
viruses		4 hrs		See **Table 2-26** for virus-based insecticides effective against specific insects. **OMRI listed.**
SPIDER MITES				
abamectin, MOA 6 (Agri-Mek) 0.7 SC	1.75 to 3.5 fl oz	12 hrs	21	Do not exceed 48 fluid ounces per acre per season, or more than two sequential applications. Active against immatures and adults.
acequinocyl, MOA 20B (Kanemite) 15 SC	31 fl oz	12 hrs	1	The use of a surfactant/adjuvant with Kanemite on tomatoes is prohibited. Active against all life stages.
bifenazate, MOA 20D (Acramite) 50 WS	0.75 to 1.0 lb	12 hrs	14	Do not make more than one application per season. Active against immatures and adults. Some egg activity.
etoxazole, MOA 10B (Zeal) 72 WSP	3 to 4 oz	12 to 78 hrs	7	Apply when mites are low because Zeal is primarily an ovicide/larvicide.
fenazaquin, MOA 21A (Magister) 1.7	24 to 36 fl oz	12 hrs	7	Do not make more than one application per year. Active against all life stages.
fenpyroximate MOA 21A (Portal) 0.4 EC	2 pts	12 hrs	15	Do not make more than two applications per season. Active against immatures and adults.
hexythiazox, MOA 10A (Savey) 50 DF	4 to 6 oz	12 hrs	—	May be applied up to burr formation in hop vines. Apply when mites are low, because Savey is primarily an ovicide, and sterilizes females.
mineral oil (TriTek) (Various brands)	1 to 2% soln.	4hrs	0	**OMRI listed.** TriTek is the only emulsified formulation of oil. No others contain an emulsifier. Active against eggs and immatures.

TABLE 2-12. INSECT CONTROL FOR LETTUCE

Insecticide, Mode of Action Code, and Formulation	Amount of Formulation Per Acre	Restricted Entry Interval (REI)	Pre harvest Interval (PHI) (Days)	Precautions and Remarks
APHID				
acetamiprid, MOA 4A (Assail) 30 SC	1.7 to 3.4 fl oz	12 hrs	7	Do not apply more than once every 7 days, and do not exceed 4- to 5-applica- tions per calendar year.
afidopyropen, MOA 9D (Versys) DC	1.5 fl oz	12 hrs	0	Do not make more than 2 sequential applications before using a different mode of action.
clothianidin, MOA 4A (Belay) 2.13 EC **Soil treatment** **Foliar treatment**	9 to 12 fl oz 3 to 4 fl oz	12 hrs 12 hrs	7	Soil application at planting only. Do not incorporate an adjuvant with foliar applications. Do not apply more than 6.4 oz per acre per season.
dimethoate 4 EC, MOA 1B	0.5 pt	48 hrs	14	For use on leaf lettuce.
flonicamid, MOA 29 (Beleaf) 50 SG	2 to 2.8 oz	12 hrs	0	
flupyradifurone, MOA 4D (Sivanto Prime) 1.67 SL	10.5 to 14.0 fl oz	4 hrs	1	

TABLE 2-12. INSECT CONTROL FOR LETTUCE (cont'd)

Insecticide, Mode of Action Code, and Formulation	Amount of Formulation Per Acre	Restricted Entry Interval (REI)	Pre harvest Interval (PHI) (Days)	Precautions and Remarks
APHID (cont'd)				
imidacloprid, MOA 4A **Soil treatment** (Admire Pro) 4.6 F (various) 1.6 F	4.4 to 10.5 fl oz 10 to 24 fl oz	12 hrs	21	Do not follow soil applications with foliar applications of any neonicotinoid insecticide. See label for soil application instructions.
Foliar treatment (Admire Pro) 4.6 F (various) 1.6 F	1.3 fl oz 3.8 fl oz	12 hrs	7	
pymetrozine, MOA 9B (Fulfill) 50 WDG	2.75 oz	12 hrs	7	Apply before aphids reach damaging levels. Do not exceed 5.5 oz per acre per season
pyrifluquinazon, MOA 9D (PQZ) 1.87 EC	2.4 to 3.2 oz	12 hrs	1	See label for rotational crop restrictions. Do not exceed 4.8 fl oz per acre per crop cycle.
spirotetramat, MOA 23 (Movento) 2 SC	4 to 5 fl oz	24 hrs	3	Do not exceed 10 fl oz per season. Requires surfactant.
thiamethoxam, MOA 4A **Soil treatment** (Platinum) 75 SG	1.66 to 3.67 oz	12 hrs	30	Do not follow applications of Platinum with foliar applications of any neonicotinoid insecticide. Platinum may be applied to direct-seeded crops in-furrow at the seeding or transplant depth, or as a narrow surface band above the seedling and followed by irrigation. Post seeding it may be applied as a transplant drench or through the drip irrigation.
Foliar treatment (Actara) 25 WD G	1.5 to 3 oz	12 hrs	7	Actara is for foliar application.
tolfenpyrad, MOA 21A (Torac) 1.29 EC	17 to 21 fl oz	12 hrs	1	Do not apply until at least 14 days after plant emergence or after transplanting to allow time for root establishment.
CABBAGE LOOPER, CORN EARWORM, ARMYWORMS				
Bacillus thuringiensis, MOA 11A (Crymax) WDG (Dipel) DF	0.5 to 1.5 lb 8 oz	4 hrs	0	Only target small armyworms with Bts. **OMRI listed (excluding Crymax).**
chlorantraniliprole, MOA 28 (Coragen eVo) 5 SC	1.2 to 2.5 oz	4 hrs	1	Foliar or drip chemigation. See label for instructions.
cyantraniliprole, MOA 28 **Soil treatment** (Verimark) 1.67 SC	5 to 13.5 oz	4 hrs	1	Verimark is for soil application only. Applications made at planting and/or via drip chemigation. Use higher rates (>10 fluid ounces) where cabbage looper is a concern. See label for application options.
Foliar treatment (Exirel) 0.83SE	7 to 17 fl oz	12 hrs	1	Exirel is for foliar application only. Use higher rates (>13.5 fluid ounces) for Cabbage loopers.
cyclaniliprole, MOA 28 (Harvanta) 50 SL	11 to 16.4 fl oz	4 hrs	1	
emamectin benzoate, MOA 6 (Proclaim) 5 WDG	3.2 to 4.8 oz	12 hrs	7	Do not make more than two sequential applications without rotating to another product with a different mode of action.
indoxacarb, MOA 22A (Avaunt eVo) 30 WDG	2.5 to 3.5 lb	12 hrs	3	Do not apply more than 14 ounces of Avaunt (0.26 lb a.i.) per acre per crop. The minimum interval between sprays is 3 days.
methomyl, MOA 1A (Lannate) 90 SP (Lannate) 2.4 LV	0.5 to 1 lb 1.5 to 3 pts	48 hrs	See label	See label for use instructions.
methoxyfenozide, MOA 18 (Intrepid) 2 F	4 to 10 fl oz	4 hrs	1	Low rates for early-season applications to young or small plants. For mid- and late-season applications, use 6 to 10 ounces.
spinetoram, MOA 5 (Radiant) 1 SC	5 to 10 fl oz	4 hrs	1	
pyrethroid, MOA 3A		12 hrs		See **Table 2-25 A** for registered pyrethroids and pre-harvest intervals. Not recommended for armyworms.
viruses		4 hrs		See **Table 2-26** for virus-based insecticides effective against specific insects **OMRI listed**.
LEAFHOPPER				
buprofezin, MOA 16 (Courier) 40 SC	9 to 13.6 fl oz	12 hrs	7	Do not apply more than 27.2 fl oz per acre per crop cycle.
dinotefuran, MOA 4A **Foliar treatment** (Venom) 70 SG (Scorpion) 35 SL	1 to 3 oz 2 to 2.25 fl oz	12 hrs	7	Do not follow soil applications with foliar applications of any neonicotinoid in- secticide. Use only one application method. Do not apply more than 6 ounces per acre (foliar) or 12 ounces per acre (soil).
Soil treatment (Venom) 70 SG (Scorpion) 35 SL	5 to 6 fl oz 9 to 10.5 fl oz		21	Soil applications may be applied by: (1) narrow band below or above the seed line at planting, (2) post seeding or transplant drench with sufficient water to ensure incorporation, or (3) drip irrigation.

TABLE 2-12. INSECT CONTROL FOR LETTUCE (cont'd)

Insecticide, Mode of Action Code, and Formulation	Amount of Formulation Per Acre	Restricted Entry Interval (REI)	Pre harvest Interval (PHI) (Days)	Precautions and Remarks
LEAFHOPPER (cont'd)				
dimethoate 4 EC, MOA 1B	0.5 pt	48 hrs	14	14-day interval for leaf lettuce.
flupyradifurone, MOA 4D **Foliar treatment** (Sivanto Prime) 1.67	7.0 to 14 fl oz	4 hrs	1	Do not apply more than 0.365 lb flupyradifurone per acre per crop per season regardless of applications method, product, or formulation.
Soil treatment	21 to 28 fl oz	4 hrs	21	Chemigation via drip, injection below the eventual seed line prior to planting, post-transplant drench following setting and covering, and planting hole drench after transplanting.
imidacloprid, MOA 4A (various) 1.6 F	3.75 fl oz	12 hrs	7	There is a 12-month plant-back restriction for several crops. Check label for restrictions.
pyrethroid, MOA 3A		12 hrs		See **Table 2-25 A** for registered pyrethroids and pre-harvest intervals.
thiamethoxam, MOA 4A (Actara) 25 WDG	1.5 to 3 oz	12 hrs	7	
tolfenpyrad, MOA 21A (Torac) 1.29 EC	14 to 21 fl oz	12 hrs	1	Do not apply until at least 14 days after plant emergence or after transplanting to allow time for root establishment.
SLUGS				
iron phosphate (Sluggo)	20 to 44 lbs	0 hrs	0	**OMRI listed.** Sluggo should be scattered around the perimeter of the crop to provide a protective barrier for slugs and snails. If slugs are inside the rows, scatter the bait on the soil around the plants and between rows. For smaller plantings use at 0.5 to 1 lb 1,000 square feet.
metaldehyde (Deadline Bullets)	25 lbs	12 hrs	0	Apply in a band to the soil between rows. Do not allow pellets to come into contact with plant parts. Do not exceed 3 applications per season or at inter- vals shorter than 14 days.

TABLE 2-13. INSECT CONTROL FOR OKRA

Insecticide, Mode of Action Code, and Formulation	Amount of Formulation Per Acre	Restricted Entry Interval (REI)	Pre harvest Interval (PHI) (Days)	Precautions and Remarks
APHID				
acetamiprid, MOA 4A (Assail) 30 SC	1.7 to 3.4 fl oz	12 hrs	7	Do not apply more than once every 7 days, and do not exceed 4 applications per season.
afidopyropen, MOA 9D (Sefina) DC	3 oz	12 hrs	0	Do not make more than 2 sequential applications before using a different mode of action.
imidacloprid, MOA 4A **Soil treatment** (Admire Pro) 4.6 F (various) 2 F	7 to 10.5 fl oz 16 to 24 fl oz	12 hrs	21	See label for soil treatment instructions.
Foliar treatment (Admire Pro) 4.6 F (various) 1.6 F	1.3 to 2.2 fl oz 3.8 fl oz	12 hrs	0	
flonicamid, MOA 29 (Beleaf) 50 SG	2 to 2.8 oz	12 hrs	0	
flupyradifurone, MOA 4D (Sivanto Prime) 1.67 SL	7.0 to 12 fl oz	4 hrs	1	
pyrifluquinazon, MOA 9B (PQZ) 1.87 EC	2.4 to 3.2 fl oz	12 hrs	1	See label for rotational crop restrictions. Do not exceed 4.8 fl oz per acre per crop cycle.
spirotetramat, MOA 23 (Movento) 2 SC	4 to 5 fl oz	24 hrs	3	Do not exceed 10 fluid ounces per season. Not for flea beetle. Requires surfactant.
sulfoxaflor, MOA 4C (Closer) 2 SC	1.5 to 2.0 fl oz	12 hrs	7	
malathion, MOA 1B (various brands) 8 F (various brands) 25 WP	1.5 pt 6 lb	12 hrs	1	

TABLE 2-13. INSECT CONTROL FOR OKRA (cont'd)

Insecticide, Mode of Action Code, and Formulation	Amount of Formulation Per Acre	Restricted Entry Interval (REI)	Pre harvest Interval (PHI) (Days)	Precautions and Remarks
BLISTER BEETLE, FLEA BEETLE, JAPANESE BEETLE				
carbaryl, MOA 1A (Sevin) XLR Plus	1 to 1.5 qt	12 hrs	3	On foliage as needed.
pyrethroid, MOA 3A		12 hrs		See **Table 2-25 A** for a list of registered pyrethroids and pre-harvest intervals.
CORN EARWORM, TOBACCO BUDWORM, EUROPEAN CORN BORER				
carbaryl, MOA 1A (Sevin) XLR Plus	1 to 1.5 qt	12 hrs	3	On foliage as needed.
chlorantraniliprole, MOA 28 (Coragen eVo) 5 SC	3 to 7.5 fl oz	4 hrs	1	Foliar or drip chemigation. Drip chemigation must be applied uniformly to the root zone. See label for instructions.
cyclaniliprole, MOA 28 (Harvanta) 50 SL	10.9 to 16.4 fl oz	4 hrs	1	Foliar applications will help suppress western flower thrips when used in a rotational program.
cyantraniliprole, MOA 28 **Soil treatment** (Verimark) 1.67 SC	5 to 10 fl oz	4 hrs	1	Verimark is for soil application only. Applications made at planting and/or via drip chemigation. See label for application options. Exirel is for foliar application only. Rates >13.5 for loopers only.
Foliar treatment (Exirel) 0.83SE	7 to 17 fl oz	12 hrs	1	
methoxyfenozide, MOA 18 (Intrepid) 2 F	8 to 16 fl oz	4 hrs	1	
novaluron, MOA 15 (Rimon) 0.83 EC	9 to 12 fl oz	12 hrs	1	
spinetoram, MOA 5 (Radiant) 1 SC	5 to 10 fl oz	4 hrs	1	For corn earworm only.
pyrethroid, MOA 3A		12 hrs		See **Table 2-25 A** for a list of registered pyrethroids and pre-harvest intervals.
viruses		4 hrs		See **Table 2-26** for virus-based insecticides effective against specific insects. **OMRI listed.**
SPIDER MITES				
bifenazate, MOA 20D (Acramite) 50 WP	0.75 to 1 lb	12 hrs	3	Do not make more than one application per season. Active against immatures and adults. Some egg activity.
fenpyroximate MOA 21A (Portal) 0.4 EC	2 pt	12 hrs	3	Do not make more than two applications per season. Active against immatures and adults.
STINK BUG, LEAFFOOTED BUG				
pyrethroid, MOA 3A		12 hrs		See **Table 2-25 A** for a list of registered pyrethroids and pre-harvest intervals.
WHITEFLY				
buprofezin, MOA 16 (Courier) 40 SC	9 to 13.6 fl oz	12 hrs	1	
chlorantraniliprole, MOA 28 (Coragen eVo) 5 SC	1.2 to 2.5 fl oz	4 hrs	1	Foliar or drip chemigation. Drip chemigation must be applied uniformly to the root zone. See label for instructions.
cyantraniliprole, MOA 28 **Soil treatment** (Verimark) 1.67 SC	6.75 to 13.5 fl oz	4 hrs	1	Apply Verimark to at planting and/or later via drip irrigation or soil injection. See label for application options.
Foliar treatment (Exirel) 0.83SE	13.5 to 20.5 fl oz	12 hrs	1	Exirel is for foliar application.
flupyradifurone, MOA 4D (Sivanto Prime) 1.67 SL	10.5 to 14.0 fl oz	4 hrs	1	
imidacloprid, MOA 4A **Soil treatment** (Admire Pro) 4.6 F (various) 2 F	7 to 14 fl oz 16 to 32 fl oz	12 hrs	21	See label for soil application instructions.
Foliar treatment (Admire Pro) 4.6 F (various) 1.6 F	1.3 to 2.2 fl oz 3.8 oz	12 hrs	0	
pyrifluquinazon, MOA 9B (PQZ) 1.87 EC	2.4 to 3.2 fl oz	12 hrs	1	
pyriproxyfen, MOA 7C (Knack) 0.86 EC	8 to 10 fl oz	12 hrs	1	Do not make more than two applications per season.
spirotetramat, MOA 23 (Movento) 2 SC	4 to 5 fl oz	24 hrs	3	Do not exceed 10 fluid ounces per season. Not for flea beetle. Requires surfactant.

TABLE 2-14. INSECT CONTROL FOR ONION

Insecticide, Mode of Action Code, and Formulation	Amount of Formulation Per Acre	Restricted Entry Interval (REI)	Pre harvest Interval (PHI) (Days)	Precautions and Remarks
ARMYWORMS, CUTWORM				
chlorantraniliprole, MOA 28 (Coragen eVo) 5SC (Vantacor) 5SC	1.2 to 2.5 fl oz 1.2 to 2.5 fl oz	4 hrs	1	
cyantraniliprole, MOA 28 (Exirel) SE	7.0 to 13.5 fl oz	12 hrs	1	
methoxyfenozide, MOA 18 (Intrepid) 2 F	4 to 8 fl oz 8 to 12 fl oz	4 hrs	1	For use against lepidopteran pests on green onion only. Use lower rates in early season on small plants; use higher rates in late season and for heavy infestations.
pyrethroid, MOA 3A		12 hrs		See **Table 2-25 A** for a list of registered pyrethroids and pre-harvest intervals.
spinetoram, MOA 5 (Radiant) 1 SC	5 to 10 fl oz	4 hrs	1	
LEAFMINER, Including allium leafminer				
cyromazine, MOA 17 (Trigard) 75 WS	2.66 oz	12 hrs	7	While cyromazine is effective against liromyza leafminers, data is lacking on allium leafminer.
Dinotefuran, MOA 4A (Venom) 70 SG (Scorpion) 35 SL	3 to 4 oz 5.75 to 7 fl oz	12 hrs	1	
chlorantraniliprole, MOA 28 (Coragen eVo) 5 SC	1.7 to 2.5 fl oz	4 hrs	1	For foliar or soil application or drip chemigation. Drip chemigation must be applied uniformly to the root zone. See label for instructions.
cyclaniliprole, MOA 28 (Harvanta) 50 SL	16.4 fl oz	12 hrs	1	
spinetoram, MOA 5 (Radiant) 1 SC	6 to 10 fl oz	4 hrs	1	Control may be improved by mixing with an adjuvant.
ONION MAGGOT, SEED CORN MAGGOT				
Onion seed pre-treated with cyromazine (Trigard) can be used to control onion and seed corn maggot.				
diazinon, MOA 1B (Diazinon) (AG 500)	2 to 4 qt	3 days		Broadcast just before planting and mix into the top 3 to 4 inches of soil.
pyrethroid, MOA 3A		12 hrs		See **Table 2-25 A** for a list of registered pyrethroids and pre-harvest intervals.
THRIPS				
abamectin, MOA 6 (Agri-Mek) SC	1.75 to 3.5 fl oz	12 hrs	30	Avoid using in combination with stick or binder product such as Bravo WeatherStik.
acetamiprid MOA 4A (Assail) 30 SC	3.4 fl oz	12 hrs	7	Control may be improved by tank mixing with an adjuvant. Do not exceed 4 applications per year.
methomyl, MOA 1A (Lannate) 2.4 LV	3 pt	48 hrs	7	May be applied by overhead sprinkler chemigation to control thrips. Add a wetting agent to improve coverage.
spinetoram, MOA 5 (Radiant) 1 SC	6 to 10 fl oz	4 hrs	1	Control may be improved by mixing with an adjuvant.
tolfenpyrad, MOA 21A (Torac) 1.29 EC	24 fl oz	12 hrs	7	Do not make more than 3 applications per crop cycle. See label restrictions for protection of pollinators.
pyrethroid, MOA 3A		12 hrs		See **Table 2-25 A** for a list of registered pyrethroids and pre-harvest intervals.

TABLE 2-15. INSECT CONTROL FOR PEA, ENGLISH, AND SNOW PEA (SUCCULENT AND DRIED)

Insecticide, Mode of Action Code, and Formulation	Amount of Formulation Per Acre	Restricted Entry Interval (REI)	Pre harvest Interval (PHI) (Days)	Precautions and Remarks
APHID				
acetamiprid MOA 4A (Assail) 30 SC	2.1 to 4.5 fl oz	12 hrs	7	Also controls leafhoppers. Succulent peas only.
pyrethroid, MOA 3A		12 hrs		See **Table 2-25 A** for a list of registered pyrethroids and pre-harvest intervals.
dimethoate, MOA 1B (Dimethoate) 400 (4E)	0.32 pt	48 hrs	0	Do not make more than one application per season, and do not feed or graze if a mobile viner is used, or for 21 days, if a stationary viner is used. Re-entry interval is 48 hours.
flupyradifurone, MOA 4D (Sivanto Prime) 1.67 SL	7.0 to 14 fl oz	4 hrs	7	Will also control leafhopper
imidacloprid, MOA 4A **Soil treatment** (Admire Pro) 4.6 F (various) 2 F	7 to 10.5 fl oz 16 to 24 fl oz	12 hrs	21	See label for soil application instructions.
Foliar treatment (Admire Pro) 4.6 F (various) 2 F	1.2 fl oz 3.5 fl oz	12 hrs	7	
ARMYWORMS, CLOVERWORM, CUTWORM, LOOPER				
chlorantraniliprole MOA 28 (Coragen eVo) 5 SC (Vantacor) 5 SC	1.2 to 2.5 fl oz 1.2 to 2.5 fl oz	4 hrs	1	
spinetoram, MOA 5 (Radiant) 1 SC	4 to 8 fl oz	4 hrs	3 (succulent) 28 (dried)	Not for cutworm.
spinosad, MOA 5 (Blackhawk)	2.2 to 3.3 oz	4 hrs	3 (succulent) 28 (dried)	
viruses		4 hrs		See **Table 2-26** for virus-based insecticides effective against specific insects. **OMRI listed.**
LEAFHOPPER, PLANT BUG, STINK BUG				
methomyl, MOA 1A (Lannate) 2.4 LV	1.5 to 3 pt	48 hrs	1 (pea) 5 (forage)	Apply to foliage as needed.
pyrethroid, MOA 3A		12 hrs		See **Table 2-25 A** for registered pyrethroids and pre-harvest intervals.
SEEDCORN MAGGOT				
See **Beans** for control				

TABLE 2-16. INSECT CONTROL FOR PEA, SOUTHERN (COWPEAS)

Insecticide, Mode of Action Code, and Formulation	Amount of Formulation Per Acre	Restricted Entry Interval (REI)	Pre harvest Interval (PHI) (Days)	Precautions and Remarks
APHID, THRIPS				
acetamiprid MOA 4A (Assail) 30SG	2 to 5.3 oz	12 hrs	7	Succulent peas only.
pyrethroid, MOA 3A		12 hrs		See **Table 2-25 A** for registered pyrethroids and pre-harvest intervals.
flupyradifurone, MOA 4D (Sivanto Prime) 1.67 SL	7.0 to 14 fl oz	4 hrs	7	Will not control thrips.
imidacloprid, MOA 4A **Soil treatment** (Admire Pro) 4.6 F (various) 2 F	7 to 10.5 fl oz 16 to 24 fl oz	12 hrs	21	See label for soil application instructions.
Foliar treatment (Admire Pro) 4.6 F (various) 2 F	1.2 fl oz 2.8 fl oz	12 hrs	7	

TABLE 2-16. INSECT CONTROL FOR PEA, SOUTHERN (COWPEAS) (cont'd)

Insecticide, Mode of Action Code, and Formulation	Amount of Formulation Per Acre	Restricted Entry Interval (REI)	Pre harvest Interval (PHI) (Days)	Precautions and Remarks
APHID, THRIPS (cont'd)				
spinetoram, MOA 5 (Radiant) 1 SC	5 to 8 fl oz	4 hrs	3 (succulent) 28 (dried)	Radiant is not effective against aphids.
spinosad, MOA 5 (Blackhawk)	2.2 to 3.3 oz	4 hrs	3 (succulent) 21 (dried)	Blackhawk is not effective against aphids.
sulfoxaflor, MOA 4C (Transform) 50 WG	0.75 to 1.0 oz	24 hrs	7	
BEAN LEAF BEETLE				
carbaryl, MOA 1A (Sevin) XLR Plus	0.5 to 1 qt	12 hrs	3 (succulent) 21 (dried)	Do not feed treated foliage to livestock for 14 days after application.
pyrethroid, MOA 3A		12 hrs		See **Table 2-25 A** for a list of registered pyrethroids and pre-harvest intervals.
CORN EARWORM, LOOPERS, EUROPEAN CORN BORER				
chlorantraniliprole MOA 28 (Coragen eVo) 5 SC (Vantacor) 5 SC	1.2 to 2.5 fl oz 1.2 to 2.5 fl oz	4 hrs	1	
cyantraniliprole, MOA 28 (Exirel) SE	10 to 20.5 fl oz	12 hrs	1 (succulent) 7 (dried)	
methoxyfenozide, MOA 18 (Intrepid) 2 F	4 to 16 fl oz	4 hrs	7	Use lower rates on smaller plants and higher rates in mid- to late season applications, against corn earworm. Do not apply more than 16 fl ounces per acre per season.
spinetoram, MOA 5 (Radiant) 1 SC	3 to 8 fl oz	4 hrs	3 (succulent) 28 (dried)	Do not apply more than 28 (succulent) or 12 fl oz (dried) per acre per season.
pyrethroid, MOA 3A		12 hrs		See **Table 2-25 A** for a list of registered pyrethroids and pre-harvest intervals.
methomyl, MOA 1A (Lannate) 90 SP	0.5 to 1 lb	48 hrs	1	Re-entry interval is 48 hr.
viruses		4 hrs		See **Table 2-26** for virus-based insecticides effective against specific insects. **OMRI listed.**
COWPEA CURCULIO				
pyrethroids, MOA 3A		12 hrs		See **Table 2-25 A** for a list of registered pyrethroids and pre-harvest intervals. Control may be poor in areas where resistant populations occur, primarily in parts of Alabama and Georgia. In areas where resistance is a problem, pyrethroid insecticides should be used at the highest labeled rate and synergized by tank mixing with 1-pint piperonyl butoxide synergist per acre. In field where resistance is a problem, applications every 3 to 5 days may be necessary to maintain control of the cowpea curculio population.
lambda-cyhalothrin, MOA 3A + chlorantraniliprole, MOA 28 (Besiege) ZC	6 to 10 fl oz	24 hrs	7 (succulent) 21 (dried)	
STINK BUG				
pyrethroid, MOA 3A		12 hrs		See **Table 2-25 A** for a list of registered pyrethroids and pre-harvest intervals. Control may be poor in areas where resistant populations occur, primarily in the Gulf Coast areas.
methomyl, MOA1A (Lannate) 90 SP	0.5 to 1 lb	48 hrs	1 (1.5 pt) 3 (>1.5 pt)	
LEAFMINER				
abamectin, MOA 6 (Agri-Mek) SC	1.75 to 3.5 fl oz	12	7 (succulent) 7 (dried)	Do not allow livestock to graze on forage.
cyromazine, MOA 17 (Trigard) 75 WP	2.66 oz	12 hrs	7 (dried)	For use on dried peas only. Allow a minimum of 17 days between successive applications.
spinetoram, MOA 5 (Radiant) 1 SC	5 to 8 fl oz	4 hrs	3 (succulent) 28 (dried)	
spinosad, MOA 5 (Blackhawk)	2.5 to 3.3 oz	4 hrs	3 (succulent) 28 (dried)	

TABLE 2-17. INSECT CONTROL FOR PEPPER

Insecticide, Mode of Action Code, and Formulation	Amount of Formulation Per Acre	Restricted Entry Interval (REI)	Pre harvest Interval (PHI) (Days)	Precautions and Remarks
APHID				
acephate, MOA 1B (Orthene) 97	8 oz	24 hrs	7	
acetamiprid, MOA 4A (Assail) 30 SC	1.7 to 3.4 fl oz	12 hrs	7	Do not apply more than once every 7 days, and do not exceed 4 applications per season.
afidopyropen, MOA 9D (Sefina) DC	3 oz	12 hrs	0	Do not make more than 2 sequential applications before using a different mode of action.
cyantraniliprole, MOA 28 (Verimark) SC	6.75 to 13.5 fl oz	4 hrs	1	Apply to soil at planting, as a transplant tray drench, in transplant water or hill drench. After planting may be applied via drip irrigation.
flonicamid, MOA 29 (Beleaf) 50 SG	2 to 4.8 oz	12 hrs	0	
flupyradifurone, MOA 4D (Sivanto Prime) 1.67 SL	7.0 to 14.0 fl oz	4 hrs	1	
imidacloprid, MOA 4A **Soil treatment** (Admire Pro) 4.6 F (various) 2 F	7 to 14 fl oz 16 to 32 fl oz	12 hrs	21	Where whitefly resistance is a concern, do not follow soil applications with foliar applications of any neonicotinoid. See label for soil application instructions. For short-term protection of transplants at planting, apply Admire Pro (0.44 oz/10,000 plants) not more than 7 days before trans- planting by 1) uniformly spraying on transplants, followed immediately by sufficient overhead irrigation to wash product into potting media; or 2) injection into overhead irrigation system using adequate volume to thoroughly saturate soil media.
Foliar treatment (Admire Pro) 4.6 F (various) 2 F	1.3 fl oz 3.8 fl oz	12 hrs	0	
pymetrozine, MOA 9B (Fulfill) 50 WDG	2.75 oz	12 hrs	0	Apply before aphids reach damaging levels. Do not exceed 5.5 ounces per acre per season. Not for flea beetle.
pyrifluquinazon, MOA 9B (PQZ) 1.87 EC	2.4 to 3.2 fl oz	12 hrs	1	See label for rotational crop restrictions. Do not exceed 4.8 fl oz per acre per crop cycle.
spirotetramat, MOA 23 (Movento) 2SC	4 to 5 fl oz	24 hrs	1	Do not exceed 10 fluid ounces per season. Requires surfactant. Will not control flea beetle.
sulfoxaflor, MOA 4C (Closer) 2 SC (Transform) WG	1.5 to 2.0 fl oz 0.75 to 1.0 oz	24 hrs	1	
tolfenpyrad, MOA 21A (Torac)	17 to 21 fl oz	12 hrs	1	
thiamethoxam, MOA 4A **Soil treatment** (Platinum) 75 SG	1.66 to 3.67 oz	12 hrs	30	Platinum may be applied to direct-seeded crops in-furrow seeding or transplant depth, post seeding or transplant as a drench, or through drip irrigation. Actara is applied as a foliar spray. Do not exceed 11 ounces per acre per season of Platinum or Actara. Check label for plant-back restrictions for several crops. Actara is for foliar application.
Foliar treatment (Actara) 25 WDG	1.5 to 3 oz	12 hrs	7	
ARMYWORMS, CORN EARWORM, LOOPER, EUROPEAN CORN BORER, HORNWORM				
Bacillus thuringiensis, MOA 11A (Dipel) DF (Xentari) DF	0.5 to 1.5 lb 0.5 to t2 lb	4 hrs	0	
chlorantraniliprole, MOA 28 (Coragen eVo) 5 SC	2 to 7.5 fl oz	4 hrs	1	Foliar or drip chemigation. Drip chemigation must be applied uniformly to the root zone. See label for instructions.
cyantraniliprole, MOA 28 **Soil treatment** (Verimark) 1.67 SC	5 to 10 fl oz	4 hrs	1	Verimark is for soil application only. Applications made at planting and/ or via drip chemigation. See label for application options.
Foliar treatment (Exirel) 0.83SE	7 to 13.5 fl oz	12 hrs	1	Exirel is for foliar application only.
cyclaniliprole, MOA 28 (Harvanta) 50 SL	10.9 to 16.4 fl oz	4 hrs	1	
emamectin benzoate, MOA 6 (Proclaim) 5 WDG	2.4 to 4.8 oz	12 hrs	7	Apply when larvae are first observed. Additional applications may be necessary to maintain control.
indoxacarb, MOA 22A (Avaunt eVo) 30 WDG	2.5 to 3.5 oz	12 hrs	3	Use only higher rate for control of armyworm and corn earworm. Do not apply more than 14 ounces of Avaunt (0.26 lb a.i. per acre per crop). Minimum interval between sprays is 5 days.
methoxyfenozide, MOA 18 (Intrepid) 2 F	4 to 16 fl oz	4 hrs	1	Apply at rates of 4 to 8 fluid ounces early in season when plants are small. Apply at rates of 8 to 16 ounces to large plants or when infestations are heavy. During periods of continuous moth flights retreatments at 7- to 14-days may be required. Do not apply more than 16 fluid ounces per application or 64 fluid ounces of Intrepid per acre per season.

TABLE 2-17. INSECT CONTROL FOR PEPPER (cont'd)

Insecticide, Mode of Action Code, and Formulation	Amount of Formulation Per Acre	Restricted Entry Interval (REI)	Pre harvest Interval (PHI) (Days)	Precautions and Remarks
ARMYWORMS, CORN EARWORM, LOOPER, EUROPEAN CORN BORER, HORNWORM (cont'd)				
novaluron, MOA 15 (Rimon) 0.83 EC	9 to 12 fl oz	12 hrs	1	The use of a surfactant/adjuvant with Rimon is prohibited on pepper.
spinetoram, MOA 5 (Radiant) 1 SC	5 to 10 fl oz	4 hrs	1	
pyrethroid, MOA 3A		12 hrs		See **Table 2-25 A** for a list of registered pyrethroids and pre-harvest intervals.
BLISTER BEETLE, STINK BUG, LEAFFOOTED BUG				
pyrethroid, MOA 3A		12 hrs		See **Table 2-25 A** for a list of registered pyrethroids and pre-harvest intervals.
dinotefuran, MOA 4A **Foliar treatment** (Venom) 70 SG (Scorpion) 35 SL	1 to 4 oz 2 to 7 fl oz	12 hrs	1	Do not combine foliar applications with soil applications, or vice versa. Use only one application method.
Soil treatment (Venom) 70 SG (Scorpion) 35 SL	5 to 6 oz 9 to 10.5 fl oz		21	
thiamethoxam, MOA 4A (Actara) 25WDG	3 to 5.5 oz	12 hrs	0	
LEAFMINER				
abamectin, MOA 6 (Agri-Mek) 0.7 SC	1.75 to 3.5 fl oz	12 hrs	7	
cyromazine, MOA 17 (Trigard) 75 WP	2.66 oz	12 hrs	0	
dimethoate 4 EC, MOA 1B	0.5 pt	48 hrs	0	Re-entry interval is 48 hr.
spinetoram, MOA 5 (Radiant) 1 SC	6 to 10 fl oz	4 hrs	1	
PEPPER MAGGOT				
acephate, MOA 1B (Orthene) 97 PE	0.75 to 1 lb	24 hrs	7	
dimethoate 4 EC, MOA 1B	0.5 to 0.67 pt	48 hrs	0	
pyrethroid, MOA 3A		12 hrs		See **Table 2-25 A** for registered pyrethroids and pre-harvest intervals
PEPPER WEEVIL				
acetamiprid, MOA 4A (Assail) 30 SC	2.1 to 3.4 fl oz	12 hrs	7	
imidacloprid, MOA 4A (Admire Pro) 4.6	2.2 fl oz	12 hrs	0	Do not exceed 6.7 fl oz per acre per crop season.
cyclaniliprole, MOA 28 (Harvanta) 50 SL	10.9 to 16.4 fl oz	4 hrs	1	
oxamyl, MOA 1A (Vydate) 2 L	2 to 4 pt	48 hrs	7	May also be applied through drip irrigation.
thiamethoxam, MOA 4A (Actara) 25 WP	3 to 4 oz	12 hrs	0	
tolfenpyrad, MOA 21A (Torac)	17 to 21 fl oz	12 hrs	1	
BROAD MITE				
abamectin, MOA 6 (Agri-Mek) 0.7 SC	1.75 to 3.5 fl oz	12 hrs	7	On foliage as needed.
fenazaquin, MOA 21A (Magister) SC	24 to 36 fl oz	12 hrs	3	Do not make more than one application per season.
fenpyroximate MOA 21A (Portal) 0.4 EC	2 pt	12 hrs	3	Do not make more than two applications per season.
spiromesifen, MOA 23 (Oberon) 2 SG	7 to 8.5 fl oz	12 hrs	7	Do not exceed 3 applications per season
spirotetramat MOA 23 (Movento) 2 SC	4 to 5 fl oz	12 hrs	1	
tolfenpyrad, MOA 21A (Torac)	17 to 21 fl oz	12 hrs	1	

TABLE 2-17. INSECT CONTROL FOR PEPPER (cont'd)

Insecticide, Mode of Action Code, and Formulation	Amount of Formulation Per Acre	Restricted Entry Interval (REI)	Pre harvest Interval (PHI) (Days)	Precautions and Remarks
THRIPS				
dinotefuran, MOA 4A (Venom) 70 SG (Scorpion) 35 SL	5 to 6 oz 9 to 10.5 fl oz	12 hrs	21	Dinotefuran is not effective against western flower thrips.
imidacloprid, MOA 4A (Admire Pro) 4.6 F	7 to 14 fl oz	12 hrs	21	See Aphids for application instructions. Treating transplants before setting in the field, followed by drip irrigation may suppress incidence of tomato spotted virus. Imidacloprid is ineffective against western flower thrips.
viruses		4 hrs		See Table 2-26 for virus-based insecticides effective against specific insects. **OMRI listed.**
cyclaniliprole, MOA 28 (Harvanta)	10.9 to 16.4 fl oz	4 hrs	1	Harvanta used in a rotational program will help suppress western flower thrips.
flonicamid, MOA 29 (Beleaf) 50 SG	2 to 4.8 fl oz	12 hrs	0	This is an option for insecticide-resistant wester flower thrips. Do not exceed 8.4 oz per acer per season.
methomyl, MOA 1A (Lannate) 2.4 LV	1.5 pt	48 hrs	3	
spinetoram, MOA 5 (Radiant) 1 SC	6 to 10 fl oz	4 hrs	1	Do not exceed 29 fluid ounces per acre per season. Control of thrips may be improved by adding a spray adjuvant. See label for instructions. This is an option for insecticide-resistant thrips.
tolfenpyrad, MOA 21A (Torac) 1.29 EC	21 fl oz	12 hrs	1	

TABLE 2-18. INSECT CONTROL FOR POTATO

Insecticide, Mode of Action Code, and Formulation	Amount of Formulation Per Acre	Restricted Entry Interval (REI)	Pre harvest Interval (PHI) (Days)	Precautions and Remarks
APHID				
acetamiprid, MOA 4A (Assail) 30 SC	2.1 to 3.4 fl oz	12 hrs	7	Do not make more than 4 applications per season. Thorough coverage is important. Assail belongs to the same class of insecticides (neonicotinoid, 4A) as Admire Pro, Belay and Platinum (soil insecticides) and Provado and Actara (foliar insecticides). Colorado potato beetle populations have developed resistance to this class.
clothianidin MOA 4A (Belay) 2.13	2 to 3 fl oz	12 hrs	7	Apply Belay 50 as foliar spray when populations reach a threshold level. Do not apply more than 3 applications. Belay belongs to the same class of insecticides (neonic- otinoid) as Admire Pro, Provado, Actara, and Platinum and Colorado potato beetle populations have the potential to become resistant to this class.
flonicamid, MOA 29 (Beleaf) 50 SG	2 to 2.8 oz	12 hrs	7	
flupyradifurone, MOA 4D (Sivanto Prime) 1.67 SL	7.0 to 12.0 fl oz	4 hrs	1	
dimethoate 4 EC, MOA 1B	0.5 to 1 pt	48 hrs	0	Do not apply more than 2 pints total per year.
imidacloprid, MOA 4A (Admire Pro) 4.6F	1.2 fl oz	12 hrs	7	To minimize selection for resistance in Colorado potato beetle, do not use acetamiprid, imidacloprid, or thiamethoxam for aphid control if either of these compounds was applied to the crop for control of Colorado potato beetle. See comments on insecticide rotation under Colorado potato beetle.
pymetrozine, MOA 9B (Fulfill) 50 WDG	2.75 oz	12 hrs	14	Allow at least 7 days between applications. Do not exceed a total of 5.5 ounces (0.17 lb a.i.) per acre per season.
thiamethoxam, MOA 4A (Actara) 25 WDG	3 oz	12 hrs	14	To minimize selection for resistance in Colorado potato beetle, do not use imidacloprid or thiamethoxam for aphid control if either of these compounds was applied to the crop for control of Colorado potato beetle.

COLORADO POTATO BEETLE

(CPB) populations in most commercial potato-growing areas have developed resistance to many insecticides. As a result, insecticides that are effective in some areas, or were effective in the past, may no longer provide control in particular areas. CPB readily develops resistance to insecticides. The following practices help to reduce the risk of resistance developing:

CROP ROTATION AND INSECTICIDE ROTATION: The use of insecticides representing different modes of action IRAC MoA class number in different years and against different generations of CPB within a year are essential if insecticide resistance is to be managed and the risks of control failures due to resistance minimized. If control failures or reduced levels of control are observed with a particular insecticide, do NOT make a second application of the same insecticide at the same or higher rate. If an additional insecticide application is necessary, a different insecticide representing a different IRAC MoA class number should be used. Because CPB adults will move between adjacent and nearby fields from one year to the next, it is important to maintain the same rotation schedule of insecticide MOA classes in adjacent fields are groups

TABLE 2-18. INSECT CONTROL FOR POTATO (cont'd)

Insecticide, Mode of Action Code, and Formulation	Amount of Formulation Per Acre	Restricted Entry Interval (REI)	Pre harvest Interval (PHI) (Days)	Precautions and Remarks
COLORADO POTATO BEETLE (cont'd)				

of nearby fields. *SCOUT FIELDS:* All insecticide applications to the potato crop, regardless of the target insect pest, have the potential to increase the resistance of the CPB to insecticides. Unnecessary insecticide applications should be avoided by scouting fields for insect pests and applying insecticides only when potentially damaging insect populations are present.

SPOT TREATMENTS: Because overwintered potato beetles invade rotated fields from sources outside the field, potato beetle infestations in rotated fields occur first along field edges early in the season. Limiting insecticide applications to infested portions of the field will provide effective control and reduce costs. Growers are advised to keep accurate records on which insecticides have been applied to their potato crop for control of CPB and on how effective those insecticides were at controlling infestations.

This will make choosing an insecticide and maintaining insecticide rotations easier. Monitoring the insecticide resistance status of local populations will also make insecticide selection easier.

Insecticide, Mode of Action Code, and Formulation	Amount of Formulation Per Acre	REI	PHI	Precautions and Remarks
abamectin, MOA 6 (Agri-Mek) 0.7 SC	1.75 to 3.5 fl oz	12 hrs	14	Apply when adults and/or small larvae are present but before large larvae appear. Do not exceed two applications per season. Apply in at least 20 gallons water per acre.
acetamiprid, MOA 4A (Assail) 30 SC	1.3 to 3.4 fl oz	12 hrs	7	Apply when most egg masses have hatched and many small, but few large larvae are present. An additional application should be used only if defoliation increases. Al- low at least 7 days between foliar applications. To minimize selection for resistance, do not use foliar applications of any IRAC MOA class 4A insecticides if any IRAC MOA class 4A insecticides were applied to the crop as soil or seed piece treatments. See comments on insecticide rotation under CPB.
chlorantraniliprole, MOA 28 (Coragen eVo) 5 SC (Vantacor) 5 SC	3.5 to 7.5 oz 1.2 to 2.5 fl oz	4 hrs	14	Do not apply more than 15.4 ounces Coragen per acre per crop season. Coragen treated insects may take several days to die but stop feeding almost immediately after treatment.
clothianidin MOA 4A (Belay) 2.13	2 to 3 fl oz	12 hrs	7	Apply Belay 50 WDG as foliar spray Apply when adults and/or small larvae are present but before large larvae appear. Do not apply more than 3 applications. Belay belongs to the same class of insecticides (neonicotinoid) as Admire Pro, Provado, Actara, and Platinum and some CPB populations have developed resistance to this class.
cyantraniliprole, MOA 28 (Verimark) 1.67 SC	6.75 to 13.5 fl oz	4 hr	NA	Apply in-furrow at planting. Do not apply any other MOA Group 28 insecticide for CPB control following an at-plant application for cyantraniliprole. When applied at 10 to 13.5 fluid ounces per acre will provide control of European corn borer in most years, except possibly in very early planted potatoes.
dinotefuran, MOA 4A (Venom) 70 SG	1 to 1.5 oz (foliar) 6.5 to 7.5 oz (soil)	12 hrs	7	Soil treatment for preplant, preemergence, or ground crack application only. To minimize selection for resistance, do not use foliar applications of any IRAC MOA class 4A insecticides if any IRAC MOA class 4A insecticides applied to crop as soil or seed piece treatments. See comments on rotation under CPB.
imidacloprid, MOA 4A **Seed Piece treatment** (Genesis) 240 g/L	0.4 to 0.6 fl oz/ 100 lb of seed tubers			Resistance has been reported and may reduce efficacy or duration of control. To minimize selection for resistance, do not use foliar applications of any IRAC MOA class 4A insecticides if any of these compounds were applied to the crop as soil or seed piece treatments. See label for specific instructions. For early-planted potatoes, control may be marginal because of the prolonged time between application and CPB emergence. Limit use to locations where CPB were a problem in the same or adjacent fields during the previous year. Do not apply other IRAC MOA class 4A insecticides to a field if seed pieces were treated with Genesis. See product label for restrictions on rotational crops.
imidacloprid, MOA 4A **Soil treatment** (Admire Pro) 4.6F (various) 2.0 F	0.74 fl oz/ 1,000 ft row	12 hrs	—	Resistance has been reported and may reduce efficacy or duration of control. See comments on insecticide rotation under CPB. Admire Pro applied in-furrow at planting time may provide season-long control. However, for early planted potatoes control may be marginal due to the prolonged time between application and CPB emergence. Use only in potato fields that have a history of potato beetle infestations. If potatoes are rotated to a field adjacent to one planted in potato last year, a barrier treatment may be effective. Admire Pro may also be applied as a seed treatment. Check label for instructions regarding this use. Check label for restrictions on planting crops follow- ing Admire Pro treated potatoes. There have been reports of low levels of resistance to imidacloprid. To minimize selection for resistance, do not use foliar applications of any IRAC MOA class 4A insecticides if any of these compounds were applied to the crop as soil or seed piece treatments. See comments on insecticide rotation under CPB.
Foliar treatment (Admire Pro) 4.6 (various) 1.6 F	1.3 fl oz 3.75 fl oz	12 hrs	7	Resistance has been reported and may reduce efficacy or duration of control. To minimize selection for resistance, do not use foliar applications of any IRAC MOA class 4A insecticides if any of these compounds were applied to the crop as soil or seed piece treatments. See comments on insecticide rotation under CPB. Apply when most of the egg masses have hatched and most larvae are small (1/8 to 3/16 in.). An additional application should be made only if defoliation increases. Allow at least 7 days between foliar applications. Do not exceed 5.6 fluid ounces of Admire Proper field per acre per season. Regardless of formulation, do NOT apply more than a total of 0.31 lb imidacloprid per season. Foliar applications of imidacloprid should not be applied If soil application was used.
imidacloprid, MOA 4A + cyfluthrin, MOA 3A Premix (Leverage) 2.7 SE	3 to 3.75 fl oz		7	There have been reports of low levels of resistance to imidacloprid. To minimize selection for resistance, do not use foliar applications of any IRAC MOA class 4A insecticides if any of these compounds were applied to the crop as soil or seed piece treatments. See comments on insecticide rotation under CPB, apply when most of the egg masses have hatched and most larvae are small (1/8 to 3/16 inch). An additional application should be made only if defoliation increases. Leverage will control Europe- an corn borer if application coincides with egg hatch and presence of small corn borer larvae. Leverage should not be used in fields treated with Admire Pro.

TABLE 2-18. INSECT CONTROL FOR POTATO (cont'd)

Insecticide, Mode of Action Code, and Formulation	Amount of Formulation Per Acre	Restricted Entry Interval (REI)	Pre harvest Interval (PHI) (Days)	Precautions and Remarks
COLORADO POTATO BEETLE (cont'd)				
ledprona (Calantha)	16 fl oz	4 hrs	0	Ledprona is most effective against small larvae. Foliar applications at 50% egg hatch followed by a second application within 7-10 days is recommended. This product will cause lethal effects over a period of several days, often larvae cease feeding but may remain on the plant. Consult label for specific application conditions and instructions.
indoxacarb, MOA 22A (Steward) EC	6.7 to 11.3 fl oz	12 hrs	7	See label for notes on use.
novaluron, MOA 15 (Rimon) 0.83 EC	9 to 12 fl oz	12 hrs	14	Novaluron is an insect growth regulator with activity against eggs and larvae. Larvae are killed as they molt to the next stage. Eggs present at the time of application are killed. Adults exposed produce few eggs. Novaluron is most effective if directed against overwintered adults when egg numbers are increasing, and small larvae are just beginning to appear. Do not apply to successive generations of CPB. Do not apply more than 24 fl oz per season.
spinosad, MOA 5 (Blackhawk) 36 WG	1.7 to 3.3 oz		3	Apply when most egg masses have hatched and both small and large larvae are present. Thorough coverage is important. Do not apply more than a total of 0.33 lb a.i. (14.4 ounces of Blackhawk or 21 ounces of Radiant) per crop. Do not apply in consecutive generations of CPB and do not make more than two applications per single generation of Colorado potato beetle. Do not make successive applications less than 7 days apart. To minimize the potential for resistance, do NOT use spinosad or spinetoram if either product was applied to a potato crop in the field or an adjacent field within the last year.
spinetoram, MOA 5 (Radiant) 1 SC	6 to 8 fl oz	4 hrs	7	See comments above under spinosad.
thiamethoxam, MOA 4A **Seed Piece treatment** (Cruiser) 5 FS	0.11 to 0.16 fl oz/ 100 lb			See label for specific instructions. Resistance to neonicotinoid insecticides has been reported and may reduce efficacy or duration of control by thiamethoxam. To minimize selection for resistance, do not use foliar applications of any IRAC MOA class 4A insecticides if any of these compounds were applied to the crop as soil or seed piece treatments. See comments on insecticide rotation under CPB. For early-planted potatoes, control may be marginal because of the prolonged time between application and CPB emergence. Limit use to locations where CPB were a problem in the same or adjacent fields during the previous year.
thiamethoxam, MOA 4A (Platinum) 75 SG (Actara) 25 WDG	1.66 to 2.67 oz	12 hrs	7	Resistance to neonicotinoid insecticides has been reported and may reduce efficacy or duration of control by thiamethoxam. To minimize selection for resistance, do not use foliar applications of any IRAC MOA class 4A insecticides if any of these compounds were applied to the crop as soil or seed piece treatments. See comments on insecticide rotation under CPB. Platinum applied in-furrow at planting time may provide season-long control. For early-planted potatoes, control may be marginal because of the prolonged time between application and CPB emergence. Limit use to locations where CPB were a problem in the same or adjacent fields in the previous year. See product label for restrictions on rotational crops.
	3 oz	12 hrs	7	Resistance to neonicotinoid insecticides has been reported and may reduce efficacy or duration of control by thiamethoxam. To minimize selection for resistance, do not use foliar applications of any IRAC MOA class 4A insecticides if any of these compounds were applied to the crop as soil or seed piece treatments. See comments on insecticide rotation under CPB. Actara is applied as foliar spray. Apply when most of the eggs have hatched and most of the larvae are small (1/8 to 3/16 inch). An additional application should be made only if defoliation increases. Allow at least 7 days between applications. Do not make more than 2 applications of Actara per crop per season.
thiamethoxam, MOA 4A + chlorantraniliprole, MOA 28 Premix (Voliam Flex)	4 oz		14	Resistance to neonicotinoid insecticides has been reported and may re- duce efficacy or duration of control by thiamethoxam. To minimize selection for resistance, do not use foliar applications of any IRAC MOA class 4A insecticides if any of these compounds were applied to the crop as soil or seed piece treatments. See comments on insecticide rotation under CPB. Voliam Flexi is applied as a foliar spray. Apply when most of the eggs have hatched and most of the larvae are small (1/8 to 3/16 inch.). An additional application should be made only if defoliation increases. Allow at least 7 days between applications. To minimize selection for resistance, do not use foliar applications of any IRAC MOA class 4A insecticides if any of these compounds were applied to the crop as soil or seed piece treatments. Do not exceed 8 ounces of Voliam Flexi. See label for rotational restrictions. Voliam Flexi can be expected to provide control of European corn borer if application is timed correctly (see European corn borer for correct timing.
EUROPEAN CORN BORER				
'Atlantic' potato is very tolerant of injury by European corn borer larvae. Consequently, control is not recommended on 'Atlantic' unless more than 30 percent of the stems are infested. Control on all other varieties is recommended when infestations reach 20 percent infested stems. Application timing is critical. Scout for eggs and treat when eggs hatch or at the first sign of larvae entering petioles. Several days of cool wet weather will kill larvae and may eliminate the need for insecticide applications. If this occurs, flag additional egg masses and apply insecticide at hatch.				
pyrethroid, MOA 3A		12 hrs		Apply when threshold is reached (usually during the first half of May). A second application may be needed if the percentage of infested stems increases substantially 7 to 10 days after the first application. Ground applications are usually more effective than aerial applications. See **Table 2-25 A** for a list of registered pyrethroids and pre-harvest intervals.

TABLE 2-18. INSECT CONTROL FOR POTATO (cont'd)

Insecticide, Mode of Action Code, and Formulation	Amount of Formulation Per Acre	Restricted Entry Interval (REI)	Pre harvest Interval (PHI) (Days)	Precautions and Remarks
EUROPEAN CORN BORER (cont'd)				
chlorantraniliprole, MOA 28 (Coragen eVo) 5 SC (Vantacor) 5 SC	3.5 to 7.5 oz 1.2 to 2.5 fl oz	4 hrs	14	Correct timing of application is important. Apply when threshold is reached (usually during the first half of May). Do not apply more than 0.2 lb ai per chlorantraniliprole per acre per crop season.
thiamethoxam, MOA 4A + chlorantraniliprole, MOA 28 Premix (Voliam Flexi)	4 oz	12 hrs	14	Voliam Flexi is applied as a foliar spray. Correct timing of application is important for control of European corn borer. Apply when threshold is reached(usually during the first half of May). Voliam Flexi can also be expected to provide control of Colorado potato beetle if most of the potato beetle eggs have hatched and most of the larvae are small (1/8 to 3/16 inch). Voliam Flexi applications targeting European corn borer will select for resistance to neonicotinoid insecticides in CPB, if present. To minimize selection for resistance to Colorado potato beetle, do not use foliar applications of any IRAC MOA class 4A insecticides if any of these compounds were applied to the crop as soil or seed piece treatments. Do not exceed 8 ounces of Voliam Flexi. See label for rotational restrictions.
spinetoram, MOA 5 (Radiant) 1 SC	6 to 8 fl oz	4 hrs	7	Do not apply more than a total of 0.25 lb a.i.(32 fl oz) per crop.
FLEA BEETLE				
imidacloprid, MOA 4A **Soil treatment** (Admire Pro) 4.6 (various) 1.6 F	0.74 fl oz/ 1,000 ft row	12 hrs	—	If imidacloprid or thiamethoxam resistant CPB occur in the field, application of imida- cloprid to control flea beetles has the potential to further increase resistance levels. Imidacloprid applied in-furrow at planting time may provide season-long control of flea beetles. However, for early-planted potatoes control may be marginal due to the prolonged time between application and crop emergence. Check label for restrictions on planting crops following Admire Pro treated potatoes.
Foliar treatment (Admire Pro) 4.6 (various) 1.6 F	1.3 fl oz 3.75 fl oz	12 hrs	7	See comments for imidacloprid resistance in CPB.
thiamethoxam, MOA 4A **Seed Piece treatment** (Cruiser) 5 FS	0.11 to 0.16 fl oz/ 100 lb	12 hrs		See label for specific instructions. For early-planted potatoes, control may be marginal because of the prolonged time between application and flea beetle emergence. Limit use to locations where Colorado potato beetles were not a problem in the same or adjacent fields during the previous year. To minimize selection for resistance, do not use foliar applications of any IRAC MOA class 4A insecticides if any of these com- pounds were applied to the crop as soil or seed piece treatments. See comments on insecticide rotation under CPB.
thiamethoxam, MOA 4A (Platinum) 2 SC	5 to 8 fl oz	12 hrs	7	Platinum applied in-furrow at planting time may provide season-long control. However, for early planted potatoes control may be marginal due to the prolonged time between application and crop emergence. Limit use to locations where CPB were not a problem in the same or adjacent fields during the previous year. See product label for restrictions on rotational crops. See comments for imidacloprid resistance in CPB.
(Actara) 25 WDG	3 oz	12 hrs	7	Actara is applied as foliar spray. See comments for imidacloprid resistance in CPB. If imidacloprid or thiamethoxam resistant CPB occur in the field, application of Voliam Flexi should control flea beetle has the potential to increase resistance levels. See comments for imidacloprid resistance in CPB.
thiamethoxam, MOA 4A + chlorantraniliprole, MOA 28 (Voliam Flexi)	4 fl oz		14	Do not exceed 8.0 fluid ounces per acre Voliam Flexi or 0.094 lb a.i/ acre of thiamethoxam-containing products or 0.2-pound a.i/acre of chlorantraniliprole-containing products per growing season. If CPB occur in the field, application of Voliam Flexi to control flea beetles has the potential to increase resistance levels. See comments for imidacloprid for imidacloprid resistance in CPB.
pyrethroid, MOA 3A		12 hrs		See **Table 2-25 A** for a list of registered pyrethroids and pre-harvest intervals.
LEAFHOPPER				
carbaryl, MOA 1A (Sevin) XLR Plus	0.5 to 1 qt	12 hrs	7	On foliage when leafhoppers first appear. Repeat every 10 days as needed. Often a problem in the mountains.
dimethoate, MOA 1B (various formulations)				Check label for rate, PHI, and REI.
imidacloprid + cyfluthrin premix, MOA 4A and 3 (Leverage) 2.7 SE (Leverage) 360	3 to 3.80 fl oz 2.8 fl oz	7 hrs	7	There have been reports of low levels of resistance to imidacloprid. To minimize selection for resistance, do not use foliar applications of any IRAC MOA class 4A insecticides if any of these compounds were applied to the crop as soil or seed piece treatments. Leverage should not be used in fields treated with Admire Pro.
methomyl, MOA 1A (Lannate) 2.4 LV	1.5 pt	48 hrs	6	
pyrethroid, MOA 3A		12 hrs		See **Table 2-25 A** for a list of registered pyrethroids and pre-harvest intervals.
LEAFMINER				
dimethoate, MOA 1B (various formulations)				Check label for rate, PHI, and REI.

TABLE 2-18. INSECT CONTROL FOR POTATO (cont'd)

Insecticide, Mode of Action Code, and Formulation	Amount of Formulation Per Acre	Restricted Entry Interval (REI)	Pre harvest Interval (PHI) (Days)	Precautions and Remarks
LEAFMINER (cont'd)				
chlorantraniliprole, MOA 28 (Coragen eVo) 5 SC	3.5 to 5 fl oz	4 hrs	14	
BLISTER BEETLE, LEAFFOOTED BUG, PLANT BUG, STINK BUG, VEGETABLE WEEVIL				
carbaryl, MOA 1A (Sevin) XLR Plus	1 to 2 qt	12 hrs	7	On foliage as needed.
pyrethroid, MOA 3A		12 hrs		See **Table 2-25 A** for a list of registered pyrethroids and pre-harvest intervals.
POTATO TUBERWORM				
chlorantraniliprole, MOA 28 (Coragen eVo) 5 SC (Vantacor) 5 SC	1.2 to 2.5 fl oz 1.2 to 2.5 fl oz	4 hrs	14	Do not exceed 4 applications per acre per crop. Do not apply more than 0.2 lbs ai/acre chlorantraniliprole containing products per acre per calendar year. Minimum interval between applications is 5 days. Performance is improved if applied via overhead chemigation (see label).
cyantraniliprole, MOA 28 (Exirel) 0.83 SE	7 to 13.5 fl oz	12 hrs	12	Apply as foliar spray. Do not apply more than 0.4 lb ai/acre (including seed treatments) of cyantraniliprole containing products per calendar year. Methylated seed oil (MSO) adjuvant at 1 gal.100 gal. spray volume (1%v/v) improves control by foliar sprays. Performance is improved if applied via overhead chemigation (see label). Do not apply more than 0.4 lb/acre (including seed treatments) of cyantraniliprole containing products per calendar year.
indoxacarb, MOA 22A (Steward) EC	5.6 to 11.3 fl oz	12 hrs	7	See label for notes on use.
methomyl, MOA 1A (Lannate) 2.4 LV	1.5 pt	48 hrs	6	Prevent late-season injury by keeping potatoes covered with soil. To prevent damage in storage, practice sanitation.
pyrethroid, MOA 3A		12 hrs		See **Table 2-25 A** for a list of registered pyrethroids and pre-harvest intervals.
THRIPS				
dimethoate 4 EC, MOA 1B	0.5 pt	48 hrs	0	
spinetoram, MOA 5 (Radiant) 1 SC	6 to 8 fl oz	4 hrs	7	
spinosad, MOA 5 (Blackhawk) 36 WG	2.25 to 3.5 oz	4 hrs	3	Control may be improved by addition of an adjuvant to the spray mixture.
WIREWORM				
bifenthrin, MOA 3A (Capture LFR)	25.5 fl oz			In furrow at planting.
broflanilide, MOA 30 (Nurizma)	0.08 to 0.16 fl oz/ 1,000 row ft	12 hrs		In-furrow at planting. Apply as a 5-7" band at planting.
clothianidin, MOA 4A (Belay) 2.13	12 fl oz	12 hrs		In-furrow at planting.
ethoprop, MOA 1B (Mocap) 15 G	1.4 lb per 1,000 row ft	48 hrs	90	In-furrow at planting.
fipronil, MOA 2B (Regent) 4 SC	3.2 fl oz	0 hrs	90	In-furrow at planting. Do NOT use T-banding over the top of a closed furrow.
phorate, MOA 1B (Thimet) 20 G	Row Treatment: 10 to 20 oz (38 in. row spacing)	12 hrs	90	Can contribute to insecticide-resistance problems with Colorado potato beetle.

TABLE 2-19. INSECT CONTROL FOR RADISH

Insecticide, Mode of Action Code, and Formulation	Amount of Formulation Per Acre	Restricted Entry Interval (REI)	Pre harvest Interval (PHI) (Days)	Precautions and Remarks
APHID, FLEA BEETLE, LEAFMINER				
pyrethroid, MOA 3A		12 hrs		Effective against flea beetles only. See **Table 2-25 A** for a list of registered pyrethroids and pre-harvest intervals.
flupyradifurone, MOA 4D (Sivanto Prime) 1.67 SL	7.0 to 14 fl oz	4 hrs	7	Will not control flea beetles or leafminer.
imidacloprid, MOA 4A **Foliar treatment** (Admire Pro) 4.6 F (various) 2 F	1.2 fl oz 2.8 fl oz	12 hrs	7	Will not control leafminer.
flonicamid, MOA 29 (Beleaf) 50 SG	2 to 2.8 oz	12 hrs	3	
thiamethoxam, MOA 4A **Soil treatment** (Platinum) 75 SG	1.7 to 2.17 oz	12 hrs	NA	Platinum is for soil application within 24 hr of planting only. See label for application instructions.
Foliar treatment (Actara) 25 WDG	1.5 to 3 oz	12 hrs	7	
ROOT MAGGOT, WIREWORM				
diazinon, MOA 1B (AG 500) (50 WP)	2 to 4 qt 4 to 8 lb	3 days		Broadcast just before planting and immediately incorporate into the upper 4 to 8 inches of soil. Do not exceed 4 qt (AG 500) or 8 lbs (50 WP) per acre per year.

TABLE 2-20. INSECT CONTROL FOR SPINACH

Insecticide, Mode of Action Code, and Formulation	Amount of Formulation Per Acre	Restricted Entry Interval (REI)	Pre harvest Interval (PHI) (Days)	Precautions and Remarks
APHID				
acetamiprid, MOA 4A (Assail) 30 SC	1.7 to 3.4 fl oz	12 hrs	7	Do not apply more than once every 7 days, and do not exceed 5 applications per calendar year.
afidopyropen, MOA 9D (Versys) DC	1.5 fl oz	12 hrs	0	Do not make more than 2 sequential applications before using a different mode of action.
clothianidin, MOA 4A (Belay) 2.13 EC **Soil treatment**	9 to 12 fl oz	12 hrs	NA	Soil application at planting only. Belay must not be applied during bloom. Do not incorporate an adjuvant with foliar applications. Do not apply more than 6.4 oz per acre per season.
Foliar treatment	3 to 4 fl oz	12 hrs	7	
cyantraniliprole, MOA 28 (Verimark) 1.67 SC	6.75 to 13.5 fl oz	4 hrs	1	Suppression only. Soil applications are made only at planting. See label for application options.
flonicamid, MOA 29 (Beleaf) 50 SG	2 to 2.8 oz	12 hrs	0	
flupyradifurone, MOA 4D (Sivanto Prime) 1.67 SL	10.5 to 12.0 fl oz	4 hrs	1	
imidacloprid, MOA 4A **Soil treatment** (Admire Pro) 4.6 F (various) 2 F	4.4 to 10.5 fl oz 10 to 24 fl oz	12 hrs	21	Do not follow soil applications with foliar applications of any neonicotinoid insecticides. See label for soil application instructions.
Foliar treatment (Admire Pro) 4.6 F (various) 1.6 F	1.2 fl oz 3.8 fl oz	12 hrs	7	
pymetrozine, MOA 9B (Fulfill) 50 WDG	2.75 oz	12 hrs	7	Apply before aphids reach damaging levels. Use sufficient water to ensure good coverage.

TABLE 2-20. INSECT CONTROL FOR SPINACH (cont'd)

Insecticide, Mode of Action Code, and Formulation	Amount of Formulation Per Acre	Restricted Entry Interval (REI)	Pre harvest Interval (PHI) (Days)	Precautions and Remarks
APHID (cont'd)				
pyrifluquinazon, MOA 9B (PQZ) 1.87 EC	2.4 to 3.2 fl oz	12 hrs	1	See label for rotational crop restrictions. Do not exceed 4.8 fl oz per acre per crop cycle.
spirotetramat, MOA 23 (Movento) 2 SC	4 to 5 fl oz	24 hrs	3	Do not exceed 10 fluid ounces per season. Requires surfactant.
thiamethoxam, MOA 4A **Soil treatment** (Platinum) 75 SG	1.7 to 3.7 oz	12 hrs	30	See label for soil application instructions.
Foliar treatment (Actara) 25 WDG	1.5 to 3 oz	12 hr	7	
tolfenpyrad, MOA 21A (Torac) 1.29 EC	17 to 21 fl oz	12 hrs	1	Do not apply until at least 14 days after plant emergence or after transplanting to allow time for root establishment.
LEAFMINER				
chlorantraniliprole, MOA 28 (Coragen eVo) 5 SC	5 to 7.5 fl oz	4 hrs	1	Foliar or drip chemigation. Drip chemigation must be applied uniformly to the root zone. See label for instructions.
cyantraniliprole, MOA 28 **Soil treatment** (Verimark) 1.67 SC	5 to 13.5 fl oz	4 hrs	N/A	Verimark is for soil application only made at planting and/or via drip chemigation. Use higher rates (>10 fluid ounces) where cabbage looper is a concern. See label for application options.
Foliar treatment (Exirel) 0.83 SE	7 to 17 fl oz	12 hrs	1	Exirel is for foliar application only. Use higher rates (>13.5 fluid ounces) for cabbage loopers. Do not apply more than 0.4 lb a.i. per acre per year of cyazypyr or cyantraniliprole containing products.
cyclaniliprole, MOA 28 (Harvanta) 50 SL	10.9 to 16.4 fl oz	4 hrs	1	
cyromazine, MOA 17 (Trigard) 75 WP	2.66 oz	12 hrs	7	
spinetoram, MOA 5 (Radiant) 1 SC	6 to 10 fl oz	4 hrs	1	Spray adjuvants may enhance efficacy against leafminers. See label for information on adjuvants.
ARMYWORMS, BEET WEBWORM, CORN EARWORM, CUTWORM, LOOPER				
chlorantraniliprole, MOA 28 (Coragen eVo) 5 SC	1.2 to 2.5 fl oz	4 hrs	3	Beet armyworm may have developed resistance to group 28 insecticides in some areas.
emamectin benzoate, MOA 6 (Proclaim) 5 WDG	3.2 to 4.8 oz	12 hrs	7	Do not make more than two sequential applications without rotating to another product with a different mode of action.
cyclaniliprole, MOA 28 (Harvanta) 50 SL	10.9 to 16.4 fl oz	4 hrs	1	Beet armyworm may have developed resistance to group 28 insecticides in some areas.
indoxacarb, MOA 22A (Avaunt eVo) 30 SG	3.5 oz	12 hrs	3	
methomyl, MOA 1A (Lannate) 90 SP (Lannate) 2.4 LV	0.5 to 1 lb 1.5 to 3 pts	48 hrs	7	Air temperature should be well above 32°F. Do not apply to seedlings less than 3 in. in diameter.
methoxyfenozide, MOA 18 (Intrepid) 2 F	4 to 10 fl oz	4 hrs	1	Use low rates for early-season applications to young or small plants and 6 to 10 oz for mid- to late-season applications.
spinetoram, MOA 5 (Radiant) 1 SC	5 to 10 fl oz	4 hrs	1	
pyrethroid, MOA 3A		12 hrs		See **Table 2-25 A** for a list of registered pyrethroids and pre-harvest intervals.
viruses		4 hrs		See **Table 2-26** for virus-based insecticides effective against specific insects. **OMRI listed.**

TABLE 2-21. INSECT CONTROL FOR SWEETPOTATO

Insecticide, Mode of Action Code, and Formulation	Amount of Formulation Per Acre	Restricted Entry Interval (REI)	Pre harvest Interval (PHI) (Days)	Precautions and Remarks
APHIDS, LEAFHOPPER, WHITEFLY				
Aphids, leafhoppers, and whiteflies are rarely a problem.				
acetamiprid, MOA 4A (Assail) 30 SC	2.1 to 3.4 fl oz	12 hrs	7	
clothianidin, MOA 4A (Belay) 2.13 SC **Soil treatment**	9 to 12 oz	12 hrs	21	Soil application as an in-furrow or sidedress application. For sidedress applica- tions, immediately cover with soil.
Foliar treatment	2 to 3 fl oz	12 hrs	14	
dinotefuran, MOA 4A (Venom) 70 SG	1 to 1.5 oz	12 hrs	7	May require tank mix for full control
flonicamid, MOA 29 (Beleaf) 50 SG	2 to 2.8 oz	12 hrs	7	
flupyradifurone, MOA 4D (Sivanto Prime) 1.67 SL	7.0 to 14.0 fl oz	4 hrs	7	For aphids and leafhopper use 7.0 to 10.5 fluid ounces, for whitefly use 10.5 to 14.0 fluid ounces.
imidacloprid, MOA 4A **Foliar treatment** (Admire Pro) 4.6 F (various) 1.6 F	1.2 fl oz 3.5 fl oz	12 hrs	7	Two applications may be needed to control heavy populations. Allow 5- to 7-days between applications.
Soil treatment (Admire Pro) 4.6 F (various) 1.6 F	4.4 fl oz 10.5 fl oz	12 hrs	60	
pymetrozine, MOA 9B (Fulfill) 50 WDG	2.75 to 5.5 oz	12 hrs	14	
pyrifluquinazon, MOA 9B (PQZ) 1.87 EC	2.4 to 3.2 fl oz	12 hrs	1	See label for rotational crop restrictions. Do not exceed 4.8 fl oz per acre per crop cycle.
sulfoxaflor, MOA 4C (Transform) 50WG	0.75 to 1 oz	24 hrs	7	Limit application to times when managed and native pollinators are least active, e.g., 2 hrs before sunset or when temperature is below 50°F
spirotetramat, MOA 23 (Movento)	4 to 5 oz	24 hrs	7	Movento must be combined with a spray adjuvant with spreader/penetrating proper- ties to maximize leaf uptake.
thiamethoxam, MOA 4A (Actara) 25 WDG	3 oz	12 hrs	14	Two applications of Actara may be needed to control heavy populations. Allow 7- to 10-days between applications. Do not exceed a total of 6 ounces of Actara per crop per season
ARMYWORMS, LOOPER, CORN EARWORM, HORNWORM				
Damaging armyworm and earworm infestations may occur in August or September. If significant infestations are present on foliage during harvest, larvae may feed on exposed roots.				
chlorantraniliprole, MOA 28 (Coragen eVo) 5 SC (Vantacor) 5 SC	1.2 to 2.5 fl oz 1.2 to 2.5 fl oz	4 hrs	1	
cyantraniliprole, MOA 28 (Exirel) 0.83 EC	10 to 20.5 fl oz	12 hrs	1	
methoxyfenozide, MOA18 (Intrepid) 2 F	6 to 10 fl oz	4 hrs	7	
novaluron, MOA 15 (Rimon) 0.83 EC	9 to 12 fl oz	12 hrs	14	Do not make more than 2 applications per crop per season.
spinetoram, MOA 5 (Radiant) 1 SC	6 to 8 fl oz	4 hrs	7	
spinosad MOA 5 (Blackhawk)	2.25 to 3.5 oz	4 hrs	7	
viruses		4 hrs		See **Table 2-26** for virus-based insecticides effective against specific insects. **OMRI listed.**
CUCUMBER BEETLE (ADULTS), JAPANESE BEETLE (ADULTS), TORTOISE BEETLE				
Cucumber beetle larvae (Diabrotica) are a serious pest of sweetpotato in LA and MS. Controlling adult cucumber beetles in areas with a history of Diabrotica damage can reduce damage to roots. Foliage feeding by beetles rarely causes economic loss, and control is not warranted unless defoliation is severe. Tor- toise beetles are frequently present but rarely reach levels requiring treatment. Treat for tortoise beetles only if significant defoliation is observed.				
carbaryl, MOA 1A (Sevin) XLR Plus	1 to 2 qt	12 hrs	7	Treat for tortoise beetles only if significant defoliation is observed. Tortoise bee- tles are frequently present but rarely reach levels requiring treatment.
pyrethroids, MOA 3A		12 hrs		See **Table 2-25 A** for a list of registered pyrethroids and pre-harvest intervals.

TABLE 2-21. INSECT CONTROL FOR SWEETPOTATO (cont'd)

Insecticide, Mode of Action Code, and Formulation	Amount of Formulation Per Acre	Restricted Entry Interval (REI)	Pre harvest Interval (PHI) (Days)	Precautions and Remarks
FLEA BEETLE, WIREWORM, WHITE GRUB, CUCUMBER BEETLE LARVAE				
bifenthrin, MOA 3A (various) 2 EC	9.6 to 19.2 fl oz	12 hrs	21	Apply as broadcast, preplant application to the soil and incorporate 4- to 6-inches prior to bed formation. This use has been demonstrated to control overwintered wireworm populations and reduce damage to roots at harvest. Post-transplant bifenthrin should be directed onto each side of the bed from the drill to the middle of the furrow and incorporated with cultivating equipment set to throw soil toward the drill. The objective is to provide a barrier of treated soil that covers the bed and furrows. Foliar sprays of various insecticides that target adults to prevent egg laying have not been shown to provide any reduction in damage to roots by wireworm larvae at harvest. **NOTE:** Broflanilide (Nurizma) must be applied as an in-furrow application behind tillage equipment (ripper bedder, bed conditioner). For best performance, consider highest labeled rate. Please see Nurizma Section 2(ee) recommendation for specific application information.
broflanilide, MOA 30 (Nurizma)	0.08 to 0.16 fl oz per 1000 ft row	12 hrs		
clothianidin MOA 4A (Belay) 2.13	12 fl oz	12 hrs		
imidacloprid, MOA 4A (Admire Pro) 4.6 SC	10.5 fl oz or 0.75 fl oz per 1,000 ft	3 days	60 days (NC, LA Only) 125 days elsewhere	
cyantraniliprole, MOA 28 (Exirel) 0.83SE	13.5 to 20.5 oz	12 hrs	1	Effective against flea beetles only.
dinotefuran, MOA 4A **Foliar treatment** (Venom) 70 SG	1 to 1.5 oz	12 hrs	7	Do not make a soil and foliar application – use one or the other. Do not exceed 6 ounces (foliar) or 12 oz (soil) of Venom per acre per season. Do not exceed 10.5 fl oz (foliar) or 21 fl oz (soil) of Scorpion per acre per season. See label for application restrictions for protection of pollinators.
Soil treatment (Venom) 70 SG	6.5 to 7.5 oz	12 hrs	7	
spirotetramat, MOA 23 (Movento)	4 to 5 fl oz	24 hrs	7	Foliar applications of Movento have shown to suppress wireworm damage to roots.
thiamethoxam, MOA 4A (Platinum) 75 SG	1.66 to 2.67 oz	12 hrs		**SEE** PRECAUTIONS AND REMARKS **ABOVE**
FRUIT FLY (VINEGAR FLY)				
pyrethrins, MOA 3A (Pyrenone)	1 gal. per 10,000 cu ft	12 hrs	—	Postharvest application in storage. Apply as a space fog with a mechanical or thermal generator. Do not make more than 10 applications.
SWEETPOTATO WEEVIL				
pyrethroid, MOA 3A		12 hrs		See **Table 2-25 A** for a list of registered pyrethroids and pre-harvest intervals.
phosmet, MOA 1B (Imidan) 70 W	1.33 lb	5 days	7	Do not make more than five applications per season.
THRIPS				
spinetoram, MOA 5 (Radiant) 1 SC	6 to 8 fl oz	4 hrs	7	
WHITEFRINGED BEETLE				
phosmet, MOA 1B (Imidan) 70 W	1.33 lb	5 days	7	Do not make more than five applications per season. Whitefringed beetle adults are active in July and August. Do not plant in fields with a recent history of white-fringed beetles. Limited to mechanical harvest only.

TABLE 2-22. INSECT CONTROL FOR TOMATO

Insecticide, Mode of Action Code, and Formulation	Amount of Formulation Per Acre	Restricted Entry Interval (REI)	Pre harvest Interval (PHI) (Days)	Precautions and Remarks
APHID, FLEA BEETLE				
acetamiprid, MOA 4A (Assail) 30 SC	1.7 to 3.4 fl oz	12 hrs	7	Do not apply more than once every 7 days, and do not exceed 5 applications per season.
afidopyropen, MOA 9D (Sefina) DC	3 oz	12 hrs	0	Will not control flea beetle. Do not make more than 2 sequential applications before using a different mode of action.
cyantraniliprole, MOA 28 (Exirel) 0.83SE	13.5 to 20.5 fl oz	12 hrs	1	
cyantraniliprole, MOA 28 (Verimark) 1.67 SC	6.75 to 13.5 fl oz	4	1	Soil applications at planting will control flea beetles and suppress aphids. See label for application options.

TABLE 2-22. INSECT CONTROL FOR TOMATO (cont'd)

Insecticide, Mode of Action Code, and Formulation	Amount of Formulation Per Acre	Restricted Entry Interval (REI)	Pre harvest Interval (PHI) (Days)	Precautions and Remarks
APHID, FLEA BEETLE (cont'd)				
dinotefuran, MOA 4A **Foliar treatment** (Venom) 70 SG	1 to 4 fl oz	12 hrs	1	
Soil treatment (Venom) 70 SG	5 to 7.5 fl oz	12 hrs	21	
dimethoate 4 EC, MOA 1B	0.5 to 1 pt	48 hrs	7	Do not exceed rate with dimethoate as leaf injury may result.
flupyradifurone, MOA 4D (Sivanto Prime) 1.67 SL	7.0 to 14 fl oz	4 hrs	1	Will not control flea beetle.
imidacloprid, MOA 4A **Soil treatment** (Admire Pro) 4.6F (various) 2F	7 to 10.5 fl oz 16 to 24 fl oz	12 hrs	21	For short-term protection at planting, apply Admire Pro transplants in the planthouse not more than 7 days before planting at the rate of 0.44 (4.6 F formulation) or 1 ounce (2 F formulation) per 10,000 plants. See label for instructions of greenhouse transplant and field soil applications.
Foliar treatment (Admire Pro) 4.6 F (various) 2 F	1.2 fl oz 3.75 fl oz	12 hrs	0	
pymetrozine, MOA 9B (Fulfill) 50 WDG	2.75 oz	12 hrs	0	For aphids only.
pyrifluquinazon, MOA 9B (PQZ) 1.87 EC	2.4 to 3.2 fl oz	12 hrs	1	See label for rotational crop restrictions. Do not exceed 4.8 fl oz per acre per crop cycle.
sulfoxaflor MOA 4C (Closer) 2 SC	1.5 to 2 fl oz	12 hrs	1	
thiamethoxam, MOA 4A (Platinum) 75 SG	1.66 to 3.67 oz	12 hrs	30	Platinum may be applied to direct-seeded crops in-furrow seeding or transplant depth, post seeding or transplant as a drench, or through drip irrigation. Do not exceed 11 ounces per acre per season of Platinum. Check label for plant-back restrictions for several crops.
(Actara) 25 WDG	2 to 3 oz	12 hrs	0	
COLORADO POTATO BEETLE				
acetamiprid, MOA 4A (Assail) 30 SC	1.3 to 3.4 fl oz	12 hrs	7	
chlorantraniliprole, MOA 28 (Coragen eVo) 5 SC	1.2 to 2.5 fl oz	4 hrs	1	Foliar or drip chemigation. Drip chemigation must be applied uniformly to the root zone. See label for instructions.
cyantraniliprole, MOA 28 **Soil treatment** (Verimark) 1.67 SC	5 to 10 fl oz	4 hrs	1	Apply Verimark to soil via drip irrigation or soil injection.
Foliar treatment (Exirel) 0.83 SE	7 to 13.5 fl oz	12 hrs	1	Exirel is for foliar application.
dinotefuran, MOA 4A (Venom) 70SG	5 to 7.5 oz (soil)	12 hrs	21	Soil application only for Colorado potato beetle.
imidacloprid, MOA 4A **Soil treatment** (Admire Pro) 4.6 F (various) 2 F	7 fl oz 16 fl oz	12 hrs	21	See application instructions above under aphid/flea beetle.
Foliar treatment (Admire Pro) 4.6 F (4 various) 2 F	1.3 to 2.2 fl oz 3.75 fl oz	12 hrs	0	
spinetoram, MOA 5 (Radiant) 1 SC	5 to 10 fl oz	4 hrs	1	
thiamethoxam, MOA 4A **Soil treatment** (Platinum) 75 SG	1.66 to 3.67 oz	12 hrs	30	Platinum may be applied to direct-seeded crops in-furrow seeding or transplant depth, post seeding or transplant as a drench, or through drip irrigation. Do not exceed 11 oz per acre per season of Platinum. Check label for plant-back restrictions for several crops.
Foliar treatment (Actara) 25 WDG	2 to 3 oz	12 hrs	0	Actara is for foliar applications.
ARMYWORM, CABBAGE LOOPER, HORNWORM, TOMATO FRUITWORM, PINWORM				
Bacillus thuringiensis, MOA 11A (Dipel) DF (Crymax) WDG	0.5 to 1 lb 0.5 to 1.5 lb	4 hrs	0	Tomato fruitworm resistance to Bts has become common in recent years. **OMRI listed (excluding Crymax).**
pyrethroid, MOA 3A				See **Table 2-25 A** for a list of registered pyrethroids and pre-harvest intervals.

TABLE 2-22. INSECT CONTROL FOR TOMATO (cont'd)

Insecticide, Mode of Action Code, and Formulation	Amount of Formulation Per Acre	Restricted Entry Interval (REI)	Pre harvest Interval (PHI) (Days)	Precautions and Remarks
ARMYWORM, CABBAGE LOOPER, HORNWORM, TOMATO FRUITWORM, PINWORM (cont'd)				
chlorantraniliprole, MOA 28 (Coragen eVo) 5 SC	1.2 to 2.5 fl oz	4 hrs	1	Foliar or drip chemigation. Drip chemigation must be applied uniformly to the root zone. See label for instructions.
cyantraniliprole, MOA 28 **Soil treatment** (Verimark) 1.67 SC	5 to 10 fl oz	4 hrs	1	Verimark is for soil application only. Applications made at planting and/or via drip chemigation after planting. See label for application options.
Foliar treatment (Exirel) 0.83SE	7 to 13.5 fl oz	12 hrs	1	Exirel is for foliar application only.
cyclaniliprole, MOA 28 (Harvanta) 50 SL	10.9 to 16.4 fl oz	4 hrs	1	
emamectin benzoate, MOA 6 (Proclaim) 5 WDG	2.4 to 4.8 oz	12 hrs	7	
indoxacarb, MOA 22A (Avaunt eVo) 30 WDG	3.5 to 6 oz	12 hrs	3	Do not apply more than 24 ounces of Avaunt (0.44 lbs a.i.) per acre per crop.
flonicamid, MOA 29 (Beleaf) 50 SG	2 to 4.8 oz	12 hrs	0	Foliar or soil applications are permissible. Soil applications should be made via drip chemigation and within 21 days of transplanting. Will not control flea beetle.
methomyl, MOA 1A (Lannate) 2.4 LV	1.5 to 3 pt	48 hrs	1	Methomyl may induce leafminer infestation.
methoxyfenozide, MOA 18 (Intrepid) 2 F	4 to 10 fl oz	4 hrs	1	Use low rates for early-season applications to young or small plants and 6 to 10 ounces for mid- and late-season applications. Intrepid provides suppression of pinworm only.
novaluron, MOA 15 (Rimon) 0.83 EC	9 to 12 fl oz	12 hrs	1	Do not make more than 3 applications per season.
spinetoram, MOA 5 (Radiant) 1 SC	5 to 10 fl oz	4 hrs	1	
viruses		4 hrs		See **Table 2-26** for virus-based insecticides effective against specific insects. **OMRI listed.**
CUTWORM				
pyrethroid, MOA 3A		12 hrs		See **Table 2-25 A** for a list of registered pyrethroids and pre-harvest intervals.
LEAFMINER				
abamectin, MOA 6 (Agri-Mek) 0.7 SC	1.75 to 3.5 fl oz	12 hrs	7	Do not exceed 48 fluid ounces per acre per season, or more than two sequential applications.
chlorantraniliprole, MOA 28 (Coragen eVo) 5 SC	5 to 7.5 fl oz	4 hrs	1	Foliar or soil chemigation. Drip chemigation must be applied uniformly to the root zone. See label for soil application instructions.
cyclaniliprole, MOA 28 (Harvanta) 50 SL	10.9 to 16.4 fl oz	4 hrs	1	
cyromazine, MOA 17 (Trigard) 75 WP	2.66 oz	12 hrs	0	See label for plant-back restrictions.
spinetoram, MOA 5 (Radiant) 1 SC	6 to 8 fl oz	4 hrs	1	Do not exceed 29 fl oz per acre per season.
SPIDER MITE				
abamectin, MOA 6 (Agri-Mek) 0.7 SC	1.75 to 3.5 fl oz	12 hrs	7	Do not make more than two sequential applications. Active against immatures and adults.
acequinocyl, MOA 20B (Kanemite) 15 SC	31 fl oz	12 hrs	1	The use of a surfactant/adjuvant with Kanemite on tomatoes is prohibited. Active against all life stages.
bifenazate, MOA 20D (Acramite) 50 WS	0.75 to 1.0 lb	12 hrs	3	Do not make more than one application per season. Active against immatures and adults. Some egg activity.
cyflumetofen, MOA 25 (Nealta) 1.67 SC	13.7 fl oz	12 hrs	3	Do not make more than one application before using an effective miticide with a different mode of action. Active against all life stages.
fenazaquin, MOA 21A (Magister) 1.7SC	24 to 36 fl oz	12 hrs	3	Do not make more than one application per year. Active against all life stages.
fenpyroximate MOA 21A (Portal) 0.4EC	2 pts	12 hrs	1	Do not make more than two applications per season. Active against immatures and adults.
spiromesifen, MOA 23 (Oberon) 2 SG	7 to 8.5 fl oz	12 hrs	1	Do not exceed 3 applications per season. Active against immatures. Some adult activity.
STINK BUGS, LEAFFOOTED BUG				
pyrethroid, MOA 3A		12 hrs		See **Table 2-25 A** for a list of registered pyrethroids and pre-harvest intervals.

TABLE 2-22. INSECT CONTROL FOR TOMATO (cont'd)

Insecticide, Mode of Action Code, and Formulation	Amount of Formulation Per Acre	Restricted Entry Interval (REI)	Pre harvest Interval (PHI) (Days)	Precautions and Remarks
STINK BUGS, LEAFFOOTED BUG (cont'd)				
dinotefuran, MOA 4A **Foliar treatment** (Venom) 70 SG (Scorpion) 35 SL	1 to 4 oz 2 to 7 oz	12 hrs	1	Do not combine foliar with soil applications, use only one method. Soil applications of Venom or Scorpion may be made in a narrow band under the plant row as a post-transplant drench, as a soil-incorporated sidedress after plants are established, or in drip irrigation water. See label for instructions. Read pollinator protection restrictions on the label.
Soil treatment (Venom) 70 SG (Scorpion) 35 SL	5 to 6 fl oz 9 to 10.5 fl oz		21	
thiamethoxam, MOA 4A (Actara) 25 WDG	3 to 5.5 oz	12 hrs	0	Do not exceed 11 ounces Actara per acre per season.
THRIPS				
Insecticide-resistant western flower thrips populations occur in certain areas, particularly those with concentrations of tomato and pepper production and where wheat is grown for grain. In some areas suppression is the best control that may be achieved.				
dimethoate 4 EC, MOA 1B	0.5 to 1 pt	48 hrs	7	
cyantraniliprole, MOA 28 (Verimark)	10 fl oz (drip chemigation)	4 hrs	1	For suppression of foliar infestation of thrips. Allow 1 to 3 days for Verimark to be translocated to leaf tissue when applied to transplants or transplant water, 2 to 5 days when applied via drip irrigation early in the season, and 7 to 10 days when applied via drip during the second half of the growing season.
cyclaniliprole, MOA 28 (Harvanta) 50 SL	10.9 to 16.4 fl oz	4 hrs	1	Harvanta will help suppress western flower thrips when used in a rotational program.
flonicamid MOA 29 (Beleaf) 50 SG	2.4 to 4.8 oz	12 hrs	0	Beleaf has shown good activity against insecticide resistant western flower thrips.
imidacloprid, MOA 4A (Admire Pro) 4.6 F **For Planthouse treatment of transplants**	0.44 fl oz per 10,000 plants	12 hrs	—	For suppression of TSWV, treatment transplants in the planthouse not more than 7 days before planting in the field. Transplants should be treated with overhead irrigation immediately after planting to ensure movement of imidacloprid into the soil media. See label for instructions. Only effective against tobacco thrips.
methomyl, MOA 1A (Lannate) 2.4 LV	1.5 to 3 pt	48 hrs	1	On foliage as needed.
novaluron, MOA 15 (Rimon) 0.83 EC	9 to 12 fl oz	12 hrs	1	Do not make more than 3 applications per season.
spinetoram, MOA 5 (Radiant) 1 SC	6 to 10 fl oz	4 hrs	1	Will control thrips on foliage, not in flowers.
tolfenpyrad, MOA 21A (Torac) 1.29 EC	21 oz	12 hrs	1	Do not make more than 2 applications per crop cycle and allow at least 14 days between applications.
WHITEFLY				
For resistance management of whiteflies, do not follow a foliar application of a neonicotinoid (MOA group 4A) with a soil application of any neonicotinoid. Use only one application method.				
acetamiprid, MOA 4A (Assail) 30 SC	2.1 to 3.4 fl oz	12 hrs	7	Do not apply more than once every 7 days, and do not exceed 5 applications per season.
afidopyropen, MOA 9D (Sefina) DC	14 oz	12 hrs	0	Do not make more than 2 sequential applications before using a different mode of action.
buprofezin, MOA 16 (Courier) 40 SC	9 to 13.6 fl oz	12 hrs	1	Use sufficient water to ensure good coverage. Do not apply more than twice per crop cycle and allow 28 days between applications.
chlorantraniliprole, MOA 28 (Coragen eVo) 5 SC	5 to 7.5 fl oz	4 hrs	1	Foliar or soil application. Drip chemigation must be applied uniformly to the root zone. See label for soil application instructions.
cyantraniliprole, MOA 28 **Soil treatment** (Verimark) 1.67 SC	6.75 to 13.5 fl oz	4 hrs	1	Apply Verimark to at planting and/or later via drip irrigation or soil injection. See label for application options.
Foliar treatment (Exirel) 0.83SE	13.5 to 20.5 fl oz	12 hrs	1	Exirel is for foliar application.
dinotefuran, MOA 4A **Soil treatment** (Venom) 70 SG (Scorpion) 35 SL	5 to 6 oz 9 to 10.5 fl oz	12 hrs	21	See soil application instructions above under stink bug. See the label for pollinator protection restrictions.
Foliar treatment (Venom) 70 SG (Scorpion) 35 SL	1 to 4 oz 2 to 7 fl oz		1	
flupyradifurone, MOA 4D (Sivanto Prime) 1.67SL **Soil treatment**	21 to 28 fl oz	12 hrs	45	Soil applications may be made through drip irrigation, at planting or post-transplant drench,
Foliar treatment	10.5 to 14 fl oz	12 hrs	1	

TABLE 2-22. INSECT CONTROL FOR TOMATO (cont'd)

Insecticide, Mode of Action Code, and Formulation	Amount of Formulation Per Acre	Restricted Entry Interval (REI)	Pre harvest Interval (PHI) (Days)	Precautions and Remarks
WHITEFLY (cont'd)				
imidacloprid, MOA 4A (Admire Pro) 4.6 F (various) 2 F	16 to 24 fl oz 7 to 10.5 fl oz	12 hrs	21	Apply through a drip irrigation system or as a transplant drench with sufficient water to reach root zone. As a sidedress, apply 2 to 4 inches to the side of the row and incorporate 1 inch or more. Residual activity will increase with increased rates. Use higher rate for late-season or continuous infestations. Trickle irrigation applications will also control aphids and suppress stinkbugs.
pyriproxyfen, MOA 7C (Knack) 0.86 EC	8 to 10 fl oz	12 hrs	1	Do not apply more than two applications per growing season, and do not make applications closer than 14 days.
pyrifluquinazon, MOA 9B (PQZ) 1.87 EC	2.4 to 3.2 fl oz	12 hrs	1	See label for rotational crop restrictions. Do not exceed 4.8 fl oz per acre per crop cycle.
spiromesifen, MOA 23 (Oberon) 2 SC	7 to 8.5 fl oz	12 hrs	1	Do not make more than 3 applications per season.
spirotetramat, MOA 23 (Movento) 2 SC	4 to 5 fl oz	24 hrs	1	Do not exceed 10 fluid ounces per season. Requires surfactant.
thiamethoxam, MOA 4A **Soil treatment** (Platinum) 75 SG	1.66 to 3.67 oz	12 hrs	30	Platinum may be applied to direct-seeded crops in-furrow at seeding or transplanting, post seeding or transplant as a drench, or through drip irrigation. Do not exceed 11 oz per acre per season of Platinum. Check label for plant-back restrictions for several crops.
Foliar treatment (Actara) 25 WDG	3 to 5.5 oz	12 hrs	0	Actara is for foliar applications.
WIREWORM				
diazinon, MOA 1B (Diazinon) AG 500 or 50 WP	2 to 4 qt	48 hrs	—	Broadcast before planting and incorporate. Wireworms may be a problem in fields previously in pasture, corn, or soybean.

TABLE 2-23. INSECT CONTROL FOR TURNIP

Insecticide, Mode of Action Code, and Formulation	Amount of Formulation Per Acre	Restricted Entry Interval (REI)	Pre harvest Interval (PHI) (Days)	Precautions and Remarks
APHID				
afidopyropen, MOA 9D (Versys) DC	1.5 fl oz	12 hrs	0	Do not make more than 2 sequential applications before using a different mode of action.
cyantraniliprole, MOA 28 (Verimark) 1.67 SC	6.75 to 13.5 fl oz	4 hrs	4	Verimark is for greens only, not root turnips. Verimark is for soil application only. Applications can be made at planting and/or later via drip chemigation. See label for application options.
dimethoate 4 EC, MOA 1B	0.5 pt	48 hrs	14	
flonicamid, MOA 29 (Beleaf) 50 SG	2 to 2.8 oz	12 hrs	0	
flupyradifurone, MOA 4D (Sivanto Prime) 1.67 SL	7 to 14 fl oz	4 hrs	7	
imidacloprid, MOA 4A **Soil treatment** (Admire Pro) 4.6 F (various) 2 F	4.4 to 10.5 fl oz 10 to 24 fl oz	12 hrs	21	See label for soil application instructions.
Foliar treatment (Admire Pro) 4.6 F (various) 2 F	1.2 fl oz 3.8 fl oz	12 hrs	7	
pymetrozine, MOA 9B (Fulfill) 50 WDG	2.75 oz	12 hrs	7	
pyrifluquinazon, MOA 9B (PQZ) 1.87 EC	2.4 to 3.2 fl oz	12 hrs	1	See label for rotational crop restrictions. Do not exceed 4.8 fl oz per acre per crop cycle.
thiamethoxam, MOA 4A **Soil treatment** (Platinum) 75 SG	1.7 to 4.01 oz	12 hrs	NA	Platinum is for soil application and Actara for foliar application.
Foliar Treatment (Actara) 25 WDG	1.5 to 3 oz	12 hrs	7	

TABLE 2-23. INSECT CONTROL FOR TURNIP (cont'd)

Insecticide, Mode of Action Code, and Formulation	Amount of Formulation Per Acre	Restricted Entry Interval (REI)	Pre harvest Interval (PHI) (Days)	Precautions and Remarks
HARLEQUIN BUG, VEGETABLE WEEVIL, YELLOWMARGINED LEAF BEETLE, FLEA BEETLE				
clothianidin, MOA 4A (Belay) 2.13 EC **Soil treatment**	4.8 to 6.4 oz	12 hrs	NA	Soil application as in in-furrow, side dress application, seed, transplant drench, or chemigation. See label for application instructions.
Foliar treatment	1.6 to 2.1 oz	12 hrs	7	
dinotefuran MOA 4A (Venom) 70 SG	2 to 3 oz	12 hrs	1	For use on turnip greens only, not turnips roots.
imidacloprid, MOA 4A **Soil treatment** (Admire Pro) 4.6 F (Various) 2 F	4.4 to 10.5 fl oz 10 to 24 fl oz	12 hrs	21	Soil applications of imidacloprid will not control harlequin bug past 20 days after application.
Foliar treatment (Admire Pro) 4.6 F (Various) 2 F	1.2 fl oz 2.4 fl oz		7	
thiamethoxam, MOA 4A **Soil treatment** (Platinum) 75 SG	1.7 to 4.0 oz	12 hrs	NA	Platinum is for soil application and Actara for foliar application.
Foliar treatment (Actara) 25 WDG	1.5 to 3 oz	12 hrs	7	
pyrethroid, MOA 3A		12 hrs		See **Table 2-25 A** for a list of registered pyrethroids and pre-harvest intervals.
spinetoram, MOA 5 (Radiant) 1 SC	5 to 10 fl oz	4 hrs	3	For yellowmargined leaf beetle only.
CABBAGE LOOPER, DIAMONDBACK MOTH				
Insecticide-resistant diamondback moth populations, widespread in the Southeast, may not be controlled with some registered insecticides. To manage resistance, avoid transplants from Georgia and Florida, where resistance is common, and avoid the repeated use of the same materials for extended periods. Repeated use of pyrethroid insecticides often aggravates diamondback moth problems. Do not allow populations to increase to large densities before treatments are initiated.				
Bacillus thuringiensis, MOA 11A (Crymax) WDG (Dipel) DF (Xentari) DF	0.5 to 1.5 lb 0.5 to 2 lb 0.5 to 2 lb	4 hrs	0	Apply when larvae are young before crop damage occurs. Applications may be repeated at 3 to 14 days to maintain control. **OMRI listed (excluding Crymax).**
chlorantraniliprole, MOA 28 (Coragen eVo) 5 SC (Vantacor) 5 SC	3.5 to 7.5.0 fl oz 1.2 to 2.5 fl oz	4 hrs	1	For turnip greens or root turnips.
cyantraniliprole, MOA 28 **Soil treatment** (Verimark) 1.67 SC	5 to 10 fl oz	4 hrs	1	Verimark and Exirel are for greens only, not root turnips Verimark is for soil application only. Applications can be made at planting and/or later via drip chemigation. See label for application options. Exirel is for foliar application only
Foliar treatment (Exirel) 0.83SE	7 to 17 fl oz	12 hrs	1	
cyclaniliprole, MOA 28 (Harvanta) 50 SL	10.9 to 16.4 fl oz	4 hrs	1	Harvanta is for foliar application only.
emamectin benzoate, MOA 6 (Proclaim) 5 WDG	2.4 to 4.8 oz	12 hrs	14	Proclaim is for use on turnip greens only. Do not make more than two sequen- tial applications.
indoxacarb, MOA 22A (Avaunt eVo) 30 WDG	2.5 to 3.5 oz	12 hrs	3	Avaunt may be applied only to turnip greens, not root turnips.
methoxyfenozide, MOA 18 (Intrepid) 2F (Intrepid Edge)	4 to 16 fl oz 4.5 to 12 fl oz	4 hrs	7 3	
spinetoram, MOA 5 (Radiant) 1 SC	5 to 10 fl oz	4 hrs	3	
viruses		4 hrs		See **Table 2-26** for virus-based insecticides effective against specific insects. **OMRI listed.**
ROOT MAGGOT				
tolfenpyrad, MOA 21A (Torac)	21 fl oz	12 hrs	1	Read soil application guidelines on label.
cyantraniliprole, MOA 28 (Verimark)	10 to 13.5 fl oz	4 hrs	At-planting only	

TABLE 2-24. FIRE ANT MANAGEMENT IN COMMERCIAL VEGETABLE FIELDS

Insecticide, Mode of Action Code, and Formulation	Amount of Formulation Per Acre	Restricted Entry Interval (REI)	Pre harvest Interval (PHI) (Days)	Precautions and Remarks
pyriproxyfen (0.5%) (Esteem)	1.5 to 2.0 lbs	12 hrs	1 day	Bait. Broadcast when ants are actively foraging. Do not overhead water-treated area for 24 hours after application. Slow acting. Esteem is labeled on most, but not all fruits and vegetables. Check label before using.
spinosad, MOA 5 (Seduce Insect Bait)	20 to 44 lbs	4 hrs	See label (varies among crops)	Scatter the bait around the perimeter of vegetable plantings to intercept pests.
abamectin, MOA 6 (Clinch Ant Bait)	1 lb	12 hrs	0	Broadcast applications to soil when ants are actively foraging. Apply after dew or rainfall on the soil surface has dried and do not apply if rainfall is anticipated within 4- to 6-hrs after application. Do not apply if rain is expected within Do not apply within 25 ft of water sources, or to areas accessible to livestock or chickens.

* Do not use Extinguish Plus in vegetable crops. This product contains methoprene + hydramethylnon and is not approved for use in vegetable crops.

Timing Bait Applications
Fire ant baits can be applied anytime during the growing season, but spring is the best time. Baits should be applied when ants are active and foraging. Fire ants are generally active on sunny days above 65°F. Foraging activity can be verified by spreading a few potato chips (fried, not baked) on the ground and checking in 20 to 30 minutes to see if they have found them.

Apply when rain is not expected for 1 to 2 days. Do not apply when leaves or soil are wet and reapply if rain falls within 6 to 12 hrs of application. A single bait treatment in the spring will reduce fire ant numbers by about 80%. For improved control subsequent applications in mid-summer and/or fall can be made. Fall treatments help reduce the number of mounds the following spring.

Insect Growth Regulator (IGR) baits (MOA 7) are slow acting and take up to two months to have their full impact. Hence, consider harvest timing when planning applications. For short season crops apply baits shortly after planting. For perennial or "U-Pick" crops, treat preventively in the fall and follow up with a spring application.

Applying Baits Properly
Rates for most granular fire ant baits range from only 1 to 2 pounds per acre, which can be easy to over-apply. Avoid using fertilizer spreaders, instead use hand-operated spreaders sold to apply fire ant baits to home lawns. Hand seeders designed to spread small seeds will also work if calibrated properly. Use a power-operated spreader for large acreages. Herd Seeder Company and Spyker Spreaders both make spreaders driven by small electric motors that can be mounted on a tractor, ATV, or other vehicle. Be sure to treat turn rows and field borders, because fire ants ae often highest in untilled areas around field edges and from where they move into fields.

TABLE 2-25. RELATIVE EFFECTIVENESS OF INSECTICIDES AND MITICIDES FOR INSECT AND MITE CONTROL ON FIELD-GROWN VEGETABLES

Not all insecticides listed below are registered on all vegetable crops. Refer to label before applying to specific crop. Ratings are based on consensus of vegetable entomologists in the SE United States.

Key: "E" Excellent; "G" Good; "F" Fair; "-" ineffective or insufficient data

Chemical class (IRAC)	Common name	Example Product	Flea Beetle	Colorado potato beetle*	Cucumber beetles	Corn earworm*	European corn borer	Fall armyworm	Cabbage looper	Imported cabbageworm	Diamondback moth*	Squash vine borer	Beet armyworm*	Stinkbugs/Harlequin bug	Squash bug	Aphids*	Thrips	Western Flower Thrips*	Leafminer	Maggots	Whiteflies*	Cutworms	Wireworms	White grubs	Spider mites*	Broad mites
1A	carbaryl	Sevin	E	F	G	F	G	F	F	G	F	F	-	-	-	-	F	-	-	-	-	F	-	-	-	-
1A	methomyl	Lannate	F	-	-	G	G	G	G	G	G	-	F	G	G	F	E	G	F	-	-	-	-	-	-	-
1A	oxamyl	Vydate	F	F	F	-	-	-	-	-	-	-	F	F	G	G	-	-	-	-	-	-	-	-	-	-
1B	malathion	Malathion	G	F	G	F	F	F	F	G	F	F	-	F	F	F	F	-	F	-	F	-	-	-	-	-
1B	acephate	Orthene	-	-	-	F	E	G	F	G	-	-	-	E	-	G	G	F	-	-	G	-	-	-	-	-
1B	diazinon	Diazinon	-	-	-	-	-	-	-	-	-	-	-	-	-	-	-	-	-	G	-	F	G	F	-	-
1B	dibrom	Dibrom	G	-	-	-	F	G	G	G	-	F	-	-	G	-	-	-	-	-	-	-	-	-	F	-
1B	dimethoate	Dimethoate	G	-	F	-	-	-	-	-	-	-	G	F	E	E	G	G	-	-	-	-	-	-	-	-
3A	permethrin	Pounce	G	F	G	F	G	F	G	E	F	E	-	F	G	F	F	-	F	-	-	G	-	-	-	-
3A	alpha cypermethrin	Fastac	G	F	G	F	G	G	G	E	F	E	-	G	G	F	F	-	F	-	-	G	-	-	-	-
3A	zeta cypermethrin	Mustang Maxx	E	F	E	F	E	G	G	E	F	E	-	G	G	F	F	-	F	-	-	E	-	-	-	-
3A	cyfluthrin	Tombstone	E	F	E	F	E	F	G	E	F	E	-	G	G	F	F	-	F	-	-	E	-	-	-	-
3A	beta cyfluthrin	Baythroid XL	E	F	E	F	E	F	G	E	F	E	-	G	G	F	F	-	F	-	-	E	-	-	-	-
3A	lambda cyhalothrin	Karate, Warrior	E	F	E	F	G	G	G	E	F	E	-	E	G	F	F	-	E	-	-	E	-	-	F	-
3A	esfenvalerate	Asana XL	G	G	G	F	G	F	G	E	F	G	-	F	G	F	F	-	F	-	-	E	-	-	-	-
3A	gamma cyhalothrin	Proaxis	E	F	E	F	G	G	G	E	F	E	-	E	G	F	F	-	E	-	-	E	-	-	-	-
3A	fenpropathrin	Danitol	G	-	G	F	F	F	E	F	G	-	E	G	F	F	-	-	-	-	-	G	-	-	F	-
3A	bifenthrin	Brigade	E	F	E	F	G	F	F	E	-	E	E	F	F	-	F	F	-	E	G	F	F	-

* Denotes that insecticide-resistant populations may occur in some areas and can affect the performance of insecticides.

TABLE 2-25. RELATIVE EFFECTIVENESS OF INSECTICIDES AND MITICIDES FOR INSECT AND MITE CONTROL ON FIELD-GROWN VEGETABLES (cont'd)

Not all insecticides listed below are registered on all vegetable crops. Refer to label before applying to specific crop. Ratings are based on consensus of vegetable entomologists in the SE United States.

Key: "E" Excellent; "G" Good; "F" Fair; "-" ineffective or insufficient data

Chemical class (IRAC)	Common name	Example Product	Flea Beetle	Colorado potato beetle*	Cucumber beetles	Corn earworm*	European corn borer	Fall armyworm	Cabbage looper	Imported cabbageworm	Diamondback moth*	Squash vine borer	Beet armyworm*	Stinkbugs/Harlequin bug	Squash bug	Aphids*	Thrips	Western Flower Thrips*	Leafminer	Maggots	Whiteflies*	Cutworms	Wireworms	White grubs	Spider mites*	Broad mites
4A	imidacloprid	Admire	F	G	E	-	-	-	-	-	-	-	-	F	E	E	G	-	-	G	G	-	F	G	-	-
4A	acetamiprid	Assail	G	E	E	-	-	-	-	-	F	-	-	F	E	E	G	-	-	-	E	-	-	-	-	-
4A	clothianidin	Belay	E	E	G	-	-	-	-	-	-	-	-	G	E	G	-	-	F	G	F	-	F	G	-	-
4A	thiamethoxam	Platinum/Actara	E	G	G	-	-	-	-	-	-	-	-	G	E	E	F	-	F	G	E	-	F	F	-	-
4A	dinotefuran	Venom/Scorpion	E	E	G	-	-	-	-	-	-	-	-	E	E	F	G	-	F	-	E	-	-	-	-	-
4C	sulfoxaflor	Closer, Transform	-	-	-	-	-	-	-	-	-	-	-	F	-	E	-	-	-	-	F	-	-	-	-	-
4D	flupyradifurone	Sivanto	-	-	-	-	-	-	-	-	-	-	-	-	G	E	-	-	-	-	E	-	-	-	-	-
5	spinosad	Blackhawk/Entrust	-	E	-	G	G	G	G	E	G	G	G	-	-	-	G	G	E	-	-	F	-	-	-	-
5	spinetoram	Radiant	-	E	-	G	E	G	G	E	G	G	G	-	-	-	E	E	E	-	-	F	-	-	-	-
6	emamectin benzoate	Proclaim	-	-	-	G	G	G	E	E	E	G	E	-	-	-	-	-	F	-	-	F	-	-	-	-
6	abamectin	Agri-Mek	-	E	-	-	-	-	-	-	-	-	-	-	-	G	F	E	-	-	-	-	-	-	E	E
7C	pyriproxyfen	Knack/Distance	-	-	-	-	-	-	-	-	-	-	-	-	-	-	-	-	-	-	G	-	-	-	-	-
9A	pyrifluquinazon	PQZ	-	-	-	-	-	-	-	-	-	-	-	-	-	E	-	-	-	-	G	-	-	-	-	-
9B	pymetrozine	Fulfill	-	-	-	-	-	-	-	-	-	-	-	-	-	E	-	-	-	-	F	-	-	-	-	-
9D	afidopyropen	Sefina, Versys	-	-	-	-	-	-	-	-	-	-	-	-	-	E	-	-	-	-	F	-	-	-	-	-
10B	etoxazole	Zeal	-	-	-	-	-	-	-	-	-	-	-	-	-	-	-	-	-	-	-	-	-	-	G	-
11A	Bt	Dipel, various	-	-	-	F	F	F	G	E	G	F	F	-	-	-	-	-	-	-	-	-	-	-	-	-
15	novaluron	Rimon	-	E	-	E	E	E	G	E	F	G	E	F	F	-	G	G	G	-	F	-	-	-	-	-
16	buprofezin	Courier	-	-	-	-	-	-	-	-	-	-	-	-	-	-	-	-	-	-	G	-	-	-	-	-
17	cyromazine	Trigard	-	G	-	-	-	-	-	-	-	-	-	-	-	-	-	-	E	-	-	-	-	-	-	-
18	methoxyfenozide	Intrepid	-	-	-	G	G	E	E	E	F	G	E	-	-	-	-	-	-	-	-	-	-	-	-	-
20B	acequinocyl	Kanemite	-	-	-	-	-	-	-	-	-	-	-	-	-	-	-	-	-	-	-	-	-	-	E	-
20D	bifenazate	Acramite/Floramite	-	-	-	-	-	-	-	-	-	-	-	-	-	-	-	-	-	-	-	-	-	-	E	-
21A	fenazaquin	Magister	-	-	-	-	-	-	-	-	-	-	-	-	-	-	-	-	-	-	F	-	-	-	G	E
21A	fenpyroximate	Portal	-	-	-	-	-	-	-	-	-	-	-	-	-	-	-	-	-	-	F	-	-	-	G	G
21A	tolfenpyrad	Torac	G	-	-	F	F	F	F	G	G	-	F	-	-	G	G	F	-	-	F	-	-	-	-	G
22A	indoxacarb	Avaunt eVo	F	G	F	E	G	G	E	E	G	G	E	-	-	-	-	-	F	-	-	F	-	-	-	-
23	spiromesifen	Oberon	-	-	-	-	-	-	-	-	-	-	-	-	-	-	-	-	-	-	F	-	-	-	G	G
23	spirotetramat	Movento	-	-	-	-	-	-	-	-	-	-	-	-	-	E	-	-	-	-	G	-	-	-	-	-
25	cyflumetofen	Nealta	-	-	-	-	-	-	-	-	-	-	-	-	-	-	-	-	-	-	-	-	-	-	G	-
28	chlorantraniliprole	Coragen/Vantacor	-	E	-	E	E	E	E	E	E	G	E	-	-	-	F	-	E	-	E	-	-	-	-	-
28	cyantraniliprole	Verimark/Exirel	G	E	F	E	E	E	E	E	E	G	E	-	-	G	F	F	E	G	E	-	-	-	-	-
28	cyclaniliprole	Harvanta	F	E	G	E	E	E	G	E	E	G	E	-	-	-	F	F	E	-	F	-	-	-	-	-
29	flonicamid	Beleaf	-	-	-	-	-	-	-	-	-	-	-	-	-	E	G	E	-	-	F	-	-	-	-	-
30	broflaniliide	Nurizma	-	-	-	-	-	-	-	-	-	-	-	-	-	-	-	-	-	-	-	-	G	G	-	-

* Denotes that insecticide-resistant populations may occur in some areas and can affect the performance of insecticides.

TABLE 2-25 A. PREHARVEST INTERVALS (IN DAYS) FOR PYRETHROID INSECTICIDES IN VEGETABLE CROPS

See TABLE 2-25 to compare relative efficacy of these products against specific insect pests. Read the pesticide label for specific rates and application instructions.

Crop		alpha cypermethrin Fastac (12 hrs)	beta cyfluthrin Baythroid XL (12 hrs)	bifenthrin Brigade eVo (12 hrs)	cypermethrin Various names (12 hrs)	cyfluthrin Tombstone (12 hrs)	esfenvalerate Asana XL (12 hrs)	fenpropathrin Danitol (24 hrs)	gamma cyhalothrin Proaxis (24 hrs)	lambda cyhalothrin Karate/Warrior (24 hrs)	permethrin Pounce (12 hrs)	zeta cypermethrin Mustang Maxx (12 hrs)
	Asparagus	NR	NR	NR	NR	NR	NR	NR	NR	NR	NR	NR
Bulb Vegetables	Onions, Green	NR	NR	NR	7	NR	NR	NR	NR	NR	NR	7
	Onions, Dry Bulb	NR	NR	NR	7	NR	NR	NR	14	14	1	7
Brassica Leafy Vegetables	Broccoli, Brussels Sprout, Cabbage, Cauliflower, Kohlrabi	1	0	7	1	0	3	7	1	1	1	1
	Collard, Mustard Green	1	0	7	1	0	7†	NR	NR	NR	1†	1
Corn	Sweet Corn	3	0	1	NR	0	1	NR	1	1	1	3
Cucurbits	Cantaloupe, Watermelon	1	0	3	NR	0	3	7	NR	1	0	1
	Cucumber, Pumpkin, Summer Squash, Winter Squash	1	0	3	NR	0	3	7	NR	1	0	1
Fruiting Vegetables	Eggplant, Pepper	1	7	7	NR	0	7	3	5	5	3	1
	Tomato	1	0	1	NR	7	1	3	5	5	0	1
	Okra	1	NR	7	NR	NR	NR	NR	NR	NR	NR	1
Legumes	Edible-podded	1	NR	3	NR	NR	3	NR	7	7	NR	1
	Succulent Shelled Pea and Bean	1	3	3	NR	3	3	7	7	7	NR	1
	Dried Shelled Pea and Bean	21	7	14	NR	7	21	NR	21	21	NR	21
Leafy Vegetables, Except Brassicas	Head and Leaf Lettuce	1	0	7	5ᴬ	0	7ᴬ	NR	1	1	1	1
	Spinach	1	0	40	NR	0	NR	NR	NR	NR	1	1
	Celery	1	0	NR	NR	0	NR	NR	NR	NR	3	1
Rot and Tuber Vegetables	Beet, Carrot, Radish, Turnip	1	0	21	NR	0	7	NR	NR	NR	1	1
	Potato	1	0	21	NR	0	NR	NR	NR	7	14	1
	Sweetpotato	1	0	21	NR	0	NR	NR	NR	7	NR	1

NR - Not registered   ᴬHead lettuce only   † Collard only

TABLE 2-26. VIRUS-BASED INSECTICIDES AND LEPIDOPTERAN LARVAE CONTROLLED

			Insect Species Managed*						
Product	Company	Virus	Fruitworm, Budworm	Cabbage Looper	Soybean Looper	Fall Armyworm	Beet Armyworm	Diamond-back moth	Tomato Leafminer
Fawligen	AgBitech	*SfMNPV*				X	X		
Gemstar	Certis	*SfMNPV*	X						
Helicovex	Andermatt	*HearNPV*	X						
Heligen	AgBitech	*HearNPV*	X						
Littovir	Andermatt	*SpliNPV*				X			
Lepigen[1]	AgBiTech	*AcMNPV*	X	X	X	X	X	X	
Loopex	Andermatt	*AcMNPV*		X					
Loopovir	Andermatt	*ChinNPV*			X				
Spexit	Andermatt	*SeNPV*					X		
Spodovir Plus	Andermatt	*SfMNPV*				X			
Spod-X	Certis	*SeNPV*					X		
Surtivo	AgBitech	*ChinNPV + HearNPV*	X	X	X				
Tutavir	Andermatt	*PhopGV*							X

* Most virus-based insecticides need to be applied when eggs or small larvae are observed; rarely are large larvae controlled when at high populations levels. DO NOT apply when most observed caterpillars are large (over ½ inch).

[1] Do not apply Lepigen with a Bacillus thuringiensis (Bt) product

TABLE 2-27. LIST OF GENERIC INSECTICIDES BY ACTIVE INGREDIENT

Active Ingredient	Original Product and Formulation (Manufacturer)	Generics and Formulation (Manufacturer)	
Abamectin	Agri-Mek 0.15 EC (Syngenta)	Abacus 0.15 EC (Rotam) Abamex 0.15EC (Nufarm) Abba Ultra 0.3 EC (AMVAC) Agri-Mek 0.7 EC (Syngenta) Averland 0.7 FC (Vive) Avow 0.15 EC (Invicitis)	Enterik 0.15 EC (Atticus) Reaper 0.15 EC (Loveland) Reaper Advance 0.15 EC (Loveland) Timectin 0.15 EC (Tide Intl.) Willowood Abamectin 0.15 LV (Willowood USA) Willowood Abamectin 0.7 SC (Willowood USA)
Acetamiprid	Assail 30 SG (UPL)	Afflict 30 SG, 70 WP, (Aceto Life Sciences) Anarchy 30 SG (Loveland) ArVida 30 SG, 70 WP (Atticus) Azomar (AgBiome Innovations)	Intruder Max (UPL) Quasar 8.5 SL (Atticus) Tristar 8.5 SL (Clearly Chemical)
Acephate	Orthene 90 SP (Valent)	Acephate 90 Prill (Adama) Acephate 90 WDG, 90 WSP (Loveland) Acephate 97 UP (UPL) Acephate 97 WDG (Adama)	Bracket 90 WDG (Winfield) Livid 90 Prill, 97 WDG (Winfield) Orthene 97 (Amvac)
Bifenazate	Acramite (Arysta)	Actuate 2 SC (Atticus) Banter 50 WDG, 4 SC (UPL) Bifenamite 50 WDG, 4 SC (Agri Star) Bifenazate 50 WDG, 4 SC (Tacoma, Willowood)	Bizate 50 WDG, 4 SC (Loveland) Enervate 4 SC, 50 WSB (Atticus) Floramite SC (OHP) Vigilant 4 SC (Arysta)
Bifenthrin	Brigade 2 EC, Capture 2 EC (FMC)	Battalion 2 EC, 10 WSP, LFC (Atticus) Bi-Dash 2E (Sharda) Bifen 2 AG Gold (Direct AG Source) Bifen 25% EC (Tacoma) Bifender FC (Vive) Bifenthrin 2 EC (Aceto) Bifenture 2 EC, 10 DF, LFC (UPL) Discipline 2 EC (Amvac) Fanfare 2 EC, ES (Adama) Frenzy Veloz (Real Farm)	Lancer 2EC (Albaugh) Nirvana RTU (Innvictis) Reveal 2EC (Innvictis) Seguro (Sharda) Sniper 2EC (Loveland) Squadron LFC (Aceto) Strict 1.5 EC (Sharda) Tundra 2 EC (Winfield) Xpedient (Amvac)
Carbaryl	Sevin 4 L, SL, XLR (Bayer)	Carbaryl 4 L (Drexel, Loveland)	
Chlorantraniliprole	Coragen 1.67 SC (FMC)	Coragen eVo 5 SC (FMC) Exceliprole 3.34 SC (Albaugh) Shenzi 3.33 SC (UPL)	Trinalor 35 WDG (Adama) Vantacor 5 SC (FMC)
Cyantraniliprole	Exirel 0.83 SE (FMC)	Kradan 0.83 SE (Altamont)	
Cyfluthrin & beta-cyfluthrin	Baythroid 1 EC (Bayer)	Cryptoid XL (Atticus) Tombstone 2 E (Loveland)	Tombstone Helios 2 E (Loveland)
Cypermethrin	Ammo (discontinued)	Holster (Loveland)	Up-Cyde 2.5EC (UPL)
Dinotefuran	Venom 70 SG (Valent)	Certador (BASF)	Scorpion 35 SL (Gowan)
Esfenvalerate	Asana XL 0.66 EC (DuPont)	S-FenvaloStar 0.66 EC (LG Life Sciences)	
Fenpyroximate	Fujimite 0.42 EC (Nichino)	Portal 0.4 EC (Nichino) Tyoga 0.42 EC (Sipcam)	
Gamma-cyhalothrin	Proaxis 0.5 EC (Loveland)	Declare Insecticide 0.5 EC (Cheminova)	Proaxis Insecticide 0.5 EC (Cheminova)
Hexithiazox	Hexygon, Savey, Onager (Gowan)	Hexamite 1 E (Albaugh) Hexcel 50 DF (Atticus) Hexy 2E (Sharda)	Proneva 1 EC (Innvictus) Ruger 1 EC (Atticus)
Imidacloprid	Admire 2 F, Pro 4.6 F, Provado 1.6 F (Bayer)	Acronyx 4 F (Atticus) Advise Four 2 FL (Winfield) Alias 2 F, 4 F (Adama) Lada 2 F (Rotam) Macho 2 FL, 4 F (Albaugh) Malice 2 F, 75 WSP (Loveland) Midash 2 SC, Forte (Sharda USA) Montana 2 F, 4 F (Rotam NA)	NuPrid 2 SC, 4 F Max, 4.6 F (Nufarm) Prey 1.6 F (Loveland) Provoke 4 F (Innvictis) Sherpa 1.6 F (Loveland) Tide Imidacloprid 2 F, 4 F (Tide) Widow 2 F (Loveland) Viloprid 1.7 FC (Vive CP) Willowood Imidacloprid 2 SC, 4 SC (Willowood) Wrangler 4 F (Loveland)
Indoxacarb	Avaunt 30 WDG, eVo 30 WDG (FMC)	Comber 30 WDG (Sharda)	

TABLE 2-27. LIST OF GENERIC INSECTICIDES BY ACTIVE INGREDIENT (cont'd)

Active Ingredient	Original Product and Formulation (Manufacturer)	Generics and Formulation (Manufacturer)	
Lambda-cyhalothrin	Karate 2 ME, Warrior 2 ME (Syngenta)	Cavalry II 2.08 SC (Growmark) Crusader 1 EC, 2 ME (Albaugh) Firestone 1 CS (Altitude Crop Innovations) Grizzly Z 1 CS, 2 CS (Winfield) Kendo 1 EC (Helm) LC Insecticide (Drexel) Labamba 1 EC (Sharda) Lambda-CY AG (Winfield United) Lambda CY 1 EC (United Phosphorous) Lambda-Cyhalothrin 1 EC (Nufarm)	Lamcap II 2.08 CS (Syngenta) Lunge 2.08 EC (UPL) Paradigm 1 EC (Winfield United) Province II 2 SC (TENKOZ) Ravage 1 EC (Innvictis) Serpent 1 EC (Atticus) Silencer 1 EC, VXN (Adama) Willowood Lambda-Cy 1 EC (Willowood)
Methomyl	Lannate (DuPont)	Lanveer LV (Invictus) Nudrin SP, LV (Rotam)	
Methoxyfenozide	Intrepid (Dow Agro)	Inspirato 2F (Atticus) Insurgent 2 F (Altamont) Invertid 2 F (Loveland) Invicar 2 SC (Albaugh) Insurgent 2 F (Altamont)	Thwartex (Agsurf) TurnStyle 2 F (UPL) Vexer 2 F (Innvictis) Zylo 2 F (UPL)
Oxamyl	Vydate C-LV, L (Corteva)	Return 2 SL (Albaugh)	
Permethrin	Pounce 3.2 EC (FMC)	Arctic 3.2 EC (Winfield) Perm-Up 3.2 EC (UPL) PermaStar 3.2 EC (LG Life Sciences)	Permethrin 3.2 EC (Loveland, Winfield, Tenkoz) Stelleto (Wilbur-Ellis)
Pymetrozine	FulFill 50 WDG (Syngenta)	Achiever 50 WDG (Aceto)	Seville 50 WDG (Atticus)
Pyriproxyfen	Distance, Knack 0.86 EC (Valent)	Cusack 0.86 EC (Atticus) Farewell (Adama)	Reemit 0.86EC (Atticus) Sever 35 WSB (Innvictis)
Thiamethoxam	Actara 25 WDG (Syngenta)	Artist 25 WDG (Sharda) Elliptica 5 F (Nufarm)	
Zeta-cypermethrin	Mustang Maxx 0.8 EC (FMC)	Cortes Maxx 0.8 EC (Atticus)	

TABLE 2-28. COMPONENTS OF INSECTICIDE MIXTURES *

Premix Trade Name	Components (Legacy trade name)
Acenthrin	acephate (Orthene) + bifenthrin (Brigade)
Athena	abamectin (Agri-Mek) + bifenthrin (Brigade)
Agri-Flex	abamectin (Agri-Mek) + thiamethoxam (Actara)
Avenger, Brigadier, IMAX Plus, Swagger, Skyraider, Tempest	bifenthrin (Brigade) + imidacloprid (Admire)
Besiege	chlorantraniliprole (Coragen) + lambda-cyhalothrin (Warrior)
Cormoran	acetamiprid (Assail) + novaluron (Rimon)
Elevest	bifenthrin (Brigade) + chlorantraniliprole (Coragen)
Enkounter	acetamiprid (Assail) + methoxyfenozide (Intrepid)
Endigo ZC	lambda-cyhalothrin (Warrior, Karate) + thiamethoxam (Actara)
Durivo, Voliam Flexi	chlorantraniliprole (Coragen) + thiamethoxam (Actara, Platinum)
Gladiator	avermectin B1 (Agri-Mek) + zeta-cypermethrin (Mustang Maxx)
Hero, Steed	bifenthrin (Brigade) + zeta-cypermethrin (Mustang Maxx)
Intrepid Edge	methoxyfenozide (Intrepid) + spinetoram (Radiant)
Killer	imidacloprid (Admire) + lambda-cyhalothrin (Warrior, Karate)
Leverage 360	imidacloprid (Admire) + beta-cyfluthrin (Baythroid XL)
Leverage 2.7	imidacloprid (Admire) + cyfluthrin (Baythroid)
Minecto Pro	cyantraniliprole (Exirel) + abamectin (Agri-Mek)
Obelisk	abamectin (Agri-Mek) + imidacloprid (Admire)
Savoy EC	acetamiprid (Assail) + bifenthrin (Brigade)
Senstar	spirotetramat (Movento) + pyriproxyfen (Knack)

* Insecticide pre-mixes usually control a wider range of insect pests than a single active ingredient and may be less costly than purchasing equivalent amounts of the premix components separately. However, pre-mixes can also have disadvantages, including unnecessary use of individual active ingredients, unnecessary selection for resistance, and application of rates that are too low or too high for specific target pests. Growers who need to apply a combination of insecticides also have the option of purchasing active ingredients separately and tank-mixing to achieve rates that are appropriate for the need. Having active ingredients available for use separately often provides greater pest management flexibility than having the same active ingredients available only as a pre-mix.

INSECT CONTROL FOR GREENHOUSE VEGETABLES

Sound cultural practices, such as sanitation and insect-free transplants, help prevent insect establishment and subsequent damage. Use separate plant production houses, use of yellow sticky traps, and timely sprays will help prevent whitefly buildup. Use of Encarsia parasites for whitefly and other biological control agents in conjunction with use of pesticides is encouraged. Unless a pesticide label specifically states that, a product. cannot be used in a greenhouse vegetable crop, the product can be used on those crops for which it is registered. However, pesticides behave differently in the field and the greenhouse, and for many products, information is not available on greenhouse crop phytotoxicity and residue retention. If unsure of the safety of a product to a crop, apply to a small area before treating the entire crop.

TABLE 2-29. INSECT CONTROL FOR GREENHOUSE VEGETABLES

Insecticide and Formulation	Amount of Formulation	Restricted Entry Interval (REI)	Pre Harvest Interval (PHI) (Days)	Precautions and Remarks
INSECT CONTROL FOR CUCUMBER				
Aphid				
flonicamid, MOA 29 (Beleaf) 50 SG	0.065 to 0.1 oz per 1000 sq ft	12 hrs	0	May be applied either to the soil as a drench or drip irrigation for preventive control or sprayed onto plans as a rescue treatment.
flupyradifurone, MOA 4D (Altus) 1.67 SL		12 hrs	1	Spray crop to wet, not to drip. Through, uniform coverage is required for good control. Use higher rates for whiteflies.
Foliar treatment	7 to 14 fl oz per 50 gal			
Soil treatment	1.4 to 1.9 fl oz per 50 gal			Apply as a soil drench using micro-irrigation, drip irrigation, overhead irrigation, or hand-held motorized calibrated equipment. Use sufficient volume to wet potting medium without loss of liquid from the bottom of the container. Irrigate carefully during the next 10 days to avoid loss of product due to leaching.
imidacloprid, MOA 4A (Admire Pro) 4.6 F	0.6 fl oz/1,000 plants	12 hrs	0	Apply in a minimum of 21 gallons water using soil drenches, microirrigation, or drip irrigation. Do not apply to immature plants as phytotoxicity may occur. Make only one application per crop per season.
insecticidal soap (M-Pede) 49 EC	1 to 2% soln.	12 hrs	0	
Cabbage looper				
Bacillus thuringiensis, MOA 11 (various)	0.5 to 1 lb OR 3 pt/100-gal water	4 hrs	—	Most formulations are **OMRI listed**
spinosad, MOA 5 (Entrust) SC	3 fl oz/100 gal	4 hrs	1	Do not make more than two consecutive applications. **OMRI listed.**
Spider mite				
insecticidal soap (M-Pede) 49 EC	1 to 2% soln.	12 hrs		Use predatory mites. **OMRI listed.**
mineral oil (TriTek)	1 to 2 gal/100 gal	4 hrs	0	Begin applications when mite populations are low and repeat at weekly intervals.
acequinocyl, MOA 20B (Kanemite) 15SC (Shuttle O) 1.25SC	31 fl oz per 43,560 sq ft, or per 100 gal	12 hrs	1	Will control spider mites and broadmites. Active against all spider mite life stages.
fenpyroximate, MOA 21A (Akari) 5 SC	1 to 2 pts per 100 gal	12 hrs	7	Active against immatures and adults.
chlorfenapyr, MOA 13 (Pylon) 2SC	9.8 to 13 fl oz/100-gal water or per acre area	12 hrs	0	Do not make more than two applications at 5- to 10-day intervals before rotating to an insecticide with a different mode of action.
Whitefly, Leafminer				
acetamiprid, MOA 4A (Assail) 30 SC	0.1 oz per 1000 sq ft	12	0	
cyantraniliprole MOA 28 (Exirel) SE	13.5 to 20.5 fl oz/ acre or per 100 gal	12 hrs	0	For best performance, use an effective adjuvant.
flonicamid, MOA 29 (Beleaf) 30 SG	0.065 to 0.1 oz per 1000 sq ft	12 hrs	0	

TABLE 2-29. INSECT CONTROL FOR GREENHOUSE VEGETABLES (cont'd)

Insecticide and Formulation	Amount of Formulation	Restricted Entry Interval (REI)	Pre Harvest Interval (PHI) (Days)	Precautions and Remarks
INSECT CONTROL FOR CUCUMBER (cont'd)				
Whitefly, Leafminer (cont'd)				
flupyradifurone, MOA 4D (Altus) 1.67 SL	—	—	1	See rates and application instructions under aphids.
imidacloprid, MOA 4A (Admire Pro) 4.6 F	0.6 fl oz/1,000 plants	12 hrs	0	Apply in a minimum of 21 gallons water using soil drenches, microirrigation, or drip irrigation. Do not apply to immature plants as phytotoxicity may occur. Make only one application per crop per season.
insecticidal soap (M-Pede) 49 EC	1 to 2% soln.	12 hrs	0	May be used alone or in combination. Acts as an exciter. **OMRI listed.**
Beauveria bassiana (Mycotrol WP)	0.25 lb/20-gal water		0	Apply when whiteflies observed. Repeat in 4- to 5-day intervals. **OMRI listed.**
INSECT CONTROL FOR LETTUCE				
Aphid, Leafminer, Whitefly				
flupyradifurone, MOA 4D (Altus) 1.67 SL		12 hrs	1	Spray crop to wet, not to drip. Through, uniform coverage is required for good control. Use higher rates for whiteflies.
Foliar treatment	7 to 14 fl oz per 50 gal			
Soil treatment	1.4 to 1.9 fl oz per 50 gal			Apply as a soil drench using micro-irrigation, drip irrigation, overhead irrigation, or hand-held motorized calibrated equipment. Use sufficient volume to wet potting medium without loss of liquid from the bottom of the container. Irrigate carefully during the next 10 days to avoid loss of product due to leaching.
pymetrozine, MOA 9B (Fulfill) 50 WG	0.063 oz per 1000 sq ft	12 hrs	0	Will not control leafminer.
pyrethrins, MOA 3A (Pyganic) 5 EC	0.25 to 0.5 fl oz per gal water	12 hrs	0	May be used alone or tank mixed with a companion insecticide (see label for details). **OMRI listed.**
malathion, MOA 1B (various) 57 EC 25 WP	1qt/100-gal water 4 lb/100-gal water	24 hrs	14 14	Will not control whitefly
insecticidal soap (M-Pede) 49 EC	1 to 2% soln.	12 hrs	0	May be used alone or in combination. Acts as an exciter. Insecticidal soaps can cause phytotoxicity under high temperatures or slow drying conditions. If unsure, apply to a small area before treating the entire crop. **OMRI listed.**
Beauveria bassiana (BotaniGard) 22 WP (Mycotrol WP)	1 lb/100-gal water 0.25 lb/20-gal water	4 hrs	0	Under high aphid or whitely pressure, apply at 2- to 5-day intervals. **OMRI listed.**
Cabbage looper				
Bacillus thuringiensis, MOA11 (Javelin) WG	0.5 to 1.25/100-gal water	4 hrs	0	
cyantraniliprole MOA 28 (Exirel) SE	10 to 20.5 fl oz per acre or per 100 gal	12 hrs	0	For best performance, use an effective adjuvant.
spinosad, MOA 5 (Entrust) SC	3 fl oz/100 gal	4 hrs	1	Do not make more than two consecutive applications.
Slugs				
iron phosphate (Sluggo)	0.5 to 1 lb/1,000 sq ft	4 hrs	1	Scatter the bait around the perimeter of the greenhouse to provide a protective barrier. If slugs are within the crop, then scatter the bait on the ground around the plants. Do not make more than 3 applications within 21 days. Sluggo will control slugs and snails, while Bug-N-Sluggo will also control earwigs, cutworms, sowbugs, and pillbugs. Sluggo is **OMRI listed.**
iron phosphate + spinosad (Bug-N-Sluggo)	0.5 to 1 lb/1,000 sq ft	4 hrs	1	
Spider mite				
insecticidal soap (M-Pede) 49 EC	1 to 2% soln.	12 hrs	0	Begin applications when mite populations are low and repeat at weekly intervals.
mineral oil (TriTek)	1 to 2 gal/100 gal	4 hrs	0	
INSECT CONTROL FOR TOMATO AND PEPPER				
Aphid				
acetamiprid, MOA 4A (Tristar) 8.5 SL	8.5 oz per 100 gal	12 hrs	3	Do not apply more than two times per crop, and do not apply more than once every 7 days.

TABLE 2-29. INSECT CONTROL FOR GREENHOUSE VEGETABLES (cont'd)

Insecticide and Formulation	Amount of Formulation	Restricted Entry Interval (REI)	Pre Harvest Interval (PHI) (Days)	Precautions and Remarks
INSECT CONTROL FOR TOMATO AND PEPPER (cont'd)				
Aphid (cont'd)				
flonicamid, MOA 29 (Beleaf) 50 SG	0.1 oz per 1000 sq ft	12 hrs	0	May be applied to the soil as a drench or drip irrigation for preventive control, or as a spray for rescue treatments. Will also control whiteflies.
flupyradifurone, MOA 4D (Altus) 1.67 SL **Foliar treatment**	7 to 14 fl oz per 50 gal	12 hrs	1 (tomato) 3 (pepper)	Spray crop to wet, not to drip. Through, uniform coverage is required for good control. Use higher rates for whiteflies.
Soil treatment	1.4 to 1.9 fl oz per 50 gal			Apply as a soil drench using micro-irrigation, drip irrigation, overhead irrigation, or hand-held motorized calibrated equipment. Use sufficient volume to wet potting medium without loss of liquid from the bottom of the container. Irrigate carefully during the next 10 days to avoid loss of product due to leaching.
imidacloprid, MOA 4A (Admire Pro) 4.6	0.6 fl oz/1,000 plants	12 hrs	0	Apply in a minimum of 16 gallons water. Apply only to plants grown in field-type soils, potting media, or mixtures thereof. Do not apply to plants grown in non-soil media such as perlite, vermiculite, rock wool, or other soilless media, or plants growing hydroponically. Do not apply to peppers. Do not exceed one application per crop. Also controls whiteflies.
malathion, MOA 1B (various) 10 A 57 EC 25 WP	1 lb/50,000 cu ft 1qt/ 100-gal water 4 lb/ 100-gal water	12 hrs	15 hrs 1 1	
insecticidal soap (M-Pede) 49 EC	1 to 2% soln.	12 hrs	0	May be used alone or in combination. Acts as an exciter.
Beauveria bassiana (Mycotrol WP)	0.25 lb/20-gal water		0	Apply when whiteflies are observed. Repeat in 4-to 5-day intervals.
Armyworm, Fruitworm, Cabbage looper, Pinworm				
Bacillus thuringiensis, MOA 11 (Javelin) WG	0.5 lb to 1.25 lb/ 100-gal of water	4 hrs	0	
(Agree) WP (Dipel) DF (Xentari) DF	1 to 2 lb 0.5 to 1.25 lb 0.5 to 1.5 lb			
chlorfenapyr, MOA 13 (Pylon) 2 SC	6.5 to 13 fl oz/100-gal water, or per acre area	12 hrs	0	Do not make more than two applications at 5- to 10-day intervals before rotat- ing to an insecticide with a different mode of action
cyantraniliprole, MOA 28 (Exirel) SE	7 to 13.5 fl oz per acre, or per 100 gal	12 hrs	1	
spinosad, MOA 5 (Entrust) SC	3 fl oz/100 gal	4 hrs	1	Do not make more than two consecutive applications. Do not apply to seedling tomatoes or peppers grown for transplants.
Leafminer				
cyantraniliprole, MOA 28 (Exirel) SE	13.5 to 20.5 fl oz/acre or per 100 gal	12 hrs	1	
chlorfenapyr, MOA 13 (Pylon) 2SC	9.8 to 13 fl oz/ 100-gal water or per acre	12 hrs	0	Do not make more than two applications at 5- to 10-day intervals before rotat- ing to a different mode of action.
spinosad, MOA 5 (Entrust) SC	10 fl oz/100 gal	4 hrs	1	Do not apply to seedlings grown for transplants
Slug				
metaldehyde (various) bait	Follow label directions	12 hrs		Apply to soil surface around plants. Do not contaminate fruit.
iron phosphate (Sluggo)	½ teaspoon per 9-inch pot		0	
Spider mite, Broad mite, Russet mite				
acequinocyl, MOA 20B (Kanemite) 15 SC (Shuttle O) 1.25SC	31 fl oz per 43,560 sq ft or per 100-gal water	12 hrs	1	Active against all life stages.

TABLE 2-29. INSECT CONTROL FOR GREENHOUSE VEGETABLES (cont'd)

Insecticide and Formulation	Amount of Formulation	Restricted Entry Interval (REI)	Pre Harvest Interval (PHI) (Days)	Precautions and Remarks
INSECT CONTROL FOR TOMATO AND PEPPER (cont'd)				
Spider mite, Broad mite, Rust mite (cont'd)				
bifenazate, MOA 25 (Floramite) SC	4 to 8 fl oz/100-gal water (1/4 to 1/2 tsp/gal)	12 hr	3	For use on tomatoes more than 1 inch in diameter at maturity. Not registered on pepper. Not for Rust mite. Active against immatures and adults. Some egg activity.
mineral oil (TriTek)	1 to 2 gal/100 gal	4 hr	0	Begin applications when mite populations are low and repeat at weekly intervals.
chlorfenapyr, MOA 13 (Pylon) 2 SC	9.8 to 13 fl oz/100- gal water or per acre area	12 hr	0	Do not make more than two applications at 5- to 10-day intervals before rotat- ing to an insecticide with a different mode of action.
cyflumetofen, MOA 25 (Sultan) 1.67SC	13.7 fl oz/100 gal	12 hrs	1	Do not make more than 2 applications. Active against all life stages.
fenpyroximate, MOA 21A (Akari) 5 SC	1 to 2 pts per 100 gal	12 hrs	1	Active against immatures and adults.
insecticidal soap (M-Pede) 49 EC	1 to 2% soln.	12 hrs	0	
Thrips, including western flower				
Beauveria bassiana (Mycotrol WP)	0.25 lb/20-gal water		0	Use screens on intake vents. Apply when whiteflies observed. Repeat in 4- to 5-day intervals.
cyantraniliprole, MOA 28 (Exirel) SE	13.5 to 20.5 fl oz per acre or per 100 gal	12 hrs	1	For foliage-feeding thrips only, not those in flowers.
flonicamid, MOA 29 (Beleaf) 50 SG	0.1 oz per 1000 sq ft	12 hrs	1	For use on tomato only.
spinosad, MOA 5 (Entrust) SC	5.5 fl oz/100 gal	4 hrs	1	Do not make more than two consecutive applications, and do not apply more than 6 times in a 12-month period against thrips. Do not apply to seedlings grown from transplants.
Whitefly				
imidacloprid, MOA 4A (Admire Pro) 4.6 F	0.6 fl oz/1,000 plants	12 hrs	0	Apply in a minimum of 16 gal of water. Apply only to plants grown in field-type soils, potting media, or mixtures thereof. Do not apply to plants grown in non- soil media such as perlite, vermiculite, rock wool, or other soil-less media, or plants growing hydroponically. Do not apply to peppers. Do not exceed one application per crop. Also controls aphids.
acetamiprid, MOA 4A (Tristar) 8.5 SL	1.25 fl oz/1000 plants	12 hrs	1	Apply only to plants growing in rock wool, perlite, or other soilless growing media. Do not apply to crops that have already been treated with imidaclo-prid, dinotefuran, or another neonicotinoid.
cyantraniliprole, MOA 28 (Exirel) SE	13.5 to 20.5 fl oz per acre, or per 100 gal	12 hrs	1	
flonicamid, MOA 29 (Beleaf) 50 SG	0.1 oz per 1000 sq ft	12 hrs	0	For use on tomato only.
flupyradifurone, MOA 4D (Altus) 1.67 SL	—	—	1 (tomato) 3 (pepper)	See rates and application instructions under aphids.
insecticidal soap (M-Pede) 49 EC	1 to 2% soln.	12 hrs	0	
pyrethrins and PBO, MOA 3 (Pyganic) 5 EC	0.25 to 0.5 fl oz per gal	12 hrs	0	May be used alone or tank mixed with a companion insecticide. (See label for details).
Beauveria bassiana (Mycotrol WP)	0.25 lb/20-gal water	4 hrs	0	Apply when whiteflies are observed. Repeat in 4- to 5-day intervals. **OMRI listed**.
buprofezin, MOA 16 (Talus) 40 SC	9 to 13.6 oz/ 100-gal water or per acre area	12 hrs	1	Insect growth regulator that affects immature stages of whiteflies. Will not kill adults. For use on tomatoes only.
pyriproxyfen, MOA 7C (Distance) 0.86 EC	6 fl oz/100-gal water	12 hrs	<1	Do not use on tomatoes less than 1 inch in diameter. Insect growth regulator that affects immature stages of whiteflies. Will not kill adults. Do not use on tomatoes more than 1 inch in diameter. Do not apply on non-bell peppers

TABLE 2-30. ALTERNATIVE IPM & BIOINSECTICIDE RECOMMENDATIONS IN VEGETABLE CROPS

NOTE: Many organic insecticides have not been tested thoroughly in the commercial vegetable production systems. Use systems-based and pest exclusion practices to reduce the overall infestation levels before using approved insecticides in organic farming systems. Targeted insecticide applications at correct rate can protect natural enemies. Always read the insecticide label before application and purchase bioinsecticides from reliable sources.

Target Pest	Cropping System	Systems-based Practices (for pest prevention)	Mechanical tactics (for pest prevention)	Biorational Insecticides [#]
Aphid	Multiple crops	Timely planting and harvest reduce water stress on crops. Use appropriate plant spacing to reduce spread and improve air movement	Use of reflective mulches to protect transplants or use insect barrier fabric immediately after transplanting	Insecticidal soap, oil blends (Pyola & others), paraffinic oil, pyrethin, *Chromobacterium* (Grandevo), Azera (insecticide premix). Do not spray oils in high heat
Armyworms	Multiple crops	Weed control, field sanitation, control soil organic residue (larvae hide under thick organic residue), timely planting & harvest to avoid late-season infestation	Remove & destroy egg masses, Insect netting for gardens or short rows*, High tunnel pest exclusion for commercial producers^	*Bacillus thuringiensis* (Xentari, Dipel, or premix), , Spinosad, Neem, *Chromobacterium* (Grandevo), Leap (insecticide premix), Spear-Lep (Synergist), Viruses
Bean leaf beetle	Snap, lima pole beans	Sanitation (removal of crop debris), timely planting	Row cover or pest exclusion (Super Light Insect Barrier, AgroFabric)	Pyrethrins (single a.i. or premix), Spinosad
Blister beetle	Multiple crops	—	Row cover or pest exclusion (Super Light Insect Barrier, AgroFabric)	Spinosad, Pyrethrin (single a.i. or premix)
Cabbage looper (small caterpillars)	Multiple crops	Remove alternate host plants (wild mus- tard, shepherd's purse), sanitation	Insect netting for gardens or short rows*. High tunnel pest exclusion for commercial producers^	*Bacillus thuringiensis* (single a.i. or premix), Neem, Pyrethrins (single a.i. or premix), Spinosad, *Chromobacterium* (Grandevo), Spear-Lep (Synergist), Viruses
Colorado potato beetle	Multiple crops	Crop rotation, planting tolerant varieties	Insect netting for gardens or short rows* Hand remove larvae & adults	Neem, Pyrethrins (use for larval control), Spinosad (watch for insecticide resistance)
Corn earworm/ Tomato fruitworm	Multiple crops	Field sanitation, removal of weedy hosts. Use pheromone traps for monitoring moth activity	Insect netting for gardens or short rows*, High tunnel pest exclusion for commercial producers^	*Bacillus thuringiensis* (single a.i. or premix), Spinosad, Neem, *Chromobacterium* (Grandevo), Spear-Lep (tank-mix with Bt, not OMRI listed), Viruses
Cowpea curculio	Cowpeas, Snap, Lima, Pole beans	Crop rotation, sanitation, early harvest of crop, tillage in fall or early spring	Row cover or pest exclusion (Super Light Insect Barrier, AgroFabric).	Pyrethrins (single a.i. or premix), Spinosad (adults are difficult to kill due to insecticide resistance)
Cucumber beetle	Multiple crops	Perimeter trap cropping with Hubbard squash is highly effective. Use organic insecticides on trap crops	Row cover or pest exclusion (Super Light Insect Barrier, AgroFabric) for early season protection	Pyrethrins (single a.i. or premix), Neem, Parasitic nematodes (weekly soil drench for caterpillars) as needed
Cutworm	Multiple crops	Need-based soil tillage and organic matter management	Row cover or pest exclusion (Super Light Insect Barrier, AgroFabric Pro) for early season protection	*Bacillus thuringiensis* (single a.i. or premix with directed spray at plant base); Spinosad foliar and stem spray; spinosad-based bait (Seduce, Bug-N- Sluggo)
Diamondback moth & Imported cabbage worm	Collard & Mustard greens	Use pheromone traps to monitor moths. Collards and Yellow Rocket can be used as trap crop (planted before main crop)	Insect netting over short rows to block moths* (soon after transplanting), destroy caterpillar clusters on leaves before dispersal	*Bacillus thuringiensis* (single a.i. or premix), Neem, Pyrethrins (single a.i. or premix), Azera (insecticide premix), Spear-Lep (tank-mix with Bt, not OMRI listed)
European corn borer	Multiple crops	Use tolerant cultivars when possible	Remove plant stalks	*Bacillus thuringiensis* (single a.i. or premix), Neem, Pyrethrins
Flea beetle	Multiple crops	Timely planting of crops, Perimeter trap cropping with eggplants	Insect netting (Super Light Insect Barrier or AgroFabric) immediately after transplanting*	Spinosad, Parasitic nematodes (drench in soil), Pyrethrins (single a.i. or premix), Azera (premix insecticide)
Grasshopper	Multiple crops	Maintain grassy patch (non-crop habitat) and use nematode insecticidal bait	Row cover or pest exclusion (Super Light Insect Barrier, AgroFabric).	Pyrethrins or Spinosad (multiple applications needed), Nolo Bait (*Nosema locustae*) for use in grassy areas or near fences (non-crop areas)
Hornworm	Tomato	Timely harvesting of fruits, remove crop debris. *Cotesia* parasitizes caterpillars naturally	Row cover or pest exclusion (Super Light Insect Barrier, AgroFabric). High tunnel pest exclusion system for commercial producers^	Spinosad, *Bacillus thuringiensis*, Pyrethrins (single a.i. or premix), Neem, Spear-Lep (tank-mix with Bt, not OMRI approved)
Japanese beetle	Multiple crops	Sunflower and sorghum trap crops may deter feeding on main crop	Row cover or pest exclusion (Super Light Insect Barrier, AgroFabric)	Pyrethrins (single a.i. or premix), Neem (multiple sprays), Milky spore disease, *Bacillus thuringiensis subsp. galleriae* (beetleJUS!)
Leaffooted bug	Fruiting vegetables (tomato, okra, eggplant)	Trap cropping with *Peredovik* sunflower & *NK300* sorghum provides significant reduction. Plant extra rows of trap crop in drought year. Plant okra away from other susceptible crops	High tunnel pest exclusion for commercial producers^	Pyrethrins and spinosad for killing nymphs in trap crops (adults are difficult to control). Kaolin clay may be effective temporarily (reapply after rain).

^ High tunnel pest exclusion using 50 percent woven shade cloth on side- and end-walls can exclude large moths and reduce caterpillar pressure. Details at https://southern.sare.org/resources/high-tunnel-pest-exclusion-system/

[#] Organic insecticides are currently available as single active ingredient (AI) or in the form of premixes, such as Azera (neem + pyrethrins), Leap (BT + methyl salicylate), and BotaniGard Maxx (pyrethrins + *Beauveria bassiana*).

* Relatively low-cost, temporary insect netting products include the Super-Light Insect Barrier (GardensAlive.com), AgroFabric Pro (https://www.7springsfarm.com) or other lightweight exclusion material are suitable for early season pest prevention.

INSECT CONTROL

TABLE 2-30. ALTERNATIVE IPM & BIOINSECTICIDE RECOMMENDATIONS IN VEGETABLE CROPS (cont'd)

NOTE: Many organic insecticides have not been tested thoroughly in the commercial vegetable production systems. Use systems-based and pest exclusion practices to reduce the overall infestation levels before using approved insecticides in organic farming systems. Targeted insecticide applications at correct rate can protect natural enemies. Always read the insecticide label before application and purchase bioinsecticides from reliable sources.

Target Pest	Cropping System	Systems-based Practices (for pest prevention)	Mechanical tactics (for pest prevention)	Biorational Insecticides #
Leafhopper	Multiple crops	Use varieties tolerant to the insect feeding	Row cover or pest exclusion system (Super Light Insect Barrier, AgroFabric)	Insecticidal soap, Pyrethrins (single a.i. or premix), Neem
Leafminer	Multiple crops	Select vigorous hybrid plant varieties	Pick and destroy mined leaves	Neem with azadirachtin, Spinosad
Onion maggot, seed corn maggot	Multiple crops	Use well-composted manure, soil tillage exposes maggots to natural enemies, field preparation ahead of planting	Soil tillage may expose maggots to predators	Spinosad-based insecticides (Seduce, Bug-N-Sluggo) may help reduce maggot buildup
Pepper & vegetable weevil	Pepper	Crop rotation, sanitation	Insect netting (Super Light Insect Barrier) for small areas	Pyrethrins (single a.i. or premix). *Bacillus thuringiensis* subsp. *galleriae* (beetleJUS!). Diatomaceous earth (DE) as abrasive & deterrent
Spider mite	Multiple crops	Plant and harvest timely; provide irrigation to plants; problem could be severe in drought years; excessive use of pyrethroid insecticides can cause spider mite outbreak; release predatory mites before an outbreak	Reduce traffic in infested crops to prevent spreading to new areas	Paraffinic oil, Neem oil, Sulfur dust or spray (check label before use), *Chenopodium* Terpene Extract (Requiem), *Isaria fumosoroseus* (PFR97), Soluble silica (Sil-Matrix). Many oil blends are available in the market; use with caution to avoid crop burn in hot dry weather. To protect beneficial mites, do *not use pyrethrins*.
Squash vine borer (SVB), Pickleworms (PW)	Pumpkin, squash	Plant & harvest cucurbit crops timely, trap cropping with Hubbard squash (Baby Blue and New England) may reduce SVB but not PW, sanitation or removing crop debris reduces build-up of pests	Temporary pest exclusion system with Super Light Insect Barrier may help reduce early infestation. High tunnel pest exclusion with 50% shade cloth can significantly reduce moth numbers	Pyrethrins and Spinosad may provide limited control; perform repeated spraying (every 3 d) on fruits with Bt-based products to stop small PW caterpillars
Squash bug	Pumpkin, squash	Trap cropping with Hubbard squash; planting tolerant varieties, sanitation (remove crop debris), use of plastic mulch may increase pest pressure	Insect netting for gardens*, High tunnel pest exclusion for commercial producers can slow down pest migration^	Pyrethrins (single a.i. or premix) and Spinosad may provide control of nymphs better than the adults on trap crops or main crop
Harlequin bug	Multiple crops	These are types of stink bugs with sucking mouthparts. Timely planting and harvest of crops	Insect netting for garden crops. High tunnel pest exclusion for commercial producers^	Pyrethrins and Spinosad may reduce infestation, but thorough coverage is needed
Thrips	Multiple crops	Timely planting, avoid planting ornamental close to vegetable crops, use blue sticky cards for monitoring and trap out in small areas	Thrips exclusion netting material is available for greenhouse producers	Spinosad, Insecticidal soap, Paraffinic oil, *Chenopodium* Terpene Extract (Requiem Prime Entomopathogenic nematodes (e.g. *Steinernema feltiae*).
Whitefly	Multiple crops	Crop rotation, avoid planting ornamental close to vegetable crops, problem intensifies in drought; release parasitoids in greenhouses	Thrips exclusion netting may also reduce whiteflies	*Metarhizium anisopliae* (Met 52, Met Master), Insecticidal soap, Neem oil, *Beauveria bassiana*, *Chenopodium* Terpene Extract (products work well for nymphs, adults are difficult to kill in outbreak conditions)
Yellowmargined leaf beetle (YMLB)	Brassica crops	Turnips and Napa Cabbage are defoliated rapidly compared to other brassicas. Trap cropping with turnips and insecticidal treatment can reduce build up. Note the occurrence of caterpillar infestations distinct from YMLB. Remove crop debris after harvest to prevent pest buildup	Insect netting (temporary exclusion methods) for blocking migratory adults*, soil tillage may reduce eggs in soil	Spinosad, *Chromobacterium* (Grandevo), Pyrethrins (use in rotation with frequent foliar applications), Azera (premix) for preventing YMLB outbreaks.
Snails & Slugs	Brassica crops	Frequent rainfall and high organic matter in soil increases snails & slugs. Do not over-irrigate. Soil tillage may destroy eggs. Remove crop debris to reduce buildup	—	Iron phosphate (e.g., Sluggo by Monterey, OMRI-certified. Bonide Slug Magic for Gardens), Volatile oil blend (Monterey All-Natural Snail & Slug Spray), Diatomaceous earth with amorphous silica (PermaGuard Crawling Insect Control), Bait premix of iron phosphate and spinosad (Bug-N-Sluggo by Certis, Monterey Sluggo Plus), Sulfur (Bug-Geta by Ortho), Metaldehyde (Southern Ag Snail & Slug Bait, Deadline Mini-Pellets), Nemaslug (endopathogenic nematode formulation)

^ High tunnel pest exclusion using 50 percent woven shade cloth on side- and end-walls can exclude large moths and reduce caterpillar pressure. Details at https://southern.sare.org/resources/high-tunnel-pest-exclusion-system/

Organic insecticides are currently available as single active ingredient (AI) or in the form of premixes, such as Azera (neem + pyrethrins), Leap (BT + methyl salicylate), and BotaniGard Maxx (pyrethrins + *Beauveria bassiana*).

* Relatively low-cost, temporary insect netting products include the Super-Light Insect Barrier (GardensAlive.com), AgroFabric Pro (https://www.7springsfarm.com) or other lightweight exclusion material are suitable for early season pest prevention.

Disease Control for Commercial Vegetables

Caution: At the time these table were prepared; the entries were believed to be useful and accurate. However, labels change rapidly, and errors are possi- ble, so the user must follow all directions on the product labels. Federal toler- ances for fungicides may be canceled or changed at any time. Information in the following tables must be used in the context of an integrated disease management program. Many diseases are successfully managed by combined strategies—using resistant varieties, crop rotation, deep-turn plowing, sani- tation, seed treatments, cultural practices, and fungicides. Always use top quality seed and plants obtained from reliable sources. Seeds are ordinarily treated by commercial producers for control of decay and damping-off diseases. Preplant fumigation of soils, nematode control chemicals, and greenhouse disease control products are provided in separate tables following the crop tables. The efficacy tables will help you select the appropriate disease control materials for some vegetable crops. These tables are located at the end of each crop table.

Rates: Some foliar rates are based on mixing a specified amount of product in 100 gal of water and applying the finished spray for complete cover- age of foliage just to the point of run off with high-pressure (over 250 psi) drop nozzle sprayers. The actual amount of product and water applied per acre will vary de- pending on plant size and row spacing. Typically, 25 to 75 gallons (gal.) per acre of finished spray are used. Concentrate spray (air blast, aircraft, etc.) rates are based on the amount of product per acre.

Caution: With concentrate sprays, it is easy to apply too much product. Some fungicides are adversely affected by pH of water; adjust pH of water if specified on label. Some fungicides will cause damage to the plant if applied at temperatures above 90°F. Do not feed treated foliage to livestock unless allowed by the label. Do not reenter fields until sprays have dried; some fungi- cides may have a reentry requirement of one to several days. Read the label. Do not exceed maximum number of applications on the label. Do not exceed the maximum limit of fungicide per acre per application or per year as stated on the label. See label for rotational crops. In all cases, follow directions on the label. The label is the law.

THE FOLLOWING ONLINE DATABASES PROVIDE CURRENT PRODUCT LABELS AND OTHER RELEVANT INFORMATION:

Database	Web Address
Crop Data Management Systems	http://www.cdms.net/Label-Database
EPA Pesticide Product and Label System	https://ordspub.epa.gov/ords/pesticides/f?p=PPLS:1
Greenbook Data Solutions	https://www.greenbook.net/
Kelly Registration Systems[1]	http://www.kellysolutions.com
National Pesticide Information Retrieval System [2]	https://www.npirs.org/state
TELUS Agronomy Label Database	https://home.agrian.com/labelcenter/results.cfm

[1] Available for AR, FL, GA, LA, MS, NC, OK, SC, TX, and VA in the southeastern US.
[2] Available for AL, AR, FL, KY, LA, TX, and VA in the southeastern US.

PESTICIDES AND THE ENDANGERED SPECIES ACT: WHAT YOU NEED TO KNOW

The following description has been endorsed by the Weed Science Society of America, Entomological Society of America, and American Phytopathological Society.

1. **What is the Endangered Species Act (ESA)?**

   The Endangered Species Act is a long-standing federal law, first passed in 1973, which requires government agencies to ensure any actions they take do not jeopardize a species that has been federally listed as endangered or threatened. When an agency has a proposed action that might affect a listed species or its habitat, they consult with one or both of the agencies that helps enforce the ESA, the U.S. Fish and Wildlife Services or the National Marine Fisheries Service (this is known as "a consultation" with "the Services"). The Services then may recommend changes to the project or action to protect listed species or habitats.

2. **How does the ESA affect pesticide use?**

   The Environmental Protection Agency (EPA) Office of Pesticide Programs (OPP) is the federal agency that regulates pesticide use. Because the use of pesticides can affect animals and plants (or their habitat), pesticide registrations are considered "actions" that would trigger an endangered species consultation.

3. **Why am I hearing about the ESA and pesticide use now?**

   Due to the complex nature of the process, the EPA has not fully completed the required endangered species consultations with the Services for pesticide registrations in the past, which has left many of those pesticides vulnerable to lawsuits. Courts have annulled pesticide registrations which has led to their removal from market. To make pesticide registrations more secure from litigation, ultimately all pesticide registrations will comply with the Endangered Species Act (https://www.epa.gov/endangered-species).

4. **How will this affect the pesticide I use today?**

   Many pesticide labels will likely have changes that could include:
   - Requirement to check the EPA's Bulletins Live! Two website and follow current ESA restrictions for the pesticide product in the bulletin (https://www.epa.gov/endangered-species/bulletins-live-two-view-bulletins)
   - Measures to reduce spray drift
   - Measures to reduce runoff/erosion
   - Other measures to reduce pesticide exposure to listed species and their habitat.

   In short, farmers and applicators should expect to see some new application requirements on their pesticide labels. But there is no need to panic. To date, no pesticide has ever been fully removed from the market based solely on endangered species risks, and that remains an unlikely scenario in the future.

5. **Why does complying with the ESA matter?**

   By starting to fully comply with the ESA, **EPA anticipates that this will give farmers and applicators more stable, reliable access to the pesticides they need.** Furthermore, the ESA has been successful at bringing back some species Americans care about – such as the bald eagle or the Eggert sunflower – and restoring them to healthy populations, which has benefited the natural and cultivated ecosystems that agriculture (and society) rely on.

TABLE 3-1. DISEASE CONTROL PRODUCTS FOR ASPARAGUS

E. Sikora, Plant Pathologist, Auburn University

Disease/Material	FRAC Code	Rate of Material Formulation	Minimum Days Harv.	Minimum Days Reentry	Method, Schedule, and Remarks
GRAY MOLD					
fenhexamid (Elevate)	17	1.5 lb/acre	180	0.5	Apply at fern stage only. Make up to four applications. Repeat at 7- to 14-day intervals if conditions favor disease development.
PHYTOPHTHORA CROWN ROT, SPEAR SLIME					
mefenoxam (various)	4	1 pt/acre	1	2	Apply over beds after seeding or covering crowns, 30- to 60-days before first cutting, and just before harvest.
fosetyl-AL (Aliette)	33	5 lb/acre	110	0.5	Apply ALIETTE WDG once per season. ALIETTE WDG should be applied to fully expanded asparagus ferns. Do not apply to ferns that are beginning to senesce. Thorough coverage is required.
Oxathiapiprolin (Orondis Gold 200)	48	4.8 to 9.6 fl oz/acre	0	4 hr	Can be applied to new planting or established plantings. See label for use directions. Make no more then 2 sequential applications of Orondis Gold 200 before rotating to a fungicide with a different mode of action.
RUST					
myclobutanil (various)	3	5 oz/acre	180	1	Begin applications to developing ferns after harvest has taken place. Repeat on a schedule not to exceed 14 days. Do not apply to harvestable spears.
sulfur (various)	M02	See label	0	1	See label for applications directions.
tebuconazole (various)	3	4 to 6 fl oz/acre	180	0.5	Apply to developing ferns at first sign of rust and repeat on a 14-day interval, no more than 3 applications per season.
copper oxychloride/ hydroxide (Badge SC)	M01	1 to 2.5 pints/acre	0	48 hr	Recommended for tank mixture with other registered products. For disease suppression only. Addition of spread/sticker is recommended.
RUST, CERCOSPORA LEAF SPOT					
chlorothalonil (various)	M05	2 to 4 lb/acre	190	0.5	Repeat applications at 14- to 28-day intervals depending on disease pressure. Do not apply more than 12 pints/acre during each growing season.
mancozeb (various)	M	See label	180	1	Apply to ferns after harvest; spray at first appearance of disease at 7-to 10-day intervals. Do not exceed 8 lb product per acre per crop.
mancozeb + azoxystrobin (Dexter MAX)	M03+11	2 to 2.2 lb/acre	180	1	Apply only on ferns after spears have been harvested. Applications should begin prior to disease development. Do not apply more than 8.5 lbs. of product per acre per season.
PURPLE SPOT					
azoxystrobin (various)	11	6 to 15.5 fl oz/acre	100	4 hr	Do not apply more than 1 foliar application of Quadris (or other group 11 fungicide) before alternating with a fungicide with a different mode of action.
chlorothalonil (various)	M05	2 to 4 lb/acre	190	0.5	Repeat applications at 14- to 28-day intervals depending on disease pressure. Do not apply more than 12 pints/acre during each growing season.
mancozeb + azoxystrobin (Dexter MAX)	M03+11	2 to 2.2 lb/acre	180	1	Apply only on ferns after spears have been harvested. Applications should begin prior to disease development. Do not apply more than 8.5 lbs. of product per acre per season.
trifloxystrobin (Flint Extra 500 SC)	11	3 to 3.8 oz/acre	180	12 hr	Apply on a 14-day interval as needed. Make applications to the fern stage only. Mow down the asparagus ferns (or allow the ferns to senesce) between the last fungicide application and harvest.

TABLE 3-2. IMPORTANCE OF ALTERNATIVE MANAGEMENT PRACTICES FOR DISEASE CONTROL IN ASPARAGUS

E. Sikora, Plant Pathologist, Auburn University; A. Keinath, Plant Pathologist, Clemson University

Scale: "E" excellent; "G" good; "F" fair; "P" poor; "NC" no control; "ND" no data.

Strategy	Rust	Cercospora blight	Stemphylium blight	Fusarium root rot	Phytophthora crown/ spear rot
Avoid overhead irrigation	F	F	F	NC	NC
Crop rotation (4 years or more)	NC	NC	NC	G	P
Clip and bury infected ferns	G	G	G	NC	NC
Destroy infected ferns	E	E	E	NC	NC
Encourage air movtement/wider row spacing	G	P	G	NC	NC
Plant in well-drained soil	NC	NC	NC	F	G
Destroy volunteer asparagus	G	NC	NC	NC	NC
Pathogen-free planting material	NC	NC	NC	E	E
Resistant/tolerant cultivars	G	G	NC	E	NC

TABLE 3-3. DISEASE CONTROL PRODUCTS FOR BASIL

L. Quesada-Ocampo, Plant Pathologist, North Carolina State University (LAST UPDATED 2025)

Disease/Material	FRAC Code	Rate of Material Formulation	Minimum Days Harv.	Minimum Days Reentry	Method, Schedule, and Remarks
DAMPING OFF (*PYTHIUM* SPP.)					
mefenoxam (Ridomil Gold SL)	4	1.0 to 2.0 pt/acre	21	2	Limit of 2 soil applications per season. Basal direct spray at 28 days after planting or after first cutting.
LEAF SPOTS, FUNGAL (*BOTRYTIS, ALTERNARIA, ANTHRACNOSE, FUSARIUM*), POWDERY MILDEW					
azoxystrobin (Aframe, Quadris)	11	6.0 to 15.5 fl oz/acre	3	4 hr	Apply as needed on a 7-day schedule following resistant management guidelines. Use a minimum of 30-gal water per acre. Do not apply more than 2 sequential applications of this active ingredient or a similar group 11.
Bacillus amyloliquefaciens strain D747 (Double Nickel LC)	BM02	0.25 to 3.0 lb/acre	0	4 hr	Apply using sufficient water to achieve full coverage. Repeat application every 7 to 10 days or as needed.
Bacillus subtilis strain IAB/BD03 (Aviv)	BM02	10 to 30 fl oz/ 100-gal water	0	4 hr	Apply with a minimum of 5 gallons of water per acre. Reapply every 7 to 14 days.
cerevisane (Romeo)	BM02	0.23 to 0.68 lb/acre	0	4 hr	Spray interval of 7 to 10 days. In high-pressure scenarios program with other fungicides.
cinnamon oil (Cinnerate)	BM01	13 to 32 fl oz/100-gal	0	0	10 to 100 gallons per acre. Repeat as needed (5-day interval).
copper octanoate (Cueva)	M01	16.8 gal/acre	0	4 hr	Reapply every 10 to 14 days as needed. Do not apply more than 0.53 lb/ Cueva per acre.
cyprodinil (Vango)	9	5.5 to 7 oz/acre	7	12 hr	Apply at 7-to-10 days interval. After 2 applications, alternate with another fungicide with a different mode of action.
cyprodinil + fludioxonil (Switch 62.5WG)	9+12	11 to 14 oz/acre	7	0.5	Limit of 56 fl oz per acre per season. Make no more than two consecutive applications before rotating to another effective fungicide with a different mode of action. Apply in a minimum spray volume of 30 gal/acre to obtain thorough coverage.
extract of *Swinglea glutinosa* (EcoSwing)	BM01	1.5 to 2 pt/acre	0	4 hr	Sufficient water recommended for foliar application, 5 to 20 gallons of water per acre.
fluopyram (Luna Privilege)	7	4.0 to 6.84 fl oz/acre	3	0.5	Limit of 13.7 fl oz per acre per season. Apply as needed on a 7- to 10-day interval. When disease pressure is severe, use the higher rates and/or shorter intervals.
fluopyram + trifloxystrobin (Luna Sensation)	7+11	5.0 to 7.6 fl oz/acre	7	0.5	Limit of 15.3 fl oz per acre per season. Apply as needed on a 7- to 10-day interval. When disease pressure is severe, use the higher rates and/or shorter intervals.
garlic oil (Brandt Organics Aleo)	BM01	3 to 12 fl oz/minimum of 20 gals per acre	0	0	The use of adjuvants is highly recommended for improving performance. Repeat as need (5-day intervals).
hydrogen peroxide + peroxyacetic acid (OxiDate 2.0)	NC	1:500 to 1:1000	0	1 hr	Apply 3 to 20 gallons of spray per acre. For best results apply at first sign of disease. Use 12.8 fl oz to 25.65 fl oz of product per 100 gallons of water. Spray 3 to 5 days intervals until control is achieved.
polyoxin D zinc salt (Affirm WDG)	19	6.2 oz/acre	0	4 hr	Apply as a full coverage foliar spray with sufficient water (50 to 300 gal per acre).
potassium bicarbonate (Carb-O-Nator)	BM01	2.5 to 5 lb/ 100 gallons of water	0	4 hr	Minimum application of 20 gallons per acre. Repeat application 10 to 14 days interval or as needs. Do not exceed the 5 pounds per 100 gallons of water.
potassium phosphite (Rampart)	P07	1 to 3 qt/ 20 to a minimum 20-gal water/acre	0	4 hr	Apply 2 to 3 weeks intervals. Do not apply at intervals less than 3 days. To avoid undesirable copper phytotoxicity, do not apply product to already treated plants with copper-based products at less than 20-days intervals.
sodium chloride (Amicos KPM)	NC	2.5 to 3.0 lb/acre	0	0	Repeat application at 10- to 14-days intervals or as needed. Do not exceed a mix rate of 5 pounds per 100 gallons of water.
Streptomyces lydicus WYEC 108 (Actinovate AG)	BM02	3 to 12 oz/acre	0	4 hr	Re-apply every 7 to 10 days. For best results use with a spreader-sticker.
DOWNY MILDEW (*PERONOSPORA BELBAHRII*)					
Bacillus amyloliquefaciens strain D747 (Double Nickel LC)	BM02	0.25 to 3.0 lb/acre	0	4 hr	Apply using sufficient water to achieve full coverage. Repeat application every 7 to 10 days or as needed.
Bacillus subtilis strain IAB/BD03 (Aviv)	BM02	10 to 30 fl oz/ 100-gal water	0	4 hr	Apply with a minimum of 5 gallons of water per acre. Reapply every 7 to 14 days.
cerevisane (Romeo)	BM02	0.23 to 0.68 lb/acre	0	4 hr	Spray interval of 7 to 10 days. In high-pressure scenarios program with other fungicides.
cyazofamid (Ranman 400SC)	21	2.75 to 3 fl oz/acre	0	0.5	Limit of 27 fl oz per acre per season. Alternate with a fungicide with a different mode of action. May be applied through sprinkler irrigation system. Can be applied in a greenhouse.
extract of *Swinglea glutinosa* (EcoSwing)	BM01	1.5 to 2 pts/acre	0	4 hr	Sufficient water recommended for foliar application, 5 to 20 gallons of water per acre.

TABLE 3-3. DISEASE CONTROL PRODUCTS FOR BASIL (cont'd)

L. Quesada-Ocampo, Plant Pathologist, North Carolina State University (LAST UPDATED 2025)

Disease/Material	FRAC Code	Rate of Material Formulation	Minimum Days Harv.	Minimum Days Reentry	Method, Schedule, and Remarks
DOWNY MILDEW (*PERONOSPORA BELBAHRII*) (cont'd)					
fenamidone (Reason 500 SC)	11	6.0 lbs/acre	2	12 hr	Maximum application per year 24.0 fl oz/acre. Minimum of 7-day intervals.
fluopicolide (Presidio)	43	4 fl oz/acre	1	12 hr	Limit of 12 fl oz per acre per year. Make no more than two sequential applications. Alternate with a fungicide with a different mode of action.
mandipropamid (Revus)	40	8 fl oz/acre	1	4 hr	Limit of 32 fl oz per acre per season. Make no more than two consecutive applications before rotating to another effective fungicide with a different mode of action.
phosphorous acid (Confine Extra, K-Phite)	P07	1 to 3 qt/20-to 100-gal water/acre	0	4 hr	Do not apply at less than 3-day intervals. Minimum 10 gallons of water/acre.
oxathiapiprolin (Segovis)	49	1.1 to 2.4 fl oz/acre	0	4 hr	Apply in at least 15 gallons of water per acre. Use higher application rate when disease is present. Minimum application intervals of 5 days. No more than 2 application.
oxathiapiprolin + mandipropamid (Orondis Ultra)	49 + 40	5.5 to 8.0 fl oz/acre	1	4 hr	Begin foliar application prior disease and continue a 7- to 10-day interval.
potassium bicarbonate (Carb-O-Nator)	BM01	2.5 to 5 lb/100 gallons of water	0	4 hr	Minimum application of 20 gallons per acre. Repeat application to 7– to 14-days interval or as needs. Do not exceed the 5 pounds per 100 gallons of water.
potassium phosphite (Rampart)	P07	1 to 3 qt/20-to minimum 20-gal water/acre	0	4 hr	Apply 2 to 3 weeks intervals. Do not apply at intervals less than 3 days. To avoid undesirable copper phytotoxicity, do not apply product to already treated plants with copper-based products at less than 20-days intervals.
FUSARIUM WILT, PYTHIUM, VERTICILLIUM AND RHIZOCTONIA ROOT ROTS					
azoxystrobin + extract of *Reynoutria sachalinensis* (Azterknot)	11 + P5	7.4 to 18.4 fl oz/acre	0	4 hr	Apply on a 7 to 14 days interval throughout the season following the resistance management guidelines. Minimum of 30-gallon pr acre. Ground application only.
Bacillus amyloliquefaciens strain D747 (Double Nickel LC)	BM02	0.125 to 1 lbs/acre	0	4hr	For soil application, spray directly onto the soil surface or lower plant parts.
garlic oil (Brandt Organics Aleo)	BM01	3 to 12 fl oz/ minimum of 20 gallons per acre	0	0	The use of adjuvants is highly recommended for improving performance. Repeat as need (5-day intervals).
hydrogen peroxide + peroxyacetic acid (OxiDate 2.0)	NC	1:500 to 1:1000	0	1 hr	Apply previous planting for better control. Post planting treat soil as needed.
phosphorous acid (Confine Extra, K-Phite)	P07	1 to 3 qt/20-to 100-gal water/acre	0	4 hr	Do not apply at less than 3-day intervals.
polyoxin D zinc salt (Affirm WDG)	19	8.0 oz/acre	0	4 hr	Apply as a soil drench every 14 to 28 days.
potassium phosphite (Rampart)	P07	1 to 3 qt/20 to a minimum 20-gal water/acre	0	4 hr	Apply 2 to 3 weeks intervals. Do not apply at intervals less than 3 days.
Streptomyces lydicus WYEC 108 (Actinovate AG)	BM02	3 to 12 oz/acre	0	4 hr	For best results apply product with damp soil.

TABLE 3-4. DISEASE CONTROL PRODUCTS FOR BEAN

D. Higgins, Plant Pathologist, Virgina Tech

Disease/Material	FRAC Code	Rate of Material Formulation	Minimum Days Harv.	Minimum Days Reentry	Method, Schedule, and Remarks
BEAN, SNAP					
ANTHRACNOSE, BOTRYTIS, SCLEROTINIA					
azoxystrobin (various)	11	6.2 to 15.4 fl oz	0	4 hr	For anthracnose only. Do not apply more than three sequential applications.
boscalid (Endura 70WG)	7	8 to 11 oz	7	0.5	Many other dried and succulent beans on label.
chlorothalonil (various)	M05	2.7 lb/acre	7	2	Spray first appearance, 11 lb limit per acre per crop, 7-day intervals. **Not for Sclerotinia control.**
thiophanate-methyl (various)	1	1 to 2 lb/acre	14	1	Spray at 25% bloom; repeat at full bloom. Do not exceed 4 lb product per season.
fluazinam (various)	29	8 to 13.6 fl oz/acre	14	3	PHI varies by crop; see label restrictions.
fluxapyroxad + pyraclostrobin (Priaxor)	7+11	4.0 to 8.0 oz	7	0.5	Begin prior to disease development and continue on a 7- to 14-day spray sched- ule. See label for specific directions for edible-podded legumes and dried-shelled legumes.
ASCOCHYTA BLIGHT, BOTRYTIS GRAY MOLD, WHITE MOLD (*SCLEROTINIA*)					
boscalid (Endura 70 WG)	7	8 to 11 oz	7	0.5	
penthiopyrad (Fontelis)	7	14 to 30 fl oz/acre	0	0.5	Begin sprays prior to disease development. Also registered for Alternaria, Anthracnose, and rust.
ALTERNARIA, ANTHRACNOSE, ASCOCHYTA, RUST, SOUTHERN BLIGHT, WEB BLIGHT					
azoxystrobin + propiconazole (Quilt Xcel; Aframe Plus)	11+3	10.5 to 14 oz/acre	7	0.5	Apply when conditions are conducive for disease. Up to three applications may be made on 7- to 14-day intervals.
azoxystrobin + propiconazole + pydiflumetofen (Miravis Neo)	11+3+7	13.7 fl oz	14	0.5	Min. application interval is 14 days. Southern blight not listed on label. Also provides suppression for white mold.
propiconazole (Tilt)	3	4 fl oz	7	1	Up to three applications can be made on a 7-14 day interval
pydiflumetofen + fludioxonil (Miravis Prime)	7+12	9.2 fl oz	14	0.5	Labeled for Alternaria and Ascochyta. Begin applications prior to disease devel- opment.
BOTRYTIS GRAY MOLD, WHITE MOLD (*SCLEROTINIA*)					
azoxystrobin + tebuconazole + thiophanate-methyl (Trevo Packed)	11+3+1	25 to 30 fl oz	14	1	Min. application interval is 14 days. See label for specific directions for edible-podded legumes.
iprodione (various)	12	1.5 to 2 pt	See label	1	Apply as foliar spray and again 5- to 7-days later or up to peak bloom if condi- tions are favorable for disease. Do not use on cowpeas
isofetamid (Kenja 400SC)	7	17 fl oz	See label	0.5	Apply at 10 to 30% bloom and 7- to 14-days later, if needed. Do not allow livestock to graze in treated area.
cyprodinil + fludioxonil (various)	9+12	11 to 14 oz	7	0.5	Begin applications prior to onset of disease and repeat on 7-day intervals if condition remain favorable for disease development.
fludioxonil (Cannonball WG)	12	7 oz	7	0.5	For white mold control, make the first application at 10-20% bloom. For gray mold begin applications prior to the onset of disease.
polyoxin D (Oso)	19	6.5 to 13 fl oz	0	4 hr	White mold not listed on label. Low rate may be used as preventative application. Begin high-rate applications at first sign of disease, repeat on 7- to 14-day intervals if conditions remain favorable for disease.
pydiflumetofen + fludioxonil (Miravis Prime)	7+12	10.3 to 13.4 fl oz	14	0.5	For white mold suppression, make the first application at 10% bloom. For gray mold control only use highest labeled rate and begin applications prior to the onset of disease.
BACTERIAL BLIGHTS					
fixed copper (various)	M01	See labels	1	1	Spray first appearance, 10-day intervals.
POWDERY MILDEW					
sulfur (various)	M02	See labels	0	1	Spray at first appearance, 10- to 14-day intervals. Avoid days over 90°F.
DOWNY MILDEW, COTTONY LEAK					
mefenoxam + copper hydroxide (Ridomil Gold Copper)	4+M01	5 lb/2.5 acres	7	2	For succulent shelled beans. Begin foliar applications at onset of disease and con- tinue on a 7-day interval. Do not make more than 2 applications per season. Do not use an adjuvant.

TABLE 3-4. DISEASE CONTROL PRODUCTS FOR BEAN (cont'd)

D. Higgins, Plant Pathologist, Virgina Tech

Disease/Material	FRAC Code	Rate of Material Formulation	Minimum Days Harv.	Minimum Days Reentry	Method, Schedule, and Remarks
BEAN, SNAP (cont'd)					
DOWNY MILDEW, COTTONY LEAK (cont'd)					
mandipropamid (Revus)	40	8 fl oz	1	4 hr	Not for cottony leak. Use of a NIS or crop oil concentrate or blend is recommend- ed
COTTONY LEAK (*PYTHIUM* SPP.)					
fenamidone (Reason 500SC)	11	5.5 to 8.2 fl oz	3	0.5	Begin applications when conditions become favorable for disease development. Do not make more than one application before alternating to a product with a different mode of action.
COTTONY LEAK, DOWNY MILDEW, PHYTOPHTHORA BLIGHT					
cyazofamid (Ranman)	21	2.75 fl oz	0	0.5	Read label for specific directions for each disease as well as use restrictions.
phosphorous acid (Fungi-Phite)	P07	1 to 2 quarts	0	4 hr	Rate listed for is for foliar application.
fluopicolide (Presidio)	40	3 to 4 fl oz	2	0.5	Not for cottony leak or downy mildew; Phytophthora blight only.
RHIZOCTONIA ROOT ROT					
azoxystrobin (various)	11	0.4 to 0.8 fl oz/ 1,000 row feet	—	4 hr	Apply in-furrow or banded applications shortly after plant emergence.
myclobutanil (various)	3	4 to 5 oz/acre	0	1	**For Rhizoctonia only.**
RUST (*UROMYCES*)					
azoxystrobin (various)	11	6.2 to 15.4 fl oz/acre	0	4 hr	Make no more than three sequential applications.
pyraclostrobin (various)	11	6.0 to 9.0 fl oz		0.5	Make no more than two sequential applications.
myclobutanil (various)	3	4 to 5 oz/acre	0	1	Spray at first appearance.
sulfur (various)	M02	See label	0	1	Spray at 7- to 10-day intervals.
tebuconazole (various)	3	4 to 6 fl oz/acre	7	0.5	Apply before disease appears when conditions favor rust development and repeat at 14-day intervals: maximum 24 fl oz per season.
WHITE MOLD (*SCLEROTINIA*)					
dicloran (Botran 5F)	14	1.3 qt/acre	2	0.5	Begin applications when disease is anticipated.
BEAN, LIMA					
ALTERNARIA, ANTHRACNOSE, ASCOCHYTA, BEAN RUST, SOUTHERN BLIGHT, WEB BLIGHT (RHIZOCTONIA)					
azoxystrobin + propiconazole (Quilt Xcel, Aframe Plus)	11+3	10.5 to 14 fl oz	7	0.5	Apply when conditions are conducive for disease. Up to three applications may be made on a 7- to 14-day interval.
azoxystrobin + propiconazole + pydiflumetofen (Miravis Neo)	11+3+7	13.7 fl oz	14	0.5	Dry lima bean only. Southern blight not listed on label. Also provides suppression for white mold. Min. application interval is 14 days.
propiconazole (Tilt)	3	4 fl oz	7	1	Up to three applications can be made on a 7- to 14 day interval.
pydiflumetofen + fludioxonil (Miravis Prime)	7+12	9.2 fl oz	14	0.5	Labeled for Alternaria and Ascochyta. Begin applications prior to disease development.
ALTERNARIA, ASCOCHYTA, CERCOSPORA, MYCOSPHAERELLA, POWDERY MILDEW, RUST					
pydiflumetofen + difenoconazole (Miravis Top)	3+7	13.7 fl oz/acre	14	0.5	For dried shelled beans only. Begin applications prior to disease development. Continue applications on a 14-day interval. Do not make more than 4 applications per season.
BOTRYTIS, LEAF SPOTS, SCLEROTINIA					
azoxystrobin (various)	11	6.2 to 15.4 fl oz/acre	0	4 hr	Leaf spots only; do not make more than three sequential applications.
azoxystrobin + tebuconazole + thiophanate-methyl (Trevo Packed)	11+3+1	25 to 30 fl oz	14	1 3 (dry)	Min. application interval is 14 days. See label for specific directions for edible-podded legumes. 3-day REI for dry beans.

TABLE 3-4. DISEASE CONTROL PRODUCTS FOR BEAN (cont'd)

D. Higgins, Plant Pathologist, Virgina Tech

Disease/Material	FRAC Code	Rate of Material Formulation	Minimum Days Harv.	Minimum Days Reentry	Method, Schedule, and Remarks
BEAN, LIMA (cont'd)					
BOTRYTIS, LEAF SPOTS, SCLEROTINIA (cont'd)					
boscalid (Endura 70WG)	7	8 to 11 oz	7	0.5	Apply at beginning of flowering or prior to onset of disease. Apply a second time at full bloom if conditions are favorable for disease.
cyprodinil + fludioxonil (various)	9+12	11 to 14 oz	7	0.5	Begin applications prior to onset of disease and repeat on 7-day intervals if condition remain favorable for disease development.
iprodione (various)	2	1.5 to 2 lb/acre	See label	1	Apply as foliar spray and again 5- to 7-days later or up to peak bloom. If conditions are favorable for disease. Do not use on cowpeas.
isofetamid (Kenja 400SC)	7	17 fl oz	See label		Apply at 10 to 30% bloom and 7- to 14-days later, if needed. Do not allow livestock to graze in treated area.
fluazinam (various)	29	8 to 13.6 fl oz/acre	30	3	PHI varies by crop; see label restrictions.
fludioxonil (Cannonball WG)	12	7 oz	7	0.5	For white mold control, make the first application at 10-20% bloom. For gray mold begin applications prior to the onset of disease.
fluopyram + prothioconazole (Propulse)	7+3	10.0 to 13.6 fl oz	7	0.5	Use for dry lima bean only. Rate listed is for white mold. Also labeled for Gray mold, Ascochyta blight, Mycosphaerella blight and anthracnose at a rate of 8.0 to 13.6 fl oz.
fluxapyroxad + pyraclostrobin (Priaxor)	7+11	4.0 to 8.0 fl oz	21	0.5	Begin prior to disease development and continue on a 7- to 14-day spray sched- ule. See label for specific directions for edible-podded legumes and dried-shelled legumes.
penthiopyrad (Fontelis)	7	14 to 30 fl oz/acre	0	0.5	Begin sprays prior to disease development.
polyoxin D (Oso)	19	6.5 to 13 fl oz	0	4 hr	White mold not listed on label. Low rate may be used preventative applications. Be- gin high-rate applications at first sign of disease and repeat on 7- to 14-day intervals if conditions remain favorable for disease development.
pydiflumetofen + fludioxonil (Miravis Prime)	7+12	10.3 to 13.4 fl oz	14	0.5	For white mold suppression, make the first application at 10% bloom. For gray mold control only use highest labeled rate and begin applications prior to the onset of disease.
pyraclostrobin (various)	11	6.0 to 9.0 fl oz	21	0.5	Make no more than two sequential applications.
thiophanate-methyl (various)	12	7 oz	7	0.5	Begin before disease develops and continue on 7-day interval until conditions no longer favor disease development. Do not apply more than 28 oz/acre. Do not apply on cowpeas.
COTTONY LEAK (*PYTHIUM* SPP.)					
fenamidone (Reason 500SC)	11	5.5 to 8.2 fl oz	3	0.5	A spreader/sticker may be used to improve disease control. Minimum interval of 7-days between applications
COTTONY LEAK, DOWNY MILDEW PHYTOPHTHORA BLIGHT					
cyazofamid (Ranman)	21	2.75 fl oz	0	0.5	Read label for specific directions for each disease as well as use restrictions.
phosphorous acid (Fungi-Phite)	P07	1 to 2 quarts	0	4 hr	Rate listed is for foliar application.
DAMPING-OFF, PYTHIUM, RHIZOCTONIA					
azoxystrobin (various)	11	0.4 to 0.8 fl oz/ 1,000 row feet	—	4 hr	**For Rhizoctonia only**. Make in-furrow or banded applications shortly after plant emergence.
azoxystrobin + mefenoxam (Uniform)	11+4	0.34 fl oz/ 1,000 row ft	—	—	Limit of one application per season. In-furrow spray. See label directions.
mefenoxam (various)	4	0.5 to 2 pt/trt acre	—	2	**For Pythium only.** Soil incorporate. See label for row rates. Use proportionally less for banded rates.
fluopyram + prothioconazole (Propulse)	7+3	6.0 to 10.0 fl oz	7	0.5	For Rhizoctonia and Fusarium: in-furrow spray during planting directed on or below seed. Dry lima bean only.
BOTRYTIS, FUSARIUM, PHOMOPSIS, RHIZOCTONIA					
penflufen + trifloxystrobin (Evergol Prime)	7+11	See label	—	—	For seed rot and damping off caused by Rhizoctonia, Fusarium, Phomopsis, or Botrytis. **Seed treatment only.**
DOWNY MILDEW					
mefenoxam + copper hydroxide (Ridomil Gold Copper)	4+M01	5 lb/2.5 acres	3	2	For succulent shelled beans. Begin foliar applications at onset of disease and contin- ue on a 7-day interval. Do not exceed 4 applications per season.

TABLE 3-5. IMPORTANCE OF ALTERNATIVE MANAGEMENT PRACTICES FOR DISEASE CONTROL IN BEANS

D. Higgins, Plant Pathologist, Virginia Tech

Scale: E, excellent; G, good; F, fair; P, poor; NC, no control; ND, no data.

	Anthracnose	Ashy stem blight	Botrytis gray mold	Cercospora	Common bacterial blight and halo blight	Fusarium root rot	Mosaic viruses	Powdery mildew	Pythium damping-off	Rhizoctonia root rot	Root knot nematode	Rust (more on pole beans)	Southern blight (*Sclerotium rolfsii*)	White mold (*Sclerotinia*)
Avoid field operations when leaves are wet	E	NC	E	F	E	NC	NC	NC	NC	NC	NC	E	NC	NC
Avoid overhead irrigation	E	NC	E	E	E	NC	NC	NC	P	NC	NC	E	NC	G
Change planting date	F	F	NC	P	F	G	F	P	E	E	P	G (early)	NC	NC
Cover cropping with antagonist	NC	ND	NC	NC	NC	NC	NC	NC	NC	NC	G	NC	NC	NC
Crop rotation	G	P	F	F	G	F	P	P	F	F	G	NC	F	E
Deep plowing	E	F	E	P	E	F	NC	NC	F	F	F	NC	E	E
Destroy crop residue	E	F	E	F	E	NC	NC	NC	P	P	F	F	G	E
Encourage air movement	E	NC	E	F	E	NC	NC	E	P	NC	NC	F	NC	G
Increase between-plant spacing	P	NC	P	F	P	P	P	P	F	F	NC	P	F	G
Increase soil organic matter	NC	F	NC	NC	NC	F	NC	NC	NC	NC	F	NC	NC	NC
Insecticidal oils	NC	NC	NC	NC	NC	F	NC	NC	NC	NC	NC	NC	NC	NC
pH management	NC	NC	NC	NC	F	NC	NC	NC	NC	NC	NC	NC	NC	NC
Plant in well-drained soil	F	F	F	NC	F	E	NC	NC	E	E	NC	NC	P	F
Plant on raised beds	F	P	F	NC	F	E	NC	NC	E	E	NC	NC	P	F
Plastic mulch bed covers	NC	NC	NC	NC	NC	NC	NC	NC	NC	NC	NC	NC	NC	F
Postharvest temperature control	NC	NC	NC	NC	NC	NC	NC	NC	NC	NC	NC	NC	NC	E
Reflective mulch	NC	NC	NC	NC	NC	NC	G	NC	NC	NC	NC	NC	NC	P
Reduce mechanical injury	NC	NC	NC	NC	F	P	NC	NC	NC	NC	NC	NC	P	NC
Rogue diseased plants	NC	NC	P	NC	NC	NC	F	NC	NC	NC	NC	NC	P	F
Row covers	NC	NC	NC	NC	NC	NC	F	NC	NC	NC	NC	NC	NC	NC
Soil solarization	NC	NC	P	NC	NC	F	NC	NC	F	G	F	NC	F	G
Pathogen-free planting material	E	G	NC	F	E	NC	G	NC	NC	NC	NC	NC	NC	NC

TABLE 3-6. DISEASE CONTROL PRODUCTS FOR BROCCOLI, BRUSSEL SPROUT, CABBAGE, AND CAULIFLOWER (HEAD AND STEM BRASSICAS)

A. Keinath, Plant Pathologist, Clemson University

Disease/Material	FRAC Code	Rate of Material Formulation	Minimum Days Harv.	Minimum Days Reentry	Method, Schedule, and Remarks
ALTERNARIA LEAF SPOT (*ALTERNARIA BRASSICICOLA, A. JAPONICA*)					
azoxystrobin + difenoconazole (Quadris Top 2.72SC)	11+3	14 fl oz/acre	1	0.5	Apply prior to disease, but when conditions are favorable on 7- to 14-day schedule. Alternate to a non-FRAC 11 fungicide after 1 application. No more than 4 applications per season.
boscalid (Endura 70 EG)	7	6 to 9 oz/acre	0	0.5	Begin applications prior to disease development and continue on a 7- to 14-day interval. Make no more than 2 applications per season.
cyprodinil + difenoconazole (Inspire Super 2.82SC)	9+3	16 to 20 fl oz/acre	7	0.5	Begin applications prior to disease development and continue on a 7- to 10-day interval. Make no more than 2 sequential applications before rotating to another effective fungicide with a different mode of action. Do not exceed 80 fl oz per season.
cyprodinil + fludioxonil (Switch 62.5WG)	9+12	11 to 14 oz/acre	7	0.5	Apply when disease first appears and continue on a 7- to 10-day interval. Do not exceed 56 oz of product per acre per year.
flutriafol (Rhyme 2.08SC)	3	5 to 7 fl oz/acre	7	0.5	Limit of 4 applications per year. Labeled for Alternaria and Cercospora leaf spots.
flutriafol + azoxystrobin (Topguard EQ 4.29SC)	3+11	4 to 8 fl oz/acre	0	0.5	Limit of 4 applications per year.

TABLE 3-6. DISEASE CONTROL PRODUCTS FOR BROCCOLI, BRUSSEL SPROUT, CABBAGE, AND CAULIFLOWER (HEAD AND STEM BRASSICAS)

A. Keinath, Plant Pathologist, Clemson University

Disease/Material	FRAC Code	Rate of Material Formulation	Minimum Days Harv.	Minimum Days Reentry	Method, Schedule, and Remarks
ALTERNARIA LEAF SPOT (*ALTERNARIA BRASSICICOLA, A. JAPONICA*)					
fluxapyroxad + pyraclostrobin (Priaxor 500 SC)	7+11	6.0 to 8.2 fl oz/acre	3	0.5	Make no more than two sequential applications before alternating with fungicides that have a different mode of action. Maximum of 3 applications. **Do not apply to turnip greens or roots.**
pydiflumetofen + fludioxonil (Miravis Prime 3.34SC)	7+12	11.4 fl oz/acre	7	0.5	Make no more than 2 sequential applications of Miravis Prime or other Group 7 and 12 fungicides before rotating to another effective fungicide with a different mode of action. Maximum 3 applications per year.
triflumizole (Procure 480SC)	3	6 to 8 fl oz/acre	1	0.5	Apply when disease first appears and continue on a 14-day interval. Do not exceed 18 fl oz per season.
ALTERNARIA LEAF SPOT, GRAY MOLD (*BOTRYTIS CINEREA*)					
fluopyram + trifloxystrobin (Luna Sensation 500SC)	7+11	5.0 to 7.6 fl oz/acre	0	0.5	Do not apply more than 15.3 fl oz per acre per season. Make no more than 2 sequential applications before rotating to a fungicide not in Group 7 or 11.
penthiopyrad (Fontelis 1.67SC)	7	14 to 30 fl oz/acre	0	0.5	Do not exceed 72 fl oz of product per year. Make no more than 2 sequential applications per season before rotating to another effective product with a different mode of action. **See additional products listed below under Downy mildew and Alternaria leaf spot.**
BLACK LEG (*LEPTOSPHAERIA OLERICOLA*)					
iprodione (Rovral 4F)	2	2 lb/acre 2 pt/acre	0	—	Apply to base of plant at 2- to 4-leaf stage. A second application may be made up to the harvest date. Do not use as a soil drench. **For broccoli only.**
fluxapyroxad + pyraclostrobin (Priaxor 500SC)	7+11	6.0 to 8.2 fl oz/acre	3	0.5	Make no more than two sequential applications before alternating with fungicides that have a different mode of action. Maximum of 3 applications. **Do not apply to turnip greens or roots.**
BLACK ROT (*XANTHOMONAS CAMPESTRIS PV. CAMPESTRIS*), DOWNY MILDEW (*HYALOPERONOSPORA PARASITICA*)					
acibenzolar-*S*-methyl (Actigard 50WG)	P01	0.5 to 1 oz/acre	7	0.5	Begin applications 7- to 10-days after thinning, not to exceed 4 applications per a season.
fixed copper (various)	M01	See labels	0	1 to 2	Apply on 7- to 10-day intervals after transplanting or shortly after seeds have emerged. Some reddening on older broccoli leaves and flecking of cabbage wrapper leaves may occur. Check label carefully for recommended rates for each disease.
CERCOSPORA LEAF SPOT (*CERCOSPORA BRASSICICOLA*)					
fluopyram + trifloxystrobin (Luna Sensation 500 C)	7+11	7.6 fl oz/acre	0	0.5	Do not apply more than 15.3 fl oz per acre per season. Make no more than 2 sequential applications before rotating to a fungicide not in Group 7 or 11.
flutriafol (Rhyme 2.08SC)	3	5 to 7 fl oz/acre	7	0.5	Limit of 4 applications per year. Labeled for Alternaria and Cercospora leaf spots.
flutriafol + azoxystrobin (Topguard EQ 4.29SC)	3+11	4 to 8 fl oz/acre	0	0.5	Limit of 4 applications per year.
pydiflumetofen + fludioxonil (Miravis Prime 3.34SC)	7+12	11.4 fl oz/acre	7	0.5	Make no more than 2 sequential applications of Miravis Prime or other Group 7 and 12 fungicides before rotating to another effective fungicide with a different mode of action. Maximum 3 applications per year.
CLUBROOT (*PLASMODIOPHORA BRASSICAE*)					
cyazofamid (Ranman 34.5SC)	21	*Transplant:* 12.9 to 25.75 fl oz/ 100-gal water			Either apply immediately after transplanting with 1.7 fl oz of solution per trans-plant, or as a banded application with soil incorporation of 6 to 8 inches prior to transplanting. Do not apply more than 39.5 fl oz/acre/season or 6 (1 soil+ 5 foliar) applications/season. Do not make more than 3 consecutive applications without rotating to fungicide with different mode of action for 3 subsequent applications.
		Banded: 20 fl oz/acre	0.5	0	
fluazinam (Omega 500F)	29	*Transplant:* 6.45 fl oz/ 100-gal water			Apply either directly as a drench to transplants or as a banded application with soil incorporation of 6 to 8 inches prior to transplanting. Use of product can delay harvest and cause some stunting without adverse effects on final yields.
		Banded: 2.6 pts/acre	50	50	
DOWNY MILDEW (*HYALOPERONOSPORA PARASITICA*)					
ametoctradin + dimethomorph (Zampro 525SC)	40+45	14 fl oz/acre	0	0.5	Do not make more than 2 sequential applications before alternating to a fungicide with a different mode of action. Addition of an adjuvant may improve performance (see label for specifics).
cyazofamid (Ranman 400SC)	21	2.75 fl oz/acre	0	0.5	Begin applications on a 7- to 10-day schedule when disease first appears, or weather is conducive. Do not apply more than 39.5 fl oz/acre/season; or 6 (1 soil + 5 foliar) applications per season. Do not make more than 3 con-secutive applications without rotating to another fungicide with a different mode of action for 3 subsequent applications.

TABLE 3-6. DISEASE CONTROL PRODUCTS FOR BROCCOLI, BRUSSEL SPROUT, CABBAGE, AND CAULIFLOWER (HEAD AND STEM BRASSICAS) (cont'd)

A. Keinath, Plant Pathologist, Clemson University

Disease/Material	FRAC Code	Rate of Material Formulation	Minimum Days Harv.	Minimum Days Reentry	Method, Schedule, and Remarks
DOWNY MILDEW (*HYALOPERONOSPORA PARASITICA*) (cont'd)					
dimethomorph (Forum 4.16SC)	40	6 oz/acre	0	0.5	Alternate every application with a non-FRAC Group 40 fungicide. Limit of 3 applications per season.
fenamidone (Reason 500SC)	11	5.5 to 8.2 oz/acre	2	0.5	Begin applications as soon as conditions become favorable for disease development. Applications should be made on a 5- to 10-day interval. Do not make more than one application of Reason 500SC before alternating with a fungicide from a different resistance management group.
fluopicolide (Presidio 4SC)	43	3 to 4 fl oz/acre	2	0.5	Must be tank mixed with another fungicide with a different mode of action. No more than 2 sequential applications before rotating to another effective product of a different mode of action. Limited to 4 applications 12 fl oz/acre per season.
fosetyl-AL (Aliette)	P07	2 to 5 lb/acre	3	1	Apply when disease first appears; then repeat on 7- to 21-day intervals. Do not tank mix with copper fungicides. A maximum of 7 applications can be made per season.
mandipropamid (Revus 2.08SC)	40	8 fl oz/acre	1	0.5	Apply prior to disease development and continue throughout season at 7- to 10- day intervals; maximum 32 fl oz per season.
oxathiapiprolin + mandipropamid (Orondis Ultra 2.33SC)	49+40	5.5 to 8 fl oz/acre	0	4 hr	Apply prior to disease development at 10-day intervals. Make no more than 2 sequential applications before alternating with fungicides that have a different mode of action. Maximum of 4 applications per crop per year of all Orondis products.
potassium phosphite (various)	P07	2 to 4 pt/acre	0	4 hr	Apply when weather is foggy as a preventative. Do not apply to plants under water or temperature stress. Spray solution should have a pH greater than 5.5. Apply in at least 30-gal water per acre.
DOWNY MILDEW, ALTERNARIA LEAF SPOT					
azoxystrobin (Quadris 2.08F)	11	6.0 to 15.5 fl oz/acre	0	4 hr	Do not make more than 2 applications before alternating to a fungicide with a different mode of action. Do not apply more than 92.3 fl oz per acre per season.
chlorothalonil (various)	M05	See labels	7	2	Apply after transplanting, seedling emergence, or when conditions favor disease development. Repeat as needed on a 7- to 10-day interval.
fenamidone (Reason 500SC)	11	5.5 to 8.2 fl oz/acre	2	0.5	Begin applications on a 5- to 10-day schedule when disease first appears, or weather is conducive. Do not apply more than 24.6 fl oz/acre/season. Do not make more than 1 application without rotating to another fungicide with a different mode of action.
fluazinam (Omega 500F)	29	15.35 fl oz/acre	7	0.5	**Apply to cabbage only.** DO NOT apply more than 5.75 pints (6 applications) per acre per year.
mancozeb (various)	M03	1.6 to 2.1 lb/acre	10	1	Spray at first appearance of disease and continue on a 7- to 10-day interval. No more than 12.8 lbs/acre per season.
mefenoxam + chlorothalonil (Ridomil Gold/ Bravo 3.76SC)	4+M05	1.5 lb/acre	7	2	Begin applications when conditions favor disease but prior to symptoms. Under severe diseases pressure use additional fungicides between 14-day intervals. Do not make more than four applications per crop.
oxathiapiprolin + mandipropamid (Orondis Opti 3.76SC)	49+M05	1.75 to 2.5 fl oz/acre	7	0.5	Apply prior to disease development at 10-day intervals. Make no more than 2 sequential applications before alternating with fungicides that have a different mode of action. Maximum of 8 applications at the low rate or 4 applications at the high rate per year of Orondis Opti, if Ultra and Opti are both used, then the maximum is 4 applications each.
POWDERY MILDEW (*ERYSIPHE POLYGONI, E. CRUCIFERARUM*)					
azoxystrobin + difenoconazole (Quadris Top 2.72 SC)	11+3	14 fl oz/acre	1	0.5	Apply prior to disease, but when conditions are favorable, on 7-to 14-day schedule. Alternate to a non-FRAC 11 fungicide after 1 application. No more than 4 applications per season.
boscalid (Endura 70EG)	7	6 to 9 oz/acre	0	0.5	Begin applications prior to disease development and continue on a 7- to 14-day interval. Make no more than 2 applications per season; disease suppression only.
cyprodinil + difenoconazole (Inspire Super 2.82SC)	9+3	16 to 20 fl oz/acre	7	0.5	Begin applications prior to disease development and continue on a 7- to10-day interval. Make no more than 2 sequential applications before rotating to another effective fungicide with a different mode of action. Do not exceed 80 fl oz per season.
cyprodinil + fludioxonil (Switch 62.5WG)	9+12	10 to 12 oz/acre	7	0.5	Apply when disease first appears and continue on 7- to 10-day intervals. Do not exceed 56 oz of product per acre per year.
fluxapyroxad + pyraclostrobin (Priaxor 500SC)	7+11	6.0 to 8.2 fl oz/acre	3	0.5	Make no more than two sequential applications before alternating with fungicides that have a different mode of action. Maximum of 3 applications. Do not apply to turnip greens or roots.
fluopyram + trifloxystrobin (Luna Sensation 500SC)	7+11	5.0 to 7.6 fl oz/acre	0	0.5	Do not apply more than 15.3 fl oz per acre per season. Make no more than 2 sequential applications before rotating to a fungicide not in Group 7 or 11.
flutriafol (Rhyme 2.08SC)	3	5 to 7 fl oz/acre	7	0.5	Limit of 4 applications per year. Labeled for Alternaria and Cercospora leaf spots.

TABLE 3-6. DISEASE CONTROL PRODUCTS FOR BROCCOLI, BRUSSEL SPROUT, CABBAGE, AND CAULIFLOWER (HEAD AND STEM BRASSICAS) (cont'd)

A. Keinath, Plant Pathologist, Clemson University

Disease/Material	FRAC Code	Rate of Material Formulation	Minimum Days Harv.	Minimum Days Reentry	Method, Schedule, and Remarks
POWDERY MILDEW (*ERYSIPHE POLYGONI, E. CRUCIFERARUM*) (cont'd)					
flutriafol + azoxystrobin (Topguard EQ 4.29SC)	3+11	4 to 8 fl oz/acre	0	0.5	Limit of 4 applications per year.
penthiopyrad (Fontelis 1.67SC)	7	14 to 30 fl oz/acre	0	0.5	Do not exceed 72 fl oz of product per year. Make no more than 2 sequential applications per season before rotating to another effective product with a different mode of action.
pydiflumetofen + fludioxonil (Miravis Prime 3.34SC)	7+12	11.4 fl oz/acre	7	0.5	Make no more than 2 sequential applications of Miravis Prime or other Group 7 and 12 fungicides before rotating to another effective fungicide with a different mode of action. Maximum 3 applications per year.
sulfur (various)	M02	See labels	0	1	Apply when disease first appears; then repeat as needed on a 14-day interval. Avoid applying on days over 90°F. Also, for use on greens (collard, kale, and mustard), rutabaga, and turnip.
triflumizole (Procure 480SC)	3	6 to 8 fl oz/acre	1	0.5	Apply when disease first appears and continue on a 14-day interval. Do not exceed 18 fl oz per season.
PYTHIUM DAMPING OFF (*PYTHIUM* SPP.), PHYTOPHTHORA BASAL STEM ROT (*PHYTOPHTHORA MEGASPERMA*)					
fluopicolide (Presidio 4F)	43	3 to 4 fl oz/acre	2	0.5	Apply as a soil drench at transplant. As plants enlarge, use apply directly to soil by chemigation on a 10-day schedule as conditions favor disease, but prior to disease development. No more than 2 sequential applications before rotating to another effective product of a different mode of action. Limited to 4 applications, 12 fl oz/acre per season.
mefenoxam (Ridomil Gold 4SL)	4	0.25 to 0.5 pt/acre	—	2	Apply in water or liquid fertilizer and incorporate in the top 2 inches of soil. For Phytophthora basal rot, increase rates to 1.0 to 2.0 pt/acre.
metalaxyl (MetaStar 2EAG)	4	4 to 8 pt/trt acre	—	2	Preplant incorporated or surface application.
WIRESTEM, RHIZOCTONIA BOTTOM ROT (*RHIZOCTONIA SOLANI*)					
azoxystrobin (Quadris 2.08SC)	11	5.8 to 8.7 fl oz/ acre on 36-in.rows	0	4 hr	Rate is equivalent to 0.4 to 0.6 fl oz per 1000 row feet. Apply at planting as a directed spray to the furrow in a band 7 inches wide. See label for other row spacings.
boscalid (Endura 70EG)	7	6 to 9 oz/acre	0	0.5	Begin applications prior to disease development and continue on a 7- to 14-day interval. Make no more than 2 applications per season.
flutolanil (Moncut 3.8SC)	7	26 fl oz/acre	45	0.5	Apply to the row at planting as an in-furrow spray or a spray directed at the base of transplants immediately after transplanting. Limit of 2 applications per year.
penthiopyrad (Fontelis 1.67SC)	7	16 to 30 fl oz/acre	0	0.5	Do not exceed 72 fl oz of product per year. Make no more than 2 sequential applications per season before rotating to another effective product with a different mode of action.
SCLEROTINIA STEM ROT, WHITE MOLD (*SCLEROTINIA SCLEROTIORUM*)					
boscalid (Endura 70EG)	7	6 to 9 oz/acre	0	0.5	Begin applications prior to disease development and continue on a 7- to 14-day interval. Make no more than 2 applications per season.
fluopyram + trifloxystrobin (Luna Sensation 500SC)	7+11	7.6 fl oz/acre	0	0.5	Do not apply more than 15.3 fl oz per acre per season. Make no more than 2 sequential applications before rotating to a fungicide not in Group 7 or 11.
penthiopyrad (Fontelis 1.67SC)	7	16 to 30 fl oz/acre	0	0.5	Do not exceed 72 fl oz of product per year. Make no more than 2 sequential applications per season before rotating to another effective product with a different mode of action.
Coniothyrium minitans (Contans WG)	BM02	1 to 4 lb/acre	0	4 hr	**OMRI listed product.** Apply to soil surface and incorporate no deeper than 2 inches. Works best when applied prior to planting or transplanting. Do not apply other fungicides for 3 weeks after applying Contans.
WHITE RUST (*ALBUGO CANDIDA*)					
pydiflumetofen + fludioxonil (Miravis Prime 3.34SC)	7+12	11.4 fl oz/acre	7	0.5	Make no more than 2 sequential applications of Miravis Prime or other Group 7 and 12 fungicides before rotating to another effective fungicide with a different mode of action. Maximum 3 applications per year.

TABLE 3-7. EFFICACY OF PRODUCTS FOR DISEASE CONTROL IN BRASSICAS

A. Keinath, Plant Pathologist, Clemson University

Scale: E, excellent; G, good; F, fair; P, poor; NC, no control; ND, no data

Active Ingredient [1]	Product	Crop Group [2]	Fungicide group [F]	Preharvest interval (Days)	Alternaria Leaf Spot	Bacterial Soft Rot	Black Rot	Black Leg	Bottom Rot (Rhizoctonia)	Cercospora & Pseudocercosporella	Clubroot	Downy Mildew	Powdery Mildew	Pythium damping-off	White mold (Sclerotinia)	Wirestem (Rhizoctonia)
acibenzolar-S-methyl	Actigard	H&S	P01	7	NC	ND	F	NC	NC	NC	NC	G	P	ND	ND	NC
ametoctradin + dimethomorph	Zampro	B	45+40	0	NC	NC	NC	NC	NC	NC	NC	E	NC	NC	NC	NC
azoxystrobin	various	B	11	0	G[R]	NC	NC	F	ND	F	NC	G	F	NC	NC	F
azoxystrobin + difenoconazole	Quadris Top	B	11+3	1	E	NC	NC	ND	ND	G	NC	G	F	NC	NC	F
boscalid [3]	Endura	B&TR	7	0 to 14	G	NC	NC	NC	NC	NC	NC	NC	P	NC	F	F
chlorothalonil	various	H&S	M05	7	F	NC	NC	NC	P	F	NC	F	F	NC	NC	NC
fixed copper [4]	various	B	M01	0	P	NC	P	NC	NC	P	NC	F	F	NC	NC	NC
cyazofamid	Ranman	B	21	0	NC	NC	NC	NC	NC	NC	NC	G	NC	NC	NC	NC
cyprodinil + fludioxonil	Switch	B	9+12	7	F	NC	NC	NC	NC	NC	NC	F	NC	F	NC	NC
difenoconazole + cyprodinil	Inspire Super	B	3+9	7	G	NC	NC	ND	NC	G	NC	NC	F	NC	P	NC
dimethomorph	Forum	B	40	0	NC	NC	NC	NC	NC	NC	NC	G	NC	NC	NC	NC
ethaboxam	Elumin	B	22	2	NC	NC	NC	NC	NC	NC	NC	F	NC	ND	NC	NC
fenamidone	Reason	B	11	2	F	NC	NC	NC	NC	NC	NC	G	NC	NC	NC	NC
fluazinam [5]	Omega 500	B	29	20 to 50	F	NC	NC	NC	NC	F	NC	NC	NC	NC	NC	G
fluopicolide	Presidio	B	43	2	NC	NC	NC	NC	NC	NC	NC	E	NC	NC	NC	NC
fluopyram + difenoconazole	Luna Flex	LG	7+3	1	ND	NC	ND	ND	ND	ND	NC	ND	ND	NC	ND	ND
fluopyram + trifloxystrobin	Luna Sensation	H&S	7+11	0	E	NC	ND	ND	ND	ND	NC	ND	ND	NC	ND	ND
flutolanil	Moncut	LG&TG	7	45	NC	NC	NC	NC	NC	NC	NC	NC	NC	NC	NC	G
flutriafol	Rhyme	B	3	7	NC	NC	NC	ND	ND	ND	NC	ND	ND	NC	ND	ND
fluxapyroxad + pyraclostrobin	Priaxor	B	7+11	3	G	NC	ND	G	ND	G	NC	F	F	NC	ND	NC
fosetyl-Al	Aliette	B	P07	3	NC	NC	NC	NC	NC	NC	NC	F	NC	NC	NC	NC
iprodione	Rovral	Broc	2	0	F	NC	NC	F	NC	NC	NC	NC	NC	NC	P	P
mandipropamid	Revus	B	40	1	NC	NC	NC	NC	NC	NC	NC	E	NC	NC	NC	NC
mancozeb	various	H&S	M03	7	F	NC	NC	NC	NC	F	NC	F	P	NC	NC	NC
mancozeb + azoxystrobin	Dexter Max	B&C	M03+11	7	F	NC	NC	NC	NC	F	NC	F	P	NC	NC	NC
mefenoxam (pre-plant)	Ridomil Gold	B	4	—	NC	NC	NC	NC	NC	NC	NC	F	NC	F[R]	NC	NC
mefenoxam + chlorothalonil	Ridomil Gold Bravo	H&S	4+M05	7	F	NC	NC	NC	P	F	NC	F	F	NC[R]	NC	NC
oxathiapiprolin + mandipropamid	Orondis Ultra	H&S	49+40	7	NC	NC	NC	NC	NC	NC	NC	E	NC	ND	NC	NC
penthiopyrad	Fontelis	B	7	0	G	NC	NC	ND	NC	ND	NC	NC	G	NC	G	NC
phosphonates	various	B	P07	0	NC	NC	NC	NC	NC	NC	G	NC	NC	NC	NC	NC
pydiflumetofen + fludioxonil	Miravis Prime	B&TG	7+12	7	G	NC	NC	ND	ND	NC	NC	NC	F	NC	F	F
pyraclostrobin [3]	Cabrio, Pyrac	B	11	0 to 3	G	NC	NC	ND	NC	NC	NC	F	F	NC	NC	P
sulfur [O]	various	B	M02	0	P	NC	NC	NC	NC	NC	P	NC	P	F	NC	NC
tebuconazole	various	B	3	7	P	NC	NC	NC	NC	F	NC	NC	NC	ND	NC	NC
triflumizole	Procure	B	3	1	NC	NC	NC	NC	NC	NC	NC	NC	G	NC	NC	NC

[1] Efficacy ratings do not necessarily indicate a labeled use for every disease.

[2] **H&S** = fungicides registered only on head and stem brassicas (broccoli, Brussels sprouts, cabbage, and cauliflower); **B** = fungicides registered on all brassica crops except turnip greens and root turnips; see Tables 3-16 and 3-31 for products registered on turnips. **LG&TG** = fungicides registered on all brassica leafy vegetables, including turnip greens (but not turnips). **Broc** = fungicide registered on broccoli only. **B&C** = fungicide registered on broccoli and cabbage only. **LG** = fungicides registered only on leafy brassica greens except turnip greens and root turnips. **B&TG** = fungicide registered on all brassica crops, including turnip greens but not root turnips. Always refer to product labels prior to use.

[3] Shorter PHI is for head and stem brassicas (broccoli, Brussel sprout, cabbage, and cauliflower) and longer PHI is for leafy brassica greens.

[4] Phytotoxicity is seen when fosetyl-Al is tank mixed with copper.

[5] Use a 20-day PHI for Omega 500 on leafy greens and a 50-day PHI for head and stem brassicas.

[F] To prevent resistance in pathogens, alternate fungicides within a group with fungicides in another group. Fungicides in the "M" group are considered "low risk" with no signs of resistance developing, so can be rotated with any other group.

[R] Resistance reported in the pathogen.

[O] OMRI-listed product

FOR LEAFY GREENS – SEE – TABLE 3-16. DISEASE CONTROL PRODUCTS FOR GREENS, LEAFY BRASSICA (COLLARD, KALE, MUSTARD, RAPE SALAD GREENS, TURNIP GREENS)

TABLE 3-8. IMPORTANCE OF ALTERNATIVE MANAGEMENT PRACTICES FOR DISEASE CONTROL IN BRASSICAS

A. Keinath, Plant Pathologist, Clemson University

Scale: E, excellent; G, good; F, fair; P, poor; NC, no control; ND, no data.

Strategy	Alternaria leaf spot	Bacterial soft rot	Black rot	Black leg	Bottom rot (Rhizoctonia)	Cercospora	Clubroot	Downy mildew	Powdery mildew	Pythium	Sclerotinia head	Mosaic viruses (CMV, TuMV, CaMV)	Wirestem (Rhizoctonia)
Avoid field operations when leaves are wet	P	F	G	F	F	P	NC	P	NC	NC	NC	NC	NC
Avoid overhead irrigation	E	E	E	E	F	E	NC	G	P	NC	NC	NC	NC
Change planting date	P	P	NC	NC	P	NC	NC	NC	NC	P	NC	P	F
Cover cropping with antagonist	NC	NC	NC	NC	NC	NC	P	NC	NC	P	NC	NC	NC
Crop rotation	F	F	G	G	P	F	NC	F	NC	NC	P	NC	P
Deep plowing	F	F	G	G	F	F	NC	F	NC	NC	F	P	F
Destroy crop residue	F	F	G	G	F	F	NC	F	NC	NC	P	P	P
Encourage air movement	F	P	P	P	F	F	NC	F	NC	P	F	NC	NC
Increase between-plant spacing	F	P	P	P	F	F	NC	F	NC	P	F	NC	NC
Increase soil organic matter	NC	NC	NC	NC	P	NC	P	NC	NC	NC	NC	NC	P
Hot water seed treatment	P	NC	E	G	NC	NC	NC	NC	NC	NC	NC	NC	NC
pH management	NC	NC	NC	NC	NC	NC	E	NC	NC	NC	NC	NC	NC
Plant in well-drained soil	P	F	P	P	G	P	E	P	NC	F	F	NC	G
Plant on raised beds	NC	F	P	NC	G	NC	E	P	NC	F	F	NC	G
Plastic mulch bed covers	P	NC	NC	NC	F	NC	NC	NC	NC	NC	NC	NC	NC
Postharvest temperature control	NC	E	NC	NC	NC	NC	NC	NC	NC	NC	NC	NC	NC
Reflective mulch	NC	NC	NC	NC	NC	NC	NC	NC	NC	NC	NC	F	NC
Reduce mechanical injury	NC	E	G	NC	NC	NC	NC	NC	NC	NC	F	NC	P
Rogue diseased plants	P	NC	NC	F	P	NC	NC	NC	NC	NC	NC	P	NC
Row covers	NC	P	NC	NC	NC	NC	NC	NC	NC	NC	NC	F	NC
Soil solarization	NC	NC	NC	P	F	NC	NC	NC	NC	P	P	NC	F
Pathogen-free planting material	F	NC	E	E	F	NC	G	NC	NC	NC	P	G	F
Resistant cultivars	F	NC	E	NC	NC	NC	P	F	F	NC	NC	NC	P
Weed control	F	NC	F	F	NC	F	F	F	F	NC	F	F	NC

CANTALOUPE – SEE – ***CUCURBITS***

CUCUMBERS – SEE – ***CUCURBITS***

TABLE 3-9. DISEASE CONTROL PRODUCTS FOR CORN, SWEET

D. Higgins, Extension Plant Pathologist, Virginia Tech

Disease/Material	FRAC Code	Rate of Material Formulation	Minimum Days Harv.	Minimum Days Reentry	Method, Schedule, and Remarks
FUSARIUM ROOT ROT (*FUSARIUM* SPP.) PYTHIUM ROOT ROT (*PYTHIUM* SPP.), RHIZOCTONIA CROWN AND BRACE ROOT ROT (*RHIZOCTONIA SOLANI*)					
azoxystrobin (various)	11	0.4 to 0.8 fl oz/ 1000 row feet	-	4 hr	Under cool, wet conditions, crop injury from soil directed applications can occur. See label for banded or in-furrow spray instructions. Use the higher rate in fields with a history of Pythium problems. **Not registered for Fusarium root rot.**
fluopyram + prothioconazole (Propulse)	7+3	13.6 fl oz	-	1	Direct in-furrow spray during planting on or below seed. **Not registered for Pythium root rot and Fusarium root rot.**
pyraclostrobin (Headline SC & EC)	11	0.1 to 0.8 fl oz/ 1000 row feet	-	0.5	See label for in-furrow spray instructions. **Not registered for Pythium root rot.**
ANTHRACNOSE STALK ROT (*COLLETOTRICHUM GRAMINICOLA*), FUSARIUM STALK ROT (*FUSARIUM* SPP.)					
flutriafol (Xyway LFR)	3	15.2 fl oz/acre	-	3	See Section 2(ee) label. Rate assumes 30" row spacing. Apply at planting but avoid direct contact with seed; see label for details.
picoxystrobin (Aproach)	11	3.0 to 6.0 oz/acre 6.0 to 12.0 oz/acre	7	0.5	Do not tank mix with an adjuvant or crop oil when spraying corn between the V8 and VT stages of growth. Use 3 to 6 fl oz rates at V4 to V7 growth stages and 6 to 12 fl oz rates at VT to R3 growth stages. **Not registered for Fusarium stalk rot.**
ANTHRACNOSE LEAF BLIGHT (*COLLETOTRICHUM GRAMINICOLA*), EYE- SPOT (*KABATIELLA ZEAE*), GRAY LEAF SPOT (*CERCOSPORA* SPP.), NORTHERN CORN LEAF BLIGHT (*EXSEROHILUM* [*HELMINTHOSPORIUM*] *TURCICUM*), NORTHERN CORN LEAF SPOT (*BIPOLARIS ZEICOLA* [*HELMINTHOSPORIUM CARBONUM*]), SOUTHERN CORN LEAF BLIGHT (*BIPOLARIS* [*HELMINTHOSPORIUM*] *MAYDIS*), COMMON RUST (*PUCCINIA SORGHI*), SOUTHERN RUST (*PUCCINIA POLYSORA*)					
azoxystrobin (various)	11	See label	7	4 hr	Maximum 2 sequential applications before alternating to a different FRAC code.
azoxystrobin + benzovindiflupyr (Elatus)	11+7	5.0 to 7.3 oz/acre	7	0.5	Limit to 2 sequential applications before alternating to a different FRAC code.
azoxystrobin + flutriafol (Topguard EQ)	11+3	5.0 to 7.0 oz/acre	7	3	Apply when disease symptoms first appear and re-apply on a 7- to 10-day interval. Maximum 2 applications per season. Do NOT apply after R4 early dough stage
azoxystrobin + propiconazole (Quilt XCEL, Avaris)	11+3	7.0 or 10.5 to 14.0 fl oz/acre	14	0.5	Must rotate every application with a non-Group 11 fungicide. Maximum 56 fl oz/acre (4 applications at high rate) per crop.
chlorothalonil (various)	M05	See label	14	2	Do NOT apply to sweet corn to be processed, grazed, or used for livestock forage. Spray at first appearance of disease, 4-to 14-day intervals. **Not registered for anthracnose, eyespot, or gray leaf spot.**
fluoxastrobin (Aftershock)	11	2.0 to 3.8 oz/acre	7 / 23	0.5	7-day PHI for ear harvest. 23-day PHI stover for feed. Maximum 2 applications per season. Do not apply after R4 early dough stage.
flutriafol (Topguard)	3	7.0 to 14.0 oz/acre	7	3	Apply preventively when conditions are favorable for disease and on a 7-day schedule thereafter. Apply no later than growth stage R4 (early dough stage).
flutriafol (Xyway LFR)	3	5.8 to 15.2 fl oz/acre	-	3	Can be applied at planting or post emergence. When applied at planting avoid direct contact with seed; see label for details. Dribble or band alongside the row for post emergence (V2-V8 growth stages) application; rainfall or irrigation is required to move fungicide into the soil. **Not registered for eyespot, anthracnose leaf blight, southern corn leaf spot.**
flutriafol + azoxystrobin + fluindapyr (Adastrio 4.0 SC)	3+11+7	7.0 to 9.0 oz/acre	0.5 / 14	0.5 / 14	14-day PHI/REI for hand harvesting and detasseling sweet corn; all other activities are 12 hrs. Apply at 10- to14-day intervals from the onset of disease through R4 growth stage. Maximum 2 sequential applications before alternating to a different FRAC code.
flutriafol + bixafen (Lucento)	3+7	3.0 to 5.5 oz/acre	10	0.5 / 5	5-day REI for detasseling sweet corn; 12 hr REI otherwise. Do NOT use an adjuvant after the V8 stage and prior to the VT stage. Maximum 2 sequential applications before alternating to a different FRAC code.
fluxapyroxad + mefentrifluconazole + pyraclostrobin (Revytek)	7+3+11	18.0 to 15.0 fl oz/acre	21	0.5	Apply at 7- to 14-day intervals. Maximum 2 sequential applications before alternating to a different FRAC code. Adjuvant damage may occur see label for details.
fluxapyroxad + pyraclostrobin (Priaxor 4.17SC)	7+11	4.0 to 8.0 oz/acre	7	0.5	Maximum 2 sequential applications before alternating to a different FRAC code. Adjuvant damage may occur see label for details.
mancozeb (various)	M03	See labels	7	1	Start applications when disease first appears and repeat at 4- to 7-day intervals. **Not registered for anthracnose, eyespot, gray leaf spot, or southern rust.**
mefentrifluconazole + fluxapyroxad (Revylok)	3+7	4.5 to 6.5 fl oz/acre	21	0.5	Apply at 7- to 10-day intervals; Adjuvant damage may occur see label for details. **Not registered for eyespot.**
mefentrifluconazole + pyraclostrobin (Veltyma)	3+11	7.0 to 10.0 fl oz/acre	21	0.5	Apply at 7- to 10-day intervals. DO NOT make more than 3 applications at 10 fl oz or 4 applications at 7 fl oz per acre per year to sweet corn.
penthiopyrad (Vertisan)	7	10.0 to 24.0 fl oz/acre	7	0.5	Maximum 2 sequential applications before alternating to a different FRAC code. **Not registered for eyespot.**
picoxystrobin (Aproach)	11	3.0 to 6.0 oz/acre 6.0 to 12.0 oz/acre	7	0.5	Do not tank-mix with an adjuvant or crop oil when spraying corn between the V8 and VT stages of growth. Use 3-6 fl oz rates at V4 to V7 growth stages and 6-12 fl oz rates at VT to R3 growth stages.

TABLE 3-9. DISEASE CONTROL PRODUCTS FOR CORN, SWEET (cont'd)

D. Higgins, Extension Plant Pathologist, Virginia Tech

Disease/Material	FRAC Code	Rate of Material Formulation	Minimum Days Harv.	Minimum Days Reentry	Method, Schedule, and Remarks
ANTHRACNOSE LEAF BLIGHT (*COLLETOTRICHUM GRAMINICOLA*), EYE- SPOT (*KABATIELLA ZEAE*), GRAY LEAF SPOT (*CERCOSPORA* SPP.), NORTHERN CORN LEAF BLIGHT (*EXSEROHILUM* [*HELMINTHOSPORIUM*] *TURCICUM*), NORTHERN CORN LEAF SPOT (*BIPOLARIS ZEICOLA* [*HELMINTHOSPORIUM CARBONUM*]), SOUTHERN CORN LEAF BLIGHT (*BIPOLARIS* [*HELMINTHOSPORIUM*] *MAYDIS*), COMMON RUST (*PUCCINIA SORGHI*), SOUTHERN RUST (*PUCCINIA POLYSORA*) (cont'd)					
propiconazole (various)	3	See label	14	1	Allow time for product to dry before rainfall. Use 4 fl oz/A rate for rusts, gray leaf spot, and eye spot. 16 fl oz per acre per crop maximum. Not registered for anthracnose.
propiconazole + azoxystrobin + benzovindiflupyr (Trivapro 2.21SE)	3+11+7	13.7 fl oz/acre	14	0.5	Begin applications prior to disease development. Continue applications through season on a 14-day interval, following the resistance management guidelines.
prothioconazole (Proline 480 SC)	3	5.7 fl oz/acre	0 14	0.5	See supplemental label for use on corn. 14-day PHI if harvested for fodder; ears and forage may be harvested the same day of application.
prothioconazole + fluopyram (Propulse)	3+7	13.6 oz/acre	0	1	Apply at 7- to 10-day intervals. Maximum 2 sequential applications before alternating to a different FRAC code.
prothioconazole + tebuconazole (Prosaro)	3+3	6.5 fl oz/acre	7 49	0.5	7-day PHI for harvest of ears or forage. 49-day PHI if harvested for fodder. Do NOT use an adjuvant after the V8 stage and prior to the VT stage. Apply at 5- to 14-day intervals.
prothioconazole + trifloxystrobin (Delaro 325SC)	3+11	8.0 fl oz/acre	0 14	0.5	0-day PHI for harvest of ears or forage. 14-day PHI if harvested for fodder. Apply at 5- to 14-day intervals. Maximum 2 sequential applications before alternating to a different FRAC code.
prothioconazole + trifloxystrobin + fluopyram (Delaro Complete)	3+11+7	8.0 fl oz/acre	0 14	0.5	0-day PHI for harvest of ears or forage. 14-day PHI if harvested for fodder. Apply at 5- to 14-day intervals. Maximum 2 sequential applications before alternating to a different FRAC code.
propiconazole + azoxystrobin + pydiflumetofen (Miravis Neo)	3+11+7	13.7 fl oz/acre	14	0.5	Begin applications prior to disease development. Continue applications through season on a 7- to 14-day interval, following the resistance management guidelines.
pyraclostrobin (Headline SC & EC)	11	6.0 to 12.0 fl oz/acre	7	0.5	Do not make more than 2 sequential applications or 6 applications of this fungicide or other group 11 fungicides per crop. **Not registered for eyespot.**
pyraclostrobin + metconazole (Headline AMP)	11+3	10.0 to 14.4 fl oz/acre	7	0.5	Maximum 2 sequential applications before alternating to a different FRAC code. Maximum 4 (high rate) or 5 applications (low rate) per crop. **Not registered for eyespot.**
tebuconazole (various)	3	4.0 to 6.0 fl oz/acre	7	19	Allow for two to four hours of drying time on corn foliage. **Most products not registered for eyespot and anthracnose.**
trifloxystrobin + propiconazole (Stratego)	11+3	10.0 fl oz/acre	14	0.5	Apply Stratego when disease first appears and continue on a 7- to 14-day interval. Alternate applications of Stratego with another product with a different mode of action than Group 11 fungicides. Maximum 3 applications per crop. **Not registered for the 3 Helminthosporium diseases.**
trifloxystrobin + prothioconazole (Stratego YLD)	11+3	4.0 to 5.0 fl oz/acre	0 14	0.5	0-day PHI for harvest of ears or forage. 14-day PHI if harvested for fodder. Alternate Stratego YLD sprays with another mode of action than a group 11 fungicide. Maximum 4 (high rate) or 5 applications (low rate) per crop. **Not registered for the 3 Helminthosporium diseases.**
trifloxystrobin + tebuconazole (Absolute Maxx)	11+3	5.0 to 6.0 fl oz/acre	7 49	19	7-day PHI for harvest of ears or forage. 49-day PHI if harvested for fodder. Apply at 10- to14-day interval if favorable conditions persist.
BROWN SPOT (*PHYSODERMA MAYDIS*), YELLOW LEAF BLIGHT (*PEYRONELLAEA ZEAMAYDIS* [*PHYLLOSTICTA MAYDIS*])					
azoxystrobin (Quadris, multiple generics)	11	See label	7	4	Maximum 2 sequential applications before alternating to a different FRAC code. **Not registered for yellow leaf blight.**
azoxystrobin + benzovindiflupyr (Elatus)	11+7	5.0 to 7.3 oz/acre	7	0.5	Maximum 2 sequential applications before alternating to a different FRAC code.
propiconazole + azoxystrobin + benzovindiflupyr (Trivapro 2.21 SE)	3+11+7	13.7 fl oz/acre	14	0.5	Begin applications prior to disease development. Continue applications through season on a 14-day interval, following the resistance management guidelines.
flutriafol (Xyway LFR)	3	5.8 to 15.2 fl oz/acre	-	3	Can be applied at planting or post emergence. When applied at planting avoid direct contact with seed; see label for details. Dribble or band alongside the row for post emergence (V2-V8 growth stages) application; rainfall or irrigation is required to move fungicide into the soil. **Not registered for yellow leaf blight.**
flutriafol + azoxystrobin + fluindapyr (Adastrio 4.0 SC)	3+11+7	7.0 to 9.0 fl oz/acre	0.5 14	0.5 14	14-day PHI/REI for hand harvesting and detasseling sweet corn; all other activities are 12 hrs. Apply at 10- to14-day intervals from the onset of disease through R4 growth stage. Maximum 2 sequential applications before alternating to a different FRAC code. **Not registered for yellow leaf blight.**
flutriafol + bixafen (Lucento)	3+7	3.0 to 5.5 fl oz/acre	10	0.5 5	5-day REI for detasseling sweet corn; 12 hr REI otherwise. Do NOT use an adjuvant after the V8 stage and prior to the VT stage. Maximum 2 sequential applications before alternating to a different FRAC code. **Not registered for yellow leaf blight.**

TABLE 3-9. DISEASE CONTROL PRODUCTS FOR CORN, SWEET (cont'd)

D. Higgins, Extension Plant Pathologist, Virginia Tech

Disease/Material	FRAC Code	Rate of Material Formulation	Minimum Days Harv.	Minimum Days Reentry	Method, Schedule, and Remarks
BROWN SPOT (*PHYSODERMA MAYDIS*), YELLOW LEAF BLIGHT (*PEYRONELLAEA ZEAMAYDIS [PHYLLOSTICTA MAYDIS]*) (cont'd)					
fluxapyroxad + mefentrifluconazole + pyraclostrobin (Revytek)	7+3+11	18.0 to 15.0 fl oz/acre	21	0.5	Maximum 2 sequential applications before alternating to a different FRAC code. Adjuvant damage may occur see label for details.
fluxapyroxad + pyraclostrobin (Priaxor)	7+11	4 to 8 fl oz/acre	7	0.5	Do not make more than 2 sequential applications before alternating to a different FRAC code. Maximum 4 (high rate) or 2 applications (low rate) per crop. Crop damage may occur when an adjuvant is used; read label for specifics.
mefentrifluconazole + pyraclostrobin (Veltyma)	3+11	7.0 to 10.0 fl oz/acre	21	0.5	Apply at 7- to 10-day intervals; Adjuvant damage may occur see label for details. DO NOT make more than 3 applications at 10 fl oz or 4 applications at 7 fl oz per acre per year to sweet corn.
penthiopyrad (Vertisan)	7	16.0 to 24.0 fl oz/acre	0	0.5	Maximum 2 sequential applications before alternating to a different FRAC code. **Not registered for yellow leaf blight**
picoxystrobin (Aproach)	11	3.0 to 6.0 fl oz/acre 6.0 to 12.0 fl oz/acre	7	0.5	Do not tank-mix with an adjuvant or crop oil when spraying corn between the V8 and VT stages of growth. Use 3 to 6 fl oz rates at V4 to V7 growth stages and 6 to 12 fl oz rates at VT to R3 growth stages.
propiconazole + azoxystrobin + pydiflumetofen (Miravis Neo 2.5 SE)	3+11+7	13.7 fl oz/acre	14	0.5	Begin applications prior to disease development. Continue applications through season on a 7- to 14-day interval, following the resistance management guidelines.
pyraclostrobin (Headline SC & EC)	11	6.0 to 12.0 fl oz/acre	7	0.5	Do not exceed 2 sequential applications of this fungicide or with other group 11 fungicides.
pyraclostrobin + metconazole (Headline AMP)	11+3	10.0 to 14.4 fl oz/acre	7	0.5	Maximum 2 sequential applications before alternating to a different FRAC code.
TAR SPOT (*PHYLLACHORA MAYDIS*)					
azoxystrobin + flutriafol (Topguard EQ)	11+3	5.0 to 7.0 oz/acre	7	3	See Section 2(ee) label and confirm availability in your state. Apply no later than R4 (early dough stage). 2 applications per season.
azoxystrobin + propiconazole (Quilt Xcel 2.2 SE)	11+3	10.5 to 14 fl oz/acre	14	0.5	See Section 2(ee) label and confirm availability in your state. Use lower rate for early application timing (V4-V8); apply 2 applications under heavy disease pressure.
flutriafol + azoxystrobin + fluindapyr (Adastrio 4.0 SC)	3+11+7	7.0 to 9.0 oz/acre	0.5 14	0.5 14	See Section 2(ee) label and confirm availability in your state. Apply when disease first appears and make a second application 10-14 days later. 14-day PHI/REI for hand harvesting and detasseling sweet; all other activities are 12 hrs.
flutriafol + bixafen (Lucento)	3+7	5.0 to 5.5 fl oz/acre	10	0.5 5	5-day REI for detasseling sweet corn; 12 hr REI otherwise. DO NOT use an adjuvant after the V8 stage and prior to the VT stage. Maximum 2 sequential applications before alternating to a different FRAC code.
fluxapyroxad + mefentrifluconazole + pyraclostrobin (Revytek)	7+3+11	8.0 to 15 fl oz/acre	21	0.5	Maximum 2 sequential applications before alternating to a different FRAC code. Adjuvant damage may occur see label for details.
fluxapyroxad + pyraclostrobin (Priaxor Xemium)	7+11	4.0 to 8.0 fl oz/acre	7	0.5	See Section 2(ee) label and confirm availability in your state. Maximum 2 sequential applications before alternating to a different FRAC code.
mefentrifluconazole + fluxapyroxad (Revylok)	3+7	4.5 to 6.5 fl oz/acre	21	0.5	Apply at 7- to 10-day intervals; Adjuvant damage may occur see label for details.
mefentrifluconazole + pyraclostrobin (Veltyma)	3+11	7.0 to 10.0 fl oz/acre	21	0.5	Apply at 7- to 10-day intervals; Adjuvant damage may occur see label for details. DO NOT make more than 3 applications at 10 fl oz or 4 applications at 7 fl oz per acre per year to sweet corn.
propiconazole + azoxystrobin + benzovindiflupyr (Trivapro 2.21 SE)	3+11+7	13.7 fl oz/acre	14	0.5	Begin applications prior to disease development. Continue applications through season on a 14-day interval, following the resistance management guidelines.
propiconazole + azoxystrobin + pydiflumetofen (Miravis Neo 2.5 SE)	3+11+7	13.7 fl oz/acre	14	0.5	Begin applications prior to disease development. Maximum 2 sequential applications before alternating to a different FRAC code.
prothioconazole + trifloxystrobin (Delaro 325 SC)	3+11	8.0 to 12.0 fl oz/acre	0 14	1	0-day PHI for harvest of ears or forage. 14-day PHI if harvested for fodder. Apply at 5- to 14-day intervals.
prothioconazole + trifloxystrobin + fluopyram (Delaro Complete)	3+11+7	8.0 fl oz/acre	0 14	0.5	0-day PHI for harvest of ears or forage. 14-day PHI if harvested for fodder. Apply at 5- to 14-day intervals. Maximum 2 sequential applications before alternating to a different FRAC code.
picoxystrobin (Aproach)	11	6.0 to 12.0 fl oz/acre	7	0.5	See Section 2(ee) label and confirm availability in your state. Maximum 2 sequential applications before alternating to a different FRAC code.
pyraclostrobin + metconazole (Headline AMP)	11+3	10.0 to 14.4 fl oz/acre	7	0.5	See Section 2(ee) label for control/suppression and confirm availability in your state. Do not exceed 2 sequential applications of this fungicide or with other group 11 fungicides.

TABLE 3-10. DISEASE CONTROL PRODUCTS FOR CUCURBITS

L. Quesada-Ocampo, Plant Pathologist, North Carolina State University

Disease/Material	FRAC Code	Rate of Material Formulation	Minimum Days Harv.	Minimum Days Reentry	Method, Schedule, and Remarks
ANGULAR LEAF SPOT (*PSEUDOMONAS*)					
fixed copper (various)	M01	See label	See label	See label	See label. Rates vary depending on the formulation. Repeated use may cause leaf yellowing.
acibenzolar-*S*-methyl (Actigard) 50 WP	P01	0.5 to 1 oz/acre	0	0.5	Apply to healthy, actively growing plants. Do not apply to stressed plants. Apply no more than 8 oz per acre per season.
BACTERIAL LEAF SPOT (*XANTHOMONAS*)					
acibenzolar-*S*-methyl (Actigard) 50 WP	P01	0.5 to 1 oz/acre	0	0.5	Apply to healthy, actively growing plants. Do not apply to stressed plants. Apply no more than 8 oz per acre per season.
fixed copper (various)	M01	See label	See label	See label	See label. Rates vary depending on the formulation. Repeated use may cause leaf yellowing.
BACTERIAL FRUIT BLOTCH (*ACIDOVORAX*)					
fixed copper (various)	M01	See label	0	0	See label. Rates vary depending on the formulation. Start applications at first bloom; ineffective once fruit reaches full size. Repeated use may cause leaf yellowing.
acibenzolar-*S*-methyl (Actigard) 50 WP	P01	0.5 to 1 oz/acre	0	0.5	Apply to healthy, actively growing plants. Do not apply to stressed plants. Apply no more than 8 oz per acre per season.
BACTERIAL WILT (*ERWINIA*)					
NA	NA	NA	NA	NA	See Insect Control section for Cucumber Beetles.
BELLY (FRUIT) ROT (*RHIZOCTONIA*)					
azoxystrobin (various)	11	See label	1	4 hr	Make banded application to soil surface or in-furrow application just before seed are covered.
azoxystrobin + chlorothalonil (Quadris Opti)	11+M05	3.2 pints/acre	1	0.5	Do not apply more than one foliar application before alternating with a fungicide with a different mode of action. Do not make more than 4 applications of QoI group 11 fungicides per crop per acre per year.
difenoconazole + benzovindiflupyr (Aprovia Top)	7+3	8.5 to 13.5 fl oz/acre	0	0.5	For belly rot control, the first application should be made at the 1- to 3-leaf crop stage with a second application just prior to vine tip or 10 to 14 days later, whichever occurs first.
fluopyram + tebuconazole (Luna Experience 3.3 F)	7+3	17 fl oz/acre	7	0.5	**ONLY APPLY TO WATERMELON**. Make no more than 2 applications before alternating to a fungicide with different active ingredients. **Do not rotate with tebuconazole. Not labeled for use in Louisiana.**
flutriafol (Rhyme)	3	7 fluid oz/acre	0	12 hrs	Apply every 30 days. Do not apply more than 4 applications per year. See labels for watermelon restrictions
thiophanate-methyl (Topsin M 70 WP, Miromar)	1	0.5 lb/acre 10.5 oz/acre	0	0.5	Apply in sufficient water to obtain runoff to soil surface.
COTTONY LEAK (*PYTHIUM* SPP.)					
metalaxyl (MetaStar 2 E)	4	4 to 8 pt/treated acre	0	2	Soil surface application in 7 in. band.
DAMPING OFF (*PYTHIUM* SPP.) AND FRUIT ROT (*PYTHIUM* SPP.)					
mefenoxam (Ridomil Gold 4 SL) (Ultra Flourish 2 EL)	4	1 to 2 pt/acre 2 to 4 pt/acre	0	2	Preplant incorporated (broadcast or band); soil spray (broadcast or band) or injection (drip irrigation).
metalaxyl (MetaStar 2 E)	4	4 to 8 pt/acre	0	2	Preplant incorporated or surface application.
propamocarb (Previcur Flex 6 F)	28	12.8 fl oz/100 gal	2	0.5	Rates based on rock wool cube saturation in the greenhouse. See label for use in seedbeds, drip system, and soil drench.
DOWNY MILDEW (*PSEUDOPERONOSPORA CUBENSIS*)					
ametoctradin + dimethomorph (Zampro 4.38SC)	45+40	14 oz/acre	0	0.5	Make no more than 2 applications before alternating to a fungicide with different active ingredients. Do not rotate with Forum. Maximum of 3 applications per crop per season.
chlorothalonil (various)	M05	See label	See label	See label	See labels. Rates vary depending on the formulation. Spray at first appearance and then at 7- to 14-day interval. Avoid late-season application after plants have reached full maturity.
chlorothalonil + cymoxanil (Cymbol Advance)	M05+27	1.9 to 3.0 pt/acre	3	0.5	Repeat application at 7-day intervals. Alternate applications with different MOA fungicide. Maximum of 17.5 pints of product per acre per year. Maximum of 15.75 lb a.i. chlorothalonil per acre per year.
chlorothalonil + zoxamide (Zing!)	M05+22	36 fl oz/acre	0	0.5	May cause sunburn in watermelon fruit, see label for details.
cyazofamid (Ranman 400SC)	21	2.1 to 2.75 fl oz/acre	0	0.5	Do not apply more than 6 sprays per crop. Make no more than 3 consecutive applications followed by 3 applications of fungicides from a different resistance management group.

TABLE 3-10. DISEASE CONTROL PRODUCTS FOR CUCURBITS (cont'd)

L. Quesada-Ocampo, Plant Pathologist, North Carolina State University

Disease/Material	FRAC Code	Rate of Material Formulation	Minimum Days Harv.	Minimum Days Reentry	Method, Schedule, and Remarks
DOWNY MILDEW (*PSEUDOPERONOSPORA CUBENSIS*) (cont'd)					
cymoxanil + propamocarb (Cymbol Balance)	27+28	28.5 fl oz/acre	3	0.5	Repeat application at 5- to 7-day intervals. Maximum of five (5) applications per year. Maximum of 142.5 fl oz of product per acre per year. Maximum of 4.5 lb a.i. propamocarb per acre per year. Maximum of 1.125 lb a.i. cymoxanil per acre per year.
ethaboxam (Elumin)	22	8 fl oz/acre	2	0.5	Do not make more than two applications per year. Do not apply at intervals of less than 14 days.
fixed copper (various)	M01	See label	See label	See label	See label. Rates vary depending on the formulation. Repeated use may cause leaf yellowing.
fluazinam (Omega 500F)	29	0.75 to 1.5 pints/acre	7 or 30	0.5	Initiate applications when conditions are favorable for disease development or when disease symptoms first appear. Repeat applications on a 7- to 10-day schedule. PHI is 7 days for cucumber squash, pumpkin (subgroup 9A). PHI is 30 days for melon, watermelon, cantaloupe (subgroup 9B).
fluopicolide (Presidio 4F)	43	3 to 4 fl oz/acre	2	0.5	Tank-mix with another downy mildew fungicide with a different mode of action.
fosetyl-AL (Aliette 80 WDG)	P07	2 to 5 lb/acre	0.5	0.5	Do not tank-mix with copper-containing products. Mixing with surfactants or foliar fertilizers is not recommended.
mandipropamid (Revus 2.08F)	40	8 fl oz/acre	1	0.5	For disease suppression only. Resistance reported.
mancozeb (various)	M03	See label	See label	See label	See label. Rates vary depending on the formulation. Labeled on all cucurbits.
mefenoxam + chlorothalonil (Ridomil Gold Bravo, Flouronil 76.5 WP)	4+M05	2 to 3 lb/acre	7	2	Spray at first appearance and repeat at 14-day intervals. Apply full rate of protectant fungicide between applications. Avoid late-season application when plants reach full maturity. Resistance reported.
oxathiapiprolin + chlorothalonil (Orondis Opti SC)	49+M05	1.7 to 2.5 pt/acre	0	0.5	Limit to 10 pt per acre per year. Limit of six foliar applications per acre per year for the same crop. Do not follow soil applications of Orondis with foliar applications of Orondis. Begin foliar applications prior to disease development and continue on a 5- to 14-day interval. Use the higher rates when disease is present.
oxathiapiprolin + mandipropamid (Orondis Ultra)	49+40	5.5 to 8 fl oz/acre	0	4 hr	Limit to 32 fl oz per acre per year. Limit of six foliar applications per acre per year for the same crop. Do not follow soil applications of Orondis with foliar applications of Orondis. Begin foliar applications prior to disease development and continue on a 5- to 14-day interval. Use the higher rates when disease is present.
propamocarb (Previcur Flex 6F)	28	1.2 pt/acre	2	0.5	Begin applications before infection; continue on a 7- to 14-day interval. Do not apply more than 6 pt per growing season. Always tank mix with another Downy mildew product.
zoxamide + mancozeb (Gavel 75 DF)	22+M03	1.5 to 2 lb/acre	5	2	Begin applications when plants are in 2-leaf stage and repeat at 7- to 10-day intervals. Labeled on all cucurbits. Maximum 8 applications per season.
FUSARIUM WILT (*FUSARIUM*)					
prothioconazole (Proline 480 SC)	3	5.7 fl oz/acre	7	0.5	One soil and two foliar applications allowed by either ground or chemigation application equipment (including drip irrigation). Do not use in water used for hand transplanting. Not for use in greenhouses or transplant houses.
Pseudomonas chlororaphis strain AFS0009 (Howler)	BM02	5 to 15 lb/acre	0	0.17	In furrow or foliar spray, Repeat at 7- to 21-day intervals. Thoroughly cover plant foliage and soil surfaces.
pydiflumetofen + fludioxonil (Miravis Prime)	7+12	11.4 fl oz/acre	1	0.5	**For Suppression Only.** Do not make more than two applications before alternating with a non-Group 7 or 12 fungicide. Follow label application methods and timing.
GUMMY STEM BLIGHT, BLACK ROT (*STAGONOSPOROPSIS*)					
chlorothalonil + cymoxanil (Cymbol Advance)	M05+27	3.0 pt/acre	3	0.5	Repeat application at 7-day intervals. Alternate applications with different MOA fungicide. Maximum of 17.5 pints of product per acre per year. Maximum of 15.75 lb a.i. chlorothalonil per acre per year.
prothioconazole (Proline 480 SC)	3	5.7 fl oz/acre	7	0.5	One soil and two foliar applications allowed by either ground or chemigation application equipment (including drip irrigation). Do not use in water used for hand transplanting. Not for use in greenhouses or transplant houses.
fluopyram + difenoconazole (Luna Flex)	7+3	12.8 to 13.6 fl oz/acre	0	0.5	Please see label.
tebuconazole (Monsoon)	3	8.0 oz/acre	7	0.5	**Suppression only.** Maximum 3 applications per season. Apply as a protective spray at 1- to 14-day intervals. Add a surfactant.

TABLE 3-10. DISEASE CONTROL PRODUCTS FOR CUCURBITS (cont'd)

L. Quesada-Ocampo, Plant Pathologist, North Carolina State University

Disease/Material	FRAC Code	Rate of Material Formulation	Minimum Days Harv.	Minimum Days Reentry	Method, Schedule, and Remarks
LEAF SPOTS: *ALTERNARIA*, ANTHRACNOSE (*COLLETOTRICHUM*), CERCOSPORA, GUMMY STEM BLIGHT (*STAGONOSPOROPSIS*), TARGET SPOT (*CORYNESPORA*)					
azoxystrobin (Quadris 2.08F)	11	11 to 15.4 fl oz/acre	1	4 hr	Make no more than one application before alternating with a fungicide with a different mode of action. Apply no more than 2.88 qt per crop per acre per season. **Do not use for gummy stem blight where resistance to group 11 (QoI) fungicides exists.**
azoxystrobin + chlorothalonil (Quadris Opti)	11+M05	3.2 pints/acre	1	0.5	Do not apply more than one foliar application before alternating with a fungicide with a different mode of action. Do not make more than 4 applications of QoI group 11 fungicides per crop per acre per year. **Do not use for gummy stem blight where resistance to group 11 (QoI) fungicides exists.**
fluopyram + difenoconazole (Luna Flex)	7+3	8.0 fl oz/acre	0	0.5	Please see label.
azoxystrobin + difenoconazole (Quadris Top 1.67SC)	11+3	12 to 14 fl oz/acre	1	0.5	Not for Target spot. Make no more than one application before alternating with fungicides that have a different mode of action. Apply no more than 56 fl oz per crop per acre per season. **Do not use for gummy stem blight where resistance to group 11 (QoI) fungicides exists.**
chlorothalonil (various)	M05	See label	See label	See label	See labels. Rates vary depending on the formulation.
chlorothalonil + cymoxanil (Cymbol Advance)	M05+27	3.0 pt/acre	3	0.5	Repeat application at 7-day intervals. Alternate applications with different MOA fungicide. Maximum of 17.5 pints of product per acre per year. Maximum of 15.75 lb a.i. chlorothalonil per acre per year.
cyprodinil + fludioxonil (Switch 62.5WG)	9+12	11 to 14 oz/acre	1	0.5	**Only for Alternaria and gummy stem blight.** Make no more than 2 applications before alternating to a different fungicide. Maximum of 4 to 5 applications at high and low rates.
difenoconazole + benzovindiflupyr (Aprovia Top)	7+3	10.5 to 13.5 fl oz/acre	0	0.5	Make no more than 2 applications before alternating to a fungicide with different active ingredients. Apply no more than 53.6 fl oz per acre per year.
difenoconazole + cyprodinil (Inspire Super 2.82SC)	3+9	16 to 20 fl oz/acre	7	0.5	**Not for Target spot.** Make no more than two sequential applications before alternating with fungicides that have a different mode of action. Apply no more than 80 fl oz per crop per acre per season.
famoxadone + cymoxanil (Tanos 50WP)	11+27	8 oz/acre	3	0.5	Only for Alternaria and Anthracnose, do not make more than one application before alternating with a fungicide that has a different mode of action; must be tank mixed with contact fungicide with a different mode of action.
fixed copper (various)	M01	See label	See label	See label	See labels. Rates vary depending on the formulation. Repeated use may cause leaf yellowing.
fluazinam (Omega 500F)	29	0.75 to 1.5 pts/acre	30	0.5	Initiate applications when conditions are favorable for disease development or when disease symptoms first appear. Repeat applications on a 7- to 10-day schedule.
fluopyram + difenoconazole (Luna Flex)	7+3	8 fl oz/acre	0	0.5	Apply at the critical time for disease control and continue as needed on a 7- to 14-day interval
fluopyram + tebuconazole (Luna Experience 3.3F)	7+3	8 to 17 fl oz/acre	7	0.5	Not for Cercospora or target spot. Make no more than 2 applications before alternating to a fungicide with different active ingredients. Do not rotate with tebuconazole.
fluopyram + trifloxystrobin (Luna Sensation 1.67F)	7+11	7.6 fl oz/acre	0	0.5	**ONLY APPLY TO WATERMELON** and only to control Alternaria and Anthracnose. Make no more than 2 applications before alternating with fungicides with different active ingredients. Maximum 4 applications per season.
fluxapyroxad + pyraclostrobin (Merivon 500SC)	7+11	4 to 5.5 fl oz/acre	0	0.5	Make no more than two sequential applications before alternating with fungicides that have a different mode of action. Maximum of 3 applications per crop.
mancozeb (various)	M03	See label	See label	See label	See labels. Rates vary depending on the formulation. Labeled on all cucurbits.
pydiflumetofen + fludioxonil (Miravis Prime)	7+12	9.2 to 11.4 fl oz/acre	1	0.5	**Begin applications prior to disease development. Follow resistance management guidelines.**
pyraclostrobin (Cabrio 20 WG)	11	12 to 16 oz/acre	0	0.5	**Do not use for Gummy stem blight where resistance to group11 (QoI) fungicides exists.** Make no more than one application before alternating to a fungicide with a different mode of action.
pyraclostrobin + boscalid (Pristine 38 WG)	11+7	12.5 to 18.5 oz/acre	0	1	Not for target spot. Do not use for gummy stem blight where resistance to group 7 and group 11 fungicides exists. Use the highest rate for anthracnose. Make no more than 4 applications per season.
thiophanate-methyl (Topsin M 70WP)	1	0.5 lb/acre 10.9 oz/acre	0	0.5	Spray at first appearance and then at 7- to 10-day intervals. Resistance reported in gummy stem blight fungus.
zoxamide + mancozeb (Gavel 75 DF)	22+M03	1.5 to 2 lb/acre	5	2	Cercospora and Alternaria only. Begin applications when plants are in 2-leaf stage and repeat at 7- to 10-day intervals. Now labeled on all cucurbits. Maximum 8 applications per season.

TABLE 3-10. DISEASE CONTROL PRODUCTS FOR CUCURBITS (cont'd)

L. Quesada-Ocampo, Plant Pathologist, North Carolina State University

Disease/Material	FRAC Code	Rate of Material Formulation	Minimum Days Harv.	Minimum Days Reentry	Method, Schedule, and Remarks
PHYTOPHTHORA BLIGHT (FOLIAGE AND FRUIT) (*PHYTOPHTHORA CAPSICI*)					
ametoctradin + dimethomorph (Zampro 4.38SC)	45+40	14 oz/acre	0	0.5	Make no more than 2 applications before alternating to a fungicide with different active ingredients. Do not rotate with Forum. Maximum of 3 applications per crop per season. Apply at planting as a preventative drench treatment. Addition of a spreading or penetrating adjuvant is recommended.
cyazofamid (Ranman 400SC)	21	2.75 fl oz/acre	0	0.5	Do not apply more than 6 sprays per crop. Make no more than 3 consecutive applications followed by 3 applications of fungicides from a different resistance management group. Resistant isolates have been found.
dimethomorph (Forum 4.17SC)	40	6 fl oz/acre	0	0.5	Must be applied as a tank-mix with another fungicide with a different mode of action. Do not make more than two sequential applications.
ethaboxam (Elumin)	22	8 fl oz/acre	2	0.5	Make no more than 2 applications before alternating to a fungicide with different active ingredients. Apply no more than 16 fl oz/acre per year.
fluazinam (Omega 500F, Lektivar 40 SC)	29	0.75 to 1.5 pt/acre	7 or 30	0.5	Initiate applications when conditions are favorable for disease development or when disease symptoms first appear. Repeat applications on a 7- to 10-day schedule. PHI is 7 days for cucumber squash, pumpkin (subgroup 9A). PHI is 30 for melon, watermelon, cantaloupe (subgroup 9B).
fluopicolide (Presidio 4F)	43	3 to 4 fl oz/acre	2	0.5	Tank-mix with another Phytophthora fungicide with a different mode of action. May be applied through drip irrigation to target crown rot phase.
mandipropamid (Revus 2.08F)	40	8 fl oz/acre	0	0.5	For disease suppression only. Apply as a foliar spray with a copper-based fungicide.
oxathiapiprolin + mefenoxam (Orondis Gold 200)	49+4	4.8 to 9.6 fl oz/acre	0	4 hr	Limit to 38.6 fl oz per acre per year. Limit of 6 applications per acre per year for the same crop. Do not follow soil applications of Orondis with foliar applications of Orondis. Apply at planting in furrow, by drip, or in transplant water. Use the higher rates for heavier soils, for longer application intervals, or for susceptible varieties.
oxathiapiprolin + mandipropamid (Orondis Ultra)	49+40	5.5 to 8 fl oz/acre	0	4 hr	Limit of 6 applications per acre per year for the same crop. Do not follow soil applications of Orondis with foliar applications of Orondis. Use the higher rates when disease is present.
PLECTOSPORIUM BLIGHT (*PLECTOSPHAERELLA*)					
azoxystrobin (Quadris 2.08F, Arius 250)	11	11 to 15.4 fl oz/acre	1	4 hr	Make no more than one application before alternating with a fungicide with a different mode of action. Apply no more than 2.88 qt per crop per acre per season, and do not make more than 4 applications of Group 11 products.
azoxystrobin + difenoconazole (Quadris Top 1.67SC)	11+3	12 to 14 fl oz/acre	1	0.5	Make no more than one application before alternating with fungicides that have a different mode of action. Apply no more than 56 fl oz per crop per acre per season.
fluxapyroxad + pyraclostrobin (Merivon 500SC)	7+11	4 to 5.5 fl oz/acre	0	0.5	Make no more than two sequential applications before alternating with fungicides that have a different mode of action. Maximum of 3 applications per crop.
trifloxystrobin (Flint 50WDG)	11	1.5 to 2 oz/acre	0	0.5	Make no more than one application before alternating with fungicides that have a different mode of action. Begin applications preventively when conditions are favorable for disease and continue on as needed on a 7- to 14-day interval.
pyraclostrobin (Cabrio 20WG)	11	12 to 16 oz/acre	0	0.5	Make no more than 1 application before alternating to a fungicide with a different mode of action.
POWDERY MILDEW (*PODOSPHAERA, GOLOVINOMYCES*)					
acibenzolar-*S*-methyl (Actigard 50WP)	P01	0.5 to 1 oz/ac	0	0.5	Make no more than 1 application before alternating to a fungicide with a different mode of action.
azoxystrobin + chlorothalonil (Quadris Opti)	11 + M05	3.2 pt/acre	1	0.5	Do not apply more than one foliar application before alternating with a fungicide with a different mode of action. Do not make more than 4 applications of Group 11 fungicides per crop per acre per year.
azoxystrobin + difenoconazole (Quadris Top 1.67SC)	11+3	12 to 14 fl oz/acre	1	0.5	Make no more than one application before alternating with fungicides that have a different mode of action. Apply no more than 56 fl oz per crop per acre per season.
chlorothalonil (various)	M05	See label	See label	See label	Spray at first appearance and then at 7- to 14-days intervals. Avoid late-season application after plants have reached full maturity. Does not control PM on leaf undersides.
chlorothalonil + cymoxanil (Cymbol Advance)	M05+27	3.0 pt/acre	3	0.5	Repeat application at 7-day intervals. Alternate applications with different MOA fungicide. Maximum of 17.5 pints of product per acre per year. Maximum of 15.75 lb a.i. chlorothalonil per acre per year.
difenoconazole + benzovindiflupyr (Aprovia Top)	7+3	10.5 to 13.5 fl oz/acre	0	0.5	Make no more than 2 applications to a fungicide with different active ingredients. Apply no more than 53.6 fl oz per acre per year.
difenoconazole + cyprodinil (Inspire Super 2.82SC)	3+9	16 to 20 fl oz/acre	7	0.5	Make no more than two sequential applications before alternating with fungicides that have a different mode of action. Apply no more than 80 fl oz per crop per acre per season.

TABLE 3-10. DISEASE CONTROL PRODUCTS FOR CUCURBITS (cont'd)

L. Quesada-Ocampo, Plant Pathologist, North Carolina State University

Disease/Material	FRAC Code	Rate of Material Formulation	Minimum Days Harv.	Minimum Days Reentry	Method, Schedule, and Remarks
POWDERY MILDEW (*PODOSPHAERA, GOLOVINOMYCES*) (cont'd)					
cyflufenamid (Torino 0.85SC)	U06	3.4 oz/acre	0	4 hr	Do not make more than 2 applications per crop.
cyprodinil + fludioxonil (Switch 62.5WG)	9+12	11 to 14 oz/acre	1	0.5	Make no more than 2 applications before alternating to a different fungicide. Maximum of 4 to 5 applications at high and low rates. Not for target spot, anthracnose, or Cercospora.
fixed copper (various)	M01	See label	See label	See label	See label. Rates vary depending on the formulation. Repeated use may cause leaf yellowing.
fluopyram + difenoconazole (Luna Flex)	7+3	8 fl oz/acre	0	0.5	Apply at the critical time for disease control and continue as needed on a 7- to 14-day interval
fluopyram + tebuconazole (Luna Experience 3.3F)	7+3	8 to 17 fl oz/acre	7	0.5	Make no more than 2 applications before alternating to a fungicide with different active ingredients. Do not rotate with tebuconazole.
flutianil (Gatten)	U13	6 to 8 fl oz/acre	0	0.5	**Not labeled on watermelon.** Do not make more than five applications per year.
fluxapyroxad + pyraclostrobin (Merivon 500SC)	7+11	4 to 5.5 fl oz/acre	0	0.5	Make no more than two sequential applications before alternating with fungicides that have a different mode of action. Maximum of 3 applications per crop.
metrafenone (Vivando)	50	15.4 fl oz/acre	0	0.5	Begin applications prior to disease and continue on in a 7- to 10- day interval.
myclobutanil (Rally 40WP)	3	2.5 to 5 oz/acre	0	1	Apply no more than 1.5 lb per acre per crop. Observe a 30-day plant-back interval.
penthiopyrad (Fontelis 1.67SC)	7	12 to 16 fl oz/acre	1	0.5	Make no more than 2 sequential applications before switching to another fungicide. Do not rotate with Pristine or Luna Experience.
pyraclostrobin + boscalid (Pristine 38WG)	11+7	12.5 to 18.5 oz/acre	0	1	Make no more than 4 applications per season
pyriofenone (Prolivo)	50	4 to 5 fl oz/acre	0	0.17	Make fungicide applications prior to disease on a 7- to 10-day interval. Do not apply more than 16 fl oz/acre/year.
quinoxyfen (Quintec 2.08SC)	13	4 to 6 fl oz/acre	3	0.5	Make no more than 2 applications before alternating to a different fungicide. Maximum of 24 fl oz/acre per year. **DO NOT USE ON SUMMER SQUASH or CUCUMBER**; labeled on winter squashes, pumpkins, gourds, melon, and watermelon.
sulfur (various)	M02	See label	See label	See label	See labels. Rates vary depending on the formulation. Do not use when temperature is over 90°F or on sulfur-sensitive varieties.
tebuconazole (Monsoon 3.6F)	3	4 to 6 fl oz/acre	7	0.5	Apply before disease appears when conditions favor development and repeat at 10- to 14-day intervals; max 24 fl oz per season.
triflumizole (Procure 50WS)	3	4 to 8 oz/acre	0	0.5	Begin applications at vining or first sign of disease and repeat at 7- to 10-day intervals.
SCAB (*CLADOSPORIUM*)					
acibenzolar-*S*-methyl (Actigard 50WP)	P01	0.5 to 1 oz/acre	0	0.5	Apply to healthy, actively growing plants. Do not apply to stressed plants. Apply no more than 8 oz per acre per season.
chlorothalonil (various)	M05	See label	See label	See label	See labels. Rates vary depending on the formulation.
WHITE MOLD (*SCLEROTINIA SCLEROTIORUM*)					
Coniothyrium minitans strain CON/M/91-08 (Contans WG)	BM02	1 to 4 lb/acre	0	4 hrs	Broadcast application and incorporate into top 2 inches of soil immediately after each application. Rotation with other fungicides allowed after 3 weeks following application. See label for different application method.
VINE DECLINE (*MONOSPORASCUS*)					
fludioxonil (Cannonball)	12	4 to 8 oz/acre	14	0.5	**APPLY ONLY TO MELONS.**

TABLE 3-11. EFFICACY OF PRODUCTS FOR DISEASE CONTROL IN CUCURBITS

L. Quesada-Ocampo, Plant Pathologist, North Carolina State University

Scale: E, excellent; G, good; F, fair; P, poor; NC, no control; ND, no data.

Active Ingredient [1,2]	Product	Fungicide group [F]	Preharvest interval (Days)	Alternaria Leaf Blight	Angular Leaf Spot	Anthracnose	Bacterial Fruit Blotch	Belly Rot	Cercospora Leaf Spot	Cottony Leak	Damping off (Pythium)	Downy Mildew [DM]	Fusarium wilt	Gummy Stem Blight	Phytophthora Blight (foliage and fruit)	Phytophthora Blight (crown and root)	Plectosporium Blight	Powdery Mildew	Target Spot
acibenzolar-*S*-methyl	Actigard	P01	0	NC	ND	NC	F	NC	NC	ND	ND	P	ND	NC	ND	ND	NC	ND	NC
ametoctradin + dimethomorph	Zampro	45+40	0	ND	NC	NC	NC	NC	NC	ND	ND	F	ND	NC	F	F	NC	NC	NC
azoxystrobin [2]	Quadris/Arius 250	11	1	G	NC	G[R]	NC	F	G	NC	NC	NC[R]	ND	NC[R]	NC	NC	F	NC [R]	G
azoxystrobin + chlorothalonil	Quadris Opti/ Arius Advance	11+ M05	0	G	NC	P[R]	NC	F	G	NC	NC	NC[R]	ND	F-P	NC	NC	F	F	F
azoxystrobin + difenoconazole	Quadris Top	11+3	1	ND	NC	P[R]	NC	ND	ND	ND	ND	ND	ND	F-P	ND	ND	F	F	ND
boscalid	Endura	7	0	ND	NC	NC	NC	ND	NC	NC	NC	NC	ND	NC[R]	NC	NC	ND	F[R]	ND
chlorothalonil [5]	various	M05	0	F	NC	G	NC	G-F	NC	NC	NC	P	ND	F	NC	NC	F	F	G
cyazofamid	Ranman	21	0	NC	NC	NC	NC	NC	NC	ND	ND	G-F	NC	F	NC	NC	NC	NC	NC
cyflufenamid	Torino	U06	0	NC	NC	NC	NC	NC	NC	NC	NC	NC	NC	NC	NC	NC	NC	F[R]	NC
cymoxanil	Curzate	27	3	NC	NC	NC	NC	NC	NC	ND	ND	P[R]	NC	NC	F	NC	NC	NC	NC
cyprodinil + fludioxonil	Switch	9+12	1	ND	NC	F	NC	ND	ND	NC	NC	NC	ND	F	NC	NC	F	F	NC
difenoconazole + benzovindiflupyr	Aprovia Top	3+7	0	ND	NC	F	NC	ND	NC	NC	NC	NC	ND	G	NC	NC	ND	G	ND
difenoconazole + cyprodinil	Inspire Super	3+9	7	F	NC	P	NC	NC	F	NC	NC	NC	ND	F	NC	NC	ND	G	ND
dimethomorph	Forum	40	0	NC	NC	NC	NC	NC	NC	NC	NC	P	ND	NC	P	NC	NC	NC	NC
ethaboxam	Elumin	22	2	NC	NC	NC	NC	NC	NC	NC	NC	G-F	NC	NC	G-F	ND	NC	NC	NC
famoxadone [2] + cymoxanil	Tanos	11+27	3	ND	NC	P	NC	ND	ND	NC	NC	F	NC	ND	NC	NC	NC	NC	NC
fenamidone	Reason	11	14	F	NC	ND	NC	NC	NC	NC	NC	P[R]	ND	NC	P	NC	NC	NC	NC
fixed copper	various [P,5]	M01	1	P	F	P	F	NC	P	NC	NC	P	ND	P	ND	NC	P	P	P
fluazinam	Omega	29	30,7	ND	NC	NC	NC	NC	NC	NC	NC	NC	ND	F	NC	NC	ND	NC	NC
fluopicolide	Presidio	43	2	NC	NC	NC	NC	NC	NC	NC	NC	P[R]	NC	NC	F[R]	F[R]	NC	NC	NC
fluopyram + difenoconazole	Luna Flex	7+3	0	ND	NC	NC	ND	NC	NC	NC	NC	NC	ND	G-F	NC	NC	NC	G	NC
fluopyram + tebuconazole	Luna Experience	7+3	7	ND	NC	NC	ND	NC	NC	NC	NC	NC	ND	G-F	NC	NC	NC	G	NC
fluopyram + trifloxystrobin	Luna Sensation	7+11	0	ND	NC	F	NC	NC	NC	NC	NC	NC	ND	F	NC	NC	NC	F	NC
fluoxastrobin	Evito	11	1	G	NC	G	NC	F	NC	NC	NC	NC[R]	ND	NC[R]	NC	NC	NC	F	NC[R]	F
flutianil	Gatten	U13	0	NC	NC	NC	NC	NC	NC	NC	NC	NC	ND	NC	NC	NC	NC	G	NC
flutriafol	Rhyme, Topguard [P]	3	0	ND	NC	NC	NC	ND	NC	NC	NC	NC	F	F	NC	NC	NC	P	NC
flutriafol + azoxystrobin	Topguard EQ	3+11	0	ND	NC	P	NC	ND	NC	NC	NC	NC	ND	P	NC	NC	P	P	F
fluxapyroxad + pyraclostrobin	Merivon	7+11	0	G	NC	F	NC	ND	ND	NC	NC	NC	ND	F	NC	NC	F	ND	ND
kresoxim-methyl	Sovran	11	0	ND	NC	ND	ND	NC	NC	NC	NC	ND	ND	NC[R]	ND	ND	NC[R]	ND
mancozeb	Various [5]	M03	5	F	NC	G	NC	G	NC	NC	NC	P	ND	F	P	NC	P	P	G
mancozeb + azoxystrobin	Dexter Max	M03+11	5	F	NC	G	NC	G	NC	NC	NC	P	ND	F	P	NC	P	P	G
mandipropamid	Revus	40	0	NC	NC	NC	NC	NC	NC	NC	NC	P	ND	NC	F	P	NC	NC	NC
mefenoxam [2]	Ridomil Gold EC, Ultra Flourish	4	0	NC	NC	NC	NC	NC	NC	F[R]	G[R]	NC	ND	NC	F[R]	F[R]	NC	NC	NC
mefenoxam [2] + chlorothalonil [5]	Ridomil Gold Bravo, Flouronil	4+M05	0	F	NC	F	NC	F	F[R]	F[R]	F[R]	ND	F	F[R]	NC	F	F	F

[1] Efficacy ratings do not necessarily indicate a labeled use for every disease
[2] Curative activity; locally systemic
[3] Systemic
[4] When used in combination with chlorothalonil or mancozeb, gives control
[5] Contact control only; no systemic control
[6] Check manufacturers label for compatibility with other products
[P] Can be phytotoxic at temperatures above 90°F; read label carefully
[F] To prevent resistance in pathogens, alternate fungicides within a group with fungicides in another group. Fungicides in the "M" group are generally considered "low risk" with no signs of resistance developing to most fungicides.
[R] Resistance reported in the pathogen
[DM] Ratings are based on efficacy and resistance on cucumber

TABLE 3-11. EFFICACY OF PRODUCTS FOR DISEASE CONTROL IN CUCURBITS (cont'd)

L. Quesada-Ocampo, Plant Pathologist, North Carolina State University

Scale: E, excellent; G, good; F, fair; P, poor; NC, no control; ND, no data.

Active Ingredient [1,2]	Product	Fungicide group [F]	Preharvest interval (Days)	Alternaria Leaf Blight	Angular Leaf Spot	Anthracnose	Bacterial Fruit Blotch	Belly Rot	Cercospora Leaf Spot	Cottony Leak	Damping off (Pythium)	Downy Mildew [DM]	Fusarium wilt	Gummy Stem Blight	Phytophthora Blight (foliage and fruit)	Phytophthora Blight (crown and root)	Plectosporium Blight	Powdery Mildew	Target Spot
mefenoxam [2] + copper [5]	Ridomil Gold Copper	4+M01	5	P	P	NC	P	NC	P	F[R]	F[R]	F[R]	ND	NC	F[R]	NC	P	NC	P
mefenoxam [2] + mancozeb [5]	Ridomil Gold MZ	4+M03	5	F	NC	F	NC	NC	F	F[R]	F[R]	F[R]	ND	P	F[R]	NC	F	NC	F
metrafenone	Vivando	50	0	NC	NC	NC	NC	NC	NC	NC	NC	NC	ND	NC	NC	NC	NC	G	NC
myclobutanil [2]	Rally	3	0	NC	NC	NC	NC	NC	NC	NC	NC	NC	ND	NC	NC	NC	NC	F	NC
oxathiapiprolin + chlorothalonil	Orondis Opti	49+M05	0	F	NC	F	NC	ND	F	NC	NC	G	ND	F	G	G	F	P	F
oxathiapiprolin + mandipropamid	Orondis Ultra	49+40	0	ND	NC	ND	NC	ND	ND	NC	NC	G	ND	ND	G	G	ND	ND	ND
oxathiapiprolin + mefenoxam	Orondis Gold	49+4	0	NC	NC	ND	NC	ND	ND	F[R]	NC	G	ND	ND	G	G	ND	ND	ND
penthiopyrad	Fontelis	7	1	ND	NC	F	NC	ND	F	NC	NC	NC	ND	NC[R]	NC	NC	NC	F	NC
phosphonate [6]	various	P07	0.5	NC	NC	NC	NC	NC	NC	NC	NC	P	ND	NC	NC	F	NC	NC	NC
potassium phosphite + tebuconazole	Viathon	P07+3	7	ND	NC	ND	NC	ND	ND	ND	ND	P	ND	F	ND	ND	NC	F	NC
propamocarb	Previcur Flex/Bruin	28	2	NC	NC	NC	NC	NC	NC	NC	NC	F[R]	ND	NC	F	NC	NC	NC	NC
prothioconazole	Proline	3	7	ND	NC	G	NC	ND	F	NC	NC	NC	G	G	NC	NC	NC	F	ND
pydiflumetofen + fludioxonil	Miravis Prime	7+12	1	ND	ND	ND	ND	ND	ND	ND	ND	F	G	ND	ND	ND	ND	ND	ND
pyraclostrobin [2]	Cabrio, Pyrac	11	0	G	NC	F	NC	ND	NC	ND	NC	NC[R]	ND	NC[R]	P	NC	G	NC[R]	E
pyraclostrobin [2] + boscalid [2]	Pristine	11+7	0	G	NC	P	NC	ND	G	NC	NC	NC[R]	ND	NC[R]	P	NC	F	F	E
quinoxyfen	Quintec	13	3	NC	NC	NC	NC	NC	NC	NC	NC	NC	ND	NC	NC	NC	NC	G[R]	NC
sulfur	various [P,5]	M02	0	NC	NC	NC	NC	NC	NC	NC	NC	NC	ND	NC	NC	NC	NC	F	NC
tebuconazole	Monsson	3	7	ND	NC	NC	NC	NC	F	NC	NC	NC	F	F[R]	NC	NC	NC	F	NC
thiophanate-methyl [3]	Topsin M	1	1	F	NC	G	NC	F	F	NC	NC	NC	NC	NC[R]	NC	NC	F	NC[R]	P
trifloxystrobin [2]	Flint	11	0	G	NC	F	NC	ND	ND	NC	NC	NC[R]	ND	NC[R]	NC	NC	G	NC[R]	G
triflumizole	Procure	3	0	NC	NC	NC	NC	NC	NC	NC	NC	NC	NC	NC	NC	NC	NC	F	NC
zoxamide + chlorothalonil	Zing!	22+ M05	5	F	NC	F	NC	F	NC	NC	NC	P	ND	F	P	NC	F	P	F
zoxamide + mancozeb	Gavel	22+ M03	5	F	NC	F	NC	NC	F	NC	NC	P	ND	F	P	NC	F	P	F

[1] Efficacy ratings do not necessarily indicate a labeled use for every disease
[2] Curative activity; locally systemic
[3] Systemic
[4] When used in combination with chlorothalonil or mancozeb, gives control
[5] Contact control only; no systemic control
[6] Check manufacturers label for compatibility with other products
[P] Can be phytotoxic at temperatures above 90°F; read label carefully
[F] To prevent resistance in pathogens, alternate fungicides within a group with fungicides in another group. Fungicides in the "M" group are generally considered "low risk" with no signs of resistance developing to most fungicides.
[R] Resistance reported in the pathogen
[DM] Ratings are based on efficacy and resistance on cucumber

TABLE 3-12. IMPORTANCE OF ALTERNATIVE MANAGEMENT PRACTICES FOR DISEASE CONTROL IN CUCURBITS

L Quesada-Ocampo, Plant Pathologist, North Carolina State University

Scale: E, excellent; G, good; F, fair; P, poor; NC, no control; ND, no data.

Strategy	Alternaria leaf blight	Angular leaf spot	Anthracnose	Bacterial fruit blotch	Bacterial wilt	Belly rot	Cercospora leaf spot	Choanephora fruit rot	Cottony leak	Downy mildew	Fusarium wilt	Gummy stem blight	Mosaic virus	Phytophthora blight	Plectosporium blight	Powdery mildew	Pythium damping off	Root knot nematode	Target spot
Avoid field operations when leaves are wet	P	F	P	F	F	NC	NC	P	NC	P	NC	P	NC	NC	ND	NC	NC	NC	NC
Avoid overhead irrigation	F	F	F	F	P	NC	P	NC	NC	F	NC	F	NC	F	P	P	NC	NC	P
Change planting date from Fall to Spring [1]	G	P	G	P	P	F	G	F	F	G	P	G	F	F	F	F	G	G	G
Cover cropping with antagonist	NC	NC	NC	NC	NC	NC	NC	NC	NC	NC	F	NC	NC	NC	NC	NC	NC	F	NC
Crop rotation with non-host (2 to 3 years)	F	F	F	F	NC	P	F	NC	NC	NC	F	F	NC	F	F	NC	P	F	F
Deep plowing	P	NC	P	NC	NC	F	P	NC	NC	NC	F	F	NC	P	P	NC	P	F	P
Destroy crop residue immediately	F	P	F-G	P	P	P	P	NC	P	F	F	F	F	P	P	F	NC	F	P
Encourage air movement [2]	F	P	F	P	NC	NC	F	F	F	F	NC	F	NC	NC	P	NC	NC	NC	F
Soil organic amendments [3]	ND	NC	ND	NC	NC	P	ND	NC	F	NC	F	NC	ND	NC	P	ND	NC	F	F	ND
Insecticidal/horticultural oils [4]	NC	NC	NC	NC	F	NC	NC	NC	NC	NC	NC	NC	F	NC	NC	F	NC	NC	NC
pH management (soil)	NC	NC	NC	NC	NC	NC	NC	ND	NC	NC	NC	NC	NC	ND	NC	NC	ND	ND	NC
Plant in well-drained soil	NC	NC	NC	NC	NC	F	NC	P	F	NC	NC	NC	NC	F	NC	NC	F	P	NC
Plant on raised beds	NC	NC	NC	NC	NC	P	NC	P	F	NC	NC	F	NC	F	NC	NC	F	P	NC
Plastic mulch bed covers	NC	NC	NC	NC	NC	F	NC	P	F	NC	G	NC	NC	F	P	NC	F	NC	NC
Postharvest temperature control (fruit)	NC	NC	F	P	NC	F	NC	F	F	NC	NC	NC	NC	F	F	NC	NC	NC	NC
Reflective mulch (additional effect over plastic mulch)	NC	NC	NC	NC	NC	NC	NC	NC	NC	NC	NC	NC	F	NC	NC	NC	NC	NC	NC
Reduce mechanical injury	P	P	P	P	F	P	P	P	NC	P	P	P	P	P	NC	NC	NC	P
Rogue diseased plants/fruit (home garden)	F	P	P	P	P	NC	P	P	P	P	P	P	F	F	NC	NC	P	F	P
Row covers (insect exclusion)	NC	NC	NC	NC	G	NC	NC	NC	NC	NC	NC	NC	G	NC	NC	NC	NC	NC	NC
Soil solarization (reduce soil inoculum)	P	NC	P	NC	NC	F	P	NC	P	NC	F	P	NC	P	P	NC	F	P	P
Pathogen-free planting material	P	E	F	E	NC	NC	F	NC	NC	NC	G	E	NC	NC	NC	NC	F	NC	NC
Resistant cultivars [5]	ND	ND	E	ND	ND	E	ND	ND	ND	G	E	ND	E	ND	ND	E	ND	ND	ND
Grafting [5]	NC	NC	NC	NC	NC	NC	NC	NC	NC	NC	E	NC	NC	NC	NC	NC	NC	G	NC
Destroy volunteer plants	F	F	F	F	F	NC	F	NC	NC	F	G	F	F	F	NC	F	NC	P	F

[1] Early planting reduces risk

[2] Air movement can be encouraged by increasing plant spacing, orienting beds with prevailing wind direction and increasing exposure of field to prevailing wind

[3] Soil organic amendments = cover crops; composted organic wastes

[4] Insecticidal/Horticultural oil = Sunspray Ultra-Fine Spray Oil (Sun Company, Inc.), JMS Stylet oil; Safe-T-Side (Brandt Consolidated, Inc.); PCC 1223 (UnitedAg Products

[5] Resistance available in some cucurbits

TABLE 3-13. EXAMPLE SPRAY PROGRAM FOR FOLIAR DISEASE CONTROL IN WATERMELON PRODUCTION

A. Keinath, Plant Pathologist, Clemson University

This spray program is based on research conducted at the Clemson Coastal Research and Education Center, Charleston, SC, and on a survey of watermelon fields in South Carolina in 2015 and 2016. The most common diseases in both survey years were gummy stem blight and powdery mildew. The spring program is designed to manage bacterial fruit blotch, various bacterial leaf spots, gummy stem blight, powdery mildew, anthracnose, and downy mildew. The fall program is designed to manage gummy stem blight, downy mildew, powdery mildew, and anthracnose.

- Protectants (chlorothalonil and mancozeb) are effective against anthracnose all season, but other disease-specific fungicides must be used against gummy stem blight, downy mildew, and powdery mildew. See Tables 3-10 and 3-11. Products rated as good (G) in Table 3-11 may be substituted for products in the spray program below.
- Start spraying when vines start to run or no later than when the first blooms (the male ones) open.
- Check http://cdm.ipmpipe.org to see where and when downy mildew has been reported on watermelon. Rotate Orondis Ultra with Ranman if downy mildew is present in your field.
- From vine run until mid-May, spray every 10 days. After mid-May or when powdery and downy mildew typically show up in your area, spray every week through harvest regardless of the weather. Weekly sprays are needed to protect watermelon from powdery and downy mildew. Dry weather limits gummy stem blight but promotes powdery mildew; dry weather does not limit downy mildew or anthracnose if they are already present in a field.
- Do not stop spraying until you stop harvesting. Downy and powdery mildew can attack a crop any time it goes more than one week without a fungicide spray. Fungicides with a 7-day PHI are not recommended during the harvest period (usually after week 5); note that mancozeb and Gavel have a 5-day PHI.
- If this spray schedule is used to select fungicides for other cucurbits (vine crops), note that cucumber does not need to be sprayed for powdery mildew, and most hybrid cantaloupe cultivars also are resistant to powdery mildew. For cantaloupe, substitute a FRAC 11 fungicide for mancozeb in weeks 5 and 7 to manage Alternaria leaf spot.

Spray	Fungicide Program for Spring Watermelon*	Comments on Spring Program	Fungicide Program for Fall Watermelon*	Comments on Fall Program
1 (vine run)	mancozeb + fixed copper	For prevention of bacterial leaf spots and fruit blotch.	chlorothalonil	
2	chlorothalonil	If fruit blotch or bacterial leaf spot is a concern, use mancozeb + fixed copper instead. Do not tank-mix copper with chlorothalonil.	chlorothalonil	
3a**	chlorothalonil or mancozeb	If fruit blotch or bacterial leaf spots are a concern, substitute mancozeb + fixed copper.	mancozeb + **Ranman**	Apply Ranman if downy mildew has been reported on watermelon in your state.
3b**	OR Quadris Top	Use this fungicide if if anthracnose was found the previous year.	OR mancozeb + Gatten	Apply Gatten if weather is unusually dry to prevent powdery mildew.
4	chlorothalonil (or mancozeb)	If fruit blotch or bacterial leaf spot is a concern, substitute mancozeb + fixed copper.	Quadris Top	Quadris Top protects against anthracnose and gummy stem blight.
5a**	mancozeb + Gatten	Starting week 5, use mancozeb to avoid injury to fruit on hot, sunny days. Note 5-day PHI on mancozeb.	**Gavel**	Gavel protects against anthracnose, gummy stem blight, and downy mildew.
5b**	OR Gatten + Luna Flex or Miravis Prime	Use this program if gummy stem blight is present.	(same as 5a)	
6	Gavel	Note 5-day PHI.	Miravis Prime	
6b	OR mancozeb plus **Orondis Ultra**	Apply Orondis Ultra if downy mildew has been reported on watermelon in your state or a neighboring state. Tank-mix Orondis Ultra with mancozeb to protect against gummy stem blight and anthracnose.		
7a**	mancozeb + Vivando	Note 5-day PHI on mancozeb.	mancozeb + **Ranman**	If downy mildew is present. Note 5-day PHI on mancozeb.
7b**	Miravis Prime + Vivando	Use if gummy stem blight is present.	Miravis Prime	If gummy stem blight is present.
8	mancozeb + **Ranman**		chlorothalonil	
9-12	If more sprays are needed after spray 8 until the last harvest, apply sprays 5 to 8 again, BUT check maximum number of applications allowed per crop			

* Fungicides for downy mildew are in bold and should be used if downy mildew has been reported on watermelon in the current season (visit *http://cdm.ipmpipe.org/* to access the downy mildew forecast). Fungicides for powdery mildew are underlined and should be applied in spring and during dry periods in fall

** Option "a" is a lower cost treatment that may be less effective. Option "b" is a more expensive systemic fungicide that is more effective when disease is already in the field or when weather conditions favor disease getting worse

TABLE 3-14. DISEASE CONTROL PRODUCTS FOR EGGPLANT

A. Keinath, Plant Pathologist, Clemson University

Disease/Material	FRAC Code	Rate of Material Formulation	Minimum Days Harv.	Minimum Days Reentry	Method, Schedule, and Remarks
azoxystrobin (various)	11	See labels	0	4 hr	Apply at flowering to prevent green fruit rot. Limit of 61.5 fl oz per acre per season. Make no more than **one** application before alternating with fungicides that have a different mode of action.
chlorothalonil + cymoxanil (Cymbol Advance)	M05+27	2 to 2.4 pint/acre	3	0.5	Make no more than 8 applications at the low rate or 7 applications at the high rate per year.
difenoconazole + benzovindiflupyr (Aprovia Top 1.62EC)	3+7	10.5 to 13.5 fl oz/acre	14	0.5	Make no more than 2 consecutive applications before switching to a non-Group 7 fungicide. Make no more than 5 applications at the low rate or 4 applications at the high rate per year.
flutriafol (Rhyme 2.08SC)	3	7 fl oz/acre	0	0.5	Limit of 4 applications per year.
flutriafol + azoxystrobin (Topguard EQ 4.29SC)	3+11	4 to 8 fl oz/acre	0	0.5	Limit of 4 applications per year.

ANTHRACNOSE FRUIT ROT (*COLLETOTRICUM* SPP.), EARLY BLIGHT (*ALTERNARIA* SPP.), GRAY MOLD (*BOTRYTIS CINEREA*), GRAY LEAF SPOT (*STEM- PHYLIUM* SPP.), SEPTORIA LEAF SPOT (*SEPTORIA LYCOPERSICI*)

Note: these leaf spots are not common on eggplant in the southeastern U.S., so preventative applications are not needed.

Disease/Material	FRAC Code	Rate of Material Formulation	Harv.	Reentry	Method, Schedule, and Remarks
boscalid (Endura 70 WG)	7	2.5 to 3.5 oz/acre	0	0.5	Limit of 21 oz per acre per season. Make no more than two sequential applications before alternating with fungicides that have a different mode of action. Labeled for early blight and gray mold ONLY.
chlorothalonil (various)	M05	1.5 pt/acre	3	1	Limit of 12 pt per acre per season. **Labeled for anthracnose and gray mold ONLY.**
chlorothalonil + cymoxanil (Cymbol Advance)	M05+27	2 to 2.4 pint/acre	3	0.5	Make no more than 8 applications at the low rate or 7 applications at the high rate per year.
fenamidone (Reason 500SC)	11	5.5 to 8.2 fl oz/acre	14	0.5	Limit of 24.6 fl oz per growing season. Make no more than **one** application before rotating to another effective fungicide with a different mode of action. Labeled for early blight only.
fludioxonil + pydiflumetofen (Miravis Prime 3.34SC)	12+7	9.2 to 11.4 fl oz/acre	0	0.5	Use high rate for gray mold. Limit of 2 applications per crop per year. Make no more than 2 sequential applications of Miravis Prime or other Group 7 and 12 fungicides before rotating to another effective fungicide with a different mode of action **Do not use in enclosed structures.**
fluopyram + difenoconazole (Luna Flex 375SC)	7+3	8.0 to 13.6 fl oz/acre	0	0.5	Make no more than 2 sequential applications of Luna Flex or other Group 7 fungicides before rotating to another effective fungicide with a different mode of action. Maximum of 2 applications per year. **Also labeled for powdery mildew and target spot.**
fluopyram + trifloxystrobin (Luna Sensation 500SC)	7+11	5.0 to 7.6 fl oz/acre	7	0.5	Limit of 27.1 fl oz (3 and 5 applications at high and low rates, respectively) per acre per year. Make no more than 2 consecutive applications before switching to a non- Group 7 and non-Group 11 fungicide.
fluoxastrobin (Aftershock, Evito 280SC)	11	2 to 5.7 fl oz/acre	3	0.5	Limit of 22.8 fl oz per acre per season. Make no more than **one** application before alternating with fungicides that have a different mode of action. **NOTE: Do not overhead irrigate for 24 hours following a spray application.** Labeled for early blight only.
mefentrifluconazole (Cevya)	3	3 to 5 fl oz/acre	0	0.5	Limit of 15 fl oz (5 and 3 applications at low and high rates, respectively) per acre per year.
penthiopyrad (Fontelis1.67SC)	7	16 to 24 fl oz/acre	0	0.5	Limit of 72 fl oz per acre per year. Make no more than 2 consecutive applications before rotating to another effective fungicide with a different mode of action.
pyraclostrobin (various)	11	8 to 12 oz/acre	0	4 hr	Apply at flowering to manage green fruit rot. Limit of 96 oz per acre per season. Make no more than **one** application before alternating with fungicides that have a different mode of action.
pyraclostrobin + fluxapyroxad (Priaxor 500SC)	11+7	4.0 to 8.0 fl oz/acre	0	0.5	Limit of 24 fl oz per acre per season. Make no more than two consecutive applications before rotating to another effective fungicide with a different mode of action. Labeled for anthracnose and early blight ONLY.
tetraconazole (Mettle 125ME)	3	6 to 8 fl oz/acre	0	0.5	Limit of 16 fl oz per acre per season. Make no more than two consecutive applications before rotating to another effective fungicide with a different mode of action.
trifloxystrobin (Flint Extra 42.6SC)	11	3.0 to 3.8 fl oz/acre	3	0.5	Limit of 16 fl oz or 5 applications per acre per season. Alternate every application with a non-FRAC Group 11 fungicide.
zoxamide + chlorothalonil (Zing! 4.9SC)	22+M05	34 fl oz/acre	3	0.5	Limit of 8 applications per year. Do not use in greenhouses.

PHOMOPSIS BLIGHT (*DIAPORTHE VEXANS*)

Disease/Material	FRAC Code	Rate of Material Formulation	Harv.	Reentry	Method, Schedule, and Remarks
copper (various)	M01	See labels	See labels	2	Make the first application at flowering. If disease is present, make additional applications at 7- to 10-day intervals. Do not spray copper when temperatures are above 90°F. Phomopsis fruit rot may develop in postharvest storage.

TABLE 3-14. DISEASE CONTROL PRODUCTS FOR EGGPLANT (cont'd)

A. Keinath, Plant Pathologist, Clemson University

Disease/Material	FRAC Code	Rate of Material Formulation	Minimum Days Harv.	Minimum Days Reentry	Method, Schedule, and Remarks
PHYTOPHTHORA BLIGHT (*PHYTOPHTHORA CAPSICI*)					
ametoctradin + dimethomorph (Zampro 525SC)	45+40	14 fl oz/acre	4	0.5	Limit of 3 applications per acre per season. Make no more than two sequential applications before rotating to another effective fungicide with a different mode of action.
cyazofamid (Ranman 400SC)	21	2.75 fl oz/acre	0	0.5	Limit of 16.5 fl. oz per acre per season. Apply to the base of the plant at transplanting or in the transplant water. Make no more than three consecutive applications followed by three consecutive applications of another effective fungicide with a different mode of action.
copper (various)	M01	See labels	0	2	Begin applications when conditions first favor disease development and repeat at 3-to10-day intervals if needed depending on disease severity. Use the higher rates when conditions favor disease. Do not spray copper when temperatures are above 90°F.
dimethomorph (Acrobat, Forum)	40	6 fl oz/acre	0	0.5	**SUPPRESSION ONLY.** Limit of 30 fl oz per acre per season. Make no more than two sequential before alternating with fungicides that have a different mode of action. **NOTE: Must tank-mix with another fungicide with a different mode of action.**
ethaboxam (Elumin 4SC)	22	8 fl oz/acre	2	0.5	Soil spray or foliar applications. 2 applications per crop per year. Must rotate with a non-FRAC Group 22 fungicide in between applications.
famoxadone + cymoxanil (Tanos 50DF)	11+27	8 to 10 oz/acre	3	0.5	**SUPPRESSION ONLY.** Make no more than **one** application before alternating with a fungicide with a different mode of action. **NOTE: Must tank-mix with another fungicide with a different mode of action (i.e., copper).**
fluazinam (Omega 500 F, Ventana 500F)	29	1 to 1.5 pt/acre	30	0.5	Apply as a soil drench at 1.5 pt per acre. For foliar applications, use 1 pt per acre. Limit of 9 pt per acre per season.
fluopicolide (Presidio 4SC)	43	3 to 4 fl oz/acre	2	0.5	Limit of 4 applications at the low rate or 3 applications at the high rate per season. Apply no more than two times sequentially before alternating with fungicides that have a different mode of action. **NOTE: Must be tank mixed with another mode of action product.**
mefenoxam + copper hydroxide (Ridomil Gold + Copper)	4+M01	2 lb/acre	7	2	See label for an optimal spray program. Limit of four applications per crop per year. Do not exceed 0.4 lb a.i. per acre per season of mefenoxam + metalaxyl (MetaStar).
oxathiapiprolin + mefenoxam (Orondis Gold DC 1.17DC) (Orondis Gold 1.67SC)	U15+4	28.0-55.0 fl oz/acre 4.8 to 9.6 fl oz/acre	7 0	2 hr 4 hr	DC: Soil applications at planting, primarily for Phytophthora blight (not *Pythium*). 200:Make no more than two sequential applications before alternating with fungicides that have a different mode of action. Maximum 9.2 fl oz/acre per year in each field.
oxathiapiprolin + mandipropamid (Orondis Ultra 2.33SC)	U15+40	5.5 to 8.0 fl oz/acre	1	4 hr	Make no more than two sequential applications before alternating with fungicides that have a different mode of action. Maximum 32 fl oz/acre per year.
mandipropamid (Revus 2.08F, Micora)	40	8 fl oz/acre	1	0.5	**SUPPRESSION ONLY.** Limit of 4 applications per acre per season. **NOTE: Must tank-mix with another fungicide with a different mode of action (i.e., copper).**
POWDERY MILDEW (*LEVEILLULA TAURICA*)					
Powdery mildew has not been reported on eggplant in the U.S. If you see it, contact your county Extension agent.					
PYTHIUM ROOT ROT (*PYTHIUM SPP.*)					
mefenoxam (various)	4	See labels	—	2	Apply in a 12 to 16 in. band or in 20-to 50-gal water per acre in transplant water. Mechanical incorporation or 0.5 to 1 in. irrigation water is needed for movement into root zone if rain in not expected. After initial application, 2 supplemental applications (1 pt per treated acre) can be applied.
metalaxyl (MetaStar 2E)	4	4 to 8 pt/treated acre	7	2	Limit of 12 pt per acre per season. Preplant (soil incorporated), at planting (in water or liquid fertilizer), or as a basil-directed spray after planting. See label for the guidelines for supplemental applications.
cyazofamid (Ranman 400SC)	21	2.75 fl oz/acre	0	0.5	Limit of 16.5 fl. oz per acre per season. Apply to the base of the plant at transplanting or in the transplant water. Make no more than three consecutive applications followed by three consecutive applications of another effective fungicide with a different mode of action.
pyraclostrobin (various)	11	12 to 16 oz/acre	0	4 hr	**SUPPRESSION ONLY.** Apply at flowering to manage green fruit rot. Limit of 4 applications per acre per season. Make no more than **one** application before alternating with fungicides that have a different mode of action.
pyraclostrobin + fluxapyroxad (Priaxor 500SC)	11+7	4.0 to 8.0 fl oz/acre	0	0.5	Limit of 2 applications per season. Best option based on tests on tomato in SC.
RHIZOCTONIA SEEDLING AND ROOT ROT (*RHIZOCTONIA SOLANI*)					
azoxystrobin (various)	11	0.4 to 0.8 fl oz/ 1,000 row feet	—	4 hr	Make in-furrow or banded applications shortly after plant emergence. Under cool, wet conditions, crop injury from soil directed applications may occur.
difenoconazole + benzovindiflupyr (Aprovia Top 1.62EC)	3+7	10.5 to 13.5 fl oz/acre	14	0.5	Make no more than 2 consecutive applications before switching to a non-Group 7 fungicide. Make no more than 5 applications at the low rate or 4 applications at the high rate per year.

TABLE 3-14. DISEASE CONTROL PRODUCTS FOR EGGPLANT (cont'd)

A. Keinath, Plant Pathologist, Clemson University

Disease/Material	FRAC Code	Rate of Material Formulation	Minimum Days Harv.	Minimum Days Reentry	Method, Schedule, and Remarks
SOUTHERN BLIGHT (*AGROATHELIA ROLFSII*)					
fluoxastrobin (Aftershock, Evito 280SC)	11	2 to 5.7 fl oz/acre	3	0.5	Limit of 22.8 fl oz per acre per season. Make no more than **one** application before alternating with fungicides that have a different mode of action. **NOTE: Do not overhead irrigate for 24 hours following a spray application.**
penthiopyrad (Fontelis 1.67SC)	7	16 to 24 fl oz/acre	0	0.5	Apply 5- to 10-days after transplanting and again 14 days later. Limit of two applications per crop. Follow with a FRAC Group 11 fungicide if additional protection is needed.
VERTICILLIUM WILT (*VERTICILLIUM DAHLIAE*)					
polyoxin D (OSO 5%)	19	6.5 to 13 fl oz/acre	0	4 hr	**SUPPRESSION ONLY.** Can be applied using banded or irrigation water applications. Limit of 6 applications at maximum rate per acre per season.

TABLE 3-15. DISEASE CONTROL PRODUCTS FOR GARLIC

N. Gauthier, Plant Pathologist, University of Kentucky (Last updated in 2023)

Disease/Material	FRAC Code	Rate of Material Formulation	Minimum Days Harv.	Minimum Days Reentry	Method, Schedule, and Remarks
BOTRYTIS BLIGHT (*BOTRYTIS* SPP.), CLADOSPORIUM LEAF BLOTCH (*CLADOSPORIUM ALLII*), PURPLE BLOTCH (*ALTERNARIA PORRI*), DOWNY MILDEW (*PERONO- SPORA DESTRUCTOR*), RUST (*PUCCINIA ALLII*)					
azoxystrobin (various)	11	6.2 to 15.4 fl oz/acre	0	4 hr	High risk for resistance; use premix products when possible. Use higher rate for downy mildew and Botrytis. Do not make more than two sequential applications.
azoxystrobin + difenoconazole (Quadris Top)	11+3	14 fl oz/acre	7	0.5	Begin sprays prior to disease onset and spray on a 7- to 14-day schedule. Do not rotate with Group 11 fungicides.
azoxystrobin + chlorothalonil (Quadris Opti)	11+M05	1.6 to 3.2 pt/acre	14	0.5	Make no more than one application before alternating with a fungicide with a different mode of action. Use higher rates for downy mildew.
azoxystrobin + mancozeb (various)	11+M03	3.2 lb/acre	7	0.5	Follow a protective 7-day schedule. Observe season limit for azoxystrobin applications.
azoxystrobin + propiconazole (various)	3+11	14 to 26 oz/acre	0 to 14	0.5	Also labeled for white rot. Apply preventatively on a 7- to 14-day schedule. Season limit 56 oz/acre.
azoxystrobin + tebuconazole (various)	11+3	8.6 to 32 oz/acre	7	0.5	Also labeled for drip-irrigation or banded applications for white rot. Apply preventatively on a 10- to 14-day schedule. Season limit 70 oz/acre.
benzovindiflupyr + difenoconazole (Aprovia Top)	7+3	10.5 oz/acre	7	0.5	Not Botrytis. Includes Stemphylium and powdery mildew. Spreading/penetrating type adjuvant recommended.
boscalid (Endura) 70 WG	7	6.8 oz/acre	7	0.5	Not for downy mildew. Do not make more than 2 sequential applications or more than 6 applications per season.
chlorothalonil (various)	M05	See label	7	2	Spray at first appearance; 7- to 14-day intervals.
chlorothalonil + cymoxanil (Ariston)	M05+27	1.6 to 2.4 pt/acre	7	0.5	No efficacy on Botrytis blight; limited efficacy on downy mildew. Apply prior to favorable infection periods; continue on 7- to 9-day interval; alternate with a different mode of action.
chlorothalonil + oxathiapiprolin (Orondis Opti)	M05+49	1.75 to 2.5 pt/acre	7	0.5	Do not combine with other products containing oxathiapiprolin (any Orondis product).
chlorothalonil + tebuconazole (Muscle ADV)	M05+3	1.1 to 1.6 pt/acre	7 to 14	0.5	Rust and purple blotch only.
chlorothalonil + zoxamide (Zing!)	M05+22	30 fl oz/acre	7	0.5	Follow protective spray schedule when diseases are in the area; continue on 7-day interval. Moderate efficacy on downy mildew.
difenoconazole + cyprodinil (Inspire Super)	3+9	16 to 20 fl oz/acre	14	0.5	Make no more than two applications before alternating with a fungicide with a different mode of action.
famoxadone + cymoxanil (Tanos)	11+27	8 oz/acre	3	0.5	No efficacy on Botrytis blight; low to moderate efficacy on downy mildew.
fenamidone (Reason)	11	5.5 oz/acre	7	0.5	No efficacy on Botrytis blight; low to moderate efficacy on downy mildew.
fixed copper (various)	M01	various	1	1	Also effective against foliar bacterial diseases.
fluazinam (various)	29	1.0 pt/acre	7	1	Initiate sprays when conditions are favorable for disease at disease onset. Spray on a 7- to 10-day or schedule.

TABLE 3-15. DISEASE CONTROL PRODUCTS FOR GARLIC (cont'd)

N. Gauthier, Plant Pathologist, University of Kentucky (Last updated in 2023)

Disease/Material	FRAC Code	Rate of Material Formulation	Minimum Days Harv.	Minimum Days Reentry	Method, Schedule, and Remarks
BOTRYTIS BLIGHT (*BOTRYTIS* SPP.), CLADOSPORIUM LEAF BLOTCH (*CLADOSPORIUM ALLII*), PURPLE BLOTCH (*ALTERNARIA PORRI*), DOWNY MILDEW (*PERONOSPORA DESTRUCTOR*), RUST (*PUCCINIA ALLII*) (cont'd)					
fluopyram + tebuconazole (Luna Experience)	7+3	8 to 12.8 oz/acre	7	0.5	Not for downy mildew but labeled for white rot suppression. Apply preventatively on 10- to 14-day schedule.
fluopyram + pyrimethanil (Luna Tranquility)	7+9	16 to 27 oz/acre	7	0.5	Not for downy mildew but labeled for white rot suppression. Apply preventatively on 10-to 14-day schedule.
fluxapyroxad + pyraclostrobin (Merivon)	7+11	4 to 11 fl oz/acre	7	0.5	Use higher rates for downy mildew suppression. Apply at disease onset; continue on a 7- to 14-day schedule. No more than 3 applications per season.
mancozeb + zoxamide (Gavel)	M03+22	1.5 to 2 lb/acre	7	2	Apply on a protective spray schedule. Do not apply to exposed bulbs.
mancozeb + azoxystrobin (Dexter Max)	M03+11	3.2 lb/acre	7	1	Do not apply to exposed bulbs.
mefenoxam + chlorothalonil (Ridomil Gold/Bravo)	4+M05	2.5 pt/acre	7	2	Spray at first appearance; 7- to 14-day intervals.
penthiopyrad (Fontelis)	7	16 to 24 oz/acre	3	0.5	Not for downy mildew, but also labeled for white rot in-furrow, drenched, or drip irrigated. Use preventatively on a 7- to 14-day schedule.
propiconazole (various)	3	2 to 8 oz/acre	0 to 14	0.5	Purple blotch and Botrytis suppression. Apply preventatively in no less than 15 gal/acre, on a 7- to 10-day schedule (16 oz/acre season limit).
pyraclostrobin (Cabrio)	11	8 to 12 oz/acre	7	0.5	Not for *Botrytis* blight. QOIs have lost their efficacy against downy mildew.
mancozeb + zoxamide (Gavel)	M03+22	1.5 to 2 lb/acre	7	2	Make no more than 2 sequential applications and no more than 6 applications per season.
mancozeb + azoxystrobin (Dexter Max)	M03+11	3.2 lb/acre	7	1	Do not apply to exposed bulbs.
pyraclostrobin +boscalid (Pristine 38 WG)	11+7	10.5 to 18.5 oz/acre	7	1	Use the highest rate for suppression only on downy mildew. Make no more than 6 appli- cations per season.
pyrimethanil (Scala 5F)	9	9 or 18 fl oz/acre	7	0.5	Not for downy mildew. Use lower rate in a tank-mix with broad-spectrum fungicide and higher rate when applied alone. Do not apply more than 54 fl oz per crop.
tebuconazole (various)	3	4 to 6 oz/acre	7	0.5	Rust and purple blotch only. Also labeled for white rot. Apply preventatively on a 10- to 14-day schedule. Season limit 12 oz/acre.
DOWNY MILDEW (*PERONOSPORA DESTRUCTOR*)					
acibenzolar-*S*-methyl (Actigard 50 WG)	P01	0.75 to 1 oz/acre	7	0.5	Downy mildew, iris yellow spot virus, and bacterial leaf streak. Apply preventative- ly; avoid usage during periods of plant stress.
ametoctradin + dimethomorph (Zampro)	45+40	14.0 fl oz/acre	0	12 hr	Tank-mix with a broad-spectrum fungicide like chlorothalonil or mancozeb.
cyazofamid (Ranman)	21	2.75 to 3 oz/acre	0		
dimethomorph (Forum 50 WP)	40	6.4 oz/acre	0	0.5	Must be applied as a tank-mix with another fungicide active against downy mildew; apply every 7- to 10-days. Do not make more than two sequential applications.
fluopicolide (Presidio)	43	3 to 4 oz/acre	2	0.5	Tank-mix with a nonionic surfactant and apply on preventative schedule.
mandipropamid (Revus)	40	8.0 fl oz/acre	7	0.5	Apply as a tank-mix with another fungicide active against downy mildew. Apply with a silicone-based adjuvant. 7- to 10-day schedule.
mandipropamid + oxathiapiprolin (Orondis Ultra)	40+49	5.5 to 8 oz/acre	7	4 hr	Do not combine with other products containing oxathiapiprolin (any Orondis product).
mefenoxam + mancozeb (Ridomil Gold MZ)	4+M03	2.5 lb/acre	7	2	Use with a suitable adjuvant.
WHITE ROT (*SCLEROTIUM CUPARTUM*)					
azoxystrobin (various)	11	See labels	0	4 hr	Do not make more than two sequential applications.
azoxystrobin + chlorothalonil (Quadris Opti)	11+M05	1.6 to 3.2 pt/acre	7	0.5	Make no more than one application before alternating with a fungicide with a different mode of action.
boscalid (Endura)	7	6.8 oz/acre	7	0.5	Apply at planting in a 4- to 6-inch banded spray. Under high disease pressure, apply as a foliar spray.
dicloran (Botran)	14	2 to 3.2 qt/acre	14	0.5	Also, for Botrytis diseases.

TABLE 3-15. DISEASE CONTROL PRODUCTS FOR GARLIC (cont'd)

N. Gauthier, Plant Pathologist, University of Kentucky (Last updated in 2023)

Disease/Material	FRAC Code	Rate of Material Formulation	Minimum Days Harv.	Minimum Days Reentry	Method, Schedule, and Remarks
WHITE ROT (*SCLEROTIUM CUPARTUM*) (cont'd)					
fludioxonil (Cannonball)	12	0.5 oz/1000 row ft	7	0.5	In-furrow application only.
iprodione (Rovral 50 WP)	2	4 lb/acre	—	1	Spray cloves as they are being covered by soil (38 to 40 in. bed spacing). One applica- tion per year.
PCNB (Blocker 4F)	14	29 oz/1000 row ft	—	0.5	In-furrow, at-planting treatment only.
thiophanate-methyl (various)	1	43.6 oz/acre	—	3	Spray directly into open furrow at planting.

TABLE 3-16. DISEASE CONTROL PRODUCTS FOR GREENS, LEAFY BRASSICA (COLLARD, KALE, MUSTARD, RAPE SALAD GREENS, TURNIP GREENS)

A. Keinath, Plant Pathologist, Clemson University

Note: For turnips harvested for roots, see Remarks and Table 3-31 Root Vegetables to find the correct rates to apply; some fungicides have higher rates for turnip greens than for turnip roots.

Disease/Material	FRAC Code	Rate of Material Formulation	Minimum Days Harv.	Minimum Days Reentry	Method, Schedule, and Remarks
SEEDLING BLIGHT, DAMPING OFF, ROOT ROT (*PYTHIUM SPP., RHIZOCTONIA SOLANI*)					
azoxystrobin + mefenoxam (Uniform 3.72SC)	11+4	0.34 fl oz/1,000 row ft	--	0	Apply as an in-furrow spray in 5 gal of water per acre prior to covering seed. Make only 1 application per season. **May be applied to turnip grown for roots.**
azoxystrobin (Quadris 2.08SC)	11	0.4 to 0.8 fl oz per 1000 row feet	0	4 hr	Apply at planting as a directed spray to the furrow in a band 7 inches wide.
ethaboxam (Elumin 4SC)	22	8 fl oz/acre	2	0.5	Apply as soil drench or through drip irrigation at seeding or transplanting. Maximum 2 applications. Do not use in greenhouses.
ALTERNARIA LEAF SPOT (*ALTERNARIA* SPP.), CERCOSPORA LEAF SPOT (*CERCOSPORA* SPP.), ANTHRACNOSE (*COLLETOTRICHUM HIGGINSIANUM*), WHITE SPOT (*PSEUDOCERCOSPORELLA CAPSELLAE*), AND VARIOUS FOLIAR FUNGAL DISEASES (SEE SPECIFIC LABELS)					
azoxystrobin (various)	11	See labels	0	4 hr	Make no more than two sequential applications before alternating with fungicides that have a different mode of action. **May be applied to turnip grown for roots.** Note that resistance has been observed in Georgia in *Alternaria japonica*.
azoxystrobin + difenoconazole (Quadris Top 2.72SC)	11+3	12 to 14 fl oz/acre	1	0.5	Make no more than one application before alternating to another fungicide with a different mode of action (NOT FRAC 11).
boscalid (Endura 70 WG)	7	6 to 9 oz/acre	14	0.5	Begin applications prior to disease development and continue on a 7- to 14-day interval. Make no more than 2 applications per season. **Do not apply to turnip greens.**
cyprodonil + fludioxonil (Switch 62.5WG)	9+12	11 to 14 oz/acre	7	0.5	Apply when disease first appears and continue on a 7- to 10-day intervals. See label for complete list of greens.
difenoconazole + cyprodinil (Inspire Super 2.82SC)	3+9	16 to 20 fl oz/acre	7	0.5	Make no more than 2 sequential applications before alternating to a fungicide with a different mode of action.
flutriafol (Rhyme 2.08SC)	3	5 to 7 fl oz/acre	7	0.5	Limit of 4 applications per year. Labeled for Alternaria and Cercospora leaf spots.
flutriafol + azoxystrobin (Topguard EQ 4.29SC)	3+11	4 to 8 fl oz/acre	0	0.5	Limit of 4 applications per year.
fluopyram + tebuconazole (Luna Experience 400SC)	7+3	6 to 8.6 fl oz/acre	7	0.5	Make no more than 2 sequential applications before alternating to a fungicide with a different mode of action. Limit of 34 fl oz (4 applications at the high rate) per acre per year. Not labeled for white spot. **Do not apply to turnip grown for roots.**
fluopyram + difenoconazole (Luna Flex 375SC)	7 + 3	10.0 to 13.6 fl oz/acre	1	0.5	Make no more than 2 sequential applications of Luna Flex or other Group 7 fungi- cides before rotating to another effective fungicide with a different mode of action. Maximum of 2 applications per year.
fluopyram + trifloxystrobin (Luna Sensation 500SC)	7+11	5 to 7.6 fl oz/acre	7	0.5	Make no more than 2 sequential applications before alternating to a fungicide with a dif- ferent mode of action. Limit of 15.3 fl oz (2 applications at the high rate) per acre per year. Not labeled for white spot. **May be applied to turnip grown for roots.**
fluxapyroxad + pyraclostrobin (Priaxor 500SC)	7+11	6.0 to 8.2 fl oz/acre	3	0.5	Make no more than two sequential applications before alternating with fungicides that have a different mode of action. Maximum of 3 applications. **Do not apply to turnip greens or roots.**

TABLE 3-16. DISEASE CONTROL PRODUCTS FOR GREENS, LEAFY BRASSICA (COLLARD, KALE, MUSTARD, RAPE SALAD GREENS, TURNIP GREENS) (cont'd)

A. Keinath, Plant Pathologist, Clemson University

Note: For turnips harvested for roots, see Remarks and Table 3-31 Root Vegetables to find the correct rates to apply; some fungicides have higher rates for turnip greens than for turnip roots.

Disease/Material	FRAC Code	Rate of Material Formulation	Minimum Days Harv.	Reentry	Method, Schedule, and Remarks
ALTERNARIA LEAF SPOT (*ALTERNARIA* SPP.), CERCOSPORA LEAF SPOT (*CERCOSPORA* SPP.), ANTHRACNOSE (*COLLETOTRICHUM HIGGINSIANUM*), WHITE SPOT (*PSEUDOCERC OSPORELLA CAPSELLAE*), AND VARIOUS FOLIAR FUNGAL DISEASES (SEE SPECIFIC LABELS) (cont'd)					
mefentrifluconazole (Cevya)	3	3 to 5 fl oz/acre	0	0.5	**For Alternaria leaf spot only.** Limit of 15 fl oz (5 and 3 applications at high and low rates, respectively) per acre per year. **Do not apply to turnip grown for roots.**
penthiopyrad (Fontelis 1.67SC)	7	14 to 30 fl oz/acre	0	0.5	Make no more than two sequential applications before alternating with fungicides that have a different mode of action. **May be applied to turnips grown for roots.**
pydiflumetofen + fludioxonil (Miravis Prime 3.34SC)	7+12	10.3 to 13.4 fl oz/acre	7	0.5	Make no more than 2 sequential applications of Miravis Prime or other Group 7 and 12 fungicides before rotating to another effective fungicide with a different mode of action. Maximum of 26.8 fl oz/acre per year or 2 applications at the high rate. **Do not apply to turnip grown for roots.**
pyraclostrobin (Cabrio 20EG) (Pyrac 2EC)	11	12 to 16 oz/acre 8 to 12 oz/acre (turnip greens)	3	0.5	Begin applications prior to disease development and continue on a 7- to 10-day interval. Make no more than 2 sequential applications before alternating to a fungicide with a different mode of action.
tebuconazole (various)	3	3 to 4 oz/acre	7	0.5	For optimum results use as a preventative treatment. Folicur 3.6 F must have 2 to 4 hours of drying time on foliage for the active ingredient to move systemically into plant tissue before rain or irrigation occurs
BACTERIAL BLIGHT (*PSEUDOMONAS*), XANTHOMONAS LEAF BLIGHT					
none					Based on field trials in SC, no fungicides, bactericides, or biopesticides are effective against these diseases. Use a 1-yr crop rotation away from all brassicas and early or once-over harvesting if disease appears.
GRAY MOLD (*BOTRYTIS CINEREA*)					
difenoconazole + cyprodinil (Inspire Super 2.82SC)	3+9	16 to 20 fl oz/acre	7	0.5	Make no more than 2 sequential applications before alternating to a fungicide with a different mode of action (FRAC Code).
fluopyram + tebuconazole (Luna Experience 400SC)	7+3	6 to 8.6 fl oz/acre	7	0.5	Make no more than 2 sequential applications before alternating to a fungicide with a different mode of action. Limit of 34 fl oz (4 applications at the high rate) per acre per year. Not labeled for white spot. **Do not apply to turnip grown for roots.**
fluopyram + difenoconazole (Luna Flex 375SC)	7 + 3	10.0 to 13.6 fl oz/acre	1	0.5	Make no more than 2 sequential applications of Luna Flex or other Group 7 fungicides before rotating to another effective fungicide with a different mode of action. Maximum of 2 applications per year.
penthiopyrad (Fontelis 1.67SC)	7	14 to 30 fl oz/acre	0	0.5	Make no more than two sequential applications before alternating with fungicides that have a different mode of action. **Maybe applied to turnips grown for roots.**
CLUBROOT (*PLASMODIOPHORA BRASSICAE*)					
cyazofamid (Ranman 34.5SC)	21	*Transplant:* 12.9 to 25.75 fl oz/ 100-gal water *Banded:* 20 fl oz/A	0.5	0	Either apply immediately after transplanting with 1.7 fl oz of solution per transplant, or as a banded application with soil incorporation of 6 to 8 inches prior to transplanting. Do not apply more than 39.5 fl oz/acre/year, including foliar sprays made for downy mildew.
fluazinam (Omega 500F, Ventana 500F)	29	*Transplant:* 6.45 fl oz/100 gal. *Soil incorporation:* 2.6 pints/acre	20	0.5	Transplant soil drench: Immediately after transplanting, apply 3.4 fl oz of transplant solution per plant. Soil incorporation: Apply in a 9-in. band and incorporate 6 to 8 in. deep before transplanting. Note: Omega may delay harvest; see label. **Do not apply to turnips grown for roots.**
DOWNY MILDEW (*HYALOPERONOSPORA PARASITICA*)					
ametoctradin + dimethomorph (Zampro 525SC)	45+40	14 fl oz/acre	0	0.5	Do not make more than 2 sequential applications before alternating to a fungicide with a different mode of action. Addition of an adjuvant may improve performance (see label for specifics).
cyazofamid (Ranman 400SC)	21	2.75 fl. oz/acre	0	0.5	Make applications on a 7- to 10-day schedule. Do not apply more than 39.5 fl. oz/acre per crop growing season, including soil applications made for clubroot.
dimethomorph (Forum 4.16SC)	40	6.4 oz/acre	0	0.5	Must be tank mixed with another fungicide active against Phytophthora blight. Do not make more than 2 sequential applications before alternating to another effective fungicide with a different mode of action. Do not make more than 5 apps/season. **Do not apply to turnip greens or roots.**
ethaboxam (Elumin 4SC)	22	8 fl oz/acre	2	0.5	Must rotate with a non-FRAC Group 22 fungicide after every application. Maximum 2 applications. Do not use in greenhouses.
fenamidone (Reason 500SC)	11	5.5 to 8.2 oz/acre	2	0.5	Begin applications as soon as conditions become favorable for disease development. Applications should be made on a 5- to 10-day interval. Do not make more than one application of Reason 500SC before alternating with a fungicide from a different resistance management group.

TABLE 3-16. DISEASE CONTROL PRODUCTS FOR GREENS, LEAFY BRASSICA (COLLARD, KALE, MUSTARD, RAPE SALAD GREENS, TURNIP GREENS) (cont'd)

A. Keinath, Plant Pathologist, Clemson University

Note: For turnips harvested for roots, see Remarks and Table 3-31 Root Vegetables to find the correct rates to apply; some fungicides have higher rates for turnip greens than for turnip roots.

Disease/Material	FRAC Code	Rate of Material Formulation	Minimum Days Harv.	Minimum Days Reentry	Method, Schedule, and Remarks
DOWNY MILDEW (*HYALOPERONOSPORA PARASITICA*) (cont'd)					
fluopicolide (Presidio 4SC)	43	3 to 4 fl. oz/acre	2	0.5	Make applications on a 7- to 10-day schedule. Presidio must be tank mixed with another fungicide with a different mode of action. Make no more than 2 sequential applications before rotating to a fungicide with different mode of action. Apply no more than12 oz per acre and make no more than 4 applications per season.
fluopyram + trifloxystrobin (Luna Sensation 500SC)	7+11	5 to 7.6 fl oz/acre	7	0.5	Make no more than 2 sequential applications before alternating to a fungicide with a different mode of action (FRAC Code). Limit of 15.3 fl oz (2 applications at the high rate) per acre per year. **May be applied to turnip grown for roots.**
fosetyl-Al (Aliette 80WDG)	P07	2 to 5 lb/acre	3	1	Apply when disease first appears; then repeat o, 7- to 21-day intervals. Do not tank-mix with copper fungicides. A maximum of 7 applications can be made per season. **Do not apply to turnip greens or roots.**
potassium phosphite (various)	P07	2 to 4 pt/acre	0	4 hr	Apply when weather is foggy as a preventative. Do not apply to plants under water or temperature stress. Spray solution should have a pH greater than 5.5. Apply in at least 30-gal water per acre.
pyraclostrobin (Cabrio 20EG)	7	12 to 16 oz/acre	3	0.5	Begin applications prior to disease development and continue on a 7- to 10-day inter- val. Make no more than 2 sequential applications before alternating to a fungicide with a different mode of action.
POWDERY MILDEW (*ERYSIPHE* SPP.)					
azoxystrobin + difenoconazole (Quadris Top 2.72SC)	11+3	12 to 14 fl oz/acre	1	0.5	Make no more than one application before alternating to another fungicide with a different mode of action (NOT FRAC 11).
boscalid (Endura 70WG)	7	6 to 9 oz/acre	14	0.5	Begin applications prior to disease development and continue on a 7- to 14-day interval. Make no more than 2 applications per season: disease suppression only. **Do not apply to turnip greens or roots.**
cyprodonil + fludioxonil (Switch 62.5WG)	9+12	11 to 14 oz/acre	7	0.5	Apply when disease first appears and continue on 7- to 10-day intervals. See label for complete list of greens. May be used on turnip where leaves only will be harvested. **Do not apply to turnip grown for roots.**
difenoconazole + cyprodinil (Inspire Super 2.82EW)	3+9	16 to 20 fl oz/acre	7	0.5	Make no more than 2 sequential applications before alternating to a fungicide with a different mode of action.
fluopyram + difenoconazole (Luna Flex 375SC)	7 + 3	10.0 to 13.6 fl oz/acre	1	0.5	Make no more than 2 sequential applications of Luna Flex or other Group 7 fungicides before rotating to another effective fungicide with a different mode of action. Maximum of 2 applications per year.
fluopyram + trifloxystrobin (Luna Sensation 500SC)	7+11	5 to 7.6 fl oz/acre	7	0.5	Make no more than 2 sequential applications before alternating to a fungicide with a different mode of action. Limit of 15.3 fl oz (2 applications at the high rate) per acre per year. **May be applied to turnip grown for roots.**
flutriafol (Rhyme 2.08SC)	3	5 to 7 fl oz/acre	7	0.5	Limit 4 applications per year. Labeled for Alternaria and Cercospora leaf spots.
fluxapyroxad + pyraclostrobin (Priaxor 500SC)	7+11	6.0 to 8.2 fl oz/acre	3	0.5	Make no more than two sequential applications before alternating with fungicides that have a different mode of action. Maximum of 3 applications. **Do not apply to turnip greens or roots.**
mefentrifluconazole (Cevya)	3	3 to 5 fl oz/acre	0	0.5	Limit of 15 fl oz (5 and 3 applications at high and low rates, respectively) per acre per year.
penthiopyrad (Fontelis 1.67SC)	7	14 to 30 fl oz/acre	0	0.5	Make no more than 2 sequential applications before alternating with fungicides that have a different mode of action (FRAC Code). **May be applied to turnips grown for roots.**
pydiflumetofen + fludioxonil (Miravis Prime 3.34SC)	7+12	10.3 to 13.4 fl oz/acre	7	0.5	Make no more than 2 sequential applications of Miravis Prime or other Group 7 and 12 fungicides before rotating to another effective fungicide with a different mode of action. Maximum of 26.8 fl oz/acre per year or 2 applications at the high rate. **Do not apply to turnip grown for roots.**
pyraclostrobin (Cabrio 20EG, Pyrac 2EC)	11	12 to 16 oz/acre	3	0.5	Begin applications prior to disease development and continue on a 7-to 10-day interval. Make no more than 2 sequential applications before alternating to a fungicide with a different mode of action (FRAC Code).
triflumizole (Procure 480SC)	3	6 to 8 oz/acre	1	0.5	Make no more than two sequential applications before rotating with fungicide with different mode of action. Do not rotate with Rally or Nova.
tebuconazole (various)	3	3 to 4 oz/acre	7	0.5	For optimum results use as a preventative treatment. Folicur 3.6 F must have 2 to 4 hours of drying time on foliage for the active ingredient to move systemically into plant tissue before rain or irrigation occurs. **May be applied to turnip grown for roots.**
PYTHIUM DAMPING OFF (*PYTHIUM* SPP.), PHYTOPHTHORA BASAL STEM ROT (*PHYTOPHTHORA MEGASPERMA*)					
mefenoxam (Ridomil Gold 4SL)	4	0.25 to 0.5 pt/acre	—	2	Apply in water or liquid fertilizer and incorporate in the top 2 inches of soil. For Phytophthora basal rot, increase rates to 1.0 to 2.0 pt/acre.

TABLE 3-16. DISEASE CONTROL PRODUCTS FOR GREENS, LEAFY BRASSICA (COLLARD, KALE, MUSTARD, RAPE SALAD GREENS, TURNIP GREENS) (cont'd)

A. Keinath, Plant Pathologist, Clemson University

Note: For turnips harvested for roots, see Remarks and Table 3-31 Root Vegetables to find the correct rates to apply; some fungicides have higher rates for turnip greens than for turnip roots.

Disease/Material	FRAC Code	Rate of Material Formulation	Minimum Days Harv.	Reentry	Method, Schedule, and Remarks
RHIZOCTONIA BOTTOM ROT (*RHIZOCTONIA SOLANI*)					
boscalid (Endura 70 WG)	7	6 to 9 oz/acre	14	0.5	Begin applications prior to disease development and continue on a 7-to 14-day interval. Make no more than 2 applications per season. **Do not apply to turnip greens.**
fluopyram + tebuconazole (Luna Experience 400SC)	7+3	8.6 fl oz/acre	7	0.5	Make no more than 2 sequential applications before alternating to fungicide with different mode of action. Limit of 34 fl oz (4 applications at high rate) per acre per year. **Do not apply to turnip grown for roots.**
WHITE MOLD, SCLEROTINIA STEM ROT (*SCLEROTINIA SCLEROTIORUM*)					
boscalid (Endura 70EG)	7	6 to 9 fl oz/acre	14	0.5	Make no more than 2 sequential applications before alternating to a fungicide with a different mode of action. Limit of 2 applications at the high rate or 3 applications at the low rate per acre per year. **Do not apply to turnip greens.**
fluopyram + tebuconazole (Luna Experience 400SC)	7+3	8.6 fl oz/acre	7	0.5	Make no more than 2 sequential applications before alternating to a fungicide with a different mode of action. Limit of 4 applications per acre per year. **Do not apply to turnip grown for roots.**
fluopyram + difenoconazole (Luna Flex 375SC)	7 + 3	10.0 to 13.6 fl oz/ acre	1	0.5	Make no more than 2 sequential applications of Luna Flex or other Group 7 fungicides before rotating to another effective fungicide with a different mode of action. Maximum of 2 applications per year.
fluopyram + trifloxystrobin (Luna Sensation 500SC)	7+11	5.0 to 7.6 fl oz/acre	7	0.5	Make no more than 2 sequential applications before alternating to a fungicide with a different mode of action. Limit of 15.3 fl oz (2 applications at the high rate) per acre per year. **May be applied to turnip grown for roots.**
penthiopyrad (Fontelis1.67SC)	7	16 to 30 fl oz/acre	0	0.5	Do not exceed 72 fl oz of product per year. Make no more than 2 sequential applications per season before rotating to another effective product with a different mode of action.
pydiflumetofen + fludioxonil (Miravis Prime 3.34SC)	7+12	13.4 fl oz/acre	7	0.5	This FIFRA Section 2(ee) Registration is legal in the following states: AL, FL, GA. NC, SC, and VA. Make no more than 2 sequential applications of Miravis Prime or other Group 7 and 12 fungicides before rotating to another effective fungicide with a different mode of action. Maximum 2 applications per year.
Coniothyrium minitans (Contans WG)	BM02	1 to 4 lb/acre	0	4 hr	**OMRI listed product.** Apply to soil surface and incorporate no deeper than 2 inches. Works best when applied prior to planting or transplanting. Do not apply other fungi- cides for 3 weeks after applying Contans.
WHITE RUST (*ALBUGO CANDIDA*)					
azoxystrobin (Quadris 2.08SC)	11	6.2 to 15.4 fl oz/acre	0	4 hr	Make no more than 2 sequential applications.
fenamidone (Reason 500SC)	11	8.2 oz/acre	2	0.5	Begin applications as soon as conditions become favorable for disease development. Applications should be made on a 5- to 10-day interval. Do not make more than 1 application of Reason 500SC before alternating with a fungicide from a different resis- tance management group.
fluxapyroxad + pyraclostrobin (Priaxor 500SC)	7+11	6.0 to 8.2 oz/acre	3	0.5	Make no more than two sequential applications before alternating with fungicides that have a different mode of action. Maximum of 3 applications. **Do not apply to turnip greens or roots.**
WIRESTEM (*RHIZOCTONIA SOLANI*)					
flutolanil (Moncut 3.8SC)	7	26 fl oz/acre	45	0.5	Apply to the row at planting as an in-furrow spray or a spray directed at the base of transplants immediately after transplanting. Limit of 2 applications per year.

FOR PRODUCT EFFICACY – SEE – TABLE 3-7. EFFICACY OF PRODUCTS FOR DISEASE CONTROL IN BRASSICAS

TABLE 3-17. DISEASE CONTROL PRODUCTS FOR LETTUCE AND ENDIVE

M.O. Jibrin, Plant Pathologist, Oklahoma State University

Disease/Material	FRAC Code	Rate of Material Formulation	Minimum Days Harv.	Minimum Days Reentry	Method, Schedule, and Remarks
BOTTOM ROT (*RHIZOCTONIA*)					
2,6-dichloro-4-nitroaniline (Botran 5F)	29	0.6 qt/acre	14	12 hr	Spray at time of planting in a 4-to-6-inch band over seedlings before or after transplanting. Do not apply more than 3.2 qt formulated product per acre per year.
azoxystrobin (various)	11	0.4 to 0.8 fl oz/ 1,000 row feet	—	4 hr	Rhizoctonia only. Make in-furrow or banded applications shortly after plant emergence.
azoxystrobin + mefenoxam (Uniform)	4 + 11	0.34 fl oz/ 1000 ft row	0	0	Apply uniform as an in-furrow spray in a minimum of 5 gal of water per acre at planting. Make only one application per crop season.
Bacillus amyloliquefaciens strain MBI 600 (Serifel)	44	4 to 16 oz/acre	0	4 hr	See label for soil application instructions. Apply a high enough water volume to soak the root zone.
Streptomyces lydicus WYEC 108 (Actinovate AG)	BM02	3 to 12 oz/acre	0	4 hr	Soil application only except for watercress. This product might be used in overhead drip or other irrigation systems.
trifloxystrobin + fluopyram (Luna Sensation)	7 +11	7.6 fl. oz/acre	0	12 hr	Apply with ground, aerial, or chemigation equipment. apply at critical timings for disease control and continue as needed at 14-day intervals. Can be applied in a band. 20 PHI for banded applications.
SEED DECAY, SEEDLING BLIGHT, DAMPING-OFF					
fludioxonil (Spirato 480 FS) (Maxim 4 FS)	12	0.08 to 0.16 fl oz/ 100 lb of seed	—	12	Used to control diseases of seed such as Aspergillus, Fusarium, and Rhizoctonia, among others. Does NOT control Pythium or Phytophthora.
Streptomyces lydicus WYEC 108 (Actinovate AG)	BM02	3 to 12 oz/acre	0	4 hr	Soil application only except for watercress. This product might be used in overhead drip or other irrigation systems.
DOWNY MILDEW					
acibenzolar-S-methyl (Actigard 50 WG)	P01	0.75 to 1 oz/acre	7	0.5	Do not apply prior to thinning or within 5 days after transplanting. Apply preventatively every 7- to 10-days, not to exceed 4 applications (4 oz) per season.
ametoctradin + dimethomorph (Zampro 525 SC)	45+40	14 fl oz/acre	0	12 hr	Do not make more than 2 sequential applications before alternating to a fungicide with a different mode of action. Addition of an adjuvant may improve performance (see label for specifics). Do not apply more than 42 fl oz per acre per season.
azoxystrobin (various)	11	6.2 to 15.4 fl oz/ acre	0	4 hr	Make no more than two sequential applications before alternating with a fungicide with a different mode of action.
cyazofamid (Ranman 400SC)	21	2.75 fl oz/acre	0	0.5	Apply on a 7- to 10-day interval when disease first appears or when conditions favorable for disease development. Do not make subsequent applications, and limit applications to six per year.
cymoxanil (Curzate)	27	3.2 to 5.0 oz/acre	3	0.5	Curzate is only labeled for lettuce and spinach. Use only in combination with a protectant fungicide. Apply on a 5- to 7-day schedule.
cymoxanil + famoxadone (Tanos)	27+11	8.0 oz	1	0.5	See label for directions.
dimethomorph (various)	40	6.4 oz/acre	0	0.5	Must be applied as a tank-mix with another fungicide active against downy mildew. Do not make more than two sequential applications.
fenamidone (Reason 500SC)	11	5.5 to 8.2 fl oz	2	0.5	Alternate with fungicides with a different mode of action.
fluopicolide (Presidio)	43	3 to 4 fl oz/acre	2	0.5	Tank-mix with another downy mildew fungicide with a different mode of action.
mandipropamid (various)	40	See label	See label	See label	Begin applications as soon as crop and/or environmental conditions become favorable for disease development. Apply on a 7- to 10-day interval depending upon disease conditions.
mono- and di-potassium salts of phosphorous acid (Alude, K-Phite)	P07	1 to 4 quarts in a minimum of 10 gal/acre	0	4 hr	Do not apply at a less than 3-day interval.
oxathiapiprolin + mandipropamid (Orondis Ultra)	49+40	5.5 to 8 fl oz/acre	0	4 hr	Limit of six applications per acre per year for the same crop. Do not follow soil applications of Orondis with foliar applications of Orondis. Use the higher rates when disease is present.
oxathiapiprolin + mefenoxam (Orondis Gold 200)	49+4	4.8 to 9.6 fl oz/acre	0	4 hr	Limit to 38.6 fl oz per acre per year. Limit of six applications per acre per year for the same crop. Do not follow soil applications of Orondis with foliar applications of Orondis. Apply at planting in furrow, by drip, or in transplant water. Use the higher rates for heavier soils, for longer application intervals, or for susceptible varieties.
propamocarb (Previcur Flex)	28	2 pt/acre	2	0.5	Previcur Plus is only labeled for head and leaf lettuce. Do not apply more than 8 pt per growing season; begin applications before infection and continue on 7- to 10-day interval.

TABLE 3-17. DISEASE CONTROL PRODUCTS FOR LETTUCE AND ENDIVE (cont'd)

M.O. Jibrin, Plant Pathologist, Oklahoma State University

Disease/Material	FRAC Code	Rate of Material Formulation	Minimum Days Harv.	Minimum Days Reentry	Method, Schedule, and Remarks
DOWNY MILDEW, LEAF SPOTS					
azoxystrobin (various)	11	6.2 to 15.4 fl oz/acre	7	4 hr	Use the highest rate for downy mildew. Make no more than 2 sequential applications before alternating with fungicides that have a different mode of action. Apply no more than 2.88 qt per crop per acre per season.
fixed copper (various)	M01	See label	See label	See label	See label. Rates vary depending on the formulation.
pyraclostrobin (various)	11	12 to 16 oz/acre	0	0.5	Begin applications prior to disease development and continue on 7- to 14-day intervals.
fluxapyroxad + pyraclostrobin (Merivon 500 SC)	7+11	4 to 11 fl oz/acre	1	0.5	Make no more than two sequential applications before alternating with fungicides that have a different mode of action. Suppression only for downy mildew.
mancozeb (various)	M03	See labels	See label	See label	Rates vary depending on the formulation. Spray at first appearance of disease and continue on a 7- to 10-day intervals.
LEAF SPOTS					
Bacillus amyloliquefaciens strain MBI 600 (Serifel)	44	4 to 16 oz/acre	0	4 hr	Apply at 7- to 10 days interval. Apply product with enough water to coverage the foliage.
cyprodinil + fludioxonil (Switch 62.5 WDG)	9+12	11 to 14 oz/acre	0	0.5	Switch also has activity against basal rot, Sclerotinia, and Gray mold. Alternate with a fungicide with a different mode of action after 2 applications.
flutriafol (Rhyme)	3	5 to 7 oz/acre	7	0.5	Apply preventatively or when conditions are favorable for disease development.
penthiopyrad (Fontelis)	7	14 to 24 fl oz/acre	3	0.5	Begin applications before disease development. **DO NOT** make more than two consecutive applications before switching to a fungicide with a different mode of action.
trifloxystrobin + fluopyram (Luna Sensation)	7+11	7.6 fl. oz/acre	0/20	12 hr	Apply with ground, aerial, or chemigation equipment. apply at the critical timings for disease control and continue as needed at 14-day intervals. Can be applied in a band. 20 PHI for banded applications
GRAY MOLD					
2,6-dichloro-4-nitroaniline (Botran 5F)	29	1.8 to 3.2 qt/acre	14	12 hr	Post-thinning and established transplants: apply as a basal drench in 50-100 gallons. Do not apply more than 3.2 qt formulated product per acre per year.
azoxystrobin + flutriafol (Topguard EQ)	3+11	6.0 to 8.0 fl oz/acre	7	12 hr	Apply preventatively at a higher rate in 7 days interval. For multiple applications refer to the guidelines under resistance management.
boscalid (Endura)	7	7 to 9 oz/acre	14	0.5	Begin applications prior to the onset of disease and continue on a 7-day interval.
penthiopyrad (Fontelis)	7	14 to 24 fl oz/acre	3	0.5	Begin applications before disease development. **DO NOT** make more than two consecutive applications before switching to a fungicide with a different mode of action.
trifloxystrobin + fluopyram (Luna Sensation)	7+11	7.6 fl. oz/acre	0/20	12 hr	Apply with ground, aerial, or chemigation equipment. apply at the critical timings for disease control and continue as needed at 14-day intervals. Can be applied in a band. 20 PHI for banded applications
SEED DECAY, SEEDLING BLIGHT, DAMPING-OFF					
fludioxonil (Spirato 480FS) (Maxim 4FS)	12	0.08 to 0.16 fl oz/ 100 lb of seed	—	12	Used to control diseases of seed such as Aspergillus, Fusarium, and Rhizoctonia, among others. Does NOT control Pythium or Phytophthora.
POWDERY MILDEW					
azoxystrobin (various)	11	6.2 to 15.4 fl oz/acre	0	4 hr	Make no more than two sequential applications before alternating with a fungicide with a different mode of action.
azoxystrobin + flutriafol (Topguard EQ)	3+11	6.0 to 8.0 fl oz /acre	7	12 hr	Apply preventatively at a higher rate in 7 days interval. For multiple applications refer to the guidelines under resistance management.
fluxapyroxad + pyraclostrobin (Merivon 500 SC)	7+11	4 to 11 fl oz/acre	1	0.5	Make no more than two sequential applications before alternating with fungicides that have a different mode of action.
myclobutanil (Rally) 40WSP	3	5 oz/acre	3	1	For use on lettuce only. Apply when disease first appears and continue on a 14-day interval.
penthiopyrad (Fontelis)	7	14 to 24 fl oz/acre	0	0.5	Begin applications before disease development. **DO NOT** make more than two sequential applications before alternating with fungicide with a different mode of action.
quinoxyfen (Quintec)	13	6 fl oz	1	1	Alternate with a fungicide with a different mode of action.
sulfur (various)	M02	See label	See label	See label	Apply at early leaf stage and repeat every 10- to 14-days or as needed. Do not apply if temperatures are expected to exceed 90°F within 3 days of application due to the risk of crop injury.

TABLE 3-17. DISEASE CONTROL PRODUCTS FOR LETTUCE AND ENDIVE (cont'd)

M.O. Jibrin, Plant Pathologist, Oklahoma State University

Disease/Material	FRAC Code	Rate of Material Formulation	Minimum Days Harv.	Minimum Days Reentry	Method, Schedule, and Remarks
POWDERY MILDEW (cont'd)					
triflumizole (various)	3	6 to 8 fl oz/acre	0	0.5	Applications should begin prior to disease development. Repeat on a 14-day schedule. Do not apply more than 18 fl oz per acre per season.
trifloxystrobin + fluopyram (Luna Sensation)	7+11	7.6 fl. oz/acre	0/20	12 hr	Apply with ground, aerial, or chemigation equipment. apply at the critical timings for disease control and continue as needed at 14-day intervals. Can be applied in a band. 20 PHI for banded applications
PYTHIUM DAMPING-OFF					
cyazofamid (Ranman 400SC)	21	2.75 fl oz/acre	0	0.5	Apply on a 7- to 10-day interval when disease first appears or when conditions favorable for disease development. Do not make subsequent applications, and limit applications to six per year.
mefenoxam (various)	4	See label	—	2	Apply preplant incorporated or surface application at planting.
metalaxyl (various)	4	See label	—	2	Banded over the row, preplant incorporated, or injected with liquid fertilizer
propamocarb (Previcur Flex)	28	2 pt/acre	2	0.5	Previcur Plus is only labeled for head and leaf lettuce. Various application methods; see label.
RUST, WHITE RUST					
cyazofamid (Ranman 400SC)	21	2.75 fl oz/acre	0	0.5	Apply on a 7-to-10-day interval when disease first appears or when conditions favorable for disease development. Do not make subsequent applications, and limit applications to six per year.
penthiopyrad (Fontelis)	7	14 to 24 fl oz/acre	3	0.5	Begin applications before disease development. **DO NOT** make more than two sequential applications before switching to a fungicide with a different mode of action.
sulfur (various)	M02	See labels	14	1	Apply at early leaf stage and repeat every 10- to 14-days or as needed. Do not apply if temperatures are expected to exceed 90°F within 3 days of application due to the risk of crop injury.
SCLEROTINIA					
boscalid (Endura)	7	See label	14	0.5	Begin applications prior to onset of disease. Use higher rate when disease pressure is high.
cyazofamid (Ranman 400SC)	21	2.75 fl oz/acre	0	0.5	Apply on a 7- to 10-day interval when disease first appears or when conditions favorable for disease development. Do not make subsequent applications, and limit applications to six per year.
Coniothyrium minitans (Contans WG)	BM02	1 to 4 lb/acre	0	4 hr	**OMRI listed product.** Apply to soil surface and incorporate no deeper than 2 inches. Works best when applied prior to planting or transplanting. Do not apply other fungicides for 3 weeks after applying Contans.
dicloran (Botran)	14	See label	14	0.5	Rate depends on the specific crop and timing of application. See label.
fludioxonil (Cannonball WP)	12	7 oz/acre	0	0.5	Ground applications only. Do not apply more than 28 oz/acre per year.
fluazinam (Orbus 4F)	29	16 to 24 fl oz/acre	30	12 hr	Apply as either foliar, band or broadcast spray or as soil drench application at thinning. Use at least 50 gal/water per acre. Do not apply after thinning. Apply every 14 days.
isofetamid (Kenja)	7	12.3 fl oz/acre	14	0.5	Application timing depends on the planting method. Make no more than two sequential applications.
trifloxystrobin + fluopyram (Luna Sensation)	7 +11	7.6 fl. oz/acre	0/20	12 hr	Apply with ground, aerial, or chemigation equipment. apply at the critical timings for disease control and continue as needed at 14-day intervals. Can be applied in a band. 20 PHI for banded applications
iprodione (Rovral)	2	1.5 to 2 lb/acre	14	1	Only for use on lettuce. Also effective for bottom rot and Botrytis. Use higher rate when disease pressure is high.
penthiopyrad (Fontelis)	7	16 to 24 fl oz/acre	3	0.5	Begin applications before disease development. Continue on 7- to 14-day intervals. Do not make more than 2 consecutive applications before switching to a fungicide with a different mode of action.
Pseudomonas chlororaphis strain AFS0009 (Howler)	BM02	5 to 15 lb/acre	0	0.17	In furrow or foliar spray, Repeat at 7- to 21-day intervals. Thoroughly cover plant foliage and soil surfaces.
pydiflumetofen + fludioxonil (Miravis Prime)	7+12	13.4 fl oz/acre	0	0.5	Application timing depends on the planting method. Apply no closer than a 7-day interval. For best results, use a soil-directed spray.

MUSKMELON (CANTALOUPE) SEE CUCURBITS

TABLE 3-18. DISEASE CONTROL PRODUCTS FOR OKRA
E. Sikora, Plant Pathologist, Auburn University

Disease/Material	FRAC Code	Rate of Material Formulation	Minimum Days Harv.	Reentry	Method, Schedule, and Remarks
ALTERNARIA, GRAY MOLD, POWDERY MILDEW					
cyprodinil + fludioxonil (various)	9+12	11 to 14 oz/acre	0	0.5	Begin applications before disease development and continue on 7- to 10-day interval. Make no more than 2 consecutive applications before alternating to a fungicide with a different. Do not apply more than 56 oz per acre per season.
ANTHRACNOSE, BACTERIAL LEAF SPOT, LEAF SPOTS, POD SPOTS, POWDERY MILDEW					
fixed copper (various)	M01	See label	0	2	Begin applications when conditions favor disease development and repeat on 5-to10-day intervals.
ANTHRACNOSE, BOTRYTIS LEAF MOLD, POWDERY MILDEW, CERCOSPORA LEAF SPOT					
chlorothalonil; cymoxanil (various)	M05+27	2 to 2.4 pints/acre	3	0.5	Begin applications before disease development and continue on a 7-day interval.
ANTHRACNOSE, GRAY LEAF SPOT, POWDERY MILDEW, CERCOSPORA LEAF SPOT					
difenoconazole; azoxystrobin (Quadris Top)	3+11	8 to 14 fl oz/acre	0	0.5	Begin applications before disease development and continue on 7- to 10-day interval. Make no more than 2 consecutive applications before alternating to a fungicide with a different mode of action. Do not apply more than 55 fl oz per acre per season
difenoconazole + cyprodinil (Inspire Super)	3+9	16 to 20 fl oz/acre	0	0.5	Begin applications before disease development and continue 7- to 10-day interval. Make no more than 2 consecutive applications before alternating to a fungicide with a different mode of action. Do not apply more than 80 fl oz per acre per season.
fludioxonil (various)	12	5 to 7 fl oz/acre	0	0.5	Begin applications before disease development and continue on 7-day interval.
ANTHRACNOSE, GRAY LEAF SPOT, POWDERY MILDEW, CERCOSPORA LEAF SPOT, RHIZOCTONIA STEM ROT					
difenoconazole + benzovindiflupyr (Aprovia Top)	3+7	10.5 to 13.5 oz/acre	0	0.5	Begin applications before disease development and continue on 7- to 10-day interval. Make no more than 2 consecutive applications before alternating to a fungicide with a different mode of action. Refer to label for information on addition of an adjuvant.
ANTHRACNOSE, POWDERY MILDEW					
azoxystrobin (various)	11	6.0 to 15.5 fl oz/acre	0	4	Do not apply more than two sequential applications before alternating with a fungicide with a different mode of action. Do not make more than 4 applications of strobilurin-type fungicides per acre per season.
chlorothalonil (various)	M05	2 to 2.4 pt/acre	3	0.5	Begin applications when disease is expected. Repeat every 7- to 10-days.
fluoxastrobin (Aftershock)	11	3 to 5.7 fl oz/acre	1	0.5	Begin application preventively and continue as needed on a 7- to 14-day interval. Alternate after each application with another registered non-group 11 product.
tetraconazole (Mettle 125 ME)	3	6 to 8 fl oz/acre	0	0.5	Begin application before onset of disease and continue on at 7- to 14-day intervals. Make no more than 2 consecutive applications before switching to a fungicide with a different mode of action.
sulfur (Microthiol Disperss)	M02	3 to 10 lb/acre		1	Apply at early leaf stage and repeat every 14 days or as needed.
ANTHRACNOSE, BLACK MOLD, EARLY BLIGHT, POWDERY MILDEW					
mefentrifluconazole (Cevya)	3	3 to 5 oz/acre	0	0.5	Apply before onset of disease on a minimum interval of 7 days.
ANTHRACNOSE, BLACK MOLD, EARLY BLIGHT, CERCOSPORA					
zoxamide + chlorothalonil	22+M05	34 fl oz/acre	3	0.3	Begin applications when seedlings emerge or after transplants are set and repeat at 7- to 10-day intervals or when weather conditions favor disease development.
BLACK MOLD, EARLY BLIGHT, GRAY LEAFSPOT, LEAF MOLD, POWDERY MILDEW, SEPTORIA LEAFSPOT, TARGET SPOT					
azoxystrobin + tebuconazole (Affiance)	3+11	9 to 12 fl oz/acre	0	0.5	Begin applications prior to disease appearance when conditions are favorable for disease development. Reapply on a 7- to 14-day interval when conditions remain favorable. Make no more than 2 consecutive applications before switching to a fungicide with a different mode of action.
boscalid (Bonafide)	7	2.5 to 3.5 oz/acre	0	0.5	Apply Bonafide before disease development and continue on a 7- to 14-day interval for early blight, Botrytis gray mold, and black mold. Use the higher rate and the shorter interval when disease pressure is high
difenoconazole+cyprodinil (Inspire Super)	3+9	16 to 20 fl oz/acre	0	0.5	Begin applications prior to disease development and continue throughout the season on a 7- to 10-day interval. Make no more than 2 consecutive applications before switching to another effective fungicide with a different mode of action.

TABLE 3-18. DISEASE CONTROL PRODUCTS FOR OKRA (cont'd)

E. Sikora, Plant Pathologist, Auburn University

Disease/Material	FRAC Code	Rate of Material Formulation	Minimum Days Harv.	Minimum Days Reentry	Method, Schedule, and Remarks
BLACK MOLD, EARLY BLIGHT, GRAY LEAFSPOT, LEAF MOLD, POWDERY MILDEW, SEPTORIA LEAFSPOT, TARGET SPOT (cont'd)					
fludioxanil (Emblem)	12	5.5 to 7 fl oz/acre	1	4	Begin applications prior to disease onset and continue on a 7- to 10-day interval if conditions remain favorable for disease development.
fluopyram + trifloxystrobin (Luna Sensation)	7+11	5 to 7.6 fl oz/acre	3	0.5	Apply when needed on a 7- to 14-day schedule.
fluopyram + difenoconazole (Luna Flex)	3+7	8.0 to 13.6 fl oz/acre	0	0.5	Apply at critical timings for disease control. Continue as needed on a 7 to 14-day interval.
mefentrifluconazole + pyraclostrobin (Veltyma)	3+11	7 to 10 fl oz/acre	0	4	Apply before the onset of disease and on a minimum interval of 7 days.
penthiopyrad (Fontelis)	7	16 to 24 fl oz/acre	0	0.5	Begin applications prior to disease onset and continue on a 7- to 14-day interval.
pydiflumetofen + fludioxonil (Miravis Prime)	7+12	9.2 to 11.4 fl oz/acre	0	0.5	Begin applications before disease development. If disease pressure is high, use the shortest interval and highest rate. Minimum application interval is 7 days. Do not make more than 2 applications at the maximum rate per year.
CERCOSPORA LEAF SPOT					
chlorothalonil (various)	M05	2 to 2.4 pt/acre	3	0.5	Begin applications when disease is expected. Repeat every 7- to 10-days.
tebuconazole (various)	3	4 to 6 fl oz/acre	3	0.5	DO NOT apply more than 24 fl oz per acre per season.
DOWNY MILDEW					
Mandipropamid (Revus)	40	8 fl oz/acre	1	4 hr	Begin applications prior to disease development and continue throughout the season on a 7-10 day interval. Make no more than 2 consecutive applications before switching to another effective non-Group 40 fungicide.
PHYTOPHTHORA BLIGHT					
cyazofamid (Ranman)	21	2.1 to 2.75 fl oz/acre	0	0.5	See label for instructions.
ethaboxam (Elumin)	22	8 fl oz/acre	2	0.5	For best results, begin application at planting/transplanting. Do not apply at intervals of less than 14 days.
fluazinam (various)	29	16 to 24 fl oz/acre	30	0.5	Can be applied as a foliar drench at transplanting followed by foliar applications thereafter. See label.
fluopicolide (Presidio)	43	3 to 4 fl oz/acre	2	0.5	See label.
oxathiapiprolin + mandipropamid (Orondis Ultra)	49+40	5.5 to 8 fl oz/acre	1	4	Apply at planting, in furrow, by drip, or in transplant water. Disease suppression only. Do not make more than 2 applications before switching to a different mode of action.
oxathiapiprolin (Orondis Gold 200)	49	4.8 to 9.6 fl oz/acre	0	4	Make no more than 2 sequential applications of Orondis Gold before rotating to fungicide with a different mode of action. DO NOT follow soil applications of Orondis Gold with foliar applications of Orondis Gold. Use either soil application or foliar applications of this product, but not both.
POWDERY MILDEW					
cyflufenamid (Fastback)	U6	3.4 oz/acre	0	4	Begin applications at first sign of disease.
pyriofenone (Prolivo 300 SC)	50	4 to 5 fl oz/acre	0	4	Make applications on a 7- to 14-day interval when conditions favor disease development. See label.
sulfur (various)	M2	3 to 10 lb/acre	—	—	Apply at early leave stage and repeat every 14 days or as needed.
metrafenone (Vivando)	U8	15.4 fl oz/acre	12	0	Begin applications prior to disease development and continue on a 7- to 14-day interval.
myclobutanil (various)	3	2.5 to 5 fl oz/acre	0	1	Do not make more than 4 applications per season. Minimum retreatment interval: 10- to 14-days.
quinoxyfen (Quintec)	13	4 to 6 fl oz/acre	0	0.5	Apply Quintec before visible symptoms of powdery mildew appear. If powdery mildew infection is established, apply Quintec in tank-mix combination with a curative fungicide.
POWDERY MILDEW, ANTHRACNOSE, CERCOSPORA LEAF SPOT					
azoxystrobin + flutriafol (Topguard EQ)	3+11	4 to 8 fl oz/acre	0	0.5	Apply preventatively or when conditions are favorable for disease development.
flutriafol (various)	3	14 fl oz/acre	0	0.5	Apply preventatively or when conditions are favorable for disease development.

TABLE 3-18. DISEASE CONTROL PRODUCTS FOR OKRA (cont'd)

E. Sikora, Plant Pathologist, Auburn University

Disease/Material	FRAC Code	Rate of Material Formulation	Minimum Days Harv.	Minimum Days Reentry	Method, Schedule, and Remarks
PYTHIUM ROOT ROT					
propamocarb hydrochloride (Previcur Flex)	28	1.2 pints/acre	5	0.5	Can be applied by directed nozzles to the lower portions of the plants and surrounding soil, or via drip irrigation, transplant/setting water, or by sprinklers.
PYTHIUM, PHYTOPHTHORA, FUSARIUM, ANTHRACNOSE,					
mono- and di-potassium salts of phosphorous acid (various)	P07	See label		4	Can be applied as a foliar application, also root dip or through irrigation system. See label for application instructions.
RHIZOCTONIA SEEDLING ROT					
azoxystrobin (various)	11	0.4 to 0.8 fl oz/ 1,000 row feet	—	4	Make in-furrow or banded applications shortly after plant emergence.

TABLE 3-19. DISEASE CONTROL PRODUCTS FOR ONION

B. Dutta, Plant Pathologist, University of Georgia (LAST UPDATED IN 2025)

Disease/Material	FRAC Code	Rate of Material Formulation	Minimum Days Harv.	Minimum Days Reentry	Method, Schedule, and Remarks
ONION (GREEN)					
DAMPING OFF (*PYTHIUM* SPP.)					
mefenoxam (Ridomil Gold 4 SL)	4	0.5 to 1 pt/trt acre	—	2	See label for low rates. Also, for dry onion.
metalaxyl (MetaStar)	4	2 to 4 pt/trt acre	—	2	Preplant incorporated or soil surface spray.
DOWNY MILDEW (*PERONOSPORA DESTRUCTOR*)					
ametoctradin + dimethomorph (Zampro)	45+40	14.0 fl oz/acre	0	0.5	Begin applications prior to disease development and continue on a 5- to 7-day spray interval.
chlorothalonil (various)	M05	See labels	14	2	**For suppression only.** Maximum of three sprays.
chlorothalonil + cymoxanil (Ariston)	M05+27	2.0 to 2.4 pt/acre	14	0.5	Apply prior to favorable infection periods; continue on 7- to 9-day interval; alternate with a different mode of action.
fixed copper (various)	M01	See labels	1	1	May also reduce bacterial rots.
famoxadone + cymoxanil (Tanos)	11+27	8 oz/acre	3	0.5	Must be tank mixed with a contact fungicide such as mancozeb. Do not make more than one sequential application before rotating to a different mode of action.
fenamidone (Reason 500SC)	11	5.5 fl oz/acre	7	0.5	Begin applications when conditions favor disease development and continue on 5- to 10-day interval. Do not apply more than 22 fl oz per growing season. Alternate with fungicide from different resistance group.
fluazinam (Omega 500)	29	1.0 pt/acre	7	1	Initiate sprays when conditions are favorable for disease or at disease onset. Spray on a 7- to 10-day schedule.
mandipropamid (Revus 2.08F)	40	8 fl oz/acre	7	0.5	Apply prior to disease development and continue throughout season at 7- to 10-day intervals; maximum 24 fl oz per season.
mefenoxam + chlorothalonil (Ridomil Gold/Bravo)	4+M05	2.5 lb/acre	14	2	
mefenoxam + copper (Ridomil Gold/copper)	4+M01	5 lb/2.5 acre	7	-	Apply 1 pack of Ridomil Gold Copper (5 lb product)/2.5 acres in sufficient water to obtain thorough coverage. Begin applications when conditions are favorable for disease, but before infection, and continue at 14-day intervals until he threat of disease is over. Use a suitable spreader sticker at rates recommended on the product label.
oxathiapiprolin + mandipropamid (Orondis Ultra)	49+40	5.5 to 8.0 fl oz/acre	7	4hr	Use higher rate if disease is present. For the best results, begin the disease resistance program with an initial treatment at planting or transplanting with a fungicide registered for its use. Apply Orondis Ultra as a foliar spray in a mixture with copper-based fungicide beginning at first appearance of symptoms.
potassium phosphite + tebuconazole (Viathon)	49+3	2 to 3 pt/acre	7	0.5	Use as a preventative treatment.

TABLE 3-19. DISEASE CONTROL PRODUCTS FOR ONION (cont'd)

B. Dutta, Plant Pathologist, University of Georgia (LAST UPDATED IN 2025)

Disease/Material	FRAC Code	Rate of Material Formulation	Minimum Days Harv.	Reentry	Method, Schedule, and Remarks
ONION (GREEN) (cont'd)					
LEAF BLIGHT (*BOTRYTIS* SPP.)					
azoxystrobin (Quadris)	11	9.0 to 15.5 fl oz/acre	0	4 hr	Resistance reported in the Southeast; use premix products when possible. Make no more than two sequential applications before alternating with fungicides that have a different mode of action. Apply no more 90 fl oz Quadris per crop per acre per season. See individual labels for application instructions and rates.
azoxystrobin + difenoconazole (Quadris Top)	11+3	12 to 14 oz/acre	7	0.5	Make no more than one application before alternating with a fungicide with a different mode of action.
azoxystrobin + chlorothalonil (Quadris Opti)	11+M05	1.6 to 3.6 pts/acre	14	2	Applications should begin prior to disease onset and subsequent applications should be made on a 7- to 14-day interval.
azoxystrobin + propiconazole (Quilt Xcel)	11+3	17.5 to 21 fl oz	0	0.5	Make only one application before rotating to a non-group 11 fungicide.
azoxystrobin + tebuconazole (Custodia)	11+3	8.6 to 12.9 fl oz	7	0.5	Use higher rate and shorter interval when disease conditions are severe.
benzovindiflupyr + difenoconazole (Aprovia Top)	7+3	10.5 oz/acre	7	0.5	Cladosporium, powdery mildew, purple blotch, rust, and Stemphylium. Spreading/penetrating type adjuvant recommended.
boscalid (Endura 70 WG)	7	6.8 oz/acre	7	0.5	Do not make more than 2 sequential applications or more than 6 applications per season.
chlorothalonil (various)	M05	See labels	14	0.5	Spray at first appearance. Maximum of three sprays.
cyprodinil (Vangard WG)	9	10 oz/acre	7	0.5	Do not make more than 2 sequential applications before alternating to a different mode of action.
cyprodinil + fludioxonil (Switch)	9+12	11 to 14 oz/acre	7	0.5	Do not plant rotational crops other than onions or strawberries for 12 months following the last application.
difenoconazole + cyprodinil (Inspire Super)	3+9	16 to 20 fl oz/acre	14	0.5	Make no more than two applications before alternating with a fungicide with a different mode of action.
fluopyram + pyrimethanil (Luna Tranquility)	7+9	16 to 27 fl oz/acre	7	0.5	When disease pressure is high, use higher rates and shorter intervals.
fluopyram + difenoconazole (Luna Flex)	7+3	10 to 13.6 fl oz	7	0.5	Please see label.
fluopyram + tebuconazole (Luna Experience)	7+3	8.0 to 12.8 fl. oz/acre	7	0.5	Observe seasonal application limits for both group 7 and group 3 fungicides.
fluxapyroxad + pyraclostrobin (Merivon)	7+11	4 to 11 fl oz/acre	7	0.5	Apply at disease onset; continue on 7- to 14-day schedule. No more than 3 applications/season.
propiconazole (Quilt)	3	14 to 27.5 oz/acre	0	0.5	Alternate with a different mode of action.
pyraclostrobin + boscalid (Pristine)	11+7	14.5 to 18.5 oz/acre	7	1	Make a maximum of 6 applications per season.
pyrimethanil (Scala)	9	9 or 18 fl oz/acre	7	0.5	Resistance reported in the Southeast. Use lower rate in a tank-mix with broad-spectrum fungicide and higher rate when applied alone. Do not apply more than 54 fl oz per crop.
PURPLE BLOTCH (*ALTERNARIA PORRI*)					
azoxystrobin (Quadris)	11	6 to 12 fl oz/acre	0	4 hr	Make no more than two sequential applications before alternating with fungicides that have a different mode of action. Apply no more 90 fl oz Quadris per crop per acre per season. See individual labels for application instructions and rates
azoxystrobin + difenoconazole (Quadris Top)	11+3	12 to 14 oz/acre	7	0.5	Make no more than one application before alternating with a fungicide with a different mode of action.
azoxystrobin + chlorothalonil (Quadris Opti)	11+M05	1.6 to 3.2 pt/acre	14	0.5	Make no more than one application before alternating with a fungicide with a different mode of action.
azoxystrobin + propiconazole (Quilt Xcel)	11+3	14 to 21 fl oz	0	0.5	Make only one application before rotating to a non-group 11 fungicide.
azoxystrobin + tebuconazole (Custodia)	11+3	8.6 to 12.9 fl oz	7	0.5	Use higher rate and shorter interval when disease conditions are severe.
chlorothalonil (various)	M05	See labels	14	2	Spray at first appearance. Maximum of three sprays.
chlorothalonil + cymoxanil (Ariston)	M05+27	2.0 to 2.4 pt/acre	14	0.5	Apply prior to favorable infection periods; continue on 7- to 9-day interval; alternate with a different mode of action.

TABLE 3-19. DISEASE CONTROL PRODUCTS FOR ONION (cont'd)

B. Dutta, Plant Pathologist, University of Georgia (LAST UPDATED IN 2025)

Disease/Material	FRAC Code	Rate of Material Formulation	Minimum Days Harv.	Minimum Days Reentry	Method, Schedule, and Remarks
ONION (GREEN) (cont'd)					
PURPLE BLOTCH (*ALTERNARIA PORRI*) (cont'd)					
chlorothalonil + tebuconazole (Muscle ADV)	M05+3	1.1 to 1.6 pt/acre	7	0.5	Apply in a protective schedule or when weather is favorable for disease.
cyprodinil (Vangard WG)	9	10 oz/acre	7	0.5	Do not make more than 2 sequential applications before alternating to a different mode of action.
cyprodinil + fludioxonil (Switch)	9+12	11 to 14 oz/acre	7	0.5	Do not plant rotational crops other than onions or strawberries for 12 months following the last application.
difenoconazole + benzovindiflupyr (Aprovia Top)	3+7	10.5 oz/acre	7	0.5	Also, for Stemphylium leaf blight. Use preventatively with a penetrating spreader.
difenoconazole + cyprodinil (Inspire Super)	3+9	16 to 20 fl oz/acre	14	0.5	Make no more than two applications before alternating with a fungicide with a different mode of action.
fluopyram + pyrimethanil (Luna Tranquility)	7+9	16 to 27 fl oz/acre	7	0.5	When disease pressure is high, use higher rates and shorter intervals.
fluopyram + tebuconazole (Luna Experience)	7+3	8.0 to 12.8 fl. oz/acre	7	0.5	Observe seasonal application limits for both group 7 and group 3 fungicides.
fluopyram + difenoconazole (Luna Flex)	7+3	10 to 13.6 fl oz	7	0.5	Please see label.
fluxapyroxad + pyraclostrobin (Merivon)	7+11	4 to 11 fl oz/acre	7	0.5	Apply at disease onset; continue on 7- to 14-day schedule. No more than 3 applications/season.
potassium phosphite + tebuconazole (Viathon)	49+3	2 to 3 pt/acre	7	0.5	Use as a preventative treatment.
propiconazole (Quilt)	3	14 to 27.5 fl oz	0	0.5	Alternate with a different mode of action.
pyraclostrobin (Cabrio)	11	8 to 12 oz/acre	7	0.5	Make no more than 2 sequential applications and no more than 6 applications per season.
pyraclostrobin + boscalid (Pristine)	11+7	10.5 to 18.5 oz/acre	7	1	Make a maximum of 6 applications per season.
pyrimethanil (Scala)	9	9 or 18 oz/acre	7	0.5	Use lower rate in a tank-mix with broad-spectrum fungicide and higher rate when applied alone. Do not apply more than 54 fl oz per crop.
STEMPHYLIUM LEAF BLIGHT (*STEMPHYLIUM VESICARIUM*)					
azoxystrobin (Quadris)	11	6 to 12 fl oz/acre	0	4 hr	Make no more than two sequential applications before alternating with fungicides that have a different mode of action. Apply no more 90 fl oz Quadris per crop per acre per season. See individual labels for application instructions and rates.
azoxystrobin + difenoconazole (Quadris Top)	11+3	12 to 14 oz/acre	7	0.5	Make no more than one application before alternating with a fungicide with a different mode of action.
azoxystrobin + propiconazole (Quilt Xcel)	11+3	14 to 26 fl oz	0	0.5	Make only one application before rotating to a non-group 11 fungicide.
fluopyram + pyrimethanil (Luna Tranquility)	7+9	16 to 27 oz/acre	7	0.5	See label
difenoconazole + cyprodinil (Inspire Super)	3+9	16 to 20 fl oz/acre	14	0.5	Make no more than two applications before alternating with a fungicide with a different mode of action.
fluopyram + difenoconazole (Luna Flex)	7+3	10 to 13.6 fl oz	7	0.5	Please see label.
fluopyram + pyrimethanil (Luna Tranquility)	7+9	16 to 27 fl oz/acre	7	0.5	When disease pressure is high, use higher rates and shorter intervals.
fluxapyroxad + pyraclostrobin (Merivon)	7+11	4 to 11 fl oz/acre	7	0.5	Apply at disease onset; continue on 7- to 14-day schedule. No more than 3 applications/season.
pyraclostrobin + boscalid (Pristine)	11+7	10.5 to 18.5 oz/acre	7	1	Make no more than 6 applications per season.
ONION (DRY)					
DAMPING OFF (*PYTHIUM SPP.*)					
mefenoxam (Ridomil Gold)	4	0.5 to 1 pt/trt acre	—	2	See label for row rates. Also, for green onion.

TABLE 3-19. DISEASE CONTROL PRODUCTS FOR ONION (cont'd)

B. Dutta, Plant Pathologist, University of Georgia (LAST UPDATED IN 2025)

Disease/Material	FRAC Code	Rate of Material Formulation	Minimum Days Harv.	Minimum Days Reentry	Method, Schedule, and Remarks
ONION (DRY) (cont'd)					
DAMPING OFF (*PYTHIUM* SPP.) (cont'd)					
metalaxyl (MetaStar)	4	2 to 4 pt/trt acre	—	2	Preplant incorporated or soil surface spray.
azoxystrobin + mefenoxam (Uniform)	11+4	0.34 fl oz/1000 ft	—	0	In-furrow treatment.
DOWNY MILDEW (*PERONOSPORA DESTRUCTOR*)					
ametoctradin + dimethomorph (Zampro)	45+40	14.0 fl oz/acre	0	0.5	Begin applications prior to disease development and continue on a 5- to 7-day spray interval.
chlorothalonil + cymoxanil (Ariston)	M05+27	1.6 to 2.4 pt/acre	7	0.5	Apply prior to favorable infection periods; continue on a 7- to 9-day interval; alternate with a different mode of action.
chlorothalonil + zoxamide (Zing!)	M05+22	30 fl oz	7	0.5	Do not apply to exposed bulbs.
cyazofamid (Ranman)	21	2.75 to 3.0 oz/acre	0	0.5	Use a surfactant for best results.
dimethomorph (Forum)	40	6 fl oz	0	0.5	Do not make more than one sequential application
famoxadone + cymoxanil (Tanos)	11+27	8.0 oz/acre	3	0.5	Must be tank mixed with a contact fungicide such as mancozeb. Do not make more than one sequential application before rotating to a different mode of action.
fenamidone (Reason)	11	5.5 fl oz/acre	7	0.5	Use as soon as environmental conditions become favorable.
fluazinam (Omega 500F)	29	1.0 pt/acre	7	1	Initiate sprays when conditions are favorable for disease or at disease onset. Spray on a 7- to 10-day schedule.
mancozeb + zoxamide (Gavel 75 DF)	M03+22	1.5 to 2 lb/acre	7	2	Do not make more than 8 applications per season. Do not apply to exposed bulbs.
mancozeb + azoxystrobin + tebuconazole (Dexter Xcel)	M03+11+3	56 to 72 fl oz/acre	14	1	Do not make more than 2 sequential applications before rotating with a product other than FRAC 11.
mefenoxam + copper hydroxide (Ridomil Gold Copper)	4+M01	5 lb/2.5 acre	7	2	Use as a foliar spray for preventative control.
mefenoxam + chlorothalonil (Ridomil Gold/Bravo)	4+M05	2.5 lb/acre	14	2	Use with a suitable adjuvant.
oxathiapiprolin + mandipropamid (Orondis Ultra)	49+40	5.5 to 8.0 fl oz/acre	7	4hr	Use higher rate if disease is present. For the best results, begin the disease resistance program with an initial treatment at planting or transplanting with a fungicide registered for its use. Apply Orondis Ultra as a foliar spray in a mixture with copper-based fungicide beginning at first appearance of symptoms.
potassium phosphite + tebuconazole (Viathon)	49+3	2 to 3 pt/acre	7	0.5	Use as a preventative treatment.
LEAF BLIGHT (*BOTRYTIS* SPP.)					
azoxystrobin (Quadris)	11	9 to 15.5 fl oz/acre	7	4 hr	Make no more than two sequential applications before alternating with fungicides with a different mode of action. Apply no more than 90 fl oz of Quadris per crop per acre per season. See individual labels for application instructions and rates.
azoxystrobin + chlorothalonil (Quadris Opti)	11+M05	1.6 to 3.2 pt/acre	14	0.5	Make no more than one application before alternating with a fungicide with a different mode of action.
azoxystrobin + tebuconazole (Custodia)	11+3	12.9 fl oz	7	0.5	Use higher rate and shorter interval when disease conditions are severe.
chlorothalonil + zoxamide (Zing!)	M05+22	30 fl oz	7	0.5	Do not apply to exposed bulbs.
cyprodinil (Vangard WG)	9	10 oz/acre	7	0.5	Do not make more than 2 sequential applications before alternating to a different mode of action.
cyprodinil + fludioxonil (Switch)	9+12	11 to 14 oz/acre	7	0.5	Do not plant rotational crops other than onions or strawberries for 12 months following the last application.
difenoconazole + cyprodinil (Inspire Super)	3+9	16 to 20 fl oz/acre	7	0.5	Make no more than two applications before alternating with a fungicide with a different mode of action.
fluopyram + difenoconazole (Luna Flex)	7+3	10 to 13.6 fl oz	7	0.5	Please see label.

TABLE 3-19. DISEASE CONTROL PRODUCTS FOR ONION (cont'd)

B. Dutta, Plant Pathologist, University of Georgia (LAST UPDATED IN 2025)

Disease/Material	FRAC Code	Rate of Material Formulation	Minimum Days Harv.	Minimum Days Reentry	Method, Schedule, and Remarks
ONION (DRY) (cont'd)					
LEAF BLIGHT (*BOTRYTIS* SPP.) (cont'd)					
fluopyram + pyrimethanil (Luna Tranquility)	7+9	16 to 27 fl oz/acre	7	0.5	When disease pressure is high, use higher rates and shorter intervals.
mancozeb + azoxystrobin + tebuconazole (Dexter Xcel)	M03+11+3	56 to 72 fl oz/acre	14	1	Do not make more than 2 sequential applications before rotating with a product other than FRAC 11.
mancozeb + zoxamide (Gavel 75 DF)	M03+22	1.5 to 2 lb/acre	7	2	Do not make more than 8 applications per season. Do not apply to exposed bulbs.
penthiopyrad (Fontelis)	7	16 to 24 fl oz/acre	3	0.5	Begin sprays prior to disease development and continue on a 7- to 14-day schedule.
pyraclostrobin (Cabrio)	11	12 oz/acre	7	0.5	Make no more than 2 sequential applications and no more than 6 applications per season.
pyrimethanil (Scala)	9	9 or 18 fl oz/acre	7	0.5	Use lower rate in a tank-mix with broad-spectrum fungicide and higher rate when applied alone. Do not apply more than 54 fl oz per crop.
NECK ROT (*BOTRYTIS* SPP.), PURPLE BLOTCH (*ALTERNARIA PORRI*), DOWNY MILDEW (*PERONOSPORA DESTRUCTOR*)					
azoxystrobin (Quadris and various)	11	9 to 15.5 fl oz/acre	7	4 hr	Make no more than two sequential applications before alternating with fungicides with a different mode of action. Apply no more than 90 fl oz of Quadris per crop per acre per season. See individual labels for application instructions and rates.
azoxystrobin + chlorothalonil (Quadris Opti)	11+M05	1.6 to 3.2 pt/acre	14	0.5	Make no more than one application before alternating with a fungicide with a different mode of action.
azoxystrobin + mancozeb (Dexter Max)	11+M05	3.2 lb/acre	7	0.5	Follow a protective 7-day schedule. Observe season limit for azoxystrobin applications.
azoxystrobin + propiconazole (various)	11+3	14 to 26 oz /acre	14	0.5	
azoxystrobin + tebuconazole (various)	11+3	See labels	7	0.5	See labels for specific rates and application instructions.
boscalid (Endura)	7	6.8 oz/acre	7	0.5	Not for downy mildew. Do not make more than 2 sequential applications or more than 6 applications per season.
chlorothalonil (various)	M05	0.9 to 1 lb/acre	7	0.5	Will only suppress neck rot and downy mildew.
chlorothalonil + tebuconazole (Muscle ADV)	M05+3	1.1 to 1.6 pt/acre	7 to 14	0.5	Rust and purple blotch only.
fluopyram + difenoconazole (Luna Flex)	7+3	10 to 13.6 fl oz	7	0.5	Please see label.
chlorothalonil + zoxamide (Zing)	M05+22	30 fl oz/acre	7	0.5	Follow protective spray schedule when diseases are in the area.
cyprodinil (Vanguard)	12	10 oz/acre	7	0.5	Suppressive only on neck rot.
difenoconazole + benzovindiflupyr (Aprovia Top)	3+7	10.5 oz/acre	7	0.5	Purple blotch and Stemphylium leaf blight. Use preventatively with a tank mixed containing a penetrating spreader. Cladosporium, powdery mildew, purple blotch, rust, and Stemphylium. Spreading/penetrating type adjuvant recommended.
difenoconazole + cyprodinil (Inspire Super)	3+9	16 to 20 oz/acre	7	0.5	Make no more than two applications before alternating with a fungicide with a different mode of action.
fluazinam (Omega 500)	29	1.0 pt/acre	7	1	Initiate sprays when conditions are favorable for disease or at disease onset. Spray on a 7- to 10-day schedule.
fluopyram + tebuconazole (Luna Experience)	7+3	8 to 12.8 oz/acre	7	0.5	Not for downy mildew. Suppresses *Sclerotium* spp.
fluopyram + pyrimethanil (Luna Tranquility)	7+9	16 to 27 oz/acre	7	0.5	Not for downy mildew. Suppresses *Sclerotium* spp.
fluxapyroxad + pyraclostrobin (Merivon)	7+11	4 to 11 fl oz/acre	7	0.5	Use higher rates for downy mildew suppression. Apply at disease onset; continue on a 7- to 14-day schedule. No more than 3 applications per season.
iprodione (Rovral 4F)	2	1.5 lb/acre	7	0.5	Not for downy mildew. Apply when conditions are favorable; 14-day intervals.
mancozeb + zoxamide (Gavel 75 DF)	M03+22	1.5 to 2 lb/acre	7	2	Do not make more than 8 applications per season. Do not apply to exposed bulbs.
mancozeb + azoxystrobin (Dexter Max)	M03+11	3.2 lb/acre	7	1	Do not apply to exposed bulbs.

TABLE 3-19. DISEASE CONTROL PRODUCTS FOR ONION (cont'd)

B. Dutta, Plant Pathologist, University of Georgia (LAST UPDATED IN 2025)

Disease/Material	FRAC Code	Rate of Material Formulation	Minimum Days Harv.	Reentry	Method, Schedule, and Remarks
ONION (DRY) (cont'd)					
NECK ROT (*BOTRYTIS* SPP.), PURPLE BLOTCH (*ALTERNARIA PORRI*), DOWNY MILDEW (*PERONOSPORA DESTRUCTOR*) (cont'd)					
mancozeb + azoxystrobin + tebuconazole (Dexter Xcel)	M03+11+3	48 to 72 fl oz/acre	14	1	Do not make more than 2 sequential applications before rotating with a product other than FRAC 11.
penthiopyrad (Fontelis)	7	16 to 24 fl oz/acre	3	12 hr	Begin sprays prior to disease development and continue on a 7- to 14-day schedule.
propiconazole (various)	3	4 to 8 oz/acre	14	0.5	Not for downy mildew. Alternate with a different mode of action.
pyraclostrobin + boscalid (Pristine)	11+7	14.5 to 18.5 oz/acre	7	1	Make no more than 6 applications per season.
tebuconazole + chlorothalonil (Muscle)	3+M05	1.1 to 1.6 pt/acre	7 to 14	0.5	Not for downy mildew or Botrytis.
tebuconazole + potassium phosphate (Viathon)	3+P07	2 to 3 pts/acre	7	0.5	
zoxamide + mancozeb (Zing!)	22+M	1.5 to 2 lb/acre	7	0.5	Use preventatively.
penthiopyrad (Fontelis)	7	24 oz/acre	3	0.5	**GA only.** Apply as a broadcast or banded spray over seeds or seedlings.
mancozeb (various)	M03	3 lb/29,000 ft row	—	—	
azoxystrobin + difenoconazole (Quadris Top)	11+3	14 fl oz/acre	7	0.5	Begin sprays prior to disease onset and spray on a 7- to 14-day schedule. Do not rotate with Group 11 fungicides.
difenoconazole + cyprodinil (Inspire Super)	3+9	16 to 20 fl oz/acre	7	0.5	Make no more than two applications before alternating with a fungicide with a different mode of action.
fluxapyroxad + pyraclostrobin (Merivon)	7+11	4 to 11 fl oz/acre	7	0.5	Apply at disease onset; continue on 7- to 14-day schedule. No more than 3 applications/season.
PINK ROOT (*PHOMA* SPP.)					
iprodione (various)	2	1.5 lb/acre 50 to 100 gal/acre	7	0	Start 7-day foliar sprays at first appearance of favorable conditions.
SMUT (*UROCYSTIS* SPP.)					
pyraclostrobin + boscalid (Pristine)	11+7	10.5 to 18.5 oz/acre	7	1	Make no more than 6 applications per season.
STEMPHYLIUM LEAF BLIGHT (*STEMPHYLIUM VESICARIUM*)					
azoxystrobin (Quadris and various)	11	9 to 15.5 fl oz/acre	7	4 hr	Make no more than two sequential applications before alternating with fungicides with a different mode of action. Apply no more than 90 fl oz of Quadris per crop per acre per season. See individual labels for application instructions and rates.
fluopyram + difenoconazole (Luna Flex)	7+3	10 to 13.6 fl oz	7	0.5	Please see label.
iprodione (Rovral 4F)	2	1.5 lb/acre	7	0.5	Not for downy mildew. Apply when conditions are favorable; 14-day intervals.
WHITE ROT (*SCLEROTIUM CEPIVORUM*)					
azoxystrobin + chlorothalonil (Quadris Opti)	11+M05	1.6 to 3.2 pt/acre	14	0.5	Make no more than one application before alternating with a fungicide with a different mode of action.
azoxystrobin + tebuconazole (Custodia)	11+3	32 fl oz	7	0.5	Make one application at 32 fl oz per acre in furrow at planting. Additional control may be obtained by including two foliar applications at 8.6 to 12.9 fl oz/acre.
dicloran (Botran)	14	5.3 lb/acre	14	0.5	Apply 5-in. band over seed row and incorporate in top 1.5 to 3 in. of soil, 1 to 2 weeks before seeding.
fludioxonil (Cannonball WG)	12	7 oz/acre	7	0.5	In furrow treatment only.
penthiopyrad (Fontelis)	7	16 to 24 fl oz/acre	3	0.5	Begin sprays prior to disease development and continue on a 7- to 14-day schedule.
tebuconazole (Toledo 36F and various)	3	20.5 fl oz/acre	7	0.5	Make one application in furrow at time of planting.
thiophanate-methyl (various)	1	See label			Spray into open furrow at time of seeding or planting in a row.

TABLE 3-20. EFFICACY OF PRODUCTS FOR DISEASE CONTROL IN ONION

Scale: E, excellent; G, good; F, fair; P, poor; NC, no control; ND, no data

Active Ingredient [1]	Product [1]	Fungicide group [F]	Preharvest interval (Days)	Bacterial Streak (*Pseudomonas viridiflava*)	Black mold (*Aspergillus niger*)	Botrytis Leaf Blight (*B. squamosa*)	Botrytis Neck Rot (*B. allii*)	Damping off (*Pythium* spp.)	Downy Mildew (*P. destructor*)	Fusarium Basal Rot (*F. oxysporum*)	Onion Smut (*Urocystis colchici*)	Center Rot (*Pantoea ananatis*)	Pink Root (*Phoma terrestris*)	Purple Blotch (*Alternaria porri*)	Stemphylium Leaf Blight and Stalk Rot	White Rot (*Sclerotium cepivorum*)
ametoctradin + dimethomorph	Zampro	40+45	0	NC	NC	NC	NC	NC	F	NC	NC	NC	NC	NC	NC	NC
azoxystrobin	Quadris	11	7	NC	G	P	NC	NC	ND	NC	ND	NC	NC	F	P	ND
azoxystrobin + difenoconazole	Quadris Top	11+3	1	NC	NC	P	NC	NC	ND	NC	NC	NC	NC	G-F	F	NC
boscalid	Endura	7	7	ND	ND	F-P	ND	ND	ND	ND	ND	ND	F	ND	P	ND
chlorothalonil	Bravo	M05	14	NC	NC	F	NC	NC	F-P	NC	NC	NC	NC	F	F	NC
chlorothalonil + zoxamide	Zing!	M05+22	7	ND	ND	ND	ND	ND	F-P	ND	ND	ND	ND	ND	ND	ND
chlorothalonil + cymoxanil	Ariston	M05+27	7	ND	ND	ND	ND	ND	F-P	ND	ND	ND	ND	ND	ND	ND
cyprodinil + fludioxonil	Switch	9+12	7	NC	NC	F	ND	NC	NC	ND	NC	NC	NC	F	F	NC
cyprodinil + difenoconazole	Inspire Super	9+3	7	ND	ND	G	ND	ND	ND	ND	ND	ND	ND	ND	G-F	ND
dichloropropene + chloropicrin, fumigant	Telone C-1	—	—	NC	NC	NC	NC	P	NC	F	NC	NC	F	NC	NC	F
difenoconazole + benzovindiflupyr	Aprovia Top	3+7	7	ND	ND	G	ND	ND	ND	ND	ND	ND	ND	ND	F	ND
dimethomorph	Forum	40	0	NC	NC	NC	NC	NC	P	NC	NC	NC	NC	NC	NC	NC
fenamidone	Reason	11	7	NC	NC	P	NC	NC	F-P	NC	NC	NC	NC	P	P	NC
famoxadone + cymoxanil	Tanos	11+27	3	NC	NC	P	NC	NC	P	NC	NC	NC	NC	F-P	P	NC
fixed copper	various	M01	1	F-G	NC	F	NC	NC	F	NC	NC	F	NC	F	NC	NC
fluazinam	Omega 50	29	2	NC	NC	G	NC	NC	G-F	NC	NC	NC	NC	E-G	G	NC
fluopyram + difenoconazole	Luna Flex	7+3	7	ND	ND	G-F	ND	ND	ND	ND	ND	ND	ND	ND	G-F	ND
fluopyram + pyrimethanil	Luna Tranquility	7+9	7	ND	ND	G-F	ND	ND	ND	ND	ND	ND	ND	ND	G-F	ND
fluopyram + tebuconazole	Luna Experience	7+3	7	ND	ND	F	ND	ND	ND	ND	ND	ND	ND	ND	F	ND
fluxapyroxad + pyraclostrobin	Merivon	7+11	7	ND	ND	F	ND	ND	ND	ND	ND	ND	ND	F	F	ND
iprodione	Rovral	2	7	NC	NC	F	P	NC	NC	NC	NC	NC	NC	F	F-P	F
mancozeb	various	M03	7	NC	NC	F	NC	NC	F	NC	F	NC	NC	F	F-P	NC
mandipropamid	Revus	40	7	NC	NC	NC	ND	F	P	NC	NC	NC	NC	NC	NC	NC
mefenoxam	Ridomil Gold EC	4	7	NC	NC	NC	NC	F	ND	NC	NC	NC	NC	NC	NC	NC
mefenoxam + chlorothalonil	Ridomil Gold Bravo	4+M01	14	NC	NC	P	NC	F	F	NC	NC	NC	NC	P	P	NC
mefenoxam + copper	Ridomil Gold/Copper	4+M01	7	F	NC	NC	NC	P	G-F	NC	NC	F	NC	NC	NC	NC
mefenoxam + mancozeb	Ridomil Gold MZ	4+M03	7	NC	NC	P	NC	P	F	NC	F	NC	NC	P	P	NC
metam sodium, fumigant	Vapam	—	—	NC	NC	NC	NC	F	NC	F	NC	NC	NC	E	NC	NC	F
penthiopyrad	Fontelis	7		ND	ND	F	ND	ND	ND	ND	ND	ND	ND	G	ND	ND
potassium phosphite + tebuconazole	Viathon	P07+3	7	ND	ND	ND	ND	ND	ND	ND	ND	ND	ND	G	ND	ND
pyraclostrobin	Cabrio	11	7	NC	ND	P	NC	NC	P	NC	ND	NC	NC	G-F	P	ND
pyraclostrobin + boscalid	Pristine	11+7	7	NC	ND	F	F	NC	P	NC	ND	NC	NC	F	F	ND
pyrimethanil	Scala	9	7	NC	ND	F-P	NC	NC	NC	ND	NC	NC	NC	F	F	NC
tebuconazole	Toledo	3	7	ND	ND	F	ND	ND	ND	ND	ND	ND	ND	F	ND	ND
oxathiapiprolin + mandipropamid	Orondis Ultra	49+40	7	NC	NC	NC	ND	ND	G-F	ND	ND	ND	ND	ND	ND	ND

[1] Efficacy ratings do not necessarily indicate a labeled use for every disease
[F] To prevent resistance in pathogens, alternate fungicides within a group with fungicides in another group. Fungicides in the "M" group are generally considered "low-risk" with no signs of resistance developing to most fungicides
[R] Resistance reported in the pathogen

TABLE 3-21. DISEASE CONTROL PRODUCTS FOR PARSLEY AND CILANTRO

A. Keinath, Plant Pathologist, Clemson University

Note: Some fungicides are registered on parsley but not on cilantro; check the Remarks section before applying any product to cilantro.

Disease/Material	FRAC Code	Rate of Material Formulation	Minimum Days Harv.	Minimum Days Reentry	Method, Schedule, and Remarks
DAMPING OFF AND ROOT ROT (*PYTHIUM, PHYTOPHTHORA*)					
fenamidone (Reason 500SC)	11	8.2 fl oz/acre	14	0.5	Limit of 24.6 fl oz/acre per year. Alternate each application with a non- FRAC Group 11 fungicide.
mefenoxam (Ridomil Gold 4 SL) (Ultra Flourish 2 EC)	4	1 to 2 pt/treated acre 2 to 4 pt/treated acre	0	0.5	Apply preplant incorporated or surface application at planting.
metalaxyl (MetaStar 2 E)	4	2 to 8 pt/treated acre	0	2	Banded over the row, preplant incorporated, or injected with liquid fertilizer.
oxathiapiprolin + mefenoxam (Orondis Gold DC 1.17DC)	49+4	13.9-27.8 fl ox/acre	7	2	Important note: this product is primarily for *Phytophthora*. For *Pythium*, use mefenoxam. Do not make more than 2 applications at the maximum rate per year, 1 application per crop.
SEEDLING BLIGHT, DAMPING OFF, ROOT ROT (*PYTHIUM SPP., RHIZOCTONIA SOLANI*)					
azoxystrobin + mefenoxam (Uniform 3.72SC)	11+4	0.34 fl oz/ 1,000 row ft	—	0	Apply as an in-furrow spray in 5 gal of water per acre prior to covering seed. Make only 1 application per season. **Do not use on cilantro.**
ALTERNARIA LEAF SPOT (*ALTERNARIA SPP.*), CERCOSPORA LEAF SPOT, EARLY BLIGHT, (*CERCOSPORA SPP.*) POWDERY MILDEW (*ERYSIPHE HERACLEI*), SEPTORIA LEAF SPOT, LATE BLIGHT (*SEPTORIA PETROSELINI*)					
azoxystrobin (various)	11	See labels	0	4 hr	Make no more than two sequential applications before alternating with fungicides that have a different mode of action. Apply no more than 1.88 lb per crop per acre per season.
azoxystrobin + flutriafol (Topguard EQ 4.3SC)	11+3	6 to 8 fl oz/acre	7	0.5	Make no more than 4 applications per crop per year. **Do not use on cilantro. Do not use in greenhouses.**
boscalid (Endura 70WG)	7	4.5 to 9 oz/acre	14	0.5	Make no more than 2 sequential applications before alternating with fungicides that have a different mode of action. Limit of 18 oz/acre per year. **Do not use in green- houses**
cyprodinil + fludioxonil (Switch 62.5WG)	9+12	11 to 14 oz/acre	0	0.5	Make no more than two sequential applications before alternating with fungicides that have a different mode of action for two applications. Apply no more than 56 oz per crop per acre per season.
fenamidone (Reason 500SC)	11	5.5 to \8.2 fl oz/acre	14	0.5	Limit of 24.6 fl oz/acre per year. Alternate each application with a non- FRAC Group 11 fungicide. Not labeled to control Septoria or powdery mildew. Use higher rate on parsley.
fixed copper (generic)	M01	See label	0	0	Spray at first disease appearance, 7- to 10-day intervals.
fludioxonil + pydiflumetofen (Miravis Prime 2.09SC)	12+7	9.2 to 13.4 fl oz/acre	0	0.5	Limit of 2 applications per crop per year. Not labeled for Cercospora leaf spot. **Do not use in greenhouses. Do not use on cilantro.**
flutriafol (Rhyme 2.08SC)	3	5 to 7 fl oz	7	0.5	Make no more than 4 applications (28 fl oz) per crop per year. **Do not use in green- houses. Do not use on cilantro.**
fluxapyroxad + pyraclostrobin (Merivon 500SC)	7+11	4 to 11 fl oz/acre	3	0.5	Make no more than two sequential applications before alternating with fungicides that have a different mode of action. Maximum of 3 applications per crop.
penthiopyrad (Fontelis 1.67F)	7	14 to 24 fl oz	3	0.5	Do not make more than two sequential applications. Maximum of 72 fl oz/ acre per year.
propiconazole (various)	3	3 to 4 fl oz/acre	14	0.5	Begin at first sign of disease and repeat at 14-day intervals. Make no more than two consecutive applications before rotating to another fungicide with a different mode of action.
pydiflumetofen + fludioxonil (Miravis Prime 2.09SC)	7+12	9.2 to 13.4 fl oz/acre	7	0.5	Make no more than 2 sequential applications of Miravis Prime or other Group 7 and 12 fungicides before rotating to another effective fungicide with a different mode of action. Maximum 2 applications at high rate and 3 applications at the low rate per year. **Do not use in greenhouses. Do not use on cilantro.**
pyraclostrobin (Cabrio 20EG, Pyrac 2EC)	11	12 to 16 oz/acre	0	0.5	Make no more than two sequential applications before alternating with fungicides that have a different mode of action. Apply no more than 64 oz per crop per acre per season.
trifloxystrobin (Flint Extra 42.6SC)	11	3.0 to 3.8 fl oz/acre	0 (broadcast) 20 (banded)	0.5	Limit of 7.6 fl oz or 2 applications per acre per season. Not labeled for Cercospora or Septoria.
triflumizole (Procure 480SC)	3	6 to 8 fl oz/acre	0	0.5	Limit of 2 applications per crop per year. Not labeled for Cercospora or Septoria.
GRAY MOLD (*BOTRYTIS CINEREA*), POWDERY MILDEW (*ERYSIPHE HERACLEI*)					
fluopyram + trifloxystrobin (Luna Sensation 500SC)	7+11	5.0 to 7.6 fl oz/acre	7	0.5	Limit of 15.3 fl oz (2 to 3 applications) at low and high rate, respectively) per acre per year. Use higher rate for gray mold. **Do not use on cilantro.**
pydiflumetofen + fludioxonil (Miravis Prime 2.09SC)	7+12	9.2 to 13.4 fl oz/acre (use high rate for Gray mold)	7	0.5	Make no more than 2 sequential applications of Miravis Prime or other Group 7 and 12 fungicides before rotating to another effective fungicide with a different mode of action. Maximum 3 applications at high rate and 4 applications at the low rate per year. **Do not use in greenhouses. Do not use on cilantro.**

TABLE 3-21. DISEASE CONTROL PRODUCTS FOR PARSLEY AND CILANTRO (cont'd)

A. Keinath, Plant Pathologist, Clemson University

Note: Some fungicides are registered on parsley but not on cilantro; check the Remarks section before applying any product to cilantro.

Disease/Material	FRAC Code	Rate of Material Formulation	Minimum Days Harv.	Reentry	Method, Schedule, and Remarks
ROOT ROT (*PHYTOPHTHORA* SPP.)					
potassium phosphite (various)	P07	2.5 to 5.0 pints/acre	0	4 hr	Limit of 7 applications per season. Do not treat plants during dormancy or when plants are under stress due to heat or inadequate moisture.
WEB BLIGHT AND ROOT ROT (*RHIZOCTONIA SOLANI*)					
azoxystrobin (Quadris 2.08F)t	11	0.125 to 0.25 oz/ 1000 row ft soil application or 6.0 to 15.5 fl oz/acre foliar application	0	4 hr	Apply as banded spray to the lower stems and soil surface. Make no more than two sequential applications. Soil applications are included in this maximum.
fluopyram + trifloxystrobin (Luna Sensation 500SC)	7+11	7.6 fl oz/acre	7	0.5	Limit of 15.3 fl oz (2 applications) at low and high rate, respectively, per acre per year. **Do not use on cilantro.**
WHITE MOLD (*SCLEROTINIA SCLEROTIORUM*)					
boscalid (Endura 70WG)	7	4.5 to 9 oz/acre	14	0.5	Make no more than 2 sequential applications before alternating with fungicides that have a different mode of action. Limit of 18 oz/acre per year. **Do not use on cilantro. Do not use in greenhouses.**
cyprodinil + fludioxonil (Switch 62.5WG)	9+12	11 to 14 oz/acre	0	0.5	Make no more than two sequential applications before alternating with fungicides with different mode of action for two applications. Apply no more than 56 oz per crop per acre per season. First application at thinning & second application 2 weeks later.
fludioxonil (various)	12	7 oz/acre	0	0.5	Make no more than 4 applications per crop per year. Appling excessive water after application may decrease efficacy. **Do not use on cilantro.**
fludioxonil + pydiflumetofen (Miravis Prime 2.09SC)	12+7	13.4 fl oz/acre	0	0.5	Make no more than 2 sequential applications of Miravis Prime or other Group 7 and 12 fungicides before rotating to another effective fungicide with a different mode of action Limit of 2 applications per crop per year. **Do not use in greenhouses. Do not use on cilantro.**
penthiopyrad (Fontelis1.67F)	7	16 to 30 fl oz	3	0.5	Do not make more than two sequential applications. Maximum of 72 fl oz/acre per year.
Coniothyrium minitans (Contans WG)	BM02	1 to 4 lb/acre	0	4 hr	**OMRI listed product.** Apply to soil surface and incorporate no deeper than 2 inches. Works best when applied prior to planting or transplanting. Do not apply other fungicides for 3 weeks after applying Contans.

TABLE 3-22. IMPORTANCE OF ALTERNATIVE MANAGEMENT PRACTICES FOR DISEASE CONTROL IN PARSLEY

A. Keinath, Plant Pathologist, Clemson University

Scale: E, excellent; G, good; F, fair; P, poor; NC, no control; ND, no data.

Strategy	Alternaria leaf spot	Cercospora leaf spot	Powdery mildew	Pythium damping off and root rot	Rhizoctonia damping off and root rot	Root knot nematode	Sclerotinia white mold	Septoria blight
Avoid field operations when leaves are wet	G	G	NC	NC	NC	NC	P	G
Avoid overhead irrigation	G	G	NC	NC	NC	NC	G	G
Biofungicide	ND	ND	F	ND	ND	ND	F	ND
Change planting date	NC	NC	NC	NC	E (early)	E (early)	G (late)	NC
Suppressive cover crops	NC	NC	NC	NC	NC	F	NC	NC
Crop rotation with non-host	E	E	NC	P	P	P	F	E
Deep plowing	G	G	P	NC	F	P	F	G
Destroy crop residue	G	G	P	NC	F	P	P	G
Encourage air movement	G	G	P	P	NC	NC	E	G
Flooding(where feasible)	NC	NC	NC	NC	F	G	G	NC
Increase soil organic matter	NC	NC	F	P	P	F	NC	NC
Hot water seed treatment	ND	ND	NC	NC	NC	NC	NC	E
Plant in well-drained soil	P	P	NC	E	G	NC	F	P
Plant on raised beds	NC	NC	NC	E	G	NC	F	NC
Plastic mulch bed covers	NC	NC	F	F	F	NC	P	NC
Postharvest temperature control	NC	NC	NC	NC	NC	NC	E	NC
Reduce mechanical injury	NC	NC	NC	NC	P	NC	G	NC
Soil solarization	F	F	NC	P	F	F	P	F
Pathogen-free seed	E	E	P	NC	NC	NC	P	E
Resistant/tolerant cultivars	NC	NC	NC	NC	P	NC	NC	F
Weed control	P	P	F	NC	NC	F	F	P

TABLE 3-23. DISEASE CONTROL PRODUCTS FOR PEA

E. Sikora, Plant Pathologist, Auburn University

Disease/Material	FRAC Code	Rate of Material Formulation	Minimum Days Harv.	Minimum Days Reentry	Method, Schedule, and Remarks
PEA (ENGLISH, GARDEN; *Pisum* **spp.)**					
ANTHRACNOSE					
azoxystrobin (various)	11	6.2 to 15.4 fl oz/acre	0	4 hr	Do not make more than two sequential applications.
penthiopyrad (Fontelis) 1.67 F	7	14 to 30 fl oz	0	0.5	Do not make more than two sequential applications. Maximum of 72 fl oz/ acre per crop.
pyraclostrobin + fluxapyroxad (Priaxor) 500 SC	11+7	4.0 to 8.0 fl oz/acre	7	0.5	Do not make more than two sequential applications. Maximum of 16 fl oz/ acre per crop. See label for specific directions for edible-podded legumes and dried-shelled legumes.
fluoxastrobin (Evito 480 SC)	11	2.0 to 4.75 fl oz/acre	7	0.5	Begin applications preventively and continue as needed on a 7- to 14-day interval.
ANTHRACNOSE, BLACK MOLD, EARLY BLIGHT, POWDERY MILDEW					
mefentrifluconazole (Provysol)	3	2.5 to 5 oz/acre	21	0.5	Apply at 7-to-14-day intervals.
azoxystrobin (various)	11	6.2 to 15.4 fl oz/acre	0	4 hr	Do not make more than two sequential applications.
ASCOCHYTA LEAF SPOT AND BLIGHT					
azoxystrobin (various)	11	6.2 to 15.4 fl oz/acre	0	4 hr	Do not make more than two sequential applications.
fluoxastrobin (Evito 480 SC)	11	2.0 to 4.75 fl oz/acre	7	0.5	Begin applications preventively and continue as needed on a 7- to 14-day interval. To be grown for pea and bean, dry seed only.
boscalid (various)	7	8 to 11 oz/acre	7	0.5	Maximum of 2 applications per crop.
penthiopyrad (Fontelis) 1.67 F	7	14 to 30 fl oz	0	0.5	Do not make more than two sequential applications. Maximum of 72 fl oz/ acre per crop.
pyraclostrobin + fluxapyroxad (Priaxor) 500 SC	11+7	4.0 to 8.0 fl oz/acre	7	0.5	Do not make more than two sequential applications. Maximum of 16 fl oz/ acre per crop. See label for specific directions for edible-podded legumes and dried-shelled legumes.
ASCOCHYTA, POWDERY MILDEW, RUST					
mefentrifluconazole + fluxapyroxad (Revylok)	3+7	4.5 to 6.5 fl oz/acre	21	0.5	Begin applications before prior to disease development and continue on 7- to 14-day interval if conditions favor disease.
mefentrifluconazole + fluxapyroxad + pyraclostrobin (Revytek)	3+7+11	8 to 13 fl oz/acre	21	0.5	Begin applications before prior to disease development and continue on 7- to 14-day interval if conditions favor disease.
mefentrifluconazole + pyraclostrobin (Veltyma)	3+11	7 to 10 fl oz/acre	21	0.5	Apply on 7 to 14-day intervals.
GRAY MOLD (*BOTRYTIS*), WHITE MOLD (*SCLEROTINIA*)					
boscalid (various)	7	8 to 11 oz/acre	7	0.5	Maximum of 2 applications per crop.
fluazinam (various)	29	8 to 13.6 fl oz/acre	30	0.5	PHI varies by crop; see label restrictions.
isofetamid (Kenja 400SC)	7	17 fl oz/acre	See label	0.5	Apply at 10 to 30% bloom and 7- to 14- days later, if needed. Do not allow livestock to graze in treated areas.
penthiopyrad (Fontelis) 1.67 F	7	14 to 30 fl oz	0	0.5	Do not make more than two sequential applications. Maximum of 72 fl oz/ acre per year.
pyraclostrobin + fluxapyroxad (Priaxor) 500 SC	11+7	4.0 to 8.0 fl oz/acre	7	12 hr	Do not make more than two sequential applications. Maximum of 16 fl oz/ acre per crop. See label for specific directions for edible-podded legumes and dried-shelled legumes.
WHITE MOLD (*SCLEROTINIA*)					
Coniothyrium minitans (Contans WG)	BM02	1 to 4 lb/acre	0	4 hr	**OMRI listed product.** Apply to soil surface and incorporate no deeper than 2 inches. Works best when applied prior to planting or transplanting. Do not apply other fungicides for 3 weeks after applying Contans.
POWDERY MILDEW					
boscalid (various)	7	8 to 11 oz/acre	7	0.5	Maximum of 2 applications per crop.
fixed copper (various)	M01	See label	0	1 to 2	See label

TABLE 3-23. DISEASE CONTROL PRODUCTS FOR PEA (cont'd)

E. Sikora, Plant Pathologist, Auburn University

Disease/Material	FRAC Code	Rate of Material Formulation	Minimum Days Harv.	Minimum Days Reentry	Method, Schedule, and Remarks
PEA (ENGLISH, GARDEN; *Pisum spp.***) (cont'd)**					
GRAY MOLD (*BOTRYTIS*), WHITE MOLD (*SCLEROTINIA*) (cont'd)					
penthiopyrad (Fontelis) 1.67 F	7	14 to 30 fl oz	0	0.5	Do not make more than two sequential applications. Maximum of 72 fl oz/ acre per year.
pyraclostrobin + fluxapyroxad (Priaxor) 500 SC	11+7	4.0 to 8.0 fl oz/acre	7	12 hr	Do not make more than two sequential applications. Maximum of 16 fl oz/ acre per crop. See label for specific directions for edible-podded legumes and dried-shelled legumes.
sulfur (various)	M02	See labels	0	1	Spray at first appearance, 10- to 14-day intervals. Do not use sulfur on wet plants or on hot days (more than 90°F).
picoxystrobin (Approach)	11	6 to 12 fl oz/acre	0	0.5	Begin applications prior to disease development and make a second application on a 5- to 14-day interval. Make a third application only after having applied a fungicide with a different mode of action. Use higher specified rate and shorter interval with high disease pressure.
PYTHIUM DAMPING OFF					
mefenoxam (Ridomil Gold) 4 EC	4	0.5 to 1 pt/trt acre	—	2	Incorporate in soil. See label for row rates.
RHIZOCTONIA ROOT ROT					
pyraclostrobin + fluxapyroxad (Priaxor) 500 SC	11+7	4.0 to 8.0 fl oz/acre	7	0.5	Purchase treated seed for control of *Rhizoctonia solani* only.
RHIZOCTONIA SEED DECAY AND SEEDLING BIGHT					
sedaxane (Vibrance)	7	See label	—	—	Seed treatment for seed Decay, Seedling Blight and Damping-off caused by Rhizoctonia solani
inpyrfluxam (Zeltera)	7	See label	—	0.5	Seed treatment for seed Decay, Seedling Blight and Damping-off caused by Rhizoctonia solani.
RUST (*UROMYCES*)					
azoxystrobin (various)	11	6.2 fl oz/acre	0	4 hr	Do not make more than two sequential applications.
penthiopyrad (Fontelis) 1.67 F	7	14 to 30 fl oz	0	0.5	Do not make more than two sequential applications. Maximum of 72 fl oz/ acre per year.
pyraclostrobin + fluxapyroxad (Priaxor) 500 SC	11+7	4.0 to 8.0 fl oz/acre	7	12 hr	Do not make more than two sequential applications. Maximum of 16 fl oz/ acre per crop. See label for specific directions for edible-podded legumes and dried shelled legumes.
PEA (SOUTHERN; *Vigna spp.***)**					
SEED DECAY (*PHOMOPSIS, FUSARIUM*); SEEDLING BLIGHT (*FUSARUM, RHIZOCTONIA*)					
thiophanate-methyl (various)	1	0.14 to 0.28 fl oz/ 100 lbs seed			Seed treatment.
ANTHRACNOSE, RUST					
azoxystrobin (various)	11	2 to 5 oz/acre	14 (dry) 0 (succulent)	4 hr	Make no more than 2 sequential applications before alternating with a fungicide with a different mode of actions. Use no more than 1.5 lb a.i. per acre per season.
ANTHRACNOSE, GRAY MOLD, WHITE MOLD (*SCLEROTINIA*)					
thiophanate-methyl (Topsin M WSB)	1	1.5 to 2 lbs/acre	28 (dry) 14 (succulent)	0.5	See label for application instructions.
ASCOCHYTA BLIGHT, GRAY MOLD, WHITE MOLD					
boscalid (various)	7	8 to 11 oz/acre	21 (dry) 7 (succulent)	0.5	Maximum of 2 applications per season.
ASCOCHYTA BLIGHT, RUST, WHITE MOLD					
prothioconazole (various)	3	5.7 fl oz /acre	7	0.5	For dried shelled peas and beans only. Maximum of 3 applications per year. Use no more than17.1 fl oz per acre per year.
ASCOCHYTA BLIGHT, WHITE MOLD					
metconazole (Quash)	3	4 oz/acre	21 (dry)	0.5	For dried shelled pea and beans only. Do not make more than 2 applications per year, but applications may be sequential. Do not apply to cowpea and field pea used for livestock feed. For suppression of white mold only.

TABLE 3-23. DISEASE CONTROL PRODUCTS FOR PEA (cont'd)

E. Sikora, Plant Pathologist, Auburn University

Disease/Material	FRAC Code	Rate of Material Formulation	Minimum Days Harv.	Minimum Days Reentry	Method, Schedule, and Remarks
PEA (SOUTHERN; *Vigna spp.***) (cont'd)**					
DOWNY MILDEW, BACTERIAL BLIGHTS					
fixed copper (various)	M01	See label	See label	See label	See label
DOWNY MILDEW, *CERCOSPORA*, ANTHRACNOSE, RUST					
chlorothalonil (various)	M05	1.4 to 2 pt/acre	14	2	Spray early bloom; repeat at 7- to 10-day intervals; for dry beans only.
ALTERNARIA, ANTHRACNOSE, ASCOCHYTA, POWDERY MILDEW, RUST, CERCOSPORA					
difenoconazole + benzovindiflupyr (Aprovia Top)	3+7	10.5 to 11 fl oz	14	0.5	Begin prior to disease development and continue on 14-day schedule.
difenoconazole + cyprodinil (Inspire Super)	3+9	16 to 20 fl oz/acre	14	0.5	Begin applications prior to disease onset when conditions are conducive for disease. Apply on a 14-day schedule making no more than 2 sequential applications before alternating to another fungicide with a different mode of action.
pyraclostrobin (Headline)	11	6 to 9 fl oz/acre	21	0.5	Begin applications preventively and continue as needed on a 7- to 14-day interval when conditions are favorable for disease development.
ALTERNARIA, ASCOCHYTA, CERCOSPORA, POWDERY MILDEW, MYCOSPHAERELLA, RUST					
pydiflumetofen + difenoconazole (Miravis Top)	3+7	13.7 fl oz/acre	14	0.5	For dried shelled peas only. Begin applications prior to disease development. Continue applications on a 14-day interval. Do not make more than 4 applications per season. **DO NOT** feed or harvest cowpeas for forage and hay.
mefentriflufluconazole (Provysol)	3	2.5 to 5 fl oz/acre	21	0.5	Apply to 7- to 14- day intervals.
pydiflumetofen + propiconazole+ azoxystrobin (Miravis Neo)	3+7+11	13.7 fl oz/acre	14	0.5	Make the first application before disease is established and no later than the onset of flowering. Continue applications through season on a 14-day interval.
pydiflumetofen + fludioxonil (Miravis Prime)	7+12	9.2 fl oz/acre	14	0.5	Make the first application before disease is established and no later than the onset of flowering. Continue applications through season on a 14-day interval.
mefentrifluconazole + fluxapyroxad (Revylok)	3+7	4.5 to 6.5 fl oz/acre	21	0.5	Begin applications before prior to disease development and continue on 7- to 14-day interval if conditions favor disease.
mefentrifluconazole + fluxapyroxad + pyraclostrobin (Revytek)	3+7+11	8 to 13 fl oz/acre	21	0.5	Begin applications before prior to disease development and continue on 7- to 14-day interval if conditions favor disease.
mefentrifluconazole + pyraclostrobin (Veltyma)	3+11	7 to 10 fl oz/acre	21	0.5	Apply at 7 to 14-day intervals.
ALTERNARIA, ANTHRACNOSE, ASCOCHYTA, RUST, SOUTHERN BLIGHT, WEB BLIGHT					
azoxystrobin + propiconazole (various)	3+11	10.5 to 14 oz/acre	7 (dry) 14 (succulent)	0.5	Apply when conditions are conducive for disease. Up to three applications may be made on 7- to 14-day intervals
azoxystrobin + difenoconazole (Quadris Top)	3+11	12 to 14 fl oz/acre	3	0.5	For dried, shelled pea and beans only. Begin applications prior to disease on- set when conditions are conducive for disease. Apply on a 14-day schedule making no more than 2 sequential applications before alternating to another fungicide with a different mode of action.
azoxystrobin + difenoconazole (various)	3+11	14 fl oz/acre	14	0.5	For dried, shelled pea and beans only. Begin applications prior to disease on- set when conditions are conducive for disease. Apply on a 14-day schedule making no more than 2 sequential applications before alternating to another fungicide with a different mode of action.
ALTERNARIA, ANTHRACNOSE, ASCOCHYTA, DOWNY MILDEW, POWDERY MILDEW, RUST, CERCOSPORA, WHITE MOLD					
picoxystrobin (Approach)	11	6 to 12 fl oz	14	0.5	Do not apply more than 3 sequential applications. For white mold, use higher rates.
fluoxastrobin (Evito 480 SC)	11	2.0 to 4.75 fl oz/acre	14	0.5	Begin applications preventively; continue as needed on 7- to 14-day interval.
penthiopyrad (Fontelis) 1.67 F	7	14 to 30 fl oz	0	0.5	Do not make more than two sequential applications. Maximum of 72 fl oz/ acre per year.
DOWNY MILDEW, CERCOSPORA, ANTHRACNOSE, RUST, POWDERY MILDEW					
pyraclostrobin (various)	7	6 to 9 fl oz	21	0.5	Make no more than 2 sequential applications before alternating with a fungicide with a different mode of action. Use no more than 18 fl oz/acre per season.
POWDERY MILDEW, RUST					
sulfur (various)	M02	See label	0	1	Spray at first appearance; 7- to 10-day interval.

TABLE 3-23. DISEASE CONTROL PRODUCTS FOR PEA (cont'd)

E. Sikora, Plant Pathologist, Auburn University

Disease/Material	FRAC Code	Rate of Material Formulation	Minimum Days Harv.	Minimum Days Reentry	Method, Schedule, and Remarks
PEA (SOUTHERN; *Vigna spp.***) (cont'd)**					
POWDERY MILDEW					
picoxystrobin (Approach)	11	6 to 12 fl oz/acre	0	0.5	Begin applications prior to disease development; make second application on a 5- to 14-day interval. Make a third application only after having applied fungicide with different mode of action. Use higher specified rate and shorter interval with high disease pressure.
PYTHIUM DAMPING OFF					
mefenoxam (various)	4	0.5 to 1 pt/treated acre	—	0.5	Broadcast or banded over the row as a soil spray at planting or preplant incorporation into the top 2 inches of soil.
metalaxyl (various)	4	2 to 4 pt/treated acre	—	2	Broadcast or banded over the row as a soil spray at planting or preplant incorporation into the top 2 inches of soil.
RHIZOCTONIA ROOT ROT					
azoxystrobin (various)	11	0.4 to 0.8 fl oz/ 1,000 row feet	—	4 hr	Make in-furrow or banded application shortly after plant emergence.
penflufen (Evergol Prime)	7	0.05 to 0.1 fl oz of product/ 100,000 seeds	—	0.5	Apply using commercial slurry or mist-type seed treatment equipment.
inpyrfluxam (Zeltera)	7	See label	—	0.5	Seed treatment. See label for application instructions.
RHIZOCTONIA AND FUSARIUM SEED AND SEEDLING DECAY					
fluxapyroxad (various)	7	0.24 to 0.47 fl oz/ 100 lbs seed	—	—	Seed treatment
sedaxane (Vibrance)	7	0.08 0 0.16 fl oz/ 100 lbs seed	—	—	Seed treatment for seed Decay, Seedling Blight and Damping-off caused by *Rhizoctonia solani*
DOWNY MILDEW					
mefenoxam + copper hydroxide (Ridomil Gold Copper)	4+M01	5 lb/2.5 acres	3	2	For black-eyed, southern and cowpea. Begin foliar applications at disease onset; continue on 7-day interval. Do not exceed 4 applications/season.
RHIZOCTONIA, AND FUSARIUM SEED ROT, DAMPING-OFF, BOTRYTIS SEEDLING BLIGHT, PHOMOPSIS SEED DECAY					
penflufen + trifloxystrobin (various)	11	Apply 0.25 to 0.5 fl oz/100 lbs seed	—	—	Apply using commercial slurry or mist-type seed treatment equipment.
WHITE MOLD (SCLEROTINIA)					
Coniothyrium minitans (Contans WG)	BM02	1 to 4 lb/acre	0	4 hr	**OMRI listed product.** Apply to soil surface and incorporate no deeper than 2 inches. Works best when applied prior to planting or transplanting. Do not apply other fungicides for 3 weeks after applying Contans.
COTTONY LEAK (*PYTHIUM* **SPP.), POD ROT (***PHYTOPHTHORA CAPSICI***)**					
fenamidone (Reason 500 C)	11	5.5 to 8.2 fl oz/acre	3	0.5	Begin applications as soon as crop and/or environmental conditions become favorable for disease development. Will also suppress pod rot caused by *Phytophthora capsici*. **DO NOT** use on COWPEA.
COTTONY LEAK, DOWNY MILDEW, *PHYTOPHTHORA CAPSICI*					
cyazofamid (Ranman)	21	2.75 fl oz/acre	0	0.5	Application instructions vary by disease; please follow label directions. **DO NOT** apply to cowpeas used for livestock feed.
SCLEROTINIA WHITE MOLD AND BOTRYTIS GRAY MOLD					
fluazinam (Omega 500F)	29	0.5 to 0.85 pt/acre	30	0.5	**DO NOT** use more than 1.75 pints of per acre. PHI varies by crop; see label restrictions.

TABLE 3-24. DISEASE CONTROL PRODUCTS FOR PEPPER

L. Quesada-Ocampo, Plant Pathologist, North Carolina State University

Scale: E, excellent; G, good; F, fair; P, poor; NC, no control; ND, no data

Disease/Material	FRAC Code	Rate of Material Formulation	Minimum Days Harv.	Minimum Days Reentry	Method, Schedule, and Remarks
APHID-TRANSMITTED VIRUSES: PVY, TEV, WMV, CMV					
JMS Stylet-Oil	NC	3 qt/100-gal water	0	Dry	Use in 50 to 200 gal per acre depending on plant size. Spray weekly when winged aphids first appear.
ANTHRACNOSE FRUIT ROT					
azoxystrobin (various)	11	See label	0	4 hr	Apply at flowering to manage green fruit rot. Limit of 61.5 fl oz per acre per season. Make no more than **one** application before alternating with fungicides that have a different mode of action.
azoxystrobin + difenoconazole (Quadris Top)	11+3	8 to 14 fl oz/acre	0	0.5	Begin application prior to disease development and continue thought the year on a 7- to 10-day interval. Limit of 55.3 fl oz per acre per season. Make no more than two consecutive applications before rotating to another effective fungicide with a different mode of action.
azoxystrobin + flutriafol (Topguard EQ)	11+3	See label	0	12 hr	Apply preventatively or when conditions are favorable for disease development. Repeat on a 7- to 14-day interval as necessary if conditions are favorable for disease development.
Bacillus amyloliquefaciens (Serifel)	44	4 to 16 oz/acre	0	4 hr	Begin applications shortly after emergence or transplanting and continue on 2- to 7-day intervals if conditions conducive to disease development.
chlorothalonil (various)	M05	See labels	7	1	See labels. Rates vary depending on the formulation.
cyprodinil + difenoconazole (Inspire Super)	9+3	16 to 20 fl oz/acre	0	0.5	Limit of 80 fl oz per acre per season.
difenoconazole + benzovindiflupyr (Aprovia Top)	11+3	10.5 to 13.5 fl oz/acre	0	0.5	Limit of 53.6 fl oz per acre per year. Not labeled for greenhouse use. No more than 2 applications of Aprovia Top may be applied on 7-day interval.
difenoconazole + tea tree oil (Regev)	3 + BM01	4 to 8.5 fl oz/acre	2	0.5	Limit to 34 fl oz/acre per year. Begin application in early plant stages and repeat on a 7- to 14-day intervals. Do not make more than 2 sequential applications before alternating to a fungicide with a different mode of action
famoxadone + cymoxanil (Tanos)	11+27	8 to 10 oz/acre	3	0.5	Make no more than **one** application before alternating with a fungicide with a different mode of action. 72 oz/acre maximum per crop cycle. Minimum of 20 gal of water of spray volume/acre. **NOTE: Must tank mix with another fungicide with a different mode of action (i.e., copper).**
fluopyram + difenoconazole (Luna Flex)	7+3	8 to 13.6 fl oz/acre	0	0.5	Limit to 27.2 fl oz/acre per year. Begin applications prior to disease development. Continue applications through season on a 7- to 14-day interval.
fluopyram + trifloxystrobin (Luna Sensation)	7+11	7.6 fl oz/acre	3	0.5	**Suppression only.** Limit to 27.1 fl oz/acre per year. Begin applications prior to disease development. Continue applications through season on a 7- to 14-day interval.
flutriafol (Rhyme)	7	7 fl oz/acre	0	0.5	Limit to 28 fl oz/acre per year. Apply when conditions are favorable for disease. Repeat as necessary on a 7-day interval.
mancozeb (various)	M03	See labels	7	1	See labels. Rates vary depending on the formulation.
mancozeb + azoxystrobin (Dexter Max)	M03+11	1.7 to 3.4 lb/acre	7	1	For states East of the Mississippi and including Mississippi, do not exceed 20.5 lbs of product/acre/season. States West of Mississippi use 1.7 to 2.25 lbs of product/acre/season; do not exceed 13.7 lbs of product/acre/season. Do not make more than one application before alternating with a fungicide not in Group 11.
mefentrifluconazole (Cevya)	3	3 to 5 fl oz/acre	0		Limit to 15 fl oz/acre per year. Begin application before onset of disease and repeat on a 7-day interval.
oxathiapiprolin + chlorothalonil (Orondis Opti premix)	49+M05	1.75 to 2.5 pt/acre	3	0.5	Limit to 10 pt/acre per year. Begin applications when disease is expected. Minimum application interval of 7- to 10-days. Make no more than 2 sequential applications before alternating with a fungicide with a different mode of action.
penthiopyrad (Fontelis)	7	24 fl oz/acre	0	0.5	**SUPPRESSION ONLY.** Limit of 72 fl oz per acre per year. Make no more than two consecutive applications before rotating to another effective fungicide with a different mode of action.
pyraclostrobin (Cabrio EG)	11	8 to 12 oz/acre	0	4 hr	Apply at flowering to manage green fruit rot. Limit of 96 oz per acre per season. Make no more than **one** sequential application before alternating with fungicide that have a different mode of action.
pyraclostrobin + fluxapyroxad (Priaxor Xemium)	11+7	4.0 to 8.0 fl oz/acre	0	0.5	Apply prior to disease development and continue on a 7- to 14-day interval. Make no more than three applications per year.
trifloxystrobin (Flint)	11	3 to 4 oz/acre	3	0.5	**SUPPRESSION ONLY.** Limit of 16 oz per acre per year. Make no more than **one** application before alternating with fungicides that have a different mode of action.
BACTERIAL SPOT (FIELD)					
acibenzolar-S-methyl (Actigard 50WG)	21	0.33 oz to 0.75 oz/acre	14	0.5	**FOR HOT PEPPERS ONLY: 0.75 oz /100gal.** Begin applications within one week of transplanting or emergence. Make up to six weekly, consecutive applications.

TABLE 3-24. DISEASE CONTROL PRODUCTS FOR PEPPER (cont'd)

L. Quesada-Ocampo, Plant Pathologist, North Carolina State University

Scale: E, excellent; G, good; F, fair; P, poor; NC, no control; ND, no data

Disease/Material	FRAC Code	Rate of Material Formulation	Minimum Days Harv.	Minimum Days Reentry	Method, Schedule, and Remarks
BACTERIAL SPOT (FIELD) (cont'd)					
Bacillus amyloliquefaciens (Serifel)	44	4 to 16 oz/acre	0	4 hr	Begin applications shortly after emergence or transplanting and continue on 2- to 7-day intervals if conditions conducive to disease development. For improved suppression of bacterial spot and speck, tank-mix or rotate with labeled copper-based bactericides.
difenoconazole + tea tree oil (Regev)	3 + BM01	4 to 8.5 fl oz/acre	2	0.5	Limit to 34 fl oz/acre per year. Begin application in early plant stages and repeat on a 7- to 14-day intervals. Do not make more than 2 sequential applications before alternating to a fungicide with a different mode of action.
famoxadone + cymoxanil (Tanos)	11+27	8 to 10 oz/acre	3	0.5	**SUPPRESSION ONLY.** 72 oz/acre maximum per crop cycle. Make no more than **one** application before alternating with a fungicide with a different mode of action. Minimum of 20 gal of water of spray volume/acre. **NOTE: Must tank mix with another fungicide with a different mode of action (i.e., copper).**
fixed copper (various)	M01	See labels	0	2	See label. Rates vary depending on the formulation. Make first application 7- to 10-days after transplanting. Carefully examine field for disease to determine need for additional applications. If disease is present, make additional applications at 5-day intervals. Applying mancozeb with copper significantly enhances bacterial spot control. **Do not spray copper when temperatures are above 90°F.**
mancozeb (various)	M03	See labels	7	1	See label. Rates vary depending on the formulation.
methyl salicylate + *Bacillus thuringiensis* subsp. *kurstaki* (Leap ES)	BM02	0.5 to 2.0 qt/acre	See label	0.5	Apply preventatively on a 5- to 10-day schedule. For best disease control, Leap ES should be used in tank-mix or rotation with other registered pathogen control products, especially if disease already observed in crop.
quinoxyfen (Quintec)	13	6.0 fl oz/acre	3	0.5	Use 6 oz of product per acre in no less than 30 gallons of water per acre. NOTE: May only be used to manage bacterial spot in Florida, Georgia, North Carolina, and South Carolina (Section 2 (ee)).
BACTERIAL SPOT (TRANSPLANTS)					
fixed copper (various)	M01	See labels	0	2	See labels. Rates vary depending on the formulation. Begin applications when conditions first favor disease development, repeat at 3- to 10-day intervals if needed depending on disease severity. Use higher rates when conditions favor disease. Do not spray copper when temperatures are > 90°F.
streptomycin sulfate (Agri-Mycin 17, Firewall, Streptrol)	25	1 lb/100 gal	—	1	**MAY ONLY BE APPLIED TO TRANSPLANTS.** Spray when seedlings are in the 2-leaf stage, continue on 5-day intervals until transplanted into field. **NOTE: Some strains resistant to streptomycin sulfate.**
BACTERIAL SPOT (SEED)					
sodium hypochlorite (Clorox 5.25%, regular formulation)	—	1 pt + 4 pt water	—	—	Add 1 TSP of surfactant (Tween-20 or 80, Silwet) to improve seed coverage.
CERCOSPORA LEAF SPOT					
azoxystrobin + difenoconazole (Quadris Top)	11+3	8 to 14 fl oz/acre	0	0.5	Limit of 55.3 fl oz per acre per season. Make no more than two consecutive applications before rotating to another effective fungicide with different mode of action. The addition of non-ionic based surfactant or oil concentrate is recommended.
azoxystrobin + flutriafol (Topguard EQ)	11+3	See label	0	0.5	Apply preventatively or when conditions are favorable for disease development. Repeat on a 7- to 14-day interval as necessary if conditions favorable for disease development.
benzovindiflupyr + difenoconazole (Aprovia Top)	7+3	10.5 to 13.5 fl oz/acre	0	0.5	Begin applications prior to disease development and continue throughout the season on a 7- to 14-day interval. For resistance management, do not apply more than two consecutive applications before switching to a non-Group 7 fungicide.
fenbuconazole (Enable)	3	6 to 12 fl oz/acre	7	0.5	Limit to 48 fl oz/acre per year. Begin application when disease is first observed and repeat on 10- to 14-day intervals. Minimum spray volume of 30 gal/acre. A surfactant should be tank mixed for optimum performance. Do not make more the 2 sequential applications with rotating to a fungicide with a different mode of action.
fixed copper (various)	M01	See labels	0	2	See labels. Rates vary depending on formulation. Begin applications when conditions first favor disease development, repeat at 3- to 10-day intervals if needed depending on severity. Use higher rates when conditions favor disease. Do not spray when temperature > 90 °F.
fluopyram + difenoconazole (Luna Flex)	7+3	8 to 13.6 fl oz/acre	0	0.5	Limit to 27.2 fl oz/acre per year. Begin applications prior to disease development. Continue applications through season on a 7- to 10-day interval.
flutriafol (Rhyme)	3	7 fl oz/acre	0	0.5	Limit to 28 fl oz/acre per year. Apply when conditions are favorable for disease. Repeat as necessary on a 7-day interval.
mancozeb (various)	M03	See labels	7	1	See labels. Rates vary depending on the formulation.

TABLE 3-24. DISEASE CONTROL PRODUCTS FOR PEPPER (cont'd)

L. Quesada-Ocampo, Plant Pathologist, North Carolina State University

Scale: E, excellent; G, good; F, fair; P, poor; NC, no control; ND, no data

Disease/Material	FRAC Code	Rate of Material Formulation	Minimum Days Harv.	Minimum Days Reentry	Method, Schedule, and Remarks
CERCOSPORA LEAF SPOT (cont'd)					
mancozeb + azoxystrobin (Dexter Max)	M03+11	1.7 to 3.4 lb/acre	7	1	For states East of the Mississippi and including Mississippi, do not exceed 20.5 lbs of product/acre/season. States West of Mississippi use 1.7 to 2.25 lbs of product/acre/season and do not exceed 13.7 lbs of product/acre/season. Do not make more than one application before alternation with a fungicide not in Group 11.
oxathiapiprolin + chlorothalonil (Orondis Opti premix)	49+M05	1.75 to 2.5 pt/acre	3	0.5	Limit to 10 pt/acre per year. Begin applications when disease is expected. Minimum application interval of 7 days. Make no more than 2 sequential applications before alternating with a fungicide with a different mode of action.
pyraclostrobin (Cabrio EG)	11	8 to 12 fl oz/acre	0	0.5	Limit of 96 fl oz per acre per season. Do not make more than one application of product before alternating to a labeled fungicide with different mode of action.
PHYTOPHTHORA FOLIAR BLIGHT AND FRUIT ROT (PHYTOPHTHORA CAPSICI)					
ametoctradin + dimethomorph (Zampro)	45+40	14 fl oz/acre	4	0.5	Limit of 42 fl oz per acre per season. Make no more than two sequential applications before rotating to another effective fungicide with a different mode of action.
cyazofamid (Ranman 400SC)	21	2.75 fl oz/acre	0	0.5	Limit to 16.5 fl oz/acre per year. Apply in transplant water at the time of transplant. Additional foliar applications can be made on a 7- to 10-day interval. For foliar sprays, a surfactant should be tank-mixed.
dimethomorph (Forum)	40	6 fl oz/acre	0	0.5	**SUPPRESSION ONLY.** Limit of 30 fl oz per acre per season. Make no more than two sequential before alternating with fungicides that have a different mode of action. **NOTE: Must tank-mix with another fungicide with a different mode of action.**
ethaboxam (Elumin)	22	8 fl oz/acre	2	0.5	**ONLY EFFECTIVE ON PHYTOPHTHORA.** Make soil spray or foliar fungicide applications beginning when conditions are favorable for disease development and prior to disease onset; continuing throughout the season. For best results, begin application at planting/transplanting. Inject (via drip irrigation) for soilborne diseases: Inject Elumin into the irrigation water at the listed application rate (see label).
famoxadone + cymoxanil (Tanos)	11+27	8 to 10 oz/acre	3	0.5	**SUPPRESSION ONLY.** 72 oz/acre maximum per crop cycle. Make no more than **one** application before alternating with a fungicide with a different mode of action. Minimum of 20 gal of water of spray volume/acre. **NOTE: Must tank mix with another fungicide with a different mode of action (i.e., copper).**
fenamidone (Reason 500SC)	11	8.2 fl oz/acre	14	0.5	**SUPRESSION ONLY.** Limit of 24.6 fl oz per growing season. Make no more than one application before rotating to another effective fungicide with a different mode of action.
fluazinam (Omega 500F, Lektivar 40SC)	29	1 to 1.5 pt/acre	30	0.5	**SUPPRESSION ONLY.** Apply as a soil drench at 1.5 pt per acre. For foliar applications, use 1 pt per acre. Limit of 9 pt per acre per season.
fluopicolide (Presidio)	43	3 to 4 fl oz/acre	2	0.5	Limit of 12 fl oz per acre per season. Make no more than two times sequentially before alternating with fungicides that have a different mode of action. **NOTE: Must be tank mixed with another mode of action product. Recently, insensitivity to this fungicide has been observed in southeastern US.**
mandipropamid (Revus, Micora)	40	8 fl oz/acre	1	0.5	**SUPPRESSION ONLY.** Limit of 32 fl oz per acre per season. **NOTE: Must tank-mix with another fungicide with a different mode of action (i.e., copper).**
oxathiapiprolin + mandipropamid (Orondis Ultra; premix)	49+40	5.5 to 8.0 fl oz/acre	See label	4hr	Use higher rate if disease is present. For best results, begin disease resistance program with initial treatment at planting or transplanting with fungicide registered for its use. Apply Orondis Ultra as foliar spray in mixture with copper-based fungicide beginning 1st appearance of symptoms.
PHYTOPHTHORA OR PYTHIUM ROOT ROT (FIELD)					
mefenoxam (Ridomil Gold, Ultra Flourish)	4	1 pt/acre	7	2	**MAY ONLY BE APPLIED AT PLANTING.** Apply in a 12 to 16 in. band or in 20- to 50-gal water per acre in transplant water. Mechanical incorporation or 0.5 to 1 in. irrigation water needed for movement into root zone if rain is not expected. After initial application, 2 supplemental applications (1 pt per treated acre) can be applied. **NOTE: Strains of *Phytophthora capsici* insensitive to Ridomil Gold have been detected in some Louisiana and North Carolina pepper fields.**
metalaxyl (MetaStar 2E)	4	4 to 8 pt/treated acre	7	2	Limit of 12 pt per acre per season. Preplant (soil incorporated), at planting (in water or liquid fertilizer), or as a basal-directed spray after planting. See label for the guidelines for supplemental applications.
oxathiapiprolin + mefenoxam (Orondis Gold)	49+4	28 to 55 fl oz/acre	7	2	See labels
POWDERY MILDEW					
azoxystrobin + difenoconazole (Quadris Top)	11+3	8 to 14 fl oz/acre	0	0.5	Begin application prior to disease development and continue thought the year on a 7– to 10-day interval. Limit of 55.3 fl oz per acre per season. Make no more than two consecutive applications before rotating to another effective fungicide with a different mode of action.

TABLE 3-24. DISEASE CONTROL PRODUCTS FOR PEPPER (cont'd)

L. Quesada-Ocampo, Plant Pathologist, North Carolina State University

Scale: E, excellent; G, good; F, fair; P, poor; NC, no control; ND, no data

Disease/Material	FRAC Code	Rate of Material Formulation	Minimum Days Harv.	Minimum Days Reentry	Method, Schedule, and Remarks
POWDERY MILDEW (cont'd)					
Bacillus amyloliquefaciens (Serifel)	44	4 to 16 oz/acre	0	4 hr	Begin applications shortly after emergence or transplanting and continue on 2- to 7-day intervals if conditions conducive to disease development.
benzovindiflupyr + difenoconazole (Aprovia Top)	7+3	10.5 to 13.5 fl oz/acre	0	0.5	Limit of 53.6 fl oz per acre per year. Make more than 2 applications before alternating to fungicide with a non-Group 7 mode of action.
chlorothalonil + cymoxanil (Ariston)	M05+27	2 to 2.44 pt/acre	3	0.5	Limit of 17.5 pt per acre per year. Repeat application at 7 days intervals.
cyprodinil + difenoconazole (Inspire Super)	9+3	16 to 20 fl oz/acre	0	0.5	Limit of 80 fl oz per acre per season.
cyprodinil + fludioxonil (Switch 62.5 WG)	9+12	11 to 14 oz/acre	0	0.5	Limit to 56 oz/acre per year. Begin application at or before onset of disease and repeat on a 7- to 10-day interval. Do not make more than 2 sequential applications with alternating to fungicide with different MOA.
cyflufenamid (Torino)	U06	3.4 oz/acre	0	4 hr	Limit to 10.2 oz/acre per year. Begin application at onset of disease. Minimum application interval of 14 days.
difenoconazole + tea tree oil (Regev)	3 + BM01	4 to 8.5 fl oz/acre	2	0.5	Limit to 34 fl oz/acre per year. Begin application early plant stages and repeat on 7- to 14-day intervals. Do not make more than 2 sequential applications before alternating to fungicide with different MOA.
fenbuconazole (Enable)	3	6 to 12 fl oz/acre	7	0.5	Limit to 48 fl oz/acre per year. Begin application when disease is first observed and repeat on 10- to 14-day intervals. Minimum spray volume of 30 gal/acre. A surfactant should be tank mixed for optimum performance. Do not make more the 2 sequential applications with rotating to a fungicide with a different mode of action.
fluopyram (Velum Prime)	7	5 to 6.84 fl oz/acre	0	0.5	Limit 13.7 fl oz/acre per year. Soil applications should be at planting/transplanting. A second application may be made 7 days later.
fluopyram + difenoconazole (Luna Flex)	7+3	8 to 13.6 fl oz/acre	0	0.5	Limit to 27.2 fl oz/acre per year. Begin applications prior to disease development. Continue applications through season on a 7- to 10-day interval.
fluopyram + trifloxystrobin (Luna Sensation)	7+11	5 to 7.6 fl oz/acre	3	0.5	Limit to 27.1 fl oz/acre per year. Begin applications prior to disease development. Continue applications through season on 7- to 14-day interval.
flutriafol (Rhyme)	3	7 fl oz/acre	0	0.5	Limit to 28 fl oz/acre per year. Apply when conditions are favorable for disease. Repeat as necessary on a 7-day interval.
mancozeb + azoxystrobin (Dexter Max)	M03+11	1.7 to 3.4 lb/acre	7	1	**Suppression only.** For states East of the Mississippi and including Mississippi, do not exceed 20.5 lbs of product/acre/season. States West of Mississippi use 1.7 to 2.25 lbs of product/acre/season and do not exceed 13.7 lbs of product/acre/season. Do not make more than one application before alternation with a fungicide not in Group 11.
mefentrifluconazole (Cevya)	3	3 to 5 fl oz/acre	0	0.5	Limit to 15 fl oz/acre per year. Begin application before onset of disease and repeat on a 7-day interval.
metrafenone (Vivando)	50	15.4 fl oz/acre	0	0.5	Limit to 46.2 fl oz/acre per year. Begin application prior to disease development. Make no more than 2 sequential applications before alternating with a fungicide with a different mode of action.
oxathiapiprolin + chlorothalonil (Orondis Opti premix)	49+M05	1.75 to 2.5 pt/acre	3	0.5	Limit to 10 pt/acre per year. Begin applications when disease is expected. Minimum application interval of 7 days. Make no more than 2 sequential applications before alternating with fungicide with different MOA.
penthiopyrad (Fontelis)	7	16 to 24 fl oz/acre	0	0.5	Limit of 72 fl oz per acre per year. Make no more than two consecutive applications before rotating to another fungicide with different MOA.
pyraclostrobin + fluxapyroxad (Priaxor Xemium)	11+7	6.0 to 8.0 fl oz/acre	0	0.5	Apply prior disease development and continue on a 7- to 14-day interval. Make no more than three applications per year.
pydiflumetofen + fludioxonil (Miravis Prime)	7+12	9.2 to 11.4 fl oz/acre	0	0.5	Limit to 22.8 fl oz/acre per year. Begin applications prior to disease development. Continue applications through season on a 7- to 10-day interval.
pyriofenone (Prolivo 300SC)	50	4 to 5 fl oz	0	4 hr	Do not exceed 16 fl oz per acre per year. Do not make more than 2 sequential applications of Prolivo or of another FRAC 50-containing fungicide before alternating to a fungicide with a different mode of action. Do not exceed 4 applications/year.
quinoxyfen (Quintec)	13	4.0 to 6.0 fl oz/acre	3	0.5	Limit of 24 fl oz per acre per year. Make no more than two consecutive applications before alternating with fungicides that have a different mode of action. **NOTE: Under certain environmental conditions leaf spotting or chlorosis may occur after application; discontinue use if symptoms occur.**
sulfur (various)	M02	See label	See label	See label	See labels. Rates vary depending on the formulation. Apply at first appearance and repeat at 14-day intervals as needed.
trifloxystrobin (Flint)	11	1.5 to 2 oz/acre	3	0.5	Limit of 16 oz per acre per year. Make no more than **one** application before alternating with fungicides that have a different mode of action.

TABLE 3-24. DISEASE CONTROL PRODUCTS FOR PEPPER (cont'd)

L. Quesada-Ocampo, Plant Pathologist, North Carolina State University

Scale: E, excellent; G, good; F, fair; P, poor; NC, no control; ND, no data

Disease/Material	FRAC Code	Rate of Material Formulation	Minimum Days Harv.	Minimum Days Reentry	Method, Schedule, and Remarks
SOUTHERN BLIGHT (*ATHELIA ROLFSII = SCLEROTIUM ROLFSII*)					
Bacillus amyloliquefaciens (Serifel)	44	4 to 16 oz/acre	0	4 hr	See label for Soil Application Instructions for In-Furrow, Drench, Shanked-In and Injected Applications.
fluopyram + trifloxystrobin (Luna Sensation)	7+11	7.6 fl oz/acre	3	0.5	**Suppression only.** Limit to 27.1 fl oz/acre per year. Begin applications prior to disease development. Continue applications through season on a 7- to 14-day interval.
fluoxastrobin (Aftershock, Evito 280SC)	11	2 to 5.7 fl oz/acre	3	0.5	Limit of 22.8 fl oz per acre per season. Make no more than **one** application before alternating with fungicides that have a different mode of action. **NOTE: Do not overhead irrigate for 24 hours following a spray application.**
benzovindiflupyr + difenoconazole (Aprovia Top)	7+3	10.5 to 13.5 fl oz/acre	0	0.5	**Suppression only.** Begin applications prior to disease development and continue throughout the season on a 7- to 14-day interval. For resistance management, do not apply more than two consecutive applications before switching to a non-Group 7 fungicide.
penthiopyrad (Fontelis)	7	16 to 24 fl oz/acre	0	0.5	Limit of 19.2 fl oz per acre per season. Make no more than two sequential applications of Fontelis before switching to a fungicide with different mode of action. **For non-bell peppers only.**
pyraclostrobin (Cabrio EG)	11	12 to 16 oz/acre	0	4 hr	**SUPPRESSION ONLY.** Apply at flowering to manage green fruit rot. Limit of 96 oz per acre per season. Make no more than one sequential application before alternating with fungicides that have a different mode of action.
pyraclostrobin + fluxapyroxad (Priaxor Xemium)	11+7	4.0 to 8.0 fl oz/acre	0	0.5	**SUPPRESSION ONLY.** Limit of 24 fl oz per acre per season. Make no more than two consecutive applications before rotating to another effective fungicide with a different mode of action.
TARGET SPOT (*CORYNESPORA CASSIICOLA*)					
boscalid (Endura)	7	3.5 oz/acre	0	0.5	Limit of 21 oz per acre per season. Make no more than two sequential applications before alternating with fungicides with different MOA.
cyprodinil + difenoconazole (Inspire Super)	9+3	16 to 20 fl oz/acre	0	0.5	Limit of 80 fl oz per acre per season.
difenoconazole + tea tree oil (Regev)	3 + BM01	4 to 8.5 fl oz/acre	2	0.5	Limit to 34 fl oz/acre per year. Begin application in early plant stages and repeat on a 7- to 14-day intervals. Do not make more than 2 sequential applications before alternating to a fungicide with a different mode of action
fluopyram + difenoconazole (Luna Flex)	7+3	10 to 13.6 fl oz/acre	0	0.5	Limit to 27.2 fl oz/acre per year. Begin applications prior to disease development. Continue applications through season on a 7- to 10-day interval.
fluopyram + trifloxystrobin (Luna Sensation)	7+11	7.6 fl oz/acre	3	0.5	Limit to 27.1 fl oz/acre per year. Begin applications prior to disease development. Continue applications through season on a 7- to 14-day interval.
fluoxastrobin (Aftershock, Evito 480SC)	11	2 to 5.7 fl oz/acre	3	0.5	Limit of 22.8 fl oz per acre per season. Make no more than one application before alternating with fungicides that have a different mode of action. **NOTE: Do not overhead irrigate for 24 hours following a spray application.**
penthiopyrad (Fontelis)	7	16 to 24 fl oz/acre	0	0.5	**SUPPRESSION ONLY.** Limit of 72 fl oz per acre per year. Make no more than two consecutive applications before rotating to another effective fungicide with a different mode of action.
pyraclostrobin (Cabrio EG)	11	8 to 12 oz/acre	0	4 hr	Apply at flowering to manage green fruit rot. Limit of 96 oz per acre per season. Make no more than one sequential application before alternating with fungicides that have a different mode of action.
pyraclostrobin + fluxapyroxad (Priaxor Xemium)	11+7	4.0 to 8.0 fl oz/acre	0	0.5	Limit of 24 fl oz per acre per season. Make no more than two consecutive applications before rotating to another effective fungicide with a different mode of action.
pydiflumetofen + fludioxonil (Miravis Prime)	7+12	9.2 to 11.4 fl oz/acre	0	0.5	Limit to 22.8 fl oz per acre per year. Begin applications prior to disease development. Continue applications through season on a 7- to 10-day interval.
GRAY LEAF SPOT (*Stemphylium solani*)					
azoxystrobin + difenoconazole (Quadris Top)	11+3	8 to 14 fl oz/acre	0	0.5	Begin application prior to disease development and continue thought the year on a 7– 10-day interval. Limit of 55.3 fl oz per acre per season. Make no more than two consecutive applications before rotating to another effective fungicide with a different mode of action.
benzovindiflupyr + difenoconazole (Aprovia Top)	7+3	10.5 to 13.5 fl oz/acre	0	0.5	Begin applications prior to disease development and continue throughout the season on a 7- to 14-day interval. For resistance management, do not apply more than two consecutive applications before switching to a non-Group 7 fungicide.
fluopyram + difenoconazole (Luna Flex)	7+3	10 to 13.6 fl oz/acre	0	0.5	Limit to 27.2 fl oz/acre per year. Begin applications prior to disease development. Continue applications through season on a 7- to 14-day interval.

TABLE 3-25. EFFICACY OF PRODUCTS FOR DISEASE CONTROL IN PEPPER

L. Quesada-Ocampo, Plant Pathologist, North Carolina State University

Scale: E, excellent; G, good; F, fair; P, poor; NC, no control; ND, no data

Active Ingredient [1]	Product	Fungicide group [F]	Preharvest interval (Days)	Anthracnose (immature fruit rot)	Bacterial spot	Phytophthora blight (root and crown)	Phytophthora blight (fruit and foliage)	Pythium damping-off	Southern blight
azoxystrobin	Quadris	11	0	F	NC	NC	NC	NC	ND
chlorothalonil	Various	M05	3	P	NC	NC	P	NC	NC
difenoconazole + benzovindiflupyr	Aprovia Top	7+3	0	G-F	NC	NC	NC	NC	NC
cyazofamid	Ranman	21	0	NC	NC	P	P	NC	NC
oxathiapiprolin + mefenoxam	Orondis Gold 200	49+4	0	NC	NC	G	F	NC	NC
oxathiapiprolin + chlorothalonil	Orondis Opti	49+M05	0	NC	NC	G-F	G	NC	NC
dimethomorph	Forum	40	4	NC	NC	NC	P	NC	NC
dimethomorph + ametoctradin	Zampro	40+45	4	NC	NC	F	F	ND	NC
famoxadone + cymoxanil	Tanos	11+27	3	NC	NC	NC	P	NC	ND
fixed copper	Various	M01	See label	P	F	NC	P	NC	NC
fluopicolide	Presidio	43	2	NC	NC	F_R	F_R	NC	NC
fluoxastrobin	Evito	11	3	F	NC	NC	NC	NC	ND
fluxapyroxad + pyraclostrobin	Priaxor Xemium	11+7	7	F	NC	NC	NC	NC	P
mancozeb [2]	Dithane, Manzate	M03	5	P	P	P	P	NC	NC
mandipropamid	Revus	40	1	NC	NC	F	F	NC	NC
mefenoxam [R]	Ridomil Gold EC, Ultra Flourish	4	0	NC	NC	P	NA	G	NC
mefenoxam [R] + copper	Ridomil Gold + copper	4+M01	14	P	F-P	NA	F	NC	NC
methyl salicylate + Bacillus thuringiensis subsp. kurstaki	Leap ES	BM02	See label	NC	G-F	NC	NC	NC	NC
penthiopyrad	Fontelis	7	0	ND	NC	NC	NC	NC	G-F
propamocarb (greenhouse use)	Previcur Flex	28	5	NC	NC	NC	NC	F	NC
pyraclostrobin	Cabrio EG	11	0	G-F	NC	NC	NC	NC	ND
quinoxyfen	Quintec	13	3	NC	P	NC	NC	NC	NC
streptomycin sulfate [3]	Agri-Mycin, Streptrol, Firewall	25	Not for field use	NC	F	NC	NC	NC	NC

TABLE 3-26. IMPORTANCE OF ALTERNATIVE MANAGEMENT PRACTICES FOR DISEASE CONTROL IN PEPPER

L. Quesada-Ocampo, Plant Pathologist, North Carolina State University

Scale: E, excellent; G, good; F, fair; P, poor; NC, no control; ND, no data

Strategy	Anthracnose (immature fruit)	Aphid-transmitted viruses (PVX,CMV,TEV, AMV,PVY)	Bacterial soft rot of fruit	Bacterial spot	Blossom-end rot	Phytophthora blight (fruit and foliage)	Phytophthora blight (root and crown)	Pythium damping off	Root-knot nematode	Southern blight	Tomato spotted wilt virus
Avoid field operations when foliage is wet	F	NC	NC	G	NC	F	P	NC	NC	NC	NC
Avoid overhead irrigation	G	NC	F	G	NC	G	G	P	NC	NC	NC
Change planting date within a season	NC	F (early)	NC	F (early)	NC	NC	NC	P (late)	F (early)	P (early)	Variable
Cover cropping with antagonist	NC	NC	NC	NC	NC	NC	NC	NC	F	NC	NC
Rotation with non-host (2 to 3 years)	G	NC	NC	NC	NC	P	P	NC	F	P	NC
Deep plowing	F	NC	NC	NC	NC	NC	NC	NC	NC	P	F	NC
Prompt destruction of crop residue	F	F	NC	P	NC	P	P	NC	F	P	NC
Promote air movement	P	NC	NC	F	NC	P	P	NC	NC	NC	NC

TABLE 3-26. IMPORTANCE OF ALTERNATIVE MANAGEMENT PRACTICES FOR DISEASE CONTROL IN PEPPER (cont'd)

L. Quesada-Ocampo, Plant Pathologist, North Carolina State University

Scale: E, excellent; G, good; F, fair; P, poor; NC, no control; ND, no data

Strategy	Anthracnose (immature fruit)	Aphid-transmitted viruses (PVX,CMV,TEV, AMV,PVY)	Bacterial soft rot of fruit	Bacterial spot	Blossom-end rot	Phytophthora blight (fruit and foliage)	Phytophthora blight (root and crown)	Pythium damping off	Root-knot nematode	Southern blight	Tomato spotted wilt virus
Use of soil organic amendments	NC	NC	NC	NC	NC	P	P	P	F	P	NC
Application of insecticidal/horticultural oils	NC	F	NC	NC	NC	NC	NC	NC	NC	NC	NC
pH management (soil)	NC	NC	NC	NC	F	NC	NC	NC	F	NC	NC
Plant in well-drained soil/raised beds	NC	NC	NC	NC	NC	NC	G	G	NC	NC	NC
Eliminate standing water/saturated areas	NC	NC	NC	NC	NC	NC	G	G	NC	NC	NC
Postharvest temp control (fruit)	NC	NC	G	NC	NC	NC	NC	NC	NC	NC	NC
Use of reflective mulch	NC	F	NC	NC	NC	NC	NC	NC	NC	NC	G
Reduce mechanical injury	NC	NC	NC	NC	NC	NC	NC	NC	NC	NC	NC
Rogue diseased plants and/or fruit	NC	NC	NC	NC	NC	F	F	NC	NC	NC	NC
Soil solarization	NC	NC	NC	NC	NC	NC	P	NC	F	NC	NC
Use of pathogen-free planting stock	F	NC	NC	G	NC	NC	NC	NC	NC	NC	NC
Use of resistant cultivars	NC	NC	NC	G	F	F	F	NC	G	NC	G
Weed management	P	F	NC	NC	NC	P	P	NC	F	NC	P

TABLE 3-27. DISEASE CONTROL PRODUCTS FOR POTATO

D. Higgins, Plant Pathologist, Virginia Tech

Disease/Material	FRAC Code	Rate of Material Formulation	Minimum Days Harv.	Minimum Days Reentry	Method, Schedule, and Remarks
BLACK SCURF (*RHIZOCTONIA SOLANI*) AND SILVER SCURF (*HELMINTHOSPORIUM SOLANI*)					
azoxystrobin (various)	11	See label	See label	See label	See labels. Rates may vary depending on the product. Apply in furrow at planting according to label direction. Do not apply more than one application without alternating away from fungicides in Group 11.
azoxystrobin + benzovindiflupyr (Elatus)	11+7	0.34 to 0.5 oz/ 1000 linear row ft	See label	0.5	Limit 9.5 oz/acre per application.
fludioxonil (Maxim PSP)	12	0.5 lb/ 100 lb seed pieces	—	0.5	Ensure thorough coverage of each seed piece.
fludioxonil + mancozeb (Maxim MZ)	12+M03	0.5 lb/ 100 lb seed pieces	—	0.5	Ensure thorough coverage of each seed piece.
fludioxonil + thiamethoxam (Cruiser Maxx Potato)	12+ insecticide	0.19 to 0.27 fl oz/ 100 lb seed pieces	—	0.5	Rate depends on seeding rate – see label. See label for additional restrictions.
fludioxonil + difenoconazole + sedazane + thiamethoxam (Cruiser Maxx Vibrance Potato)	12+3+7+ insecticide	0.5 fl oz/100 lb seed pieces	—	0.5	See label for additional restrictions.
fluopyram (Luna Privilege)	7	5.47 fl oz/acre (ground) 2.82 oz/acre (aerial)	7	0.5	Use on a 5- to 7-day interval. Do not apply more than 10.95 oz/acre/ season for ground application and no more than 8.46 oz/acre/season for aerial application. Do not make more than 2 applications before alternating with a fungicide with a different mode of action. Labeled for **silver scurf only.**
fluoxastrobin (Aftershock, Evito 480 SC)	11	0.16 to 0.24 fl oz/ 1,000 ft of row	7	0.5	Apply in furrow at planting according to label directions. Do not apply more than 22.8 fl oz of product per acre per year including seed treatment use. Alternate with fungicide from different resistance management group.
flutolanil (Moncut 70DF, Moncut SC)	7	0.71 to 1.1 lb/acre 16.0 to 25.0 fl oz/acre	—	0.5	For **black scurf only**. Apply as an in-furrow spray by directing spray uniformly around and over the seed-piece inch a 4 to 8 in band prior to covering with soil.

TABLE 3-27. DISEASE CONTROL PRODUCTS FOR POTATO (cont'd)

D. Higgins, Plant Pathologist, Virginia Tech

Disease/Material	FRAC Code	Rate of Material Formulation	Minimum Days Harv.	Minimum Days Reentry	Method, Schedule, and Remarks
BLACK SCURF (*RHIZOCTONIA SOLANI*) AND SILVER SCURF (*HELMINTHOSPORIUM SOLANI*) (cont'd)					
flutolanil + mancozeb (MonCoat MZ)	7+M03	0.75 lb to 1.0 lb/ 100 lb seed piece	—	1	Apply to seed-pieces immediately after cutting. Ensure thorough coverage.
mancozeb (various)	M03	See label	—	1	For **black scurf only.**
penflufen + prothioconazole (Emesto Sliver)	7+3	0.31 fl oz per 100 lb of seed pieces	—	0.5	For disease control, good coverage of the seed piece is required.
penthiopyrad (Vertisan)	7	0.7 to 1.6 fl oz/ 1,000 ft of row	7	0.5	Maximum rate is 24 fl oz per acre per year. No more than 2 applications before switching to a different mode of action. Provides suppression of **black scurf only.**
polyoxin D (Oso 5 SC)	19	6.5 to 13.0 fl oz/acre	0	4 hr	Apply as banded spray in-furrow at planting, either just before placement of seed pieces or over seed pieces before covering with soil. See label for addi- tional instructions.
thiophanate-methyl (various)	1	0.5 to 0.7 fl oz/ 100 lb seed pieces	—	0.5	Ensure thorough coverage of each seed piece.
FUSARIUM SEEDPIECE DECAY (*FUSARIUM SPP.*), RHIZOCTONIA STEM CANKER (*RHIZOCTONIA SOLANI*), STREPTOMYCES COMMON SCAB (*STREPTOMYCES SCABIES*)					
fludioxonil (various)	12	See label	—	0.5	Label rates may vary depending on the product.
fludioxonil + mancozeb (Maxim MZ)	12+M03	0.5 lb/100 lb seed	—	1	Do not use treated seedpieces for feed or food. **NOT** labeled for Streptomyces common scab. See label for treatment instructions.
mancozeb (various)	M03	See label	—	1	Label rates may vary depending on the product.
penthiopyrad (Vertisan)	7	0.7 to 1.6 oz/ 1,000 ft of row	7	0.5	Maximum rate is 24 fl oz per application. Labeled for **Rhizoctonia stem canker only.**
thiabendazole (Mertect 340-F)	1	0.42 fl oz/ 2,000 lb of tuber	—	0.5	Fusarium Tuber Rot Only. Mist unwashed tubers on a conveyor line, with tumbling action, entering storage in sufficient water for complete coverage. Additional treat- ments may be needed before shipping. Do not treat seed potatoes after cutting.
EARLY BLIGHT (*ALTERNARIA SOLANI*), WHITE MOLD (*SCLEROTINIA SCLEROTIORUM*)					
azoxystrobin + difenoconazole (Quadris Top)	11+3	8.0 to 14.0 fl oz/acre	14	0.5	Apply at 7- to 14-day intervals. Apply no more than 2 sequential applications without alternating with a fungicide with a different mode of action. Limit of 55.3 lb product per acre per year. Limit of 0.46lb a.i./acre/year of difeno- conazole-con- taining products; limit of 2.0 lb a.i. per acre/year of azoxystrob- in-containing products. **Not registered for white mold.**
boscalid (Endura)	7	3.5 to 10 oz/acre	10	0.5	For control of Sclerotinia white mold, use 5.5 to 10 oz rate and begin applica- tions prior to row closure or at the onset of disease. Make a second application 14 days later if conditions favor disease development. Do not exceed 2 applica- tions per season. For Early blight control, use 3.5 to 4.5 oz rate. Do not exceed four applications per season. Limit of 20.5 oz of product per acre per season. Limit of 2 applications before alternating with a fungicide with a different mode of action.
boscalid + mefentrifluconazole (Endura PRO)	7+3	18.5 to 20 fl oz	10	0.5	Begin applications prior to the onset of disease. Do not make more than 3 appli- cations at 20 fl ozs per acre per year. Do not make more than two (2) sequential applications of Endura PRO before alternating to a labeled non-Group 3 or non- Group 7 fungicide.
cyprodinil + fludioxonil	9+12	11 to 14.0 oz/acre	14	0.5	**Not registered in all states – check manufacture's website. Not registered for white mold.**
fluazinam (Omega 500 F)	29	5.5 to 8.0 fl oz/acre	14	0.5	Begin applications when plants are 6 to 8 in. tall or when conditions favor dis- ease development. Repeat applications at 7- to 10-day intervals. For late blight, use the 5.5 fl oz rate. **DO NOT** apply more than 3.5 pt per acre during each growing season. **Not registered for early blight.**
fludioxonil + pydiflumetofen (Miravis Prime)	7+12	9.2 to 11.4 fl oz/acre	14		For white mold suppression, use highest labeled rate and apply at of before row closure followed by a second application 14 days later. **DO NOT** make more than two consecutive applications before alternation with another FRAC group.
fluopyram (Velum Prime)	7	6.0 to 6.84 fl oz/acre	7	0.5	Apply specified dosage as an in-furrow spray during planting directed on or below seed or using overhead chemigation equipment. Maximum 2 applica- tions per year.
fluopyram + penflufen (Velum Rise)	7+7	13.0 fl oz/acre	—	0.5	Apply as an in-furrow spray at planting. One application per year.
fluopyram + prothioconazole (Luna Pro)	7+3	10.0 fl oz/acre	14	0.5	Maximum 2 applications per year. **DO NOT** apply more than 0.446 lbs fluopy- ram or 0.267 lbs prothioconazole per acre per year from all uses, including seed treatment, soil and foliar applications.

TABLE 3-27. DISEASE CONTROL PRODUCTS FOR POTATO (cont'd)

D. Higgins, Plant Pathologist, Virginia Tech

Disease/Material	FRAC Code	Rate of Material Formulation	Minimum Days Harv.	Minimum Days Reentry	Method, Schedule, and Remarks
EARLY BLIGHT (*ALTERNARIA SOLANI*), WHITE MOLD (*SCLEROTINIA SCLEROTIORUM*) (cont'd)					
fluopyram + pyrimethanil (Luna Tranquility)	7+9	11.2 oz/acre	7	0.5	Apply at 7- to 14-day intervals. Do not make more than 2 sequential applications without switching to fungicide outside of Group 7 or Group 9.
iprodione (various)	2	See label	14	1	Rates may vary depending on the product.
mefentrifluconazole (Provysol)	3	3 to 5 fl oz/acre	7	0.5	**Not registered for white mold.** Max 3 applications for early blight.
metconazole (Quash)	3	2.5 to 4.0 oz/acre	1	0.5	Limit 16 oz/acre/season. Make no more than 2 applications before changing modes of action. Limit to 4 applications per year. Use the 4 oz rate for white mold.
metiram + pyraclostrobin (Cabrio Plus)	M03+11	2.0 to 2.9 lb/acre	14	1	Apply at 7- to 14-day intervals. Do not apply more than 17.4 lb/acre product per season. Do not apply more than 2 sequential applications before alternating with a fungicide with a different mode of action. Use at 2.9 lb/acre rate for **SUPPRESSION of white mold.**
penthiopyrad (Vertisan)	7	10.0 to 24.0 oz/acre	7	0.5	Apply at 7- to 14-day intervals. Make no more than 2 applications before alternating with a fungicide with a different mode of action. For **SUPPRESSION of white mold**, use at 14 to 24 oz/acre. Do not exceed 72 oz per acre per year. Do not apply more than 11.25 oz a.i. per acre per year in total from any combination of seed, soil, or foliar applications.
pyraclostrobin (Headline, Headline SC)	11	6.0 to 12.0 fl oz/acre	3	0.5	**DO NOT** exceed more than six foliar applications or 72 total oz of product per acre per season. For early blight, use 6 to 9 oz rate; for **SUPPRESSION of white mold**, use 6 to 12 oz rate, depending on weather conditions and disease pressure. Do not apply more than one time before alternating with a fungicide with a different mode of action.
pyrimethanil (Scala SC)	9	7.0 fl oz/acre	7	0.5	Apply at 7- to 14-day intervals. Do not apply more than 35 fl oz per acre per season. **Not registered for white mold.**
triphenyltin hydroxide	30	4.0 to 6.0 oz/acre	7	2	Full coverage of the foliage is necessary for best results. Do not exceed 18.0 fluid ounces per acre per season.
thiophanate-methyl (various)	1	See label	See label	0.5	Rates may vary depending on the product.
EARLY BLIGHT (*ALTERNARIA SOLANI*) AND LATE BLIGHT (*PHYTOPHTHORA INFESTANS*)					
azoxystrobin (various)	11	See label	14	4 hr	Rates may vary depending on the product. Do not apply more than one application without alternating away from fungicides in Group 11. See label for limits of active ingredients
azoxystrobin + chlorothalonil (Quadris Opti)	11+M05	1.6 pt/acre	14	0.5	Apply at 5- to 7-day intervals. Do not apply more than one application without alternating away from fungicides in Group 11. See label for limits of active ingredients.
chlorothalonil (various)	M05	See label	7	0.5	Rates may vary depending on the product.
chlorothalonil + cymoxanil (Ariston)	M05+27	2.0 pt/acre	14	0.5	Apply at 5- to 7-day intervals. Do not exceed 17.5 pt of product per acre per 12-month period.
chlorothalonil + zoxamide (Zing!)	M05+22	24.0 to 34.0 fl oz/acre	7	0.5	Apply at 5- to 7-day intervals. Do not make more than 2 sequential applications before alternating with a fungicide that has a different mode of action. Do not make more than 8 applications per acre per season. Use 30 to 34 fl oz rate for late blight.
fixed copper (various)	M01	See label	0	1	See label. Rates vary depending on the formulation.
cymoxanil + famoxadone (Tanos)	27+11	6.0 to 8.0 oz/acre	14	0.5	Use rate of 6 fl oz only for early blight. Do not apply more than 48 oz/acre per crop season and no more than 72 oz/acre per 12 months. Do not make more than one application before alternating with a fungicide with a different mode of action.
dimethomorph (Forum)	40	4.0 to 6.0 fl oz/acre	4	1	Must tank-mix if using less than 6 fl oz rate; if used alone, use 6 oz rate. **DO NOT** make more than 5 applications per season. Limit 30 fl oz/acre/season.
fenamidone (Reason 500SC)	11	5.5 to 8.2 oz/acre	14	0.5	Begin applications when conditions favor disease development and continue on 5- to 10-day interval. Do not apply more than 24.6 fl oz per growing season. Alternate with fungicide from different resistance management group.
fluoxastrobin (Aftershock, Evito 480 SC)	11	2.0 to 3.8 fl oz/acre	7	0.5	Begin applications when conditions favor disease development on 7- to 10-day intervals. Do not apply more than once before alternating with fungicides that have a different mode of action. Do not apply more than 22.8 fl oz per acre per season. For late blight, apply at full label rate.
fluxapyroxad + pyraclostrobin (Priaxor Xemium)	7	4.0 to 8.0 fl oz/acre	7	0.5	Apply at 7- to 14-day intervals. Do not apply more than 24 oz per acre per season including in furrow and foliar uses.
mancozeb + azoxystrobin (Dexter Max)	M03+11	1.6 to 2.1 lb/acre	14	1	Do not exceed 16 lbs product/acre/crop. Season limits apply for azoxystrobin.

TABLE 3-27. DISEASE CONTROL PRODUCTS FOR POTATO (cont'd)

D. Higgins, Plant Pathologist, Virginia Tech

Disease/Material	FRAC Code	Rate of Material Formulation	Minimum Days Harv.	Minimum Days Reentry	Method, Schedule, and Remarks
EARLY BLIGHT (*ALTERNARIA SOLANI*) AND LATE BLIGHT (*PHYTOPHTHORA INFESTANS*) (cont'd)					
mancozeb + chlorothalonil (Elixir)	M03+M05	1.8 to 2.4 lb/acre	14	1	Do not apply more than 18 lbs product/crop/year.
metiram (Polyram 80DF)	M03	1.5 to 2.0 lb/acre	14	1	Do not apply more than 14 lb product/crop/year.
mandipropamid + difenoconazole (Revus Top)	40+3	5.5 to 7.0 fl oz/acre	14	0.5	After 2 applications, switch to a different mode of action. Do not apply more than 28 fl oz/acre/season.
mefenoxam + chlorothalonil (Ridomil Gold Bravo SC)	4+M05	2.5 pints/acre	14	2	See label for limits on application limits per season and application interval.
mefenoxam + mancozeb (Ridomil Gold MZ WG)	4+M03	2.5 lb/acre	14	2	Apply at 14-day intervals for up to 3 applications.
mefentrifluconazole + pyraclos- trobin (Veltyma)	3+11	5 to 10 fl oz/acre	7	0.5	Late blight: Follow application with a labeled non-Group 11 late blight specific fungicide 5 to 7 days later.
polyoxin D (Oso 5 SC)	19	6.5 to 13.0 fl oz/acre	0	4 hr	Tank mix with other fungicides for resistance management.
propamocarb hydrochloride (Previcur Flex)	28	0.7 to 1.2 pints/acre	14	0.5	Tank-mix with a protectant fungicide such as mancozeb or chlorothalonil. Do not exceed 6 pints of product/acre/season.
pyraclostrobin (Headline, Headline SC)	11	6.0 to 12.0 fl oz/acre	3	1	**DO NOT** exceed more than six foliar applications or 72 total oz of product per acre per season. For early blight, use 6 to 9 oz rate. Do not apply more than once before alternating with fungicide with different MOA.
pyraclostrobin + chlorothalonil (Cabrio Plus)	11+M05	2.0 to 2.9 lb/acre	14	1	Do not apply more than 2 applications before switching to a different mode of action. Do not exceed 17.4 lbs/acre/season. For late blight, use 12 lb/acre rate.
pyrimethanil (Scala 5F)	9	7.0 fl oz/acre	7	0.5	**Only labeled for early blight**. Do not apply more than 35 fluid ounces crop.
trifloxystrobin (Gem 500SC)	11	2.9 to 3.8 fl oz/acre	7	0.5	Must tank-mix with a non-Group 11 fungicide for late blight. Use the 3.8 oz rate for late blight. Do not make more than 1 application without switching to a different mode of action. Do not exceed 6 applications or 23 fl oz product/acre/ season.
triphenyltin hydroxide (Super Tin 4L) (Super Tin 80WP, Agri Tin)	30	4.0 to 6.0 fl oz/acre 2.5 to 3.75 oz/acre	7	2	For Super Tin 4L, the 3.0 fl oz rate may be used if tank mixed. Add to 3 to 15 gallons of water depending on method of application. Season application limit apply–see label.
zoxamide + chlorothalonil (Zing!)	22+M05	30.0 to 34.0 fl oz/A 24 to 34 fl oz/A	7	0.5	Do not make more than 2 consecutive applications before switching to another effective fungicide with a different mode of action. Check label for disease specific rates.
zoxamide + mancozeb (Gavel 75DF)	22+M03	1.5 to 2.0 lb/acre	14	2	Do not make more than 6 applications or apply more than 12 lbs product/acre/season.
LATE BLIGHT (*PHYTOPHTHORA INFESTANS*)					
ametoctradin + dimethomorph (Zampro)	45+40	11.0 to 14.0 fl oz/acre	4		Do not make more than 2 applications without switching to a different mode of action. Do not exceed 42 fl oz/acre/season and 3 applications/season.
cyazofamid (Ranman 400SC)	21	1.4 to 2.75 fl oz/acre	7	0.5	Do not apply more than 10 sprays per crop. Make no more than 3 consecutive applications then follow with 3 applications of another MOA.
cymoxanil (Curzate 60DF)	27	3.2 oz/acre	14	0.5	**USE ONLY WITH A PROTECTANT FUNGICIDE** such as mancozeb or chlorothalonil. No more than 7 applications/crop/year.
dimethomorph (Forum)	40	4.0 to 6.0 fl oz/acre	4	0.5	If applying less than 6 fl oz rate, must tank-mix with non-Group 40 fungicide. Do not exceed 5 applications or 30 fl oz product/acre/season.
fluazinam (Omega 500F, Omega Top MP)	29	5.5 to 8.0 fl oz/acre	14	0.5	Begin applications when plants are 6 to 8 in. tall or when conditions favor disease development. Repeat applications at 7- to 10-day intervals. For late blight, use the 5.5 fl oz rate. **DO NOT** apply more than 3.5 pt per acre during each growing season.
mefenoxam + copper hydroxide (Ridomil Gold/Copper)	4+M01	2.0 lb/acre	14	2	**MUST** tank-mix with a protectant fungicide. Apply at 14-day intervals up to 3 applications; alternate and follow with full rate of a protectant.
Mono- and di-potassium salts of phosphorous acid (various)	P07	See label	0	4 h	Mix with a fungicide labeled for control of late blight. See label for in-furrow application or foliar application rates.
oxathiapiprolin + chlorothalonil (Orondis Opti)	49+M05	1.75 to 2.5 pints/acre	7	0.5	Do not make more than 2 sequential applications without switching to a differ- ent mode of action. Maximum 4 applications per season at the highest labeled rate. Do not mix soil applications and foliar applications. See label for addition resistance management requirements.

TABLE 3-27. DISEASE CONTROL PRODUCTS FOR POTATO (cont'd)

D. Higgins, Plant Pathologist, Virginia Tech

Disease/Material	FRAC Code	Rate of Material Formulation	Minimum Days Harv.	Minimum Days Reentry	Method, Schedule, and Remarks
EARLY BLIGHT (*ALTERNARIA SOLANI*) AND LATE BLIGHT (*PHYTOPHTHORA INFESTANS*) (cont'd)					
oxathiapiprolin + mandipropamid (Orondis Ultra)	49+40	5.5 to 8.0 fl oz/acre	5 / 14	4 hr	Do not make more than 2 sequential applications without switching to a different mode of action. Maximum 4 applications per season at the highest labeled rate. Do not mix soil applications and foliar applications. See label for addition resistance management requirements.
azoxystrobin + mefenoxam (Quadris Ridomil Gold SL)	11+4	0.82 fl oz/ 1,000 ft of row	—	0	Apply as an in-furrow spray in 3 to 15 gal of water per acre at planting.
cyazofamid (Ranman 400SC)	21	1.4 to 2.75 fl oz/ acre (foliar) 0.42 fl oz/1,000 ft (in-furrow)	7	0.5	For pink rot and Pythium leak, apply at the high rate. Do not apply more than 10 sprays per crop or more than 27.5 fl oz/acre/season. Make no more than 3 consecutive applications followed by 3 applications from a different resistance management group.
PINK ROT (*PHYTOPHTHORA ERYTHROSEPTICA*), PYTHIUM LEAK (*PYTHIUM SPP.*)					
azoxystrobin + mefenoxam (Quadris Ridomil Gold SL)	11+4	0.82 fl oz/ 1,000 ft of row	—	0	Apply as an in-furrow spray in 3 to 15 gal of water per acre at planting.
cyazofamid (Ranman 400SC)	21	1.4 to 2.75 fl oz/ acre (foliar) 0.42 fl oz/1,000 ft (in-furrow)	7	0.5	For pink rot and Pythium leak, apply at the high rate. Do not apply more than 10 sprays per crop or more than 27.5 fl oz/acre/season. Make no more than 3 consecutive applications followed by 3 applications from a different resistance management group.
ethaboxam (Elumin)	22	8 fl oz	N/A	0.5	Apply using a 6" to 8" band directly over seedpiece or in furrow where seed-piece will be dropped, prior to furrow closure. Apply as a side dressing between hilling and tuber initiation. Do not exceed 2 applications/year or 16 fl oz/acre/ year.
mefenoxam (Ridomil Gold SL, Ultra Flourish)	4	0.42 fl oz/ 1,000 ft of row 0.84 fl oz/ 1,000 ft of row	7	2	See labels for maximum amount of product allowable per season. PHI is based on foliar application for Ultra Flourish.
mefenoxam + chlorothalonil (Ridomil Gold/Bravo)	4+M05	2.5 pt/acre	14	2	Apply at flowering and then continue on a 14-day interval. Do not exceed more than four applications per crop.
mefenoxam + copper hydroxide (Ridomil Gold/Copper)	4+M01	2.0 lb/acre	14	2	Apply at 14-day intervals for up to 3 applications. Alternate with a protectant fungicide.
mefenoxam + mancozeb (Ridomil Gold MZ)	4+M03	2.5 lb/acre	14	2	Apply at 14-day intervals for up to 4 applications.
metalaxyl (Metalaxyl 2E AG, MetaStar 2E)	4	12.8 fl oz/acre	14	2	Preplant incorporated or soil surface spray
mono- and di-potassium salts of phosphorous acid (various)	P07	2.5 to 10 pints/acre	0	4h	See label for in-furrow application or foliar application rates.
POWDERY MILDEW (*ERYSIPHE CICHORACEARUM; LEVEILLULA TAURICA*)					
azoxystrobin (various)	11	See label	14	4h	See label. Rates may vary depending on the product. Apply in furrow at planting according to label direction. Do not apply more than one application without alternating away from fungicides in Group 11.
azoxystrobin + chlorothalonil (Quadris Opti)	11+M03	1.6 pt/acre	14	0.5	Do not apply more than 1.5 lb a.i./acre/year of azoxystrobin; do not apply more than 11.25 lb a.i./acre/year of chlorothalonil. Do not make more than 1 application before alternating with a fungicide with a different mode of action. Do not apply this product or other fungicides in Group 11 more than 6 times in a season.
azoxystrobin + difenoconazole (Quadris Top)	11+3	8.0 to 14.0 fl oz/acre	14	0.5	Apply at 7- to 14-day intervals. Apply no more than 2 sequential applications without alternating with a fungicide with a different mode of action. Do not apply more than 55.3 lb product per acre per year. Do not apply more than 0.46 lb a.i./acre/year of difenoconazole-containing products; do not apply more than 2.0 lb a.i./acre /year of azoxystrobin-containing products.
fludioxonil + pydiflumetofen (Miravis Prime)	7+12	9.2 to 11.4 fl oz/acre	14	0.5	Do not make more than two consecutive applications before alternation with another FRAC group.
fluopyram + pyrimethanil (Luna Tranquility)	7+9	11.2 fl oz/acre	7	0.5	Do not make more than 2 sequential applications without switching to a fungicide outside of Group 7 or Group 9. Limit 54.7 fl oz/acre/season.
fluxapyroxad + pyraclostrobin (Priaxor Xemium)	7+11	6.0 to 8.0 fl oz/acre	7	0.5	Limit 3 applications per season and no more than 2 applications before switching to a different mode of action. Do not apply more than 24 fl oz/ acre/season including in furrow and foliar uses.
mancozeb + azoxystrobin (Dexter Max)	M03+11	1.6 to 2.1 lb/acre	14	1	Do not exceed 16 lbs product/acre/crop. Season limits apply for azoxystrobin. For suppression of powdery mildew.
mandipropamid + difenoconazole (Revus Top)	40+3	5.5 to 7.0 fl oz/acre	14	0.5	Begin applications when conditions favor disease development, on 7- to 10-day intervals. Do not apply more than twice before alternating with fungicides that have a different mode of action. Do not apply more than 28 fl oz per acre per season.

TABLE 3-27. DISEASE CONTROL PRODUCTS FOR POTATO (cont'd)

D. Higgins, Plant Pathologist, Virginia Tech

Disease/Material	FRAC Code	Rate of Material Formulation	Minimum Days Harv.	Minimum Days Reentry	Method, Schedule, and Remarks
POWDERY MILDEW (*ERYSIPHE CICHORACEARUM*; *LEVEILLULA TAURICA*) (cont'd)					
metconazole (Quash)	3	2.5 to 4.0 oz/acre	1	0.5	Limit 16 oz/acre/season. Make no more than 2 applications before changing modes of action. Limit to 4 applications per year. Use the 4 oz rate for white mold.
metiram + pyraclostrobin (Cabrio Plus)	M03+11	2.9 lb/acre	14	1	Apply at 7- to 14-day intervals. Do not apply more than 17.4 lb/acre product per season. Do not apply more than 2 sequential applications before alternating with a fungicide with a different mode of action.
penthiopyrad (Vertisan)	7	10.0 to 24.0 fl oz/acre	7	0.5	Apply at 7- to 14-day intervals. Make no more than 2 applications before alternating with fungicide with different mode of action. Do not exceed 72 oz per acre per year. Do not apply more than 11.25 oz a.i. per acre per year in total from any combination of seed, soil, or foliar applications.
pyraclostrobin (Headline; Headline SC)	11	6.0 to 12.0 fl oz/acre	3	0.5	**DO NOT** exceed more than six foliar applications or 72 total oz of product per acre per season. Do not apply more than one time before alternating with a fungicide with a different mode of action.
sulfur (various)	M02	See label	—	1	Rates vary among products; see label.

PUMPKIN SEE *CUCURBITS*

RADISH SEE *ROOT VEGETABLES*

RHUBARB SEE *LEAFY PETIOLE VEGETABLES*

SCALLION SEE *ONION, GREEN*

SHALLOT SEE *ONION, DRY*

SUMMER SQUASH SEE *CUCURBITS*

WINTER SQUASH SEE *CUCURBITS*

TABLE 3-28. DISEASE CONTROL PRODUCTS FOR ROOT VEGETABLES (EXCEPT SUGAR BEET)

N. Gauthier, Plant Pathologist, University of Kentucky

Disease/Material	FRAC Code	Rate of Material Formulation	Minimum Days Harv.	Minimum Days Reentry	Method, Schedule, and Remarks
BEET (RED, GARDEN, TABLE), CARROT, PARSNIP, RADISH, TURNIP – HARVESTED FOR ROOTS ONLY (UNLESS NOTED)					
ALTERNARIA LEAF BLIGHT, CERCOSPORA LEAF SPOT					
azoxystrobin (Quadris Flowable)	11	6 to 15.5 fl oz/acre	0	4 hr	No more than 1 application before alternating with a fungicide with a different mode of action. Apply no more than 120 fl oz per acre per year. Also labeled for soilborne diseases. Generics are available, refer to labels.
azoxystrobin + chlorothalonil (Quadris Opti)	11+M05	2.4 pt/acre	0	0.5	**FOR USE ON CARROTS ONLY.**
azoxystrobin + difenoconazole (Quadris Top)	11+3	12 to 14 fl oz/acre	7	0.5	**FOR USE ON CARROTS ONLY.** Apply no more than 56 fl oz per acre per year.
azoxystrobin + propiconazole (Quilt Xcel)	11+3	14 fl oz	14	0.5	**FOR USE ON CARROTS ONLY.** No more than 1 application before alternating with a non-Group 11 fungicide. Make no more than 56 fl oz per acre per year. Generics are available, refer to labels.
boscalid (Endura)	7	4.5 oz/acre	0	0.5	Not for *Cercospora*. Do not make more than 2 consecutive applications or more than 5 applications per season. Also labeled for use on harvested radish leaves for *Alternaria* control. Also labeled for white mold/cottony rot.

TABLE 3-28. DISEASE CONTROL PRODUCTS FOR ROOT VEGETABLES (EXCEPT SUGAR BEET) (cont'd)

N. Gauthier, Plant Pathologist, University of Kentucky

Disease/Material	FRAC Code	Rate of Material Formulation	Minimum Days Harv.	Minimum Days Reentry	Method, Schedule, and Remarks
BEET (RED, GARDEN, TABLE), CARROT, PARSNIP, RADISH, TURNIP – HARVESTED FOR ROOTS ONLY (UNLESS NOTED)					
chlorothalonil (Bravo WeatherStik)	M05	1.5 to 2 pt/acre	See label	0.5	**FOR USE ON CARROTS AND PARSNIPS ONLY.** Spray at first appearance, 7- to 10-day intervals. Apply no more than 20 pts per acre per year. Generics are available, refer to labels.
fixed copper (various)	M01	See label	0	1 to 2	**CHECK LABEL FOR SPECIFIC CROP LISTINGS.** Make sure product is labeled in state prior to use. Make sure crop on label. Various formulations are available; labels vary by product, refer to label.
cyprodinil + fludioxonil (Switch)	9+12	11 to 14 oz/acre	7	0.5	Not for *Cercospora*. Apply when disease first appears and continue on 7- to 10-day intervals if conditions remain favorable for disease development. Do not exceed 56 oz of product per acre per year. For radish, make no more than 2 applications per year. For other root crops, do not exceed 56 oz of product per acre per year. Also labeled for use on harvested garden beet, turnip, and radish leaves.
cyprodinil + difenoconazole (Inspire Super)	3+9	16 to 20 fl oz/acre	7	0.5	**FOR USE ON CARROTS ONLY.** Make no more than 2 consecutive applications before alternating with a fungicide with a different mode of action. Apply more than 80fl oz/acre per year. Generics are available, refer to labels.
difenoconazole + fluopyram (Luna Flex)	3+7	11 to 13.6 fl oz/acre	7	0.5	**FOR USE ON CARROTS ONLY.** Do not make more than 2 consecutive applications before rotating to a labeled non-Group 3 or non-Group 7 fungicide. Apply no more than 27.2 fl oz per acre per year. Also labeled for white mold/cottony rot in carrots.
fenamidone (Reason 500SC)	11	8.2 fl oz/acre	14	0.5	**NOT LABELED FOR RADISH GROWN AS ROOT VEGETABLES OR FOR CARROT FOLIAR DISEASES.** Apply no more than 24.6 fl oz per season. Apply with sprayer or in sprinkler irrigation. Do not use a spreader/sticker on carrots. Also labeled for Pythium and Phytophthora control only in carrot. Also labeled for use on harvested turnip greens.
fluazinam (Omega)	29	16 fl oz/acre	7	0.5	**FOR USE ON CARROTS AND TURNIP GREENS ONLY.** Turnip roots treated with Omega are not for human or animal consumption. Apply no more than 128 fl oz per acre per year. Also labeled for white mold/cottony rot and southern blight in carrots.
fluopyram + trifloxystrobin (Luna Sensation)	7+11	See label	7	0.5	Not for *Cercospora* except on carrots. Do not make more than 2 consecutive applications before rotating to a labeled non-Group 7 or non-Group 11 fungicide. Carrot rate is 4.0 to 7.6 fl oz/acre. Other root crop rates are 5 to 5.8 fl oz/acre. Also labeled for white mold/cottony rot and southern blight.
fluxapyroxad + pyraclostrobin (Merivon Xemium)	7+11	4 to 5.5 fl oz/acre	7	0.5	Do not make more than 2 consecutive applications before rotating to a labeled on-Group 7 or non-Group 11 fungicide. Make no more than 3 applications per season. Use maximum rate for Cercospora leafspot.
iprodione (Rovral)	2	1 to 2 pt/acre	0	1	**FOR USE ON CARROTS ONLY.** Not for *Cercospora*. Make no more than 4 applications per season. Generics are available, refer to labels.
mefentrifluconazole (Cevya)	3	3 to 5 fl oz/acre	7	0.5	Do not apply more than 15 fl oz/acre/season. Also labeled for use on harvested radish and turnip greens.
penthiopyrad (Fontelis)	7	16 to 30 fl oz/acre	0	0.5	Make no more than 2 consecutive applications before alternating with fungicide with different mode of action. Apply no more than 61 fl oz/acre/year. Also labeled for harvested garden beet, turnip, and radish leaves. Labeled for white mold/cottony rot and southern blight.
propiconazole (Tilt)	3	3 to 4 fl oz/acre	14	1	Use higher rate for carrots. Make no more than 2 consecutive applications before alternating with a fungicide with a different mode of action. Apply no more than 16 fl oz/acre/season. Also labeled for use on harvested radish and turnip greens. Generics are available, refer to labels.
pydiflumetofen + fludioxonil (Miravis Prime)	7+12	6.8 fl oz/acre	7	0.5	Make no more than 2 consecutive applications before alternating with a different mode of action. Use no more than 27.2 fl oz/acre/season. Also labeled for use on harvested radish and turnip greens.
pyraclostrobin (Cabrio)	11	8 to 12 oz/acre	0	0.5	Alternate with a fungicide with a different mode of action. Apply no more than 48 oz/acre/season. Also labeled for use on harvested garden beet, turnip, and radish leaves.
pyraclostrobin + boscalid (Pristine)	11+7	8 to 10.5 oz/acre	0	0.5	Make no more than 2 consecutive applications before alternating with a different mode of action. Use no more than 63 oz or make no more than 6 applications per season.
tebuconazole (Monsoon)	3	3 to 7.2 fl oz/acre	7	0.5	**FOR USE ON TURNIPAND GARDEN BEETS ONLY.** Repeat applications at 12- to 14-day intervals. Apply no more than 28.8 fl oz/acre/season. Turnip root rate: 4 to 7.2 fl oz/acre. Turnip greens: 3 to 4 fl oz/acre. Garden beet root & greens: 3 to 7.2 fl oz/acre. Generics are available, refer to labels.
trifloxystrobin (Flint)	11	2 to 2.9 oz/acre	7	0.5	**NOT FOR RADISHES.** Make no more than 1 application before alternating with a fungicide with another mode of action. Make no more than 4 applications of trifloxystrobin or other strobilurin fungicides per season. Flint rate for radish is 2 to 4 oz/acre; other crops use 2 to 3 oz/acre. Gem 500SC is not registered for radish.

TABLE 3-28. DISEASE CONTROL PRODUCTS FOR ROOT VEGETABLES (EXCEPT SUGAR BEET) (cont'd)

N. Gauthier, Plant Pathologist, University of Kentucky

Disease/Material	FRAC Code	Rate of Material Formulation	Minimum Days Harv.	Minimum Days Reentry	Method, Schedule, and Remarks
BEET (RED, GARDEN, TABLE), CARROT, PARSNIP, RADISH, TURNIP – HARVESTED FOR ROOTS ONLY (UNLESS NOTED) (cont'd)					
CERCOSPORA LEAF SPOT OR BLIGHT, POWDERY MILDEW					
azoxystrobin (Quadris)	11	6 to 15.5 fl oz/acre	0	4 hr	No more than 1 application before alternating with a fungicide with a different mode of action. Apply no more than 120 fl oz/acre/year. Also labeled for soilborne diseases. Generics available, refer to labels.
azoxystrobin + difenoconazole (Quadris Top)	11+3	12 to 14 fl oz/acre	7	0.5	FOR USE ON CARROTS ONLY. Apply no more than 56 fl oz per acre per year.
azoxystrobin + propiconazole (Quilt Xcel)	11+3	14 fl oz	14	0.5	FOR USE ON CARROTS ONLY. No more than 1 application before alternating with a non-Group 11 fungicide. Make no more than 42 fl oz/acre/year.
boscalid (Endura)	7	4.5 oz/acre	0	0.5	Not for *Cercospora*. Do not make more than 2 consecutive applications or more than 5 applications per season. Also labeled for use on harvested radish leaves.
fixed copper (various)	M01	See labels	0	1 to 2	CHECK LABEL FOR SPECIFIC CROP LISTINGS. Make sure product is labeled in state prior to use. Various formulations are available; labels vary by product, refer to labels.
cyprodinil + fludioxonil (Switch)	9+12	11 to 14 oz/acre	7	0.5	Not for *Cercospora*. Apply when disease first appears and continue on 7- to 10-day intervals if conditions remain favorable for disease development. For radish, make no more than 2 applications per year. For other root crops, do not exceed 56 oz of product per acre per year. Also labeled for use on harvested garden beet, turnip, and radish leaves.
cyprodinil + difenoconazole (Inspire Super)	3+9	14 to 20 fl oz/acre	7	0.5	FOR USE ON CARROTS ONLY. Make no more than 2 consecutive applications before alternating with a fungicide with a different mode of action. Apply more than 80 fl oz/acre per year.
difenoconazole + fluopyram (Luna Flex)	3+7	11 to 13.6 fl oz/acre	7	0.5	FOR USE ON CARROTS ONLY. Do not make more than 2 consecutive applications before rotating to a labeled non-Group 3 or non-Group 7 fungicide. Apply no more than 27.2 fl oz per acre per year.
fluopyram + trifloxystrobin (Luna Sensation)	7+11	See label	7	0.5	Not for *Cercospora* except on carrot. Do not make more than 2 consecutive applications before rotating to a labeled non-Group 7 or non-Group 11 fungicide. Carrot rate is 4.0 to 7.6 fl oz/acre.
fluxapyroxad + pyraclostrobin (Merivon)	7+11	4 to 5.5 fl oz/acre	7	0.5	Do not make more than 2 consecutive applications before rotating to labeled non-Group 7 or non-Group 11 fungicide. Make no more than 3 applications per season. Use max rate for Cercospora leaf spot.
mefentrifluconazole (Cevya)	3	3 to 5 fl oz/acre	7	0.5	Do not apply more than 15 fl oz/acre/season.
penthiopyrad (Fontelis)	7	16 to 30 fl oz/acre	0	0.5	Make no more than 2 consecutive applications before alternating with a fungicide with a different mode of action. Apply no more than 61 fl oz/acre per year. Also labeled for use on harvested garden beet, turnip, and radish leaves.
propiconazole (Tilt)	3	3 to 4 fl oz/acre	14	0.5	Use higher rate for carrots. Make no more than 2 consecutive applications before alternating with a fungicide with a different mode of action. Apply no more than 16 fl oz/acre/season. Also labeled for use on harvested radish and turnip greens. Generics available, refer to labels.
pydiflumetofen + fludioxonil (Miravis Prime)	7+12	6.8 fl oz/acre	7	0.5	Make no more than 2 consecutive applications before alternating with a fungicide with a different mode of action. Use no more than 27.2 fl oz/acre/season. Also labeled for use on harvested radish and turnip greens.
pyraclostrobin (Cabrio)	11	8 to 12 oz/acre	0	0.5	Alternate with a fungicide with a different mode of action. Apply no more than 48 oz/acre/season. Also labeled for use on harvested garden beet, turnip, and radish leaves.
pyraclostrobin + boscalid (Pristine)	11+7	8 to 10.5 oz/acre	0	0.5	Make no more than 2 consecutive applications before alternating with a fungicide with a different mode of action. Use no more than 63 oz or make no more than 6 applications per season.
sulfur (various)	M02	3 to 10 lb/acre		1	POWDERY MILDEW ONLY. Spray at first appearance. Avoid applying on days > 90°F. Also labeled for harvested garden beet and turnip leaves. Various formulations available; labels vary by product, refer to labels.
trifloxystrobin (Flint)	11	2 to 2.9 fl oz/acre	7	0.5	Make no more than 1 application before alternating with a fungicide with another mode of action. Make no more than 4 applications of trifloxystrobin or other strobilurin fungicides per season. Flint rate for radish is 2 to 4 oz/acre, other crops use 2 to 3 oz/acre. Gem 500SC not labeled for radish. Flint not labeled for powdery mildew on radish but may be used for Cercospora.
PHYTOPHTHORA BASAL STEM ROT					
mefenoxam (Ridomil Gold 4 SL) (Ultra Flourish 2 EC)	4	See label	—	2	May be applied preplant incorporated or as a soil surface spray after planting. See labels for applications and rates. Generics available.

TABLE 3-28. DISEASE CONTROL PRODUCTS FOR ROOT VEGETABLES (EXCEPT SUGAR BEET) (cont'd)

N. Gauthier, Plant Pathologist, University of Kentucky

Disease/Material	FRAC Code	Rate of Material Formulation	Minimum Days Harv.	Minimum Days Reentry	Method, Schedule, and Remarks
BEET (RED, GARDEN, TABLE), CARROT, PARSNIP, RADISH, TURNIP – HARVESTED FOR ROOTS ONLY (UNLESS NOTED) (cont'd)					
PHYTOPHTHORA BASAL STEM ROT (cont'd)					
metalaxyl (MetaStar 2E)	4	4 to 8 pt/trt acre	—	2	May be applied preplant incorporated or as soil surface spray after planting. See labels for applications and rates. Generics available.
fenamidone (Reason)	11	8.2 fl oz/acre	14	0.5	**NOT LABELED FOR RADISH OR CARROT ROOTS.** Make no more than 1 application before alternating with a mefenoxam-containing fungicide. Apply no more than 24.6 fl oz per growing season. Applied with sprayer or in sprinkler irrigation.
fluopicolide (Presidio)	43	3 to 4 fl oz/acre	7	0.5	Can be applied with a sprayer or in sprinkler irrigation. Regardless of method, must be applied in combination with fungicide with different mode of action and labeled for that method. No more than 2 consecutive applications before alternating with a Pythium fungicide with different mode of action. Maximum of 12 fl oz/ acre/year. May be applied preplant incorporated. Do not use on turnips intended for livestock. Not labeled for Phytophthora on carrot.
phosphorus acid (Fungi-Phite)	P07	1 to 3 qt/acre	0	0.3	Various formulations available; labels vary by product, refer to labels.
PYTHIUM ROOT ROT, ROOT DIEBACK, CAVITY SPOT (*PYTHIUM* SPP.)					
azoxystrobin (Quadris)	11	0.4 to 0.8 fl oz/ 1000 row ft	0	4 hr	Make one application, applied either in-furrow at planting, in a 7-inch band over the row prior to or shortly after planting, or in drip irrigation. Generics are also available; refer to labels.
azoxystrobin + mefenoxam (Uniform)	11+4	0.34 fl oz/ 1000 row ft	—	0	**NOT FOR CARROTS.** In-furrow treatment only at planting.
cyazofamid (Ranman)	21	6 fl oz/acre	14	0.5	**FOR USE ON CARROTS ONLY.** May be applied preplant incorporated, as a pre-emergent surface band, or in sprinkler irrigation. Applications can be repeated at 14-day intervals but must alternate with a Pythium fungicide with a different mode of action. Do not apply more than 30 fl oz per season. Do not use any adjuvant.
fluopicolide (Presidio)	43	3 to 4 fl oz/acre	7	0.5	Can be applied with a sprayer or in sprinkler irrigation. Regardless of method, must be applied in combination with a fungicide with a different mode of action and labeled for that method. No more than 2 consecutive applications before alternating with a Pythium fungicide with a different mode of action. Maximum of 12 fl oz/acre/ year. May also be applied preplant incorporated. Do not use on turnips intended for livestock. Use with highest rate for carrot.
fenamidone (Reason 500SC)	11	8.2 fl oz/acre	14	0.5	**NOT LABELED FOR RADISH ROOTS.** Make no more than 1 application before alternating with a mefenoxam-containing fungicide. Apply no more than 24.6 fl oz per season. Apply with sprayer or in sprinkler irrigation. Do not use a spreader/ sticker on carrots. Pythium control only for carrot.
mefenoxam (Ridomil Gold 4 SL) (Ultra Flourish 2 EC)	4	See label	—	2	May be applied preplant incorporated or as a soil surface spray after planting. See labels for applications and rates. Generics available.
metalaxyl (MetaStar 2E)	4	See label	—	2	May be applied preplant incorporated or to soil surface spray after planting. Seed treatment formulations are also available, refer to labels.
phosphorus acid (Fungi-Phite) (Confine Extra)	P07	1 to 3 qt/acre	0	0.3	Various formulations are available; labels vary by product, refer to labels.
RUST (*PUCCINIA* SPP.)					
fluopyram + trifloxystrobin (Luna Sensation)	7+11	See label	7	0.5	Not for *Cercospora* except on carrots. Do not make more than 2 consecutive applications before rotating to a labeled non-Group 7 or non-Group 11 fungicide. Carrot rate is 4.0 to 7.6 fl oz/acre. Other root crop rates are 5 to 5.8 fl oz/acre.
RUST (*PUCCINIA* SPP.)					
penthiopyrad (Fontelis)	7	16 to 30 fl oz/acre	0	0.5	Make no more than 2 consecutive applications before alternating with a fungicide with a different mode of action. Apply no more than 61 fl oz/acre per year.
sulfur (various)	M02	See label		1	Spray at first appearance. Avoid applying on days over 90°F. Also labeled for use on harvested garden beet and turnip leaves. Various formulations are available; labels vary by product, refer to labels
trifloxystrobin (Flint)	11	2 to 2.9 oz/acre	7	0.5	Make no more than 1 application before alternating with a fungicide with another mode of action. Make no more than 4 applications of trifloxystrobin or other strobilurin fungicides per season. Flint is not labeled for rust on radish. Gem 500SC not labeled for radish.
WHITE MOLD (*SCLEROTINIA* SPP.) AND GRAY MOLD (*BOTRYTIS* SPP.)					
boscalid (Endura)	7	7.8 oz	0	0.5	No more than 2 applications before alternating with a fungicide with a different mode of action. Limit of 3 applications per season. Generics available, refer to labels.

TABLE 3-28. DISEASE CONTROL PRODUCTS FOR ROOT VEGETABLES (EXCEPT SUGAR BEET) (cont'd)

N. Gauthier, Plant Pathologist, University of Kentucky

Disease/Material	FRAC Code	Rate of Material Formulation	Minimum Days Harv.	Reentry	Method, Schedule, and Remarks
BEET (RED, GARDEN, TABLE), CARROT, PARSNIP, RADISH, TURNIP – HARVESTED FOR ROOTS ONLY (UNLESS NOTED) (cont'd)					
WHITE MOLD (*SCLEROTINIA* SPP.) AND GRAY MOLD (*BOTRYTIS* SPP.) (cont'd)					
chlorothalonil (Bravo WeatherStik)	M05	1.5 to 2 pt/acre	See label	0.5	**FOR USE ON CARROTS AND PARSNIPS ONLY.** For gray mold on parsnip only. Spray at first appearance, 7- to 10-day intervals. Various formulations and generics available; labels vary by product, refer to labels.
difenoconazole + fluopyram (Luna Flex)	3+7	11 to 13.6 fl oz/acre	7	0.5	**FOR USE ON CARROTS ONLY.** Do not make more than 2 consecutive applications before rotating to a labeled non-Group 3 or non-Group 7 fungicide. Apply no more than 27.2 fl oz per acre per year.
fluazinam (Omega)	29	16 fl oz/acre	7	0.5	**FOR USE ON CARROTS ONLY.** For white mold only. Apply no more than 128 fl oz per acre per year.
fluopyram + trifloxystrobin (Luna Sensation)	7+11	See label	7	0.5	Not for *Cercospora* except on carrots. Do not make more than 2 consecutive applications before rotating to a labeled non-Group 7 or non-Group 11 fungicide. Carrot rate is 4.0 to 7.6 fl oz/acre. Other root crop rates are 5 to 5.8 fl oz/acre.
WHITE MOLD/COTTONY ROT (*SCLEROTINIA* SPP.) AND GRAY MOLD (*BOTRYTIS* SPP.), (POSTHARVEST) SOUTHERN BLIGHT (*AGROATHELIA ROLFSII*)					
azoxystrobin (Quadris)	11	0.4 to 0.8 fl oz/ 1000 row ft	0	4 hr	Make one application, applied either in-furrow at planting, in a 7-inch band over the row prior to or shortly after planting, or in drip irrigation. Generics available, refer to labels.
fluazinam (Omega)	29	16 fl oz/acre	7	0.5	**FOR USE ON CARROTS AND TURNIP GREENS ONLY.** Turnip roots treated with Omega are not for human or animal consumption. Apply no more than 128 fl oz per acre per year.
fluopyram + trifloxystrobin (Luna Sensation)	7+11	See label	7	0.5	Do not make more than 2 consecutive applications before rotating to a labeled non-Group 7 or non-Group 11 fungicide. Carrot rate for *Sclerotinia* suppression is 7.6 fl oz/acre.
penthiopyrad (Fontelis)	7	16 to 30 fl oz/acre	0	0.5	Make no more than 2 consecutive applications before alternating with a fungicide with a different mode of action. Apply no more than 61 fl oz/acre per year. Also labeled for use on harvested garden beet, turnip, and radish leaves.
thiabendazole (Mertect)	1	41 fl oz/100 gal	—	0.5	**FOR USE ON CARROTS ONLY.** Dip harvested roots for 5 to 10 seconds. Do not rinse.
SOUTHERN BLIGHT (*AGROATHELIA ROLFSII*), WHITE RUST (*ALBUGO* SPP.)					
azoxystrobin (Quadris)	11	6 to 15.5 fl oz/acre	0	4 hr	No more than 1 application before alternating with a fungicide with a different mode of action. Apply no more than 123 fl oz per acre per season. Generics available, refer to labels.
azoxystrobin + difenoconazole (Quadris Top)	11+3	12 to 14 fl oz/acre	7	0.5	**FOR USE ON CARROTS ONLY.** Apply no more than 56 fl oz per acre per year.
fluazinam (Omega)	29	1 pt/acre	7	0.5	**FOR USE ON CARROTS ONLY.**
penthiopyrad (Fontelis)	7	16 to 30 fl oz/acre	0	0.5	Make no more than 2 consecutive applications before alternating with a fungicide with a different mode of action. Apply no more than 61 fl oz/acre per year. Not registered for white rust.
pyraclostrobin + boscalid (Pristine)	11+7	8 to 10.5 oz/acre	0	0.5	Make no more than 2 consecutive applications before alternating with a different mode of action. Use no more than 63 oz or make no more than 6 applications per season. Not registered for white rust.
WHITE RUST (*ALBUGO* SPP.)					
mefenoxam + copper hydroxide (Ridomil Gold/Copper)	4+M01	1 to 2 pt/acre	7	1	**FOR USE ON CARROTS AND RADISHES ONLY.** Preplant or at-planting application.
pyraclostrobin (Cabrio EG)	11	8 to 16 oz/acre	0	0.5	Alternate with a fungicide with a different mode of action. Apply no more than 48 oz/acre/season. Also labeled for use on harvested garden beet, turnip, and radish leaves. Also labeled for use on harvested garden beet, turnip, and radish leaves. Generics available, refer to labels.

TABLE 3-29. IMPORTANCE OF ALTERNATIVE MANAGEMENT PRACTICES FOR DISEASE CONTROL IN CARROT

N. Gauthier, Plant Pathologist, University of Kentucky

Scale: E, excellent; G, good; F, fair; P, poor; NC, no control; ND, no data

Strategy	Alternaria blight	Cercospora blight	Powdery mildew	Pythium cavity spot	Pythium damping off	Southern root rot	Rhizoctonia cavity spot	Sclerotinia postharvest	Botrytis postharvest	Bacterial leaf blight	Root-knot nematode
Avoid field operations when leaves wet	P	P	NC	NC	NC	NC	NC	NC	NC	F	NC
Avoid overhead irrigation	F	F	NC	NC	NC	NC	NC	F	NC	F	NC
Change planting date	P	P	NC	F	F	F	NC	NC	NC	NC	F
Cover cropping with antagonist	NC	NC	NC	NC	NC	NC	NC	NC	NC	NC	F
Crop rotation	F	F	NC	P	P	F	F	F	NC	F	F
Deep plowing	G	G	P	NC	NC	F	F	F	P	G	NC
Destroy crop residue	E	E	P	NC	NC	NC	P	NC	P	E	P
Encourage air movement	F	F	NC	NC	NC	NC	NC	F	NC	NC	NC
Plant in well-drained soil	NC	NC	NC	G	G	P	F	F	NC	NC	NC
Plant on raised beds	NC	NC	NC	F	F	NC	F	P	NC	NC	NC
Postharvest temperature control	NC	NC	NC	NC	NC	NC	NC	E	E	NC	NC
Reduce mechanical injury	NC	NC	NC	NC	NC	NC	NC	F	G	NC	NC
Destroy volunteer carrots	F	F	P	NC	NC	F	F	NC	NC	NC	F
Pathogen-free planting material	E	E	NC	NC	NC	NC	NC	NC	NC	E	NC
Resistant cultivars	G	G	F	NC	NC	NC	NC	NC	NC	NC	NC

TABLE 3-30. DISEASE CONTROL PRODUCTS FOR SPINACH

K. Cochran, Extension Specialist, Texas A&M University

Disease/Material	FRAC Code	Rate of Material Formulation	Minimum Days Harv.	Minimum Days Reentry	Method, Schedule, and Remarks
DAMPING OFF (*PYTHIUM* SPP.)					
cyazofamid (Ranman 400SC)	21	2.75 fl oz/acre	0	0.5	Apply to soil in a banded application or in transplant water at time of transplanting. For control of oomycete diseases.
mefenoxam (Ridomil Gold SL) (Ultra Flourish)	4	1 to 2 pt/acre 2 to 4 pt/acre	21	2	Broadcast or banded over the row as a soil spray or pre-plant incorporation into the top two inches of soil. For Ultra Flourish, PHI = 3 days only if soil applications do not exceed 4 pt/acre and foliar mefenoxam applications do not exceed 0.25 lb a.i./acre, otherwise PHI = 21 days. For control of oomycete diseases.
metalaxyl (various)	4	4 to 8 pt/acre	21	2	Broadcast or banded over the row as a soil spray or pre-plant incorporation into the top two inches of soil.
oxathiapiprolin + mefenoxam (Orondis Gold)	49+4	13.9 to 27.8 fl oz/acre	21	2	Apply using any of the following methods: in-furrow, transplant water, banded spray, and drip irrigation. If using drip irrigation, delay until after emergence. PHI is 3 days if soil application does not exceed 1.0 lb mefenoxam/acre/year and foliar application does not exceed 0.25 lb mefenoxam/acre/year. For control of oomycete diseases.
phosphorous acid (mono- and di-potassium salts) (Reliant, Fungi-Phite)	P07	1 to 2 qt/acre	0	4 hr	Do not apply more than 6 times per crop cycle.
propamocarb hydrochloride (Previcur Flex)	28	2 pt/acre	2	0.5	Do not exceed two applications per crop.
SEEDLING BLIGHT, DAMPING OFF, ROOT ROT (*PYTHIUM* SPP., *RHIZOCTONIA SOLANI*, *FUSARIUM* SPP., *VERTICILLIUM* SPP.)					
azoxystrobin + mefenoxam (Uniform)	11+4	0.34 fl oz/ 1,000row ft	--	0	Apply as an in-furrow spray in 5 gal of water per acre prior to covering seed. Make only application per season.
Bacillus amyloliquefaciens strain MBI 600* (Serifel, biofungicide)	BM02	4 to 16 ozs/A (0.25 to 1.0 lb/A)	00	0.5	For control of soilborne pathogens, apply in furrow, shanked-in, injection, or soil drench with adequate water volume to soak soil through root zone. Apply in furrow, shanked-in, injection, or soil drench with adequate water volume to soak soil through root zone. For management of *Rhizoctonia solani*, *Fusarium* spp., *Verticillium* spp. per label.

TABLE 3-30. DISEASE CONTROL PRODUCTS FOR SPINACH (cont'd)

K. Cochran, Extension Specialist, Texas A&M University

Disease/Material	FRAC Code	Rate of Material Formulation	Minimum Days Harv.	Minimum Days Reentry	Method, Schedule, and Remarks
SEEDLING BLIGHT, DAMPING OFF, ROOT ROT (*PYTHIUM* SPP., *RHIZOCTONIA SOLANI*, *FUSARIUM* SPP., *VERTICILLIUM* SPP.) (cont'd)					
Bacillus subtilis (Serenade ASO)	BM02	2 to 4 qt/acre Or 1 to 4 qt/acre when in tank mix	0	4 hr	*Fusarium* spp. *Phytophthora* spp. *Pythium* spp. *Rhizoctonia* spp. *Sclerotinia* spp. *Verticillium* spp.; Apply per label as soil surface drench, shanked-in, side-dress, injected, or in-furrow.
fluopyram + trifloxystrobin (Luna Sensation)	7+11	7.6 fl oz/acre	0	0.5	Use a rotation partner outside of groups 7 and 11.
Streptomyces sp. strain K61 (Mycostop biofungicide)	BM02	Seed treatment: 0.04 to 4 oz / lb seed 3.5 to 16 oz / acre (soil spray or drench) 1.75 to 16 oz / acre (banded, in-furrow, or side dress)	0	4 hr	Do not combine as a tank mix. Apply as a seed treatment [applied on farm at planting or as a commercial liquid or dry seed treatment], by spraying or drenching the growth substrate or soil, by incorporation into a growth substrate or soil, as a dip, as a foliar spray, or by hydroponic or chemigation application in the field or greenhouse as an aqueous suspension Apply with adequate water. Follow with regular irrigation to move product into soil & root zone. Repeat every 2-6 wks as needed. In banded, in-furrow, or side-dress applications, apply in a minimum of 10 gal water or nutrient solution/acre before seeding.
Trichoderma harzianum Rifai strain T-22 + *Trichoderma virens* strain G-41 (Rootshield Plus WP or granules)	BM02	Greenhouse drench 3.0 to 8.0 oz/100 gal In-furrow spray or transplant starter solution 16.0 to 32.0 oz/acre Greenhouse chemigation 3.0 to 8.0 oz/100 gal Field chemigation 3.0 to 8.0 oz/100 gal	0	4 hr	Reported to control *Phytophthora*, *Pythium*, *Fusarium*, *Rhizoctonia*, *Cylindrocladium*, and *Thielaviopsis* species. Verify tank mix compatibility, many fungicides are not compatible (ie : imazalil, propiconazole, tebuconazole, and triflumizole). Do not apply immediately before these products are used. Soil temperatures must be over 50F to be active.
DOWNY MILDEW (*PERONOSPORA FAINOSA* F. SP. *SPINACIAE*)					
ametoctradin + dimethomorph (Zampro)	45+40	14 fl oz/acre	0	0.5	Do not apply with or in rotation with mandipropamid or dimethomorph.
cymoxanil (Curzate)	27	5 oz/acre	1	0.5	Apply with a protectant fungicide. Apply no more than 30 oz per acre in a 12-month period. **Not labeled for use in Louisiana.**
dimethomorph (Forum)	40	6 fl oz/acre	0	0.5	Do not apply with or in rotation with mandipropamid.
fluopyram + trifloxystrobin (Luna Sensation)	7+11	7.6 fl oz/acre	0	0.5	Use a rotation partner outside of groups 7 and 11.
mandipropamid (Revus)	40	8 fl oz/acre	1	4 hr	Make no more than 2 consecutive applications before alternating with a fungicide with a different mode of action. Apply no more than 32 fl oz/acre/season. Do not apply with or in rotation with dimethomorph.
oxathiapiprolin (Orondis Gold 200)	49+4	4.8 to 9.6 fl oz/acre	0	4 hr	Apply at planting in furrow or by drip, or in subsequent drip irrigation. Make no more than 2 consecutive applications before alternating with a fungicide with a different mode of action. Do not apply more than 19.2 fl oz/acre per year.
phosphorous acid (mono- and di-potassium salts) (Reliant, Fungi-Phite)	P07	1 to 2 qt/acre	0	4 hr	Do not apply more than 6 times per crop cycle.
propamocarb hydrochloride (Previcur Flex)	28	2 pt/acre	2	0.5	Use 1.33 to 2 pints per acre if mixing with another fungicide. Do not exceed two applications per crop.
DOWNY MILDEW (*PERONOSPORA FAINOSA* F. SP. *SPINACIAE*), WHITE RUST (*ALBUGO OCCIDENTALIS*)					
acibenzolar-*S*-methyl (Actigard)	P01	0.5 to 0.75 oz/acre	7	0.5	Do not apply to young seedlings or plants stressed due to drought, excessive moisture, cold weather, or herbicide injury.
aluminum tris (O-ethyl phosphonate) (Aliette WDG)	P07	2 to 5 lbs/acre	3	0.5	Do not mix with copper. The lower rate (2 to 3 lbs/acre) can be used when mixed with another fungicide labeled for downy mildew. Otherwise, 3 to 5 lbs/acre should be used. **Not labeled for use in MS.**
azoxystrobin + flutriafol (Topguard EQ)	11+3	6 to 8 fl oz/acre	7	0.5	Use a rotation partner outside of groups 3 and 11. Do not apply more than 4 applications per year.
cyazofamid (Ranman)	21	2.75 fl oz/acre	0	0.5	Do not make more than 3 consecutive applications before alternating to a fungicide with a different mode of action.
famoxadone + cymoxanil (Tanos)	11+27	8 to 10 oz/acre	1	0.5	Must be tank mixed with a contact downy mildew fungicide with a different mode of action. Make no more than 1 application before alternating with a fungicide with a different mode of action. Apply no more than 84 oz/acre per cropping season.

TABLE 3-30. DISEASE CONTROL PRODUCTS FOR SPINACH (cont'd)

K. Cochran, Extension Specialist, Texas A&M University

Disease/Material	FRAC Code	Rate of Material Formulation	Minimum Days Harv.	Minimum Days Reentry	Method, Schedule, and Remarks
DOWNY MILDEW (*PERONOSPORA FAINOSA* F. SP. *SPINACIAE*), WHITE RUST (*ALBUGO OCCIDENTALIS*) (cont'd)					
fenamidone (Reason 500SC)	11	5.5 to 8.2 fl oz/acre	2	0.5	Make no more than 1 application before alternating with a fungicide with a different mode of action. Apply no more than 24.6 fl oz/acre per cropping season.
fixed copper (various)	M01	See labels	0	2	Some formulations of copper may cause leaf flecking.
fluopicolide (Presidio)	43	3 to 4 fl oz/acre	2	0.5	Tank-mix with another downy mildew fungicide with a different mode of action. Apply as a foliar spray or in drip irrigation.
fluxapyroxad + pyraclostrobin (Merivon)	7+11	4 to 11 fl oz/acre	1	0.5	**Do not tank-mix Merivon with any pesticides, adjuvants, fertilizers, nutrients, or any other additives.** Do not make more than 2 consecutive applications before rotating to a labeled non-Group 7 or non-Group 11 fungicide. Make no more than 3 applications per season.
laminarin (Vacciplant)	P04	14 to 22 fl oz/acre	0	4 hr	Use preventatively. Tank-mix with another registered fungicide under moderate to severe disease pressure.
mefenoxam (Ridomil Gold) (Ultra Flourish)	4	0.25 pt/acre 0.25 to 0.5 pt/acre	21	2	Shank application 21 days after planting or after first cutting. Two shanked applications may be made on a 21-day interval.
polyoxin D zinc salt (OSO 5%SC)	19	6.5 to 13 fl oz/acre	0	4 hr	Do not make more than 6 applications per season at the maximum-labeled rate.
pyraclostrobin (Cabrio)	11	8 to 16 oz/acre	0	0.5	Make no more than 2 consecutive applications before alternating with a fungicide with a different mode of action. Apply no more than 64 oz per acre per growing season.
mefenoxam + copper hydroxide (Ridomil Gold/Copper)	4+M01	2 lb/acre	3	2	Spray to foliage. Use with preplant Ridomil Gold soil application.
metalaxyl (various)	4	1 pt/trt acre	21	2	Shank in 21 days after planting. Apply no more than 2-shanked applications on a 21-day interval.
VARIOUS LEAF SPOTS					
azoxystrobin + flutriafol (Topguard EQ)	11+3	6 to 8 fl oz/acre	7	0.5	Use a rotation partner outside of groups 3 and 11. Do not apply more than 4 applications per year.
Bacillus amyloliquefaciens strain MBI 600* (Serifel, biofungicide)	BM02	4 to 16 ozs/A (0.25 to 1.0 lb/A)	0	0.5	Apply prior to infection and continue on 7- to 10-day intervals as needed. Use the higher rate and shorter interval when disease pressure is high. For Bacterial leaf spot, begin applications prior to disease onset and continue on 2- to 10-day intervals if conditions are conducive to disease development. Ensure application is performed with adequate spray volume to ensure coverage of plant tissues.
Bacillus mycoides isolate J* (LifeGard WG)	P06	4.5 oz/ 100 gal water	0	4 hr	4.5 oz/100 gal water; amount applied per acre will depend on finished spray volume required to adequately cover the crop. Per label, do not apply less than 1 oz or more than 4.5 oz product/acre. For control of bacterial, fungal & oomycete (watermold) foliar pathogens, make preventative applications every 7-14d to maintain control. For control of Stemphyllium leaf spot, apply preventatively (min 3 days prior to anticipated favorable conditions), with repeat application at 3-7d. If disease pressure is high, rotation or mix with other fungicide products is warranted to provide control.
Bacillus subtilis (Serenade ASO)	BM02	2-4 qt/acre Or 1-4 qt/acre when used in tank mix	0	4 hr	Begin applications when environmental conditions are favorable for disease development of fungal, bacterial, and oomycete foliar diseases.
cyprodinil + fludioxonil (Switch 62.5WG)	9+12	11 to 14 fl oz/acre	0	0.5	Make no more than 1 application before alternating with a fungicide with a different mode of action. Apply no more than 24.6 fl oz/acre per growing season.
fenamidone (Reason 500SC)	11	5.5 to 8.2 fl oz/acre	2	0.5	Make no more than 1 application before alternating with a fungicide with a different mode of action. Apply no more than 24.6 fl oz/acre per growing season.
Mefentrifluconazole (Ceyva)	3	3 - 5 fl oz/acre	0	0.5	Apply as foliar spray before onset of disease, minimum interval of 7 days, Max rate/acre of 15 fl oz/year
fixed copper (various)	M01	See labels	0	2	Some formulations of copper may cause flecking on the leaves.
fludioxonil (Cannonball WG)	12	7 fl oz/acre	7	0.5	Make no more than 2 consecutive applications before alternating with a fungicide with a different mode of action.
fluopyram + trifloxystrobin (Luna Sensation)	7+11	7.6 fl oz/acre	0	0.5	Use a rotation partner outside of groups 7 and 11.
flutriafol (Rhyme)	3	5 to 7 fl oz/acre	7	0.5	Do not apply more than 28 fl oz/acre, or 4 applications, per year. 7 days application interval.

TABLE 3-30. DISEASE CONTROL PRODUCTS FOR SPINACH (cont'd)

K. Cochran, Extension Specialist, Texas A&M University

Disease/Material	FRAC Code	Rate of Material Formulation	Minimum Days Harv.	Minimum Days Reentry	Method, Schedule, and Remarks
VARIOUS LEAF SPOTS (cont'd)					
fluxapyroxad + pyraclostrobin (Merivon)	7+11	4 to 11 fl oz/acre	1	0.5	**Do not tank-mix Merivon with any pesticides, adjuvants, fertilizers, nutrients, or any other additives.** Do not make more than 2 consecutive applications before rotating to a labeled non-Group 7 or non-Group 11 fungicide. Make no more than 3 applications per season.
laminarin (Vacciplant)	P04	14 to 22 fl oz/acre	0	4 hr	Use preventatively. Tank-mix with another registered fungicide under moderate to severe disease pressure.
penthiopyrad (Fontelis)	7	16 to 24 fl oz/acre	3	0.5	Make no more than 2 consecutive applications before alternating with a fungicide with a different mode of action. Apply no more than 72 fl oz/acre/year.
phosphorous acid (mono- and di- potassium salts) (Reliant, Fungi-Phite)	P07	1 to 2 qt/acre	0	4 hr	Do not apply more than 6 times per crop cycle.
polyoxin D zinc salt (OSO 5%SC)	19	3.75 to 13 fl oz/acre	0	4 hr	Do not make more than 6 applications per season at the maximum-labeled rate. Not registered for use in KY, MS, AR, OK, or KS.
pyraclostrobin (Cabrio EG)	11	12 to 16 oz/acre	0	0.5	Make no more than 2 consecutive applications before alternating with a fungicide with a different mode of action. Apply no more than 64 oz per acre per growing season.
pydiflumetofen + fludioxonil (Miravis Prime)	7+12	9.2 to 13.4 fl oz/acre	0	0.5	Make no more than 2 consecutive applications before alternating with a fungicide with a different mode of action. Apply no more than 36.5 fl oz per acre per growing season.
trifloxystrobin (Flint Extra)	11	3 to 3.8 fl oz/acre	See label	0.5	REI is 0 days for broadcast foliar uses and 20 days for banded applications. Apply no more than 7.6 fl oz per acre per growing season.

TABLE 3-31. DISEASE CONTROL PRODUCTS FOR SWEETPOTATO

L. Quesada-Ocampo and A. Gorny, Plant Pathologists, North Carolina State University

Disease/Material	FRAC Code	Rate of Material Formulation	Minimum Days Harv.	Minimum Days Reentry	Method, Schedule, and Remarks
BLACK ROT (*CERATOCYSTIS FIMBRIATA*), SCURF (*MONILOCHAETES INFUSCANS*), AND FOOT ROT (*PLENODOMUS DESTRUENS*)					
thiabendazole (Mertect 340 F)	1	107 fl oz/100 gal	0.5	0.5	Dip seed roots 1 to 2 minutes and plant immediately.
POSTHARVEST BLACK ROT (*CERATOCYSTIS FIMBRIATA*)					
azoxystrobin + fludioxonil + difenoconazole (Stadium)	11+12 +3	1 fl oz per 2,000 lbs of roots	—	—	Section 2 (ee) label for NC and CA expires 12/31/2025. Ensure proper coverage, use tumbling, and mix the fungicide solution in sufficient water volume. Do not make more than one postharvest application.
thiabendazole (Mertect 340 F)	1	0.42 fl oz per 2,000 lb of roots or 0.42 fl oz/gal	0.5	0.5	Postharvest treatment of sweetpotato for control of black rot. Limit to one application during packing. Mist washed roots on a conveyor line, with tumbling action, before packing with 0.42 fl oz of Mertect to each 2,000 lb of roots in sufficient water for complete coverage. Alternatively, dip the roots for 20 seconds in 0.42 fl oz of Mertect per gal of water. Ensure roots are dry before packing.
fludioxonil + azoxystrobin (Archive)	11 + 12	0.6 to 1.0 fl oz/ 2000 lbs roots	-	-	In line aqueous spray. Ensure proper coverage of the tubers. Tubers most be tumbling as they are treated. Use T-jet CDA or similar application system. No more than 1 post-harvest application.
CIRCULAR SPOT (*SCLEROTIUM ROLFSII*), SCLEROTIAL BLIGHT (*SCLEROTIUM ROLFSII*), RHIZOCTONIA STEM CANKER (*RHIZOCTONIA SOLANI*), PYTHIUM ROOT ROT (*PYTHIUM*)					
azoxystrobin (Quadris) 2.08 F	11	0.4 to 0.8 fl oz/ 1,000 row feet	—	4 hr	Make in-furrow or banded applications shortly after transplanting.
dichloran (Botran 5F)	14	0.6 qt/7.5 gal (Seed Dip) 5.73 oz in 14 gal/ 1000 linear feet of plant bed (Plant bed spray)	—	0.5	Labeled for Southern blight (*Sclerotium rolfsii*). **Seed dip:** Dip seed sweetpotatoes 10 to 15 seconds in a well-agitated fungicide suspension. Drain sweetpotatoes and bed promptly. Prepare fresh fungicide suspension daily. **Plant bed spray:** Spray or sprinkle over bedded sweetpotatoes before covering them with soil. **Note:** Not for use in Virginia, Tennessee, or South Carolina.

TABLE 3-31. DISEASE CONTROL PRODUCTS FOR SWEETPOTATO (cont'd)

L. Quesada-Ocampo and A. Gorny, Plant Pathologists, North Carolina State University

Disease/Material	FRAC Code	Rate of Material Formulation	Minimum Days Harv.	Reentry	Method, Schedule, and Remarks
CIRCULAR SPOT (*SCLEROTIUM ROLFSII*), SCLEROTIAL BLIGHT (*SCLEROTIUM ROLFSII*), RHIZOCTONIA STEM CANKER (*RHIZOCTONIA SOLANI*), PYTHIUM ROOT ROT (*PYTHIUM*) (cont'd)					
fluazinam (Omega 500F)	29	5.5 to 8 fl oz/acre	14	0.5	Labeled for control of white mold (Sclerotinia). Begin applications when plants are 6 to 8 inches tall. Repeat applications at intervals of 7 to 10 days. See label for rate. Do not apply more than 3.5 pints/acre/year.
SEED-BORNE AND SOILBORNE FUNGI THAT CAUSE DECAY, DAMPING OFF OR SEEDLING BLIGHT					
fludioxonil (Maxim 4 FS)	12	0.08 to 0.16 fl oz/ 100 lb of propagating roots	—	0.5	Apply uniformly to seed roots as a water-based slurry.
DAMPING OFF (*PYTHIUM* SPP.)					
azoxystrobin + mefenoxam (Uniform)	11+4	0.34 fl oz/ 1000 row ft	36	0	Apply as in-furrow spray at planting. Make one application per crops season.
cyazofamid (Ranman 400SC)	21	6.1 fl oz/acre	7	0.5	Apply at planting. Refer to label for details.
ethaboxam (Elumin)	22	8 fl oz/acre	—	0.5	Apply in-furrow or as a side dressing over seed piece. Do not make more than two applications per year or apply more than 16 fl oz/acre/year.
Fluopicolide (Presidio)	43	3 to 4 fl oz/acre	7	0.5	Must be tank mixed with a labeled rate of another fungicide active against the target pathogen, but with a different moved of action. Repeat applications at 10-day intervals.
mefenoxam (Ridomil Gold) 4 SL	4	1 to 2 pt/treated acre	—	2	Incorporate in soil. See label for row rate.
mefenoxam + oxathiapiprolin (Orondis Gold)	4+49	27.8 fl oz/acre	14	2	Apply in-furrow at planting.
metalaxyl (MetaStar) 2 E	4	4 to 8 pt/treated acre	7	2	Preplant incorporated or soil surface spray.
FOLIAR DISEASES (*ALTERNARIA*)					
azoxystrobin (Aframe, generic)	11	6 to 15.5 fl oz/acre	0	4 hr	Limit to 123 fl oz per acre per season. For soilborne disease control, refer to label. Begin foliar applications prior to disease and continue on a 5- to 7-day interval.
azoxystrobin+ difenoconazole (Quadris Top)	11+3	8 to 14 fl oz/acre	14	0.5	Begin foliar applications prior to disease and continue on a 7- to 14-day interval.
cyprodinil + fludioxonil (Switch 62.5WG)	9+12	11 to 14 oz/acre	7	0.5	Begin foliar applications prior to disease and continue on a 7- to 10-day interval.
difenoconazole + benzovindiflupyr (Aprovia Top)	7+3	10.5 to 13.5 fl oz/acre	14	0.5	No more than two applications can be made at a 7-day interval; all other applications must be made at a 14-day interval. Apply no more than 27 fl oz per acre per year.
difenoconazole + tea tree oil (Regev)	3+BM01	4 to 8.5 fl oz/acre	14	0.5	Limit to 34 fl oz/acre per year. Apply in early plant stages and repeat on 7-to 14-day intervals. Do not make more than 2 sequential applications.
fenamidone (Reason 500SC)	11	5.5 to 8.2 fl oz/acre	14	0.5	Begin foliar applications prior to disease and continue on a 5- to 10- day interval.
fluoxastrobin (Aftershock)	11	2 to 3.8 fl oz/acre	7	0.5	Limit to 22.8 fl oz per acre per year. For soilborne disease control, refer to label. Begin foliar applications prior to disease and continue on a 7- to 10-day interval.
picoxystrobin (Aproach)	11	6 to 12 fl oz/acre	7	0.5	Apply at 100% bloom of primary inflorescence, or prior to row closure. Apply again 14 days later. Use higher rate with high disease pressure.
pydiflumetofen + difenoconazole (Miravis Top)	3+7	13.7 fl oz/acre	14	0.5	Apply a sufficient volume of water to ensure good coverage. Do not make more than two applications of Miravis Top or other Group 3 and 7 fungicides before alternation with a fungicide that is not in Group 3 or 7. Do not make more than 4 applications at the maximum application rate per year.
pydiflumetofen + fludioxonil (Miravis Prime)	7+12	6.8 fl oz/acre	7	0.5	Limit to 20.4 fl oz/acre per year. Begin applications prior to disease develop- ment. Continue applications through season on a 7- to 10-day interval.
pyraclostrobin (Cabrio EG)	11	8 to 12 oz/acre	0	0.5	Do not apply more than 48 fl oz per acre per season. Alternate with a fungicide with a different mode of action after each use.
pyrimethanil (Scala SC)	9	7 fl oz/acre	7	0.5	Begin foliar applications prior to disease and continue on a 7- to 14- day interval.
trifloxystrobin (Flint)	11	3 to 3.8 oz/acre	7	0.5	Apply on a 7- to 10-day interval as needed. Do not make more than six applications per year or apply more than 23 fl oz per acre per year.
POSTHARVEST FUSARIUM ROT (*FUSARIUM*)					
azoxystrobin + fludioxonil + difenoconazole (Stadium)	11+12 +3	1.25 fl oz per 2,000 lbs of roots	—	—	Ensure proper coverage, use tumbling, and mix the fungicide solution in suffi- cient water volume. Do not make more than one postharvest application.

TABLE 3-31. DISEASE CONTROL PRODUCTS FOR SWEETPOTATO (cont'd)

L. Quesada-Ocampo and A. Gorny, Plant Pathologists, North Carolina State University

Disease/Material	FRAC Code	Rate of Material Formulation	Minimum Days Harv.	Minimum Days Reentry	Method, Schedule, and Remarks
MOTTLE NECROSIS (PYTHIUM POSTHARVEST)					
potassium phosphite (Alude)	P07	1 1/4 quarts/acre	0	4 hr	Foliar spray at 5- to 14-day intervals depending on disease incidence.
POWDERY MILDEW					
azoxystrobin + difenoconazole (Quadris Top)	11+3	8 to 14 fl oz/acre	14	0.5	Begin foliar applications prior to disease and continue on a 7- to 14-day interval.
cyprodinil + fludioxonil (Switch 62.5WG)	9+12	11 to 14 oz/acre	7	0.5	Begin foliar applications prior to disease and continue on a 7- to 10-day interval.
difenoconazole + benzovindiflupyr (Aprovia Top)	7+3	10.5 to 13.5 fl oz/acre	14	0.5	No more than two applications can be made at a 7-day interval; all other applications must be made at a 14-day interval. Apply no more than 27 fl oz per acre per year.
fluopyram + pyrimethanil (Luna Tranquility)	7+9	11.2 fl oz/acre	7	0.5	Limit to 54.7 fl oz per acre per year. Do not make more than two sequential applications of Group 7-containing fungicides. Labeled for Alternaria and Sclerotinia. Apply at 7- or 14-day intervals.
mefentrifluconazole	3	3 to 5 fl oz/acre	7	0.5	Limit to 15 fl oz/acre per year. Apply on 7-to 14-day intervals.
mefentrifluconazole + pyraclostrobin (Veltyma)	3+11	5 to 10 fl oz/acre	7	0.5	Limit to 30 fl oz/acre per year. Apply before onset of disease and repeat on a minimum interval of 14 days.
metconazole (Quash)	3	2.5 to 4 fl oz/acre	1	0.5	Begin foliar applications prior to disease and continue on a 7- to 10-day interval. Do not apply more than 16 fl oz per year or four times per year. Do not make more than two sequential applications before alternating with products with different modes of action.
penthiopyrad (Vertisan)	7	0.7 to 24 oz/acre	7	0.5	For soilborne disease control, refer to label. Begin foliar applications prior to disease and continue on a 7- to 14-day interval.
pyraclostrobin (Cabrio EG)	11	8 to 12 oz/acre	0	0.5	Do not apply more than 48 fl oz per acre per season. Alternate with a fungicide with a different mode of action after each use.
pydiflumetofen + difenoconazole (Miravis Top)	3+7	13.7 fl oz/acre	14	0.5	Apply a sufficient volume of water to ensure good coverage. Do not make more than two applications of Miravis Top or other Group 3 and 7 fungicides before alternation with a fungicide that is not in Group 3 or 7. Do not make more than 4 applications at the maximum application rate per year.
pydiflumetofen + fludioxonil (Miravis Prime)	7+12	6.8 fl oz/acre	7	0.5	Limit to 20.4 fl oz/acre per year. Begin applications prior to disease development. Continue applications through season on a 7- to 10-day interval.
POSTHARVEST RHIZOPUS SOFT ROT (*RHIZOPUS*)					
azoxystrobin + fludioxonil + difenoconazole (Stadium)	11+12 +3	1.0 fl oz per 2,000 lbs of roots	—	—	Section 2 (ee) label for NC and CA expires 12/31/2025. Ensure proper coverage, use tumbling, and mix the fungicide solution in sufficient water volume. Do not make more than one postharvest application.
dicloran (Botran 5F)	14	0.6 qt/100 gal	—	—	Spray or dip. Dip for 5 to 10 seconds in well-agitated suspension. Add ½-pound Botran to 100 gallons of treating suspension after 500 bushels treated. Do not rinse.
fludioxonil (Scholar 1.9 SC)	12	16 to 32 fl oz/100 gal	—	—	Dip for approximately 30 seconds in well-agitated solution and allow sweetpotatoes to drain. Add 8 fl oz to 100 gals after 500 bushels are treated. ALTERNATIVELY, mix 16 fl oz in 7 to 25 gal of water, wax/emulsion, or aqueous dilution of wax/ oil emulsion. Can also be used to disinfest tanks, refer to label.
WHITE MOLD (*SCLEROTINIA*)					
Bacillus amyloliquefaciens (Serifel)	BM02	4 to 16 oz/acre	0	4 hr	Begin foliar applications shortly after emergence or transplanting and continue on 7- to 10-day intervals
boscalid (Endura)	7	5.5 to 10 oz/acre	10	0.5	Begin applications prior to disease development and apply again at a 7- to 14-day interval. Do not apply more than 20 fl oz per year. Do not make more than two sequential applications before alternating with products with different modes of action.
fluazinam (Ventana)	29	5.5 to 8 fl oz/acre	14	12	Limit to 56 fl oz/acre per year. Begin applications when plants 6 to 8 inches tall and continue on a 7-to-14-day interval.
pydiflumetofen + difenoconazole (Miravis Top)	3+7	13.7 fl oz/acre	14	0.5	Apply a sufficient volume of water to ensure good coverage. Do not make more than two applications of Miravis Top or other Group 3 and 7 fungicides before alternating with fungicide not in Group 3 or 7. Do not make more than 4 applications at maximum application rate per year.
Coniothyrium minitans (Contans WG)	BM02	1 to 4 lb/acre	0	4 hr	**OMRI listed product.** Apply to soil surface and incorporate no deeper than 2 inches. Works best when applied prior to planting or transplanting. Do not apply other fungicides for 3 weeks after applying Contans.
fluazinam (Omega 500F, Lektivar 40SC)	29	5.5 to 8 fl oz/acre	14	0.5	Initiate applications when conditions are favorable for disease development or when disease symptoms first appear. Repeat applications on a 7- to 10-day schedule. Do not apply more than 3.5 pints per year.

TABLE 3-31. DISEASE CONTROL PRODUCTS FOR SWEETPOTATO (cont'd)

L. Quesada-Ocampo and A. Gorny, Plant Pathologists, North Carolina State University

Disease/Material	FRAC Code	Rate of Material Formulation	Minimum Days Harv.	Minimum Days Reentry	Method, Schedule, and Remarks
WHITE MOLD (*SCLEROTINIA*) (cont'd)					
metconazole (Quash)	3	4 oz/acre	1	0.5	Make an application prior to disease development and apply again 14 days later. Do not apply more than 16 fl oz per year or four times per year. Do not make more than two sequential applications before alternating with products with different modes of action.
pydiflumetofen + difenoconazole (Miravis Top)	3+7	13.7 fl oz/acre	14	0.5	Apply a sufficient volume of water to ensure good coverage. Do not make more than two applications of Miravis Top or other Group 3 and 7 fungicides before alternation with a fungicide that is not in Group 3 or 7. Do not make more than 4 applications at the maximum application rate per year.
SCURF (*MONILOCHAETES INFUSCANS*) AND SCLEROTIAL BLIGHT (*SCLEROTIUM ROLFSII*)					
dicloran (Botran 5F)	14	0.6 qt/100 gal	—	—	**Seed dip:** Dip seed sweetpotatoes 10 to 15 seconds in a well-agitated fungicide suspension. Drain sweetpotatoes and bed promptly. Prepare fresh fungicide suspension daily. **Plant bed spray:** For a 42 in plant bed mix 5.73 oz in 14 gal water per 1000 linear ft/plant bed. Do not exceed 3.2 qt/acre.
thiabendazole (Mertect 340 F)	1	107 fl oz/100 gal	0.5	0.5	Dip seed roots 1 to 2 minutes and plant immediately.
SOUTHERN BLIGHT (*SCLEROTIUM ROLFSII*)					
dicloran (Botran) 75 W	14	1 lb/100 gal	—	—	**Seed dip:** Dip seed sweetpotatoes 10 to 15 seconds in a well-agitated fungicide suspension. Drain sweetpotatoes and bed promptly. Prepare fresh fungicide suspension daily. **Plant bed spray:** Spray or sprinkle over bedded sweetpotatoes before covering them with soil.
difenoconazole + benzovindiflupyr (Aprovia Top)	7+3	10.5 to 13.5 fl oz/acre	14	0.5	No more than two applications can be made at a 7-day interval; all other applications must be made at a 14-day interval. Apply no more than 27 fl oz per acre per year.
WHITE RUST (*ALBUGO IPOMOEA-PANDURATAE*)					
azoxystrobin (Quadris)	11	6.2 to 15.4 fl oz/acre	7	4 hr	Make no more than two sequential applications before alternating with fungicides that have a different mode of action. Apply no more than 2.88 quarts per crop per acre per season.
fenamidone (Reason 500SC)	11	5.5 to 8.2 fl oz/acre	14	0.5	Begin applications when conditions favor disease development and continue on 5- to 10-day interval. Do not apply more than 16.4 fluid ounces per growing season. Alternate with a fungicide from different resistance management group.
pyraclostrobin (Cabrio EG)	11	8 to 16 oz/acre	0	0.5	Do not apply more than 48 ounces per acre per season. Alternate with a fungicide with a different mode of action after each use.

TABLE 3-32. SWEETPOTATO STORAGE HOUSE SANITATION

L. Quesada-Ocampo, Plant Pathologist, North Carolina State University

Material	Rate per 1,000 Cubic Feet of Space	Methods and Remarks
Heat	140°F 4 to 8 hr/day for 7 days or 180°F for 30 min	See remarks under water, produce, and equipment sanitation. The storage house, ventilation system, and equipment must be exceptionally clean and moist during the procedure. *Caution:* rot-causing organisms inside a drain will probably not be exposed to lethal temperature.

TABLE 3-33. EFFICACY OF PRODUCTS FOR DISEASE CONTROL IN SWEETPOTATO

L. Quesada-Ocampo, Plant Pathologist, North Carolina State University

Scale: E, excellent; G, good; F, fair; P, poor; NC, no control; ND, no data.

Active Ingredient	Product	Fungicide Group	Nematicide (N) or Fungicide (F)	Alternaria leaf spot	Black rot (C. fimbriata)	Fusarium	Java black rot (D. gossypina)	Nematodes	Pythium	Rhizopus soft rot (R. stolonifer)	Southern blight (S. rolfsii)	Sclerotinia	Scurf (M. infuscans)	Soil rot/Pox (S. ipomoea)
azoxystrobin + fludioxonil + difenoconazole	Stadium	11+12+3	F	ND	E	ND	ND	ND	ND	E	ND	ND	ND	ND
boscalid	Endura	7	F	ND	ND	ND	ND	ND	ND	ND	ND	E	ND	ND
chlorine sanitizer postharvest		N/A	F	ND	F	ND	P	ND	ND	F	ND	ND	P	NC
chloropicrin		N/A	N, F	ND	P	F	F	F	ND	ND	F	ND	ND	F
Coniothyrium minitans	Contans WG	BM02	F	ND	ND	ND	ND	ND	ND	ND	F	ND	ND	ND
dicloran	Botran 75W	14	F	ND	P	ND	P	ND	ND	F	P	G	F	NC
1,3-dichloropropene	Telone II	N/A	N	ND	ND	P	ND	G	ND	ND	ND	ND	ND	ND
ethaboxam	Elumin	22	F	ND	ND	ND	ND	ND	G	ND	ND	ND	ND	ND
ethoprop	Mocap	N/A	N	ND	ND	ND	ND	P	ND	ND	ND	ND	ND	ND
fludioxonil	Scholar	12	F	ND	F	ND	ND	NC	ND	F	ND	ND	ND	NC
fluopicolide	Presidio	43	F	ND	ND	ND	ND	ND	G	ND	ND	ND	ND	ND
fluopyram	Velum Prime	7	N, F	ND	ND	ND	ND	G	ND	ND	ND	ND	ND	ND
mefenoxam	Ridomil Gold	4	F	ND	ND	ND	ND	ND	G	ND	ND	ND	ND	ND
metconazole	Quash	3	F	ND	E	ND	ND	ND	ND	ND	ND	ND	ND	ND
metam sodium	Vapam	N/A	N	ND	P	F	ND	F	ND	ND	ND	ND	ND	ND
metalaxyl	MetaStar	4	F	ND	ND	ND	ND	ND	F	ND	ND	ND	ND	ND
oxamyl	Vydate	N/A	N	ND	ND	ND	ND	F	ND	ND	ND	ND	ND	ND
Pseudomonas syringae	Bio-Save	N/A	F	ND	ND	ND	ND	ND	ND	P	ND	ND	ND	ND
thiabendazole	Mertect 340-F	1	F	ND	E	P	F	ND	ND	E	F	ND	P	NC

TABLE 3-34. IMPORTANCE OF ALTERNATIVE MANAGEMENT PRACTICES FOR DISEASE CONTROL IN SWEETPOTATO

L. Quesada-Ocampo, Plant Pathologist, North Carolina State University

Scale: E, excellent; G, good; F, fair; P, poor; NC, no control; ND, no data

Strategy	Alternaria leaf spot	Black rot	Fusarium	Java black rot (D. gossypina)	Nematodes	Pythium	Rhizopus soft rot (R. stolonifer)	Sclerotinia	Southern blight	Scurf (M. infuscans)	Soil rot/Pox (S. ipomoea)	Sweetpotato Feathery Mottle virus
Crop rotation (3 to 4 years)	P	F	F	F	F	P	NC	F	F	P	F	NC
Disease-free planting stock	NC	E	G	G	F	P	NC	NC	P	E	P	G
Resistant cultivars	F	NC	F	F[s]	F[s]	P	F	F	F	P	G	F
Careful handling to reduce mechanical injury	NC	F	F	NC	NC	P	E	F	NC	NC	NC	NC
Cutting plants (in beds) above soil line	NC	G	F	F	G[x]	P	NC	NC	NC	G	P	NC
Soil sample for nematode analysis	NC	NC	NC	NC	E	P	NC	NC	NC	NC	NC	NC
Sanitation (equipment, fields, storage houses)	F	E	P	F	NC	P	E	NC	NC	P	NC	NC
Manage insects that transmit pathogens	NC	P	NC	NC	NC	P	NC	NC	NC	NC	NC	E
Sulfur added to soil to reduce pH	NC	NC	NC	NC	NC	P	NC	NC	NC	NC	F	NC
Prompt curing and proper storage conditions	NC	E	F	F	NC	P	E	NC	NC	NC	NC	NC
Site selection (drainage)	P	NC	F	F	NC	E	F	G	P	NC	P	NC
Manage insects that cause feeding injuries to roots	NC	G	NC	P	NC	P	P	NC	NC	NC	NC	NC
Avoid harvesting when soils are wet	F	G	NC	F	NC	G	F	F	NC	NC	NC	NC

[s] Resistant cultivars for root knot nematode are susceptible to reniform nematode. Cultivars resistant to Southern root-knot nematode (*Meloidogyne incognita*) may not be resistant to Guava root-knot nematode (*Meloidogyne enterolobii*).

[x] Cutting plants above soil line provides good reduction in movement in nematodes but will not control nematodes already established in soil.

TABLE 3-35. DISEASE CONTROL PRODUCTS FOR TOMATILLO

I. Meadows, Extension Plant Pathologist, North Carolina State University

Disease/Material	FRAC Code	Rate of Material Formulation	Minimum Days Harv.	Minimum Days Reentry	Method, Schedule, and Remarks
EARLY BLIGHT (*ALTERNARIA*)					
azoxystrobin (various)	11	5 to 6.2 fl oz/acre	0	4 h	Limit of 37 fl oz per crop/acre/season. Make no more than **one** application before alternating with fungicides that have a different mode of action. A tank mixture with Dimethoate or some adjuvants may cause crop injury—see label.
azoxystrobin + chlorothalonil (Quadris Opti)	11+M05	1.6 pt/acre	0	0.5	Make no more than **one** application before alternating with fungicides that have a different mode of action. Do not apply until 21 days after transplant or within +/- 6 days of a postemergence broadcast application of Sencor.
azoxystrobin + difenoconazole (Quadris Top 2.72F)	11+3	8 fl oz/acre	0	0.5	Limit of 47 fl oz/acre/season. Do not apply until 21 days after transplanting or 35 days after seeding.
azoxystrobin + flutriafol (Topguard EQ)	11+3	4 to 8 fl oz/acre	0	0.5	Use of some adjuvants may cause crop injury – see label. Do not exceed 4 applications/year.
boscalid (Endura 70WDG)	7	2.5 to 3.5 oz/acre	0	0.5	Limit of 21 oz/acre/season. Do not exceed 6 applications/season. Make no more than two sequential applications before alternating with fungicides that have a different mode of action.
cyprodinil + difenoconazole (Inspire Super 2.82F)	9+3	16 to 20 fl oz/acre	0	0.5	Limit of 80 fl oz/acre/season.
cyprodinil + fludioxonil (Alterity 62.5WG, Switch 62.5WG)	9+12	11 to 14 oz/acre	0	0.5	Limit of 56 oz/acre/year. After two applications, rotate to another fungicide with a different mode of action for two applications.
difenoconazole + benzovindiflupyr (Aprovia Top)	3+7	10.5 to 13.5 fl oz/acre	0	0.5	Apply on a 7- to 14-day interval. No more than 2 consecutive applications allowed before switching to a non-FRAC 7. Use of a spreading adjuvant is recommended.
difenoconazole + mandipropamid (Revus Top 4.16F)	3+40	5.5 to 7 fl oz/acre	1	0.5	Limit of 28 fl oz/acre/season. Make no more than 2 consecutive applications/season before alternating with fungicides that have a different mode of action.
difenoconazole + tea tree oil (Regev)	3+BM01	4.0 to 8.5 fl oz/acre	2	0.5	Minimum application interval is 7 days. Do not make more than 2 sequential applications before alternating to a fungicide with a different mode of action. Maximum applications is 4/year.
fenamidone (Reason 500SC)	11	5.5 to 8.2 fl oz/acre	14	0.5	Limit of 24.6 fl oz/growing season. Make no more than **one** application before rotating to another effective fungicide with a different MOA.
fluopyram + difenoconazole (Luna Flex)	7+3	8.0 to13.6 fl oz/acre	0	0.5	Maximum 2 applications/year at high rate. Maximum interval between applications is 7 days.
fluopyram + trifloxystrobin (Luna Sensation)	7+11	7.6 fl oz/acre	3	0.5	Apply on a 7- to 14-day interval. Do not exceed 27.1 fl oz/acre/year or 5 applications.
fluoxastrobin (Aftershock, Evito 480SC)	11	2.0 to 5.7 fl oz/acre	3	0.5	Limit of 22.8 fl oz/acre/season. Make no more than **one** application before alternating with fungicides that have a different mode of action. **NOTE: Do not overhead irrigate for 24 hours following a spray application.**
flutriafol (Rhyme)	3	3.5 to 7 fl oz/acre	0	0.5	Tank mix with mancozeb for improved disease control. Limit to 4 applications/year.
mancozeb + zoxamide (Gavel 75DF)	M03+22	1.5 to 2 lb/acre	5	2	Limit to 8 applications/year.
mefentrifluconazole (Cevya)	3	3 to 5 fl oz/acre	0	0.5	Maximum of 15 fl oz/acre/year.
penthiopyrad (Fontelis 1.67F)	7	16 to 24 fl oz/acre	0	0.5	Do not exceed 72 fl oz of product/year. Make no more than two sequential applications/season before alternating with fungicides that have a different mode of action.
polyoxin D zinc salt (Ph-D; OSO 5% SC) (OSO)	19	6.2 oz/acre (Ph-D) 3.75 to 13.0 fl oz/acre (OSO)	0	4 h	Limit of five applications/season. Make no more than **one** application before alternating with fungicides that have a different mode of action.
pyraclostrobin (Cabrio 20EG)	11	8 to 12 oz/acre	0	0.5	Limit of 96 oz/acre/season. Make no more than **one** application before alternating with fungicides that have a different mode of action.
pyrimethanil (Scala)	9	7 fl oz/acre	1	0.5	Do not apply more than 35 fl oz/acre/crop.
tetraconazole (Mettle125 ME)	3	6 to 8 fl oz/acre	0	0.5	Make no more than 2 sequential applications before switching to a fungicide with a different mode of action. Can be applied by ground, air, or chemigation. Do not apply more than 16 fl oz or exceed 5 applications of Mettle/acre/year. Do not apply more than 0.125 lb/acre/year of tetraconazole containing products.
trifloxystrobin (various)	11	See label	3	0.5	Application limits apply – see label. Make no more than **one** application before alternating with a fungicide that has a different mode of action.

TABLE 3-35. DISEASE CONTROL PRODUCTS FOR TOMATILLO (cont'd)

I. Meadows, Extension Plant Pathologist, North Carolina State University

Disease/Material	FRAC Code	Rate of Material Formulation	Minimum Days Harv.	Minimum Days Reentry	Method, Schedule, and Remarks
POWDERY MILDEW (*LEVEILULLA*, *OIDIUM*)					
azoxystrobin (various)	11	5 to 6.2 fl oz/acre	0	4 hr	Limit of 37 fl oz/crop/acre/season. Make no more than **one** application before alternating with fungicides with a different MOA.
azoxystrobin + chlorothalonil (Quadris Opti)	11+M05	1.6 pt/acre	0	0.5	Make no more than **one** application before alternating with fungicides that have a different mode of action. Do not apply until 21 days after transplant or within +/- 6 days of a postemergence broadcast application of Sencor. Do not use an adjuvant.
azoxystrobin + difenoconazole (Quadris Top 2.72F)	11+3	8 fl oz/acre	0	0.5	Limit of 47 fl oz/acre/season. Do not apply until 21 days after transplanting or 35 days after seeding. Do not use an adjuvant or tank mix with any EC product. Tank mixing with Dimethoate may cause crop injury.
azoxystrobin + flutriafol (Topguard EQ)	11+3	4 to 8 fl oz/acre	0	0.5	Use of some adjuvants may cause crop injury – see label.
chlorothalonil (Bravo Weather Stick 6F)	M05	1.5 pt/acre	3	0.5	Limit of 12 pints/acre/season.
chlorothalonil + cymoxanil (Ariston 4.34F, Cymbol Advance)	M05+27	2 to 2.44 pt/acre	3	0.5	Limit of 17.5 pt/acre/year.
cyflufenamid (Fastback)	U06	4 oz/acre	0	4 hr	Make no more than 3 applications/year. Minimum application interval is 14 days.
cyprodinil + difenoconazole (Inspire Super 2.82F)	9+3	16 to 20 fl oz/acre	0	0.5	Limit of 80 fl oz/acre/season.
cyprodinil + fludioxonil (Alterity 62.5WG, Switch 62.5WG)	9+12	11 to 14 oz/acre	0	0.5	Limit 56 oz/acre/year. After two applications, rotate to another fungicide with a different mode of action for two applications.
difenoconazole + benzovindiflupyr (Aprovia Top)	3+7	10.5 to 13.5 fl oz/acre	0	0.5	Apply on a 7- to 14-day interval. No more than 2 consecutive applications allowed before switching to a non-FRAC 7. Use of a spreading adjuvant is recommended.
difenoconazole + tea tree oil (Regev)	3+BM01	4.0 to 8.5 fl oz/acre	2	0.5	Minimum application interval is 7 days. Do not make more than 2 sequential applications before alternating to a fungicide with a different mode of action. Maximum applications is 4/year.
fluopyram + difenoconazole (Luna Flex)	7+3	8.0 to 13.6 fl oz/acre	0	0.5	Maximum 2 applications/year at high rate. Maximum interval between applications is 7 days.
fluopyram + trifloxystrobin (Luna Sensation)	7+11	5 to 7.6 fl oz/acre	3	0.5	Apply on a 7- to 14-day interval. Rotate away from FRAC 7 and 11 after 2 applications. Limit of 5 applications/year.
flutriafol (Rhyme)	3	5 to 7 fl oz/acre	0	0.5	Tank mix with mancozeb for improved disease control. Limit to 4 applications/year.
mefentrifluconazole (Cevya)	3	3 to 5 fl oz/acre	0	0.5	Maximum of 15 fl oz/acre/year.
oxathiapiprolin + chlorothalonil (Orondis Opti)	49+M05	1.75 to 2.5 pt/acre	3	0.5	Do not follow soil applications with foliar applications of any oxathiapiprolin containing products. See label for season and application limits.
penthiopyrad (Fontelis 1.67F)	7	16 to 24 fl oz/acre	0	0.5	Do not exceed 72 fl oz of product/year. Make no more than two consecutive applications/season before rotating to a fungicide with a different mode of action.
polyoxin D (Ph-D 11.3WDG)	19	6.2 oz/acre	0	4 h	Limit of five applications/season. Make no more than **one** application before alternating with fungicides that have a different mode of action.
pydiflumetofen + fludioxonil (Miravis Prime)	7+12	9.2 to 11.4 fl oz/acre	0	0.5	Do not make more than two applications of Miravis Prime or other Group 7 and 12 fungicides before alternation with a fungicide that is not in Group 7 or 12. Do not exceed 22.8 fl oz/acre/season. Apply by ground, air, or chemigation.
pyraclostrobin (Cabrio 20EG)	11	8 to 16 oz/acre	0	0.5	Limit of 96 oz/acre/season. Make no more than **one** application before alternating with fungicides that have a different mode of action.
mandipropamid + difenoconazole (Revus Top 4.16F)	40+3	5.5 to 7 fl oz/acre	0	0.5	Limit of 28 fl oz/acre/season. Make no more than two consecutive applications before alternating with fungicides that have a different mode of action.
tetraconazole (Mettle125 ME)	3	6 to 8 fl oz/acre	0	0.5	Make no more than 2 sequential applications before switching to a fungicide with a different mode of action. Can be applied by ground, air, or chemigation. Do not apply more than 16 fl oz or exceed 5 applications of Mettle/acre/year. Do not apply more than 0.125 lb/acre/year of tetraconazole containing products.
tolfenpyrad (Torac)	39	21 fl oz/acre	1	0.5	**DISEASE SUPPRESSION ONLY.** See label for application intervals and season limits.
trifloxystrobin (Flint Extra)	11	3 fl oz/acre	3	0.5	**DISEASE SUPPRESSION ONLY.** Application limits apply – see label. Make no more than **one** application before alternating with a fungicide that has a different mode of action.

TABLE 3-36. DISEASE CONTROL PRODUCTS FOR TOMATO

I. Meadows, Extension Plant Pathologist, North Carolina State University; R. Singh, Plant Pathologist, Louisiana State University Agricultural Center

Disease/Material	FRAC Code	Rate of Material Formulation	Minimum Days Harv.	Minimum Days Reentry	Method, Schedule, and Remarks
TOMATO (TRANSPLANTS produced in a greenhouse or other controlled environment)					
It is strongly recommended to treat seed to eliminate plant pathogens on or within the seed. See section on SEED TREATMENTS for information on treating raw(naked) seed.					
BACTERIAL CANKER (CLAVIBACTER)					
bacteriophage (AgriPhage-CMM)	No code	1 pt/12 to 25 gal of water/9,600 sq ft	NA	4 hr	Consult your vegetable Extension Specialist for information on requirements needed to use bacteriophage. Bacteriophages are most effective when applied during or after last watering of the day.
sodium hypochlorite (CPPC Ultra Bleach 2; 6.15%)	NC	1 qt+4 qt water	NA	0	Wash seed for 40 min in solution with continuous agitation; air dry promptly. Use 1 gal of solution/1 lb seed. **NOTE: Ultra Bleach 2 seed treatment is not compatible with pelleted (coated) seed.**
streptomycin sulfate (various)	25	1 lb/100 gal	NA	0	Begin application at first true leaf stage; repeat weekly until transplant.
BACTERIAL SPOT (XANTHOMONAS), BACTERIAL SPECK (PSEUDOMONAS)					
bacteriophage (AgriPhage)	No code	1 pt/12 to 25 gal of water/9,600 sq ft	NA	4 hr	Consult your vegetable Extension Specialist for information on requirements needed to use bacteriophage. Bacteriophages are most effective when applied during or after last watering of the day.
copper (various)	M01	See label	NA	2	Begin application at first true leaf stage, repeat at 3- to 7-day intervals until transplanting. Alternating with streptomycin sulfate is recommended.
mancozeb (various)	M03	See label	NA	1	See label for state restrictions and rates. **NOTE: Use a full rate of fixed copper in combination with mancozeb. Mancozeb alone does not control bacteria.**
streptomycin sulfate (various)	25	1 lb/100 gal	NA	0.5	Begin application at first true leaf stage, repeat weekly until transplanting. For plant bed use only.
EARLY BLIGHT (ALTERNARIA), GRAY MOLD (BOTRYTIS), LATE BLIGHT (PHYTOPHTHORA INFESTANS)					
mancozeb (various)	M03	See label	NA	1	See label for state restrictions and rates. **NOTE: Use full rate of fixed copper in combination with mancozeb if bacteria control required.**
GRAY MOLD (BOTRYTIS), BOTRYTIS STEM CANKER, EARLY BLIGHT (ALTERNARIA), POWDERY MILDEW (ERYSIPHE, OIDIUM)					
cyprodinil + fludioxonil (Alterity 62.5WG, Switch 62.5WG)	9+12	11 to 14 oz/acre	NA	0.5	**DO NOT APPLY TO GRAPE OR CHERRY TOMATO.** After 2 applications, switch to a different mode of action for 2 applications.
fluopyram + pyrimethanil (Luna Tranquility)	7+9	11.2 fl oz/acre	NA	0.5	See label for limits on application amounts/season. Do not make more than 2 applications of Group 7 or 9 fungicides without switching to a different mode of action. **NOT REGISTERED FOR USE IN LA.**
penthiopyrad (Fontelis 1.67SC)	7	0.5 to 0.75 fl oz/gal	NA	0.5	Use 1 gallon of spray/1,360 ft^2. Do not make more than 2 applications before switching to a different mode of action.
LATE BLIGHT (PHYTOPHTHORA INFESTANS)					
mandipropamid (Micora)	40	5.5 to 8.0 fl oz/acre (5,000 ft^2)	NA	4 h	Apply no more than 2 applications before switching to another mode of action.
propamocarb (Previcur Flex 6F)	28	0.7 to 1.5 pt/acre	NA	0.5	Can be used as a drench before or after transplanting.
PYTHIUM DAMPING-OFF (PYTHIUM)					
cyazofamid (Ranman 400SC)	21	3.0 fl oz/100 gal	NA	0.5	Apply as a soil drench to seedling tray or at the time of transplant.
propamocarb (Previcur Flex 6F)	28	1.5 pt/acre	NA	0.5	Limit of 7.5 pt/acre/season. Do not make more than 1 application before alternating to a fungicide with a different mode of action.
TOMATO (FIELD)					
ANTHRACNOSE (COLLETOTRICHUM)					
azoxystrobin (various)	11	See label	0	4 h	See label. Do not make more than 1 application before alternating to a fungicide with a different mode of action. A tank mixture with Dimethoate or some adjuvants may cause crop injury—see label.
azoxystrobin + chlorothalonil (Quadris Opti)	11+M05	1.6 pt/acre	0	0.5	Do not make more than 1 application before alternating to a fungicide with a different mode of action. Do not apply within 21 days after transplanting or 35 days after seeding. Do not apply within +/- 6 days of a postemergence application of Sencor. Do not use an adjuvant.
azoxystrobin + difenoconazole (Quadris Top)	11+3	8 fl oz/acre	0	0.5	Do not make more than 1 application before alternating to a fungicide with a different mode of action. Limit 47 fl oz/acre/season. Do not apply within 21 days after transplanting or 35 days after seeding. Do not use an adjuvant or tank mix with any EC product. Tank mixing with Dimethoate may cause crop injury.
azoxystrobin + flutriafol (Topguard EQ)	11+3	4 to 8 fl oz	0	0.5	Do not use adjuvants or EC formulated tank-mix partners on fresh market tomatoes. Do not exceed 4 applications/year or 8 fl oz of product/acre. Limits of each a.i. apply – see label.

TABLE 3-36. DISEASE CONTROL PRODUCTS FOR TOMATO (cont'd)

I. Meadows, Extension Plant Pathologist, North Carolina State University; R. Singh, Plant Pathologist, Louisiana State University Agricultural Center

Disease/Material	FRAC Code	Rate of Material Formulation	Minimum Days Harv.	Minimum Days Reentry	Method, Schedule, and Remarks
TOMATO (FIELD) (cont'd)					
ANTHRACNOSE (*COLLETOTRICHUM*) (cont'd)					
chlorothalonil (various)	M05	See label	0	0.5	Refer to individual labels for rates and restrictions.
copper (various)	M01	See label	0	2	Refer to individual labels for rates and restrictions.
cymoxanil + chlorothalonil (Ariston, Cymbol Advance)	27+ M05	1.9 pt/acre	3	0.5	Check copper labels for specific precautions and limitations for mixing with this product.
difenoconazole + benzovindiflupyr (Aprovia Top)	7+3	10.5 to 13.5 fl oz/acre	0	0.5	Do not make more than 2 consecutive applications before alternating to a non-Group 7 fungicide. Limit of 53.6 fl oz/acre/season. Limits of each a.i. application – see label. Use of a spreading adjuvant is recommended.
difenoconazole + cyprodinil (Inspire Super)	3+9	16 to 20 fl oz/acre	0	0.5	Limit of 80 fl oz/acre/season. Do not make more than 2 consecutive applications before alternating to a fungicide with a different mode of action. Limits of each a.i. application – see label.
difenoconazole + tea tree oil (Regev)	3+BM01	4.0 to 8.5 fl oz/acre	2	0.5	Minimum application interval is 7 days. Do not make more than 2 sequential applications before alternating to a fungicide with a different mode of action. Do not exceed 4 applications/year.
famoxadone + cymoxanil (Tanos)	11+27	8 oz/acre	3	0.5	Limit of 72 fl oz/acre/12-month period. Do not make more than one application before alternating to a fungicide with a different mode of action. **NOTE: Must be tank mixed with a contact fungicide that has a different mode of action.**
fluopyram + difenoconazole (Luna Flex)	7+3	8.0 to 13.6 fl oz/acre	0	0.5	Maximum 2 applications/year at high rate. Maximum interval between applications is 7 days.
fluopyram + trifloxystrobin (Luna Sensation)	7+11	7.6 fl oz/acre	3	0.5	**DISEASE SUPPRESSION ONLY.** Do not exceed 5 applications or 27.1 fl oz/ acre/season. Do not make more than 2 consecutive applications before alternating to a fungicide with a different mode of action.
flutriafol (Rhyme)	3	5 to 7 fl oz/acre	0	0.5	Do not exceed 4 applications or 28 fl oz of product/acre/season.
fluxapyroxad + pyraclostrobin (Priaxor Xemium)	7+11	4 to 8 fl oz/acre	0	0.5	Limit of 24 fl oz/acre/season. Do not make more than 2 consecutive applications before alternating to fungicide with different mode of action.
mancozeb (various)	M03	See label	5	1	See label for rates.
mancozeb + azoxystrobin (Dexter Max)	M03+11	0.8 to 1.6 lb/acre	5	1	For states East of the Mississippi including Mississippi, do not exceed 12 lb of product/acre/season. States West of Mississippi use 0.8 to 1.1 lb/acre and do not exceed 9.14 lb of product/acre/season. Do not exceed 12 lb of product/acre/ season. Do not make more than 1 application before alternating with a fungicide not in Group 11. On fresh market tomato, do not tank-mix with an adjuvant or an EC formulation. **Tank mixture with dimethoate may cause crop injury.**
mandipropamid + difenoconazole (Revus Top)	40+3	5.5 to 7 fl oz/acre	1	0.5	Limit of 28 fl oz/acre/season. Do not make more than 2 consecutive applications before alternating to a fungicide with a different mode of action. Limits of each a.i. apply – see label.
mefenoxam + chlorothalonil (Ridomil Gold Bravo SC)	4+M05	2.5 to 3.25 pt/acre	7	2	Apply a protectant between applications of Ridomil Gold Bravo SC.
mefentrifluconazole (Cevya)	3	3 to 5 fl oz/acre	0	0.5	Maximum of 15 fl oz/acre/year.
oxathiapiprolin + chlorothalonil (Orondis Opti)	49+M05	1.75 to 2.5 pt/acre	3	0.5	Do not follow soil applications with foliar applications of any oxathiapiprolin containing products. See label for season and application limits.
penthiopyrad (Fontelis 1.67SC)	7	24 fl oz/acre	0	0.5	**DISEASE SUPPRESSION ONLY.** Limit of 72 fl oz/acre/season. Do not make more than 2 consecutive applications before alternating to a fungicide with a different mode of action.
pydiflumetofen + fludioxonil (Miravis Prime)	7+12	9.2 to 11.4 fl oz/acre	0	0.5	Do not make more than 2 applications of Miravis Prime or other Group 7 and 12 fungicides before alternating to a fungicide that is not in Groups 7 or 12. Do not exceed 22.8 fl oz/acre/season. Apply by ground, air, or chemigation.
pyraclostrobin (Cabrio EG)	11	8 to 12 oz/acre	0	0.5	Limit of 96 fl oz/acre/season. Do not make more than 2 applications before alternating to a fungicide with a different mode of action.
tetraconazole (Mettle125ME)	3	6 to 8 fl oz/acre	0	0.5	Make no more than 2 sequential applications before switching to a fungicide with a different mode of action. Can be applied by ground, air, or chemigation. Do not apply more than 16 fl oz or exceed 5 applications of Mettle/acre/year. Do not apply more than 0.125 lb/acre/year of tetraconazole containing products.
trifloxystrobin (various)	11	See label	3	0.5	**DISEASE SUPPRESSION ONLY.** Seasonal limits apply – see label. Do not make more than 1 application before alternating to a fungicide with a different mode of action.

TABLE 3-36. DISEASE CONTROL PRODUCTS FOR TOMATO (cont'd)

I. Meadows, Extension Plant Pathologist, North Carolina State University; R. Singh, Plant Pathologist, Louisiana State University Agricultural Center

Disease/Material	FRAC Code	Rate of Material Formulation	Minimum Days Harv.	Minimum Days Reentry	Method, Schedule, and Remarks
TOMATO (FIELD) (cont'd)					
BACTERIAL SPOT *(XANTHOMONAS)*, BACTERIAL SPECK *(PSEUDOMONAS)*					
acibenzolar-*S*-methyl (Actigard 50WG)	P01	0.33 to 0.75 oz/acre	14	0.5	Should only be applied to healthy, actively growing plants. Do not exceed 8 applications/crop/season.
Bacillus mycoides isolate J (LifeGard WG)	P06	4.5 oz/100 gal	0	4 h	Apply before or immediately after transplant. Repeat on a 7-day interval.
bacteriophage (AgriPhage)	NC	3 to 8 oz/9,600 sq ft	0	0	Consult your vegetable Extension Specialist for information on requirements needed to use bacteriophage. Bacteriophages are most effective when applied during or after last watering of the day.
copper (various)	M01	See label	0	0	Use a full rate of fixed copper in combination with mancozeb for best results.
mancozeb (various)	M03	See label	5	1	For states East of the Mississippi, use 1.5 to 3 lb of product/acre. States West of the Mississippi use 1.5 to lb of product/acre. **NOTE: Use a full rate of fixed copper in combination with mancozeb. Mancozeb alone does not control bacteria.**
mancozeb + azoxystrobin (Gavel 75DF)	M03+11	2.0 lb/acre	5	2	Limit to 8 applications/year.
GRAY MOLD *(BOTRYTIS)*					
boscalid (Endura 70 WDG)	7	9 to 12.5 oz/acre	0	0.5	Limit of 25 oz/acre/season. Do not make more than 2 consecutive applications and no more than 2 applications/crop/year.
chlorothalonil (various)	M05	See label	0	0.5	Refer to individual labels for rates and restrictions.
chlorothalonil + cymoxanil (Ariston, Cymbol Advance)	M05+27	1.9 pt/acre	3	0.5	Limit of 17.5 pt/acre/12-month period. If mixing with copper, check copper label for limitations.
cyprodinil + fludioxonil (Alterity 62.5WG, Switch 62.5WG)	9+12	11 to 14 oz/acre	0	0.5	Limit of 56 oz/acre/season. Do not make more than 2 consecutive applications before alternating to a fungicide with a different mode of action for 2 applications. Limits of each a.i. application – see label.
difenoconazole + cyprodinil (Inspire Super)	3+9	16 to 20 fl oz/acre	0	0.5	Limit of 80 fl oz/acre/season. Do not make more than 2 consecutive applications before alternating to a fungicide with a different mode of action.
difenoconazole + tea tree oil (Regev)	3+BM01	4.0 to 8.5 fl oz/acre	2	0.5	Minimum application interval is 7 days. Do not make more than 2 sequential applications before alternating to a fungicide with a different mode of action. Maximum applications are 4/year.
fluopyram + difenoconazole (Luna Flex)	7+3	8.0 to 13.6 fl oz/acre	0	0.5	Maximum 2 applications/year at high rate. Maximum interval between applications is 7 days.
fluopyram + pyrimethanil (Luna Tranquility)	7+9	11.2 fl oz/acre	1	0.5	See label for limits on application amounts/season. Do not make more than 2 applications of Group 7 or 9 fungicides without switching to a different mode of action. **NOT REGISTERED FOR USE IN LA.**
fluopyram + trifloxystrobin (Luna Sensation)	7+11	7.6 fl oz/acre	3	0.5	Do not exceed 5 applications or 27.1 fl oz/acre/season. Do not make more than 2 consecutive applications before alternating to fungicide with different mode of action. Limits of each a.i. application – see label.
fluxapyroxad + pyraclostrobin (Priaxor Xemium)	7+11	4 to 8 fl oz/acre	0	0.5	**DISEASE SUPPRESSION ONLY.** Limit of 24 fl oz and 3 applications/acre/season. Do not make more than 2 consecutive applications before alternating to a fungicide with a different MOA.
mancozeb (various)	M03	See label	5	1	See label for rates.
mefenoxam + chlorothalonil (Ridomil Gold Bravo SC)	4+M05	2.5 to 3.25 pt/acre	7	2	Apply a protectant between applications of Ridomil Gold Bravo SC.
oxathiapiprolin + chlorothalonil (Orondis Opti)	49+M05	1.75 to 2.5 pt/acre	3	0.5	Do not follow soil applications with foliar applications of any oxathiapiprolin containing products. See label for season and application limits.
penthiopyrad (Fontelis 1.67SC)	7	16 to 24 fl oz/acre	0	0.5	Limit of 72 fl oz/acre/season. Do not make more than 2 consecutive applications before alternating to a fungicide with a different MOA.
pydiflumetofen + fludioxonil (Miravis Prime)	7+12	11.4 fl oz/acre	0	0.5	**DISEASE SUPPRESSION ONLY.** Do not make more than 2 applications of Miravis Prime or other Group 7 and 12 fungicides before alternating to a fungicide that is not in Groups 7 or 12. Do not exceed 22.8 fl oz/acre/season. Apply by ground, air, or chemigation.
pyraclostrobin (Cabrio EG)	11	12 to 16 oz/acre	0	0.5	**DISEASE SUPPRESSION ONLY.** Do not make more than 2 consecutive applications before alternating to a fungicide with a different mode of action. Do not exceed 96 oz/acre/season.
pyrimethanil (Scala SC)	9	7 fl oz/acre	1	0.5	Limit of 35 fl oz/acre/season.

TABLE 3-36. DISEASE CONTROL PRODUCTS FOR TOMATO (cont'd)

I. Meadows, Extension Plant Pathologist, North Carolina State University; R. Singh, Plant Pathologist, Louisiana State University Agricultural Center

Disease/Material	FRAC Code	Rate of Material Formulation	Minimum Days Harv.	Reentry	Method, Schedule, and Remarks
TOMATO (FIELD) (cont'd)					
BUCKEYE ROT AND PHYTOPHTHORA FRUIT ROT (*PHYTOPHTHORA* SPP.)					
azoxystrobin (various)	11	See label	0	4 h	Do not make more than one application before alternating to a fungicide with a different mode of action. A tank mixture with Dimethoate or some adjuvants may cause crop injury—see label.
azoxystrobin + chlorothalonil (Quadris Opti)	11+M05	1.6 pt/acre	0	0.5	Limit of 5 applications of any Group 11 fungicide. Do not make more than 1 application before alternating to a fungicide with a different mode of action. Do not apply earlier than 21 days after transplant or within +/- 6 days of a postemer- gence application of Sencor. Do not use an adjuvant.
famoxadone + cymoxanil (Tanos)	11+27	8 oz/acre	3	0.5	**DISEASE SUPPRESSION ONLY.** Do not make more than 1 application before alternating to a fungicide with a different mode of action. **NOTE: Must be tank-mixed with a contact fungicide that has a different mode of action.**
fluopicolide (Presidio 4F)	43	3 to 4 fl oz/acre	2	0.5	Do not make more than 2 consecutive applications before alternating to a fungicide with a different mode of action. Use only in combination with a labeled rate of another fungicide product with a different mode of action.
mancozeb + azoxystrobin (Dexter Max)	M03+11	0.8 to 1.6 lb/acre	5	1	For states East of the Mississippi and including Mississippi, do not exceed 12 lb of product/acre/season. States West of Mississippi use 0.8 to 1.1 lb/acre and do not exceed 9.14 lb of product/acre/season. Do not make more than 1 application before alternating to a fungicide not in Group 11. On fresh market tomato, do not tank-mix with an adjuvant or EC formulation. **Tank mixture with dimethoate may cause crop injury.**
mancozeb + zoxamide (Gavel 75DF)	M03+22	1.5 to 2 lb/acre	5	2	For states East of Mississippi, do not exceed 16 lb/acre/year. States West of Mississippi, do not exceed 8 lb/acre/year.
mefenoxam (Ridomil Gold SL) (Ultra Flourish)	4	1 pt/acre 2pt/acre	7	2	Apply uniformly to soil at time of planting. Incorporate mechanically if rainfall is not expected before seeds germinate. See labels for application limits.
mefenoxam + chlorothalonil (Ridomil Gold Bravo SC)	4+M05	2.5 pt/acre	7	2	Apply a protectant between applications of Ridomil Gold Bravo SC.
phosphorous acid, mono- and dipotassium salts (Resist 57)	P07	1 to 3 qt/acre	0	4 h	Do not apply to plants that are heat or moisture stressed or dormant. Avoid applying to plants that had been treated with copper within 20 days.
oxathiapiprolin + chlorothalonil (Orondis Opti)	49+M05	1.75 to 2.5 pt/acre	0	0.5	Do not make more than 2 consecutive applications before alternating to a fungicide with different mode of action and no more than 6 total applications/season. Do not exceed 10 pints of product/acre/year. Do not mix soil applications and foliar applications. See labels for application limits.
oxathiapiprolin + mefenoxam (Orondis Gold DC)	49+4	28 to 55 fl oz/acre	7	4 h	Apply at-plant by in-furrow, transplant water, banded surface spray, or drip irrigation. Do not follow soil applications with foliar applications with any product containing FRAC 49 fungicide. Do not make more than one soil application/crop. Not for use in nursery production of transplants. **NOT REGISTERED FOR USE in FL, GA, or MS.**
oxathiapiprolin + mandipropamid (Orondis Ultra)	40+49	5.5 to 8.0 fl oz/acre	1	4 h	Do not make more than 2 consecutive applications before alternating to a fungicide with different mode of action and no more than 6 applications/season. Do not exceed 32 fl oz of product/acre/year. Do not mix soil applications and foliar applications. See labels for application limits.
DAMPING-OFF (*PYTHIUM*)					
fosetyl-Al (Aliette WDG, Linebacker WDG)	P07	2.5 to 5 lb/acre	14	0.5	Do not tank-mix with copper. Do not exceed 20 lb product/season. Check label for specific counties in each state where use is prohibited
mefenoxam (Ridomil Gold SL) (Ultra Flourish)	4	1 pt/acre 2pt/acre	7	2	Apply uniformly to soil at time of planting. Incorporate mechanically if rainfall is not expected before seeds germinate. A 2nd application may be made up to 4 weeks before harvest. See labels for application limits.
Oxathiapiprolin + mefenoxam (Orondis Gold DC)	49+4	28 to 55 fl oz/acre	7	4 h	Apply at-plant by in-furrow, transplant water, banded surface spray, or drip irrigation. Do not follow soil applications with foliar applications with any product containing FRAC 49 fungicide. Do not make more than one soil application/crop. Not for use in nursery production of transplants. See label for season limits.
phosphorous acid, mono- and dipotassium salts (Resist 57)	P07	2 to 4 qt/acre (chemigation, soil application) See label for other application types	0	4 h	Do not apply at less than 3-day intervals. Do not apply to plants when plants will remain wet for longer than 4 hours.
propamocarb (Previcur Flex)	28	1.5 pt/acre	5	0.5	Limit of 7.5 pt/acre/season. Do not make more than 1 application before alternating to a fungicide with a different mode of action.

TABLE 3-36. DISEASE CONTROL PRODUCTS FOR TOMATO (cont'd)

I. Meadows, Extension Plant Pathologist, North Carolina State University; R. Singh, Plant Pathologist, Louisiana State University Agricultural Center

Disease/Material	FRAC Code	Rate of Material Formulation	Minimum Days Harv.	Minimum Days Reentry	Method, Schedule, and Remarks
TOMATO (FIELD) (cont'd)					
GRAY LEAF SPOT (*STEMPHYLIUM* SPP.)					
azoxystrobin + difenoconazole (Quadris Top)	11+3	8 fl oz/acre	0	0.5	Do not apply until 21 days after transplanting or 35 days after seeding. Limit of 47 fl oz/acre/season. Do not make more than 2 consecutive applications before alternating to a fungicide with a different mode of action. See label for tank-mix cautions. Do not use an adjuvant or tank mix with any EC product. Tank mixing with Dimethoate may cause crop injury.
chlorothalonil (various)	M05	See label	0	0.5	Refer to individual labels for rates and restrictions.
chlorothalonil + cymoxanil (Ariston, Cymbol Advance)	M05+27	1.9 to 3 pt/acre	3	0.5	Limit of 17.5 pt/acre/12-month period. If mixing with copper, check copper label for limitations.
difenoconazole + benzovindiflupyr (Aprovia Top)	7+3	10.5 to 13.5 fl oz/acre	0	0.5	Do not make more than 2 applications before alternating to a non-Group 7 fungi- cide. See label for application intervals and limits/season. Use of a spreading adjuvant is recommended.
difenoconazole + cyprodinil (Inspire Super)	3+9	16 to 20 fl oz/acre	0	0.5	Limit of 80 fl oz/acre/season. Do not make more than 2 consecutive applications before alternating to a fungicide with a different mode of action.
difenoconazole + tea tree oil (Regev)	3+BM01	4.0 to 8.5 fl oz/acre	2	0.5	Minimum application interval is 7 days. Do not make more than 2 sequential applications before alternating to a fungicide with a different mode of action. Do not exceed 4 applications/year.
fluopyram + difenoconazole (Luna Flex)	7+3	10 to 13.6 fl oz/acre	0	0.5	Maximum 2 applications/year at high rate. Maximum interval between applications is 7 days.
fluopyram + trifloxystrobin (Luna Sensation)	7+11	7.6 fl oz/acre	3	0.5	Do not exceed 5 applications or 27.1 fl oz/acre/season. Do not make more than 2 consecutive applications before alternating to a fungicide with a different mode of action.
fluopyram + pyrimethanil (Luna Tranquility)	7+9	11.2 fl oz/acre	1	0.5	See label for limits on application amounts/season. Do not make more than 2 applications of Group 7 or 9 fungicides without switching to a different mode of action. **NOT REGISTERED FOR USE IN LA.**
mancozeb (various)	M03	See label	5	1	See label for limits on application amounts/season.
mancozeb + azoxystrobin (Dexter Max)	M03+11	0.8 to 1.6 lb/acre	5	1	For states East of the Mississippi including Mississippi, do not exceed 12 lb of product/acre/season. States West of Mississippi use 0.8 to 1.1 lb/acre and do not exceed 9.14 lb of product/acre/season. Do not exceed 12 lbs of product/ acre/ season. Do not make more than 1 application before alternating to a fungicide not in Group 11. On fresh market tomato, do not tank-mix with an adjuvant or an EC formulation. **Tank mixture with dimethoate may cause crop injury.**
mancozeb + copper (ManKocide)	M03+M1	1 to 3 lb/acre	5	2	Limit of 58 lb/acre/season East of the Mississippi River.
mancozeb + zoxamide (Gavel 75DF)	M3+22	1.5 to 2 lb/acre	5	2	Limit of 16 lb/acre/season East of the Mississippi River.
mandipropamid + difenoconazole (Revus Top)	40+3	5.5 to 7 fl oz/acre	1	0.5	Limit of 28 fl oz/acre/season. Do not make more than 2 consecutive applications before alternating to a fungicide with a different mode of action.
mefenoxam + chlorothalonil (Ridomil Gold Bravo SC)	4+M05	2.5 pt/acre	7	2	Apply a protectant between applications of Ridomil Gold Bravo SC.
pydiflumetofen + fludioxonil (Miravis Prime)	7+12	9.2 to 11.4 fl oz/acre	0	0.5	Do not make more than two applications of Miravis Prime or other Group 7 and 12 fungicides before alternation with a fungicide that is not in Group 7 or 12. Do not exceed 22.8 fl oz/acre/season. Apply by ground, air, or chemigation.
tetraconazole (Mettle125 ME)	3	6 to 8 fl oz/acre	0	0.5	Make no more than 2 sequential applications before switching to a fungicide with a different mode of action. Can be applied by ground, air, or chemigation. Do not apply more than 16 fl oz or exceed 5 applications of Mettle/acre/year. Do not apply more than 0.125 lb/acre/year of tetraconazole containing products.
trifloxystrobin (various)	11	See label	3	0.5	Season limits apply – see label. Do not make more than 1 application before alternating to a fungicide with a different mode of action.
EARLY BLIGHT (*ALTERNARIA*), SEPTORIA LEAF SPOT (*SEPTORIA*), AND TARGET SPOT (*CORYNESPORA*)					
azoxystrobin (various)	11	5 to 6.2 fl oz/acre	0	4 h	Limit of 37 fl oz/acre/season. Do not make more than 1 application before alternating to a fungicide with a different mode of action. A tank mixture with Dimethoate or some adjuvants may cause crop injury—see label.
azoxystrobin + chlorothalonil (Quadris Opti)	11+M05	1.6 pt/acre	0	0.5	Make no more than **1** application before alternating with fungicides that have a different mode of action. Do not apply until 21 days after transplant or within +/- 6 days of a postemergence broadcast application of Sencor. Do not use an adjuvant.
azoxystrobin + difenoconazole (Quadris Top)	11+3	8 fl oz/acre	0	0.5	Do not apply until 21 days after transplanting or 35 days after seeding. Limit of 47 fl oz/acre/season. Do not make more than 2 consecutive applications before alternating to a fungicide with a different mode of action.

TABLE 3-36. DISEASE CONTROL PRODUCTS FOR TOMATO (cont'd)

I. Meadows, Extension Plant Pathologist, North Carolina State University; R. Singh, Plant Pathologist, Louisiana State University Agricultural Center

Disease/Material	FRAC Code	Rate of Material Formulation	Minimum Days Harv.	Minimum Days Reentry	Method, Schedule, and Remarks
TOMATO (FIELD) (cont'd)					
EARLY BLIGHT (*ALTERNARIA*), SEPTORIA LEAF SPOT (*SEPTORIA*), AND TARGET SPOT (*CORYNESPORA*) (cont'd)					
azoxystrobin + flutriafol (Topguard EQ)	11+3	4 to 8 fl oz/acre	0	0.5	Do not use adjuvants or EC formulated tank-mix partners on fresh market tomatoes. Do not exceed 4 applications/year. Limits on both a.i.'s apply - see label.
boscalid (Endura WDG)	7	2.5 to 3.5 oz/acre	0	0.5	Limit of 21 oz/acre/season. Do not make more than 2 consecutive applications and no more than 2 applications/crop/year.
chlorothalonil (various)	M05	See label	0	0.5	Refer to individual labels for rates and restrictions.
chlorothalonil + cymoxanil (Ariston, Cymbol Advance)	M05+27	1.9 to 3 pt/acre	3	0.5	Limit of 17.5 pt/acre/12-month period. If mixing with copper, check copper label for limitations.
copper (various)	M01	See label	0	0	Labeled for Septoria leaf spot only.
copper + mancozeb (ManKocide)	M01+M03	1 to 3 lb/acre	5	2	Apply every 3 to 7 days.
cyprodinil + fludioxonil (Alterity 62.5WG, Switch 62.5WG)	9+12	11 to 14 oz/acre	0	0.5	Limit of 56 oz/acre/season. Do not make more than 2 consecutive applications before alternating to a fungicide with different mode of action for two applications. **For early blight control only.**
difenoconazole + benzovindiflupyr (Aprovia Top)	7+3	10.5 to 13.5 fl oz/acre	0	0.5	Do not make more than 2 consecutive applications before alternating with a non-Group 7 fungicide. See label for application intervals and limits/season. Use of a spreading adjuvant is recommended.
difenoconazole + cyprodinil (Inspire Super)	3+9	16 to 20 fl oz	0	0.5	Limit of 80 fl oz/acre/season. Do not make more than 2 consecutive applications before alternating to a fungicide with a different mode of action.
difenoconazole + tea tree oil (Regev)	3+BM01	4.0 to 8.5 fl oz/acre	2	0.5	Minimum application interval is 7 days. Do not make more than 2 sequential applications before alternating to a fungicide with a different mode of action. Maximum applications are 4/year.
famoxadone + cymoxanil (Tanos)	11+27	6 to 8 oz/acre	3	0.5	Limit of 72 fl oz/acre/season. Do not make more than 1 application before alternating to a fungicide with a different mode of action. **Must be tank mixed with a contact fungicide that has a different mode of action. For Septoria leaf spot and target spot use 8 oz/acre.**
fenamidone (Reason 500SC)	11	5.5 to 8.2 fl oz/acre	14	0.5	Limit of 24.6 fl oz/acre/season. Do not make more than 1 application before alternating to a fungicide with a different mode of action. **NOT labeled for target spot control.**
fluopyram + difenoconazole (Luna Flex)	7+3	8.0 to 13.6 fl oz/acre	0	0.5	Maximum 2 applications/year at high rate. Maximum interval between applications is 7 days. **For target spot use a minimum of 10.0 fl oz/acre.**
fluopyram + trifloxystrobin (Luna Sensation)	7+11	5 to 7.6 fl oz/acre	3	0.5	Do not exceed 5 applications or 27.1 fl oz/acre/season. Do not make more than 2 consecutive applications before alternating to a fungicide with a different mode of action. **Use 7.6 fl oz rate for gray leaf spot and target spot.**
fluopyram + pyrimethanil (Luna Tranquility)	7+9	11.2 fl oz/acre	1	0.5	See label for limits on application amounts/season. Do not make more than 2 applications of Group 7 or 9 fungicides before alternating to a fungicide with a different mode of action. **NOT REGISTERED FOR USE IN LA.**
fluoxastrobin (Aftershock, Evito 480SC)	11	2.0 to 5.7 oz/acre	3	0.5	Limit of 22.8 fl oz/acre/season. Do not make more than 1 application before alternating to a fungicide with a different mode of action. **Controls target spot and early blight only.**
flutriafol (Rhyme)	3	3.5 to 7 fl oz/acre	0	0.5	Tank-mix with mancozeb for improved early blight control. Do not exceed more than 4 applications or 28 fl oz product/acre/season. Use a minimum of 5 fl oz/acre for target spot control. **Not labeled for Septoria leaf spot control.**
fluxapyroxad + pyraclostrobin (Priaxor Xemium)	7+11	4 to 8 oz/acre	0	0.5	Limit of 24 fl oz/acre/season. Do not make more than 2 consecutive applications before alternating to a fungicide with a different mode of action.
mancozeb (various)	M03	See label	5	1	See label.
mancozeb + azoxystrobin (Dexter Max)	M03+11	0.8 to 1.6 lb/acre	5	1	For states East of the Mississippi and including Mississippi, do not exceed 12 lb of product/acre/season. States West of Mississippi use 0.8 to 1.1 lb/acre and do not exceed 9.14 lb of product/acre/season. Do not exceed 12 lb of product/acre/season. Do not make more than 1 application before alternating to a fungicide not in Group 11. On fresh market tomato, do not tank-mix with an adjuvant or an EC formulation. Tank mixture with dimethoate may cause crop injury. **For target spot control East of the Mississippi, use highest rate.**
mancozeb + zoxamide (Gavel 75DF)	M03+22	1.5 to 2 lb/acre	5	2	For states East of the Mississippi, do not exceed 16 lb/acre/year. States West of the Mississippi do not exceed 8 lb/acre/year. **Not labeled for target spot.**
mandipropamid + difenoconazole (Revus Top)	40+3	5.5 to 7 fl oz/acre	1	0.5	Limit of 28 fl oz/acre/season. Do not apply more than 2 consecutive applications before alternating to a fungicide with a different mode of action.

TABLE 3-36. DISEASE CONTROL PRODUCTS FOR TOMATO (cont'd)

I. Meadows, Extension Plant Pathologist, North Carolina State University; R. Singh, Plant Pathologist, Louisiana State University Agricultural Center

Disease/Material	FRAC Code	Rate of Material Formulation	Minimum Days Harv.	Minimum Days Reentry	Method, Schedule, and Remarks
TOMATO (FIELD) (cont'd)					
EARLY BLIGHT (*ALTERNARIA*), SEPTORIA LEAF SPOT (*SEPTORIA*), AND TARGET SPOT (*CORYNESPORA*) (cont'd)					
mefenoxam + chlorothalonil (Ridomil Gold Bravo SC)	4+M05	2.5 pt/acre	7	2	Apply a protectant between applications of Ridomil Gold Bravo SC.
mefentrifluconazole (Cevya)	3	3 to 5 fl oz/acre	0	0.5	Maximum of 15 fl oz/acre/year.
penthiopyrad (Fontelis 1.67SC)	7	16 to 24 fl oz/acre	0	0.5	Limit of 72 fl oz/acre/season. Do not make more than 2 consecutive applications before alternating to a fungicide with a different mode of action.
propamocarb (Previcur Flex)	28	0.7 to 1.5 pt/acre	5	0.5	Limit of 7.5 pt/acre/season. Do not make more than 1 application before alternating to a fungicide with a different mode of action. Tank-mix with a compatible fungicide for optimal early blight control. **For early blight control only.**
pydiflumetofen + fludioxonil (Miravis Prime)	7+12	9.2 to 11.4 fl oz/acre	0	0.5	Do not make more than two applications of Miravis Prime or other Group 7 and 12 fungicides before alternation with a fungicide that is not in Group 7 or 12. Do not exceed 22.8 fl oz/acre/season. Apply by ground, air, or chemigation.
pyraclostrobin (Cabrio EG)	11	8 to 12 oz/acre	0	0.5	Limit of 96 fl oz/acre/season. Do not make more than 2 applications before alternating to a fungicide with a different mode of action.
pyrimethanil (Scala SC)	9	7 fl oz/acre	1	0.5	Limit of 35 fl oz/acre/season. Use only in a tank-mix with another fungicide recommended for early blight. **For early blight control only.**
tetraconazole (Mettle125 ME)	3	6 to 8 fl oz/acre	0	0.5	Make no more than 2 sequential applications before switching to a fungicide with a different mode of action. Can be applied by ground, air, or chemigation. Do not apply more than 16 fl oz or exceed 5 applications of Mettle/acre/year. Do not apply more than 0.125 lb/acre/year of tetraconazole containing products.
trifloxystrobin (various)	11	See label	3	0.5	Limit of 16 fl oz or 5 applications/acre/season. Do not make more than 1 application before alternating to a fungicide with a different mode of action. **Not labeled for target spot. Disease suppression for Septoria leaf spot.**
zinc dimethyldithiocarbamate (Ziram 76DF)	M03	3 to 4 lb/acre	7	2	Limit of 24 lb/acre/season. **DO NOT USE ON CHERRY TOMATOES. For early blight and Septoria leaf spot only.**
zoxamide + chlorothalonil (Zing!)	22+M03	36 fl oz/acre	5	0.5	Do not make more than 2 consecutive applications before alternating to a fungicide with a different mode of action. See label for application limits. **For early blight and Septoria leaf spot only.**
FUSARIUM WILT (*FUSARIUM OXYSPORUM F.SP. LYCOPERSICI*)					
pydiflumetofen + fludioxonil (Miravis Prime)	7+12	11.4 fl oz/acre	0	0.5	**SUPPRESSION ONLY.** Make one application after transplanting or within 7-to 14-days later. Make a 2nd application 14-to 21-days later. Do not make more than 2 applications. See label for application methods.
POWDERY MILDEW (*LEVEILULLA, OIDIUM*)					
azoxystrobin (various)	11	5 to 6.2 fl oz/acre	0	4 h	Limit of 37 fl oz/acre/season. Do not make more than 1 application before alternating to a fungicide with a different mode of action. A tank mixture with Dimethoate or some adjuvants may cause crop injury—see label.
azoxystrobin + chlorothalonil (Quadris Opti)	11+M03	1.6 pt/acre	0	0.5	Make no more than **one** application before alternating with fungicides that have a different mode of action. Do not apply until 21 days after transplant or within +/- 6 days of a postemergence broadcast application of Sencor. Do not use an adjuvant
azoxystrobin + difenoconazole (Quadris Top)	11+3	8 fl oz/acre	0	0.5	Do not apply until 21 days after transplanting or 35 days after seeding. Limit of 47 fl oz/acre/season. Make no more than 2 consecutive applications before alternating to fungicide with a different mode of action. See label for tank-mix cautions. Do not use adjuvant or tank mix with any EC product. Tank mixing with Dimethoate may cause crop injury.
azoxystrobin + flutriafol (Topguard EQ)	11+3	4 to 8 fl oz/acre	0	0.5	Do not use adjuvants or EC formulated tank-mix partners on fresh market tomatoes. Do not exceed 4 applications/year. Limits on both a.i. apply-see label.
cyflufenamid (Fastback)	U06	4 oz/acre	0	4 h	Make no more than 3 applications/year. Minimum application interval is 14 days.
cyprodinil + fludioxonil (Alterity 62.5WG, Switch 62.5WG)	9+12	11 to 14 oz/acre	0	0.5	Limit of 56 oz/acre/season. Do not make more than 2 consecutive applications before alternating to a fungicide with different mode of action for 2 applications
difenoconazole + cyprodinil (Inspire Super)	3+9	16 to 20 fl oz	0	0.5	Limit of 80 fl oz/acre/season. Do not make more than 2 consecutive applications before alternating to fungicide with different mode of action.
difenoconazole + benzovindiflupyr (Aprovia Top)	7+3	10.5 to 13.5 fl oz/acre	0	0.5	Do not make more than 2 consecutive applications before alternating a non- Group 7 fungicide. See label for application intervals and limits/season. Use of a spreading adjuvant is recommended.
difenoconazole + tea tree oil (Regev)	3+BM01	4.0 to 8.5 fl oz/acre	2	0.5	Minimum application interval is 7 days. Do not make more than 2 sequential applications before alternating to a fungicide with a different mode of action. Maximum applications are 4/year.

TABLE 3-36. DISEASE CONTROL PRODUCTS FOR TOMATO (cont'd)

I. Meadows, Extension Plant Pathologist, North Carolina State University; R. Singh, Plant Pathologist, Louisiana State University Agricultural Center

Disease/Material	FRAC Code	Rate of Material Formulation	Minimum Days Harv.	Minimum Days Reentry	Method, Schedule, and Remarks
TOMATO (FIELD) (cont'd)					
POWDERY MILDEW (*LEVEILULLA, OIDIUM*) (cont'd)					
fluopyram + difenoconazole (Luna Flex)	7+3	8.0 to 13.6 fl oz/acre	0	0.5	Maximum 2 applications/year at high rate. Maximum interval between applications is 7 days.
fluopyram + pyrimethanil (Luna Tranquility)	7+9	11.2 fl oz/acre	1	0.5	DISEASE SUPPRESSION ONLY. See label for limits on application amounts/season. Do not make more than 2 applications of Group 7 or 9 fungicides before alternating to a fungicide with a different mode of action. **NOT REGISTERED FOR USE IN LA.**
fluopyram + trifloxystrobin (Luna Sensation)	7+11	5 to 7.6 fl oz/acre	3	0.5	Do not exceed 5 applications or 27.1 fl oz/acre/season. Do not make more than 2 consecutive applications before alternating to a fungicide with a different mode of action. **Use 7.6 fl oz rate for gray leaf spot and target spot.**
flutriafol (Rhyme)	3	5 to 7 fl oz/acre	0	0.5	Tank-mix with mancozeb for improved disease control. Use lower rate if tank mixed. Do not exceed 4 applications/year or more than 28 fl oz of product/acre/year.
fluxapyroxad + pyraclostrobin (Priaxor Xemium)	7+11	6 to 8 fl oz/acre	0	0.5	Limit of 24 fl oz/acre/season. Do not make more than 2 consecutive applications before alternating to a fungicide with a different mode of action.
mancozeb + azoxystrobin (Dexter Max)	M03+11	1.6 lb/acre	5	1	Do not exceed 12 lb of product/acre/season. Do not make more than 1 application before alternating to a fungicide not in Group 11. On fresh market tomato, do not tank-mix with an adjuvant or an EC formulation. **Tank mixture with dimethoate may cause crop injury.**
mandipropamid + difenoconazole (Revus Top)	40+3	5.5 to 7 fl oz/acre	1	0.5	Limit of 28 fl oz/acre/season. Do not apply more than 2 consecutive applications before alternating to a fungicide with a different mode of action.
mefentrifluconazole (Cevya)	3	3 to 5 fl oz/acre	0	0.5	Maximum of 15 fl oz/acre/year.
myclobutanil (various)	3	See label	1	0	Spray weekly beginning at first sign of disease. Do not apply more than 1.25 lb/acre. Observe a 30-day plant back interval between last application and planting new crop.
penthiopyrad (Fontelis 1.67SC)	7	16 to 24 fl oz/acre	0	0.5	Limit of 72 fl oz/acre/season. Do not make more than 2 consecutive applications before alternating to fungicide with different mode of action.
pyraclostrobin (Cabrio EG)	11	8 to 16 oz/acre	0	0.5	Limit of 96 fl oz/acre/season. Do not make more than 2 applications before alternating to a fungicide with a different mode of action.
pyriofenone (Prolivo 300SC)	50	4 to 5 fl oz	0	4 h	Do not exceed 16 fl oz/acre/year. Do not make more than 2 sequential applications of Prolivo or of another FRAC 50-containing fungicide before alternating to a fungicide with a different mode of action. Do not exceed 4 applications/year.
pydiflumetofen + fludioxonil (Miravis Prime)	7+12	9.2 to 11.4 fl oz/acre	0	0.5	Do not make more than two applications of Miravis Prime or other Group 7 and 12 fungicides before alternation with a fungicide that is not in Group 7 or 12. Do not exceed 22.8 fl oz/acre/season. Apply by ground, air, or chemigation.
tetraconazole (Mettle125 ME)	3	6 to 8 fl oz/acre	0	0.5	Make no more than 2 sequential applications before switching to a fungicide with a different mode of action. Can be applied by ground, air, or chemigation. Do not apply more than 16 fl oz or exceed 5 applications of Mettle/acre/year. Do not apply more than 0.125 lb/acre/year of tetraconazole containing products.
trifloxystrobin (various)	11	See label	3	0.5	DISEASE SUPPRESSION ONLY. Do not exceed 16 fl oz/acre/crop or 5 applications/acre/season. Do not exceed 1 application before switching to a fungicide with a different mode of action.
tolfenpyrad (Torac)	39	21 fl oz/acre	1	0.5	DISEASE SUPPRESSION ONLY. Do not exceed 42 fl oz/acre/crop. Do not exceed 2 applications/crop cycle and do not exceed 4 applications/year. **Provides SUPPRESSION of powdery mildew.**
sulfur (various)	M02	See label	See label	1	Follow labels. **May cause leaf burn if used under high temperatures.**
LATE BLIGHT (*PHYTOPHTHORA INFESTANS*)					
azoxystrobin (various)	11	6.2 fl oz/acre	0	4 h	Limit of 37 fl oz/acre/season. Do not make more than 1 application before alternating to a fungicide with a different mode of action. A tank mixture with Dimethoate or some adjuvants may cause crop injury—see label. **NOTE: Apply at 5- to 7-day intervals for effective late blight management.**
chlorothalonil (various)	M05	See label	0	0.5	Refer to individual labels for rates and restrictions.
copper (various)	M01	See label	0	0	
azoxystrobin + chlorothalonil (Quadris Opti)	11+M05	1.6 pts/acre	0	0.5	Make no more than **one** application before alternating with fungicides that have a different mode of action. Do not apply until 21 days after transplant or within +/- 6 days of a postemergence broadcast application of Sencor. Do not use an adjuvant.

2026 Southeastern U.S. Vegetable Crop Handbook

TABLE 3-36. DISEASE CONTROL PRODUCTS FOR TOMATO (cont'd)

I. Meadows, Extension Plant Pathologist, North Carolina State University; R. Singh, Plant Pathologist, Louisiana State University Agricultural Center

Disease/Material	FRAC Code	Rate of Material Formulation	Minimum Days Harv.	Minimum Days Reentry	Method, Schedule, and Remarks
TOMATO (FIELD) (cont'd)					
LATE BLIGHT (*PHYTOPHTHORA INFESTANS*) (cont'd)					
azoxystrobin + flutriafol (Topguard EQ)	11+3	4 to 8 fl oz/acre	0	0.5	Do not use adjuvants or EC formulated tank-mix partners on fresh market tomatoes. Do not exceed 4 applications/year. Limits on both a.i.'s apply-see label.
chlorothalonil (various)	M05	See label	0	0.5	Refer to individual labels for rates and restrictions.
cymoxanil + chlorothalonil (Ariston, Cymbol Advance)	27+M05	1.9 to 3.0 pts/acre	3	0.5	Limit of 17.5 pints /acre/12-month period. Check copper labels for specific pre-cautions and limitations for mixing with this product.
cyazofamid (Ranman 400SC)	21	2.1 to 2.75 fl oz/acre	0	0.5	Limit of 16.5 fl oz/acre/season. Do not make more than 1 application before alternating to a fungicide with a different mode of action.
cymoxanil (Curzate 60DF)	27	3.2 to 5 oz/acre	3	0.5	Limit of 30 oz/acre/12-month period. Use only in combination with a labeled rate of a protectant fungicide. If late blight is present, use 5 oz/acre on a 5-day schedule. **NOT REGISTERD FOR USE IN LA.**
dimethomorph (Forum 4.18F)	40	6 fl oz/acre	4	0.5	Limit of 30 fl oz and 5 applications/acre/season. Performance is improved if tank-mixed with another fungicide with a different mode of action.
dimethomorph + ametoctradin (Zampro)	40+45	14 fl oz/acre	4	0.5	Limit of 42 fl oz/acre/season. Do not make more than 2 consecutive applications before alternating to a fungicide with a different mode of action. The addition of a spreading or penetrating adjuvant is recommended to improve product performance.
fenamidone (Reason 500SC)	11	5.5 to 8.2 fl oz/acre	14	0.5	Limit of 24.6 fl oz/acre/season. Do not make more than 1 application before alternating to a fungicide with a different mode of action.
fluopicolide (Presidio 4F)	43	3 to 4 fl oz/acre	2	0.5	Do not make more than 2 consecutive applications before alternating to a fungicide with a different mode of action. Use only in combination with a labeled rate of another fungicide product with a different mode of action.
fluoxastrobin (Aftershock, Evito 480 SC)	11	5.7 fl oz/acre	3	0.5	**DISEASE SUPPRESSION ONLY**. Limit of 22.8 fl oz/acre/season. Do not make more than 1 application before alternating to a fungicide with a different mode of action.
fluxapyroxad + pyraclostrobin (Priaxor Xemium)	7+11	8 fl oz/acre	7	0.5	**DISEASE SUPPRESSION ONLY**. Limit of 24 fl oz/acre/season. Do not make more than 2 consecutive applications before alternating to a fungicide with a different mode of action.
mancozeb (various)	M03	See label	5	1	
mancozeb + azoxystrobin (Dexter Max)	M03+11	0.8 to 1.6 lb/acre	5	1	For states East of the Mississippi and including Mississippi, do not exceed 12 lb of product/acre/season. States West of Mississippi use 0.8 to 1.1 lb/ acre and do not exceed 9.14 lb of product/acre/season. Do not exceed 12 lb of product/acre/season. Do not make more than 1 application before alternating to a fungicide not in Group 11. On fresh market tomato, do not tank-mix with an adjuvant or an EC formulation. **Tank mixture with dimethoate may cause crop injury.**
mancozeb + copper hydroxide (ManKocide)	M03+M01	1 to 3 lb/acre	5	2	Apply at 7- to 10-day interval.
mancozeb + zoxamide (Gavel 75DF)	M03+22	1.5 to 2 lb/acre	5	2	For states East of the Mississippi River, do not exceed 16 lb/acre/year. States West of the Mississippi River, do not exceed 8 lb/acre/year.
mandipropamid (Revus)	40	8 fl oz/acre	1	4 h	Limit of 32 fl oz/acre/season. Do not make more than 2 consecutive applications before alternating to fungicide with different mode of action.
mandipropamid + difenoconazole (Revus Top)	40+3	5.5 to 7 oz/acre	1	0.5	Limit of 28 fl oz/acre/season. Do not make more than 2 consecutive applications before alternating to fungicide with different mode of action.
mefenoxam (Ridomil Gold SL) (Ultra Flourish)	4	1 pt/acre 2pt/acre	7	2	Apply uniformly to soil at time of planting. Incorporate mechanically if rainfall is not expected before seeds germinate. See labels for application limits.
mefenoxam + chlorothalonil (Ridomil Gold Bravo)	4+M05	2.5 pt/acre	5	2	See label for application limits.
mefenoxam + mancozeb (Ridomil Gold MZ WG)	4+M03	2.5 lb/acre	5	2	Do not make more than 3 applications or 7.5 lb/acre/season of Ridomil Gold MZ.
oxathiapiprolin + chlorothalonil (Orondis Opti)	49+M05	1.75 to 2.5 pints/acre	0	0.5	Do not make more than 2 sequential applications before alternating to a fungicide with a different mode of action and no more than 6 total applications on multiple crops in the same year. Do not mix soil applications and foliar applications. See label for application limits.
oxathiapiprolin + mandipropamid (Orondis Ultra)	49+40	5.5 to 8.0 fl oz/acre	1	4 h	Do not make more than 2 sequential applications before alternating to a fungicide with a different mode of action and no more than 6 total applications on multiple crops/season. Limit applications apply—see label. Do not mix soil applications and foliar applications.

TABLE 3-36. DISEASE CONTROL PRODUCTS FOR TOMATO (cont'd)

I. Meadows, Extension Plant Pathologist, North Carolina State University; R. Singh, Plant Pathologist, Louisiana State University Agricultural Center

Disease/Material	FRAC Code	Rate of Material Formulation	Minimum Days Harv.	Minimum Days Reentry	Method, Schedule, and Remarks
TOMATO (FIELD) (cont'd)					
LATE BLIGHT (*PHYTOPHTHORA INFESTANS***) (cont'd)**					
phosphorous acid, mono- and dipotassium salts (Resist 57)	P07	1 to 3 qt/acre	0	4 h	Do not apply to plants that are heat or moisture stressed or dormant. Avoid applying to plants that had been treated with copper within 20 days.
pyraclostrobin (Cabrio EG)	11	8 to 16 oz/acre	0	0.5	Do not make more than 2 consecutive applications before alternating to a fungicide with a different mode of action. Do not exceed 96 oz/acre/season.
propamocarb (Previcur Flex)	28	0.7 to 1.5 pt/acre	5	0.5	Limit of 7.5 pt/acre/season. Do not make more than 1 application before alternating to a fungicide with a different mode of action.
trifloxystrobin (various)	11	See label	3	0.5	Season limits apply – see label. Must tank-mix and alternate Flint Extra with a protectant for late blight control.
zoxamide + chlorothalonil (Zing!)	22+M05	36 fl oz/acre	5	0.5	Do not make more than 2 consecutive applications before alternating to a fungicide with a different mode of action. Do not tank-mix with another fungicide if the target pest is only late blight. Tank-mix only if a partner product is required to control other diseases.
LEAF MOLD (*FULVIA FULVA* = *PASSALORA FULVA***)**					
azoxystrobin + difenoconazole (Quadris Top)	11+3	8 fl oz/acre	0	0.5	Do not apply until 21 days after transplanting or 35 days after seeding. Limit of 47 fl oz/acre/season. Do not make more than 2 consecutive applications before alternating to fungicide with different mode of action. Do not use an adjuvant or tank mix with any EC product. Tank mixing with Dimethoate may cause crop injury.
chlorothalonil (various)	M05	See label	0	0.5	Refer to individual labels for rates and restrictions.
chlorothalonil + cymoxanil (Ariston, Cymbol Advance)	M05+27	1.9 pt/acre	3	0.5	Limit of 17.5 pt/acre/12-month period. If mixing with copper, check copper label for limitations.
difenoconazole + benzovindiflupyr (Aprovia Top)	7+3	10.5 to 13.5 fl oz/acre	0	0.5	Do not make more than 2 applications before alternating to a non-Group 7 fungicide. See label for application intervals and limits/season. Use of a spreading adjuvant is recommended.
difenoconazole + cyprodinil (Inspire Super)	3+9	16 to 20 fl oz/acre	0	0.5	Limit of 80 fl oz/acre/season. Do not make more than 2 consecutive applications before alternating to fungicide with different mode of action.
difenoconazole + mandipropamid (Revus Top)	3+40	5.5 to 7 fl z/acre	1	0.5	Make no more than 2 consecutive applications before switching to another fungicide with a different mode of action. See label for application limits
difenoconazole + tea tree oil (Regev)	3+BM01	4.0 to 8.5 fl oz/acre	2	0.5	Minimum application interval is 7 days. Do not make more than 2 sequential applications before alternating to a fungicide with a different mode of action. Maximum applications are 4/year.
famoxadone + cymoxanil (Tanos)	11+27	8 oz/acre	3	0.5	Do not make more than 1 application before alternating to a fungicide with a different mode of action. **NOTE: Must be tank mixed with a contact fungicide that has a different mode of action.** See label for application limits.
fluopyram + difenoconazole (Luna Flex)	7+3	8.0 to 13.6 fl oz/acre	0	0.5	Maximum 2 applications/year at high rate. Maximum interval between applications is 7 days.
mancozeb (various)	M03	See label	5	1	
mancozeb + azoxystrobin (Dexter Max)	M03+11	0.8 to 1.6 lb/acre	5	1	For states East of the Mississippi and including Mississippi, do not exceed 12 lb of product/acre/season. States West of Mississippi use 0.8 to 1.1 lb/ acre and do not exceed 9.14 lb of product/acre/season. Do not exceed 12 lb of product/acre/season. Do not make more than 1 application before alternation to a fungicide not in Group 11. On fresh market tomato, do not tank-mix with an adjuvant or an EC formulation. **Tank mixture with dimethoate may cause crop injury**.
mancozeb + copper hydroxide (ManKocide 61DF)	M03+ M01	1 to 3 lb/acre	5	2	Apply at 7- to 10-day intervals.
mancozeb + zoxamide (Gavel 75DF)	M03+22	1.5 to 2 lb/acre	5	2	For states East of the Mississippi River, do not exceed 16 lb/acre/year. States West of the Mississippi River, do not exceed 8 lb/acre/year.
mefenoxam + chlorothalonil (Ridomil Gold Bravo SC)	4+M05	2.5 pt/acre	7	2	Apply a protectant between applications of Ridomil Gold Bravo SC.
pydiflumetofen + fludioxonil (Miravis Prime)	7+12	9.2 to 11.4 fl oz/acre	0	0.5	Do not make more than two applications of Miravis Prime or other Group 7 and 12 fungicides before alternation with a fungicide that is not in Group 7 or 12. Do not exceed 22.8 fl oz/acre/season. Apply by ground, air, or chemigation.
SOUR ROT (*GEOTRICHUM CANDIDUM***)**					
fludioxonil + propiconazole (Chairman)	3+12	See label	0	0	Use as a post-harvest dip, drench, or high-volume spray to control certain post-harvest rots. See label for details.

TABLE 3-36. DISEASE CONTROL PRODUCTS FOR TOMATO (cont'd)

I. Meadows, Extension Plant Pathologist, North Carolina State University; R. Singh, Plant Pathologist, Louisiana State University Agricultural Center

Disease/Material	FRAC Code	Rate of Material Formulation	Minimum Days Harv.	Minimum Days Reentry	Method, Schedule, and Remarks
TOMATO (FIELD) (cont'd)					
SOUR ROT (*GEOTRICHUM CANDIDUM*) (cont'd)					
propiconazole (various)	3	See label	0	0	Use as a post-harvest dip, drench, or high-volume spray to control certain post-harvest rots. See label for details.
fludioxonil (Scholar SC)	12	See label	0	0	Use as a post-harvest dip, drench, or high-volume spray to control certain post-harvest rots. See label for details.
SOUTHERN BLIGHT (*AGROATHELIA ROLFSII* = *ATHELIA ROLFSII* = *SCLEROTIUM ROLFSII*)					
difenoconazole + benzovindiflupyr (Aprovia Top)	7+3	10.5 to 13.5 fl oz/acre	0	0.5	**DISEASE SUPPRESSION ONLY.** Do not make more than 2 applications before alternating to a non-Group 7 fungicide. See label for application intervals and limits/season.
fluazinam (Omega 500F)	29	16 to 24 fl oz/acre	7	12	Apply the initial application as a soil directed spray at or within the day of transplanting at 24 fl oz (0.782 lb ai)/acre. Follow the soil directed spray at transplant with up to five foliar applications at 7- to 14-day retreatment intervals. Do not make more than 6 applications at 24 fl oz/acre/year. Do not apply more than 144 fl oz of the product/acre/year
fluopyram + trifloxystrobin (Luna Sensation)	7+11	7.6 fl oz/acre	3	0.5	**DISEASE SUPPRESION ONLY.** Do not make more than 2 applications before alternating to a non-Group 7 or 11 fungicide. See label for application intervals and limits/season.
fluoxastrobin (Aftershock, Evito 480SC)	11	2.0 to 5.7 fl oz/acre	3	0.5	**DISEASE SUPPRESSION ONLY.** Begin applications when conditions favor disease development, on 7- to 10-day intervals. Do not make more than 1 application before alternating to a fungicide with a different mode of action. Do not apply more than 22.8 fl oz/acre/season.
fluxapyroxad + pyraclostrobin (Priaxor Xemium)	7+11	4 to 8 fl oz/acre	0	0.5	**DISEASE SUPPRESSION ONLY.** Limit of 24 fl oz/acre/season. Do not make more than 2 consecutive applications before alternating to a fungicide with a different mode of action. Do not mix Priaxor with any other products, adjuvants, additives, or nutrients for application to fresh market tomatoes at less than 20 gal/acre spray volume.
penthiopyrad (Fontelis 1.67SC)	7	1 to 1.6 fl oz/ 1000 row ft	0	0.5	Apply as a soil drench to seedling tray or at the time of transplant. See label for application limits.
pyraclostrobin (Cabrio EG)	11	12 to 16 oz/acre	0	0.5	**DISEASE SUPPRESSION ONLY.** Limit of 96 fl oz/acre/season. Do not make more than 2 applications before alternating to a fungicide with a different mode of action.
TIMBER ROT, WHITE MOLD, OR SCLEROTINIA STEM ROT (*SCLEROTINIA*)					
boscalid (Endura WDG)	7	12.5 oz/acre	0	0.5	Limit 2 applications/year and a maximum of 25 oz/acre/year.
Coniothyrium minitans strain CON/M/91-08 (Contans WG)	BM02	1 to 4 lb/acre	0	4 h	Apply at crop emergence or crop transplant. Do not tank-mix or apply other fungicides within 7 days before or after use.
fluxapyroxad + pyraclostrobin (Priaxor Xemium)	7+11	4 to 8 fl oz/100 gal	0	0.5	**DISEASE SUPPRESSION ONLY.** Limit of 24 fl oz/acre/season. Do not make more than 2 consecutive applications before alternating to a fungicide with a different mode of action. See label for application limits. Do not mix Priaxor with any other products, adjuvants, additives, or nutrients for application to fresh market tomatoes at less than 20 gal/acre spray volume.
pyraclostrobin (Cabrio EG)	11	12 to 16 oz/acre	0	0.5	**DISEASE SUPPRESSION ONLY.** Limit of 96 fl oz/acre/season. Do not make more than 2 applications before alternating to a fungicide with a different mode of action.

TABLE 3-37. IMPORTANCE OF ALTERNATIVE MANAGEMENT PRACTICES FOR DISEASE CONTROL IN TOMATO

I. Meadows, Extension Plant Pathologist, North Carolina State University; R. Singh, Plant Pathologist, Louisiana State University Agricultural Center

Scale: E, excellent; G, good; F, fair; P, poor; NC, no control; NA, not applicable; ND, no data

* Bacterial canker (foliar or systemic) is rarely observed on open field grown tomatoes in deep Southern states.
** Tomato spotted wilt virus is transmitted by thrips.
*** Race specific.

Strategy	Bacterial canker*	Bacterial speck	Bacterial spot	Bacterial wilt	Buckeye rot	Alternaria stem canker/ Early blight	Fusarium wilt	Gray mold (Botrytis)	Late blight	Leaf mold (greenhouse or open field)	Powdery mildew	Septoria leaf spot	Southern blight	Target spot (greenhouse or open field)	Tomato spotted wilt virus **	Tomato yellow leaf curl virus	Verticillium wilt
Use of resistant cultivars	NA	P	P	NA	NA	F	G***	NA	G	P	F	NR	NA	P	G	G	G***
Use of disease-free seed or transplants	G	G	G	NA	NC	NC	NA	NC	NC	F	NC	P	NA	F	NA	G	NA
Use of seed treatments	G	G	G	NA	NC	P	NA	NC	P	F	NA	P	NA	ND	NA	NA	NA
Use of sanitation practices at transplant stage	G	G	G	NA	NC	G	NA	G	NC	F	NC	NC	NA	F	NC	F	NA
Use of grafted rootstocks	NC	NC	NC	G***	NC	NC	G***	NC	NC	NC	NC	NC	G	NC	NC	NA	NC
Crop rotation (3-4 years) or tomato-free period	F	P	P	F	F	F	F	NC	NC	F	NC	P	F	P	NC	G	F
Control of solanaceous weeds	F	NC	NC	NC	F	F	NC	F	F	F	F	F	F	NC	F	F	NA	NC
Fertility	NC	NC	NC	NC	NC	F	NC	F	NC	ND	NC	NC	NC	ND	NC	NA	NC
Deep plow	NC	NC	NC	NC	NC	NC	NC	NC	NC	ND	NC	NC	NC	F	ND	NC	NA	NC
Use of cover crops	NC	NC	NC	NC	F	P	NC	NC	NC	ND	NC	NC	NC	ND	NC	NA	NC
Destroy crop residue	F	NC	NC	P	NC	P	F	NC	NC	F	NC	F	F	F	ND	NA	F
Rogue plants	F	NC	NC	NC	NC	NC	NC	NC	NC	NC	NC	NC	NC	NC	NC	NA	NC
Promote air movement	F	F	F	NC	P	P	NC	F	F	F	P	F	NC	F	NA	NA	NC
Use of plastic or reflective mulches	NC	NC	NC	NC	F	F	NC	NC	NC	NC	NC	F	NC	NC	G	G	NC
Do not handle plants when wet	G	G	G	NC	NC	P	NC	NC	P	F	NC	P	NC	F	NC	NA	NC
Use drip irrigation (avoid overhead irrigation)	F	F	F	NC	P	F	NC	F	F	F	NC	F	NC	F	NC	NA	NC
Use of biological control or biorational products	P	P	F	NC	NC	P	P	P	P	P	P	NC	P	P	NC	F	NC
Use of foliar fungicides/ bactericides	F	F	F	NC	G	G	NC	G	G	G	G	G	NC	F	NA	NA	NC
Use of insecticides	NC	NC	NC	NA	NC	NC	NC	NC	NC	NC	NC	NC	NC	NC	F	G	NC
Soil fumigation	NC	NC	NC	NC	F	P	G	NC	NC	NC	NC	NC	G	NC	NC	NA	F

TABLE 3-38. EFFICACY OF PRODUCTS FOR DISEASE CONTROL IN TOMATO [1]

I. Meadows, Extension Plant Pathologist, North Carolina State University; R. Singh, Plant Pathologist, Louisiana State University Agricultural Center

Scale: E, excellent; G, good; F, fair; P, poor; NC, no control; NA, not applicable; ND, no data

Active ingredient	Product	Fungicide Group [F]	Preharvest Interval (days)	Anthracnose	Bacterial canker (foliar)	Bacterial speck	Bacterial spot	Early blight	Gray mold (Botrytis)	Gray leaf spot (Stemphylium)	Late blight and Buckeye Rot	Leaf mold (Fulvia fulva)	Powdery mildew	Septoria leaf spot	Southern blight	Target spot
azoxystrobin [2]	Quadris	11	0	G	NC	NC	ND	E[R]	NC	E[R]	F	NC	E	G	ND	P[R]
azoxystrobin + chlorothalonil	Quadris Opti	11+M05	0	G	NC	NC	ND	E[R]	NC	E	F-G	NC	E	E	ND	P[R]
azoxystrobin + difenoconazole	Quadris Top	11+3	0	ND	NC	NC	NC	G	ND	G	F	ND	G	G	ND	ND
azoxystrobin + flutriafol	Topguard EQ	11+3	0	G	NC	NC	NC	G	ND	G	ND	ND	ND	ND	P	G
acibenzolar-S-methyl [9]	Actigard	P01	14	P	ND	F	F	NC	NC	NC	NC	NC	NC	NC	NC	NC
bacteriophage	AgriPhage	NC	0	NC	NC	P	P	NC	NC	NC	NC	NC	NC	NC	NC	NC
bacteriophage	AgriPhage-CMM	NC	0	NC	F	NC	NC	NC	NC	NC	NC	NC	NC	NC	NC	NC
benzovindiflupyr + difenoconazole	Aprovia Top	7+3	0	ND	NC	NC	NC	G	ND	G	ND	ND	F	P	NC	ND
boscalid	Endura	7	0	ND	NC	NC	NC	G	G	G	NC	ND	ND	ND	NC	P[R]
chlorothalonil	Bravo, Chloronil, Echo, Equus, Initiate	M05	0	F	NC	NC	NC	F	F	F	G	F	P	F	NC	F
chlorothalonil + cymoxanil	Ariston	M05+27	3	ND	NC	NC	NC	F	F	F	G	F	P	F	NC	F
copper [4]	(various)	M01	0	P	F	F	P[R]	F	NC	F	F	F	P	F	NC	NC
cyazofamid	Ranman	21	0	NC	NC	NC	NC	NC	NC	NC	F	NC	NC	NC	NC	NC
cymoxanil	Curzate	27	3	NC	NC	NC	NC	NC	NC	NC	F	NC	NC	NC	NC	NC
cyprodinil + fludioxonil	Switch, Alterity	9+12	0	ND	NC	NC	NC	F	F	F	NC	NC	F	NC	ND	NC
dimethomorph	Forum	40	4	NC	NC	NC	NC	NC	NC	NC	F	NC	NC	NC	NC	NC
difenoconazole + cyprodinil	Inspire Super	3+9	0	ND	NC	NC	NC	G	G	G	NC	G	G	F	ND	F
difenoconazole + tea tree oil	Regev	3+BM01	2	NC	NC	NC	NC	G	ND	ND	ND	ND	ND	ND	ND	ND
dimethomorph + ametoctradin	Zampro	40+45	4	NC	NC	NC	NC	NC	NC	NC	G	NC	NC	NC	NC	NC
famoxadone + cymoxanil	Tanos	11+27	3	ND	P	P	P	F	NC	F	F	F	ND	F	NC	F[R]
fenamidone	Reason	11	14	ND	NC	NC	NC	F	NC	F	F	NC	NC	P	ND	NC
fluazinam	Omega	29	7	ND	NC	NC	NC	ND	ND	ND	ND	ND	ND	ND	NZ_D	ND
fluopicolide	Presidio	43	2	ND	NC	NC	NC	NC	NC	NC	G	NC	NC	NC	NC	NC
fluoxastrobin	Evito 480SC, Aftershock	11	3	ND	NC	NC	NC	E[R]	P	E	F	NC	E[R]	G	F	F[R]
fluopyram + difenoconazole	Luna Flex	7+3	0	G	NC	NC	NC	E	E	E	NC	G	G	E	ND	G
fluopyram + trifloxystrobin	Luna Sensation	7+11	3	ND	NC	NC	NC	G	NC	G	ND	ND	G	ND	ND	ND
fluopyram + pyrimethanil	Luna Tranquility	7+9	1	NC	NC	NC	NC	G	ND	ND	ND	G	F	ND	ND	ND
flutriafol	Rhyme	3	0	ND	NC	NC	NC	G	ND	G	ND	ND	ND	ND	P	G
fluxapyroxad + pyraclostrobin	Priaxor	7+11	0	ND	NC	NC	NC	G	P	G	P	G	G	F	F	F
mancozeb	Dithane, Koverall, Manzate, Penncozeb	M03	5	F	NC	NC	P	F	P	F	F	F	NC	F	NC	F

[1] Efficacy ratings do not necessarily indicate a labeled use for every disease.

[2] Contact control only; not systemic.

[3] Biological control products consisting of a virus that attacks pathogenic bacteria.

[4] Fixed coppers include: Basicop, Champ, Champion, Citcop, Copper-Count-N, Kocide, Nu-Cop, Super Cu, Tenn-Cop, Top Cop with Sulfur, and Tri-basic copper sulfate.

[5] Streptomycin may only be used on transplants; not registered for field use.

[6] Sulfur may be phytotoxic; follow label carefully.

[7] Curative activity; not systemic.

[8] Curative activity; systemic.

[9] Systemic activated resistance.

[10] Do not use on cherry tomatoes.

[11] Can cause stunting on tomatoes.

[F] To prevent resistance in pathogens, alternate fungicides within a group with fungicides in another group. Fungicides in the "M" group are generally considered "low-risk" with no signs of resistance developing to most fungicides. "NC" indicates that the product has not been classified into a group.

[R] Resistance reported in the pathogen.

[O] OMRI-listed product.

TABLE 3-38. EFFICACY OF PRODUCTS FOR DISEASE CONTROL IN TOMATO [1] (cont'd)

I. Meadows, Extension Plant Pathologist, North Carolina State University; R. Singh, Plant Pathologist, Louisiana State University Agricultural Center

Scale: E, excellent; G, good; F, fair; P, poor; NC, no control; NA, not applicable; ND, no data

Active ingredient	Product	Fungicide Group [F]	Preharvest Interval (days)	Anthracnose	Bacterial canker (foliar)	Bacterial speck	Bacterial spot	Early blight	Gray mold (Botrytis)	Gray leaf spot (Stemphylium)	Late blight and Buckeye Rot	Leaf mold (Fulvia fulva)	Powdery mildew	Septoria leaf spot	Southern blight	Target spot
mancozeb + fixed copper	ManKocide	M03+M01	5	F	NC	F	F	F	NC	F	F	F	NC	F	NC	NC
mancozeb + azoxystrobin	Dexter Max	M03+11	5	ND	ND	ND	ND	G	ND	G	ND	ND	ND	ND	ND	ND
mancozeb + zoxamide	Gavel	M03+22	3	ND	NC	P	P	F	NC	F	F	F	NC	F	NC	NC
mandipropamid + difenoconazole	Revus Top	40+3	1	ND	NC	NC	NC	F	NC	F	G	F	F	F	NC	G
mefenoxam [8]	Ridomil Gold SL	4	7	NC	NC	NC	NC	P	P	P	E[R]	F	NC	F	NC	NC
mefenoxam [8] + chlorothalonil	Ridomil Gold Bravo	4+M05	5	NC	NC	NC	NC	P	P	P	E[R]	F	NC	F	NC	NC
mefenoxam + copper	Ridomil Gold/Copper	4+M01	14	NC	NC	NC	NC	NC	NC	NC	E[R]	NC	NC	NC	NC	NC
mefenoxam + mancozeb	Ridomil Gold MZ	4+M03	5	NC	NC	NC	NC	NC	NC	NC	G[R]	NC	NC	NC	NC	NC
myclobutanil	(various)	3	0	ND	NC	NC	NC	NC	NC	NC	NC	NC	G	NC	NC	ND
oxathiapiprolin + chlorothalonil	Orondis Opti	49+ M05	0	NC	NC	NC	NC	P	NC	P	E	NC	NC	NC	NC	NC
oxathiapiprolin + mefenoxam	Orondis Gold	49+4	7	NC	NC	NC	NC	NC	NC	NC	E	NA	NC	NC	NC	NC
oxathiapiprolin + mandipropamid	Orondis Ultra	49+40	1	NC	NC	NC	NC	NC	NC	NC	E	NC	NC	NC	NC	NC
penthiopyrad	Fontelis	7	0	ND	NC	NC	NC	G	F	G	NC	ND	F	F	F	F
phosphorous acid, mono- and dipotassium salts of	Resist 57	P07	0	NC	NC	NC	NC	NC	NC	NC	F	NC	NC	NC	NC	NC
polyoxin D zinc salt	Ph-D: OSO 5% SC	19	0	ND	ND	ND	ND	F	F	F	ND	ND	F	ND	NC	F
propamocarb	Previcur Flex	28	5	NC	NC	NC	NC	P	NC	P	F	NC	NC	NC	NC	NC
pydiflumetofen + fludioxonil	Miravis Prime	7+12	0	ND	NC	NC	F	E	ND	E	NC	ND	ND	E	ND	ND
pyraclostrobin	Cabrio	11	0	ND	NC	NC	NC	E[R]	P	E	F	NC	E	G	F	E[R]
pyrimethanil	Scala	9	1	ND	NC	NC	NC	F	F	F	NC	ND	ND	ND	NC	F
streptomycin [5]	Agri-Mycin 17, Ag-Streptomycin, Harbour	25	0	NC	NC	F	F	NC	NC	NC	NC	NC	NC	NC	NC	NC
sulfur [6,O]	(various)	M02	0	ND	NC	NC	NC	NC	NC	NC	NC	F	NC	NC	NC	NC
tetraconazole	Mettle	3	0.5	ND	NC	NC	NC	G	ND	G	NC	ND	E	G	ND	G
tolfenpyrad	Torac	39		ND	NC	NC	NC	ND	ND	ND	NC	E	ND	ND	ND	ND
trifloxystrobin	Flint, Flint Extra	11	3	ND	NC	NC	NC	E[R]	P	E	NC	E	G	F	F[R]	
zinc dimethyldithiocarbamate [10]	Ziram	M03	7	ND	NC	NC	NC	F	NC	F	ND	NC	ND	F	NC	ND
zoxamide + chlorothalonil	Zing!	22+M05	5	NC	NC	NC	NC	F	ND	F	F	F	ND	F	NC	ND

[1] Efficacy ratings do not necessarily indicate a labeled use for every disease.
[2] Contact control only; not systemic.
[3] Biological control products consisting of a virus that attacks pathogenic bacteria.
[4] Fixed coppers include: Basicop, Champ, Champion, Citcop, Copper-Count-N, Kocide, Nu-Cop, Super Cu, Tenn-Cop, Top Cop with Sulfur, and Tri-basic copper sulfate.
[5] Streptomycin may only be used on transplants; not registered for field use.
[6] Sulfur may be phytotoxic; follow label carefully.
[7] Curative activity; not systemic.
[8] Curative activity; systemic.
[9] Systemic activated resistance.
[10] Do not use on cherry tomatoes.
[11] Can cause stunting on tomatoes.

[F] To prevent resistance in pathogens, alternate fungicides within a group with fungicides in another group. Fungicides in the "M" group are generally considered "low-risk" with no signs of resistance developing to most fungicides. "NC" indicates that the product has not been classified into a group.

[R] Resistance reported in the pathogen.

[O] OMRI-listed product.

TABLE 3-39. EXAMPLE SPRAY PROGRAM FOR FOLIAR DISEASE CONTROL IN FRESH-MARKET TOMATO PRODUCTION

I. Meadows, Extension Plant Pathologist, North Carolina State University; R. Singh, Plant Pathologist, Louisiana State University Agricultural Center

Week	Program 1	Program 2	Program 3	Program 4 [1]
		BEFORE HARVEST (weeks 1 to 8)		
1	mancozeb (M) + copper (M) + Actigard (21)	mancozeb (M) + copper (M) + Actigard (21)	mancozeb (M) + copper (M) + Actigard (21)	mancozeb (M) + copper (M) + Actigard (21)
2	mancozeb (M) + copper (M)	mancozeb (M) + copper (M)	mancozeb (M) + copper (M)	mancozeb (M) + copper (M)
3	mancozeb (M) + Inspire Super (3+9) + Actigard (21)	mancozeb (M) + Priaxor (7+11) + Actigard (21)	mancozeb (M) + Aprovia Top (7+3) OR Luna Tranquility (7+9) + Actigard (21)	mancozeb (M) + strobilurin [1] (11) + Actigard (21)
4	mancozeb (M) + copper (M)	mancozeb (M) + copper (M)	mancozeb (M) + copper (M)	mancozeb (M) + copper (M)
5	mancozeb (M) + Fontelis (7) OR Endura (7) + Actigard (21)	mancozeb (M) + Switch (9+12) + Actigard (21)	mancozeb (M) + Aprovia Top (7+3) OR Luna Tranquility (7+9) + Actigard (21)	mancozeb (M) + Fontelis (7) OR Endura (7) + Actigard (21)
6	mancozeb (M) + copper (M)	mancozeb (M) + copper (M)	mancozeb (M) + copper (M)	mancozeb (M) + copper (M)
7 [2]	mancozeb (M) + Inspire Super (3+9) + Actigard (21)	mancozeb (M) + Priaxor (7+11) + Actigard (21)	mancozeb (M) + Aprovia Top (7+3) OR Luna Tranquility (7+9) + Actigard (21)	mancozeb (M) + strobilurin [1] (11) + Actigard (21)
8 [2]	mancozeb (M) + copper (M)	mancozeb (M) + copper (M)	mancozeb (M) + copper (M)	mancozeb (M) + copper (M)
		DURING HARVEST (weeks 9 to 15)		
9	Fontelis (7) OR Endura (7) + chlorothalonil (M)	Switch (9+12) + chlorothalonil (M)	Aprovia Top (7+3) OR Luna Tranquility (7+9) + chlorothalonil (M)	Fontelis (7) OR Endura (7) + chlorothalonil (M)
10	For late blight: Presidio (43) OR Ranman (21) OR Orondis Ultra (49+40) OR Zampro (45+40) OR Revus Top (40+3) For early blight: chlorothalonil (M)			
11 [3]	For late blight: chlorothalonil (M) For early blight: Fontelis (7) OR Endura (7) OR Switch (9+3)			
12 [3]	For late blight: Presidio (43) OR Ranman (21) OR Orondis Ultra (49+40) OR Zampro (45+40) OR Revus Top (40+3) For early blight: chlorothalonil (M)			
13 [3]	For late blight: chlorothalonil (M) For early blight: Fontelis (7) OR Endura (7) OR Switch (9+3)			
14 [3]	Finish season with chlorothalonil			

[1] Resistance to strobilurins is known to occur in the early blight pathogen in NC; if resistance is suspected, use alternate program.
[2] For late season plantings: If late blight is in the area, consider chlorothalonil for late blight control.
[3] Continue early blight program or use Revus Top if early blight pressure is high for weeks 11-14.

TURNIP GREENS SEE *GREENS AND LEAFY BRASSICAS*

TURNIP ROOTS SEE *ROOT VEGETABLES*

WATERMELONS SEE *CUCURBITS*

TABLE 3-40. NEMATODE CONTROL IN VEGETABLE CROPS

J. Desaeger, Nematologist, University of Florida

Crop losses due to nematodes can be avoided or reduced by using the following management tactics.

1. Practice crop rotation using poor or non-host (cover) crops.
2. Plow out and expose roots immediately after the last harvest.
3. Plow or disk the field two to four times before planting.
4. Use nematode-free planting material.
5. Sample soil and have it assayed for nematodes, preferably at the end of each crop cycle. There is a fee for each sample. Ship sample via DHL, FedEx, or UPS to: State Agency.
6. Where warranted, fumigate, or use other nematicides according to guidelines listed on the label. (For fumigation, soil should be warm, well worked, and free from undecomposed plant debris and have adequate moisture for proper diffusion of chemical into soil pores.)
7. For in-row application, insert chisels 6 to 8 inches deep and throw a high, wide bed up over it; do not rework rows after fumigating. In deep sand soils, deep shanking (up to 18 inches deep) may give better results.
8. For broadcast treatments, insert chisels 6 to 8 inches deep, and space chisels 12 inches apart for most fumigants; use 5-inch spacing for Vapam.
9. Row rates in this section are stated for rows on 40-inch spacing. For other row spacings, multiply the stated acre rate by the appropriate conversion factor to determine the amount of material applied per acre (Do not alter stated amount per 100-foot row).

This will be a guide to the amount of material to purchase for the acreage you want to treat:

Your Row Spacing (inches)	Conversion Factor	Your Row Spacing (inches)	Conversion Factor
24	1.67	42	0.952
26	1.54	44	0.909
28	1.43	46	0.870
30	1.33	48	0.833
32	1.25	5 ft	0.667
34	1.18	6 ft	0.556
36	1.11	7 ft	0.476
38	1.05	8 ft	0.417
40	1.00		

For example, if 10 gallons per acre are used on 40-inch rows, for 36-inch rows, it will take 11.1 gallons to treat an acre.

CAUTION: Read labels carefully. Some products have restrictive crop rotations.

FUMIGANTS

New labels require extensive risk mitigation measures including fumigant management plans (FMPs), buffer restrictions, Worker Protection Safety (WPS) standards and other measures. Details are on the labels and see http://www2.epa.gov/soil-fumigants. Some fumigants are registered on many vegetable crops but with crop-or soil-type-specific rates; others are registered for specific crops and/or in certain states only. Follow all labels carefully. The label is the law.

TABLE 3-41. EFFICACY OF FUMIGANTS OR FUMIGANT COMBINATIONS FOR MANAGING SOILBORNE NEMATODES, DISEASES, AND WEEDS

J. Desaeger, Nematologist, University of Florida; A. Gorny, Nematologist, North Carolina State University; I. A. Chowdhury, University of Georgia

Scale: ranges from "—" = not effective to "+++++" = most efficacious.

Product	Rate per Treated Acre [2] Volume (gal)	Weight (lb)	Nematodes	Disease	Nutsedge	Weeds: Annual [1]
Telone II (1,3-D)	15 to 27	153 to 275	+++++	+	—	—
Telone EC [3]	9 to 24 [4]	91 to 242 [4]	+++++	+	—	—
Telone C17 (1,3-D + chloropicrin)	32.4 to 42	343 to 445	+++++	+++	+	+
Telone C35 (1,3-D + chloropicrin)	39 to 50	437 to 560	+++++	+++++	+	++
InLine (1,3-D + chloropicrin) [3]	29 to 57.6 (See Label)	325 to 645 (See Label)	+++++	+++++	+	+++
Pic-Clor 60 (chloropicrin + 1,3-D)	48.6	588	+++++	+++++	+	+++
Pic-Clor 80 (chloropicrin + 1,3-D)	32.4 to 42	343 to 445	++++	+++++	+	+++
Pic-Clor 60 EC [3]	42.6	503	+++++	+++++	+	+++
Metam potassium [5]	30 to 62	318 to 657	+++	+++	+	++++
Metam sodium [5] (MS)	37.5 to 75	379 to 758	+++	+++	+	++++
Chloropicrin + MS [5]	19.5 to 31.5 + 37.5 to 75	275 to 444 + 379 to 758	+++	+++++	++	++++
Chloropicrin	48.6	150 to 350	++	+++++	++	—
Tri-Pic 100 [3]	8 to 24	100 to 300	+	+++++	—	—
Dominus (allyl isothiocyanate) [6]	25 to 40 [4]	212 to 340 [4]	++	+++	+	+++

[1] Fumigants with lower efficacy against weeds may require a complementary herbicide or hand-weeding program, although use of virtually impermeable film (VIF) or impermeable film (TIF) may increase weed control, particularly with Telone C35 or Paladin. Refer to the Herbicide Recommendation section of this guide for directions pertaining to herbicide applications. To prevent phytotoxicity, soil should be undisturbed and unplanted for 7- to 14-days after Telone application. However, the planting interval depends upon soil temperature and moisture. Telone can persist for more than 21 days under cool or wet soil conditions.

[2] Rates can sometimes be reduced if products are applied with VIF or TIF.

[3] Product is formulated for application through drip lines under a plastic mulch; efficacy is dependent on good distribution of the product in the bed profile.

[4] Labelled rates are per *broadcast-equivalent* acre, NOT per treated acre.

[5] Metam potassium can be Metam KLR, K-Pam, Sectagon K54 or other registered formulations, and should be used in soils with high sodium content. Metam sodium can be Vapam, Sectagon 42, Metam CLR or other registered formulations.

[6] Dominus is registered in a few states in the Southeast, but there is limited experience with the product through University or independent trials in our region; growers may want to consider this on an experimental basis. Planting interval is 10 days. The active ingredient allyl isothiocyanate is like the active ingredient in metam sodium products (methyl isothiocyanate) and is likely to behave in a similar manner with a similar pest control profile.

TABLE 3-42. MANAGEMENT OF SOILBORNE NEMATODES WITH NON-FUMIGANT NEMATICIDES

J. Desaeger, Nematologist, University of Florida; A. Gorny, Nematologist, North Carolina State University; I. A. Chowdhury, University of Georgia

Nematodes are best managed through an integrated program (IPM). Key management options may include securing advisory/predictive soil samples, crop rotation, fallow periods, host resistance, soil amendments, flooding, soil solarization, suppressive cover crops, & other options.

Material (Product)	Vegetable Crop	Application Method	Rate/Acre	Rate/ 1000 Ft Row	Schedule and Remarks
ethoprop (Mocap 15G)	Bean (snap & lima)	Band	13 to 20 lb	0.9 to 1.4 lb	Do not place in-furrow or allow granules to contact seed. Incorporate 2 to 4" deep in 12" to 15" band, at planting. Use higher rates for higher nematode populations
		Broadcast	40 to 54 lb	N/A	Do not place in-furrow or allow granules to contact seed. Incorporate 2 to 4" deep no more than 3 days before planting. Use higher rates for higher nematode populations.
	Cabbage	Band	13 lb	0.9 lb	Do not place in-furrow or allow granules to contact seed. Incorporate 2 to 4" deep in 15" band at planting.
		Broadcast	34 lb	N/A	Do not place in-furrow or allow granules to contact seed. Incorporate 2 to 4" deep no more than 1 week before planting.
	Corn (field and sweet)	Band	—	0.75 to 1.0 lb	Incorporate 2 to 4" deep in 12 to 15" band.
		Broadcast	40 lb	N/A	Incorporate 2 to 4" deep no more than 3 days before, to at planting.

TABLE 3-42. MANAGEMENT OF SOILBORNE NEMATODES WITH NON-FUMIGANT NEMATICIDES

J. Desaeger, Nematologist, University of Florida; A. Gorny, Nematologist, North Carolina State University; I. A. Chowdhury, University of Georgia

Nematodes are best managed through an integrated program (IPM). Key management options may include securing advisory/predictive soil samples, crop rotation, fallow periods, host resistance, soil amendments, flooding, soil solarization, suppressive cover crops, & other options.

Material (Product)	Vegetable Crop	Application Method	Rate/Acre	Rate/ 1000 Ft Row	Schedule and Remarks
ethoprop (Mocap 15G) (cont'd)	Cucumber	Band	13 lb	2.1 lb	Do not place in-furrow or allow granules to contact seed. Incorporate 2 to 4" deep in 12 to 15" band (7 ft row spacing) at planting.
	Potatoes	Band	20 lb	1.4 lb	For suppression of stubby root nematode populations. Incorporate 2 to 4" deep; Band should be 12" to 15" wide (36" row spacing) at planting; do not apply once seedlings have begun to emerge.
		Broadcast	40 to 60 lb	N/A	For suppression of moderate to heavy stubby root nematode populations; apply within 2 wk before planting; do not apply once seedlings have begun to emerge.
	Sweetpotato	Band only	20 to 26 lb	1.6 to 2.1 lb	Incorporate 2 to 4" deep in centered 12 to 15" band (at least 42" row spacing) 2 to 3 wk before planting. Use no more than one application per season.
ethoprop (Mocap EC)	Potatoes	Band	63.9 fl. oz	4.4 fl. oz	For suppression of stubby root nematode populations. Incorporate 2 to 4" deep; Band should be 12' wide (36" row spacing) at planting or before crop emergence.
		Broadcast	1 to 1.5 gal	N/A	For suppression of moderate to heavy stubby root nematode populations; apply and immediately incorporate no more than 2 wk before planting or crop emergence.
	Sweetpotato	Band only	63.5 to 85.9 fl. oz	5.1 to 6.9 fl. oz.	Incorporate 2 to 4" deep in centered 12 to 15" band (at least 42" row spacing) 2 to 3 wk before planting. Use no more than one application per season.
oxamyl (Vydate L)	Carrot, Cucumber, Cantaloupe, Honeydew melon, Watermelon, Squash, Pumpkin, Eggplant, Pepper (bell and non-bell), Sweetpotatoes, Tomatoes	Preplant Foliar, Drip/ soil injection; Chemigation In-furrow	See label	—	Note: Vydate L is labeled for these crops to manage nematodes and certain insects. However, the label varies by crop, method of application and state. Therefore, follow the label carefully for the best method, rate, and timing.
oxamyl (Vydate C-LV)	Potatoes	Foliar ground; Chemigation Aerial; At-plant In-furrow	See label	—	Note: Vydate C-LV is labeled for potatoes to manage nematodes and certain insects. However, the label varies by crop, method of application, timing, crop rotation program, and state. Therefore, follow the label carefully for the best method, rate, and timing.
fluopyram (Velum Prime)	Brassica (Cole) Leafy Vegetables (group 5) Cucurbits (group 9) Fruiting vegetables (tomato subgroup)	Chemigation	6.5 to 6.84 fl oz	—	See label for specific labeled crops within the subgroups. Chemigation into root zone through low-pressure drip, trickle, micro-sprinkler, or equivalent equipment. Can be applied the day of harvest. Do not make more than 2 sequential applications with this product or any other product in FRAC group 7. Velum is also labeled for powdery mildew control. The first foliar fungicide application after Velum Prime should be a product from a different FRAC group.
	Potato	Chemigation	6.5 to 6.84 fl oz	—	Apply specified dosage using overhead chemigation equipment. May offer suppression only if root-knot pressure is high and other methods of suppression should also be employed. Velum Prime is also registered to suppress early blight (*Alternaria solani*) and suppress white mold (*Sclerotinia sclerotiorum*).
	Sweetpotato	Chemigation	6.0 to 6.84 fl oz	—	Apply as post-planting drench, or hill drench. May offer suppression only if root-knot pressure is high and other methods of suppression should also be employed. Velum Prime is also registered to suppress white mold (*Sclerotinia sclerotiorum*).
fluensulfone (Nimitz)	Cucurbits (group 9); Fruiting Vegetables (groups 8-10); Brassica Leafy Vegetables (group 5) Celery, lettuce, spinach (group 4) Potato, Sweetpotato	Broadcast soil	3.5 to 7.0 pt	N/A	Apply and incorporate 6 to 8" deep at least 7 days before transplanting. Irrigate with 0.5 to 1.0" of water 2 to 5 days after application.
		Banded soil	See label	See label	Table 2 on label specifies rate based on row spacing. Incorporate 6 to 8" deep at least 7 days before transplanting. Irrigate with 0.5 to 1.0 in. water 2 to 5 days after application.
		Drip irrigation	—	—	Table 3 on label specifies rate based on bed width. Uniformly wet entire bed width and root zone 6 to 8 in. deep at least 7 days before trans- planting. Irrigate with 0.5 to 1.0" of water 2 to 5 days after application.
terbufos (Counter 20G)	Sweet corn	Band	—	4.5 to 6 fl oz	Place granules in 4-5" band over the open seed furrow and incorporate thoroughly into top 1" of the soil. Apply no more than 6.5 lb. per acre.
		In-furrow	—	4.5 to 6 fl oz	Place granules directly into the open seed furrow behind the planter seed opener.
spirotetramat (Movento)	Brassica (Cole) Leafy Vegetables (group 5) Fruiting Vegetables (groups 8-10) Leafy vegetables (group 4, except Brassica) Potato and other tubers	Foliar Chemigation	—	4.0 to 5.0 fl oz	Note: Movento is labeled for these crops to suppress or control nematode and insect pests. Must be tank mixed with a spray adjuvant/ additive to maximize leaf uptake. Follow the label carefully for the application method, interval, and timing.

TABLE 3-42. MANAGEMENT OF SOILBORNE NEMATODES WITH NON-FUMIGANT NEMATICIDES

J. Desaeger, Nematologist, University of Florida; A. Gorny, Nematologist, North Carolina State University; I. A. Chowdhury, University of Georgia

Nematodes are best managed through an integrated program (IPM). Key management options may include securing advisory/predictive soil samples, crop rotation, fallow periods, host resistance, soil amendments, flooding, soil solarization, suppressive cover crops, & other options.

Material (Product)	Vegetable Crop	Application Method	Rate/Acre	Rate/ 1000 Ft Row	Schedule and Remarks
fluazaindolizine (Salibro)	Carrot	Pre-plant incorporated; in-season drip chemigation	See label	See label	Pre-plant incorporate within 7 days prior to planting for maximum residual efficacy. Do not apply more than 61.4 fl oz of product per acre per year. Do not apply within 65 days of harvest. See label for single and split application rates.
	Cucurbits (group 9), Fruiting Vegetables (groups 8-10)	Pre-plant incorporate; pre-plant drip; at-plant drip; in-season drip chemigation	See label	See label	Pre-plant incorporate within 7 days prior to planting for maximum residual efficacy. Do not apply more than 23 fl oz (for cucurbits) or 61.4 fl oz (for fruiting vegetables) of product per acre per year. Do not apply within 1 days of harvest. See label for single and split application rates.
	Tuber and Corm Vegetables (group 1C)	Pre-plant incorporate; in-furrow treatment; in-season drip chemigation	See label	See label	Pre-plant incorporate within 7 days prior to planting for maximum residual efficacy. Do not apply more than 61.4 fl oz of product per acre per year. Do not apply within 40 days of harvest. See label for single and split application rates. Do not apply to sweetpotato, ginger, or taro outside of the United States.

GREENHOUSE VEGETABLE CROP DISEASE CONTROL

Note: Follow manufacturer's directions on label in all cases.

Caution: At the time this table was prepared, the entries were believed to be useful and accurate, however, labels change rapidly, and errors are possible, so the user must follow all directions on the pesticide container. See product labels for application limits per crop/ season.

Information in the following table must be used in the context of a total disease control program. For example, many diseases are controlled using resistant varieties, crop rotation, sanitation, seed treatment, and cultural practices. Always use top-quality seed or plants obtained from reliable sources. Seeds are ordinarily treated by the seed producer for the control of seed decay and damping-off. Most foliar diseases can be reduced or controlled by maintaining relative humidity below 90%, by keeping the air circulating in the house with a large overhead polytube, and by avoiding water on the leaves.

Caution: The risk of pesticide exposure in the greenhouse is high. Use protective clothing laundered daily or after each exposure. Ventilate during application and use appropriate personal protective equipment (PPE).

TABLE 3-43. GREENHOUSE DISEASE CONTROL FOR VARIOUS VEGETABLE CROPS

R. A. Melanson, Extension Plant Pathologist, Mississippi State University; I. Meadows, Extension Plant Pathologist, North Carolina State University

Crop/Disease	Product [1]	FRAC	Rate of Material	Minimum Days Harv.	Minimum Days Reentry	Method, Schedule, and Remarks
GREENHOUSE						
Sanitation	Solarization	NA	140°F, 4 to 8 hr For 7 days	—	—	Close greenhouse during hottest and sunniest part of summer for at least one week. Greenhouse must reach at least 140°F each day. Remove debris, heat sensitive materials, and keep greenhouse and contents moist. Will not control pests 0.5 inches or deeper in soil. Not effective against tobacco mosaic virus (TMV).
	Added heat	NA	180°F for 30 min	—	—	Remove all debris and heat-sensitive materials. Keep house and contents warm.

[1] Products registered for field use may be used on greenhouse crops (but not transplants) unless excluded on the label. Always check the label before applying a product. Additional products not prohibited for greenhouse use on basil, cucurbits, and lettuce not listed in this table are available.

[2] Former names of diseases and pathogens listed in this table that may be still be listed on fungicide labels are as follows: *Acidovorax citrulli* (formerly *Acidovorax avenae* subsp. *citrulli*) *Alternaria linariae* (formerly *A. solani* and *A. tomatophila*); *Agroathelia rolfsii* (formerly *Athelia rolfsii* and *Sclerotium rolfsii*); *Golovinomyces* spp. (formerly *Erysiphe* spp.) or *Golovinomyces cichoracearum* (formerly *Erysiphe cichoracearum*); *Fulvia fulva* (formerly *Cladosporium fulvum* and *Passalora fulva*); *Boeremia exigua* (formerly *Phoma exigua*); *Plectosphaerella cucumerina* (formerly *Plectosporium tabacinum*); *Podosphaera xanthii* (formerly *Sphaerotheca fuliginea*); *Stagonosporopsis* spp. (formerly *Didymella*); Xanthomonas leaf spot (formerly bacterial leaf spot); and (*Xanthomonas cucurbitae* (formerly *X. campestris* pv. *cucurbitae*).

TABLE 3-43. GREENHOUSE DISEASE CONTROL FOR VARIOUS VEGETABLE CROPS (cont'd)

R. A. Melanson, Extension Plant Pathologist, Mississippi State University; I. Meadows, Extension Plant Pathologist, North Carolina State University

Crop/Disease	Product [1]	FRAC	Rate of Material	Minimum Days Harv.	Minimum Days Reentry	Method, Schedule, and Remarks
SOIL						
Soilborne diseases and weeds		—	See soil fumigants table and check soil fumigant label if registered for greenhouse use.			Preplant soil treatment.
BASIL (LAST UPDATED IN 2025)						
Anthracnose, Alternaria blight, bacterial blight, Botrytis, downy mildew, leaf spot, Rhizoctonia leaf blight	copper, fixed (various)	M01	See label	See label	See label	**Some products are OMRI-listed.** See product labels for complete application instructions, specific crop and disease labels, and greenhouse usage. **Check state registration status prior to use.**
Alternaria leaf spot (*Alternaria* spp.), Botrytis leaf blight (*Botrytis* spp.), Fusarium blight (*Fusarium* spp.)	cyprodinil + fludioxonil (Switch 62.5WG)	9+12	11 to 14 oz/acre	7	0.5	*Not prohibited for greenhouse use.* After two applications, alternate with another fungicide with a different mode of action for two applications. Do not exceed 56 oz of product per acre per year. See label for application limits of active ingredients.
Downy mildew (*Peronospora belbahrii*)	cyazofamid (Ranman 400SC, Segway O)	21	2.75 to 3.0 fl oz/acre	0	0.5	Do not exceed 9 applications of product per crop. Alternate applications with fungicides that have a different mode of action. Do not make more than three consecutive applications before switching to products that have a different mode of action for three applications before returning to Ranman 400SC or Segway O. Do not exceed 27 fl oz of product per acre per year. See label for surfactant recommendations. Segway O is also labeled for use against Phytophthora root rot (*Phytophthora* spp.).
Alternaria leaf spot (*Alternaria* spp.), Botrytis leaf blight (*Botrytis* spp.), Fusarium blight (*Fusarium* spp.)	fludioxonil (Spirato GHN)	12	5.5 to 7 fl oz/acre	7	0.5	After two applications of product, alternate to a fungicide with a different mode of action for two applications. Do not exceed 28 fl oz of product per acre per year.
Gray mold (*Botrytis cinerea*), powdery mildew (*Golovinomyces* spp.[2])	fluopyram + trifloxystrobin (Luna Sensation)	7+11	5.0 to 7.6 fl oz/acre (powdery mildew) 7.6 fl oz/acre (gray mold)	7	0.5	*Not prohibited for greenhouse use.* Do not exceed 15.3 fl oz of product per acre per year. Do not apply more than 0.446 lb fluopyram or 0.25 lb trifloxystrobin per acre per year. Do not make more than two sequential applications of product or of any FRAC 7- or FRAC 11-containing fungicide before alternating with a fungicide from a different FRAC group.
Labels may include: Anthracnose, Downy mildew, powdery mildew	hydrogen peroxide + peroxyacetic acid (OxiDate 2.0, OxiDate 5.0, ZeroTol 2.0)	NG	See label	0	See label	**OMRI-listed.** See labels for additional instructions and labeled diseases, and precautions, including those related to the use and application of metal-based chemicals. Do not store spray solution. **OxiDate 2.0:** *Not prohibited for greenhouse use.* **OxiDate 5.0:** *Not prohibited for greenhouse use. Determine if product can be used safely on greenhouse crops prior to application.*
Downy mildew (*P. belbahrii*)	mandipropamid (Micora)	40	8.0 fl oz/acre (0.9 fl oz/5,000 sq ft)	1	4 hr	*For basil transplants grown in enclosed greenhouses with permanent flooring for resale to consumers.* Do not make more than two applications per crop before switching to a fungicide with a different mode of action. Do not apply within 1 day of shipping. See label for additional restrictions and recommendations.
Labels may include: Downy mildew, root rots	phosphonates (mono- and dibasic salts of phosphites, phosphorous acid, potassium phosphite) (various)	P07	See label	See label	See label	See product labels for complete application instructions, labeled diseases, restrictions, and crop and greenhouse usage. **Check state registration status prior to use.**
Labels may include: Botrytis, downy mildew, powdery mildew	potassium bicarbonate (Carb-O-Nater, MilStop SP)	NC	2.5 to 5.0 lb/100 gal (Carb-O-Nator) 1.25 to 5.0 lb/100 gal (MilStop)	0	4 hr (Carb-O-Nator) 1 hr (MilStop)	**OMRI-listed.** Do not store unused spray solution. See labels for additional labeled diseases and instructions. **Carb-O-Nator:** Final spray solution should not be below pH 7.0. Some tank-mixes may be incompatible and may cause phytotoxicity. **MilStop:** Do not exceed 0.5 lb of product per 4,350 sq ft or 1.15 lb product per 10,000 sq ft per application. Do not adjust pH after mixing. Do not mix with other soluble pesticide or fertilizers not compatible with mild alkaline solutions.

[1] Products registered for field use may be used on greenhouse crops (but not transplants) unless excluded on the label. Always check the label before applying a product. Additional products not prohibited for greenhouse use on basil, cucurbits, and lettuce not listed in this table are available.

[2] Former names of diseases and pathogens listed in this table that may be still be listed on fungicide labels are as follows: *Acidovorax citrulli* (formerly *Acidovorax avenae* subsp. *citrulli*) *Alternaria linariae* (formerly *A. solani* and *A. tomatophila*); *Agroathelia rolfsii* (formerly *Athelia rolfsii* and *Sclerotium rolfsii*); *Golovinomyces* spp. (formerly *Erysiphe* spp.) or *Golovinomyces cichoracearum* (formerly *Erysiphe cichoracearum*); *Fulvia fulva* (formerly *Cladosporium fulvum* and *Passalora fulva*); *Boeremia exigua* (formerly *Phoma exigua*); *Plectosphaerella cucumerina* (formerly *Plectosporium tabacinum*); *Podosphaera xanthii* (formerly *Sphaerotheca fuliginea*); *Stagonosporopsis* spp. (formerly *Didymella*); Xanthomonas leaf spot (formerly bacterial leaf spot); and (*Xanthomonas cucurbitae* (formerly *X. campestris* pv. *cucurbitae*).

TABLE 3-43. GREENHOUSE DISEASE CONTROL FOR VARIOUS VEGETABLE CROPS (cont'd)

R. A. Melanson, Extension Plant Pathologist, Mississippi State University; I. Meadows, Extension Plant Pathologist, North Carolina State University

Crop/Disease	Product [1]	FRAC	Rate of Material	Minimum Days Harv.	Minimum Days Reentry	Method, Schedule, and Remarks
BASIL (LAST UPDATED IN 2025) (cont'd)						
Powdery mildew (*Golovinomyces* spp.[2])	trifloxystrobin (Flint Extra)	11	3.8 fl oz/acre	0	0.5	**Not prohibited for greenhouse use.** Do not exceed 7.6 fl oz per acre per year. Do not apply more than 2 sequential applications of another FRAC 11-containing fungicide before rotating with fungicide from different FRAC group.
CUCURBITS – *TRANSPLANT PRODUCTION*						
Alternaria leaf blight (*Alternaria cucumerina*), Alternaria leaf spot (*A. alternata*), anthracnose (*Colletotrichum orbiculare*), Cercospora leaf spot (*C. citrullina*), downy mildew (*Pseudoperonospora cubensis*), gummy stem blight (*Stagonosporopsis* spp.[2]), Myrothecium canker (*Paramyrothecium roridum*[2]); Phoma blight (*Boeremia exigua*[2]), Phyllosticta leaf spot (*P. cucurbitacearum*), Plectosporium blight (*Plectosphaerella cucumerina*[2]), powdery mildew (*Podosphaera xanthii*, *Golovinomyces cichoracearum*)[2], Septoria leaf blight (*S. cucurbitacearum*)	azoxystrobin + benzovindiflupyr (Mural)	11+7	0.6 to 0.8 oz/ 5,000 sq ft	—	0.5	**Allowed for use on transplants grown for resale to consumers.** Do not exceed two applications of product per crop. See label for restrictions for product use when multiple crops are grown in the same area. Do not apply more than one application of a FRAC 11-containing fungicide before alternating to another fungicide with a different mode of action. See label for additional restrictions when multiple crops are grown in the same area.
Gray mold (*Botrytis cinerea*)	fenhexamid (Decree 50 WDG)	17	1.5 lb/acre (stand-alone) 1.0 to 1.5 lb/acre (tank-mix)	—	0.5	**Labeled for cucumbers ONLY.** Do not make more than two consecutive applications. See label for tank-mixing instructions. Do not exceed 6.0 lb product per acre per crop cycle to transplants. Do not exceed six applications per acre per crop cycle at the lowest use rate or four applications per acre per crop cycle at the highest use rate.
Alternaria leaf blight (*A. cucumerina*), Alternaria leaf spot (*A. alternata*), gummy stem blight (*Stagonosporopsis*[2]), powdery mildew (*P. xanthii, G. cichoracearum*)[2]	fludioxonil (Spirato GHN)	12	5.5 to 7 fl oz/acre	—	0.5	After two applications of product, alternate to a fungicide with a different mode of action for two applications. Do not exceed 28 fl oz of product per acre per year. See label for restrictions regarding applications made in the vicinity of aquatic areas.
Corynespora leaf spot (*Corynespora cassiicola*), early blight (*Alternaria* sp.), gray mold (*Botrytis*), gummy stem blight (*Stagonosporopsis* spp.[2]), powdery mildew (*Podosphaera* sp.[2]), scab (*Cladosporium* sp.)	polyoxin D zinc salt (Affirm WDG)	19	6.2 oz/acre	—	4 hr	**Allowed for use in greenhouses ONLY in the production of seedlings and transplants; product is not for use in the production of edible commodities.** Do not apply more than five applications per season. See product label for maximum limits of active ingredient and applications per season. See product labels for additional application instructions and restrictions.
Powdery mildew	triflumizole (Terraguard SC)	3	2 to 4 fl oz/100 gal	—	0.5	**For use on cucumber only.** For use only as a foliar spray. For use in commercial greenhouse production only. Do not exceed two applications per crop when applying to transplants. See label for additional application instructions.

[1] Products registered for field use may be used on greenhouse crops (but not transplants) unless excluded on the label. Always check the label before applying a product. Additional products not prohibited for greenhouse use on basil, cucurbits, and lettuce not listed in this table are available.

[2] Former names of diseases and pathogens listed in this table that may be still be listed on fungicide labels are as follows: *Acidovorax citrulli* (formerly *Acidovorax avenae* subsp. *citrulli*) *Alternaria linariae* (formerly *A. solani* and *A. tomatophila*); *Agroathelia rolfsii* (formerly *Athelia rolfsii* and *Sclerotium rolfsii*); *Golovinomyces* spp. (formerly *Erysiphe* spp.) or *Golovinomyces cichoracearum* (formerly *Erysiphe cichoracearum*); *Fulvia fulva* (formerly *Cladosporium fulvum* and *Passalora fulva*); *Boeremia exigua* (formerly *Phoma exigua*); *Plectosphaerella cucumerina* (formerly *Plectosporium tabacinum*); *Podosphaera xanthii* (formerly *Sphaerotheca fuliginea*); *Stagonosporopsis* spp. (formerly *Didymella*); *Xanthomonas* leaf spot (formerly bacterial leaf spot); and (*Xanthomonas cucurbitae* (formerly *X. campestris* pv. *cucurbitae*).

TABLE 3-43. GREENHOUSE DISEASE CONTROL FOR VARIOUS VEGETABLE CROPS (cont'd)

R. A. Melanson, Extension Plant Pathologist, Mississippi State University; I. Meadows, Extension Plant Pathologist, North Carolina State University

Crop/Disease	Product [1]	FRAC	Rate of Material	Minimum Days Harv.	Minimum Days Reentry	Method, Schedule, and Remarks
CUCURBITS – *AFTER TRANSPLANTING IN A GREENHOUSE*						
Alternaria leaf blight (*A. cucumerina*), Alternaria leaf spot (*A. alternata*), anthracnose (*C. orbiculare*), Cercospora leaf spot (*C. citrullina*), downy mildew (*P. cubensis*), gummy stem blight (*Stagonosporopsis* spp.[2]), Phoma blight (*P. exigua*[2]), Phyllosticta leaf spot (*P. cucurbitacearum*), Plectosporium blight (*P. cucumerina*[2]), powdery mildew (*P. xanthii, G. cichoracearum*[2]), Septoria leaf blight (*S. cucurbitacearum*)[2]	azoxystrobin + difenoconazole (Quadris Top)	11+3	12 to 14 fl oz/acre	1	0.5	*Not prohibited for greenhouse use.* **Do not use for transplant production.** Make no more than one application of a QoI-containing fungicide (FRAC Group 11) before alternating to another fungicide with a different mode of action. Do not exceed 56 fl oz of product per acre per year. See label for application limits of active ingredients. The addition of a spreading/penetrating type adjuvant is recommended. See label for additional instructions.
Labels may include: Alternaria leaf spot, angular leaf spot, anthracnose, bacterial fruit blotch (suppression), downy mildew, gray mold, gummy stem blight, powdery mildew, scab, Ulocladium leaf spot	copper, fixed (various)	M01	See label	See label	See label	**Some products are OMRI-listed.** See product labels for complete application instructions, specific crop and disease labels, and greenhouse usage. **Check state registration status prior to use.**
Powdery mildew	cyflufenamid (Torino)	U06	3.4 oz/acre	0	4 hr	*Not prohibited for greenhouse use.* Do not make more than two applications per year. Do not exceed 6.8 oz of product per acre per calendar year. See label for other application instructions and restrictions. **Use with caution as resistance has been reported in NY.**
Downy mildew (*P. cubensis*)	cymoxanil (Curzate 60DF)	27	3.2 to 5.0 oz/acre	3	0.5	Always apply in a tank-mix with the labeled rate of a protectant fungicide. Do not exceed 30 oz of product or 6 applications per year. **Curzate is not registered for use in LA. Resistance has been observed.**
Alternaria leaf blight and spot (*A. cucumerina* and *A. alternata*), gummy stem blight (*Stagonosporopsis* spp.[2]), powdery mildew (*P. xanthii, G. cichoracearum*)[2]	cyprodinil + fludioxonil (Switch 62.5WG)	9+12	11 to 14 oz/acre	1	0.5	*Not prohibited for greenhouse use.* After two applications, alternate with another fungicide with a different mode of action for two applications. Do not exceed 56 oz of product per acre per year. See label for application limits of active ingredients.
Alternaria leaf blight and spot (*A. cucumerina* and *A. alternata*), anthracnose (*C. orbiculare*), Cercospora leaf spot (*C. citrullina*), gummy stem blight (*Stagonosporopsis* spp.[2]), powdery mildew (*P. xanthii, G. cichoracearum*)[2], Septoria leaf blights (*S. cucurbitacearum*)	difenoconazole + cyprodinil (Inspire Super)	3+9	0.37 to 0.46 fl oz/ 1,000 sq ft	7	0.5	*Only allowed for use in the greenhouse on cucumber.* Do not apply more than 80 fl oz of product per acre per season. Do not apply product for more than 50% of sprays per crop. See label for application limits of active ingredients. Make no more than two consecutive applications per season before alternating with fungicides that have a different mode of action.
Alternaria leaf blight (*A. cucumerina*), anthracnose (*Colletotrichum* spp.), downy mildew (*P. cubensis*) **Suppression:** Phytophthora blight (*Phytophthora capsici*)	famoxadone + cymoxanil (Tanos)	11+27	8 oz/acre 8 to 10 oz/acre (for diseases listed under suppression)	3	0.5	*Not prohibited for greenhouse use.* Product must be tank-mixed with a contact fungicide with a different mode of action. Do not exceed 32 oz of product per acre per crop cycle or 72 oz per acre per 12-month period. Do not make more than one application of product before alternating with a fungicide that has a different mode of action. See label for tank-mixing instructions. Do not exceed four applications of Tanos or FRAC Group 11 fungicides per cropping cycle. For suppression of foliar and fruit phases ONLY of Phytophthora blight.

[1] Products registered for field use may be used on greenhouse crops (but not transplants) unless excluded on the label. Always check the label before applying a product. Additional products not prohibited for greenhouse use on basil, cucurbits, and lettuce not listed in this table are available.

[2] Former names of diseases and pathogens listed in this table that may be still be listed on fungicide labels are as follows: *Acidovorax citrulli* (formerly *Acidovorax avenae* subsp. *citrulli*) *Alternaria linariae* (formerly *A. solani* and *A. tomatophila*); *Agroathelia rolfsii* (formerly *Athelia rolfsii* and *Sclerotium rolfsii*); *Golovinomyces* spp. (formerly *Erysiphe* spp.) or *Golovinomyces cichoracearum* (formerly *Erysiphe cichoracearum*); *Fulvia fulva* (formerly *Cladosporium fulvum* and *Passalora fulva*); *Boeremia exigua* (formerly *Phoma exigua*); *Plectosphaerella cucumerina* (formerly *Plectosporium tabacinum*); *Podosphaera xanthii* (formerly *Sphaerotheca fuliginea*); *Stagonosporopsis* spp. (formerly *Didymella*); *Xanthomonas* leaf spot (formerly bacterial leaf spot); and (*Xanthomonas cucurbitae* (formerly *X. campestris* pv. *cucurbitae*).

TABLE 3-43. GREENHOUSE DISEASE CONTROL FOR VARIOUS VEGETABLE CROPS (cont'd)

R. A. Melanson, Extension Plant Pathologist, Mississippi State University; I. Meadows, Extension Plant Pathologist, North Carolina State University

Crop/Disease	Product [1]	FRAC	Rate of Material	Minimum Days Harv.	Minimum Days Reentry	Method, Schedule, and Remarks
CUCURBITS – *AFTER TRANSPLANTING IN A GREENHOUSE* (cont'd)						
Gray mold (*B. cinerea*)	fenhexamid (Decree 50WDG)	17	1.5 lb/acre (stand-alone) 1.0 to 1.5 lb/acre (tank-mix)	0	0.5	**Labeled for cucumbers ONLY.** For use in transplant production and greenhouse production. Do not make more than two consecutive applications. See label for additional tank-mixing instructions. Do not exceed 6.0 lb product per acre per crop cycle (transplants, greenhouse production). See label for additional restrictions.
Alternaria leaf spot (*A. cucumerina*), downy mildew (*P. cubensis*), gummy stem blight (*Stagonosporopsis* spp.[2]), Phytophthora blight (*P. capsici*)	fluazinam (Omega 500F, Orbus 4F)	29	12 to 24 fl oz/acre	7	0.5	*Not prohibited for greenhouse use.* See labels for specific rate instructions for Phytophthora blight and gummy stem blight and for additional restrictions. **PHI and yearly limits are for cucurbit crops in subgroup 9B**, which includes cucumbers; see labels for additional cucurbits included in this subgroup and for application information for cucurbits in subgroup 9A, which includes melon. **Omega:** Do not exceed 120 fl oz of product per acre per year. **Orbus:** Do not exceed 96 fl oz of product per acre per year. **Orbus is not registered for use in AL, AR, MS, or WV; it was pending approval in TN.**
Alternaria leaf blight (*A. cucumerina*), Alternaria leaf spot (*A. alternata*), gummy stem blight (*Stagonosporopsis*[2]), powdery mildew (*P. xanthii, G. cichoracearum*)[2]	fludioxonil (Spirato GHN)	12	5.5 to 7 fl oz/acre	1	0.5	After two applications of product, alternate to a fungicide with a different mode of action for two applications. Do not exceed 28 fl oz of product per acre per year. See label for restrictions regarding applications made in the vicinity of aquatic areas.
Alternaria leaf spot (*A. cucumerina*), anthracnose (*Colletotrichum* spp.), Botrytis gray mold (*B. cinerea*), gummy stem blight (*Stagonosporopsis* spp.[2]), powdery mildew (*P. xanthii, G. cichoracearum*)[2] **Suppression:** gummy stem blight (*Stagonosporopsis* spp.[2]),	fluopyram + trifloxystrobin (Luna Sensation)	7+11	4.0 to 7.6 fl oz/acre (powdery mildew) 7.6 fl oz/acre (other listed diseases)	0	0.5	*Not prohibited for greenhouse use.* Do not exceed 27.1 fl oz of product per acre per year. Do not apply more than 0.446 lb fluopyram or 0.5 lb trifloxystrobin per acre per year. Do not make more than four applications per year. Do not make more than two sequential applications of product or of any FRAC 7- or FRAC 11-containing fungicide before alternating with a fungicide from a different FRAC group. *See label for information regarding potential foliar discoloration after product application.*
Downy mildew (*P. cubensis*), Phytophthora root and fruit rots (*Phytophthora* spp.)	fosetyl-Al (Aliette WDG, Linebacker WDG)	P07	2.0 to 5.0 lb/acre	0.5	1	*Not prohibited for greenhouse use.* For foliar application. Phytotoxicity may occur if products are tank mixed with copper products, applied to plants with copper residues, or mixed with adjuvants. Do not tank-mix products with copper products. See labels for additional restrictions, season limits, and application instructions. **Aliette WDG:** Do not make more than seven applications per season. **See label for county restrictions in some states. Linebacker WDG:** Do not make more than seven applications per year. Do not exceed 35 lb of product per acre per year. **Check state registration status prior to use. See label for county restrictions in some states.**
Labels may include: Alternaria, anthracnose, downy mildew, powdery mildew, root rots	hydrogen peroxide + peroxyacetic acid (OxiDate 2.0, OxiDate 5.0, ZeroTol 2.0)	NC	See label	See label	See label	**OMRI-listed.** See labels for list of labeled cucurbits. See labels for additional instructions, labeled diseases, and precautions, including those related to the use and application of metal-based chemicals. Do not store spray solution. *Determine if products can be used safely on greenhouse crops prior to application.*
Labels may include: Alternaria leaf spot, anthracnose, Cercospora leaf spot, downy mildew, gummy stem blight, scab	mancozeb (various)	M03	See label	See label	See label	See product labels for complete application instructions, specific crop and disease labels, and greenhouse usage. **Check state registration status prior to use.**

[1] Products registered for field use may be used on greenhouse crops (but not transplants) unless excluded on the label. Always check the label before applying a product. Additional products not prohibited for greenhouse use on basil, cucurbits, and lettuce not listed in this table are available.

[2] Former names of diseases and pathogens listed in this table that may be still be listed on fungicide labels are as follows: *Acidovorax citrulli* (formerly *Acidovorax avenae* subsp. *citrulli*) *Alternaria linariae* (formerly *A. solani* and *A. tomatophila*); *Agroathelia rolfsii* (formerly *Athelia rolfsii* and *Sclerotium rolfsii*); *Golovinomyces* spp. (formerly *Erysiphe* spp.) or *Golovinomyces cichoracearum* (formerly *Erysiphe cichoracearum*); *Fulvia fulva* (formerly *Cladosporium fulvum* and *Passalora fulva*); *Boeremia exigua* (formerly *Phoma exigua*); *Plectosphaerella cucumerina* (formerly *Plectosporium tabacinum*); *Podosphaera xanthii* (formerly *Sphaerotheca fuliginea*); *Stagonosporopsis* spp. (formerly *Didymella*); Xanthomonas leaf spot (formerly bacterial leaf spot); and (*Xanthomonas cucurbitae* (formerly *X. campestris* pv. *cucurbitae*).

TABLE 3-43. GREENHOUSE DISEASE CONTROL FOR VARIOUS VEGETABLE CROPS (cont'd)

R. A. Melanson, Extension Plant Pathologist, Mississippi State University; I. Meadows, Extension Plant Pathologist, North Carolina State University

Crop/Disease	Product [1]	FRAC	Rate of Material	Minimum Days Harv.	Minimum Days Reentry	Method, Schedule, and Remarks
CUCURBITS – *AFTER TRANSPLANTING IN A GREENHOUSE* (cont'd)						
Alternaria leaf spot, Cercospora leaf spot, downy mildew, Phytophthora rot (*P. capsici*)	mancozeb + zoxamide (Gavel 75DF)	M03+ 22	1.5 to 2.0 lb/acre	5	2	*Not prohibited for greenhouse use.* See label for application limits and instructions Do not exceed 16 lb per acre per year. Do not make more than eight applications per acre per year. See label for additional instructions and precautions.
Powdery mildew (*Golovinomyces* spp. and *Podosphaera* spp.)[2]	myclobutanil (Albaugh Sonoma 20 EWAG)	3	4.75 to 9.5 fl oz/acre	0	1	*Not prohibited for greenhouse use.* Do not exceed 46 fl oz of product per acre per crop. See product labels for additional instructions and restrictions. **Product is not registered for use in AL, FL, KY, MS, OK, SC, TN, or TX.**
Alternaria leaf spot and blight (*Alternaria* spp.), gray mold (*B. cinerea*), gummy stem blight (*Stagonosporopsis* spp.)[2], powdery mildew (*P. xanthii, G. cichoracearum*)[2], Sclerotinia stem rot (*S. sclerotiorum*)	penthiopyrad (Fontelis)	7	0.375 to 0.5 fl oz/ gal of spray per 1,360 sq ft	1	0.5	*Allowed for use in greenhouse production of edible-peel cucurbits (cucumbers, summer squash).* Do not exceed 67 fl oz of product per acre per year. Make no more than two consecutive applications per season before alternating with fungicides that have a different mode of action. **Resistance in *Stagonosporopsis*[2] (gummy stem blight) is widespread.** See label for additional instructions for gummy stem blight management with Fontelis.
Labels may include: Downy mildew, root rots	phosphonates (mono- and dibasic salts of phosphites, phosphorous acid, potassium phosphite) (various)	P07	See label	See label	See label	See product labels for complete application instructions, labeled diseases, restrictions, and crop and greenhouse usage. **Check state registration status prior to use.**
Anthracnose (*C. orbiculare*), early blight (*Alternaria* spp.), gray mold (*Botrytis*), gummy stem blight (*Stagonosporopsis* spp.[2]), powdery mildew (*Podosphaera* spp.[2]), scab (*Cladosporium* spp.), target leaf spot (*C. cassiicola*)	polyoxin D zinc salt (OSO 5%SC)	19	6.5 to 13.0 fl oz/acre	0	4 hr	*Not prohibited for greenhouse use.* See product label for maximum limits of active ingredient and applications per season. See product label for additional application instructions and labeled diseases.
Labels may include: Alternaria leaf spot, anthracnose, Botrytis, downy mildew, powdery mildew, Septoria leaf spot	potassium bicarbonate (Kaligreen, MilStop SP)	NC	2.5 to 5.0 lb/acre (Kaligreen) 1.25 to 5.0 lb/100 gal (MilStop)	1 (Kaligreen) 0 (MilStop)	4 hr (Kaligreen) 1 hr (MilStop)	*OMRI-listed.* Do not store unused spray solution. See labels for additional labeled diseases and instructions. **Kaligreen:** *Not prohibited for greenhouse use.* Do not allow mixture to stand; use within 24 hours. Do not mix with highly acidic products. **MilStop:** Do not exceed 0.5 lb of product per 4,350 sq ft or 1.15 lb product per 10,000 sq ft per application. Do not adjust pH after mixing. See label for compatibility cautions with other products.
Damping off and root rots (*Phytophthora* spp., *Pythium* spp.)	propamocarb hydrochloride (Previcur Flex)	28	See label	See label	0.5	See label for application uses. Product can be applied through a drip system or as a soil drench. Do not apply more than six total applications (preseeding and/or seedling treatment and after transplanting). Do not apply more than two preseeding and/or seedling treatment applications or four applications of product after transplanting per cropping cycle. Do not apply more than two foliar applications per cropping cycle. Do not mix with other products. Phytotoxicity may occur if applied to dry growing media. See label for other application instructions.
Gray mold (*B. cinerea*)	pyrimethanil (Scala SC)	9	18 fl oz/acre	1	0.5	*Only allowed for use in the greenhouse on cucumber.* Use only in well-ventilated plastic tunnel or glass houses; ventilate for at least two hours after application of product. Do not exceed 36 fl oz of product per acre per crop. Do not exceed two applications per crop(maximum of three crop cycles per year). See label for cautions.

[1] Products registered for field use may be used on greenhouse crops (but not transplants) unless excluded on the label. Always check the label before applying a product. Additional products not prohibited for greenhouse use on basil, cucurbits, and lettuce not listed in this table are available.

[2] Former names of diseases and pathogens listed in this table that may be still be listed on fungicide labels are as follows: *Acidovorax citrulli* (formerly *Acidovorax avenae* subsp. *citrulli*) *Alternaria linariae* (formerly *A. solani* and *A. tomatophila*); *Agroathelia rolfsii* (formerly *Athelia rolfsii* and *Sclerotium rolfsii*); *Golovinomyces* spp. (formerly *Erysiphe* spp.) or *Golovinomyces cichoracearum* (formerly *Erysiphe cichoracearum*); *Fulvia fulva* (formerly *Cladosporium fulvum* and *Passalora fulva*); *Boeremia exigua* (formerly *Phoma exigua*); *Plectosphaerella cucumerina* (formerly *Plectosporium tabacinum*); *Podosphaera xanthii* (formerly *Sphaerotheca fuliginea*); *Stagonosporopsis* spp. (formerly *Didymella*); Xanthomonas leaf spot (formerly bacterial leaf spot); and (*Xanthomonas cucurbitae* (formerly *X. campestris* pv. *cucurbitae*).

TABLE 3-43. GREENHOUSE DISEASE CONTROL FOR VARIOUS VEGETABLE CROPS (cont'd)

R. A. Melanson, Extension Plant Pathologist, Mississippi State University; I. Meadows, Extension Plant Pathologist, North Carolina State University

Crop/Disease	Product [1]	FRAC	Rate of Material	Minimum Days Harv.	Minimum Days Reentry	Method, Schedule, and Remarks
CUCURBITS – AFTER TRANSPLANTING IN A GREENHOUSE (cont'd)						
Powdery mildew	pyriofenone (Prolivo 300SC)	50	4 to 5 fl oz	0	4 hr	Do not exceed 16 fl oz per acre per year. Do not make more than two applications of Prolivo or of another FRAC 50-containing fungicide in a row before rotating to a fungicide from a different FRAC group.
	sulfur (various)	M02	See label	See label	See label	**Some products are OMRI-listed.** See product labels for complete application instructions, specific crop and disease labels, and greenhouse usage. **Check *C. melo* cultivars for phytotoxicity prior to use. Check state registration status prior to use.**
Anthracnose (*Colletotrichum* spp.), gummy stem blight (*Stagonosporopsis* spp.[2]), powdery mildew (*Golovinomyces* and *Podosphaera*)[2], target spot (*Corynespora*)	thiophanate methyl (various)	1	See label	See label	See label	*Not prohibited for greenhouse use.* See product labels for complete application instructions, specific crop and disease labels, and greenhouse usage. **Check state registration status prior to use. **Resistance in *Stagonosporopsis*[2] (gummy stem blight) is widespread.****
Powdery mildew **Suppression:** Downy mildew	tolfenpyrad (Torac)	39	21.0 fl oz/acre	1	0.5	*Not prohibited for greenhouse use.* Do not exceed 42 fl oz or two applications of product per acre per crop cycle. Do not exceed two applications per crop cycle or four applications of product per year. Allow at least 14 days between applications. IRAC 21A.
Plectosporium blight (*P. cucumerina*)[2], powdery mildew (*P. xanthii, G. cichoracearum*)[2] **Suppression:** Downy mildew (*P. cubensis*)	trifloxystrobin (Flint Extra)	11	2.0 to 3.8 fl oz/acre (others) 3.8 fl oz/acre (downy mildew)	7	0.5	*Not prohibited for greenhouse use.* Alternate every FRAC 11 fungicide application with at least one application of a fungicide from a different FRAC group to reduce the potential for resistance. Do not exceed four applications or 15.2 fl oz of product per acre per year. The minimum interval between applications is 7 days. ****Resistance in *P. cubensis* (downy mildew) is widespread.****
Powdery mildew	triflumizole (Procure 480SC, Terraguard SC)	3	4 to 8 fl oz/acre (Procure) 2 to 4 fl oz/100 gal (Terraguard)	0 (Procure) 1 (Terraguard)	0.5	**Procure:** *Not prohibited for greenhouse use.* Do not exceed 24 fl oz of product per crop per year. If applying to a crop grown from seed, do not exceed four applications per crop per year at the 4 to 6 oz/acre rate or three applications per crop per year at the 8 oz/acre rate. If applying to a transplanted crop, do not apply more than three applications of product per crop per year. Label specifies the following powdery mildew species: *G. cichoracearum*[2], *P. xanthii*[2]. **Terraguard SC: For use on cucumber only.** For use only as a foliar spray. For use in commercial greenhouse production only. Do not exceed 16 fl oz of product per acre per cropping system. See label for additional application instructions. Do not exceed four applications per crop.
EGGPLANT AND PEPPER - *TRANSPLANT PRODUCTION*						
Anthracnose (*Colletotrichum* spp.), Cercospora leaf spot (*C. capsici*), gray leaf spot (*Stemphylium solani*), powdery mildew (*Oidiopsis sicula*), Rhizoctonia stem rot (*R. solani*) **Suppression:** Southern blight (*Agroathelia rolfsii*[2])	azoxystrobin + benzovindiflupyr (Mural)	11+7	0.6 to 0.8 oz/ 5,000 sq ft	—	0.5	*Allowed for use on transplants grown for resale to consumers.* Do not exceed two applications of product per crop. See label for restrictions for product use when multiple crops are grown in the same area. Do not apply more than two consecutive applications of product before alternating to another fungicide from a different FRAC group. See label for additional restrictions when multiple crops are grown in the same area.
Gray mold (*Botrytis cinerea*)	fenhexamid (Decree 50 WDG)	17	1.5 lb/acre (stand-alone) 1.0 to 1.5 lb/acre (tank-mix)	—	0.5	Do not make more than two consecutive applications. See label for additional tank-mixing instructions. Do not exceed 6.0 lb product per acre per crop cycle to transplants. Do not exceed six applications per acre per crop cycle at the lowest use rate or four applications per acre per crop cycle at the highest use rate.
Botrytis rot (*Botrytis* spp.), early blight (*A. linariae*[2]), powdery mildew (*L. taurica, O. sipula*) **Suppression:** Anthracnose (*C. coccodes*)	polyoxin D zinc salt (Affirm WDG)	19	6.2 oz/acre	—	4 hr	*Allowed for use in greenhouses ONLY in the production of seedlings and transplants; product is not for use in the production of edible commodities.* See product labels for maximum limits of active ingredient and applications per season. See product labels for additional application instructions and labeled diseases.

[1] Products registered for field use may be used on greenhouse crops (but not transplants) unless excluded on the label. Always check the label before applying a product. Additional products not prohibited for greenhouse use on basil, cucurbits, and lettuce not listed in this table are available.

[2] Former names of diseases and pathogens listed in this table that may be still be listed on fungicide labels are as follows: *Acidovorax citrulli* (formerly *Acidovorax avenae* subsp. *citrulli*) *Alternaria linariae* (formerly *A. solani* and *A. tomatophila*); *Agroathelia rolfsii* (formerly *Athelia rolfsii* and *Sclerotium rolfsii*); *Golovinomyces* spp. (formerly *Erysiphe* spp.) or *Golovinomyces cichoracearum* (formerly *Erysiphe cichoracearum*); *Fulvia fulva* (formerly *Cladosporium fulvum* and *Passalora fulva*); *Boeremia exigua* (formerly *Phoma exigua*); *Plectosphaerella cucumerina* (formerly *Plectosporium tabacinum*); *Podosphaera xanthii* (formerly *Sphaerotheca fuliginea*); *Stagonosporopsis* spp. (formerly *Didymella*); Xanthomonas leaf spot (formerly bacterial leaf spot); and (*Xanthomonas cucurbitae* (formerly *X. campestris* pv. *cucurbitae*).

TABLE 3-43. GREENHOUSE DISEASE CONTROL FOR VARIOUS VEGETABLE CROPS (cont'd)

R. A. Melanson, Extension Plant Pathologist, Mississippi State University; I. Meadows, Extension Plant Pathologist, North Carolina State University

Crop/Disease	Product [1]	FRAC	Rate of Material	Minimum Days Harv.	Minimum Days Reentry	Method, Schedule, and Remarks
LETTUCE (LAST UPDATED IN 2025)						
Downy mildew (*Bremia lactucae*)	acibenzolar-*S*-methyl (Actigard 50WG)	21	0.75 to 1 oz/acre	7	0.5	*Not prohibited for greenhouse use.* For use on head and leaf lettuce. Product should be applied to healthy, actively growing plants. Product should not be applied to stressed plants. Do not exceed 9.5 oz per acre per year. Do not apply more than one application of product on head lettuce intended for bag purposes. Do not apply prior to thinning or within 5 days of transplanting. See label for other instructions, restrictions, and precautions.
Anthracnose, downy mildew, leaf spot, powdery mildew, Septoria leaf spot	copper, fixed (various)	M01	See label	See label	See label	*Some products are OMRI-listed.* See product labels for complete application instructions, specific crop and disease labels, and greenhouse usage. **Check state registration status prior to use.**
Downy mildew (*B. lactucae*), Pythium damping-off (*Pythium* spp.), white rust (*Albugo occidentalis*)	cyazofamid (Ranman 400SC)	21	2.75 fl oz/acre	0	0.5	*Not prohibited for greenhouse use.* For use on head and leaf lettuce. Do not exceed 16.5 fl oz of product or 0.43 lb cyazofamid per acre per year. Do not apply more than six applications of product per crop. Alternate applications with a fungicide from a different FRAC group. See label for additional application instructions specific to each disease and regarding resistance management and tank-mixing.
Downy mildew (*B. lactucae*)	cymoxanil (Curzate 60 DF)	27	3.2 to 5.0 oz/acre (head) 5 oz/acre (leaf)	3 (head) 1 (leaf)	0.5	*Not prohibited for greenhouse use.* For use on head and leaf lettuce. Use only with the labeled rate of a protectant fungicide. Do not exceed 30 oz or six applications of product per year. **Curzate not registered for use in LA.**
Alternaria leaf spot (*Alternaria* spp.), gray mold (*Botrytis cinerea*), Sclerotinia rot (*Sclerotinia* spp.), basal rot (*B. exigua*), Septoria leaf spot (*Septoria lactucae*) **Suppression**: Powdery mildew (*Golovinomyces cichoracearum* [2])	cyprodinil + fludioxonil (Switch 62.5WG)	9+12	11 to 14 oz/acre	0	0.5	*Not prohibited for greenhouse use.* For use on head and leaf lettuce. After two applications, alternate with another fungicide with a different mode of action for two applications. Do not exceed 56 oz of product per acre per year. See label for application limits of active ingredients.
Botrytis gray mold rot (*B. cinerea*), drop rot, *Sclerotinia minor*, watery soft rot (*Sclerotinia sclerotiorum*)	dicloran (Botran 5F)	14	See label	14	0.5	For use on head and leaf lettuce. Product has different application rates for at planting, pre-thinning, and post-thinning and established transplants. See label for detailed application instructions and tank-mix precautions. Do not exceed 3.2 qt of product per acre per year. **Not registered for use in VA or WV.**
Downy mildew (*B. lactucae*), white rust (*A. occidentalis*)	famoxadone + cymoxanil (Tanos)	11+27	8 to 10 oz/acre	1	0.5	*Not prohibited for greenhouse use.* For use on head and leaf lettuce. Product must be tank mixed with a contact fungicide with a different mode of action. Do not exceed 48 oz of product per acre per crop season. Do not make more than one application of product before alternating with a fungicide that has a different mode of action. See label for tank mixing instructions. No more than 50% of total applications in cropping season should contain Tanos or other FRAC Group 11 fungicides.
Gray mold (*B. cinerea*)	fenhexamid (Decree 50 WDG)	17	1.5 lb/acre (stand-alone) 1.0 to 1.5 lb/acre (tank-mix)	3	0.5	For use in transplant production and greenhouse production. Do not make more than two consecutive applications. See label for additional tank-mixing instructions. Do not exceed 3.0 lb product per acre per crop cycle.
Alternaria leaf spot (*Alternaria* spp.), basal rot (*B. exigua*), gray mold (*B. cinerea*), Sclerotinia rot (*Sclerotinia* spp.), Septoria leaf spot (*S. lactucae*) **Suppression**: Powdery mildew (*G. cichoracearum* [2])	fludioxonil (Spirato GHN, Cannonball WG)	12	5.5 to 7 fl oz/acre (Spirato GHN) 7.0 oz/acre (Cannonball)	0	0.5	For use on head and leaf lettuce. **Spirato GHN:** After two applications of product, alternate to a fungicide with a different mode of action for two applications. Do not exceed 28 fl oz of product per acre per year. **Cannonball WG:** *Not prohibited for greenhouse use.* Products are labeled for use against basal rot, gray mold, and Sclerotinia rot. Do not exceed 28 oz of product per acre per year. After two applications of product, alternate to a fungicide with a different mode of action for two applications. See product labels for additional instructions and restrictions.

[1] Products registered for field use may be used on greenhouse crops (but not transplants) unless excluded on the label. Always check the label before applying a product. Additional products not prohibited for greenhouse use on basil, cucurbits, and lettuce not listed in this table are available.

[2] Former names of diseases and pathogens listed in this table that may be still be listed on fungicide labels are as follows: *Acidovorax citrulli* (formerly *Acidovorax avenae* subsp. *citrulli*) *Alternaria linariae* (formerly *A. solani* and *A. tomatophila*); *Agroathelia rolfsii* (formerly *Athelia rolfsii* and *Sclerotium rolfsii*); *Golovinomyces* spp. (formerly *Erysiphe* spp.) or *Golovinomyces cichoracearum* (formerly *Erysiphe cichoracearum*); *Fulvia fulva* (formerly *Cladosporium fulvum* and *Passalora fulva*); *Boeremia exigua* (formerly *Phoma exigua*); *Plectosphaerella cucumerina* (formerly *Plectosporium tabacinum*); *Podosphaera xanthii* (formerly *Sphaerotheca fuliginea*); *Stagonosporopsis* spp. (formerly *Didymella*); *Xanthomonas* leaf spot (formerly bacterial leaf spot); and (*Xanthomonas cucurbitae* (formerly *X. campestris* pv. *cucurbitae*).

TABLE 3-43. GREENHOUSE DISEASE CONTROL FOR VARIOUS VEGETABLE CROPS (cont'd)

R. A. Melanson, Extension Plant Pathologist, Mississippi State University; I. Meadows, Extension Plant Pathologist, North Carolina State University

Crop/Disease	Product [1]	FRAC	Rate of Material	Minimum Days Harv.	Minimum Days Reentry	Method, Schedule, and Remarks
LETTUCE (LAST UPDATED IN 2025) (cont'd)						
Downy mildew (*Bremia* spp.), gray mold (*Botrytis* spp.), lettuce drop (*S. minor, S. sclerotiorum*), powdery mildew (*G. cichoracearum*)	fluopyram + trifloxystrobin (Luna Sensation)	7+11	7.6 fl oz/acre	See label	0.5	**Not prohibited for greenhouse use.** Do not exceed 5.3 fl oz of product per acre per year. Do not apply more than 0.446 lb fluopyram or 0.375 lb trifloxystrobin per acre per year. Do not make more than two sequential applications of product or of any FRAC7- or FRAC 11-containing fungicide before alternating with a fungicide from a different FRAC group.
Downy mildew (*B. lactucae*)	fosetyl-Al (Aliette WDG, Linebacker WDG)	P07	2.5 to 5.0 lb/acre	3	1 (Aliette) 1 (Linebacker)	**Not prohibited for greenhouse use.** For foliar application on head and leaf lettuce. Phytotoxicity may occur if products are tank mixed with copper products, applied to plants with copper residues, or mixed with adjuvants. Do not tank-mix products with copper products. See labels for additional restrictions, season limits, and application instructions not included below. **Aliette WDG:** Do not exceed seven applications per season. Speckling can occur when applied to lettuce. **Linebacker WDG:** Do not exceed 35 lb of product per acre per year. Do not make more than seven applications per year. **Check state registration status prior to use.**
Labels may include: Botrytis gray mold, downy mildew, powdery mildew	hydrogen peroxide + peroxyacetic acid (OxiDate 2.0, OxiDate 5.0, ZeroTol 2.0)	NG	See label	0	See label	**OMRI-listed.** See labels for additional instructions, labeled diseases, and precautions, including those related to the use and application of metal-based chemicals. Do not store spray solution. **OxiDate 2.0:** *Not prohibited for greenhouse use.* **OxiDate 5.0:** *Not prohibited for greenhouse use. Determine if product can be used safely on greenhouse crops prior to application.*
Bottom rot (*Rhizoctonia solani*), gray mold (*Botrytis cinerea*), lettuce drop (*Sclerotinia* spp.)	iprodione (Meteor, Rovral 4F)	2	1.5 to 2.0 pt/acre	14	1	**Not prohibited for greenhouse use.** See product labels for application instructions and additional restrictions. Do not cultivate after application. Do not drench. Use of products at residential sites in prohibited. **Meteor:** Do not make more than three applications to each crop per season. **Rovral:** Do not make more than three applications per crop. Do not exceed 6 pt per acre per crop.
Sclerotinia drop (*S. minor* and *S. sclerotiorum*)	isofetamid (Kenja 400SC)	7	12.3 fl oz	14	0.5	For use on head and leaf lettuce. **Not prohibited for greenhouse use. Do not use for commercial transplant production in the greenhouse.** Do not exceed two applications of product per acre per year. Do not make more than two applications of Kenja or of another FRAC 7-containing fungicide in a row before rotating to a fungicide from a different FRAC group.
Alternaria leaf spot, anthracnose, downy mildew, Septoria	mancozeb + azoxystrobin (Dexter Max)	M03+11	1.7 to 2.25 lb/acre	10	1	**See label for restrictions on greenhouse use.** For use on head and leaf lettuce. Remove residues from head lettuce by stripping and trimming. Do not apply more than one application of product or FRAC Group 11 fungicide before alternating to a fungicide with a different mode of action. Do not exceed 13.7 lb of product per acre per season. See label for application limits of active ingredients. See label for additional instructions and cautions regarding residues, adjuvants, and tank mixing.
Anthracnose, downy mildew	mancozeb + copper hydroxide (ManKocide)	M03+M01	1 to 2 lb/acre	10	See label	**Not prohibited for greenhouse use.** Do not exceed 26 lb of product per acre per year. Plant injury may occur in some varieties. Determine crop sensitivity prior to use. Phytotoxicity may occur when spray solution has a pH of less than 6.5 or when certain environmental conditions occur.
Downy mildew (*B. lactucae*)	mandipropamid (Micora)	40	5.5 to 8.0 fl oz/acre (0.65 to 0.9 fl oz per 5,000 sq ft)	—	4 hr	For use on head and leaf lettuce. **For lettuce transplants grown in enclosed greenhouses with permanent flooring for resale to consumers.** Do not apply more than two applications of product per crop. Do not apply consecutive applications. Apply in a tank-mix with another downy mildew fungicide with a different mode of action. The addition of a spreading/penetrating type adjuvant is recommended.

[1] Products registered for field use may be used on greenhouse crops (but not transplants) unless excluded on the label. Always check the label before applying a product. Additional products not prohibited for greenhouse use on basil, cucurbits, and lettuce not listed in this table are available.

[2] Former names of diseases and pathogens listed in this table that may be still be listed on fungicide labels are as follows: *Acidovorax citrulli* (formerly *Acidovorax avenae* subsp. *citrulli*) *Alternaria linariae* (formerly *A. solani* and *A. tomatophila*); *Agroathelia rolfsii* (formerly *Athelia rolfsii* and *Sclerotium rolfsii*); *Golovinomyces* spp. (formerly *Erysiphe* spp.) or *Golovinomyces cichoracearum* (formerly *Erysiphe cichoracearum*); *Fulvia fulva* (formerly *Cladosporium fulvum* and *Passalora fulva*); *Boeremia exigua* (formerly *Phoma exigua*); *Plectosphaerella cucumerina* (formerly *Plectosporium tabacinum*); *Podosphaera xanthii* (formerly *Sphaerotheca fuliginea*); *Stagonosporopsis* spp. (formerly *Didymella*); Xanthomonas leaf spot (formerly bacterial leaf spot); and (*Xanthomonas cucurbitae* (formerly *X. campestris* pv. *cucurbitae*).

TABLE 3-43. GREENHOUSE DISEASE CONTROL FOR VARIOUS VEGETABLE CROPS (cont'd)

R. A. Melanson, Extension Plant Pathologist, Mississippi State University; I. Meadows, Extension Plant Pathologist, North Carolina State University

Crop/Disease	Product [1]	FRAC	Rate of Material	Minimum Days Harv.	Minimum Days Reentry	Method, Schedule, and Remarks
LETTUCE (LAST UPDATED IN 2025) (cont'd)						
Powdery mildew (*G. cichoracearum* [2])	myclobutanil (AgriStar Sonoma 20 EW AG)	3	9.5 fl oz/acre	3	1	*Not prohibited for greenhouse use.* For use on head and leaf lettuce. Do not exceed 9.5 fl oz of product per acre per application or 38.0 fl oz of product per acre per season. Do not make more than four applications of product per season. See product labels for additional instructions and restrictions. **Sonoma 20EW AG is not registered for use in AL, FL, KY, MS, OK, SC, TN, or TX.**
Labels may include: Downy mildew	phosphonates (mono- and dibasic salts of phosphites, phosphorous acid, potassium phosphite) (various)	P07	See label	See label	See label	See product labels for complete application instructions, labeled diseases, restrictions, and crop and greenhouse usage. **Check state registration status prior to use.**
Alternaria leaf spot (*Alternaria* spp.); Botrytis damping off, leaf blight, and rot (*Botrytis* spp.); bottom rot (*R. solani*), lettuce drop (*Sclerotinia* spp.), powdery mildew (*G. cichoracearum*) **Suppression:** Downy mildew (*B. lactucae*)	polyoxin D zinc salt (OSO 5%SC)	19	6.5 to 13.0 fl oz/acre	0	4 hr	*Not prohibited for greenhouse use.* For use on head, leaf, iceberg, and romaine lettuce. Check products label for maximum limits of active ingredient per acre per season. See product label for additional application instructions.
Labels may include: Alternaria leaf spot, Botrytis, downy mildew, powdery mildew	potassium bicarbonate (Carb-O-Nator, MilStop SP)	NC	2.5 to 5.0 lb/ 100 gal (Carb-O-Nator) 1.25 to 5.0 lb/100 gal (MilStop)	0	4 hr (Carb-O-Nator) 1 hr (MilStop)	**OMRI-listed.** Do not store unused spray solution. See labels for additional labeled diseases and instructions. **Carb-O-Nator:** Final spray solution should not be below pH 7.0. Some tank-mixes may be incompatible and may cause phytotoxicity. **MilStop SP:** Do not exceed 0.5 lb of product per 4,350 sq ft or 1.15 lb product per 10,000 sq ft per application. Do not adjust pH after mixing. Do not mix with other soluble pesticide or fertilizers not compatible with mild alkaline solutions.
Damping off and root rots (*Phytophthora* spp., *Pythium* spp.)	propamocarb hydrochloride (Previcur Flex)	28	See label	2	0.5	**For greenhouse use on leaf lettuce only.** Product can be applied as a foliar treatment in leaf lettuce – see label for instructions. Do not apply more than six total applications (preseeding and/or seedling treatment, after transplanting, and foliar applications). Do not apply more than two preseeding and/or seedling applications, four total applications after transplanting, or two foliar applications of product per crop cycle. Do not mix with other products. Phytotoxicity may occur if applied to dry growing media. See label for other application instructions and maximum use rates.
Powdery mildew	sulfur (various)	M02	See label	See label	See label	**Some products are OMRI-listed.** See product labels for complete application instructions, specific crop and disease labels, and greenhouse usage. **Check state registration status prior to use.**
Powdery mildew **Suppression:** Downy mildew	tolfenpyrad (Torac)	39	21.0 fl oz/acre	1	0.5	*Not prohibited for greenhouse use.* Do not apply until at least 14 days after emergence or transplanting. Do not exceed 42 fl oz or two applications of product or 0.42 lb of tolfenpyrad per acre per crop cycle. Do not exceed four applications of product per year. IRAC 21A.
Alternaria leaf spot (*Alternaria* spp.), anthracnose (*Colletotrichum* spp.), powdery mildew (*G. cichoracearum* [2])	trifloxystrobin (Flint Extra)	11	3.0 to 3.8 fl oz/acre	See label	0.5	*Not prohibited for greenhouse use.* For use on head and leaf lettuce. Do not exceed 7.6 fl oz of product per acre per year. Do not apply more than two sequential applications of product or another FRAC 11-containing fungicide before rotating with a fungicide from a different FRAC group.
Alternaria leaf spot/black spot (*Alternaria* spp.), powdery mildew (*Golovinomyces* spp.[2])	triflumizole (Procure 480SC)	3	6 to 8 fl oz/acre	0	0.5	*See label for greenhouse usage allowances.* For use on head and leaf lettuce. See label for additional uses on lettuce. Do not exceed two applications of product per crop per year. Do not exceed 16 fl oz of product per crop per year.

[1] Products registered for field use may be used on greenhouse crops (but not transplants) unless excluded on the label. Always check the label before applying a product. Additional products not prohibited for greenhouse use on basil, cucurbits, and lettuce not listed in this table are available.

[2] Former names of diseases and pathogens listed in this table that may be still be listed on fungicide labels are as follows: *Acidovorax citrulli* (formerly *Acidovorax avenae* subsp. *citrulli*) *Alternaria linariae* (formerly *A. solani* and *A. tomatophila*); *Agroathelia rolfsii* (formerly *Athelia rolfsii* and *Sclerotium rolfsii*); *Golovinomyces* spp. (formerly *Erysiphe* spp.) or *Golovinomyces cichoracearum* (formerly *Erysiphe cichoracearum*); *Fulvia fulva* (formerly *Cladosporium fulvum* and *Passalora fulva*); *Boeremia exigua* (formerly *Phoma exigua*); *Plectosphaerella cucumerina* (formerly *Plectosporium tabacinum*); *Podosphaera xanthii* (formerly *Sphaerotheca fuliginea*); *Stagonosporopsis* spp. (formerly *Didymella*); Xanthomonas leaf spot (formerly bacterial leaf spot); and (*Xanthomonas cucurbitae* (formerly *X. campestris* pv. *cucurbitae*).

TABLE 3-43. GREENHOUSE DISEASE CONTROL FOR VARIOUS VEGETABLE CROPS (cont'd)

R. A. Melanson, Extension Plant Pathologist, Mississippi State University; I. Meadows, Extension Plant Pathologist, North Carolina State University

Crop/Disease	Product [1]	FRAC	Rate of Material	Minimum Days Harv.	Minimum Days Reentry	Method, Schedule, and Remarks
TOMATO — *TRANSPLANT PRODUCTION*						
Anthracnose (*Colletotrichum* spp.), black mold (*Alternaria alternata*), early blight (*Alternaria linariae* [2]), gray leaf spot (*Stemphylium* spp.), leaf mold (*Fulvia fulva* [2]), powdery mildew (*Leveilula taurica*), Septoria leaf spot (*Septoria lycopersici*), target spot (*Corynespora cassiicola*)	azoxystrobin + benzovindiflupyr (Mural)	11+7	0.6 to 0.8 oz/ 5,000 sq ft	—	0.5	**Allowed for use on transplants grown for resale to consumers. Do not apply to tomatoes grown in a greenhouse for the purpose of producing and harvesting fruit.** Do not exceed two applications of product per crop for plants grown indoors for resale to consumers. Do not apply until 21 days after transplanting or 35 days after seeding. See label for restrictions for product use when multiple crops are grown in the same area. See label for instructions and cautions regarding the use of adjuvants and tank mixing. Do not apply more than two consecutive applications of product before alternating to another fungicide from a different FRAC group.
Damping-off (*Pythium* spp.)	cyazofamid (Ranman 400SC, Segway O)	21	3 fl oz/100 gal	—	0.5	Apply as a soil drench. Do not use a surfactant. One fungicide application can be made to the seedling tray at planting or any time afterwards until one week before transplanting. See label for additional instructions. **Segway O is only allowed for use on transplants grown for resale to consumers. Do not apply to tomatoes grown in a greenhouse for the purpose of producing and harvesting fruit.**
Gray mold (*Botrytis cinerea*)	fenhexamid (Decree 50 WDG)	17	1.5 lb/acre (stand-alone) 1.0 to 1.5 lb/acre (tank-mix)	—	0.5	Do not make more than two consecutive applications. See label for additional tank-mixing instructions. Do not exceed 6.0 lb product per acre per crop cycle to transplants. Do not exceed six applications per acre per crop cycle at the lowest use rate or four applications per acre per crop cycle at the highest use rate.
Late blight (*Phytophthora infestans*)	mandipropamid (Micora)	40	5.5 to 8.0 fl oz/acre (0.65 to 0.9 fl oz/ 5,000 sq ft)	—	4 hr	**For tomato transplants grown in enclosed greenhouses with permanent flooring for re-sale to consumers. Do not apply to tomatoes grown in a greenhouse for the purpose of producing and harvesting fruit.** Do not make more than two applications of product per crop. Do not make more than two consecutive applications before switching to a fungicide with a different mode of action. The addition of a spreading/penetrating type adjuvant is recommended.
Botrytis rot (*Botrytis* spp.), early blight (*A. linariae*[2]), powdery mildew (*L. taurica*, *O. sipula*) *Suppression*: Anthracnose (*C. coccodes*)	polyoxin D zinc salt (Affirm WDG)	19	6.2 oz/acre	—	4 hr	**Allowed for use in greenhouses ONLY in the production of seedlings and transplants; product is not for use in the production of edible commodities.** See product labels for maximum limits of active ingredient and applications per season. See product labels for additional application instructions.
Damping off and root rots (*Phytophthora* spp., *Pythium* spp.)	propamocarb hydrochloride (Previcur Flex)	28	See label	—	0.5	**For pre-seeding and/or seedling treatment (before transplanting).** Do not apply more than two pre-seeding and/or seedling applications per cropping cycle. Do not mix with other products. See label for specific use directions and maximum use rates. Phytotoxicity may occur if applied to dry growing media.
Crown and basal rot (*Fusarium* spp., *Rhizoctonia solani*, *Sclerotinia* spp.), damping-off (*Pythium* spp., *Rhizoctonia* spp.), spots and blights (*Alternaria* spp., *Cercospora* spp., *Phoma* spp., *Septoria* spp.), Phytophthora blight (*Phytophthora* spp.), powdery mildew (*Leveilula* spp. and *Oidiopsis* spp.), rots and blights (*Botrytis* spp.)	pyraclostrobin + boscalid (Pageant Intrinsic)	11+7	See label	—	0.5	**For transplant production for the home consumer market. Not for use on transplants intended for agricultural production fields or tomatoes grown in a greenhouse for the purpose of producing and harvesting fruit.** Do not tank-mix with adjuvants or other agricultural products. Do not apply more than two consecutive applications in any crop production cycle or more than three applications to any crop during a growing cycle. Do not apply to consecutive transplant crops within the same production structure. Do not exceed 118 oz product per year to the same production crop. Product application rates vary depending on target disease. See label for application rates and additional restrictions and instructions.
Labels may include: Bacterial canker, speck, and/ or spot	Streptomycin sulfate (Agri-Mycin 50, Firewall 17 WP, Firewall 50 WP, Harbour)	25	See labels	—	0.5	See labels for restrictions, cautions, application instructions, and labeled diseases for each product. **Not all products are registered for use in all states. Check state registration status prior to use.**

[1] Products registered for field use may be used on greenhouse crops (but not transplants) unless excluded on the label. Always check the label before applying a product. Additional products not prohibited for greenhouse use on basil, cucurbits, and lettuce not listed in this table are available.

[2] Former names of diseases and pathogens listed in this table that may be still be listed on fungicide labels are as follows: *Acidovorax citrulli* (formerly *Acidovorax avenae* subsp. *citrulli*) *Alternaria linariae* (formerly *A. solani* and *A. tomatophila*); *Agrothelia rolfsii* (formerly *Athelia rolfsii* and *Sclerotium rolfsii*); *Golovinomyces* spp. (formerly *Erysiphe* spp.) or *Golovinomyces cichoracearum* (formerly *Erysiphe cichoracearum*); *Fulvia fulva* (formerly *Cladosporium fulvum* and *Passalora fulva*); *Boeremia exigua* (formerly *Phoma exigua*); *Plectosphaerella cucumerina* (formerly *Plectosporium tabacinum*); *Podosphaera xanthii* (formerly *Sphaerotheca fuliginea*); *Stagonosporopsis* spp. (formerly *Didymella*); Xanthomonas leaf spot (formerly bacterial leaf spot); and (*Xanthomonas cucurbitae* (formerly *X. campestris* pv. *cucurbitae*).

TABLE 3-43. GREENHOUSE DISEASE CONTROL FOR VARIOUS VEGETABLE CROPS (cont'd)

R. A. Melanson, Extension Plant Pathologist, Mississippi State University; I. Meadows, Extension Plant Pathologist, North Carolina State University

Crop/Disease	Product [1]	FRAC	Rate of Material	Minimum Days Harv.	Minimum Days Reentry	Method, Schedule, and Remarks
TOMATO — *TRANSPLANT PRODUCTION* (cont'd)						
Powdery mildew	triflumizole (Terraguard SC)	3	2 to 4 fl oz/100 gal	—	0.5	Commercial greenhouse production only. Use only as a foliar spray. Can be used on greenhouse transplants. Do not exceed 16 fl oz of product per acre per cropping system. Do not exceed 4 applications per crop.
TOMATO — *AFTER TRANSPLANTING IN A GREENHOUSE*						
Bacterial speck (*Pseudomonas syringae* pv. *tomato*), bacterial spot (*Xanthomonas* spp.)	acibenzolar-*S*-methyl (Actigard 50 WG)	P01	0.33 to 0.75 oz/acre	14	0.5	*Not prohibited for greenhouse use.* Product should be applied to healthy, actively growing plants. Product should not be applied to stressed plants. Do not apply on intervals less than 7 days. Do not exceed 6 oz of product per acre per year. See label for other instructions, restrictions, and precautions.
Anthracnose (*Colletotrichum* spp.), black mold (*A. alternata*), early blight (*A. linariae* [2]), gray leaf spot (*Stemphylium* spp.), leaf mold (*F. fulva* [2]), powdery mildew (*L. taurica*), Septoria leaf spot (*S. lycopersici*), target spot (*C. cassiicola*)	azoxystrobin + difenoconazole (Quadris Top)	11+3	8 fl oz/acre	0	0.5	*Not prohibited for greenhouse use.* **Do not use for transplant production.** Do not make more than two consecutive applications before switching to a fungicide with a different mode of action. Do not exceed 47 fl oz of product per acre per year. See label for application limits of active ingredients. Do not apply until 21 days after transplanting or 35 days after seeding. Do not use with adjuvants or tank-mix with any EC product on fresh market tomatoes. Plant injury may occur with the use of adjuvants or when product is tank mixed with dimethoate.
Bacterial spot (*Xanthomonas* spp.), bacterial spot (*Pseudomonas syringae* pv. *tomato*), bacterial canker (*Clavibacter michiganensis* subsp. *michiganensis*)	bacteriophage (Agriphage, Agriphage-CMM)	NA	1 to 2 pt/ 50 to 100 gal water	0	4 hr	See label for use in hydroponic systems or transplants.
Gray mold (*Botrytis cinerea*), late blight (*Phytophthora infestans*), leaf mold (*F. fulva* [2]), powdery mildew (*L. taurica*, *Oidium neolycopersici*), target spot (*C. cassiicola*), white mold (*S. sclerotiorum*)	Banda de Lupinus albus doce (BLAD) (ProBLAD Verde)	BM01	18.1 to 45.7 fl oz/acre	1	4 hr	*Not prohibited for greenhouse use.* **OMRI-listed.** Do not make more than two sequential applications before alternating to a fungicide with a different mode of action. Do not make more than five foliar applications of product per harvest cycle. **Check state registration status prior to use.**
Powdery mildew	cyflufenamid (Torino)	U06	3.4 oz/acre	0	4 hr	*Not prohibited for greenhouse use.* Do not make more than three applications per year. Do not exceed 10.2 oz product per acre per calendar year. See label for other application instructions and restrictions.
Late blight (*P. infestans*), Phytophthora blight (*P. capsici*)	cyazofamid (Ranman 400SC)	21	2.1 to 2.75 fl oz/acre (late blight) 2.75 fl oz/acre (Phytophthora blight)	0	0.5	Do not exceed 16.5 fl oz per acre per year. Do not exceed six applications of product per crop. See label for surfactant recommendations. Alternate applications with fungicides that have a different mode of action. Do not make more than three consecutive applications before switching to products that have a different mode of action for three applications before returning to Ranman 400SC. See label for application instructions.
Late blight (*P. infestans*)	cymoxanil (Curzate 60DF)	27	3.2 to 5.0 oz/acre	3	0.5	*Not prohibited for greenhouse use.* Use only with the labeled rate of a protectant fungicide with a different mode of action. Do not exceed 30 oz of product or six applications per year. **Curzate not registered in LA.**
Early blight (*A. linariae* [2]), gray mold (*B. cinerea*), powdery mildew (*L. taurica*)	cyprodinil + fludioxonil (Switch 62.5WG)	9+12	11 to 14 oz/acre	0	0.5	*Not prohibited for greenhouse use.* After two applications, alternate with another fungicide with a different mode of action for two applications. Do not exceed 56 oz of product per acre per year. See label for application limits of active ingredients. Do not exceed four applications per year.

[1] Products registered for field use may be used on greenhouse crops (but not transplants) unless excluded on the label. Always check the label before applying a product. Additional products not prohibited for greenhouse use on basil, cucurbits, and lettuce not listed in this table are available.

[2] Former names of diseases and pathogens listed in this table that may be still be listed on fungicide labels are as follows: *Acidovorax citrulli* (formerly *Acidovorax avenae* subsp. *citrulli*) *Alternaria linariae* (formerly *A. solani* and *A. tomatophila*); *Agroathelia rolfsii* (formerly *Athelia rolfsii* and *Sclerotium rolfsii*); *Golovinomyces* spp. (formerly *Erysiphe* spp.) or *Golovinomyces cichoracearum* (formerly *Erysiphe cichoracearum*); *Fulvia fulva* (formerly *Cladosporium fulvum* and *Passalora fulva*); *Boeremia exigua* (formerly *Phoma exigua*); *Plectosphaerella cucumerina* (formerly *Plectosporium tabacinum*); *Podosphaera xanthii* (formerly *Sphaerotheca fuliginea*); *Stagonosporopsis* spp. (formerly *Didymella*); *Xanthomonas* leaf spot (formerly bacterial leaf spot); and (*Xanthomonas cucurbitae* (formerly *X. campestris* pv. *cucurbitae*).

TABLE 3-43. GREENHOUSE DISEASE CONTROL FOR VARIOUS VEGETABLE CROPS (cont'd)

R. A. Melanson, Extension Plant Pathologist, Mississippi State University; I. Meadows, Extension Plant Pathologist, North Carolina State University

Crop/Disease	Product [1]	FRAC	Rate of Material	Minimum Days Harv.	Minimum Days Reentry	Method, Schedule, and Remarks
TOMATO —*AFTER TRANSPLANTING IN A GREENHOUSE* (cont'd)						
Anthracnose (*Colletotrichum* spp.), black mold (*A. alternata*), early blight (*A. linariae* [2]), gray mold (*B. cinerea*), gray leaf spot (*S. botryosum*), leaf mold (*F. fulva* [2]), powdery mildew (*L. taurica*), Septoria leaf spot (*S. lycopersici*), target spot (*C. cassiicola*)	difenoconazole + cyprodinil (Inspire Super)	3+9	16 to 20 fl oz/acre	0	0.5	***Not prohibited for greenhouse use.*** Do not exceed 80 fl oz of product per acre per year. See label for application limits of active ingredients. Make no more than two consecutive applications per season before alternating with fungicides that have a different mode of action.
Anthracnose (*Colletotrichum* spp.), early blight (*A. linariae* [2]), late blight (*P. infestans*), leaf mold (*P. fulva*[2]), Septoria leaf spot (*S. lycopersici*), target spot (*C. cassiicola*) **Suppression:** Bacterial canker (*Clavibacter michiganensis* subsp. *michiganensis*), bacterial speck (*Pseudomonas syringae* pv. *syringae*), bacterial spot (*Xanthomonas* spp.), buck-eye rot (*Phytophthora* spp.)	famoxadone + cymoxanil (Tanos)	11+27	6 to 8 oz/acre (early blight) 8 oz/acre (other labeled diseases)	3	0.5	***Not prohibited for greenhouse use.*** Product must be tank mixed with a contact fungicide with a different mode of action. Do not exceed 72 oz per acre per crop cycle or 12-month period. Do not make more than one application of product before alternating with a fungicide that has a different mode of action. See label for tank-mixing instructions. No more than 50% of total applications in a cropping season should contain Tanos or other FRAC Group 11 fungicides.
Botrytis gray mold (*B. cinerea*)	fenhexamid (Decree 50 WDG)	17	1.5 lb/acre (stand-alone) 1.0 to 1.5 lb/acre (tank-mix)	0	0.5	Do not make more than two consecutive applications. See label for additional tank-mixing instructions. Do not exceed 6.0 lb product per acre per crop cycle for greenhouse production. Do not exceed six applications per acre per crop cycle at the lowest use rate or four applications per acre per crop cycle at highest use rate.
Early blight (*A. linariae* [2]), gray mold (*B. cinerea*), powdery mildew (*L. taurica*)	fludioxonil (Spirato GHN)	12	5.5 to 7 fl/oz/acre	0	0.5	After two applications of product, alternate to a fungicide with a different mode of action for two applications. Do not exceed 28 fl oz of product per acre per year. Do not exceed four applications per year.
Black mold (*A. alternata*), early blight (*A. linariae* [2]), gray mold (*B. cinerea*), gray leaf spot (*Stemphylium* spp.), powdery mildew (*O. taurica, L. taurica*), Septoria leaf spot (*S. lycopersici*), target spot (*C. cassiicola*) **Suppression:** Anthracnose (*Colletotrichum* spp.), southern blight (*A. rolfsii* [2]), white mold (*Sclerotinia* spp.)	fluopyram + trifloxystrobin (Luna Sensation)	7+11	5.0 to 7.6 fl oz/acre (early blight, powdery mildew, Septoria leaf spot) 7.6 fl oz/acre (other listed diseases)	3	0.5	***Not prohibited for greenhouse use.*** Do not exceed 27.1 fl oz of product per acre per year. See label for application limits of active ingredients. Do not apply more than 0.446 lb fluopyram or 0.5 lb trifloxystrobin per acre per year. Do not make more than five applications per year. Do not make more than two sequential applications of product or of any FRAC 7- or FRAC 11-containing fungicide before alternating with a fungicide from a different FRAC group.
Black mold (*A. alternata*), early blight (*A. linariae* [2]), gray mold (*B. cinerea*), gray leaf spot (*Stemphylium* spp.), Septoria leaf spot (*S. lycopersici*), target spot (*C. cassiicola*) **Suppression:** Powdery mildew (*O. taurica, L. taurica*)	fluopyram + pyrimethanil (Luna Tranquility)	7+9	11.2 fl oz/acre	1	0.5	Do not exceed 54.7 fl oz of product per acre per year. Do not apply more than 0.446 lb fluopyram or 1.4 lbs pyrimethanil per acre per year. Do not make more than two sequential applications of product or of any FRAC 7- or FRAC 9-containing fungicide before alternating with a fungicide from a different FRAC group. Apply only in well-ventilated plastic tunnel houses or glass houses. Ventilate for at least 2 hours after application. **Luna Tranquility is not registered for use in LA.**

[1] Products registered for field use may be used on greenhouse crops (but not transplants) unless excluded on the label. Always check the label before applying a product. Additional products not prohibited for greenhouse use on basil, cucurbits, and lettuce not listed in this table are available.

[2] Former names of diseases and pathogens listed in this table that may be still be listed on fungicide labels are as follows: *Acidovorax citrulli* (formerly *Acidovorax avenae* subsp. *citrulli*) *Alternaria linariae* (formerly *A. solani* and *A. tomatophila*); *Agroathelia rolfsii* (formerly *Athelia rolfsii* and *Sclerotium rolfsii*); *Golovinomyces* spp. (formerly *Erysiphe* spp.) or *Golovinomyces cichoracearum* (formerly *Erysiphe cichoracearum*); *Fulvia fulva* (formerly *Cladosporium fulvum* and *Passalora fulva*); *Boeremia exigua* (formerly *Phoma exigua*); *Plectosphaerella cucumerina* (formerly *Plectosporium tabacinum*); *Podosphaera xanthii* (formerly *Sphaerotheca fuliginea*); *Stagonosporopsis* spp. (formerly *Didymella*); Xanthomonas leaf spot (formerly bacterial leaf spot); and (*Xanthomonas cucurbitae* (formerly *X. campestris* pv. *cucurbitae*).

TABLE 3-43. GREENHOUSE DISEASE CONTROL FOR VARIOUS VEGETABLE CROPS (cont'd)

R. A. Melanson, Extension Plant Pathologist, Mississippi State University; I. Meadows, Extension Plant Pathologist, North Carolina State University

Crop/Disease	Product [1]	FRAC	Rate of Material	Minimum Days Harv.	Minimum Days Reentry	Method, Schedule, and Remarks
TOMATO —AFTER TRANSPLANTING IN A GREENHOUSE (cont'd)						
Damping-off (*Pythium* spp.), root rots (*Phytophthora* spp.)	fosetyl-Al (Aliette WDG, Linebacker WDG)	P07	2.5 to 5.0 lb/acre	14	1	*Not prohibited for greenhouse use.* For foliar application. Phytotoxicity may occur if products are tank mixed with copper products, applied to plants with copper residues, or mixed with adjuvants. Do not tank-mix products with copper products. See labels for additional restrictions and application instructions not included below. **Products are not labeled for use on tomato in certain counties in AL, KY, LA, NC, and TN.** **Aliette WDG:** Do not exceed 20 lb of product per acre per season. **Aliette is not labeled for use on tomato in certain counties in AL, KY, LA, NC, and TN. Aliette has a 2(ee) Recommendation for suppression of bacterial spot in FL;** see label for application instructions and restrictions. **Linebacker WDG:** Do not exceed 20 lb of product per acre per year. Do not exceed four applications of product per acre per year. Check state registration status prior to use. **Linebacker is not labeled for use on tomato in certain counties in AL, KY, LA, NC, and TN.**
Labels may include: Anthracnose, *Alternaria*, bacterial speck and spot, Botrytis gray mold, leaf mold, early blight, *Fusarium*, late blight, powdery mildew, *Pythium, Rhizoctonia*	hydrogen peroxide + peroxyacetic acid (OxiDate 2.0, OxiDate 5.0, ZeroTol 2.0)	NG	See label	0	See label	OMRI-listed. See labels for additional instructions, labeled diseases, and precautions, including those related to the use and application of metal-based chemicals. Do not store spray solution. **OxiDate 2.0:** Not prohibited for greenhouse use. Determine if product can be used safely on greenhouse crops prior to application. **OxiDate 5.0:** Not prohibited for greenhouse use. Determine if product can be used safely on greenhouse crops prior to application. **ZeroTol 2.0:** Determine if product can be used safely on greenhouse crops prior to application.
Labels may include: Anthracnose, bacterial speck and spot, early blight, gray leaf spot, late blight, leaf mold, Septoria leaf spot	mancozeb (various)	M03	See label	See label	See label	See product labels for complete application instructions, specific crop and disease labels, and greenhouse usage. **Check state registration status prior to use.**
Anthracnose, bacterial speck and spot, early blight, gray leaf spot, late blight, leaf mold, Septoria leaf spot *East of the Mississippi:* black mold, buckeye rot, powdery mildew, target spot	mancozeb + azoxystrobin (Dexter Max)	M03+11	See label	5	1	*See label for restrictions on greenhouse use.* **Do not use for commercial transplant production.** See product label for complete application instructions, including seasonal product limits and restrictions and/or cautions for mixing with other products, additives, or adjuvants. Product rates vary depending on location and disease.
Bacterial speck and spot, buckeye rot, early blight, gray leaf spot, late blight, leaf mold, Septoria leaf spot	mancozeb + zoxamide (Gavel 75DF)	M03+22	1.5 to 2.0 lb/acre (all others) 2.0 lb/acre (bacterial speck and spot)	5	2	*Not prohibited for greenhouse use.* Do not exceed 8 lb per acre per year (west of the Mississippi River) or 16 lb per acre per year (east of the Mississippi River). Do not make more than four applications per acre per year (west of the Mississippi River) or eight applications per acre per year (east of the Mississippi River). For bacterial speck and spot, apply the full rate of product in a tank-mix with a full rate of a fixed copper. See label for other application limits. **Product has a 2(ee) recommendation for anthracnose management in AL, AR, FL, GA, KY, LA, MS, NC, OK, SC, TN, VA, and WV.**
Late blight (*P. infestans*)	mandipropamid (Revus)	40	8.0 fl oz/acre	1	4 hr	**Do not use for transplant production.** Do not make more than two consecutive applications per season before alternating with a fungicide that has a different mode of action. Do not use for transplant production. Do not exceed 32 fl oz of product per acre per season. See label for application limits when multiple croppings are produced.

[1] Products registered for field use may be used on greenhouse crops (but not transplants) unless excluded on the label. Always check the label before applying a product. Additional products not prohibited for greenhouse use on basil, cucurbits, and lettuce not listed in this table are available.

[2] Former names of diseases and pathogens listed in this table that may be still be listed on fungicide labels are as follows: *Acidovorax citrulli* (formerly *Acidovorax avenae* subsp. *citrulli*) *Alternaria linariae* (formerly *A. solani* and *A. tomatophila*); *Agroathelia rolfsii* (formerly *Athelia rolfsii* and *Sclerotium rolfsii*); *Golovinomyces* spp. (formerly *Erysiphe* spp.) or *Golovinomyces cichoracearum* (formerly *Erysiphe cichoracearum*); *Fulvia fulva* (formerly *Cladosporium fulvum* and *Passalora fulva*); *Boeremia exigua* (formerly *Phoma exigua*); *Plectosphaerella cucumerina* (formerly *Plectosporium tabacinum*); *Podosphaera xanthii* (formerly *Sphaerotheca fuliginea*); *Stagonosporopsis* spp. (formerly *Didymella*); *Xanthomonas* leaf spot (formerly bacterial leaf spot); and (*Xanthomonas cucurbitae* (formerly *X. campestris* pv. *cucurbitae*).

TABLE 3-43. GREENHOUSE DISEASE CONTROL FOR VARIOUS VEGETABLE CROPS (cont'd)

R. A. Melanson, Extension Plant Pathologist, Mississippi State University; I. Meadows, Extension Plant Pathologist, North Carolina State University

Crop/Disease	Product [1]	FRAC	Rate of Material	Minimum Days Harv.	Minimum Days Reentry	Method, Schedule, and Remarks
TOMATO —*AFTER TRANSPLANTING IN A GREENHOUSE* (cont'd)						
Anthracnose (*Colletotrichum* spp.), black mold (*A. alternata*), early blight (*A. linariae* [2]), gray leaf spot (*S. botryosum*), late blight (*P. infestans*), leaf mold (*F. fulva* [2]), powdery mildew(*L. taurica*), Septoria leaf spot (*S. lycopersici*), target spot (*C. cassiicola*)	mandipropamid + difenoconazole (Revus Top)	40+3	5.5 to 7.0 fl oz/acre	1	0.5	*Not prohibited for greenhouse use.* Do not make more than two consecutive applications per season before alternating with a fungicide that has a different mode of action. Do not exceed 28 fl oz of product per acre per season. See label for application limits of active ingredients. The addition of a spreading/penetrat- ing type adjuvant is recommended.
Late blight (*P. infestans*)	mefenoxam + mancozeb (Ridomil Gold MZ WG)	4+ M03	2.5 lb/acre	5	2	*Not prohibited for greenhouse use.* Do not exceed 7.5 lb of product per acre per year. Do not exceed three applications per year. See label for application limits. Apply a protectant fungicide in between applications of product. See label for other restrictions.
Powdery mildew (*Leveilula* spp.)	myclobutanil (Sonoma 20 EW AG)	3	4.75 to 7.6 fl oz/acre	0	1	*Not prohibited for greenhouse use.* Do not exceed 38.0 fl oz of product or 0.5 lb myclobutanil per acre per crop. See product labels for additional instructions and restrictions. **Sonoma 20EW AG is not registered for use in AL, FL, KY, MS, OK, SC, TN, or TX.**
Buckeye rot, late blight, Phytophthora blight (foliar)	oxathiapiprolin + mandipropamid (Orondis Ultra)	49+40	See comments	1	4 hr	**Do not use in nursery production of transplanted crops.** Do not make more than two consecutive applications before alter- nating with fungicides that have a different mode of action. Use a rate range of 2 to 5 ml (0.07 to 0.167 fl oz or 0.42 to 1 tsp) per gallon of spray per 1,518 sq ft. See label for additional restrictions and yearly limits.
Black mold (*Alternaria alternata*), Botrytis gray mold (*Botrytis* spp.), early blight(*A. linariae* [2]), powdery mildew (*Leveilula, Oidiopsis*), Rhizoctonia seedling blight (*Rhizoctonia* spp.), Septoria leaf spot (*Septoria lycopersici*), southern blight (*Agroathelia rolfsii*), target spot (*C. cassiicola*) *Suppression*: Anthracnose (*Colletotrichum* spp.)	penthiopyrad (Fontelis)	7	16 to 24 fl oz/acre 24 fl oz/acre (anthracnose) 1 to 1.6 fl oz/ 1,000- row-ft	0	0.5	Basal stem rot: apply as a directed spray to the base of the plant 5 to 10 days after transplant followed by a 2nd application 14 days later.
Labels may include: Late blight, root rots	phosphonates (mono- and dibasic salts of phosphites, phosphorous acid, potassium phosphite) (various)	P07	See label	See label	See label	See product labels for complete application instructions, labeled diseases, restrictions, and crop and greenhouse usage. **Check state registration status prior to use.**
Botrytis gray mold (*Botrytis* spp.), early blight (*A. linariae* [2]), leaf mold (*F. fulva* [2]) powdery mildew (*Leveilula, Oidiopsis*) *Suppression*: Anthracnose (*Colletotrichum* spp.), late blight (*P. infestans*), target spot (*C. cassiicola*)	polyoxin D zinc salt (OSO 5% SC)	19	3.75 to 13 fl oz/acre	0	4 hr	*Not prohibited for greenhouse use.* Check products labels for maximum limits of active ingredient and applications per season.

[1] Products registered for field use may be used on greenhouse crops (but not transplants) unless excluded on the label. Always check the label before applying a product. Additional products not prohibited for greenhouse use on basil, cucurbits, and lettuce not listed in this table are available.

[2] Former names of diseases and pathogens listed in this table that may be still be listed on fungicide labels are as follows: *Acidovorax citrulli* (formerly *Acidovorax avenae* subsp. *citrulli*) *Alternaria linariae* (formerly *A. solani* and *A. tomatophila*); *Agroathelia rolfsii* (formerly *Athelia rolfsii* and *Sclerotium rolfsii*); *Golovinomyces* spp. (formerly *Erysiphe* spp.) or *Golovinomyces cichoracearum* (formerly *Erysiphe cichoracearum*); *Fulvia fulva* (formerly *Cladosporium fulvum* and *Passalora fulva*); *Boeremia exigua* (formerly *Phoma exigua*); *Plectosphaerella cucumerina* (formerly *Plectosporium tabacinum*); *Podosphaera xanthii* (formerly *Sphaerotheca fuliginea*); *Stagonosporopsis* spp. (formerly *Didymella*); Xanthomonas leaf spot (formerly bacterial leaf spot); and (*Xanthomonas cucurbitae* (formerly *X. campestris* pv. *cucurbitae*).

TABLE 3-43. GREENHOUSE DISEASE CONTROL FOR VARIOUS VEGETABLE CROPS (cont'd)

R. A. Melanson, Extension Plant Pathologist, Mississippi State University; I. Meadows, Extension Plant Pathologist, North Carolina State University

Crop/Disease	Product [1]	FRAC	Rate of Material	Minimum Days Harv.	Minimum Days Reentry	Method, Schedule, and Remarks
TOMATO —AFTER TRANSPLANTING IN A GREENHOUSE (cont'd)						
Labels may include: Alternaria diseases, anthracnose, Botrytis, powdery mildew, Septoria leaf spot	potassium bicarbonate (Carb-O-Nator, MilStop SP)	NC	2.5 to 5.0 lb/100 gal (Carb-O-Nator) 1.25 to 5.0 lb/100 gal (MilStop SP)	0	4 hr (Carb-O-Nator) 1 hr (MilStop SP)	***OMRI-listed.*** Do not store unused spray solution. See labels for additional labeled diseases and instructions. **Carb-O-Nator:** Final spray solution should not be below pH 7.0. Some tank-mixes may be incompatible and may cause phytotoxicity. Spray solution should be applied within 12 hours of preparation. **MilStop:** Do not exceed 0.5 lb of product per 4,350 sq ft or 1.15 lb of product per 10,000 sq ft per application. Do not adjust pH after mixing. Do not mix with other soluble pesticide or fertilizers not compatible with mild alkaline solutions.
Damping off and root rots (*Phytophthora* spp., *Pythium* spp.)	propamocarb hydrochloride (Previcur Flex)	28	See label	See label	0.5	Product applied through a drip system or as a soil drench. Do not apply more than four applications of product after transplanting per crop cycle. Do not mix with other products. Phytotoxicity may occur if applied to dry growing media. See label for other application instructions and maximum use rates.
Gray mold (*B. cinerea*), early blight (*A. linariae* [2])	pyrimethanil (Scala SC)	9	7 fl oz/acre	1	0.5	Use only in well-ventilated plastic tunnel or glass houses; ventilate for at least 2 hours after application of product. Use only in a tank-mix with another fungicide for early blight. Do not exceed 35 fl oz per acre per crop, with a maximum of two crop cycles per year. See label for cautions.
Powdery mildew (*Leveilula*, *Oidium* spp.)	pyriofenone (Prolivo 300 SC)	50	4 to 5 fl oz/acre	0	4 hr	***Not prohibited for greenhouse use.*** Do not make more than two sequential applications of Prolivo or another Group 50-containing fungicide before rotating to a fungicide with a different mode of action. Do not apply more than 16 fl oz of product or 0.32 lb of pyriofenone per acre per year. Do not apply more than four applications per year.
Powdery mildew	sulfur (various)	M02	See label	See label	See label	*Some products are OMRI-listed. See product labels for complete application instructions, specific crop and disease labels, and greenhouse usage. Check state registration status prior to use.*
Suppression: Powdery mildew	tolfenpyrad (Torac)	39	21.0 fl oz/acre	1	0.5	***Not prohibited for greenhouse use.*** Do not exceed 42 fl oz or two applications of product or 0.42 lb of tolfenpyrad per acre per crop cycle. Do not exceed four applications of product per year. Allow at least 14 days between applications. IRAC 21A.
Early blight (*A. linariae* [2]), gray leaf spot (*Stemphylium* spp.), late blight (*P. infestans*) *Suppression:* Anthracnose (*Colletotrichum* spp.), Septoria leaf spot (*S. lycopersici*), powdery mildew (*O. taurica*)	trifloxystrobin (Flint Extra)	11	See label 3.0 to 3.8 fl oz/acre (only for suppressed diseases)	3	0.5	***Not prohibited for greenhouse use.*** See label for application rates specific for early blight, gray leaf spot, and late blight. Alternate every FRAC 11 fungicide application with at least one application of a fungicide from a different FRAC group. Product must be tank mixed and alternated with a protectant fungicide for control of late blight. Do not exceed five applications or 16 fl oz of product per acre per year. The minimum interval between applications is 7 days.
Powdery mildew	triflumizole (Terraguard SC)	3	2 to 4 fl oz/100 gal	1	0.5	Commercial greenhouse production only. For use only as a foliar spray. Do not exceed 16 fl oz of product per acre per cropping system. Do not exceed four applications per crop.
Anthracnose, early blight, Septoria leaf spot	zinc dimethyl-dithiocarbamate (Ziram 76 DF)	M03	3 to 4 lb/acre	7	2	***Not prohibited for greenhouse use.*** Do not use on cherry tomatoes. Do not exceed 23.7 lb of product per acre per crop cycle. Product can be mixed with copper fungicides to enhance bacterial disease control.

[1] Products registered for field use may be used on greenhouse crops (but not transplants) unless excluded on the label. Always check the label before applying a product. Additional products not prohibited for greenhouse use on basil, cucurbits, and lettuce not listed in this table are available.

[2] Former names of diseases and pathogens listed in this table that may be still be listed on fungicide labels are as follows: *Acidovorax citrulli* (formerly *Acidovorax avenae* subsp. *citrulli*) *Alternaria linariae* (formerly *A. solani* and *A. tomatophila*); *Agroathelia rolfsii* (formerly *Athelia rolfsii* and *Sclerotium rolfsii*); *Golovinomyces* spp. (formerly *Erysiphe* spp.) or *Golovinomyces cichoracearum* (formerly *Erysiphe cichoracearum*); *Fulvia fulva* (formerly *Cladosporium fulvum* and *Passalora fulva*); *Boeremia exigua* (formerly *Phoma exigua*); *Plectosphaerella cucumerina* (formerly *Plectosporium tabacinum*); *Podosphaera xanthii* (formerly *Sphaerotheca fuliginea*); *Stagonosporopsis* spp. (formerly *Didymella*); Xanthomonas leaf spot (formerly bacterial leaf spot); and (*Xanthomonas cucurbitae* (formerly *X. campestris* pv. *cucurbitae*).

TABLE 3-44. EFFICACY OF PRODUCTS FOR GREENHOUSE TOMATO DISEASE CONTROL

I. Meadows, Extension Plant Pathologist, North Carolina State University

Scale: E, excellent; G, good; F, fair; P, poor; NC, no control; ND, no data.

Active Ingredient [1]	Product(s)	Fungicide (FRAC) Group [F]	Preharvest Interval (Days)	Anthracnose (*Colletotrichum coccodes*)	Bacterial Soft Rot (*Erwinia* spp.)	Bacterial Canker (*Clavibacter michiganensis* subsp. *michiganensis*)	Botrytis Gray Mold (*Botrytis cinerea*)	Buckeye Rot (*Phytophthora* spp.)	Early Blight (*Alternaria linariae*)[3]	Late Blight (*Phytophthora infestans*)	Leaf Mold (*Passalora Fulvia*)[3]	Powdery Mildew (*Leveillula taurica*)	Pythium Root Rot (*Pythium* spp.)	Rhizoctonia Root Rot (*Rhizoctonia solani*)	Septoria Leaf Spot (*Septoria lycopersici*)	Target Spot (*Corynespora cassiicola*)	Timber Rot (White mold) (*Sclerotinia sclerotiorum*)
acibenzolar-*S*-methyl	Actigard	P01	14	NC	F	F	NC	NC	NC	NC	NC	NC	NC	NC	NC	NC	NC
azoxystrobin + difenoconazole	Quadris Top	11+3	0	ND	NC	NC	ND	NC	G	F	ND	G	NC	ND	G	G	ND
bacteriophage	AgriPhage-CMM	NC	0	NC	P	F	NC	NC	NC	NC	NC	NC	NC	NC	NC	NC	NC
Bacillus subtilis	various	BM02	0	ND	ND	NC	P	NC	P	NC	ND	P	NC	NC	ND	P	ND
BLAD	Fracture	BM01	1	ND	NC	NC	F	NC	ND	NC	ND	F	NC	ND	ND	ND	ND
cyprodinil + fludioxonil	Switch 62.5WG	9+12	0	G	NC	NC	E	NC	F	NC	ND	G	NC	F	F	ND	ND
cyazofamid	Ranman	21	0	NC	NC	NC	NC	NC	NC	G	NC	NC	F	NC	NC	NC	NC
cymoxanil	Curzate	27	0	NC	NC	NC	NC	NC	NC	NC	NC	NC	F	NC	NC	NC	NC
difenoconazole + cyprodinil	Inspire Super	3+9	0	G	NC	NC	E	NC	G	NC	ND	G	NC	NC	ND	G	F
famoxadone + cymoxanil	Tanos	11+27	3	F	ND	P	ND	NC	F	NC	F	P	NC	NC	F	F	ND
fenamidone	Reason	11	14	ND	NC	NC	ND	NC	G	F	ND	G	NC	ND	G	G	ND
fenhexamid	Decree 50 WDG	17	1	NC	NC	NC	NC	NC	NC	NC	NC	F	NC	NC	NC	NC	NC
fixed copper	various	M01	0	P	F	F	P	NC	F	NC	P	P	NC	NC	P	P	NC
mancozeb	various	M03	5	G	NC	NC	F	NC	G	NC	F	NC	NC	NC	F	F	ND
mancozeb + azoxystrobin	Dexter Max	M03+11	5	ND	NC	NC	ND	NC	G	G[R]	ND	G	NC	ND	G	G	ND
mandipropamid	Revus	40	1	NC	NC	NC	NC	NC	NC	G	NC	NC	F	NC	NC	NC	NC
mandipropamid + difenoconazole	Revus Top	40+3	1	G	NC	NC	NC	G	G	G	NC	G	F	NC	G	ND	ND
neem oil	Triact 7	NC	0	ND	ND	ND	ND	ND	ND	NC	ND	ND	ND	ND	ND	ND	ND
oxathiapiprolin + mandipropamid	Orondis Ultra	49+40	1	NC	NC	NC	NC	NC	NC	G	NC	NC	F	NC	NC	NC	NC
penthiopyrad	Fontelis	7	0	P	NC	NC	F	NC	G	NC	G	NC	NC	NC	F	G	ND
polyoxin D zinc salt	Affirm WDG, OSO 5%SC	19	0	F	NC	NC	F[R]	ND	F	NC	ND	F	ND	ND	ND	F	ND
propamocarb hydrochloride	Previcur Flex	28	5	NC	NC	NC	NC	F	NC	G	NC	NC	F	NC	NC	NC	NC
pyrimethanil	Scala	9	1	NC	NC	NC	F[R]	NC	NC	NC	NC	NC	NC	NC	NC	NC	NC
Streptomyces sp. strain K61	Mycostop	BM02	0	NC	NC	NC	NC	NC	NC	NC	NC	F	F	NC	NC	NC	NC
streptomycin sulfate [2]	Agri-Mycin 50, Firewall 17 WP, Firewall 50 WP, Harbour	25	0	NC	F	F	NC	NC	NC	NC	NC	NC	NC	NC	NC	NC	NC
sulfur [P]	various	M02	0	P	NC	NC	NC	NC	NC	NC	NC	F	NC	NC	NC	NC	NC

[1] Efficacy ratings do not necessarily indicate a labeled use for every disease

[2] For use on transplants only

[3] Former names of pathogens listed in this table that may still be listed on fungicide labels are as follows: *Alternaria linariae* (formerly *A. solani* and *A. tomatophila*) and *Fulvia fulva* (formerly *Cladosporium fulvum* and *Passalora fulva*)

[P] Sulfur may be phytotoxic; follow label carefully

[F] To prevent resistance in pathogens, alternate fungicides within a group with fungicides in another group. Fungicides in the "M" groups are generally considered "low-risk" with no signs of resistance developing to most fungicides. "NC" indicates that the product has not been classified into a fungicide group

[R] Resistance reported in the pathogen

SEED TREATMENTS

Seed sanitation to eradicate bacterial or viral plant pathogens: When treating vegetable seeds, it is critical to follow the directions exactly, be- cause germination can be reduced by the treatment and/or the pathogen may not be eliminated. The effect of a treatment on germination should be determined on a small lot of seeds prior to treating large amounts of seed. Treatments should not be applied to pelleted seed, previously treated seed, or old or poor-quality seed. A protective fungicide treatment (see below) can be applied to the seed following treatment for bacterial pathogens.

Seed treatments to prevent damping off diseases: Most commercially available vegetable seeds come treated with at least one fungicide and/ or insecticide. Vegetable producers who would like to apply their own seed treatment should purchase non-treated seed. While many fungicides are labeled for use on vegetable seed, most fungicides are restricted to commercial treatment only and should not be applied by producers. Labeled fungicides can be applied to seed following treatment for bacterial pathogens. Do not use fungicide treated seed for food or feed.

HOT WATER TREATMENT

By soaking seed in hot water, seed-borne fungi and bacteria can be reduced, if not eradicated, from the seed coat. Hot water soaking will not kill pathogens associated with the embryo nor will it remove seed-borne plant viruses from the seed surface.

1. Place seed loosely in a weighted cheesecloth or nylon bag.
2. Warm the seed by soaking it for 10 minutes in 100°F (37°C) water.
3. Transfer the warmed seed into a water bath already heated to the temperature recommended for the vegetable seed being treated (Table 3-47). The seeds should be completely submerged in the water for the recommended amount of time (Table 3-47). Agitation of the water during the treatment process will help to maintain a uniform temperature in the water bath.
4. Transfer the hot water treated seed into a cold-water bath for five minutes to stop the heating action.
5. Remove seed from the cheesecloth or nylon bag and spread them evenly on clean paper towel or a sanitized drying screen to dry. Do not dry seed in areas where fungicides, pesticides or other chemi- cals are located.
6. Seed can be treated with a labeled fungicide to protect against damping off pathogens.

CHLORINE BLEACH TREATMENT

Treating seeds with a solution of chlorine bleach can effectively remove bacterial pathogens and some viruses (i.e., Tobacco Mosaic Virus) that are borne on the surface of seeds.

1. Add 1 quart (946 ml) of household bleach to 5 quarts (4.7 L) of potable water.
2. Add a drop or two of liquid dish detergent or a commercial surfactant such as Activator 90 or Silwet to the disinfectant solution. Add seed to the disinfectant solution (1 pound of seed per 4 quarts of disinfectant solution) and agitate for 1 minute.
3. Prepare fresh disinfectant solution for each batch of seeds to be treated.
4. Rinse the seed in a cold-water bath for 5 minutes to remove residual disinfectant.
5. Spread seeds evenly on clean paper towel or a sanitized drying screen to dry. Do not dry seed in area where fungicides, pesticides, or other chemicals are located.
6. Seed can be treated with a labeled fungicide to protect against damping off pathogens.

HYDROCHLORIC ACID TREATMENT

Tomato seed can be treated with a dilute solution of hydrochloric acid (HCl) solution to eliminate seed-borne bacterial pathogens such as Xanthomonas spp. (Bacterial leaf spot), Pseudomonas syringae pv. tomato (Bacterial speck) and Clavibacter michiganensis subsp. michiganensis (Bacterial canker). Hydrochloric acid can also be used to remove TMV from the surface of tomato seed. Do not use HCL-treated seed for food or animal feed.

1. Prepare a 5% solution of HCl by adding one part acid to 19 parts potable water. Prepare the acid solution in a well-ventilated area and avoid direct skin contact with the acid.
2. Soak seeds for 6 hours with gentle agitation.
3. Carefully drain the acid off the seed and rinse seed under running potable water for 30 minutes. Alternatively, rinse the seeds 10 to 12 times with potable water to remove residual acid.
4. Spread seeds evenly on clean paper towel or a sanitized drying screen to dry. Do not dry seed in area where fungicides, pesticides, or other chemicals are located.
5. Seed can be treated with a labeled fungicide to protect against damping off pathogens

TRISODIUM PHOSPHATE TREATMENT

Tomato seed can be treated with trisodium phosphate (TSP) to eradicate seed transmitted TMV. Do not use TSP-treated seed for food or animal feed.

1. Prepare a 10% solution of TSP (1 part TSP in 9 parts potable wa- ter). Trisodium phosphate is available at most home supply or paint stores. Avoid direct skin contact with the TSP solution.
2. Soak seed for 15 minutes in the disinfectant solution.
3. Rinse the seed in a cold-water bath for 5 minutes to remove residual disinfectant.
4. Spread seeds evenly on clean paper towel or a sanitized drying screen to dry. Do not dry seed in area where fungicides, pesticides, or other chemicals are located.
5. Seed can be treated with a labeled fungicide to protect against damping off pathogens.

TESTING SEED GERMINATION AFTER SEED TREATMENTS

Randomly select 100 seeds from each seed lot.

1. Treat 50 seeds using one of the sanitizers described above.
2. After the treated seed has dried and before application of a protectant fungicide, plant the treated and non-treated seed separately in flats containing planting mix according to standard practice. Label each group as treated or non-treated.
3. Allow the seeds to germinate and grow until the first true leaf ap- pears (to allow for differences in germination rates to be observed).
4. Count seedlings in each group separately.
5. Determine the percent germination for each group: # seedlings emerged ÷ # seeds planted x 100.
6. Compare percentage germination between the treated and
7. non-treated groups. Percent germination should be within 5% of each other.

TABLE 3-45. RECOMMENDED TEMPERATURES & TREATMENT TIMES FOR HOT WATER DISINFESTATION OF VEGETABLE SEED

Vegetable Crop	Water Temperature (°F/°C)	Soaking Time (Minutes)
Broccoli	122/50	20 to 25
Brussels sprout	122/50	25
Cabbage	122/50	25
Carrot	122/50	15 to 20
Cauliflower	122/50	20
Celery	122/50	25
Chinese cabbage	122/50	20
Collard	122/50	20
Cucumber [1]	122/50	20
Eggplant	122/50	25
Garlic	120/49	20
Kale, Kohlrabi	122/50	20
Lettuce	118/48	30
Mint	112/44	10
Mustard, Cress, Radish	122/50	15
Onion	115/46	60
Pepper	125/51	30
Rape, Rutabaga	122/50	20
Shallot	115/46	60
Spinach	122/50	25
Tomato	122/50	25
Turnip	122/50	20

[1] Cucurbits other than cucumbers can be severely damaged by hot water treatment and should be disinfested using chlorine bleach

CLEANING AND SANITATION

Sanitation is a broad term used in the food industry. For produce growers, this term is often defined by requirements within the PSR or standards within GAPs. Sanitation includes the basic sanitary conditions and practices within the food industry for growing and handling of food safely. Measures required to prevent equipment, tools, facilities (i.e., packing- houses), and sanitation practices from becoming a route of contamination for produce and food contact surfaces include:

- Pre and postharvest water quality
- Cleaning of tools and equipment
- Storage and maintenance of tools and equipment
- Maintenance of toilet and handwashing facilities
- Control of pests
- Maintenance of plumbing
- Disposal of sewage and waste

PREHARVEST WATER SANITATION

There are many things to consider when choosing a water treatment system, and the system is unique to each water source/situation. Selecting the appropriate water system treatment is probably the most difficult part of treating your water. The questions below highlight the basic steps you should consider prior to choosing a water treatment method. If you decide to choose a water treatment method, you should contact your local resources to help you decide the best method for your operation.

CONSIDERATIONS FOR CHOOSING A WATER TREATMENT METHOD FOR YOUR FARM OPERATION

- *Is the product EPA-approved?* Chemicals must be EPA-registered for their use based on the EPA label for the product. If the product does not have this information, you cannot use it.
- *What is the cost of set-up, continued use, and maintenance?* Some treatment methods require a larger investment than others at set-up, and all costs (including recurring costs such as chemical costs) should be assessed at the start to ensure successful implementation of the treatment.
- *What are the management and monitoring requirements?* All of these treatments require proper management and monitoring in order to treat the water effectively and consistently. Be sure you have the time and resources to invest in order to successfully accomplish these goals on your farm.
- *What infrastructure do you already have on your farm?* The infrastructure you already have on the farm may more easily lend itself to one treatment versus another (i.e., electricity, injection point, etc.).
- *What is the source of your water?* Not all sources of water are equal. Keep in mind the population of indicator microorganisms like
- *E. coli* and other factors like turbidity and alkalinity. This can drive increased chemical demand to achieve your goals. In the case of UV treatment, more turbid water might require additional filtration or longer contact time with the UV light.
- *What is the application method?* How you size and implement a water treatment depends on your irrigation system and flow rate.
- *What is the sensitivity of your crops to the treatment?* As some crops may be more sensitive to particular treatments, it is important to consider your crop tolerance.
- *Are there WPS and other restricted entry considerations?* This is another important consideration for resources that may be needed for your employees to implement the treatment.
- *Are there other EPA or local environmental considerations?* Depending on where you are located, environmental considerations for treated water may vary before water is discharged back into the environment.

Source: Buchanan, J.R., Chapin, T.K., Danyluk, M.D., Gunter, C.C., Strawn, L.K., Wszelaki, A.L., Critzer, F.J. 2019. Bridging the GAPs: Approaches for Treating Water On-Farm. Curriculum v1.0.

For more information, visit:
https://vegetables.tennessee.edu/preharvest-water/

POSTHARVEST WATER SANITATION

Produce may be washed after harvest to remove dirt and debris. According to FSMA, water used for harvest and postharvest activities must have no detectable generic *E. coli* in a 100 mL water sample. Generic *E. coli* is an indicator organism, with the idea that if there is generic *E. coli* in the water, there is a higher likelihood that there could be other pathogenic organisms.

Growers may use dump tanks or bins for washing produce, which oftentimes relies on recirculated and batch water. Single-pass systems (i.e., spray bars) may also be used. For recirculated and batch water systems, it is particularly important to maintain the sanitary quality of the water throughout its use. Potential contamination from produce in a batch water system, like a dump tank, can contaminate the water, thus spreading the microorganisms or cross-contaminating other produce in the dump tank. The use of sanitizers in wash water systems can significantly reduce this risk of cross-contamination. Table 3-49 provides a list of sanitizers that are EPA-approved for use in wash-water systems.

CLEANING AND SANITIZING FOOD CONTACT SURFACES

A critical component within a sanitation program includes procedures for cleaning and sanitizing, particularly for surfaces on which food comes into contact. The Produce Safety Rule states that growers must "inspect, maintain, and clean and, when necessary and appropriate, sanitize all food contact surfaces of equipment and tools used in covered activities as frequently as reasonably necessary to protect against contamination of covered produce."

Identify Food Contact Surfaces

Food contact surfaces are surfaces that will directly touch the produce and are priority areas for cleaning and sanitizing. Examples of food contact surfaces can include knives, harvest bins, transportation equipment, tables, conveyor belts, and packing materials. You must not forget about employees' hands – they can be a food contact surface too, emphasizing the need for proper handwashing.

Cleaning and Sanitizing Procedures

Once the priority areas for cleaning and sanitizing have been identified, implement proper cleaning and sanitizing procedures. Proper cleaning and sanitizing procedures follow four basic steps, as described in Table 3-48. Keep in mind that the PSR defines criteria for water used during harvest and postharvest activities that include no detectable generic *E. coli* in a 100 mL water sample.

The use of appropriate detergents is important to consider as they help reduce the surface tension of water and surround and lift soil from the surface. Upon rinsing the detergent, the surface may appear visually clean; however, microorganisms can still be present even after the cleaning step. Food contact surfaces should include a sanitization step to reduce the number of microorganisms to a safe level. According to the EPA definition, food contact sanitizers reduce the bacterial count on a surface by 5 logs or by 99.999%. For instance, if there are 1,000,000 bacteria on a surface before the sanitizer is applied, this number should be reduced to 10 bacterial cells after the sanitizer is applied and dried. A list of commonly used sanitizers can be found in Table 3-49. For food-contact surfaces, sanitizers are often formulated as a last step after cleaning and rinsing.

IMPLEMENTING A CLEANING AND SANITATION PROGRAM

Having a dedicated trained crew and crew leader is critical to implementing a cleaning and sanitation program in farms and packing houses. Always conduct daily housekeeping activities in packing areas regardless of the frequency of cleaning and sanitation of other areas. It is important to budget sufficient time for equipment breakdown, cleaning, sanitation, and reassembly of equipment. The use of personal protective equipment (PPE) indicated on the chemical label must be followed for safety. Additionally, label instructions must be followed when mixing and applying sanitizers to ensure their safety and effectiveness. The use of application equipment for cleaners and sanitizers (such as *foamers*) can greatly reduce the time it takes to clean and sanitize equipment and can allow for access to hard-to-reach areas, especially in packing lines. Having a dedicated set of tools for bathrooms that are never used in any other areas where produce is present is important to prevent cross-contamination. Working with a cleaning and sanitation company directly is always helpful. Some companies offer training resources for their products, enhanced labels, and provide advice on application equipment. ATP testing and environmental monitoring can be used to verify that cleaning and sanitation procedures are actually working. Consult with your food safety specialist about what options are feasible for your operation.

TRAINING EMPLOYEES

Training employees on how to clean and sanitize surfaces on the farm is a crucial way to reduce contamination on the farm. Employees must be trained to understand the difference between cleaning and sanitizing, especially that an unclean surface cannot be sanitized. Training employees with a video, like this one from the University of Maryland, https://www.you-tube.com/watch?v=mXtFlWt67z0 and pausing to demonstrate farm-specific cleaning and sanitizing procedures can be an effective training tool.

TABLE 3-46. BASIC CLEANING AND SANITIZING PROCEDURES FOR FOOD CONTACT SURFACES

E. Rogers, Area Specialized Agent, Fresh Produce Food Safety, North Carolina State University

Step	Purpose	Implementation
Remove dirt and debris	Remove all loose soil and food residue.	This could include either the use of clean water or physical removal of soil using a brush or broom; avoid the use of high-pressure sprayers to prevent spreading dirt and microorganisms through the air onto additional surfaces.
Clean	Further removal of soil and food residue to ensure effective sanitizer use and prevent the formation of biofilms.	Use an appropriate detergent for the type of soil that needs to be removed.
Rinse	Remove detergent and soil/food residue.	Use clean water.
Sanitize	Reduction of microbial levels.	Apply a sanitizer approved for use on food contact surfaces following the instructions on the label.

On some farms, employing **dry cleaning** methods like brushing, vacuuming, or using blowers might prove more advantageous and safer. This is because introducing moisture to a dry procedure can potentially promote the growth of germs, elevating the associated risks. Dry cleaning involves the physical removal of visible soil and debris from surfaces, followed by the application of an EPA-approved sanitizer for its intended use. This method is particularly suitable for packing areas where the introduction of water could promote microbial proliferation. Dry cleaning is a prevalent practice in the fields of berry and nut production. Additionally, it can be employed for equipment that lacks adequate drying time between uses, as equipment not thoroughly dried may facilitate germ growth between uses.

Note that sanitizers and disinfectants are not the same. Disinfectants will eliminate all the organisms listed on the product label, which often includes not only bacteria but viruses and fungi, as well. Disinfectants are not generally used for food-contact surfaces because they can leave harmful chemical residues and corrode equipment over time. Sanitizers are formulated to be used on food and food contact surfaces. Both sanitizers and disinfectants are considered antimicrobial pesticides and, therefore, are regulated by the U.S. Environmental Protection Agency (EPA). Following the label is required by law. The label will typically provide information on how the chemical should be handled and whether the chemical and the concentration are meant to be used on food-contact surfaces. This information can also be found on the EPA Pesticide Product Labeling System (PPLS) website: https://www.epa.gov/pesticide-la-bels/pesticide-product-label-system-ppls-more-information by searching the chemical product with the EPA registration number found on the label.

Verifying and maintaining the correct concentration of a sanitizer is key to effectively reducing microbial levels on surfaces and in the wash water. For more information on using chlorine and peroxyacetic acid for fruit and vegetable washing and packing, visit:

How to Use and Monitor Chlorine (Sodium/Calcium Hypochlorite) in Fruit and Vegetable Wash water and on Equipment and Food Contact Surfaces https://extension.tennessee.edu/publications/Documents/SP798-A.pdf

Using Peroxyacetic Acid (PAA) in Fruit and Vegetable Washing and Packing
https://extension.tennessee.edu/publications/Documents/SP798-B.pdf

TABLE 3-47. WATER, PRODUCE, AND EQUIPMENT SANITATION

The trade names listed are intended to aid in the crop identification of products and are intended neither to promote the use of specific trade names nor to discourage the use of generic products. The target rate and formulation are based on the current EPA Pesticide Product Label. Make sure to read the product label before preparing any sanitizer solution for the designated use.

E. Rogers, Area Specialized Agent, Fresh Produce Food Safety, North Carolina State University (LAST UPDATED IN 2025)

Medium/Sanitizer	Contact Time (seconds)	Rate of Material to Use		Method, Schedule, and Remarks
		Target Rate (ppm)*	Formulation	
Calcium hypochlorite (Aquafit, ECR aquachlor, PPG Calcium Hypochlorite tablets)	120	25 to 400 chlorine	Varies between commodities	Sanitizing solution: 1 oz/200 gal to make a solution of 25 ppm available chlorine. A second wash is required.
Chlorine dioxide (ProOxine, Anthium Dioxide, Adox 750)	10 to 20	3 to 5 chlorine	Varies between products; see product labels.	Maintain water pH between 6.0 and 10. Restricted to large operations. Requires automated and controlled injection systems. Wear PPE during preparation and handling. After treatment of fruits and vegetables followed by a clean water rinse. **NOTE: Chlorine dioxide is explosive.**
Chlorine gas (99.9%)	—	Contact supplier for rates.		Restricted to very large operations. Requires automated and controlled injection systems. Regulated by both the EPA (water) and FDA (food contact surfaces).
Hydrogen peroxide + peroxyacetic acid	Contact times vary depending on the governing sanitary code. Post-sanitation rinse is not necessary.			
		Target Rate (ppm)*	Formulation	
(BioSide HS)	60	80 peroxyacetic acid	1 fl oz/16.4 gal	
(PAA Sanitizer FP)	45	88 to 100 peroxyacetic acid	3 to 3.5 fl oz/16 gal	
(Perasan A)	60	30 to 300 peroxyacetic acid	0.6 to 6 fl oz/10 gal	
(SaniDate 5.0)	45	27 to 96t peroxyacetic acid	59.1 to 209.5 fl oz/1,000 gal	
(StorOx 2.0)	30 to 180	26 to 93 peroxyacetic acid	1.5 to 5.4 fl oz/10 gal	
(Tsunami 100)	90	80 peroxyacetic acid	6.7 fl oz/100 gal	
(Victory)	90	30 to 80 peroxyacetic acid (foodborne pathogens)	1 fl oz/16.4 gal	
(VigorOx 15 F&V)	—	45 peroxyacetic acid (foodborne pathogens)	0.54 fl oz/16 gal	
(Maguard 5626)	30	28 to 90 peroxyacetic acid	60 to 195 fl oz/ 1,000 gal	
Sodium hypochlorite (5.25%)	Monitor free chlorine or change the solution when it is visibly dirty. Rinse produce with potable water prior to packing. Wear goggles & rubber gloves when handling. Maintain water pH between 6.0 & 7.5. Noxious chlorine gas can be released when pH drops below 6.0.			
(Bleach labeled for fruit and vegetable washing)	120	25 chlorine	1 fl oz/15 gal	**NOTE: Household bleach that contains fragrances or anti-splash agents for household use is NOT registered for use with fresh produce.**
(Dibac)	120	25 chlorine	1 fl oz/20 gal	
(Agclor 310)	120	30 to 400	Varies between commodities	
(Dynachlor)	120	25 chlorine	5 fl oz/200 gal	
(Extract-2)	120	25 chlorine	5 fl oz/200 gal	
(JP Optimum CRS)	—	25 chlorine	0.75 fl oz/10 gal	
(Zep FS Formula 4665)	120	25 chlorine	5 fl oz/200 gal	
Calcium hypochlorite (Aquafit, ECR aquachlor, PPG Calcium Hypochlorite Tablets)	120	200 (non-porous surfaces), 600 (porous surfaces)	3 oz/20 gal	Do not rinse or soak equipment overnight.
Chlorine dioxide (ProOxine, Sanogene, Anthium Dioxide, Adox 750)	60 to 10 min	10 to 20 (porous or non-porous surfaces) 500 (ceilings, floors, & walls)	Varies between products; see product labels.	Wear PPE during preparation and handling. **NOTE: Chlorine dioxide is explosive.**
Hydrogen peroxide + peroxyacetic acid	Contact time varies depending on the governing sanitary code. Consult labels as some products require a post-application rinse with potable water.			
(BioSide HS)	60 or more	93 to 500 peroxyacetic acid	0.7 to 3.8 fl oz/10 gal	
(Oxidate 2.0)	See label	100 to 300 peroxyacetic acid	1.25 to 1.5 fl oz/gal	If treated surfaces will contact food, thoroughly rinse surfaces with clean water. If the product is applied via fogging or spraying, workers must wear appropriate PPE (see label).
(PAA Sanitizer FP)	60 or more	88 to 130 peroxyacetic acid (non-porous surfaces)	1 to 1.5 fl oz/5 gal	
(Perasan A)	60 or more	82 to 500 peroxyacetic acid	1 to 6.1 fl oz/5 gal	
(SaniDate 5.0)	60	147 to 500 peroxyacetic acid	1.6 to 5.4 fl oz/5 gal	
(StorOx 2.0)	60	86 peroxyacetic acid	0.5 fl oz/1 gal	
(VigorOx SP-15 F&V)	60	85 to 123 peroxyacetic acid and 57 to 82 hydrogen peroxide	0.31 to 0.45 fl oz/5 gal	
(Maguard 5626)	60	154 peroxyacetic acid	1 fl oz/6 gal	

* Recommendations for active/available forms

** Recommendations are for clean water only. Always wash off organic debris and soil with water prior to sanitizing. Rates and contact time are dependent on surface type.

TABLE 3-47. WATER, PRODUCE, AND EQUIPMENT SANITATION (cont'd)

The trade names listed are intended to aid in the identification of products and are intended neither to promote the use of specific trade names nor to discourage the use of generic products. The target rate and formulation are based on the current EPA Pesticide Product Label. Make sure to read the product label before preparing any sanitizer solution for the designated use.

E. Rogers, Area Specialized Agent, Fresh Produce Food Safety, North Carolina State University (LAST UPDATED IN 2025)

Medium/Sanitizer	Contact Time (seconds)	Rate of Material to Use		Method, Schedule, and Remarks
		Target Rate (ppm)*	Formulation	
Sodium hypochlorite (5.25%) (bleach labeled for sanitizing of food contact surfaces)	120	100 to 200 chlorine	1 to 1.5 fl oz/3 gal	Noxious chlorine gas can be released when the pH drops below 6.0. Porous surfaces require a thorough post-disinfection rinse with potable water. Allow all surface types to air dry prior to re-use.
Sodium hypochlorite (9.2%)	120	600 chlorine (porous surfaces)	12 fl oz/10 gal	
(Dibac)	120	100 to 200 chlorine (non-porous surfaces)	6 fl oz/5 gal	
Sodium hypochlorite (12.5%)				
(Agclor 310)	120	100 to 200 chlorine (non-porous surfaces)	1 fl oz/10 gal	
(Dynachlor)	120	100 to 200 chlorine (non-porous surfaces)	1 to 2 fl oz/10 gal	
	120	600 chlorine (porous surfaces)	6 fl oz/10 gal	
(Extract-2)	120	100 to 200 (non-porous surfaces)	1 to 2 fl oz/10 gal	
(Zep FS Formula 4665)	120	100 to 200 chlorine (non-porous surfaces)	1 to 2 fl oz/10 gal	
	120	600 chlorine (porous surfaces)	6 fl oz/10 gal	

* Recommendations for active/available forms
** Recommendations are for clean water only. Always wash off organic debris and soil with water prior to sanitizing. Rates and contact time are dependent on surface type.

VARIOUS AND ALTERNATIVE FUNGICIDES

TABLE 3-48. VARIOUS FUNGICIDES FOR USE ON VEGETABLE CROPS

R. Singh, Plant Pathologist, LSU AgCenter

Not all trade names are registered in all states; check the registration status of each product prior to use. Check product labels to confirm that a product is labeled for your intended use.

Common Name (FRAC)	Trade Name(s)		
aluminum tris (O-ethyl phosphate) (FRAC P07)	See fosetyl-Al.		
azoxystrobin (FRAC 11)	Acadia 2SC (Atticus)	Azoxyzone (LG Life Sciences)	Quadris (Syngenta)
	Aframe (Syngenta)	AZteroid FC (Vive Crop Protection)	Satori Fungicide (Loveland Products)
	Arius 250 (Sipcam Agro USA)	AZteroid FC 3.3 (Vive Crop Protection)	Tetraban (Winfield United)
	A-Zox 25SC (Sharda USA)	Gold Rush (Altitude Crop Innovations)	Tigris Azoxy 2SC (Tigris)
	Azoxy 2SC Prime (Prime Source)	Heritage (Syngenta)	Trevo (Innvictis Crop Care)
	AzoxyStar (Albaugh)	Mazolin (AgBiome Innovations)	Willowood Azoxy 2SC (Willowood USA)
chlorothalonil (FRAC M05)	Bravo Ultrex (Adama)	Chlorothalonil 720 (Drexel)	Initiate 720 (Loveland Products)
	Bravo Weather Stik (Adama)	Echo 720 Agricultural Fungicide (Sipcam Agro USA)	Praiz (Winfield United)
	Bravo Zn (Adama)	Echo 90 DF Agricultural Fungicide (Sipcam Agro USA)	RIALTO 720F (ATTICUS)
	Chloronil 720 (Syngenta)	Equus 720SST (AMVAC)	
copper diammonia diacetate complex (FRAC M01)	Copper-Count-N (Mineral Research & Dev. Corp)		

TABLE 3-48. VARIOUS FUNGICIDES FOR USE ON VEGETABLE CROPS (cont'd)
R. Singh, Plant Pathologist, LSU AgCenter

Not all trade names are registered in all states; check the registration status of each product prior to use. Check product labels to confirm that a product is labeled for your intended use.

Common Name (FRAC)	Trade Name(s)		
copper hydroxide (FRAC M01)	Americop 40DF (Industrias Quimicos del Valles, SA) Champ Dry Prill (Nufarm) Champ Formula 2F (Nufarm) Champ WG (Nufarm) ChampION++ (Nufarm) Kalmor (OHP) Kentan DF (Isagro USA)	Kocide 2000-O (Certis USA) Kocide 50DF (Certis USA) Kocide 3000-O (Certis USA) Kocide HCu (Certis USA) KOP-Hydroxide (Drexel) KOP Hydroxide 50W (Drexel) Nu-Cop 3L (Albaugh) Nu-Cop 30HB (Albaugh)	Nu-Cop 50DF (Albaugh) Nu-Cop 50WP (Albaugh) Nu-Cop HB (Albaugh) Nu-Cop XLR (Albaugh) Previsto (Gowan) SPINNAKER (SIPCAM AGRO USA)
copper octanoate (FRAC M01)	Camelot-O (SePRO) Cueva Fungicide Concentrate (Certis USA)		
copper oxychloride (FRAC M01)	COC DF (Albaugh)		
copper oxychloride + copper hydroxide (FRAC M01)	Badge SC (Gowan) Badge X2 (Gowan)		
copper(cuprous) oxide (FRAC M01)	Nordox (NORDOX Industrier AS) Nordox 75WG (NORDOX Industrier AS)		
copper sulfate (basic) (FRAC M01)	Basic Copper 53 (Albaugh)	Cuprofix Ultra 40 Disperss (UPL)	Cuproxat Flowable (NuFarm)
copper sulfate pentahydrate (FRAC M01)	KOP-5 (Drexel Chemical Company) Instill (S.T. Biologicals)	Magna-Bon CS 2005 (Magna-Bon Agricultural Control Solutions)	Mastercop (ADAMA) Phyton 35 (Phyton Corporation)
fludioxonil (FRAC 12)	Cannonball WG (Syngenta)	Cannonball WP (Syngenta)	Spirato GHN (Nufarm)
fosetyl-Al aluminum tris (O-ethyl phosphate) (FRAC P07)	Aliette WDG Fungicide (Bayer Crop Science) Linebacker WDG (NovaSource)		
iprodione (FRAC 2)	Meteor (UPL)	Nevado 4F (Adama)	Rovral 4 Flowable Fungicide (FMC Corp.)
mancozeb (FRAC M03)	Dithane F-45 Rainshield (Corteva Agriscience) Dithane M45 (Corteva Agriscience) Fortuna 75WDG (Agria Canada)	Koverall (FMC Corporation) Manzate Max (UPL) Manzate Pro-Stick (UPL)	Penncozeb 75DF (UPL) Penncozeb 80WP (UPL) Roper DF Rainshield (Loveland Products)
mefenoxam (FRAC 4)	Apron XL (Syngenta) Ridomil Gold GR (Syngenta)	Ridomil Gold SL (Syngenta) Ultra Flourish (Nufarm)	
myclobutanil (FRAC 3)	Rally 40WSP (Corteva Agriscience)	Sonoma 25EW AG (Albaugh)	Sonoma 40WSP (Albaugh)
phosphite, potassium (FRAC P07)	Helena Prophyt (Helena) Reveille (Helena)		
phosphite (mono- and dibasic salts) (FRAC P07)	Helena ProPhyt and Helena Prophyt (HAE) (Helena)	Phostrol (Nufarm)	Phostrol 500 (Nufarm)
propiconazole (FRAC 3)	AmTide Propiconazole 41.8% EC (AmTide) Fitness (Loveland Products) Mentor (Syngenta) Propi-Star EC (Albaugh)	Propicure 3.6F (United Supplies, Inc.) Shar-Shield PPZ (Sharda USA) Tide Propiconazole 41.8EC (Tide International)	Tilt (Syngenta) Topaz (Winfield Solutions) Vigil (Innvictis Crop Care)
sulfur (FRAC M02)	Cosavet-DF (Sulphur Mills Limited) CSC 80% Thiosperse (Martin Resources) CSC Dusting Sulfur (Martin Resources) CSC Thioben 90 (Martin Resources) CSC Wettable Sulfur (Martin Resources) Dusting Sulfur (Loveland Products; Wilbur-Ellis) First Choice Dusting Sulfur (Loveland Products)	IAP Dusting Sulfur (Independent Agribusiness Professionals) InteGro Magic Sulfur Dust (InteGro Inc.) Kumulus DF (Micro Flo and Wilbur-Ellis) Liquid Sulfur Six (Helena) Micro Sulf (Nufarm) Microfine Sulfur (Loveland Products) Microthiol Disperss (United Phosphorus) Special Electric Sulfur (Wilbur-Ellis)	Spray Sulfur (Wilbur-Ellis) Sulfur 6L (Arysta and Micro Flo) Sulfur 90W (Drexel) Sulfur DF (Wilbur-Ellis) THAT Flowable Sulfur (Stoller Enterprises) Thiolux (Loveland Products) Yellow Jacket Wettable Sulfur II (Georgia Gulf Sulfur)
tebuconazole (FRAC 3)	Monsoon (Loveland Products) Onset 3.6L (Winfield Solutions) Tebu-Crop 3.6F (Sharda USA)	Tebucon 3.6F (Repar Corp.) Tebuconazole 3.6F (Solera Source Dynamics) TebuStar 3.6L (Albaugh)	Tebuzol 3.6F (United Phosphorus)
thiophanate-methyl (FRAC 1)	Cercobin (Cheminova) Incognito 4.5F (Makhteshim Agan of North) Incognito 85WDG (MANA)	Thiophanate Methyl 85WDG (Makhteshim Agan of North) T-Methyl 4.5 Ag (Helena) T-Methyl 4.5F (Nufarm)	T-Methyl 70WSB (Nufarm) Topsin 4.5FL (UP) Topsin M WSB (UPL)

TABLE 3-49. BIOPESTICIDES, FUNGICIDE, AND NEMATICIDE ALTERNATIVES FOR VEGETABLES

R. A. Melanson, Extension Plant Pathologist, Mississippi State University; J. Desaeger, Nematologist, University of Florida

Active Ingredient [1]	Product [1]	Target Diseases/Pests	PHI (days)	REI	Greenhouse Use	OMRI-listed	Comments [2]
allyl isothiocyanate	Dominus (*Gowan*)	Certain soil-borne fungi and nematodes	—	5 days	Yes	No	Preplant soil biofumigant. See label for other restrictions and application instructions. **Check state registration status prior to use.**
azadirachtin	AzaGuard (*BioSafe Systems*), Ecozin Plus 1.2% ME (*Amvac*), Molt-X (*BioWorks, Inc*)	Plant parasitic nematodes	0*	4 hr	Yes	Yes	*See label for use rates that may restrict the PHI. See labels for other restrictions and application instructions. **Other products containing the active ingredient azadirachtin are available.**
Bacillus amyloliquefaciens strain D747	Convergence, Double Nickel 55, Double Nickel LC (*Certis*); Triathlon BA (*OHP*)	Various diseases (see label for crop-specific diseases)	0	4 hr*	Not Prohibited (Convergence, Double Nickel 55) Yes (Double Nickel LC, Triathlon BA)	Yes	*See labels for REI exceptions and specific greenhouse uses. Do not use highly acidic or alkaline water to mix sprays. See labels for additional application instructions and restrictions. **Convergence is only allowed for use on legumes and root and tuber vegetables.** *FRAC BM02.*
Bacillus amyloliquefaciens strain F727 cells and spent fermentation media	Amplitude, Stargus (*Marrone Bio Innovations/ProFarm*)	Various diseases (see label for crop-specific diseases)	See label	See label	See label	Yes	See label for application instructions, restrictions, and vegetable crops on which products may be used. *FRAC BM02.*
Bacillus amyloliquefaciens strain ENV503	BellaTrove Companion Maxx WP (*Douglas Plant Health*)	Root and foliar diseases (see label for crop-specific diseases)	0	4 hr*	Yes*	Yes	See product labels for instructions on various application uses, particularly those in greenhouse production. *See labels for exceptions to the REI and for specifics of greenhouse use. **Check state registration status of products prior to use.** *FRAC BM02.*
Bacillus amyloliquefaciens strain MBI 600	Serifel (*BASF*)	Various foliar and soil-borne diseases (suppression, see label for crop-specific diseases)	0	12 hr	Not prohibited	Yes	See label for application instructions, restrictions, and vegetable crops on which products may be used. *FRAC BM02.*
Bacillus amyloliquefaciens subspecies *plantarum* strain FZB42	Bexfond (*Corteva*)	Certain fungal and oomycete soil-borne diseases (see label for crop-specific diseases)	None	4 hr*	Not prohibited	Yes	*See labels for exceptions to the REI. See label for application instructions, precautions, and restrictions. **Check state registration status of products prior to use.** *FRAC BM02.*
Bacillus mycoides isolate J	LifeGard WG (*Certis*)	Various diseases (see label for crop-specific diseases)	0	4 hr	Yes	Yes	See product label for restrictions and specific application instructions. **Product has 2(ee) Recommendations for Botrytis leaf blight in onions in some states** *FRAC P06.*
Bacillus pumilus strain QST 2808	Sonata (*Bayer/Wilbur Ellis*)	Early blight, late blight, downy mildew, powdery mildew, leaf blights, rust (see label for crop-specific diseases)	0	4 hr	Yes*	No	Products are not OMRI-listed, but labels state that they can be used for organic production. *See labels for specifics of greenhouse use.
Bacillus subtilis strain AFS032321	Theia (*Certis*)	Various diseases (see label for crop-specific diseases)	0	4 hr	Yes*	Yes	See product labels for instructions on various application uses, particularly those in greenhouse production. *See labels for specifics of greenhouse use. **Check state registration status prior to use.** *FRAC BM02.*
Bacillus subtilis strain IAB/BS03	AVIV (*Seipasa/Summit Agro*)	Various diseases (see label for crop-specific diseases)	See label	4 hr*	Yes	Yes	See product label for restrictions and specific application instructions. *See label for REI exceptions and specifics for greenhouse use. **Check state registration status prior to use.**
Bacillus subtilis strain QST 713	Cease (*BioWorks Inc*); Serenade ASO, Serenade Opti (*Bayer*)	Various diseases (see label for crop-specific diseases)	0	4 hr*	Yes (Cease, Serenade ASO) Not prohibited (Serenade Opti)	Yes	See label for product-specific instructions regarding product application. See product labels for specific greenhouse crops to which products may be applied. *See label for REI exceptions. *FRAC BM02.*

[1] Data on efficacy is limited or not available for many products listed in this table. Therefore, products listed in this table are not recommended based on efficacy. Other active ingredients or products containing the listed active ingredients may also be available.

[2] Consult product labels to determine vegetables for which a particular product is labeled.

[3] Former names of diseases and pathogens listed in this table that may still be listed on fungicide labels are as follows: *Agroathelia rolfsii* (formerly *Athelia rolfsii* and *Sclerotium rolfsii*); *Fulvia fulva* (formerly *Cladosporium fulvum* and *Passalora fulva*); *Xanthomonas vesicatoria* (formerly *Xanthomonas campestris* pv. *vesicatoria*).

TABLE 3-49. BIOPESTICIDES, FUNGICIDE, AND NEMATICIDE ALTERNATIVES FOR VEGETABLES (cont'd)

R. A. Melanson, Extension Plant Pathologist, Mississippi State University; J. Desaeger, Nematologist, University of Florida

Active Ingredient [1]	Product [1]	Target Diseases/Pests	PHI (days)	REI	Greenhouse Use	OMRI-listed	Comments [2]
Bacillus subtilis var. *amyloliquefaciens* strain FZB24	Taegro 2 (*(Novonesis/ Novozymes)*)	Various seedling diseases caused by *Fusarium*, *Rhizoctonia*, *Phytophthora*, and *Pythium* and suppression of various soil-borne and foliar diseases (see label for crop-specific diseases)	See label	4 hr*	Yes*	Yes	See product label for restrictions, cautions, and specific application instructions. *See labels for exceptions for REIs and for specifics of greenhouse use. **FRAC BM02.**
Bacillus thuringiensis subsp. *kurstaki* strain ABTS-351 fermentation solids, spores, and insecticidal toxins + methyl salicylate	Leap ES (*Valent*)	*Pseudomonas* spp. and *Xan- thomonas* spp. (suppression)	See label	12 hr	Yes	No	For use on tomato and pepper. See product label for restrictions and application and mixing instructions.
bacteriophage active against *Xanthomonas vesicatoria* and *Pseudomonas syringae* pv. *tomato*	AgriPhage (*Omni- Lytics*)	Bacterial spot and speck	0	4 hr	Yes (see label)	No	Product is not OMRI-listed, but label states that it can be used for organic production. Product is strain specific (active against *Xanthomonas vesicatoria* and *Pseudomonas syringae* pv. *tomato*) and labeled for use on tomato and pepper. Do not tank-mix product with denaturing agents or copper salts. See label for application instructions. **Check state registration status prior to use.**
bacteriophage active against *Clavibacter michiganensis* pv. *michiganensis*	AgriPhage-CMM (*OmniLytics*)	Bacterial canker	0	4 hr	Yes (see label)	No	Product is not OMRI-listed, but label states that it can be used for organic production. Product is specific for *Clavibacter michiganensis* pv. *michiganensis* that causes bacterial canker in tomato and is only labeled for use on tomatoes. See label for application & tank-mixing instructions. **Check state registration status prior to use.**
Banda de Lupinus albus doce (BLAD)	ProBLAD Verde (*SymAgro*)	Botrytis gray mold, powdery mildew, and others (see label for crop-specific diseases)	1	4 hr	Not prohibited	Yes	For use on some cucurbits, fruiting vegetables and leafy greens. **Check state registration status prior to use. FRAC BM01.**
Burkholderia spp. strain A396 cells (heat-killed) and spent fermentation media	Bronte (previously Majestene (ProFarm))	Nematodes (see label for list of specific nematodes)	0	4 hr*	Yes	Yes	Bionematicide. See label for additional application instructions and restrictions. *See label for REI exceptions. **Product has a 2 (ee) Recommendation for nematodes on carrots, potatoes, and sweetpotatoes.**
Cerevisane (cell walls of *Saccharomyces cerevisiae* strain LAS117)	Romeo (*Agrauxine / Lesaffre Yeast Corporation*)	Certain fungal and oomycete diseases (see label for crop-specific diseases)	0	4 hr	Yes	Yes	Systemic resistance inducer. See label for additional application instructions, precautions, and restrictions. **Check state registration status prior to use.**
Chitosan	Kitogard (Sym Agro)	Nematode suppression	0	0 hr	Yes	No	See label for mixing instructions and restrictions. Check state registration status prior to use.
cinnamon oil	Cinnerate (*SymAgro*)	Diseases such as powdery mildew and rusts and sporulated forms of fungi as *Botrytis cinerea* & *Fulvia fulva*[3]	0	See label	Yes	Yes	See label for mixing instructions and restrictions. Check state registration status prior to use.
citric acid	FungOUT (*AEF Global*)	Various diseases (see label for crop-specific diseases)	0	4 hr	Yes	Yes	Check state registration status of products prior to use.
Coniothyrium minitans strain CON/M/91-08	LAL Stop Contans WG (*Sipcam Agr/ Lallemand Plant Care*)	Sclerotinia diseases (*Sclerotinia sclerotiorum* and *S. minor*)	See label	4 hr	Yes	Yes	Apply to soil or potting medium. Do not tank-mix products with other fungicides. Rotation with other fungicides allowed after 3 weeks following the application of this product. *See label for REI exceptions. See label for additional application instructions and restrictions. **Tomato not included in list of fruiting vegetables on this label. FRAC BM01.**

[1] Data on efficacy is limited or not available for many products listed in this table. Therefore, products listed in this table are not recommended based on efficacy. Other active ingredients or products containing the listed active ingredients may also be available.

[2] Consult product labels to determine vegetables for which a particular product is labeled.

[3] Former names of diseases and pathogens listed in this table that may still be listed on fungicide labels are as follows: *Agroathelia rolfsii* (formerly *Athelia rolfsii* and *Sclerotium rolfsii*); *Fulvia fulva* (formerly *Cladosporium fulvum* and *Passalora fulva*); *Xanthomonas vesicatoria* (formerly *Xanthomonas campestris* pv. *vesicatoria*).

TABLE 3-49. BIOPESTICIDES, FUNGICIDE, AND NEMATICIDE ALTERNATIVES FOR VEGETABLES (cont'd)

R. A. Melanson, Extension Plant Pathologist, Mississippi State University; J. Desaeger, Nematologist, University of Florida

Active Ingredient [1]	Product [1]	Target Diseases/Pests	PHI (days)	REI	Greenhouse Use	OMRI-listed	Comments [2]
copper	See disease control tables for individual crops. **FRAC M01.**						
extract of *Reynoutria sachalinensis*	Regalia, Regalia CG (*ProFarm*)	Certain bacterial and fungal diseases (see label for crop-specific diseases)	0	4 hr*	Not prohibited (Regalia) / Yes (Regalia CG)	Yes	See labels for additional application instructions and restrictions. *See label for RE I exceptions. **FRAC P05.**
extract of *Swinglea glutinosa*	EcoSwing (*Gowan*)	Fungal diseases such as powdery mildew (oidiums) and *Botrytis cinerea*	0	4 hr	Yes	Yes	Rates vary for field and enclosed space production – see label. Dilution water should have pH less than 8. Use product mixture promptly after mixing; do not allow tank-mix to sit for more than 6 hr. **FRAC BM01.**
fats and glyceridic oils of margosa	Debug On (*MGK*)	Nematodes, *Rhizoctonia solani, Sclerotinia sclerotiorum, Athelia rolfsii*	See label	4 hr	Yes	Yes	See label for application instructions and restrictions. **Check state registration status prior to use.**
fats and glyceridic oils of margosa + azadirachtin	Debug Tres, Debug Turbo (*MGK*)	Nematodes, *Rhizoctonia solani, Sclerotinia sclerotio- rum, Agroathelia rolfsii*[3], others (see label)	See label	4 hr	Yes	Yes	See label for application instructions and restrictions. **Check state registration status prior to use.**
garlic oil	Debug Optimo, Debug Tres DebugTurbo (*MGK*)	Various bacterial and fungal diseases (see label)	See label	4 hr (others) / See label (Turbo)	Yes (others) / Not prohibited (Turbo)	Yes	Do not apply when temperatures are above 90°F. See label for application instructions and additional restrictions. **Check state registration status prior to use.**
Gliocladium virens strain GL-21	SoilGard (*Certis*)	Damping-off and root rots	See label	4 hr*	Yes	Yes	Do not apply in conjunction with chemical fungicides. See label for additional application instructions, precautions, & restrictions. *See label for REI exceptions. **FRAC BM02.**
harpin protein	Employ (*Plant Health Care, Inc*)	Nematodes (suppression)	0	4 hr*	Yes	No	Product is used to suppress nematode egg production. See product labels for application instructions and restrictions. *See label for REI exceptions. **Check state registration status of product prior to use.**
hydrogen peroxide + hydrogen dioxide	Perpose Plus (*BioWorks*)	Various diseases (see label for crop-specific diseases)	See label	See label	Yes	Yes	See label for application instructions, precautions, and restrictions. **Check state registration status of product prior to use.**
hydrogen peroxide + peroxyacetic acid	OxiDate 2.0, OxiDate 5.0, ZeroTol 2.0 (*BioSafe Systems*); Jet-Ag 5% (*ProFarm*); TerraClean 5.0 (*BioSafe Systems*)	Various diseases (see label for crop-specific diseases)	See label	See label	Yes*	Yes	Use spray solution the same day it is prepared; do not store spray or reuse mixed spray solution. *See label for restrictions on greenhouse use. Determine if products can be used safely on greenhouse crops prior to application. See product labels for additional precautions, restrictions, and instructions.
inactivated *Burkholderia rinojensis* strain A396 cells and spent fermentation media	Arino, Bronte (*ProFarm*)	Nematodes (see label for crop-specific nematodes)	See label	4 hr*	Yes (Bronte) / Not prohibited (Arino)	No	Check product labels to determine which products are labeled on specific vegetables. *See labels for REI exceptions. **Check state registration status of product prior to use.**
kaolin	Surround WP (*NovaSource*)	Powdery mildew (cucurbit crops)	See label	4 hr	Yes	Yes	Suppression only. See product label for additional precautions, restrictions, and instructions.
laminarin	Vacciplant (*UPL*)	Various diseases (see label for crop-specific diseases)	See label	4 hr	Yes*	No	Do not apply product post-harvest. *Product can be used in greenhouse applications prior to transplant. See labels for additional application instructions, precautions, & restrictions. **FRAC P04.**
milk	N/A	Viruses [(tomato mosaic virus (ToMV) and tobacco mosaic virus (TMV)]	Until spray dries	0	Yes	Yes	Spray plants until runoff. Dip hands every 5 min while handlings plants. Dip tools for 1 min; do not rinse. Use in combination with seed treatments and sanitation practices. **Sooty mold may develop on treated plants.**

[1] Data on efficacy is limited or not available for many products listed in this table. Therefore, products listed in this table are not recommended based on efficacy. Other active ingredients or products containing the listed active ingredients may also be available.

[2] Consult product labels to determine vegetables for which a particular product is labeled.

[3] Former names of diseases and pathogens listed in this table that may still be listed on fungicide labels are as follows: *Agroathelia rolfsii* (formerly *Athelia rolfsii* and *Sclerotium rolfsii*); *Fulvia fulva* (formerly *Cladosporium fulvum* and *Passalora fulva*); *Xanthomonas vesicatoria* (formerly *Xanthomonas campestris* pv. *vesicatoria*).

TABLE 3-49. BIOPESTICIDES, FUNGICIDE, AND NEMATICIDE ALTERNATIVES FOR VEGETABLES (cont'd)

R. A. Melanson, Extension Plant Pathologist, Mississippi State University; J. Desaeger, Nematologist, University of Florida

Active Ingredient [1]	Product [1]	Target Diseases/Pests	PHI (days)	REI	Greenhouse Use	OMRI-listed	Comments [2]
mineral oil	SuffOil-X (*BioWorks*); TriTek (*Brandt*)	powdery mildew on certain vegetable crops	See label	4 hr	Yes	Yes	Products may also be used to control certain insects on listed crops. **Check state registration status of products prior to use. FRAC NC.**
mustard oil and capsaicin	Dazitol (*Champon Millenium Chemicals*)	Various soilborne fungi and nematodes	See label	See label	See label	No	Dazitol is a preplant soil treatment product. See label for application instructions, precautions, and restrictions. **Check state registration status of product prior to use.**
neem oil (extract)	Triact 70 (*OHP*); Trilogy (*Certis USA*)	Foliar fungal diseases (see label for specifics)	See label	4 hr	Yes (Triact) / Not prohibited (Trilogy)	Yes	May cause leaf burn; test on a small number of plants before spraying entire crop. **Toxic to bees. Check state registration status of products prior to use. FRAC NC.**
oil from cottonseed, corn, and garlic	GC-3, Mildew Cure (*JH Biotech Inc*)	Powdery mildew	See label	See label	See label	Yes	See label for application instructions, precautions, and restrictions. **Check state registration status of products prior to use.**
oil from rosemary, clove, thyme, and peppermint	Sporan EC[2] (*KeyPlex*)	Various diseases (see label for diseases listed)	None	None	Yes	Yes	See label for application instructions. The use of an adjuvant is highly recommended. **Check state registration status of product prior to use.**
paraffinic oil	JMS Stylet-Oil, Organic JMS Stylet-Oil (*JMS Flower Farms*)	Various diseases (see label for crop-specific diseases)	See label	4 hr	Yes	Yes	Do not apply to vegetables when temperatures are below 50°F. See labels for additional restrictions & for compatibility information. **Check state registration status of products prior to use.**
phosphorous acid	See disease control tables for individual crops. **FRAC P07.**						
polyoxin D zinc salt	See disease control tables for individual crops. **FRAC 19.**						
potassium bicarbonate	Kaligreen (*OAT Agrio Co/Brandt*), MilStop SP (*BioWorks*)	Various diseases (see label for crop-specific diseases)	1 (Kaligreen) / 0 (MilStop)	1 hr (MilStop) / 4 hr (Kaligreen)	Yes (MilStop) / Not prohibited (Kaligreen)	Yes	See labels for application instructions, precautions, and restrictions.
potassium salts of fatty acids	Des-X (*Certis*), M-Pede (*Gowan*)	Powdery mildew	(Des-X) 12 hr / (M-Pede) 0	12 hr / 12 hr	Yes	Yes	See product labels for notes regarding plant sensitivity, site uses, and use restrictions.
potassium silicate	Sil-MATRIX LC (*Certis USA*)	Powdery mildew	0	4 hr	Not prohibited	Yes	Avoid contact with glass. Tank-mix with a non-ionic surfactant for best results.
Pseudomonas chlororaphis strain AFS009	Howler EVO (*Certis*)	Various diseases (see label for crop-specific diseases)	0	4 hr*	Yes	Yes	See labels for application instructions, cautions, restrictions, and crops on which products may be used. *See labels for REI exceptions. **Check state registration status of products prior to use.** Howler EVO has 2 (ee) recommendations for some crops/diseases. **FRAC BM02.**
Purpureocillium lilacinum strain PL11	NemaClean 10% WP (*Certis*)	Nematodes (see label for list of specific nematodes)	0	12 hr*	Yes	Yes	Bionematicide. See product labels for mixing restrictions and application instructions. *See labels for REI exceptions. **Check state registration status of products prior to use.**
Purpureocillium lilacinum strain PL251	LALNIX ACT DC (*Lallemand*)	Nematodes (see label for list of specific nematodes)	0	12 hr*	Yes	Yes	Bionematicide. See product labels for mixing restrictions and application instructions. *See labels for REI exceptions. **Check state registration status of products prior to use.**
rhamnolipid biosurfactant (from *Pseudomonas aeruginosa*)	Zonix (*Stepan Agricultural Solutions, Jeneil Biosurfactant Co*)	Certain fungal diseases (zoosporic diseases such as blight and downy mildew; see label for specific pathogen groups)	See label	See label	See label	Yes	See label for application instructions, cautions, and restrictions. **Check state registration status prior to use.**

[1] Data on efficacy is limited or not available for many products listed in this table. Therefore, products listed in this table are not recommended based on efficacy. Other active ingredients or products containing the listed active ingredients may also be available.

[2] Consult product labels to determine vegetables for which a particular product is labeled.

[3] Former names of diseases and pathogens listed in this table that may still be listed on fungicide labels are as follows: *Agroathelia rolfsii* (formerly *Athelia rolfsii* and *Sclerotium rolfsii*); *Fulvia fulva* (formerly *Cladosporium fulvum* and *Passalora fulva*); *Xanthomonas vesicatoria* (formerly *Xanthomonas campestris* pv. *vesicatoria*).

TABLE 3-49. BIOPESTICIDES, FUNGICIDE, AND NEMATICIDE ALTERNATIVES FOR VEGETABLES (cont'd)

R. A. Melanson, Extension Plant Pathologist, Mississippi State University; J. Desaeger, Nematologist, University of Florida

Active Ingredient [1]	Product [1]	Target Diseases/Pests	PHI (days)	REI	Greenhouse Use	OMRI-listed	Comments [2]
saponins of *Quillaja saponaria*	Nema-Q (*Brandt*)	Nematodes	See label	24 hr*	Not prohibited	Yes	See label for application instructions, cautions, and restrictions. *See label for REI exceptions. **Check state registration status prior to use.**
Sesame oil	Sesamin EC (*Brandt*)	Nematodes	None	See label	Not prohibited	No	Nematicide. See label for application instructions, cautions, and restrictions. **Check state registration status prior to use.**
sodium carbonate peroxyhydrate	PerCarb (*BioSafe Systems*)	Various foliar bacterial and fungal diseases (see label for crop-specific diseases)	0	See label	Yes	Yes	Corrosive. Do not mix PerCarb with other products with an acidic pH. Do not adjust pH after mixing product. See label for additional instructions and information regarding compatibility and plant sensitivity. **Check state registration status prior to use.**
Streptomyces sp. strain K61	Mycostop (*Lallemand Plant Care*)	Seed, root, and stem rots and wilt diseases caused by certain pathogens; suppression of certain diseases (see label)	See label	4 hr	Yes	Yes	See label for additional instructions, restrictions, and cautions. See product label for crops and specific diseases on which product can be used. **Check state registration status of product prior to use.** *FRAC BM02.*
Streptomyces lydicus WYEC 108	Actinovate AG (*Novozymes*)	Foliar diseases and/or damping-off and root rots (see label for crop-specific diseases)	0	4 hr*	Not prohibited	Yes	*See product labels for REI exemptions See label for instructions on various application uses. *FRAC BM02.*
sulfur	See disease control tables for individual crops. *FRAC M02.*						
tea tree oil	Timorex Act (*Stockton/SummitAgro*)	Various foliar diseases (see label for list)	See label	4 hr / 12 hr	Yes	Yes	Do not spray when temperatures are above 95°F. See labels for additional instructions, restrictions, and cautions. **Check state registration status of products prior to use** *FRAC BM01.*
thyme oil	Guarda (*BioSafe Systems*); Promax, Proud 3 (*Huma Gro and Fertilgold/Bio Huma Netics*); Thyme Guard (*Agro Research Intl.*)	See label	See label	See label	See label	Yes (others) No (Guarda)	See specific product labels for instructions, cautions, and restrictions. **Check state registration status of products prior to use.**
Trichoderma harzianum Rifai strain T-22	RootShield WP, RootShield Granules (*BioWorks*)	Root pathogens (such as *Pythium*, *Rhizoctonia*, and *Fusarium*)	See label	4 hr*	Yes	Yes	Products are for use in soil or planting applications or seed or propagative material treatments. Product should not be applied to bok choi or chickpea. Not for use on aquatic crops. *See product labels for REI exemptions and specifics for greenhouse use. See product labels for additional instructions, uses, and restrictions on various uses and for information on compatibility. *FRAC BM02.*
Trichoderma harzianum Rifai strain KRL-AG2	Trianum-G, Trianum-P (*Koppert Biological Systems*)	Soilborne or root pathogens (such as *Pythium*, *Rhizoctonia*, and *Fusarium*)	See label	See label	See label	No	Products are for use in soil or planting mix applications or seed or propagative material treatments. Product should not be applied to chickpea. Not for use on aquatic crops. **Check product label to determine if product can be applied when aboveground harvestable food is present.** See product labels for instructions on various application uses and on compatibility with other products. **Check state registration status of products prior to use.**

[1] Data on efficacy is limited or not available for many products listed in this table. Therefore, products listed in this table are not recommended based on efficacy. Other active ingredients or products containing the listed active ingredients may also be available.

[2] Consult product labels to determine vegetables for which a particular product is labeled.

[3] Former names of diseases and pathogens listed in this table that may still be listed on fungicide labels are as follows: *Agrothelia rolfsii* (formerly *Athelia rolfsii* and *Sclerotium rolfsii*); *Fulvia fulva* (formerly *Cladosporium fulvum* and *Passalora fulva*); *Xanthomonas vesicatoria* (formerly *Xanthomonas campestris* pv. *vesicatoria*).

TABLE 3-49. BIOPESTICIDES, FUNGICIDE, AND NEMATICIDE ALTERNATIVES FOR VEGETABLES (cont'd)

R. A. Melanson, Extension Plant Pathologist, Mississippi State University; J. Desaeger, Nematologist, University of Florida

Active Ingredient [1]	Product [1]	Target Diseases/Pests	PHI (days)	REI	Greenhouse Use	OMRI-listed	Comments [2]
Trichoderma harzianum Rifai strain T-22 + *T. virens* strain G-41	RootShield PLUS Granules, RootShield PLUS*W-P (*BioWorks*)	Soilborne or root pathogens (such as *Pythium, Rhizoctonia,* and *Fusarium*)	See label (Granules) 0 (WP)	4 hr*	Yes	Yes	Products are for use in soil or planting mix applications or seed or propagative material treatments. Products should not be applied to bok choi or chickpea. Not for use on aquatic crops. *See product labels for REI exemptions and specifics for greenhouse use. See product labels for additional instructions, uses, and restrictions and for information on various application uses and on compatibility with other products. **FRAC BM02.**
Trichoderma spp. (*T. asperellum* strain ICC 012 and *T. gamsii* strain ICC 080)	BIO-TAM 2.0 (*SePRO*)	Certain fungal diseases (see label)	See label	4 hr*	Yes	Yes	See label for a list of incompatible fungicides and county restrictions in some states. See label for REI exceptions and restrictions on greenhouse use. **Check state registration status of product prior to use.**
Ulocladium oudemansii strain U3	BotryStop (*BioWorks*)	*Botrytis* spp. and *Sclerotinia* spp. (see label for crop-specific diseases)	See label	4hr	Not prohibited	Yes	Product should be stored in a cool, dry place at or below 68°F. See label for tank-mixing and application instructions.
Yeast extract hydrolysate from *Saccharomyces cerevisiae*	KeyPlex 350 OR (*KeyPlex*)	bacterial leaf spot (tomatoes)	See label	See label	Not prohibited	No	See product label for instructions on various application uses. **Check state registration status of product prior to use.**

[1] Data on efficacy is limited or not available for many products listed in this table. Therefore, products listed in this table are not recommended based on efficacy. Other active ingredients or products containing the listed active ingredients may also be available.

[2] Consult product labels to determine vegetables for which a particular product is labeled.

[3] Former names of diseases and pathogens listed in this table that may still be listed on fungicide labels are as follows: *Agroathelia rolfsii* (formerly *Athelia rolfsii* and *Sclerotium rolfsii*); *Fulvia fulva* (formerly *Cladosporium fulvum* and *Passalora fulva*); *Xanthomonas vesicatoria* (formerly *Xanthomonas campestris* pv. *vesicatoria*).

FUNGICIDE RESISTANCE MANAGEMENT

The Fungicide Resistance Action Committee (FRAC, http://www.frac.info/) has organized fungicides according to FRAC groups, which reflect chemical structure and Mode of Action (MoA). Fungicides within a given FRAC group control fungi in a comparable manner and share the same risk for fungicide resistance development. Some fungicides are referred to as high- or at-risk fungicides because of their specific MoA's and therefore have an elevated risk for resistance development.

Groups of fungicides, such as the QoI's (FRAC group 11) or phenylamides (FRAC group 4) are prone to resistance development due to extremely specific MoA's. Fungicides in high- or at-risk groups should be rotated and/or tank mixed with broad spectrum, protectant fungicides (FRAC group M03 or M05) to delay the development of resistant strains of fungi. For more information on fungicide resistance management see: http://www.frac.info/

TABLE 3-50. FUNGICIDE MODES OF ACTION FOR FUNGICIDE RESISTANCE MANAGEMENT

R. A. Melanson, Extension Plant Pathologist, Mississippi State University

FRAC Code	Fungicide Resistance Risk	Group Name	Example Active ingredients	Example Products
BM01	Unknown	Plant extract	Extract from lupine	ProBLAD Verde
			Extract from *Swinglea glutinosa*	EcoSwing
BM02	Unknown	Microbial	*Bacillus* spp.	Serifel
			Coniothyrium spp.	Contans WG
			Trichoderma spp.	RootShield products
			Streptomyces spp.	Actinovate AG
M01	Low	Inorganic copper	Fixed copper	Many; see Table 3-48
M02	Low	Inorganic sulfur	Sulfur	Many; see Table 3-48
M03	Low	Dithiocarbamates	Mancozeb	Many; see Table 3-48
M05	Low	Chloronitriles	Chlorothalonil	Many; see Table 3-48
P01	Unknown	Benzo-thiadiazole (BTH)	Acibenzolar-*S*-methyl	Actigard

TABLE 3-50. FUNGICIDE MODES OF ACTION FOR FUNGICIDE RESISTANCE MANAGEMENT (cont'd)

R. A. Melanson, Extension Plant Pathologist, Mississippi State University

FRAC Code	Fungicide Resistance Risk	Group Name	Example Active ingredients	Example Products
P05	Unknown	Plant extract	extract of *Reynoutria sachalinensis*	Regalia products
P06	Unknown	Microbial	*Bacillus mycoides*	LifeGard WG
P07	Low	Phosphonates	Fosetyl AL	Many; see Table 3-50
U06	Medium	Phenyl-acetamide	Cyflufenamid	Torino
U13	Unknown	Thiazolidines	Flutianil	Gatten
1	High	Methyl benzimidazole carbamates (MBC)	Thiophanate-methyl	Topsin M
2	Medium to high	Dicarboximides	Iprodione	Rovral
3	Medium	Demethylation inhibitors (DMI)	Triflumizole	Procure
3	Medium	Demethylation inhibitors (DMI)	Myclobutanil	Rally
4	High	Phenylamide (PA)	Mefenoxam	Ridomil Gold
7	Medium to high	Succinate dehydrogenase inhibitors (SDHI)	Boscalid	Endura
7	Medium to high	Succinate dehydrogenase inhibitors (SDHI)	Penthiopyrad	Fontelis
9	Medium	Anilino-pyrimidines (AP)	Pyrimethanil	Scala
11	High	Quinone outside inhibitors (QoI)	Pyraclostrobin	Cabrio
11	High	Quinone outside inhibitors (QoI)	Trifloxystrobin	Flint
11	High	Quinone outside inhibitors (QoI)	Azoxystrobin	Quadris
12	Low to medium	Phenylpyrroles (PP)	Fludioxinil	Switch 62.5 (also contains cyprodinil)
13	Medium	Azonaphthalenes (AZA)	Quinoxyfen	Quintec
14	Low to medium	Aromatic hydrocarbons (AH)	Dicloran	Botran 75W
19	Medium	Polyoxins	Polyoxin D	OSO 5%SC
21	Medium	Quinone inside Inhibitors (Qil)	Cyazofamid	Ranman
22	Low to medium	Benzamides (toluamides)	Zoxamide	Gavel 75DF (also contains mancozeb)
25	High	Glucopyranosyl antibiotics	Streptomycin	Agri-Mycin 17
27	Low to medium	Cyanoacetamide-oximes	Cymoxanil	Curzate 60DF
28	Low to medium	Carbamates	Propamocarb	Previcur Flex
29	Low	Dinitroanilines	Fluazinam	Omega 500F
40	Low to medium	Carboxylic acid amides (CAA)	Dimethomorph	Forum
40	Low to medium	Carboxylic acid amides (CAA)	Mandipropamid	Revus
43	High	Benzamides	Fluopicolide	Presidio
45	Medium to high	Triazolo-pyrimidylamine (QioSI)	Ametoctradin	Zampro (also contains dimethomorph)
49	Medium to high	Oxysterol binding protein inhibitors (OSBPI)	Oxathiapiprolin	Orondis
50	Medium	Aryl-phenyl ketones	Metrafenone	Vivando

Weed Control for Commercial Vegetables

THE FOLLOWING ONLINE DATABASES PROVIDE CURRENT PRODUCT LABELS AND OTHER RELEVANT INFORMATION:

Database [1]	Web Address
Agrian Label Database	https://home.agrian.com/
Crop Data Management Systems	http://www.cdms.net/Label-Database
EPA Pesticide Product and Label System	https://iaspub.epa.gov/apex/pesticides/f?p=PPLS:1
Greenbook Data Solutions	https://www.greenbook.net/
Kelly Registration Systems [2]	http://www.kellysolutions.com
Arkansas Department of Agriculture [3]	https://www.agriculture.arkansas.gov/plant-industries/pesticide-section/registration/
North Carolina Department of Agriculture [4]	https://www.ncagr.gov/SPCAP/pesticides/Index24c.htm

[1] Additional databases not included in this list may also be available. Please read the database terms of use when obtaining information from a particular website

[2] Available for AK,AL,AZ, CA, CO, CT, DE, FL, GA, IA, ID, IN, KS, MA, MD, MN, MO, MS, NC, ND, NE, NJ, NV, NY, OK, OH, OR, PA, SC, SD, VA, VT, WA, and WI. Kelly Registration Systems works with State Departments of Agriculture to provide registration and license information

[3] Arkansas products only

[4] North Carolina products only

PESTICIDES AND THE ENDANGERED SPECIES ACT: WHAT YOU NEED TO KNOW

The following description has been endorsed by the Weed Science Society of America, Entomological Society of America, and American Phytopathological Society.

1. What is the Endangered Species Act (ESA)?

The Endangered Species Act is a long-standing federal law, first passed in 1973, which requires government agencies to ensure any actions they take do not jeopardize a species that has been federally listed as endangered or threatened. When an agency has a proposed action that might affect a listed species or its habitat, they consult with one or both of the agencies that helps enforce the ESA, the U.S. Fish and Wildlife Services or the National Marine Fisheries Service (this is known as "a consultation" with "the Services"). The Services then may recommend changes to the project or action to protect listed species or habitats.

2. How does the ESA affect pesticide use?

The Environmental Protection Agency (EPA) Office of Pesticide Programs (OPP) is the federal agency that regulates pesticide use. Because the use of pesticides can affect animals and plants (or their habitat), pesticide registrations are considered "actions" that would trigger an endangered species consultation.

3. Why am I hearing about the ESA and pesticide use now?

Due to the complex nature of the process, the EPA has not fully completed the required endangered species consultations with the Services for pesticide registrations in the past, which has left many of those pesticides vulnerable to lawsuits. Courts have annulled pesticide registrations which has led to their removal from market. To make pesticide registrations more secure from litigation, ultimately all pesticide registrations will comply with the Endangered Species Act (https://www.epa.gov/endangered-species).

4. How will this affect the pesticide I use today?

Many pesticide labels will likely have changes that could include:
- Requirement to check the EPA's Bulletins Live! Two website and follow current ESA restrictions for the pesticide product in the bulletin (https://www.epa.gov/endangered-species/bulletins-live-two-view-bulletins)
- Measures to reduce spray drift
- Measures to reduce runoff/erosion
- Other measures to reduce pesticide exposure to listed species and their habitat.

In short, farmers and applicators should expect to see some new application requirements on their pesticide labels. But there is no need to panic. To date, no pesticide has ever been fully removed from the market based solely on endangered species risks, and that remains an unlikely scenario in the future.

5. Why does complying with the ESA matter?

By starting to fully comply with the ESA, **EPA anticipates that this will give farmers and applicators more stable, reliable access to the pesticides they need.** Furthermore, the ESA has been successful at bringing back some species Americans care about – such as the bald eagle or the Eggert sunflower – and restoring them to healthy populations, which has benefited the natural and cultivated ecosystems that agriculture (and society) rely on.

TABLE 4-1. CHEMICAL WEED CONTROL IN ASPARAGUS

Weed	Herbicide and Formulation	MOA*	Amount of Formulation Per Acre	Intervals** REI (Hours)	Intervals** PHI (Days)	Precautions and Remarks
ASPARAGUS (seeded and new crown plantings), Preemergence						
Contact kill of all green foliage, stale bed	paraquat Gramoxone 3SL Gramoxone 2SL	22	 1.7 to 2.7 pt 2.5 to 4 pt	24	6	Apply to emerged weeds before crop emergence as a broadcast or band treatment over a preformed row. Row should be formed several days ahead of planting and treating to allow maximum weed emergence. Use a nonionic surfactant at a rate of 16 to 32 oz per 100-gal spray mix or 1-gal approved crop oil concentrate per 100-gal spray mix. Paraquat product labels require applicators to take an EPA-approved training every 3 years to mix, load, and apply paraquat.
Annual and perennial grass and broadleaf weeds, stale bed application	glyphosate 4SL 5.5SL 5.88SL	9	 16 to 32 fl oz 11 to 22 fl oz 10 to 21 fl oz	4	N/A	Apply to emerged weeds before crop emergence. Row should be formed several days ahead of planting and treating to allow maximum weed emergence. Perennial weeds may require higher rates. The need for an adjuvant depends on brand used.
Broadleaf weeds and some annual grasses	caprylic acid + capric acid Homeplate, Suppress	26	 3 to 9% v/v	24	0	May be applied prior to planting as a burndown treatment for emerged weeds, as a preemergence application after seeding but before emergence, as a directed or shielded application between rows or as a postharvest application. Use higher spray volumes for high weed density and weeds larger than 5 in. Coverage is important for acceptable weed control. May be tank mixed with other herbicides. See label for further instructions.
Annual grasses and small-seeded broadleaf including Palmer amaranth	linuron Lorox DF 50 WDG	7	 1 to 2 lb	24	1	Plant seed 0.5 in. deep in coarse soils and 1 in. deep in fine soils. Apply to soil surface. See label for further instruction.
Annual grasses and certain broadleaf weeds	pendimethalin Prowl H2O 3.8 AS	3	 Up to 8.2 pt	24	14	Newly planted crown asparagus only. Do not apply to newly seeded asparagus. Newly planted crowns must be covered with at least 2 to 4 in. of soil prior to application. Do not apply Prowl H$_2$O at more than 2.4 pints per acre in sandy soils. See label for more information.
ASPARAGUS (seeded and new crown plantings), Postemergence						
Annual and perennial grasses	clethodim Arrow, Clethodim, Intensity, Select 2 EC Select Max, Intensity One 1 EC	1	 6 to 8 oz 9 to 16 oz	24	21	Apply to emerged grasses. Consult the manufacturer's label for best times to treat specific grasses. Refer to label for adjuvant and rate. Adding crop oil to Poast may increase the likelihood of crop injury at high air temperatures. With fluazifop, add 1 qt of nonionic surfactant or 1-gal crop oil concentrate per 100-gal of spray mix.
	fluazifop Fusilade DX 2 EC	1	 6 to 16 oz	12	1	
	sethoxydim Poast 1.5 EC	1	 1.5 to 2.5 pt	12	1	
Annual grasses and small-seeded broadleaf including Palmer amaranth	linuron Lorox DF 50 WDG	7	 1 to 2 lb	24	1	Apply when ferns are 6 to 18 in. tall. Make one or two applications, but do not exceed 2 lb active ingredient total per acre. Do not use with fertilizer, surfactant, or crop oil, as injury will occur. Use the lower rate on coarse soils. Not recommended on sand or loamy sand soils
ASPARAGUS, (established at least 2 yr. old), Preemergence						
Annual grasses and small-seeded broadleaf weeds	linuron Lorox DF 50 WDG	7	 1 to 2 lb	24	1	Apply before spear emergence or immediately after a cutting. Do not use a surfactant or fertilizer solution in spray mixture. Use the lower rates on coarse soils. Not recommended for sand or loamy sand soils. Repeat applications may be made but do not exceed 4 lb per acre per year. Lorox can also be applied as a directed spray to the base of the ferns. Lorox will also control emerged annual broadleaf weeds up to 3 in. tall.
	napropamide Devrinol DF XT50 DF Devrinol 2-XT 2 EC	15	 8 lb 2 gal	24	N/A	Apply to the soil surface in spring before weed and spear emergence. Do not exceed 8 lb per acre per year. See XT labels for information regarding delay in irrigation event.
	trifluralin Treflan 4 L, Treflan HFP 4 EC Treflan 10 G	3	 1 to 4 pt 5 to 10 lbs	12	N/A	In winter or early spring, apply to dormant asparagus after ferns are removed but before spear emergence, or apply after harvest in late spring or early summer. In a calendar year, the maximum rate is 2 pt or 10 lbs per acre for coarse soils, 3 pt or 15 lbs on medium soils and 4 pt or 20 lbs on fine soils. **See label for further restrictions on rates for soil types and on split application.** Apply at least 14 d prior to the first spear harvest or after final harvest. Do not apply over the top of emerged spears or severe injury may occur.

* Mode of action (MOA) code developed by the Weed Science Society of America with the Herbicide Resistance Action Committee (HRAC)

** REI – Restricted Entry Interval; PHI- Pre-Harvest Interval

TABLE 4-1. CHEMICAL WEED CONTROL IN ASPARAGUS (cont'd)

Weed	Herbicide and Formulation	MOA*	Amount of Formulation Per Acre	Intervals** REI (Hours)	Intervals** PHI (Days)	Precautions and Remarks
ASPARAGUS, (established at least 2 yr. old), Preemergence (cont'd)						
Annual grasses and small-seeded broadleaf weeds (cont'd)	pendimethalin Prowl H2O 3.8 AS	3	Up to 8.2 pt	24	14	Apply at least 14 days prior to the first spear harvest or after seasonal harvest is complete. Do not apply over the top of emerged spears as severe injury may occur. Do not apply at a rate of 2.4 pt per acre in sandy soils. See label for more information.
Annual broadleaf and grass weeds	diuron Karmex 80 DF Direx 4 L	7	 1 to 4 lb 0.8 to 3.2 qt	12	N/A	Apply in spring before spear emergence but no *earlier* than 4 weeks before spear emergence. A second application may be made immediately after last harvest. Diuron also controls small, emerged weeds but less effectively.
	flumioxazin Chateau 51 SW	14	 6 oz	12	N/A	Apply only to dormant asparagus no sooner than 14 days before spears emerge or after the last harvest. Do not apply more than 6 oz per acre during a single growing season. Provides residual weed control. Can be tank mixed with paraquat for control of emerged weeds. Apply in a minimum of 15-gal spray mix per acre. Add a nonionic surfactant at 1 qt per 100-gal of spray mix. A spray-grade nitrogen source (either ammonium sulfate at 2 to 2.5 lb per acre or 28 to 32 percent nitrogen solutions at 1 to 2 qt per acre) may be added to increase herbicidal activity.
	metribuzin Tricor DF, Dimetric DF 75 WDG TriCor 4F	5	 1.3 to 2.67 lb 2 to 4 pt	12	14	Make a single application to small, emerged weeds and the soil surface in early spring before spear emergence or after final cutting. Do not apply within 14 days of harvest or after spear emergence. A split application can be used. See label for rates.
	terbacil Sinbar 80 WDG	5	 See labels	12	5	Apply in spring before weed emergence and spear emergence or immediately after last clean-cut harvest. Use the lower rate on sandy soils and the higher rate on silty or clay soils. Do not use on soils containing less than 1% organic matter nor on gravelly soils or eroded areas where subsoil or roots are exposed. See label about rotation restrictions.
	norflurazon Solicam 80DF	12	2.5 to 5 lb	12	14	Rate is soil type dependent. See label for rates and tank mix information.
	mesotrione Callisto 4 F	27	 3 to 7.7 fl oz	12	N/A	Preemergence application. Apply as a spring application prior to spear emergence, after final harvest, or both. For optimum control, apply after fern mowing, disking, or other tillage operation but before spear emergence. Directed or semi directed application. Apply after final harvest with care to minimize contact with any standing asparagus spears to avoid crop injury. Do not make more than two applications per year or apply more than 7.7 oz per acre per year.
ASPARAGUS (established at least 2 yr. old), Postemergence						
Broadleaf weeds	2,4-D amine 4, others	4	 1.3 to 2.67 lb	48	3	Apply in spring before spear emergence or immediately following a clean cutting. Make no more than two applications during the harvest season and these should be spaced at least 1 month apart. Postharvest sprays should be directed under ferns, avoiding contact with ferns, stems, or emerging spears. Add a nonionic surfactant at a rate of 1 qt per 100-gal spray mix. Do not apply if sensitive crops are planted nearby or if conditions favor drift.
	dicamba Clarity 4 L, other DGA salt formulations	4	 8 to 16 oz	24	1	Apply to emerged and actively growing weeds in 40 to 60-gallons of diluted spray per treated acre immediately after cutting in the field but at least 24 hours before the next cutting. If spray contacts emerged spears, twisting of spears may occur. Discard twisted spears. See label for more information. Follow pre-cautions on label concerning drift to sensitive crops.
	carfentrazone-ethyl Aim 1.9 EW or 2 EC	14	 Up to 2 oz	12	5	Apply one to two applications. Use higher rate when weeds are under stress or are larger. See label for further instructions.
Contact kill of emerged annual weeds, suppression of emerged perennial weeds, and contact kill of volunteer ferns	paraquat Parazone 3 SL Gramoxone 2 SL	22	 1.7 to 2.7 pt 2.5 to 4 pt	24	6	Apply to control emerged weeds (including volunteer ferns). Apply in a minimum of 20-gal spray mix per acre to control weeds before spears emerge or after last harvest. Do not apply within 6 days of harvest. Use a nonionic surfactant at a rate of 1 qt per 100-gal spray mix or 1-gal approved crop oil concentrate per 100-gal spray mix. Paraquat product labels require applicators to take an EPA-approved training every 3 years to mix, load, and apply paraquat.
Volunteer ferns (seedling) and certain broadleaf weeds	linuron Lorox DF 50 WDG	7	 1 to 2 lb	24	1	Apply before cutting season or immediately after. Do not apply within 1 day of harvest. Lorox will also control emerged annual broadleaf weeds that are up to 3 inches.

* Mode of action (MOA) code developed by the Weed Science Society of America with the Herbicide Resistance Action Committee (HRAC)

** REI – Restricted Entry Interval; PHI- Pre-Harvest Interval

TABLE 4-1. CHEMICAL WEED CONTROL IN ASPARAGUS (cont'd)

Weed	Herbicide and Formulation	MOA*	Amount of Formulation Per Acre	Intervals** REI (Hours)	Intervals** PHI (Days)	Precautions and Remarks
ASPARAGUS, (established at least 2 yr. old), Preemergence (cont'd)						
Annual and perennial grass and broadleaf weeds; established volunteer ferns	glyphosate 4SL 5.5SL 5.88SL	9	 16 to 32 fl oz 11 to 22 fl oz 10 to 21 fl oz	4	7	Apply to emerged weeds up to 1 week before spear emergence or immediately after last cutting has removed all aboveground parts or as a directed spray under mature fern. Avoid contact with the stem to reduce risk of injury. Perennial weeds may require higher rates of glyphosate. For spot treatment, apply immediately after cutting but prior to emergence of new spears. Certain glyphosate formulations may require the addition of a surfactant. Adding non-ionic surfactant to glyphosate formulated with nonionic surfactant may result in reduced weed control.
Yellow and purple nutsedge, wild radish, non-ALS resistant pigweed, cocklebur, ragweed, and other broadleaf weeds	halosulfuron-methyl Profine 75, Sandea 75 DF	2	0.5 to 1.5 oz	12	1	**Postemergence and post-transplant.** Apply before or during harvesting season. Do not use nonionic surfactant or crop oil because unacceptable crop injury may occur. Without the addition of a nonionic surfactant, post emergence weed control maybe reduced. **Post-harvest.** Apply after final harvest with drop nozzles to limit contact with crop. Contact with the fern may result in temporary yellowing. Add a non-ionic surfactant at 1 qt per 100-gal of spray mixture. Under heavy nutsedge pressure, split applications will be more effective; see label for details.
	clethodim Arrow, Clethodim, Intensity, Select 2 EC Select Max, Intensity One 1 EC	1	 6 to 8 oz 9 to 16 oz	24	21	For Select Max, add 2 pt nonionic surfactant per 100-gal spray mixture.
	fluazifop Fusilade DX 2 EC	1	 6 to 16 oz	12	1	Apply to actively growing grasses not under drought stress. See label for rates for specific weeds. See label for adjuvant and rate.
	sethoxydim Poast 1.5 EC	1	 1 to 1.5 pt	12	1	Apply to emerged grasses. Consult manufacturer's label for specific rates and best times to treat. See label for adjuvant and rate. Adding crop oil to Poast may increase the likelihood of crop injury at high air temperatures and high humidity.

* Mode of action (MOA) code developed by the Weed Science Society of America with the Herbicide Resistance Action Committee (HRAC)

** REI – Restricted Entry Interval; PHI- Pre-Harvest Interval

TABLE 4-2. CHEMICAL WEED CONTROL IN BEANS

Weed	Herbicide and Formulation	MOA*	Amount of Formulation Per Acre	Intervals** REI (Hours)	Intervals** PHI (Days)	Precautions and Remarks
BEANS, Preplant and Preemergence						
Contact kill of all green foliage, stale bed application	paraquat Parazone 3 SL Gramoxone 2 SL	22	 0.8 to 1.3 pt 2 to 4 pt	24	NA	**Lima or snap beans only.** Apply in a minimum of 10-gal spray mix per acre to emerged weeds before crop emergence as a broadcast or band treatment over a preformed row. Use sufficient water to give thorough coverage. Row should be formed several days ahead of planting and treating to allow maximum weed emergence. Use a nonionic surfactant at a rate of 16 to 32 oz per 100-gal spray mix or 1-gal approved crop oil concentrate per 100-gal spray mix. Paraquat product labels require applicators to take an EPA-approved training every 3 years to mix, load, and apply paraquat.
Most broadleaf weeds less than 4 in. tall or rosettes less than 3 in. in diameter; does not control grasses	carfentrazone-ethyl Aim 1.9 EW or 2 EC	14	Up to 2 oz	12	0	**Legume vegetable group (Group 6) such as but not limited to edamame, kidney bean, lima bean, pinto bean, snap bean, soybean, and wax bean only.** Apply prior to or no later than one day after planting. Use a nonionic surfactant or crop oil with Aim. See label for rate. Coverage is essential for good weed control. Can be tank mixed with other registered burndown herbicides.
Annual and perennial grass and broadleaf weeds, stale bed application	glyphosate 4SL 5.5SL 5.88SL	9	 16 to 32 fl oz 11 to 22 fl oz 10 to 21 fl oz	4	7	Various beans are covered. Apply to emerged weeds before crop emergence. Perennial weeds may require higher rates of glyphosate. Consult the manufacturer's label for rates for specific weeds. Certain glyphosate formulations may require the addition of a surfactant. Adding nonionic surfactant to glyphosate formulated with nonionic surfactant may result in reduced weed control. See label for details.

* Mode of action (MOA) code developed by the Weed Science Society of America with the Herbicide Resistance Action Committee (HRAC)

** REI – Restricted Entry Interval; PHI- Pre-Harvest Interval

TABLE 4-2. CHEMICAL WEED CONTROL IN BEANS (cont'd)

Weed	Herbicide and Formulation	MOA*	Amount of Formulation Per Acre	Intervals** REI (Hours)	Intervals** PHI (Days)	Precautions and Remarks
BEANS, Preplant and Preemergence (cont'd)						
Broadleaf weeds and some annual grasses	caprylic acid + capric acid Homeplate, Suppress	26	3 to 9% v/v	24	0	May be applied prior to planting as a burndown treatment for emerged weeds, as a preemergence application after seeding but before emergence, as a directed or shielded application between rows. Use higher spray volumes for high weed density and weeds larger than 5 in. Coverage is important for acceptable weed control. May be tank mixed with other herbicides. See label for further instructions.
Annual grasses and small-seeded broadleaf weeds	ethalfluralin Sonalan HFP 3 EC	3	1.5 to 3 pt	24	N/A	**Dry beans only.** See label for specific bean. Apply preplant and incorporate into the soil 2 to 3 in. deep using a rototiller or tandem disk. If groundcherry or nightshade is a problem, the rate range can be increased to 3 to 4.5 pt per acre. For broader spectrum control, Sonalan may be tank mixed with Eptam or Dual Magnum. Read the combination product labels for directions, cautions, and limitations before use.
	dimethenamid Outlook 6.0 EC	15	12 to 18 oz	12	70	**Dry beans only.** See label for specific bean. Apply preplant incorporated, preemergence to the soil surface after planting, or early postemergence (first to third trifoliate stage). Dry beans may be harvested 70 or more days after Outlook application. For soils having 3% or greater organic matter, see label for rate. See label for further instructions including those for tank mixtures.
	trifluralin Treflan 4 L, Treflan HFP 4 EC Treflan 10 G	3	1 to 1.5 pt 5 to 10 lb	12	N/A	**Dry, lima, or snap beans only.** See label for specific bean. Apply preplant and incorporate into the soil 2 to 3 in. deep within 8 hours. Incorporate with a power-driven rototiller or by cross disking.
	pendimethalin Prowl H2O 3.8 AS	3	1.5 to 3 pt	24	N/A	**Edible beans: dry, lima, or snap beans and certain others.** See label for specific bean. Apply preplant and incorporate into the soil 2 to 3 in. using a power-driven rototiller or by cross disking. **DO NOT APPLY AFTER SEEDING.**
	S-metolachlor Brawl, Dual Magnum, Medal 7.62 EC Brawl II, Dual II Magnum, Medal II 7.64 EC	15	1 to 2 pt 1 to 2 pt	24	50	**Dry, lima, or snap beans and certain others.** See label for specific bean, and specific rate based on soil texture. Apply preplant incorporated or preemergence to the soil surface after planting.
Annual grasses and broadleaf weeds	clomazone Command 3 ME	13	0.4 to 0.67 pt	12	45	**Snap beans (succulent) only.** Apply to the soil surface immediately after seeding. Offers weak control of pigweed. See label for further instructions..
Yellow and purple nutsedge, grasses and some small-seeded broadleaf weeds	EPTC Eptam 7 EC	15	3.5 pt	12	45	**Dry or snap beans only.** See label for specific bean. Apply preplant and incorporate immediately to a depth of 3 in. or may be applied at lay-by as a directed application before bean pods start to form to control late season weeds. See label for instructions on incorporation. May be tank mixed with Prowl. Do not use on black-eyed beans, lima beans, or other flat-podded beans except Romano
Many broadleaf weeds	fomesafen Reflex 2 EC	14	1 to 1.5 pt	24	N/A	**Dry bean and snap beans only.** Apply preplant surface or preemergence. Total use per year cannot exceed 1.5 pt per acre. See label for further instructions and precautions.
Yellow and purple nutsedge, common cocklebur, and other broadleaf weeds	halosulfuron-methyl Profine 75, Sandea 75 DF	2	0.5 to 0.75 oz	12	N/A	**Dry beans and succulent snap beans including lima beans only.** Apply after seeding but prior to cracking. Do not apply more than 0.67 oz product per acre to dry bean. Data are lacking on runner-type snap beans. See label for other instructions.
Broadleaf weeds including morningglory, pigweed, smartweed, and purslane	imazethapyr Pursuit 2 EC	2	1.5 oz	4	30	**Dry beans and lima beans only.** See label for specific bean. Apply preemergence or preplant incorporated. Pursuit should be applied with a registered preemergence grass herbicide. Snap beans only. Apply preemergence or preplant incorporated. For preplant incorporated application, apply within 1 week of planting. May be used with a registered grass herbicide. Reduced crop growth, quality, yield, and/or delayed crop maturation may result.
BEANS, Postemergence						
Annual broadleaf weeds and yellow nutsedge	bentazon Basagran 4 SL	6	1 to 1.5 pt	48	30	**Dry, lima, or snap beans only.** Apply overtop of beans and weeds when beans have one expanded trifoliate. Two applications spaced 7 to 10 days apart may be made for nutsedge control. Do not apply more than 2 qt per season. Use of crop oil as an adjuvant will improve weed control but will likely increase crop injury. See label regarding crop oil concentrate use.

* Mode of action (MOA) code developed by the Weed Science Society of America with the Herbicide Resistance Action Committee (HRAC)
** REI – Restricted Entry Interval; PHI- Pre-Harvest Interval

TABLE 4-2. CHEMICAL WEED CONTROL IN BEANS (cont'd)

Weed	Herbicide and Formulation	MOA*	Amount of Formulation Per Acre	Intervals** REI (Hours)	Intervals** PHI (Days)	Precautions and Remarks
BEANS, Postemergence (cont'd)						
Many broadleaf weeds	fomesafen Reflex 2 EC	14	0.75 to 1 pt	24	30, 45	**Dry or snap beans only.** See label for specific bean. Apply postemergence to dry beans or snap beans that have at least one expanded trifoliate leaf. Include a nonionic surfactant at 1 qt per 100-gal spray mixture. Total use per year cannot exceed 1.5 pt per acre. Postemergence application of fomesafen can cause significant injury to the crop. See label for further information. For dry beans, PHI = 45 days. For snap beans, PHI= 30 days.
Most broadleaf weeds less than 4 in. tall or rosettes less than 3 in. in diameter. (Does not control grasses.)	carfentrazone-ethyl Aim 1.9 EW or 2 EC	14	Up to 2 oz	12	0	**Edible beans: edamame, kidney bean, lima bean, pinto bean, snap bean, and wax bean only.** Apply post-directed using hooded sprayers for control of emerged weeds. If crop is contacted, burning of contacted area will occur. Use a nonionic surfactant or crop oil with Aim. See label for rate. Coverage is essential for good weed control. Can be tank mixed with other registered herbicides.
Yellow and purple nutsedge	EPTC Eptam 7 EC	15	3.5 pt	12	45	**Green or dry beans only.** See label for specific bean. Do not use on lima bean or pea. Apply and incorporate at last cultivation as a directed spray to soil at the base of crop plants before pods start to form.
Yellow and purple nutsedge, common cocklebur, and other broadleaf weeds	halosulfuron-methyl Profine 75, Sandea 75 DF	2	0.5 to 0.66 oz	12	30	**Succulent snap beans, including lima beans.** Apply after crop has reached 2-to 4-trifoliate leaf stage but prior to flowering. Postemergence application may cause significant but temporary stunting and may delay crop maturation. Use directed spray to limit crop injury. See label for further precautions. Data lacking on runner-type snap beans.
Annual broadleaf weeds, including morningglory, pigweed, smartweed, and purslane	imazethapyr Pursuit 2 EC	2	1.5 to 3 oz	4	30, 60	**Dry beans and snap beans only.** See label for specific bean. Use only 1.5 oz EC formulation on snap bean and up to 3 oz on dry beans. Apply postemergence to 1- to 3-in. weeds (one to four leaves) when dry beans have at least one fully expanded trifoliate leaf. Add nonionic surfactant at 2 pt per 100-gal of spray mixture with all postemergence applications. For snap beans, PHI = 30 days. For dry bean, PHI = 60 days. See label for instructions on use.
Most emerged weeds	glyphosate 4SL 5.5SL 5.88SL	9	16 to 32 fl oz 11 to 22 fl oz 10 to 21 fl oz	4	14	**Row middles only.** See label for specific bean. Apply as a hooded spray in row middles, as shielded spray in row middles, as wiper applications in row middles, or post-harvest. Spot treatment is allowed in some bean crops. To avoid severe injury to crop, do not allow herbicide to contact foliage, green shoots, stems, exposed, roots, or fruit of crop.
Annual and perennial grasses	sethoxydim Poast 1.5 EC	1	1 to 1.5 pt	12	15, 30	**Dry or succulent beans only. See label for specific bean.** For succulent beans, products with quizalofop are limited to snap beans. Apply to emerged grasses. Consult manufacturer's label for specific rates and best times to treat. See label for specific adjuvant and rate. Adding crop oil to Poast may increase the likelihood of crop injury at high air temperatures. Do not apply on days that are unusually hot and humid. Do not apply within 15 days and 30 days of harvest for succulent and dry beans, respectively.
	quizalofop p-ethyl Assure II 0.88 EC Targa 0.88 EC	1	6 to 12 oz	12	15,30	**Dry or succulent beans only.** See label for specific bean. For succulent beans, products with quizalofop are limited to snap beans. Apply to emerged grasses. Consult manufacturer's label for specific rates and best times to treat. With sethoxydim, add 1 qt of crop oil concentrate per acre. With quizalofop, add 1-gal oil concentrate or 1 qt nonionic surfactant per 100-gal spray. Adding crop oil to Poast may increase the likelihood of crop injury at high air temperatures. Do not apply on days that are unusually hot and humid. Do not apply within 15 days and 30 days of harvest for succulent and dry beans, respectively.
	clethodim Arrow, Clethodim, Intensity, Select 2 EC Select Max, Intensity One 1 EC	1	6 to 8 oz 9 to 16 oz	24	21	**Dry or succulent beans.** See label for specific bean. Apply postemergence for control of emerged grasses. See label for specific rate for crop. Adding crop oil may increase the likelihood of crop injury at high air temperatures. Highly effective in controlling annual bluegrass. Apply to actively growing grasses not under drought stress.

* Mode of action (MOA) code developed by the Weed Science Society of America with the Herbicide Resistance Action Committee (HRAC)
** REI – Restricted Entry Interval; PHI- Pre-Harvest Interval

TABLE 4-3. CHEMICAL WEED CONTROL BEETS

Weed	Herbicide and Formulation	MOA*	Amount of Formulation Per Acre	Intervals** REI (Hours)	Intervals** PHI (Days)	Precautions and Remarks
BEETS (Garden or Table), Preplant						
Annual and perennial grasses and broadleaf weeds, stale bed application	glyphosate 4SL 5.5SL 5.88SL	9	 16 to 32 fl oz 11 to 22 fl oz 10 to 21 fl oz	4	N/A	**Garden beets only.** Apply to emerged weeds before seeding or after seeding but before crop emergence. Perennial weeds may require higher rates. Certain glyphosate formulations may require the addition of a surfactant. Adding nonionic surfactant to glyphosate formulated with nonionic surfactant may result in reduced weed control.
Emerged broadleaf and grass weeds	pelargonic acid Scythe 4.2 EC	26	 1 to 10% v/v	12	N/A	Apply as a preplant burndown treatment.
Emerged broadleaf weeds	pyraflufen ET Herbicide 0.208 EC	14	 0.5 to 2 oz	12	N/A	Garden beets only. Apply as a preplant burndown treatment in a minimum of 10-gallon solution per acre.
BEETS (Garden or Table), Preemergence						
Annual grasses (crabgrass spp., foxtail spp., barnyardgrass, annual ryegrass, annual bluegrass) and broadleaf weeds (Lamium spp., lambsquarters, common purslane, redroot pigweed, shepherdspurse)	cycloate Ro-Neet 6 EC	15	 0.5 to 0.67 gal	48	N/A	Use on mineral soils only. Use higher rate on heavier soils. Read label for further instructions.
BEETS (Garden or Table), Postemergence						
Broadleaf weeds including sowthistle clover, cocklebur, jimsonweed, vetch, ragweed	clopyralid Stinger 3 EC	4	 0.25 to 0.5 pt	12	30	Apply to beets having 2 to 8 leaves when weeds are small and actively growing. Will control most legumes. Do not apply within 30 days of harvest. Do not apply more than 0.5 pt per acre per year. See label for information regarding rotational restrictions. *Not registered for use in Florida*
Broadleaf weeds including wild mustard, common lambsquarters, common chickweed, purslane suppression	phenmedipham Spin-Aid 1.3EC	6	 3 to 6 pt	12	60	**Red garden beets only.** Apply to red garden beets in the 2- to 6-leaf stage. Rate is dependent on crop stage. See label for specific rate. Best control occurs when applied to weeds in cotyledon to 2-leaf stage. Minor crop stunting may be observed for approximately 10 days. Do not include spray adjuvant.
Broadleaf weeds including wild mustard, shepherd's purse	triflusulfuron Upbeet 50 DF	2	 0.5 oz	4	30	Garden beets. Apply when beets are at the 2- to 4-leaf stage. Additional applications may be made at the 4- to 6- and 6- to 8-leaf stages. Total amount must not exceed 1.5 oz per acre per growing season.
Annual and perennial grasses	sethoxydim Poast 1.5 EC	1	 1 to 1.5 pt	12	60	Apply postemergence for control of emerged grasses. Consult manufacturer's label for specific rates and best times to treat. See label for adjuvant and rate. Adding crop oil to Poast may increase the likelihood of crop injury at high air temperatures and high humidity.
	clethodim Arrow, Clethodim, Intensity, Select 2 EC Select Max, Intensity One 1 EC	1	 6 to 8 oz 9 to 16 oz	24	30	Apply postemergence for annual grasses at 6 to 8 oz per acre or bermudagrass and johnsongrass at 8 oz per acre. See label for adjuvant and rate. Adding crop oil may increase the likelihood of crop injury at high air temperatures and high humidity. Highly effective in controlling annual bluegrass. Apply to actively growing grasses not under drought stress.
BEETS (Garden or Table), Postemergence						
Most emerged weeds except for resistant pigweed	glyphosate 4SL 5.5SL 5.88SL	9	 16 to 32 fl oz 11 to 22 fl oz 10 to 21 fl oz	4	14	Apply as a hooded spray in row middles, as shielded spray in row middles, as wiper applications in row middles, or postharvest. To avoid severe injury to crop, do not allow herbicide to contact foliage, green shoots, stems, exposed, roots, or fruit of crop. The need for an adjuvant depends on brand used.
Annual broadleaf weeds including morningglory, spiderwort, and very small pigweed	carfentrazone-ethyl Aim 1.9 EW or 2 EC	14	 Up to 2 oz	12	0	Apply post-directed using hooded sprayers for control of emerged weeds. If crop is contacted, burning of contacted area will occur. Use a crop oil concentrate or a nonionic surfactant with Aim. See label for directions. Coverage is essential for good weed control. Can be tank mixed with other registered herbicides.

* Mode of action (MOA) code developed by the Weed Science Society of America with the Herbicide Resistance Action Committee (HRAC).
** REI – Restricted Entry Interval; PHI- Pre-Harvest Interval

BROCCOLI – SEE COLE CROPS

CABBAGE – SEE COLE CROPS

TABLE 4-4. CHEMICAL WEED CONTROL IN CANTALOUPE

Weed	Herbicide and Formulation	MOA*	Amount of Formulation Per Acre	Intervals** REI (Hours)	Intervals** PHI (Days)	Precautions and Remarks
CANTALOUPE (MUSKMELON), Preplant and Preemergence						
Suppression or control of most annual grasses and broadleaf weeds, full rate required for nutsedge control Cantaloupe Cucumber	metam sodium Vapam HL 42%	N/A	37.5 to 75 gal	120	N/A	For nutsedge control, use 75-gal per acre. Rates are dependent on soil type & weeds present. Apply when soil moisture is at field capacity (100 to 125%). Apply through soil injection using a rotary tiller or inject with knives no more than 4 in. apart; follow immediately with roller to smooth & compact the soil surface or with mulch. May apply through drip irrigation prior to planting 2nd crop on mulch. Plant back interval is often 14 to 21 days & can be 30 days in some environments. See label for all restrictions & additional information.
Broadleaf weeds and some annual grasses	caprylic acid + capric acid Homeplate, Suppress	26	3 to 9% v/v	24	0	May be applied prior to planting as a burndown treatment for emerged weeds, as a preemergence application after seeding but before emergence, Use higher spray volumes for high weed density and weeds larger than 5 in. Coverage is important for acceptable weed control. May be tank mixed with other herbicides. See label for further instructions.
Most broadleaf weeds less than 4 in. tall or rosettes less than 3 in. in diameter; does not control grasses	carfentrazone-ethyl Aim 1.9 EW or 2 EC	14	Up to 2 oz	12	0	**Transplanted crop.** Apply no later than one day before transplanting crop. **Seeded crop.** Apply no later than 7 days before seeding crop. Use a crop oil at up to 1-gal per 100-gal of spray solution or nonionic surfactant at 2 pt per 100-gal of spray solution. Coverage is essential for good weed control. Can be tank mixed with other registered burndown herbicides.
Contact kill of all green foliage, stale bed application	paraquat Parazone 3 SL Gramoxone 2 SL	22	1.3 to 2.7 pt 2 to 4 pt	24	N/A	Apply in a minimum of 10-gal spray mix per acre to emerged weeds before crop emerges or before transplanting as a broadcast or band treatment over a pre-formed row. Seedbeds or plant beds should be formed as far ahead of treatment as possible to allow maximum weed emergence. Use a nonionic surfactant at a rate of 16 to 32 oz per 100-gal spray mix or 1-gal approved crop oil concentrate per 100-gal spray mix. Paraquat product labels require applicators to take an EPA-approved training every 3 years to mix, load, and apply paraquat.
Annual and perennial grass and broadleaf weeds, stale bed application	glyphosate 4SL 5.5SL 5.88SL	9	16 to 32 fl oz 11 to 22 fl oz 10 to 21 fl oz	4	N/A	Apply to emerged weeds at least 3 days before seeding or transplanting. Perennial weeds may require higher rates of glyphosate. Consult manufacturer's label for rates for specific weeds. When applying Roundup before transplanting crops into plastic mulch, carefully remove residues of this product from the plastic prior to transplanting. To prevent crop injury, residues can be removed by 0.5 in. rainfall or by applying water via a sprinkler system. Certain glyphosate formulations may require the addition of a surfactant. Adding nonionic surfactant to glyphosate formulated with nonionic surfactant may result in reduced weed control.
Emerged broadleaf weeds	pyraflufen ET Herbicide 0.208 EC	14	0.5 to 2 oz	12	N/A	Apply as a preplant burndown treatment in a minimum of 10-gallons per acre. Addition of a crop oil concentrate at 1 to 2% is recommended for optimum weed control. See label for additional information.
Emerged broadleaf and grass weeds	pelargonic acid Scythe 4.2 EC	26	1 to 10% v/v	12	N/A	Apply before crop emergence and control emerged weeds. There is no residual activity. May be tank mixed with soil residual compounds. See label for instruction. May also be used as a banded spray between row middles. Use a shielded sprayer directed to the row middles to reduce drift to the crop.
Annual grasses and small-seeded broadleaf weeds	bensulide Prefar 4 EC	8	5 to 6 qt	12	N/A	Apply preplant and incorporate into the soil 1 to 2 in. (1 in. incorporation is optimum) with a rototiller or tandem disk or apply preemergence after seeding and follow with irrigation. Check re-plant restrictions for small grains and other crops on label.
Annual grasses and broadleaf weeds; weak on pigweed and morningglory	clomazone Command 3 ME	13	0.4 to 0.67 pt	12	N/A	Apply immediately after seeding or just prior to transplanting with transplanted crop. Roots of transplants must be below the chemical barrier when planting. See label for further instruction.

* Mode of action (MOA) code developed by the Weed Science Society of America with the Herbicide Resistance Action Committee (HRAC).
** REI – Restricted Entry Interval; PHI– Pre-Harvest Interval

TABLE 4-4. CHEMICAL WEED CONTROL IN CANTALOUPE (cont'd)

Weed	Herbicide and Formulation	MOA*	Amount of Formulation Per Acre	Intervals** REI (Hours)	Intervals** PHI (Days)	Precautions and Remarks
CANTALOUPE (MUSKMELON), Preplant and Preemergence (cont'd)						
Annual grasses and some small-seeded broadleaf weeds	ethalfluralin Curbit 3 EC	3	3 to 4.5 pt	24	N/A	Apply post plant to seeded crop prior to crop emergence, or as a banded spray between rows after crop emergence or transplanting. See label for timing. Shallow cultivation, irrigation, or rainfall within 5 days is needed for good weed control. Do not use under mulches, row covers, or hot caps. Under conditions of unusually cold or wet soil and air temperatures, crop stunting and injury may occur. Crop injury can occur if seeding depth is too shallow.
Annual grasses and broadleaf weeds	ethalfluralin +clomazone Strategy 2.1 L	3 + 13	2 to 6 pt	24	N/A	Apply to the soil surface immediately after seeding crop for preemergence control of weeds. **DO NOT APPLY PRIOR TO PLANTING CROP. DO NOT SOIL INCORPORATE.** May also be used as a banded treatment between rows after crop emergence or transplanting. Do not apply over or under plastic mulch.
Yellow and purple nutsedge and broadleaf weeds	halosulfuron-methyl Profine 75, Sandea 75 DF	2	0.5 to 1 oz	12	57	Apply after seeding or prior to transplanting crop. For transplanted crop, do not transplant until 7 days after application. Rate can be increased to 1 ounce of product per acre to middles between rows.
Annual grasses, some small-seeded broadleaf weeds	pendimethalin Prowl H2O 3.8 AS	3	Up to 2.1 pt	24	35	**Row middles only.** May be applied sequentially in bareground and plasticulture production systems at a minimum of 21 days apart. Refer to label for specific instructions.
Broadleaf weeds and yellow nutsedge	imazosulfuron League 75 DF	2	4 to 6.4 oz	12	48	**Row middles only.** Use a shielded sprayer directed to the row middles to reduce drift to the crop. In plasticulture, prevent the spray from contacting the plastic. Consult label for further instructions.
CANTALOUPE (MUSKMELON), Postemergence						
Annual grasses and small-seeded broadleaf weeds	trifluralin Treflan 4 L, Treflan HFP 4 EC Treflan 10 G	3	1 to 2 pt 5 to 10 lb	12	30	Apply as a directed spray to soil between rows after crop emergence when crop plants have reached three to four true leaf stage of growth. Avoid contacting foliage as slight crop injury may occur. Set incorporation equipment to move treated soil around base of crop plants. Will not control emerged weeds.
	pendimethalin Prowl H2O 3.8 AS	3	Up to 2.1 pt	24	35	May be applied sequentially in bare ground and plasticulture production systems at a minimum of 21 days apart. Refer to label for specific instructions.
Yellow and purple nutsedge and broadleaf weeds including cocklebur, galinsoga, smartweed, ragweed, wild radish, and non- ALS resistant pigweed	halosulfuron-methyl Profine 75, Sandea 75 DF	2	0.5 to 1 oz	12	57	Apply postemergence only after the crop has reached 3 to 5 true leaves but before first female flowers appear. Do not apply sooner than 14 days after transplanting. Use nonionic surfactant at 1 qt per 100-gal of spray solution with all postemergence applications. Avoid over-the-top application when temperature and humidity are high.
Most broadleaf weeds less than 4 in. tall or rosettes less than 3 in. in diameter; does not control grasses	carfentrazone-ethyl Aim 1.9 EW or 2 EC	14	Up to 2 oz	12	0	Apply post-directed using hooded sprayers for control of emerged weeds. If crop is contacted, burning of contacted area will occur. Use a crop oil concentrate or a nonionic surfactant with Aim. See label for directions. Coverage is essential for good weed control. Can be tank mixed with other registered herbicides.
Most emerged weeds	glyphosate 4SL 5.5SL 5.88SL	9	16 to 32 fl oz 11 to 22 fl oz 10 to 21 fl oz	4	14	Apply as a hooded spray in row middles, as shielded spray in row middles, as wiper applications in row middles, or postharvest. To avoid severe injury to crop, do not allow herbicide to contact foliage, green shoots, stems, exposed, roots, or fruit of crop.
Annual and perennial grasses only	sethoxydim Poast 1.5 EC	1	1 to 1.5 pt	12	3	Apply to emerged grasses. Consult manufacturer's label for specific rates and best times to treat. See label for adjuvant and rate. Adding crop oil to Poast may increase the likelihood of crop injury at high air temperatures and high humidity.
	clethodim Arrow, Clethodim, Intensity, Select 2 EC Select Max, Intensity One 1 EC	1	6 to 8 oz 9 to 16 oz	24	14	Apply postemergence for control of emerged grass in cantaloupes (muskmelons). Adding crop oil may increase the likelihood of crop injury at high air temperatures and high humidity. Highly effective in controlling annual bluegrass. Apply to actively growing grasses not under drought stress.

* Mode of action (MOA) code developed by the Weed Science Society of America with the Herbicide Resistance Action Committee (HRAC).
** REI – Restricted Entry Interval; PHI- Pre-Harvest Interval

TABLE 4-5. CHEMICAL WEED CONTROL IN CARROTS

Weed	Herbicide and Formulation	MOA*	Amount of Formulation Per Acre	Intervals** REI (Hours)	Intervals** PHI (Days)	Precautions and Remarks
CARROTS, Preplant						
Contact kill of all green foliage, stale bed application	paraquat Parazone 3 SL Gramoxone 2 SL	22	 1.3 to 2.7 pt 2 to 4 pt	24	N/A	Apply to emerged weeds before crop emergence as a broadcast or band treatment over a preformed row. Use sufficient water to give thorough coverage. Row should be formed several days ahead of planting and treating to allow maximum weed emergence. Use a nonionic surfactant at a rate of 16 to 32 oz per 100-gal spray mix or 1-gal approved crop oil concentrate per 100-gal spray mix. Paraquat product labels require applicators to take an EPA-approved training every 3 years to mix, load, and apply paraquat.
Annual and perennial grass and broadleaf weeds, stale bed application	glyphosate 4SL 5.5SL 5.88SL	9	 16 to 32 fl oz 11 to 22 fl oz 10 to 21 fl oz	4	N/A	Apply to emerged weeds before seeding or crop emergence. Perennial weeds may require higher rates. Certain glyphosate formulations require the addition of surfactant. Adding no nionic surfactant to glyphosate formulated with nonionic surfactant may result in reduced weed control.
Emerged broadleaf and grass weeds	pelargonic acid Scythe 4.2 EC	26	 1 to 10% v/v	12	N/A	Apply as a preplant burndown or prior to emergence of plants from seed. There is no residual activity. May be tank mixed with soil residual herbicides. See label for instructions. May also be used as a banded spray between row middles. Use a shielded sprayer directed to the row middles to reduce drift to the crop.
CARROTS, Preemergence						
Annual grasses and small-seeded broadleaf weeds	trifluralin Treflan 4 L, Treflan HFP 4 EC Treflan 10 G	3	 1 to 2 pt 5 to 10 lb	12	N/A	Apply preplant and incorporate into the soil 2 to 3 in. within 8 hours. Use lower rate on coarse soils with less than 2% organic matter.
Broadleaf and grass weeds	prometryn Caparol 4L	5	2 to 4 pt	12	30	Apply as preemergence and or postemergence over the top to carrot. Make POST application through the six-leaf stage of carrot. See label for application rate and crop rotation restrictions. PHI = 30 days.
	pendimethalin Prowl H2O 3.8 AS	3	2 pt	24	60	Apply post plant within 2 days after planting but prior to crop emergence. See label for instruction on layby treatment.
CARROTS, Postemergence						
Annual grasses and broadleaf weeds	linuron Lorox DF 50 WDG	7	 1.5 to 3 lb	24	14	Apply as a broadcast spray after carrots are at least 3 in. tall. If applied earlier crop injury may occur. Avoid spraying after three or more cloudy days. Repeat applications may be made, but do not exceed 4 lb of Lorox DF per acre per season. Do not use a surfactant or crop oil. Carrot varieties vary in their resistance; therefore, determine tolerance to Lorox DF before adoption as a field practice to prevent potential crop injury. See label for further directions.
Annual broadleaf weeds and some grasses	metribuzin Dimetric, TriCor DF 75 WDG TriCor 4 F	5	 0.33 lb 0.5 pt	12	60	Apply overtop when weeds are less than 1 in. tall, and carrots have 5 to 6 true leaves. A second application may be made after a time interval of at least 3 weeks. Do not apply unless 3 sunny days precede application. Do not apply within 3 days of other pesticide applications. PHI = 60 days.
Annual and perennial grasses	clethodim Arrow, Clethodim, Intensity, Select 2 EC Select Max, Intensity One 1 EC	1	 6 to 8 oz 9 to 16 oz	24	30	Apply to actively growing grasses not under drought stress. See label for adjuvant and rate. Adding crop oil may increase the likelihood of crop injury at high air temperatures and high humidity. Do not mix with other pesticides. Highly effective in controlling annual bluegrass.
	fluazifop Fusilade DX 2 EC	1	 6 to 24 oz	12	45	Apply to actively growing grasses not under drought stress. Up to 48 oz of Fusilade DX may be applied per year. See label for rates for specific weeds. See label for adjuvant and rate. Do not mix with other pesticides.
	sethoxydim Poast 1.5 EC	1	 1 to 1.5 pt	12	30	Apply to actively growing grasses not under drought stress. Consult manufacturer's label for specific rate and best times to treat. Adding crop oil may increase the likelihood of crop injury at high air temperatures and high humidity. Do not apply with other pesticides.
CARROTS, Row Middles						
Most broadleaf weeds less than 4 in. tall or rosettes less than 3 in. in diameter; does not control grasses	carfentrazone-ethyl Aim 1.9 EW or 2 EC	14	 Up to 2 oz	12	0	Apply as a hooded spray in row middles for control of emerged weeds. If crop is contacted, burning of contacted area will occur. Use a crop oil concentrate or a nonionic surfactant with Aim. See label for directions. Coverage is essential for good weed control. Can be tank mixed with other registered herbicides.
Most emerged weeds	glyphosate 4SL 5.5SL 5.88SL	9	 16 to 32 fl oz 11 to 22 fl oz 10 to 21 fl oz	4	14	Apply as a hooded spray in row middles, as shielded spray in row middles, as wiper applications in row middles, or postharvest. To avoid severe injury to crop, do not allow herbicide to contact foliage, green shoots or stems, exposed roots, or fruit of crop.

CAULIFLOWER – SEE COLE CROPS

TABLE 4-6. CHEMICAL WEED CONTROL IN CELERY

Weed	Herbicide and Formulation	MOA*	Amount of Formulation Per Acre	REI (Hours)	PHI (Days)	Precautions and Remarks
CELERY, Preplant						
Annual and perennial grass and broadleaf weeds, stale bed application	glyphosate 4SL 5.5SL 5.88SL	9	 16 to 32 fl oz 11 to 22 fl oz 10 to 21 fl oz	4	N/A	Apply to emerged weeds before crop emergence. Perennial weeds may require higher rates. Consult the manufacturer's label for rates for specific weeds. Certain glyphosate formulations require the addition of surfactant. Adding nonionic surfactant to glyphosate formulated with nonionic surfactant may result in reduced weed control.
Broadleaf weeds and some annual grasses	caprylic acid + capric acid Homeplate, Suppress	26	3 to 9% v/v	24	0	May be applied prior to planting as a burndown treatment for emerged weeds, as a preemergence application after seeding but before emergence, as a directed or shielded application between rows or as a postharvest application. Use higher spray volumes for high weed density and weeds larger than 5 in. Coverage is important for acceptable weed control. May be tank mixed with other herbicides. See label for further instructions.
Cutleaf evening primrose, Carolina geranium, henbit, and a few grasses	oxyfluorfen Goaltender 4 F Goal 2 XL	14	 Up to 1 pt Up to 2 pt	24	N/A	**Transplants only.** Apply to soil surface of pre-formed beds at least 30 days prior to transplanting.
Emerged broadleaf and grass weeds	pelargonic acid Scythe 4.2 EC	26	1 to 10% v/v	12	N/A	Apply as a preplant burndown. There is no residual activity. May be tank mixed with soil residual compounds. See label for instructions. May also be used as a banded spray between row middles. Use a shielded sprayer directed to the row middles to reduce drift to the crop.
CELERY, Preemergence						
Annual grasses and small-seeded broadleaf weeds	trifluralin Treflan 4 L, Treflan HFP 4 EC Treflan 10 G	3	 1 to 2 pt 5 to 10 lb	12	N/A	Apply incorporated to direct seeded or transplant celery before planting, at planting, or immediately after planting. Use lower rate on coarse soils with less than 2% organic matter.
Annual grasses, some small-seeded broadleaves	bensulide Prefar 4 EC	15	5 to 6 qt	12	N/A	See label for rotation restrictions.
CELERY, Postemergence						
Annual broadleaf and grass weeds	linuron Lorox DF 50 WDG	7	1.5 to 3 lb	24	45	Apply after celery is transplanted and established but before celery is 8 in. tall. Grasses should be less than 2 in. in height, and broadleaf weeds should be less than 6 in. tall. Do not tank mix with other products including surfactant or crop oil. Avoid spraying after 3 or more cloudy days or when temperature exceeds 85°F. Not recommended for sands or loamy sand soil.
	clethodim Arrow, Clethodim, Intensity, Select 2 EC	1	6 to 8 oz	24	30	Apply to actively growing grasses not under drought stress. See label for adjuvant and rate. Highly effective in controlling annual bluegrass. Adding crop oil may increase the likelihood of crop injury at high air temperature and high humidity.
Annual broadleaf and grass weeds (cont'd)	Select Max, Intensity One 1 EC		9 to 16 oz			
	sethoxydim Poast 1.5 EC	1	1 to 1.5 pt	12	30	Apply to actively growing grasses not under drought stress. See label for adjuvant and rate. Adding crop oil to Poast may increase the likelihood of crop injury at high air temperatures and high humidity.
Most broadleaf weeds less than 4 in. tall or rosettes less than 3 in. in diameter; does not control grasses	carfentrazone-ethyl Aim 1.9 EW or 2 EC	14	Up to 2 oz	12	0	Apply post-directed using hooded sprayers for control of emerged weeds. If crop is contacted, burning of contacted area will occur. Use a crop oil concentrate or a nonionic surfactant with Aim. See label for directions. Coverage is essential for good weed control. Can be tank mixed with other registered herbicides.
Most emerged weeds	glyphosate 4SL 5.5SL 5.88SL	9	 16 to 32 fl oz 11 to 22 fl oz 10 to 21 fl oz	4	14	**Row middles only.** Apply as a hooded spray in row middles, as shielded spray in row middles, as wiper applications in row middles, or postharvest. To avoid severe injury to crop, do not allow herbicide to contact foliage, green shoots, stems, exposed, roots, or fruit of crop.

* Mode of action (MOA) code developed by the Weed Science Society of America with the Herbicide Resistance Action Committee (HRAC)
** REI – Restricted Entry Interval; PHI- Pre-Harvest Interva

TABLE 4-7. CHEMICAL WEED CONTROL IN CILANTRO

Weed	Herbicide and Formulation	MOA*	Amount of Formulation Per Acre	Intervals** REI (Hours)	Intervals** PHI (Days)	Precautions and Remarks
CILANTRO, Preemergence (PRE)						
Annual grasses and broadleaf weeds	prometryn Caparol 4L	5	2 to 3.2 pt	12	30	Rates are soil-dependent. See label for more information. Do not use on sand or loamy soils. Check label for crop rotation restrictions.
Annual grasses and small-seeded broadleaf weeds	linuron Lorox DF 50 WDG	7	1 to 2 lb	24	21, 155	Some cultivars may be susceptible to injury. Do not use on sandy or loamy soils, or soils with less than 1% organic matter. Plant at least 0.5 in. deep. PHI for leaves = 21 days. PHI for coriander seed = 155 days.
CILANTRO, Postemergence						
Annual grasses and small-seeded broadleaf weeds (1 to 3 leaf stage)	linuron Lorox DF 50 WDG	7	1 to 2 lb	24	21, 155	Apply to crop plants with a minimum of 3 true leaves to avoid significant injury. Early occurring minor injury should not affect yield. If no injury occurs after the initial application, a second may be made 14 days after the first. Injury may occur under high temperatures, following cloudy periods or when mixed with other pesticides or adjuvants (see label for details). PHI for leaves = 21 days. PHI for coriander seed = 155 days.
Annual and perennial grasses	sethoxydim Poast 1.5 EC	1	1 to 1.5 pt	12	15	Maximum use rate per season is 3 pt per acre.
Most broadleaf weeds less than 4 in. tall or rosettes less than 3 in. in diameter; does not control grasses	carfentrazone-ethyl Aim 1.9 EW or 2 EC	14	Up to 2 oz	12	0	Apply post-directed using hooded sprayers for control of emerged weeds. If crop is contacted, burning of contacted area will occur. Use a crop oil concentrate or a nonionic surfactant with Aim. See label for directions. Coverage is essential for good weed control. Can be tank mixed with other registered herbicides.
Broadleaf weeds and some annual grasses	caprylic acid + capric acid Homeplate, Suppress	26	3 to 9% v/v	24	0	May be applied prior to planting as a burndown treatment for emerged weeds, as a preemergence application after seeding but before emergence, as a directed or shielded application between rows or as a post-harvest application. Use higher spray volumes for high weed density and weeds larger than 5 in. Coverage is important for acceptable weed control. May be tank mixed with other herbicides. See label for further instructions.

* Mode of action (MOA) code developed by the Weed Science Society of America with the Herbicide Resistance Action Committee (HRAC)
** REI – Restricted Entry Interval; PHI- Pre-Harvest Interval

TABLE 4-8. CHEMICAL WEED CONTROL IN COLE CROPS: BROCCOLI, CABBAGE, CAULIFLOWER

Weed	Herbicide and Formulation	MOA*	Amount of Formulation Per Acre	Intervals** REI (Hours)	Intervals** PHI (Days)	Precautions and Remarks
COLE CROPS (Broccoli, Cabbage, Cauliflower), Preplant and Preemergence						
Contact kill of all green foliage, stale seedbed application	paraquat Parazone 3 SL Gramoxone 2 SL	22	1.3 to 2.7 pt 2 to 4 pt	24	N/A	Apply in a minimum of 10-gal spray mix per acre to emerged weeds before crop emergence or transplanting as a broadcast or band treatment over a preformed row. Use sufficient water to give thorough coverage. Row should be formed several days ahead of planting and treating to allow maximum weed emergence. Use a nonionic surfactant at a rate of 16 to 32 oz per 100-gal spray mix or 1-gal approved crop oil concentrate per 100-gal spray mix. Paraquat product labels require applicators to take an EPA-approved training every 3 years to mix, load, and apply paraquat.
Most broadleaf weeds less than 4 in. tall or rosettes less than 3 in. in diameter; does not control grasses	carfentrazone-ethyl Aim 1.9 EW or 2 EC	14	Up to 2 oz	12	0	Apply no later than one day before transplanting, or seven days before seeding. See label for rate for crop oil or nonionic surfactant. Coverage is essential for good weed control. See label for more information.
Annual and perennial grass and broadleaf weeds, stale bed application	glyphosate 4SL 5.5SL 5.88SL	9	16 to 32 fl oz 11 to 22 fl oz 10 to 21 fl oz	4	N/A	Apply to emerged weeds before crop emergence or before transplanting. Perennial weeds may require higher rates of glyphosate. Consult the manufacturer's label for rates for specific weeds. When applying Roundup before transplanting crops into plastic mulch, care must be taken to remove residues of this product from the plastic prior to transplanting. To prevent crop injury, residues can be removed by 0.5 in. rainfall or by applying water via a sprinkler system applied at one time. Certain glyphosate formulations may require the addition of a surfactant. Adding nonionic surfactant to glyphosate formulated with nonionic surfactant may result in reduced weed control.

* Mode of action (MOA) code developed by the Weed Science Society of America with the Herbicide Resistance Action Committee (HRAC)
** REI – Restricted Entry Interval; PHI- Pre-Harvest Interval

TABLE 4-8. CHEMICAL WEED CONTROL IN COLE CROPS: BROCCOLI, CABBAGE, CAULIFLOWER (cont'd)

Weed	Herbicide and Formulation	MOA*	Amount of Formulation Per Acre	Intervals** REI (Hours)	Intervals** PHI (Days)	Precautions and Remarks
COLE CROPS (Broccoli, Cabbage, Cauliflower), Preplant and Preemergence (cont'd)						
Emerged broadleaf and grass weeds	pelargonic acid Scythe 4.2 EC	26	1 to 10% v/v	12	N/A	Also labeled for collards, kale, mustard/turnip greens. Apply as a pre-plant burndown or prior to emergence from seed. There is no residual activity. May be tank mixed with soil residual herbicides. See label for instructions. May also be used as a banded spray between row middles. Use a shielded sprayer directed to the row middles to reduce drift to the crop.
Broadleaf weeds and some annual grasses	caprylic acid + capric acid Homeplate, Suppress	26	3 to 9% v/v	24	0	Do not apply in cabbage except Chinese cabbage. May be applied prior to planting as a burndown treatment for emerged weeds, as a preemergence application after seeding but before emergence, as a directed or shielded application between rows or as a postharvest application. Use higher spray volumes for high weed density and weeds larger than 5 in. Coverage is important for acceptable weed control. May be tank mixed with other herbicides. See label for further instructions.
Annual grasses, some small-seeded broadleaves	bensulide Prefar 4 EC	15	5 to 6 qt	12	N/A	Also labeled for Chinese broccoli, broccoli raab, Chinese cabbage (bok choy, Napa), Chinese mustard cabbage (gai choy), and kohlrabi. Apply preplant or preemergence after planting. With preemergence application, irrigate immediately after application. See label for more directions.
	trifluralin Treflan 4 L, Treflan HFP 4 EC Treflan 10 G	3	1 to 1.5 pt 5 to 10 lb	12	N/A	Also labeled for Brussels sprouts. Apply and incorporate prior to transplanting. Caution: If soil conditions are cool and wet, reduced stands and stunting may occur. Direct seeded Cole crops exhibit marginal tolerance to higher than recommended rates.
Annual grasses and broadleaf weeds; weak on pigweed	clomazone Command 3 ME	13	0.67 pt	12	45	**Transplanted Broccoli** Apply within 48 hours of transplanting.
Annual grasses and broadleaf weeds, yellow nutsedge suppression	S-metolachlor Dual Magnum 7.62 EC	15	8 to 16 oz	24	N/A	Cabbage – direct seeded and transplanted; Chinese cabbage (Napa); Chinese cabbage (Bok Choy); broccoli, and cauliflower. Chinese cabbage may be more sensitive to injury from Dual magnum. **This is a Section 24(c) Special Local Need Label.** Growers must make sure Dual Magnum is registered for use in their state and obtain the label prior to application. Irrigation following the application of Dual Magnum will increase the risk of crop injury. Use lower rates on coarse-textured soils and higher rates on fine-textured soils. See label for more information. **Mulched Systems with Transplanted Crop.** *Option 1:* Apply 8 to 16 oz per acre to the soil surface of pre-formed beds prior to laying plastic. Ensure the plastic laying process does not incorporate or disturb the treated bed. Unless restricted by other products, crops may be transplanted immediately following Dual Magnum application. *Option 2:* Apply 8 to 16 oz per acre overtop of crop at least 10 days after transplanting to ensure root system is well developed. Does not control emerged weeds. Limited data available for NC. **Bare Ground Application for Transplanted Crop.** After transplanting, irrigate to seal the soil around the transplanted root ball. Five to ten days after transplanting and irrigating, apply Dual Magnum over the top of transplants. If soil is not sealed around the transplant root ball, crop injury may occur. **Direct Seeded Application.** May be applied over the top after the crop reaches a minimum of 3 in. tall. **Row Middle Application to Transplanted and Direct Seeded Crop.** Apply as a banded application at a rate up to 1.25 pt per acre.
Hairy galinsoga, common lambsquarters, redroot pigweed, and Palmer amaranth	sulfentrazone Spartan 4 F	14	2.25 to 4.5 oz	12	N/A	**Cabbage – Transplanted Processing only).** Early applications may be applied 60 days prior to planting up to planting time. Application rate depends on soil type. May also be applied as broadcast application prior to transplanting, or a banded application up to 72 hours after transplanting.
Annual grasses and small-seeded broadleaf weeds, including galinsoga, common ragweed, and smartweed	napropamide Devrinol DF, Devrinol DF-XT 50 DF Devrinol 2-XT 2 EC	15	4 lb 4 qt	24	N/A	Includes Brussels sprouts. Apply to weed-free soil just after seeding or transplanting as a surface application. Light cultivations, rainfall, or irrigation will be necessary within 24 hr to activate this chemical. When applied to cooler soil temperatures for a spring broccoli crop, transient stunting may occur.
Many broadleaf weeds, including galinsoga, common ragweed, and smartweed	oxyfluorfen Goal 2XL, Galigan 2EC GoalTender 4 F	14	1 to 2 pt 0.5 to 1 pt	24	N/A	**Transplants only.** Surface apply before transplanting. Do not incorporate or knock the bed off after application. ***Do not spray over the top of transplants.*** Oxyfluorfen is weak on grasses. Expect to see some temporary crop injury.

* Mode of action (MOA) code developed by the Weed Science Society of America with the Herbicide Resistance Action Committee (HRAC)
** REI – Restricted Entry Interval; PHI- Pre-Harvest Interval

TABLE 4-8. CHEMICAL WEED CONTROL IN COLE CROPS: BROCCOLI, CABBAGE, CAULIFLOWER (cont'd)

Weed	Herbicide and Formulation	MOA*	Amount of Formulation Per Acre	REI (Hours)	PHI (Days)	Precautions and Remarks
COLE CROPS (BROCCOLI, CABBAGE, CAULIFLOWER), — Postemergence						
Most emerged weeds	glyphosate 4SL 5.5SL 5.88SL	9	 16 to 32 fl oz 11 to 22 fl oz 10 to 21 fl oz	4	14	**Row middles only.** Apply as a hooded spray in row middles, as shielded spray in row middles, as wiper applications in row middles, or postharvest. To avoid severe injury to crop, do not allow herbicide to contact foliage, green shoots, stems, exposed, roots, or fruit of crop.
Most broadleaf weeds less than 4 in. tall or rosettes less than 3 in. in diameter; does not control grasses	carfentrazone-ethyl Aim 1.9 EW or 2 EC	14	 Up to 2 oz	12	0	Apply post-directed using hooded sprayers for control of emerged weeds. If crop is contacted, burning of contacted area will occur. Use crop oil concentrate at up to 1-gal per 100-gal solution or a nonionic surfactant at 2 pt per 100-gal of spray solution. Coverage is essential for good weed control. Can be tank mixed with other registered herbicides.
Broadleaf weeds including sowthistle, clover, cocklebur, jimsonweed, and ragweed	clopyralid Stinger 3 EC	4	 0.25 to 0.5 pt	12	30	Labeled for broccoli, cabbage, cauliflower, broccoli raab, Brussels sprouts, Cavolo broccoli, Chinese cabbage (bok choy), Chinese broccoli, Chinese mustard, and Chinese cabbage (Napa). Apply to crop when weeds are small and actively growing. Will control most legumes.
Annual and perennial grasses only	clethodim Arrow, Clethodim, Intensity, Select 2 EC Select Max, Intensity One 1 EC	1	 6 to 8 oz 9 to 16 oz	24	30	Apply to emerged grasses. Consult manufacturer's label for specific rates and best times to treat. See label for adjuvant and rate. Adding crop oil to Poast or Select may increase the likelihood of crop injury at high air temperature and high humidity.
	sethoxydim Poast 1.5 EC	1	 1 to 1.5 pt	12	30	

* Mode of action (MOA) code developed by the Weed Science Society of America with the Herbicide Resistance Action Committee (HRAC)
** REI – Restricted Entry Interval; PHI- Pre-Harvest Interval

TABLE 4-9. CHEMICAL WEED CONTROL IN CORN, SWEET

Weed	Herbicide and Formulation	MOA*	Amount of Formulation Per Acre	REI (Hours)	PHI (Days)	Precautions and Remarks
CORN, SWEET, Preplant Burndown						
Most broadleaf weeds less than 4 in. tall or rosettes less than 3 in. in diameter; does not control grasses	carfentrazone-ethyl Aim 1.9 EW or 2 EC	14	 Up to 2 oz	12	N/A	Apply prior to planting or within 24 hours after planting. Use a crop oil concentrate or a nonionic surfactant with Aim. For optimum performance, make applications to actively growing weeds up to 4 in. high or rosettes less than 3 in. across. Coverage is essential for good weed control. Optimum broad-spectrum control of annual and perennial weeds requires a tank mix with burndown herbicides such as glyphosate, paraquat or 2,4-D.
Contact kill of all green foliage, stale bed, and minimum tillage application	paraquat Parazone 3 SL Gramoxone 2 SL	22	 1.3 to 2.7 pt 2 to 4 pt	24	N/A	Apply in a minimum of 20-gal spray mix per acre to emerged weeds before crop emergence as a broadcast or band treatment over a preformed row. Seed-beds should be formed several days ahead of planting and treating to allow maximum weed emergence. Plant with a minimum of soil movement for best results. Use a nonionic surfactant at a rate of 16 to 32 oz per 100-gal spray mix or 1 gal approved crop oil concentrate per 100-gal spray mix. May be tank mixed with preemergence sweetcorn herbicides and herbicide combinations. Check label for directions and specific rates. Paraquat product labels require applicators to take an EPA-approved training every 3 years to mix, load, and apply paraquat.
Annual and perennial grass and broadleaf weeds, stale bed application	glyphosate 4SL 5.5SL 5.88SL	9	 16 to 32 fl oz 11 to 22 fl oz 10 to 21 fl oz	4	N/A	Apply to emerged weeds before crop emergence. Do not feed crop residue to livestock for 8 weeks following treatment. Perennial weeds may require higher rates of glyphosate. Consult manufacturer's label for rates for specific weeds. Check label for directions. Certain glyphosate formulations require addition of surfactant. Adding nonionic surfactant to glyphosate formulated with nonionic surfactant may result in reduced weed control. Glyphosate-resistant horseweed (marestail) is now common in eastern North Carolina counties. If horseweed is present at planting time, a tank mixture of paraquat and atrazine is suggested.

* Mode of action (MOA) code developed by the Weed Science Society of America with the Herbicide Resistance Action Committee (HRAC)
** REI – Restricted Entry Interval; PHI- Pre-Harvest Interval

TABLE 4-9. CHEMICAL WEED CONTROL IN CORN, SWEET (cont'd)

Weed	Herbicide and Formulation	MOA*	Amount of Formulation Per Acre	Intervals** REI (Hours)	Intervals** PHI (Days)	Precautions and Remarks
CORN, SWEET, Preemergence						
Broadleaf weeds and some annual grasses	caprylic acid + capric acid Homeplate, Suppress	26	3 to 9% v/v	24	0	May be applied prior to planting as a burndown treatment for emerged weeds, as a preemergence application after seeding but before emergence, as a directed or shielded application between rows or as a post-harvest application. Use higher spray volumes for high weed density and weeds larger than 5 in. Coverage is important for acceptable weed control. May be tank mixed with other herbicides. See label for further instructions.
Broadleaf weeds	2,4-D Amine 4, others	4	1 to 3 pt	48	45	May be tank mixed with glyphosate for broad-spectrum weed control including glyphosate-resistant horseweed (marestail). See label for planting restrictions if applied prior to planting.
Most annual grass weeds, including fall panicum, broadleaf signalgrass, and small-seeded broadleaf weeds	dimethenamid-P Outlook 6.0 EC	15	12 to 21 oz	12	50	Apply to soil surface immediately after planting. May be tank mixed with atrazine, glyphosate, or paraquat. See label for other herbicides that may be tank mixed to broaden weed control spectrum.
	metolachlor Me-Too-Lachlor II, Parallel 7.8 EC	15	1 to 2 pt	24	N/A	See comments for *S*-metolachlor products. Products containing *S*-metolachlor are more active on weeds per unit of formulated product than those containing metolachlor. See label for all instructions.
	S-metolachlor Brawl, Dual Magnum, Medal 7.62 EC Brawl II, Dual II Magnum, Medal II 7.64 EC	15	1 to 2 pt 1 to 2 pt	24	30	Apply to soil surface immediately after planting. May be tank mixed with atrazine, glyphosate, or simazine. Check label for directions. Rate is soil-texture and organic-matter dependent. See label for details. Dual II Magnum contains the corn safener benoxacor, thus is safer on sweet corn than Dual Magnum, which does not contain benoxacor.
	pyroxasulfone Zidua 85 WG	15	1.5 to 4 oz	12	37	Rate ranges based on soil texture. See label for specific rate relating to your fields. Sweetcorn seed must be planted a minimum of 1 in. deep. Provides suppression of Texas panicum, seedling johnsongrass, and shattercane. See label regarding tank mixtures for broader spectrum control and/or control of emerged weeds.
Most annual broadleaf and grass weeds	atrazine AAtrex 4L, others Various 90 WDG	5	1 to 2 qt 1.1 to 2.2 lb	12	N/A	Apply to soil surface immediately after planting. Shallow cultivations will improve control. Check label for restrictions on rotational crops. See label for reduced rate if soil coverage with plant residue is less than 30% at planting. Does not control fall panicum or smooth crabgrass. May be tank mixed with metolachlor, alachlor, glyphosate, paraquat, bentazon, or simazine. Check label for directions.
	dimethenamid-P + atrazine Guardsman Max 5 F	15 + 5	2.4 to 4.6 pt	12	N/A	Apply to soil surface immediately after planting. Does not control Texas panicum, seedling johnsongrass, or shattercane adequately. Adjust rate for soil texture and organic matter according to label. See label for reduced rate if soil coverage with plant residue is less than 30% at planting. See labels for comments on rotational crops. See label for additional instructions.
	S-metolachlor + atrazine Bicep II Magnum 5.5 F	15 + 5	1.3 to 2.6 qt	24	N/A	Apply to soil surface immediately after planting. Does not adequately control Texas panicum, seedling johnsongrass, or shattercane. May not adequately control cocklebur, morningglory, or sicklepod. Cultivation or other herbicides may be needed. See label for rates based on soil texture and organic matter and for information on setback requirements from streams and lakes. See label for reduced rate if soil coverage with plant residue is less than 30% at planting and for comments on rotational crops.
Small-seeded broadleaf weeds and some annual grass weeds	saflufenacil + dimethenamid-P Verdict 5.57 EC	14 + 15	10 to 18 oz	12		Registered for processing sweet corn only. Apply preplant surface, pre-plant incorporated or preemergence after seeding. Do not apply to emerged sweet corn.
Broadleaf weeds, including large-seeded weeds such as cocklebur and annual grass and partial control of yellow nutsedge	bicyclopyrone + mesotrione + *S*-metolachlor AcuronFlexi 2.86 L	27 + 27 + 15	2 to 2.25 qt	24	N/A	Apply preplant or preemergence to sweet corn. Severe injury will occur if applied to emerged sweet corn.
	S-metolachlor + atrazine + mesotrione + bicyclopyrone Acuron 3.44 L	15 + 5 + 27 + 27	2.5 to 3 qt	24	N/A	Apply preplant or preemergence to sweet corn. Severe injury will occur if applied to emerged sweet corn.

* Mode of action (MOA) code developed by the Weed Science Society of America with the Herbicide Resistance Action Committee (HRAC)
** REI – Restricted Entry Interval; PHI- Pre-Harvest Interval

TABLE 4-9. CHEMICAL WEED CONTROL IN CORN, SWEET (cont'd)

Weed	Herbicide and Formulation	MOA*	Amount of Formulation Per Acre	Intervals** REI (Hours)	Intervals** PHI (Days)	Precautions and Remarks
CORN, SWEET, Postemergence						
Grass and broadleaf weeds	**pendimethalin** Prowl H2O 3.8 Prowl 3.3 EC	3	2 to 4 pt 1.8 to 4.8 pt	24	60	Apply preemergence before crop germinate or postemergence until sweet corn is 20 to 24 in. tall or in the V8 growth stage whichever is more restrictive. Do not apply in reduced, minimum, or no-till sweet corn. Rate is dependent on organic matter content. See label for additional information. See label for tank mix options.
Broadleaf and grass weeds	**simazine** Princep 4L	5	1.6 to 2 qt	12	45	Apply preemergence before weeds and crop emerge. See label for tank mix options.
Most annual broadleaf and grass weeds	**atrazine** AAtrex 4L, others Various 90 WDG	5	2 qt 2.2 lb	12	N/A	Atrazine cannot exceed 2.5 lb a.i. per acre per calendar year. Apply overtop before weeds exceed 1.5 in. in height. See label for additional information in controlling larger weeds. See label for amount of oil concentrate to add to spray mix. See label on setback requirements from streams and lakes. Do not apply after corn is 12 in tall.
Annual grasses and broadleaf weeds	**dimethenamid-P + atrazine** Outlook 6.0 EC + AAtrex 4 F or 90 WDG	15 + 5	8 to 21 oz +2 qt 2.2 lb	12	50	Apply overtop corn before crop reaches 12 in. tall and before weeds exceed the two-leaf stage. Larger weeds will not be controlled. Good residual control of annual grass and broadleaf weeds. Do not apply to corn 12 in. or taller. Also available as the commercial products Guardsman or LeadOff.
	S-metolachlor + atrazine Dual II Magnum 7.64 EC + AAtrex 4 F AAtrex 90 WDG	15 + 5	1 to 1.67 pt + 1 to 2 qt 1.3 to 2.2 lb	24	30	Apply overtop corn (5 in. or less) before weeds exceed the two-leaf stage. Larger weeds will not be controlled. Good residual control of annual grass and broadleaf weeds. Also available as Bicep II or Bicep II Magnum.
Cocklebur, common ragweed, jimsonweed, Pennsylvania smartweed, velvetleaf, yellow nutsedge, and morningglory	**bentazon** Basagran 4 SL	6	1 to 2 pt	48	N/A	Apply early postemergence overtop when weeds are small, and corn has one to five leaves. See label for rates according to weed size and special directions for annual morningglory and yellow nutsedge control. Use a crop oil at a rate of 1 qt per acre.
Many broadleaf weeds	**mesotrione** Callisto 4 EC	27	3 oz	12	45	Apply overtop corn 30 in. or less or 8 leaves or less to control emerged broadleaf weeds. Use nonionic surfactant at 2 pt per 100- gal of spray solution. DO NOT add UAN or AMS when making post application in sweet corn or severe injury will occur. Most effective on small weeds, however, if weeds are greater than 5 in. or for improved control of certain weeds, certain atrazine formulations may be mixed with this herbicide. See label for further information.
Annual broadleaf weeds and some grasses	**tembotrione** Laudis 3.5 L	27	3 oz	12	N/A	Can be applied overtop or with drop nozzles to sweet corn from emergence up to V7 stage. Does not control sicklepod or prickly sida and only suppresses morningglory. Controls or suppresses some grasses. See label for weeds controlled and recommended size for treatment. Herbicide sensitivity in all hybrids and inbreds of sweet corn has not been tested. See label for information on adjuvant use. May be tank mixed with atrazine to increase weed spectrum and consistency of control. If tank mixed with atrazine, do not apply if corn is 12 in. tall or greater. See label for further restrictions and instructions.
	topramezone Impact 2.8 L	27	0.75 oz	12	45	Can be applied overtop or with drop nozzles to sweet corn from emergence until 45 days prior to harvest. Does not control sicklepod and only suppresses morningglory. Controls or suppresses some grasses. See label for weeds controlled and recommended size for treatment. This product has not been tested on all inbred line for tolerance. See label for information on adjuvant use. See label for further restrictions and instructions.
	topramezone + dimethenamid-P Armezon PRO 5.26 L	27 + 15	14 to 24 oz	12	50	Do not apply to sand textured soils with less than 3% organic matter.
Velvetleaf, spreading dayflower, morningglory species, and redroot pigweed. Will not control grasses.	**fluthiacet-methyl** Cadet 0.91 L	14	0.6 to 0.9 oz	12	40	Processing sweet corn only. Apply to small weeds, generally about 2 in. tall. Will control large velvetleaf up to 36 in. See label for information on adjuvant use. See label for further restrictions and instructions. May be applied from preplant to 48 inches tall, before tasseling.

* Mode of action (MOA) code developed by the Weed Science Society of America with the Herbicide Resistance Action Committee (HRAC)
** REI – Restricted Entry Interval; PHI- Pre-Harvest Interval

TABLE 4-9. CHEMICAL WEED CONTROL IN CORN, SWEET (cont'd)

Weed	Herbicide and Formulation	MOA*	Amount of Formulation Per Acre	Intervals** REI (Hours)	Intervals** PHI (Days)	Precautions and Remarks
CORN, SWEET, Postemergence (cont'd)						
Annual broadleaf weeds	fluthiacet-methyl + mesotrione Solstice	14 + 27	 2.5 to 3.15 oz	 12	 40	Apply up to the V8 growth stage (or 30 in. tall). See label for crop rotation restrictions. Do not include nitrogen-based adjuvants (UAN or AMS) when making postemergence application or severe injury will occur. Use nonionic surfactant at 1 qt per 100-gallons of spray. See label for further instructions.
Velvetleaf, pigweed, nightshade, morningglory, common lambsquarters	carfentrazone-ethyl Aim 1.9 EW or 2 EC	14	 0.5 to 1 oz	12	3	Apply postemergence to actively growing weeds less than 4 in. high (rosettes less than 3 in. across) up to the 14-leaf collar stage of corn. Rates above 0.5 oz will aid in controlling larger weeds and certain weeds (see label for specific rate). Directed sprays will lessen the chance of crop injury and allow later application. Coverage of weeds is essential for control. Use nonionic surfactant (2 pt per 100-gal of spray) with all applications. Under dry conditions, the use of crop oil concentrate may improve weed control. Mix with atrazine to improve control of many broadleaf weeds. Limited information is available concerning the use of this product in sweet corn. Do not apply more than 2 oz per acre per season.
Broadleaf weeds including sowthistle, clover, cocklebur, jimsonweed, ragweed, Jerusalem artichoke and thistle	clopyralid Stinger 3 EC	4	 0.25 to 0.7 pt	12	30	Processing sweet corn only. Apply to sweet corn when weeds are small (less than 5-leaf stage) and actively growing. Addition of surfactants, crop oils, or other adjuvants is not usually necessary when using Stinger. Use of adjuvants may reduce selectivity to the crop. Do not apply to sweet corn over 18 in. tall. Will control most legumes.
Cocklebur, passion-flower (maypop), pigweed, pokeweed, ragweed, smartweed (Pennsylvania), velvetleaf	halosulfuron-methyl Profine 75, Sandea 75 DF	2	 0.67 to 1 oz	12	30	Apply over the top or with drop nozzles to sweet corn from spike to lay-by for control of emerged weeds. Add nonionic surfactant at 1 to 2 qt per 100-gal of spray solution. See label for all instructions and restrictions.
Cocklebur, pigweed, lambsquarters, morningglory, sicklepod, and many other annual broadleaf weeds	2,4-D amine Various brands 3.8 SL	4	 0.5 to 1 pt	48	45	Use 0.5 pt of 2, 4-D overtop when corn is 4 to 5 in. tall, and weeds are small. Increase rate to 1 pt as corn reaches 8 in. Use drop nozzles and direct spray toward base if corn is over 8 in. tall. Do not cultivate for about 10 days after spraying, as corn may be brittle. Reduce rate of 2, 4-D if extremely hot and soil is wet. For better sicklepod and horsenettle control, add a nonionic surfactant when using a directed spray at a rate of 1 qt per 100-gal spray solution.
Annual grasses and broadleaf weeds	paraquat Parazone 3 SL Gramoxone 2 SL	22	 0.8 to 1.3 pt 1.2 to 2 pt	24	NA	DO NOT SPRAY OVER THE TOP OF CORN OR SEVERE INJURY **WILL OCCUR.** Make a postdirected application in a minimum of 20-gal spray mix per acre to emerged weeds when the smallest corn is at least 10 in. tall. Use nonionic surfactant at a rate of 16 to 32 oz per 100-gal spray mix or 1-gal approved crop oil concentrate per 100- gal spray mix. Use of a hooded or shielded sprayer will reduce crop injury. Paraquat product labels require applicators to take an EPA-approved training every 3 years to mix, load, and apply paraquat.
Certain grasses, including barnyardgrass, foxtails, Texas panicum, and johnsongrass; and broadleaf weeds, including bur cucumber, jimsonweed, pigweed, pokeweed, and smartweeds	nicosulfuron Accent 75 WDG	2	 0.67 oz	4	70	Apply to sweet corn up to 12 in. tall or up to and including 5 leaf collars. For corn 12 to 18 in. tall, apply only with drop nozzles. Sweet corn hybrids vary in their sensitivity to Accent. Do not apply to Merit sweet corn. Contact company representative for information on other local hybrids that have been evaluated with Accent. Accent may be applied to corn previously treated with Fortress, Aztec, or Force, or non-organophosphate soil insecticides regardless of soil type. See label for more information on use of soil insecticides with Accent. Label prohibits application of Accent to corn previously treated with Counter insecticide and indicates that applying Accent to corn previously treated with Counter 20 CR, Lorsban, or Thimet may result in unacceptable crop injury, especially on soils with less than 4% organic matter. See label for information on use.
	nicosulfuron + mesotrione Revulin Q 51.2 WDG	22 + 27	 3.44 to 4 oz	12	70	Apply to sweet corn up to 12 in. tall or up to and including 5 leaf collars. For corn 12 to 18 in. tall, apply only with drop nozzles. Sweet corn hybrids vary in their sensitivity to Accent. Do not apply to Merit sweet corn. Contact company representative for information on other local hybrids that have been evaluated with Accent. Accent may be applied to corn previously treated with Fortress, Aztec, or Force, or non-organophosphate soil insecticides regardless of soil type. See label for more information on use of soil insecticides with Accent. Label prohibits application of Accent to corn previously treated with Counter insecticide and indicates that applying Accent to corn previously treated with Counter 20 CR, Lorsban, or Thimet may result in unacceptable crop injury, especially on soils with less than 4% organic matter. Postemergence applications of Revulin Q may cause crop bleaching in some sweet corn hybrids. Crop bleaching is usually transient. See label for further information.

* Mode of action (MOA) code developed by the Weed Science Society of America with the Herbicide Resistance Action Committee (HRAC)
** REI – Restricted Entry Interval; PHI- Pre-Harvest Interval

TABLE 4-10. CHEMICAL WEED CONTROL IN CUCUMBER

Weed	Herbicide and Formulation	MOA*	Amount of Formulation Per Acre	Intervals** REI (Hours)	Intervals** PHI (Days)	Precautions and Remarks
CUCUMBERS, Preplant and Preemergence						
Suppression or control of most annual grasses and broadleaf weeds, full rate required for nutsedge control	metam sodium Vapam HL 42%	N/A	37.5 to 75 gal	120	N/A	Rates are dependent on soil type and weeds present. Apply when soil moisture is at field capacity (100 to 125%). Apply through soil injection using a rotary tiller or inject with knives no more than 4 in. apart; follow immediately with a roller to smooth and compact the soil surface or with mulch. May apply through drip irrigation prior to planting a second crop on mulch. Plant back interval is often 14 to 21 days and can be 30 days in some environments. See label for all restrictions and additional information.
Emerged broadleaf and grass weeds	pelargonic acid Scythe 4.2 EC	26	1 to 10% v/v	12	N/A	Apply before crop emergence and control emerged weeds. There is no residual activity. May be tank mixed with soil residual compounds. See label for further instructions. May also be used as a banded spray between row middles. Use shielded sprayer directed to the row middles to reduce drift to the crop.
Broadleaf weeds and some annual grasses	caprylic acid + capric acid Homeplate, Suppress	26	3 to 9% v/v	24	0	May be applied prior to planting as a burndown treatment for emerged weeds, as a preemergence application after seeding but before emergence, as a directed or shielded application between rows or as a postharvest application. Use higher spray volumes for high weed density and weeds larger than 5 in. Coverage is important for acceptable weed control. May be tank mixed with other herbicides. See label for further instructions.
Most broadleaf weeds less than 4 in. tall or rosettes less than 3 in. in diameter, does not control grasses.	carfentrazone-ethyl Aim 1.9 EW or 2 EC	14	Up to 2 oz	12	N/A	Aim 1.9 EW is registered for application in transplant production systems only. Aim 2 EC is registered in seeded and transplant production systems. Apply no later than one day before transplanting or no later than 7 days before seeding crop. See label for information about application timing. Use a crop oil at up to 1-gal per 100-gal of spray solution or a nonionic surfactant at 2 pt per 100-gal of spray solution. Coverage is essential for good weed control. Can be tank mixed with other registered burndown herbicides.
CUCUMBERS, Preplant and Preemergence						
Contact kill of all green foliage, stale bed application	paraquat Parazone 3 SL Gramoxone 2 SL	22	1.3 to 2.7 pt 2 to 4 pt	24	N/A	Apply in a minimum of 10-gal spray mix per acre to emerged weeds before crop emergence as a broadcast or band treatment over a preformed row. Use sufficient water to give thorough coverage. Row should be formed several days ahead of planting and treating to allow maximum weed emergence. Use a nonionic surfactant at a rate of 16 to 32 oz per 100-gal spray mix or 1-gal approved crop oil concentrate per 100-gal spray mix. Paraquat product labels require applicators to take an EPA-approved training every 3 years to mix, load, and apply paraquat.
Annual and perennial grass and broadleaf weeds, stale bed application	glyphosate 4SL 5.5SL 5.88SL	9	16 to 32 fl oz 11 to 22 fl oz 10 to 21 fl oz	4	N/A	Apply to emerged weeds at least 3 days before seeding or transplanting. Perennial weeds may require higher rates of glyphosate. Consult the manufacturer's label for rates for specific weeds. When applying Roundup before transplanting crops into plastic mulch, care must be taken to remove residues of this product from the plastic prior to transplanting. To prevent crop injury, residues can be removed by 0.5 in. rainfall or by applying water via a sprinkler system. Certain glyphosate formulations require the addition of surfactant. Adding nonionic surfactant to glyphosate formulated with nonionic surfactant may result in reduced weed control.
Annual grasses, some small-seeded broadleaves	bensulide Prefar 4 EC	15	5 to 6 qt	12	N/A	Registered for cucurbit vegetable group (Crop grouping 9). Apply preplant and incorporate into the soil 1 to 2 in. (1 in. incorporation is optimum) with a rototiller or tandem disk or apply to the soil surface after seeding and follow with irrigation within 36 hours after application. Check replant restrictions for small grains on label.
Annual grasses and broadleaf weeds; weak on pigweed	clomazone Command 3 ME	13	0.4 to 1 pt	12	30	Apply immediately after seeding. See label for further information.
Annual grasses and some small-seeded broadleaf weeds	ethalfluralin Curbit 3 EC	3	3 to 4.5 pt	24	N/A	Apply post plant to seeded crop prior to crop emergence, or as a banded spray between rows after crop emergence or transplanting. See label for timing. Shallow cultivation, irrigation, or rainfall within 5 days is needed for good weed control. Do not use under mulches, row covers, or hot caps. Under conditions of unusually cold or wet soil and air temperatures, crop stunting or injury may occur. Crop injury can occur if seeding depth is too shallow.
Annual grasses and broadleaf weeds	ethalfluralin + clomazone Strategy 2.1 L	3 + 13	2 to 6 pt	24	45	Apply to the soil surface immediately after crop seeding for preemergence control of weeds. **DO NOT APPLY PRIOR TO PLANTING CROP. DO NOT SOIL INCORPORATE.** May also be used as a banded treatment between rows after crop emergence or transplanting. Do not apply over or under plastic mulch.
Yellow and purple nutsedge and broadleaf weeds	halosulfuron-methyl Profine 75, Sandea 75 DF	2	0.5 to 1 oz	12	21	Apply after seeding or prior to transplanting crop. For transplanting, do not transplant until 7 days after application. For seeded or transplanted cucumbers in plasti-culture, do not plant within 7 days of Sandea application. Rate can be increased to 1 ounce of product per acre to middles between rows.

* Mode of action (MOA) code developed by the Weed Science Society of America with the Herbicide Resistance Action Committee (HRAC)

** REI – Restricted Entry Interval; PHI- Pre-Harvest Interval

TABLE 4-10. CHEMICAL WEED CONTROL IN CUCUMBER (cont'd)

Weed	Herbicide and Formulation	MOA*	Amount of Formulation Per Acre	Intervals** REI (Hours)	Intervals** PHI (Days)	Precautions and Remarks
CUCUMBERS, Postemergence						
Annual grasses and small-seeded broadleaf weeds	**trifluralin** Treflan 4 L Treflan HFP 4 EC Treflan 10 G	3	1 to 2 pt 5 to 10 lb	12	30	Will not control emerged weeds. Row middles only. To improve preemergence control of late emerging weeds apply as a directed spray to soil between rows after crop emergence when crop plants have reached three to four true leaf stage of growth. Avoid contacting crop foliage as slight crop injury may occur. Set incorporation equipment to move treated soil around base of crop plants. PHI = 30 days.
Yellow and purple nutsedge and broadleaf weeds including cocklebur, galinsoga, smart weed, ragweed, wild radish, and pigweed	**halosulfuron-methyl** Profine 75, Sandea 75 DF	2	0.5 to 1 oz	12	14	**Apply postemergence only** after the crop has reached 3 to 5 true leaves but before first female flowers appear. Do not apply sooner than 14 days after transplanting. Use nonionic surfactant at 1 qt per 100-gal of spray solution with all postemergence applications.
Most broadleaf weeds less than 4 in. tall or rosettes less than 3 in. in diameter; does not control grasses	**carfentrazone-ethyl** Aim 1.9 EW or 2 EC	14	Up to 2 oz	12	N/A	Apply post-directed using hooded sprayers for control of emerged weeds. If crop is contacted, burning of contacted area will occur. Use crop oil concentrate at up to 1-gal per 100-gal solution or a nonionic surfactant at 2 pt per 100-gal of spray solution. Coverage is essential for good weed control. Can be tank mixed with other registered herbicides.
Most emerged weeds	**glyphosate** 4SL 5.5SL 5.88SL	9	16 to 32 fl oz 11 to 22 fl oz 10 to 21 fl oz	4	14	**Row middles only.** Apply as a hooded spray in row middles, as shielded spray in row middles, as wiper applications in row middles, or postharvest. To avoid severe injury to crop, do not allow herbicide to contact foliage, green shoots, stems, exposed, roots, or fruit of crop.
Annual grasses	**sethoxydim** Poast 1.5 EC	1	1 to 1.5 pt	12	14	Apply to emerged grasses. Consult manufacturer's label for specific rates and best times to treat and adjuvant and rate. Adding crop oil to Poast may increase the likelihood of crop injury at high air temperatures and high humidity.
	clethodim Arrow, Clethodim, Intensity, Select 2 EC Select Max, Intensity One 1 EC	1	6 to 8 oz 9 to 16 oz	24	14	Control of emerged grasses. See label for adjuvant and rate. Adding crop oil may increase the likelihood of crop injury at high air temperatures and high humidity. Highly effective in controlling annual bluegrass. Apply to actively growing grasses not under drought stress.

* Mode of action (MOA) code developed by the Weed Science Society of America with the Herbicide Resistance Action Committee (HRAC)
** REI – Restricted Entry Interval; PHI- Pre-Harvest Interval

TABLE 4-11. CHEMICAL WEED CONTROL IN EGGPLANT

Weed	Herbicide and Formulation	MOA*	Amount of Formulation Per Acre	Intervals** REI (Hours)	Intervals** PHI (Days)	Precautions and Remarks
EGGPLANT, Preplant						
Suppression or control of most annual grasses and broadleaf weeds, full rate required for nutsedge control	**metam sodium** Vapam HL 42%	N/A	37.5 to 75 gal	120	N/A	Rates are dependent on soil type and weeds present. Apply when soil moisture is at field capacity (100 to 125%). Apply through soil injection using a rotary tiller or inject with knives no more than 4 in. apart; follow immediately with a roller to smooth and compact the soil surface or with mulch. May apply through drip irrigation prior to planting a second crop on mulch. Plant back interval is often 14 to 21 days and can be 30 days in some environments. See label for all restrictions and additional information. Chloropicrin (150 lb per acre broadcast) will also be needed when laying first crop mulch to control nutsedge.
Contact kill of all green foliage, stale bed application	**paraquat** Parazone 3 SL Gramoxone 2 SL	22	1.3 to 2.7 pt 2 to 4 pt	24	N/A	Apply in a minimum of 10-gal spray mix per acre to emerged weeds before transplanting as a broadcast or band treatment over a preformed row. Use sufficient water to give thorough coverage. Row should be formed several days ahead of planting and treating to allow maximum weed emergence. Use a nonionic surfactant at a rate of 16 to 32 oz per 100-gal spray mix or 1-gal approved crop oil concentrate per 100-gal spray mix. Paraquat product label require applicators to take an EPA-approved training every 3 years to mix, load, and apply paraquat.

* Mode of action (MOA) code developed by the Weed Science Society of America with the Herbicide Resistance Action Committee (HRAC)
** REI – Restricted Entry Interval; PHI- Pre-Harvest Interval

TABLE 4-11. CHEMICAL WEED CONTROL IN EGGPLANT (cont'd)

Weed	Herbicide and Formulation	MOA*	Amount of Formulation Per Acre	Intervals** REI (Hours)	Intervals** PHI (Days)	Precautions and Remarks
EGGPLANT, Preplant (cont'd)						
Most broadleaf weeds less than 4 in. tall or rosettes less than 3 in. in diameter; does not control grasses	carfentrazone-ethyl Aim 1.9 EW or 2 EC	14	Up to 2 oz	12	N/A	Aim 1.9 EW is registered for application in transplant production systems only. Aim 2 EC is registered in seeded and transplant production systems. Apply no later than one day before transplanting crop (Aim 1.9 EW or Aim 2 EC) or no later than 7 days before seeding crop (Aim 2 EC only). See label for information about application timing. Use a crop oil at up to 1-gal per 100-gal of spray solution or a nonionic surfactant at 2 pt per 100-gal of spray solution. Coverage is essential for good weed control.
Annual and perennial grass and broadleaf weeds, stale bed application	glyphosate 4SL 5.5SL 5.88SL	9	16 to 32 fl oz 11 to 22 fl oz 10 to 21 fl oz	4	N/A	Apply to emerged weeds at least 3 days before seeding or transplanting. Perennial weeds may require higher rates of glyphosate. When applying Roundup before transplanting crops into plastic mulch, care must be taken to remove residues of this product from the plastic prior to transplanting. To prevent crop injury, residues can be removed by 0.5 in. rainfall or by applying water via a sprinkler system. Certain glyphosate formulations require the addition of surfactant. Adding nonionic surfactant to glyphosate formulated with nonionic surfactant may result in reduced weed control.
Broadleaf weeds and some annual grasses	caprylic acid + capric acid Homeplate, Suppress	26	3 to 9% v/v	24	0	May be applied prior to planting as a burndown treatment for emerged weeds, as a preemergence application after seeding but before emergence, as a directed or shielded application between rows or as a postharvest application. Use higher spray volumes for high weed density and weeds larger than 5 in. Coverage is important for acceptable weed control.
EGGPLANT, Preemergence						
Annual grasses, some small-seeded broadleaves	bensulide Prefar 4 EC	15	5 to 6 qt	12	N/A	Apply preplant incorporated (1 in. incorporation is optimum) or preemergence after planting. With preemergence application, irrigate immediately after application. See label for more directions.
Annual grasses and some broadleaf weeds	napropamide Devrinol, Devrinol DF-XT 50 DF Devrinol 2-XT 2 EC	15	2 to 4 lb 2 to 4 qt	24	N/A	Transplanted eggplant only. Apply preplant and incorporate into soil 1 to 2 in. using a rototiller or tandem disk. Shallow cultivations or irrigation will improve control. See label for small grains replanting restrictions. May also be applied in the row middles between plastic covered beds. See label for more information. See XT labels for information regarding delay in irrigation event.
EGGPLANT, Postemergence						
Annual and perennial grasses only	sethoxydim Poast 1.5 EC	1	1 to 1.5 pt	12	20	Apply to emerged grasses. Consult manufacturer's label for specific rates and best times to treat and adjuvant and rate. Adding crop oil to Poast may increase the likelihood of crop injury at high air temperatures and high humidity.
	clethodim Arrow, Clethodim, Intensity, Select 2 EC Select Max	1	6 to 8 oz 9 to 16 oz	24	20	Apply postemergence for control of grasses. See label for adjuvant and rate. Adding crop oil may increase the likelihood of crop injury at high air temperature and high humidity. Highly effective in controlling annual bluegrass. Apply to actively growing grasses not under drought stress.
EGGPLANT, Row Middles						
Most broadleaf weeds less than 4 in. tall or rosettes less than 3 in. in diameter; does not control grasses	carfentrazone-ethyl Aim 1.9 EW or 2 EC	14	Up to 2 oz	12	0	Apply post-directed using hooded sprayers for control of emerged weeds. If crop is contacted, burning of contacted area will occur. Use crop oil concentrate at up to 1-gal per 100-gal solution or a nonionic surfactant at 2 pt per 100-gal of spray solution. Coverage is essential for good weed control.
Most emerged weeds	glyphosate 4SL 5.5SL 5.88SL	9	16 to 32 fl oz 11 to 22 fl oz 10 to 21 fl oz	4	14	Apply as a hooded spray in row middles, as shielded spray in row middles, as wiper applications in row middles, or postharvest. To avoid severe injury to crop, do not allow herbicide to contact foliage, green shoots or stems, exposed roots, or fruit of crop.
Yellow and purple nutsedge and broadleaf weeds	halosulfuron-methyl Profine 75, Sandea 75 DF	2	0.5 to 1 oz	12	30	Apply between rows as a postemergence spray. Do not allow spray to contact crop or plastic mulch. Early season application will give postemergence and preemergence control. For postemergence applications, use nonionic surfactant at 1 qt per 100-gal of spray solution.
Contact kill of all green foliage	paraquat Parazone 3 SL Gramoxone 2 SL	22	1.3 pt 2 pt	24	N/A	Apply in 10-gal spray mix as a shielded spray to emerged weeds between rows of eggplant. Use a nonionic surfactant at a rate of 16 to 32 oz per 100-gal spray mix or 1-gal approved crop oil concentrate per 100-gal spray mix. Do not allow spray to contact crop or injury will result. Paraquat product labels require applicators to take an EPA-approved training every 3 years to mix, load, and apply paraquat.

* Mode of action (MOA) code developed by the Weed Science Society of America with the Herbicide Resistance Action Committee (HRAC)

** REI – Restricted Entry Interval; PHI- Pre-Harvest Interval

TABLE 4-12. CHEMICAL WEED CONTROL IN GARLIC

Weed	Herbicide and Formulation	MOA*	Amount of Formulation Per Acre	Intervals** REI (Hours)	Intervals** PHI (Days)	Precautions and Remarks
GARLIC, Preplant						
Annual and perennial grass and broadleaf weeds	glyphosate 4SL 5.5SL 5.88SL	9	 16 to 32 fl oz 11 to 22 fl oz 10 to 21 fl oz	4	N/A	Stale bed application. Apply to emerged weeds at least 3 days before planting. Perennial weeds may require higher rates of glyphosate. Consult the manufacturer's label for rates for specific weeds. Certain glyphosate formulations require the addition of surfactant. Adding nonionic surfactant to glyphosate formulated with nonionic surfactant may result in reduced weed control.
	paraquat Parazone 3 SL Gramoxone 2 SL	22	 1.7 to 2.7 pt 2.5 to 4 pt	24	60	Apply in a minimum of 20-gal spray mix per acre to emerged weeds before crop emergence as a broadcast or band treatment over a preformed row. Row should be formed several days ahead of planting and treating to allow maximum weed emergence. Use a nonionic surfactant at a rate of 16 to 32 oz per 100-gal spray mix or 1-gal approved crop oil concentrate per 100-gal spray mix. PHI = 60 days. Paraquat product labels require applicators to take an EPA-approved training every 3 years to mix, load, and apply paraquat.
Most broadleaf weeds less than 4 in. tall or rosettes less than 3 in. in diameter; does not control grasses	carfentrazone-ethyl Aim 1.9 EW or 2 EC	14	 Up to 2 oz	12	N/A	Apply no later than 30 days before planting. See label for specific Aim rate relating to weed species and proper adjuvant and rate. Coverage is essential for good weed control. Can be tank mixed with other registered burndown herbicides.
Emerged broadleaf weeds	pyraflufen ET Herbicide 0.208 EC	14	 0.5 to 2 oz	12	N/A	Apply as a preplant burndown treatment in a minimum of 10-gallons of solution per acre. See label for information on use of adjuvant.
Emerged broadleaf and grass weeds	pelargonic acid Scythe 4.2 EC	26	 1 to 10% v/v	12	N/A	Apply as a preplant burndown treatment or use in row middles using shielded sprayer.
Emerged broadleaf weeds and some annual grasses	caprylic acid + capric acid Homeplate, Suppress	26	 3 to 9% v/v	24	0	May be applied prior to planting as a burndown treatment for emerged weeds, as a preemergence application after seeding but before emergence, as a directed or shielded application between rows or as a postharvest application. Use higher spray volumes for high weed density and weeds larger than 5 in. Coverage is important for acceptable weed control. May be tank mixed with other herbicides. See label for further instructions.
GARLIC, Preplant incorporated or Preemergence						
Annual grasses, some small-seeded broadleaves	bensulide Prefar 4 EC	15	 5 to 6 qt	12	N/A	Apply preplant incorporated (1 in. incorporation is optimum) or preemergence after planting. With preemergence application, irrigate immediately after application. See label for more directions.
Residual control of annual grasses and small-seeded broadleaf weeds	dimethenamid-P Outlook 6 EC	15	 12 to 21 oz	12	30	For preemergence, weed control. Apply after crop has reached 2 true leaves until a minimum of 30 days before harvest. If applications are made to transplanted crop, DO NOT apply until transplants are in the ground and soil has settled around transplants with several days to recover.
	flumioxazin Chateau 51 SW	14	 6 oz	12	N/A	For preemergence, weed control. Apply prior to garlic and weed emergence. Application should be made within 3 days after planting garlic.
	pendimethalin Prowl H2O 3.8 AS Prowl 3.3 EC	3	 1.2 to 3.6 pt 1.5 to 3 pt	24	45	For preemergence, weed control. Apply preemergence after planting but prior to weed and crop emergence or postemergence to garlic in the 1 to 5 true leaf stage. Prowl can be applied sequentially by applying preemergence followed by a postemergence application. Does not control emerged weeds.
GARLIC, Postemergence						
Residual control of henbit, purslane, pigweed, primrose, smartweed, and many others; controls small, emerged weeds as well	oxyfluorfen Galigan 2E, Goal 2XL GoalTender 4F	14	 0.5 pt 0.25 pt	48	60	Transplanted dry bulb only. May be used as a postemergence spray to both the weeds and crop after the garlic has at least two fully developed true leaves. Some injury to garlic may result. Injury will be more severe if the chemical is applied during cool, wet weather. Weeds should be in the 2- to 4-leaf stage for best results.
Annual and perennial grasses only	clethodim Arrow, Clethodim, Intensity, Select 2 EC Select Max, Intensity One 1 EC	1	 6 to 8 oz 9 to 16 oz	24	45	Apply to emerged grasses. Consult manufacturer's label for specific rates and best times to treat and adjuvant and rate. Adding crop oil may increase the likelihood of crop injury at high air temperatures and high humidity. Highly effective in controlling annual bluegrass.
	fluazifop Fusilade DX 2 EC	1	 6 to 24 oz	12	45	Apply to emerged grasses. Consult manufacturer's label for specific rates and best times to treat and adjuvant and rate. Do not apply on days that are unusually hot and humid.

* Mode of action (MOA) code developed by the Weed Science Society of America with the Herbicide Resistance Action Committee (HRAC)
** REI – Restricted Entry Interval; PHI- Pre-Harvest Interval

TABLE 4-12. CHEMICAL WEED CONTROL IN GARLIC (cont'd)

Weed	Herbicide and Formulation	MOA*	Amount of Formulation Per Acre	Intervals** REI (Hours)	Intervals** PHI (Days)	Precautions and Remarks
GARLIC, Postemergence (cont'd)						
Annual and perennial grasses only (cont'd)	sethoxydim Poast 1.5 EC	1	 1 to 1.5 pt	 12	 30	Apply to emerged grasses. Consult manufacturer's label for specific rates and best times to treat and adjuvant and rate. Adding crop oil to Poast may increase the likelihood of crop injury at high air temperatures and high humidity.
GARLIC, Row Middles						
Most emerged weeds	glyphosate 4SL 5.5SL 5.88SL	9	 16 to 32 fl oz 11 to 22 fl oz 10 to 21 fl oz	 4	 14	Row middles only. Apply as a hooded spray in row middles, as shielded spray in row middles, as wiper applications in row middles, or postharvest. To avoid severe injury to crop, do not allow herbicide to contact foliage, green shoots, stems, exposed, roots, or fruit of crop.
Most broadleaf weeds less than 4 in. tall or rosettes less than 3 in. in diameter; does not control grasses	carfentrazone-ethyl Aim 1.9 EW or 2 EC	14	 Up to 2 oz	 12	 0	Apply post-directed using hooded sprayers for control of emerged weeds. If crop is contacted, burning of contacted area will occur. Use a nonionic surfactant or crop oil with Aim. See label for rate. Coverage is essential for good weed control. Can be tank mixed with other registered herbicides.

* Mode of action (MOA) code developed by the Weed Science Society of America with the Herbicide Resistance Action Committee (HRAC)
** REI – Restricted Entry Interval; PHI- Pre-Harvest Interval

TABLE 4-13. CHEMICAL WEED CONTROL IN GREENS

Weed	Herbicide and Formulation	MOA*	Amount of Formulation Per Acre	Intervals** REI (Hours)	Intervals** PHI (Days)	Precautions and Remarks
GREENS (Collard, Kale, Mustard Greens, and Turnip), Preplant and Preemergence						
Emerged broadleaf and grass weeds	pelargonic acid Scythe 4.2 EC	26	 1 to 10% v/v	 12	 N/A	Apply as preplant burndown to emerged weeds. See label for instruction. May also be used as a banded spray between row middles. Use a shielded sprayer directed to the row middles to reduce drift to the crop.
Broadleaf weeds and some annual grasses	caprylic acid + capric acid Homeplate, Suppress	26	 3 to 9% v/v	 24	 0	May be applied prior to planting as a burndown treatment for emerged weeds, as a preemergence application after seeding but before emergence, as a directed shielded application between rows or as a postharvest application. Use higher spray volumes for high weed density and weeds larger than 5 in. Coverage is important for acceptable weed control. May be tank mixed with other herbicides. See label for further instructions.
Contact kill of all green foliage, stale bed application	paraquat Parazone 3 SL Gramoxone 2 SL	22	 1.3 to 2.7 pt 2 to 4 pt	 24	 N/A	**Collard and turnip only.** Apply in a minimum of 10-gal spray mix per acre to emerged weeds before crop emergence or transplanting as a broadcast or band treatment over a preformed row. Use sufficient water to give thorough coverage. Row should be formed several days ahead of planting and treating to allow maximum weed emergence. Use a nonionic surfactant at a rate of 16 to 32 oz per 100-gal spray mix or 1-gal approved crop oil concentrate per 100-gal spray mix. Paraquat product labels require applicators to take an EPA-approved training every 3 years to mix, load, and apply paraquat.
Annual and perennial grass and broadleaf weeds, stale bed application	glyphosate 4SL 5.5SL 5.88SL	9	 16 to 32 fl oz 11 to 22 fl oz 10 to 21 fl oz	 4	 N/A	Apply to emerged weeds before crop emergence. Do not feed crop residue to livestock for 8 weeks following treatment. Perennial weeds may require higher rates of glyphosate. Consult the manufacturer's label for rates for specific weeds. Certain glyphosate formulations require the addition of surfactant. Adding nonionic surfactant to glyphosate formulated with nonionic surfactant may result in reduced weed control.
Annual grasses and small-seeded broadleaf weeds	trifluralin Treflan 4 L, Treflan HFP 4 EC Treflan 10 G	3	 1 to 1.5 pt 5 to 7.5 lb	 12	 N/A	**Greens: collard, kale, mustard, and turnip (fresh or processing).** Apply pre- plant and incorporate into the soil 2 to 3 in. within 8 hr using a rototiller or tandem disk. Do not use if turnip roots are to be consumed. Georgia, North Carolina, and South Carolina have a Section 24(c) Special Local Need Label for Treflan application in turnip roots.
	bensulide Prefar 4 EC	15	 5 to 6 qt	 12	 N/A	**Brassica (Cole) leafy vegetable group.** Not labeled for turnip. Apply preplant or preemergence after planting. With preemergence application, irrigate immediately after application. See label for more directions.

* Mode of action (MOA) code developed by the Weed Science Society of America with the Herbicide Resistance Action Committee (HRAC)
** REI – Restricted Entry Interval; PHI- Pre-Harvest Interval

TABLE 4-13. CHEMICAL WEED CONTROL IN GREENS (cont'd)

Weed	Herbicide and Formulation	MOA*	Amount of Formulation Per Acre	Intervals** REI (Hours)	Intervals** PHI (Days)	Precautions and Remarks
GREENS (Collard, Kale, Mustard Greens, and Turnip), Preplant and preemergence (cont'd)						
Annual grasses and broadleaf weeds, yellow nutsedge suppression	**S-metolachlor** Dual Magnum 7.62 EC	15	Check label	24	30, 60	**Collards and Kale ONLY. This is a Section 24(c) Special Local Need Label.** Growers must obtain the label prior to making Dual Magnum application. Irrigation following the application of Dual Magnum will increase the risk of crop injury. **Use lower rates on coarse-textured soils and higher rates on fine-textured soils. See label for more information.** **Mulched Systems with Transplanted Crop.** *Option 1:* Apply 8 to 16 oz per acre to the soil surface of pre-formed beds prior to laying plastic. Ensure the plastic laying process does not incorporate or disturb the treated bed. Unless restricted by other products, crops may be transplanted immediately following Dual Magnum application. *Option 2:* Apply 8 to 16 oz per acre overtop of crop at least 10 days after transplanting to ensure root system is well developed. Does not control emerged weeds. **Bare Ground Application for Transplanted Crop.** After transplanting, irrigate to seal the soil around the transplanted root ball. Five to ten days after transplanting and irrigating, apply Dual Magnum over the top of transplants. If soil is not sealed around the transplant root ball, crop injury may occur. **Direct Seeded Application.** May be applied over the top after the crop reaches 3 in. tall. **Row Middle Application to Transplanted and Direct Seeded Crop.** Apply as a banded application at a rate up to 1.25 pt per acre. PHI = 30 days for bok choy, collard, and kale PHI = 60 days for broccoli, cabbage, cauliflower, Chinese cabbage
GREENS (Collard, Kale, Mustard Greens, and Turnip Greens or roots), Postemergence						
Broadleaf weeds including sowthistle clover, cocklebur, jimsonweed, and ragweed	**clopyralid** Stinger 3 EC	4	0.3 to 0.5 pt	12	30, 15	**Kale, collards, mustard, turnip, mizuna, mustard spinach, and rape. See label to determine if other Brassica (Cole) leafy vegetables are registered.** Apply to crop when weeds are small and actively growing. Will control most legumes. For kale, collards, mustard, and turnip (roots) PHI = 30 days. For turnip tops PHI = 15 days. Mustard green injury has been observed in some research trials.
Annual and perennial grasses only	**clethodim** Arrow, Clethodim, Intensity, Select 2 EC Select Max, Intensity One 1 EC	1	6 to 8 oz 9 to 16 oz	24	14, 30	Apply postemergence for control of grasses. See label for adjuvant and rate. Adding crop oil may increase the likelihood of crop injury at high air temperatures and high humidity. Highly effective in controlling annual bluegrass. Apply to actively growing grasses not under drought stress. For greens crops, PHI = 14 days. For turnip roots, PHI = 30 days.
	sethoxydim Poast 1.5 EC	1	1 to 1.5 pt	12	30, 14	**ALSO LABELED FOR RAPE GREENS.** Apply to emerged grasses. Consult manufacturer's label for specific rates and best times to treat. See label for adjuvant and rate. Adding crop oil to Poast may increase the likelihood of crop injury at high air temperatures and high humidity. For greens crops, PHI = 30 days. For root crops, PHI = 14 days.

* Mode of action (MOA) code developed by the Weed Science Society of America with the Herbicide Resistance Action Committee (HRAC)

** REI – Restricted Entry Interval; PHI- Pre-Harvest Interval

TABLE 4-14. CHEMICAL WEED CONTROL IN HOPS

Weed	Herbicide and Formulation	MOA*	Amount of Formulation Per Acre	Intervals** REI (Hours)	Intervals** PHI (Days)	Precautions and Remarks
HOPS, Preplant and Preemergence						
Broadleaf weeds including chickweed, wild radish, and henbit. Limited control of annual grasses such as barnyard grass and large crabgrass	flumioxazin Chateau 51 SW	14	6 oz	12	N/A	Apply to dormant hops November through February.
Broadleaf weeds and annual grasses	norflurazon Solicam 80 DF	12	2.5 to 5 lb	12	60	Apply as a directed treatment a minimum of 6 months after planting hops. Rate is soil texture dependent.
Annual grasses and small-seeded broadleaf weeds	trifluralin Treflan 4 L, Treflan HFP 4 EC Treflan 10 G	3	 1 to 1.5 pt 5 to 7.5 lb	12	N/A	See label for rate information. Shallow incorporate to established, dormant crop. Use equipment that will ensure thorough soil mixing with minimal damage to crop.
Annual grasses and small-seeded broadleaf weeds	pendimethalin Prowl H2O 3.8 AS	3	1.1 to 4.2 qt	24	90	See label for instructions.
HOPS, Postemergence						
Canada thistle	clopyralid Solix 3, Spur	4	0.3 to 0.67			Some transient minor leaf cupping may occur to lower leaves and suckers if spray is exposed to plant. PHI = 30 days.
Broadleaf weeds	2,4-D Amine 4, others	4	See labels		3	Apply with a hooded sprayer for row middles. Hop foliage is susceptible to this product.
Annual and perennial grasses	clethodim Arrow, Clethodim, Intensity, Select 2 EC Select Max, Intensity One 1 EC	1	 6 to 8 oz 9 to 16 oz	24	21	For repeat applications make on a minimum of a 14-day interval.
Emerged broadleaf and grass weeds	pelargonic acid Scythe 4.2 EC	26	1 to 10% v/v	12	N/A	Apply before crop emergence to control emerged weeds. There is no residual activity. Avoid contact with foliage and green bark. See label for further instructions. May be used to control basal sucker growth.
Most broadleaf weeds less than 4 in. tall or rosettes less than 3 in. in diameter, does not control grasses	carfentrazone-ethyl Aim 1.9 EW or 2 EC	14	Up to 2 oz	12	7	Directed hooded spray for row middles. Most effective on broadleaf weeds less than 4 in. in height.
Annual and perennial grass and broadleaf weeds	glyphosate 4SL 5.5SL 5.88SL	9	 16 to 32 fl oz 11 to 22 fl oz 10 to 21 fl oz	4	N/A	Directed hooded spray for row middles. Avoid contact with green shoots and foliage.

* Mode of action (MOA) code developed by the Weed Science Society of America with the Herbicide Resistance Action Committee (HRAC)
** REI – Restricted Entry Interval; PHI- Pre-Harvest Interval

TABLE 4-15. CHEMICAL WEED CONTROL IN LETTUCE

Weed	Herbicide and Formulation	MOA*	Amount of Formulation Per Acre	Intervals** REI (Hours)	Intervals** PHI (Days)	Precautions and Remarks
LETTUCE, Preplant						
Suppression or control of most annual grasses and broadleaf weeds, full rate required for nutsedge control	metam sodium Vapam HL 42%	N/A	37.5 to 75 gal	120	N/A	Rates are dependent on soil type and weeds present. Apply when soil moisture is at field capacity (100 to 125%). Apply through soil injection using a rotary tiller or inject with knives no more than 4 in. apart; follow immediately with a roller to smooth and compact the soil surface or with mulch. May apply through drip irrigation prior to planting a second crop on mulch. Plant back interval is often 14 to 21 days and can be 30 days in some environments. See label for all restrictions and additional information.
Contact kill of all green foliage, stale bed application	paraquat Parazone 3 SL Gramoxone 2 SL	22	 1.3 to 2.7 pt 2 to 4 pt	24	N/A	Apply in a minimum of 10-gal spray mix per acre to emerged weeds before crop emerges as a broadcast or band treatment over a preformed row. Row should be formed several days ahead of planting and treating to allow maximum weed emergence. Use a nonionic surfactant at a rate of 16 to 32 oz per 100-gal spray solution or 1-gal approved crop oil concentrate per 100-gal spray mix. Paraquat product labels require applicators to take an EPA-approved training every 3 years to mix, load, and apply paraquat.
Annual and perennial grass and broadleaf weeds, stale bed application	glyphosate 4SL 5.5SL 5.88SL	9	 16 to 32 fl oz 11 to 22 fl oz 10 to 21 fl oz	4	N/A	Apply to emerged weeds before crop emergence. Do not feed crop residue to livestock for 8 weeks following treatment. Perennial weeds may require higher rates of glyphosate. Consult the manufacturer's label for rates for specific weeds. Certain glyphosate formulations require the addition of surfactant. Adding nonionic surfactant to glyphosate formulated with nonionic surfactant may result in reduced weed control.
Broadleaf weeds and some annual grasses	caprylic acid + capric acid Homeplate, Suppress	26	 3 to 9% v/v	24	0	May be applied prior to planting as a burndown treatment for emerged weeds, as a preemergence application after seeding but before emergence, as a directed or shielded application between rows or as a post-harvest application. Use higher spray volumes for high weed density and weeds larger than 5 in. Coverage is important for acceptable weed control.
LETTUCE, Preplant or Preemergence						
Annual grasses and small-seeded broadleaf	benefin Balan 60 WDG	3	2 to 2.5 lb	12	N/A	Apply preplant and incorporate 2 to 3 in. deep with a rototiller or tandem disk before seeding or transplanting.
	bensulide Prefar 4 EC	15	5 to 6 qt	12	N/A	Apply preplant incorporated (1 in. incorporation is optimum) or preemergence after planting. With preemergence application, irrigate immediately after application. See label for more directions.
Most annual grasses and broadleaf weeds	pronamide Kerb 3.3 SC	3	1.25 to 5 pt	24	55	**Kerb 3.3 SC has a supplemental label allowing application on leaf and head lettuce.** Also labeled in endive, escarole, or radicchio greens. Can be applied preplant, post plant, or postemergence in banded, bed-topped or broadcast applications or a split application can be made. See label for more information. Consult label for planting restrictions for rotational crops.
LETTUCE, Postemergence						
Annual and perennial grasses	sethoxydim Poast 1.5 EC clethodim Arrow, Clethodim, Intensity, Select 2 EC Select Max, Intensity One 1 EC	1 1	 1 to 1.5 pt 6 to 8 oz 9 to 16 oz	12 24	30, 15 14	Arrow, Clethodim, and Select are only registered for leaf lettuce. Consult manufacturer's label for specific rates and best times to treat and adjuvant and rate. Use of Poast or clethodim with crop oil may increase the likelihood of crop injury at high air. Temperatures and high humidity. Do not apply sethoxydim within 30 days of harvest on head lettuce or within 15 days of harvest on leaf lettuce.
Annual and perennial grasses	fluazifop Fusilade DX 2 EC	1	6 to 24 oz	12	14	Registered in leaf and head lettuce.
Most annual grasses and broadleaf weeds	pronamide Kerb 3.3 SC	3	1.25 to 5 pt	24	55	**Kerb 3.3 SC has supplemental label now allowing use on leaf lettuce as well as head lettuce.** Apply before weed germination, if possible, no later than weeds in the 2-leaf stage. See label for restrictions and use patterns. Consult label for rotational restrictions and other restrictions.
Most emerged weeds	glyphosate 4SL 5.5SL 5.88SL	9	 16 to 32 fl oz 11 to 22 fl oz 10 to 21 fl oz	4	14	**Row middles only.** Apply as a hooded spray in row middles, as shielded spray in row middles, as wiper applications in row middles, or postharvest. To avoid severe injury to crop, do not allow herbicide to contact foliage, green shoots, stems, exposed, roots, or fruit of crop.
Most broadleaf weeds less than 4 in. tall or rosettes less than 3 in. in diameter; does not control grasses	carfentrazone-ethyl Aim 1.9 EW or 2 EC	14	Up to 2 oz	12	N/A	Apply post-directed using hooded sprayers for control of emerged weeds. If crop is contacted, burning of contacted area will occur. Use a nonionic surfactant or crop oil with Aim. See label for rate. Coverage is essential for good weed control. Can be tank mixed with other registered herbicides.

* Mode of action (MOA) code developed by the Weed Science Society of America with the Herbicide Resistance Action Committee (HRAC)
** REI – Restricted Entry Interval; PHI- Pre-Harvest Interval

TABLE 4-16. CHEMICAL WEED CONTROL IN OKRA

Weed	Herbicide and Formulation	MOA*	Amount of Formulation Per Acre	Intervals** REI (Hours)	Intervals** PHI (Days)	Precautions and Remarks
OKRA, Preplant and Preemergence						
Annual and perennial grass and broadleaf weeds, stale bed application	glyphosate 4SL 5.5SL 5.88SL	9	 16 to 32 fl oz 11 to 22 fl oz 10 to 21 fl oz	4	N/A	Apply to emerged weeds before crop emergence. Perennial weeds may require higher rates. Consult the manufacturer's label for rates for specific weeds. Certain glyphosate formulations require the addition of surfactant. Adding nonionic surfactant to glyphosate formulated with nonionic surfactant may result in reduced weed control.
Most broadleaf weeds less than 4 in. tall or rosettes less than 3 in. in diameter; does not control grasses	carfentrazone-ethyl Aim 1.9 EW or 2 EC	14	 Up to 2 oz	12	N/A	Apply no later than 1 day before transplanting crop. Use a nonionic surfactant or crop oil with Aim. See label for rate. Coverage is essential for good weed control. Can be tank mixed with other registered burndown herbicides.
Emerged broadleaf weeds and some annual grasses	caprylic acid + capric acid Homeplate, Suppress	26	 3 to 9% v/v	24	0	May be applied prior to planting as a burndown treatment for emerged weeds, as a preemergence application after seeding but before emergence, as a directed or shielded application between rows or as a postharvest application. Use higher spray volumes for high weed density and weeds larger than 5 in. Coverage is important for acceptable weed control. May be tank mixed with other herbicides. See label for further instructions.
Annual grasses and small-seeded broadleaf weeds	trifluralin Treflan 4 L, Treflan HFP 4 EC Treflan 10 G	3	 1 to 2 pt 5 to 10 lb	12	N/A	Apply preplant and incorporate into the soil 2 to 3 in. within 8 hours using a rototiller or tandem disk.
Broadleaf and grass weeds	prometryn Caparol 4L	5	 1.5 to 3 pt	12	14	Apply preemergence and or post-directed application. Make a single preemergence application of Caparol at 3 pt per acre after planting and before crop emergence or a sequential application (see label for further details). Do not exceed 3 pt per acre of Caparol per season. See label for crop rotation restrictions.
Annual broadleaf weeds including pigweed spp.	mesotrione Callisto 4 L	27	 6 oz	12	28	May be applied as a row middle **or** hooded POST-directed application but not both. For preemergence row middle application, apply as a banded application to the row middles prior to weed emergence. Leave 1 ft. of untreated area over the okra row or 6 in. on each side of the planted row. Do not apply Callisto directly over the planted row or severe injury may occur. Injury risk is greatest on coarse textured soils (sand, sandy loam, or loamy sand).
OKRA, Postemergence						
Annual and perennial grasses	sethoxydim Poast 1.5 EC	1	 1 to 1.5 pt	12	14	Apply to actively growing grasses not under drought stress. Do not apply on days that are unusually hot and humid.
Most broadleaf weeds less than 4 in. tall or rosettes less than 3 in. in diameter; does not control grasses	carfentrazone-ethyl Aim 1.9 EW or 2 EC	14	 Up to 2 oz	12	0	Emerged broadleaf weeds. Apply to **row middles only** with a hooded sprayer. Use crop oil concentrate or nonionic surfactant at recommended rates.
Annual and perennial grass and broadleaf weeds	glyphosate 4SL 5.5SL 5.88SL	9	 16 to 32 fl oz 11 to 22 fl oz 10 to 21 fl oz	4	14	**Row middles only.** Apply as a hooded spray in row middles, as shielded spray in row middles, as wiper applications in row middles, or postharvest. To avoid severe injury to crop, do not allow herbicide to contact foliage, green shoots or stems, exposed roots, or fruit of crop.
Annual broadleaf weeds including pigweed 3 in or less	mesotrione Callisto 4 L	27	See label	12	28	May be applied as a row middle **or** hooded POST-directed application but not both. For postemergence hooded application, okra must be at least 3 in. tall. Minimize amount of Callisto that contacts okra foliage or crop injury will occur. PHI = 28 days.
Yellow and purple nutsedge and broadleaf weeds	halosulfuron-methyl Profine 75, Sandea 75 DF	2	 0.5 to 1 oz	12	30	Apply to row middles as a postemergence shielded or hooded spray to avoid contact of herbicide with planted crop. In plasticulture, do not allow spray to contact plastic. Do not apply more than 2 oz per acre per 12-month period.

* Mode of action (MOA) code developed by the Weed Science Society of America with the Herbicide Resistance Action Committee (HRAC)
** REI – Restricted Entry Interval; PHI- Pre-Harvest Interval

TABLE 4-17. CHEMICAL WEED CONTROL IN ONION

Weed	Herbicide and Formulation	MOA*	Amount of Formulation Per Acre	REI (Hours)	PHI (Days)	Precautions and Remarks
ONIONS, Preplant and Preemergence						
Suppression or control of most annual grasses and broadleaf weeds, full rate required for nutsedge control	metam sodium Vapam HL 42%	N/A	37.5 to 75 gal	120	N/A	**Dry bulb and green onion.** Rates are dependent on soil type and weeds present. For nutsedge control, use 75-gal per acre. Apply when soil moisture is at field capacity (100 to 125%). Apply through soil injection using a rotary tiller or inject with knives no more than 4 in. apart; follow immediately with a roller to smooth and compact the soil surface or with mulch. May apply through drip irrigation prior to planting a second crop on mulch. Plant back interval is often 14 to 21 days and can be 30 days in some environments. See label for all restrictions and additional information.
Contact kill of all green foliage, stale bed application	paraquat Parazone 3 SL Gramoxone 2 SL	22	 1.7 to 2.7 pt 2 to 4 pt	24	N/A	Apply in a minimum of 20-gal spray mix per acre to emerged weeds before crop emergence or transplanting as a broadcast or band treatment over a preformed row. Row should be formed several days ahead of planting and treating to allow maximum weed emergence. Use a nonionic surfactant at a rate of 16 to 32 oz per 100-gal spray mix or 1-gal approved crop oil concentrate per 100-gal spray mix. Paraquat product labels require applicators to take an EPA-approved training every 3 years to mix, load, and apply paraquat.
Annual and perennial grass and broadleaf weeds	glyphosate 4SL 5.5SL 5.88SL	9	 16 to 32 fl oz 11 to 22 fl oz 10 to 21 fl oz	4	N/A	Apply to emerged weeds before crop emergence. Perennial weeds may require higher rates of glyphosate. Consult the manufacturer's label for rates for specific weeds. Use on direct seeded onions only. Certain glyphosate formulations require the addition of surfactant. Adding nonionic surfactant to glyphosate formulated with nonionic surfactant may result in reduced weed control.
Most broadleaf weeds less than 4 in. tall or rosettes less than 3 in. in diameter; does not control grasses	carfentrazone-ethyl Aim 1.9 EW or 2 EC	14	Up to 2 oz	12	N/A	Apply no later than 30 days before planting. See label for specific Aim rate relating to weed species and proper adjuvant and rate. Coverage is essential for good weed control. Can be tank mixed with other registered burndown herbicides.
Emerged broadleaf weeds and some annual grasses	caprylic acid + capric acid Homeplate, Suppress	26	3 to 9% v/v	24	0	May be applied prior to planting as a burndown treatment for emerged weeds, as a preemergence application after seeding but before emergence, as a directed or shielded application between rows or as a postharvest application. Use higher spray volumes for high weed density and weeds larger than 5 in. Coverage is important for acceptable weed control. May be tank mixed with other herbicides. See label for further instructions.
Annual grasses, some small-seeded broadleaves	bensulide Prefar 4 EC	15	5 to 6 qt	12	N/A	**Dry bulb only.** Apply preplant incorporated (1 in. incorporation is optimum) or preemergence after planting. With preemergence application, irrigate immediately after application. See label for more directions and rotation restrictions.
Annual broadleaf weeds	oxyfluorfen Galigan 2E, Goal 2XL GoalTender 4 F	14	 1 to 2 pt 1 pt	48	45	**Transplanted dry bulb only.** Apply as a single application immediately (within 2 days) after transplanting for preemergence control of weeds. Injury can occur if applications are made during cool, wet weather or prior to the full development of the true leaves. See label for rates and instructions for use.
Most annual grasses and some broadleaf weeds	pendimethalin Prowl H2O 3.8 AS Prowl 3.3 EC	3	See label	24	30 45	**Dry bulb and green onion (chives, leeks, spring onions, scallions, Japanese bunching onions, green shallots, and green eschalots). Prowl 3.3 EC is not registered for green onions.** For preemergence, weed control. Apply when onions have two to nine true leaves (dry bulb) and two to three leaves (green onion) but prior to weed emergence. For green onion, the soil must be a muck soil or be a mineral soil with at least 3% organic matter. See label for additional information on rate depending on soil type. For dry bulb onion, PHI = 45 days. For green onions, PHI = 30 days.
	dimethenamid-P Outlook 6 EC	15	12 to 21 oz	12	30	**Dry bulb and green onion (leeks, spring onions or scallions, Japanese bunching onions, green shallots or eschalots).** For preemergence, weed control. Apply after crop has reached 2 true leaves until a minimum of 30 days before harvest. If applications are made to transplanted crop, DO NOT apply until transplants are in the ground and soil has settled around transplants with several days to recover.
Annual grasses and broadleaf weeds, yellow nutsedge suppression	S-metolachlor Dual Magnum 7.62 EC	15	8 to 16 oz	24	21, 60	**Dry bulb onion and green onion. This is a Section 24(c) Special Local Needs Label.** Growers must obtain a label. Irrigation following the application of Dual Magnum will increase the risk of crop injury. **Use lower rates on coarse-textured soils and higher rates on fine-textured soils.** See label for more information. **Seeded Application.** Do not apply before 4-leaf stage. Once onion has reached the 4-leaf stage, apply 8 oz per acre. When onions reach 6-leaf stage, rate can be increased to 12 oz per acre. **Transplant Dry Bulb.** Transplant and then irrigate to seal soil around the root ball. Apply within 48 hours of planting. Heavy irrigation following the application of Dual Magnum will increase the risk of crop injury. PHI = 21 days for green onion. PHI = 60 days for bulb onion.

* Mode of action (MOA) code developed by the Weed Science Society of America with the Herbicide Resistance Action Committee (HRAC)
** REI – Restricted Entry Interval; PHI- Pre-Harvest Interval

TABLE 4-17. CHEMICAL WEED CONTROL IN ONION (cont'd)

Weed	Herbicide and Formulation	MOA*	Amount of Formulation Per Acre	REI (Hours)	PHI (Days)	Precautions and Remarks
ONIONS, Postemergence						
Most annual broadleaf weeds	oxyfluorfen Galigan 2E, Goal 2XL GoalTender 4 F	14	 1 to 2 pt 1 pt	48	45	**Dry bulb only.** May be used as a postemergence spray to both the weeds and crop after the onions have at least two fully developed true leaves. Some injury to onions may result. Injury will be more severe if the chemical is applied during cool, wet weather. Weeds should be in the 2- to 4-leaf stage for best results. Do not make more than four applications per year.
Common lambsquarters, common chickweed, common purslane, black nightshade, ladysthumb, Pennsylvania smartweed, redroot pigweed, and some annual grasses	ethofumesate Nortron 4 SC	8	 16 to 32 oz	12	30	Apply at planting or just after planting prior to weed emergence. Can be used postemergence at 16 oz per acre. See label for more information. Rainfall of at least 0.5 in. is needed for activation.
Most broadleaf weeds less than 4 in. tall or rosettes less than 3 in. diameter; does not control grasses	carfentrazone-ethyl Aim 1.9 EW or 2 EC	14	 Up to 2 oz	12	0	Apply post-directed using hooded sprayers for control of emerged weeds. If crop is contacted, burning of contacted area will occur. Use a nonionic surfactant or crop oil with Aim. See label for rate. Coverage is essential for good weed control. Can be tank mixed with other registered herbicides.
Most emerged weeds	glyphosate 4SL 5.5SL 5.88SL	9	 16 to 32 fl oz 11 to 22 fl oz 10 to 21 fl oz	4	14	**Row middles only.** Apply as a hooded spray in row middles, as shielded spray in row middles, as wiper applications in row middles, or postharvest. To avoid severe injury to crop, do not allow herbicide to contact foliage, green shoots, stems, exposed, roots, or fruit of crop.
Annual and perennial grasses only	fluazifop Fusilade DX 2 EC	1	 6 to 24 oz	12	14, 45	**Dry bulb and green onion.** Apply to emerged grasses. Consult manufacturer's label for specific rates and best times to treat. Do not apply on days that are unusually hot and humid. For green onions, PHI = 14 days. For dry bulb onion, PHI = 45 days.
	sethoxydim Poast 1.5 EC	1	 1 to 1.5 pt	12	30	**Dry bulb and green.** Apply to emerged grasses. Consult manufacturer's label for specific rates and best times to treat and adjuvant and rate. Adding crop oil to Poast may increase the likelihood of crop injury at high air temperatures. Do not apply Poast on days that are unusually hot and humid.
	clethodim Arrow, Clethodim, Intensity, Select 2 EC Select Max, Intensity One 1 EC	1	 6 to 8 oz 9 to 16 oz	24	14, 45	**Dry bulb only.** Apply to emerged grasses. Consult the manufacturer's label for specific rates and best times to treat. Adding crop oil may increase the likelihood of crop injury at high air temperatures and high humidity. Do not apply Select on unusually hot and humid days. Intensity One may be applied to dry bulb onions or green onions (leeks, scallions or spring onions, Japanese bunching onion, shallots or eschalots). Do not exceed 16 ounces of Intensity. For green onions, PHI = 14 days. For dry bulb onion, PHI = 45 days.

* Mode of action (MOA) code developed by the Weed Science Society of America with the Herbicide Resistance Action Committee (HRAC)
** REI – Restricted Entry Interval; PHI- Pre-Harvest Interval

TABLE 4-18. CHEMICAL WEED CONTROL IN PEAS

Weed	Herbicide and Formulation	MOA*	Amount of Formulation Per Acre	REI (Hours)	PHI (Days)	Precautions and Remarks
PEAS, GREEN/ENGLISH, Preplant and Preemergence						
Contact kill of all green foliage, stale bed application	paraquat Parazone 3 SL Gramoxone 2 SL	22	 1.3 to 2.7 pt 2 to 4 pt	24	N/A	Apply in a minimum of 10-gal spray mix per acre to emerged weeds before crop emergence as a broadcast or band treatment over a preformed row. Use sufficient water to give thorough coverage. Row should be formed several days ahead of planting and treating to allow maximum weed emergence. Use a nonionic surfactant at a rate of 16 to 32 oz per 100-gal spray mix or 1-gal approved crop oil concentrate per 100-gal spray mix. Paraquat product labels require applicators to take an EPA-approved training every 3 years to mix, load, and apply paraquat.

* Mode of action (MOA) code developed by the Weed Science Society of America with the Herbicide Resistance Action Committee (HRAC)
** REI – Restricted Entry Interval; PHI- Pre-Harvest Interval

TABLE 4-18. CHEMICAL WEED CONTROL IN PEAS (cont'd)

Weed	Herbicide and Formulation	MOA*	Amount of Formulation Per Acre	Intervals** REI (Hours)	Intervals** PHI (Days)	Precautions and Remarks
PEAS, GREEN/ENGLISH, Preplant and Preemergence (cont'd)						
Most broadleaf weeds less than 4 in. tall or rosettes less than 3 in. in diameter; does not control grasses	carfentrazone-ethyl Aim 1.9 EW or 2 EC	14	Up to 2 oz	12	N/A	Apply prior to planting or emergence of crop. Use a nonionic surfactant or crop oil with Aim. See label for rate. Coverage is essential for good weed control. Can be tank mixed with other registered burndown herbicides.
Annual and perennial grass and broadleaf weeds	glyphosate 4SL 5.5SL 5.88SL	9	 16 to 32 fl oz 11 to 22 fl oz 10 to 21 fl oz	4	N/A	Apply to emerged weeds before crop emergence. Do not feed crop residue to livestock for 8 weeks following treatment. Perennial weeds may require higher rates of glyphosate. Consult the manufacturer's label for rates for specific weeds. Certain glyphosate formulations require the addition of a surfactant. Adding nonionic surfactant to glyphosate formulated with nonionic surfactant may result in reduced weed control.
Broadleaf weeds and some annual grasses	caprylic acid + capric acid Homeplate, Suppress	26	3 to 9% v/v	24	0	May be applied prior to planting as a burndown treatment for emerged weeds, as a preemergence application after seeding but before emergence, as a directed or shielded application between rows. Use higher spray volumes for high weed density and weeds larger than 5 in. Coverage is important for acceptable weed control. May be tank mixed with other herbicides. See label for further instructions.
Broadleaf weeds	saflufenacil Sharpen 3.42 SL	14	1 oz	12	3	**Dry field pea edible pea (sugar snap, English pea, garden pea, green pea, marrow fat pea) and chickpea only.** Apply as a preplant/preemergence burndown of small actively growing broadleaf weeds. Can also be used preplant incorporated or preemergence in edible pea. See label for directions. Do not apply more than 2 ounces per acre per season. Apply as a preplant/preemergence burndown. Do not apply more than 2 oz per acre per season.
Annual grasses and small-seeded broadleaf weeds	pendimethalin Prowl H2O 3.8 AS Prowl 3.3 EC	3	 1.5 to 3 pt 1.8 to 3.6 pt	24	N/A	**Southernpeas (cowpeas) and snap beans only.** Apply preplant and incorporate into the soil 2 to 3 in. using a power-driven rototiller or by cross disking. **DO NOT APPLY AFTER SEEDING. Do not apply when air temperature is below 45°F.**
	trifluralin Treflan 4 L, Treflan HFP 4EC Treflan 10 G	3	 1 to 2 pt 5 to 10 lb	12	N/A	**English peas only.** Apply preplant and incorporate to a depth of 2 to 3 in. within 8 hr with a rototiller or tandem disk.
Annual grasses and broadleaf weeds; weak on pigweed	clomazone Command 3 ME	13	1.3 pt	12	45	Apply to the soil surface immediately after seeding. See label for further instruction.
Annual grasses, small-seeded broadleaf weeds, and suppression of yellow nutsedge	S-metolachlor Brawl, Dual Magnum, Medal 7.62 EC Brawl II, Dual II Magnum, Medal II 7.64 EC	15	 1 to 2 pt 1 to 2 pt	24	N/A	Apply to soil surface immediately after seeding. Shallow cultivations will improve control. See label for specific rate.
Annual broadleaf weeds including morningglory, pigweed, smartweed, and purslane	imazethapyr Pursuit 2 EC	2	Up to 3 oz	4	N/A	**English peas only.** Apply preplant incorporated or to soil surface immediately after planting. See label for more details.
PEAS, GREEN, Postemergence						
Annual broadleaf weeds and yellow nutsedge	bentazon Basagran 4 SL	6	1 to 2 pt	48	10	Apply overtop of peas when weeds are small, and peas have at least three pairs of leaves (four nodes). **DO NOT ADD CROP OIL CONCENTRATE TO SPRAY MIX.** Do not apply when peas are in bloom.
Most broadleaf weeds less than 4 in. tall or rosettes less than 3 in. in diameter; does not control grasses	carfentrazone-ethyl Aim 1.9 EW or 2 EC	14	Up to 2 oz	12	N/A	Apply prior to planting or emergence of crop. Use a nonionic surfactant or crop oil with Aim. See Label for rate. Coverage is essential for good weed control. Can be tank mixed with other registered burndown herbicides.
Most emerged weeds	glyphosate 4SL 5.5SL 5.88SL	9	 16 to 32 fl oz 11 to 22 fl oz 10 to 21 fl oz	4	14	**Row middles only.** Apply as a hooded spray in row middles, as shielded spray in row middles, as wiper applications in row middles, or postharvest. To avoid severe injury to crop, do not allow herbicide to contact foliage, green shoots or stems, exposed roots, or fruit of crop.

* Mode of action (MOA) code developed by the Weed Science Society of America with the Herbicide Resistance Action Committee (HRAC)
** REI – Restricted Entry Interval; PHI- Pre-Harvest Interval

TABLE 4-18. CHEMICAL WEED CONTROL IN PEAS (cont'd)

Weed	Herbicide and Formulation	MOA*	Amount of Formulation Per Acre	Intervals** REI (Hours)	Intervals** PHI (Days)	Precautions and Remarks
PEAS, GREEN, Postemergence (cont'd)						
Annual and perennial grasses	sethoxydim Poast 1.5 EC	1	 1 to 1.5 pt	12	30, 15	Apply to emerged grasses. Consult manufacturer's label for specific rates and best times to treat and adjuvant and rate. Adding crop oil may increase the likelihood of crop injury at high air temperatures and high humidity. For dry beans, PHI = 30 days. For succulent beans, PHI = 15 days.
	quizalofop p-ethyl Assure II 0.88 EC Targa 0.88 EC	1	 6 to 12 oz 6 to 12 oz	12	30, 60	For dry beans, PHI = 60 days. For succulent beans, PHI = 30 days.
Annual broadleaf weeds including morningglory, pigweed, smartweed, and purslane	imazethapyr Pursuit 2 EC	2	 Up to 3 oz	4	30	**See label for pea type.** Apply postemergence to 1- to 3-in. weeds (one to four leaves) when peas are at least 3 in. high but prior to five nodes and before flowering. Add nonionic surfactant at 2 pt per 100-gal of spray mix. See label for crop rotation restrictions.
Broadleaf and grass weeds	imazamox Raptor 1 SL	2	 4 oz		30	**Dry peas only.** Apply postemergence before bloom stage but after dry peas have at least 3 pairs of leaves. See label for further information.
PEAS, SOUTHERN (Cowpeas, Blackeyed peas, Field peas), Preplant or Preemergence						
Contact kill of all green foliage, stale bed application	paraquat Parazone 3 SL Gramoxone 2 SL	22	 1.3 to 2.7 pt 2 to 4 pt	24	N/A	Apply in a minimum of 20-gal spray solution to emerged weeds before crop emergence as a broadcast or band treatment over a preformed row. Use sufficient water to give thorough coverage. Row should be formed several days ahead of planting and treating to allow maximum weed emergence. Use a nonionic surfactant at a rate of 16 to 32 oz per 100-gal spray mix or 1-gal approved crop oil concentrate per 100-gal spray mix. Paraquat product labels require applicators to take an EPA-approved training every 3 years to mix, load, and apply paraquat.
Annual and perennial grass and broadleaf weeds, stale bed application.	glyphosate 4SL 5.5SL 5.88SL	9	 16 to 32 fl oz 11 to 22 fl oz 10 to 21 fl oz	4	N/A	Apply to emerged weeds before crop emergence. Do not feed crop residue to livestock for 8 weeks following treatment. Perennial weeds may require higher rates of glyphosate. Consult the manufacturer's label for rates for specific weeds. Certain glyphosate formulations require the addition of a surfactant. Adding nonionic surfactant to glyphosate formulated with nonionic surfactant may result in reduced weed control.
Annual grasses and small-seeded broadleaf weeds	pendimethalin Prowl H2O 3.8 AS Prowl 3.3 EC	3	 1.5 to 3 pt 1.8 to 3.6 pt	24	N/A	**NOT LABELED IN BLACKEYED PEAS.** Apply preplant and incorporate into the soil 2 to 3 in. using a power-driven rototiller or by cross disking. **DO NOT APPLY AFTER SEEDING.**
	trifluralin Treflan 4 L, Treflan HFP 4 EC Treflan 10 G	3	 1 to 2 pt 5 to 10 lb	12	N/A	Apply preplant and incorporate into the soil 2 to 3 in. deep within 8 hr with a rototiller or tandem disk.
Annual grasses and broadleaf weeds	clomazone Command 3 ME	13	 0.4 to 0.67 pt	12	45	**Succulent Southernpeas only.** Apply to the soil surface immediately after seeding. Offers weak control of pigweed. See label for further instruction.
Annual grasses, small-seeded broadleaf weeds, and suppression of yellow nutsedge	S-metolachlor Brawl, Dual Magnum, Medal 7.62 EC Brawl II, Dual II Magnum, Medal II 7.64 EC	15	 1 to 2 pt 1 to 2 pt	24	50	Apply to soil surface immediately after planting. Shallow cultivations will improve control. May also be soil incorporated before planting.
Annual grasses and broadleaf weeds including morninggglory, pigweed, and purslane	imazethapyr Pursuit 2 EC	2	 Up to 4 oz	4	N/A	Apply preemergence or preplant incorporated. See label for rate for specific pea species.
Most broadleaf weeds less than 4 in. tall or rosettes less than 3 in. in diameter; does not control grasses	carfentrazone-ethyl Aim 1.9 EW or 2 EC	14	 Up to 2 oz	12	0	Apply post-directed using hooded sprayers for control of emerged weeds. If crop is contacted, burning of contacted area will occur. Use a nonionic surfactant or crop oil with Aim. See Label for rate. Coverage is essential for good weed control. Can be tank mixed with other registered herbicides.
PEAS, SOUTHERN, Postemergence						
Annual broadleaf weeds and yellow nutsedge	bentazon Basagran 4 SL	6	 1 to 2 pt	48	30	Apply overtop of peas when weeds are small, and peas have at least three pair of leaves (four nodes). **DO NOT ADD CROP OIL CONCENTRATE TO SPRAY MIX.** See label for weeds controlled with Basagran. Do not apply when peas are in bloom.

* Mode of action (MOA) code developed by the Weed Science Society of America with the Herbicide Resistance Action Committee (HRAC)
** REI – Restricted Entry Interval; PHI- Pre-Harvest Interval

TABLE 4-18. CHEMICAL WEED CONTROL IN PEAS (cont'd)

Weed	Herbicide and Formulation	MOA*	Amount of Formulation Per Acre	Intervals** REI (Hours)	Intervals** PHI (Days)	Precautions and Remarks
PEAS, SOUTHERN, Postemergence						
Annual broadleaf weeds including morningglory, pigweed, smartweed, and purslane	imazethapyr Pursuit 2 EC	2	Up to 4 oz	4	30	**Southern peas and certain dry peas.** Apply postemergence to 1- to 3-in. weeds (one to four leaves) when peas are at least 3 in. in height but prior to five nodes and flowering. Add nonionic surfactant at 2 pt per 100-gal of spray mixture with all postemergence applications. See label for rate for specific pea species.
Most broadleaf weeds less than 4 in. tall or rosettes less than 3 in. in diameter; does not control grasses	carfentrazone-ethyl Aim 1.9 EW or 2 EC	14	Up to 2 oz	12	0	Apply post-directed using hooded sprayers for control of emerged weeds. If crop is contacted, burning of contacted area will occur. Use a nonionic sur-factant or crop oil with Aim. See label for rate. Coverage is essential for good weed control. Can be tank mixed with other registered herbicides.
Most emerged weeds	glyphosate 4SL 5.5SL 5.88SL	9	16 to 32 fl oz 11 to 22 fl oz 10 to 21 fl oz	4	14	**Row middles only.** Apply as a hooded spray in row middles, as shielded spray in row middles, as wiper applications in row middles, or postharvest. To avoid severe injury to crop, do not allow herbicide to contact foliage, green shoots, stems, exposed, roots, or fruit of crop.
Annual and perennial grasses	quizalofop p-ethyl Assure II 0.88 EC Targa 0.88 EC	1	6 to 12 oz 6 to 12 oz	12	30, 60	Apply to emerged grasses. Consult manufacturer's label for specific rates and best times to treat. With sethoxydim, add 1 qt of crop oil concentrate per acre. With quizalofop, add 1-gal oil concentrate or 1 qt nonionic surfactant per 100-gal spray. Adding crop oil to Assure II or Poast may increase the likelihood of crop injury at high air temperatures. Do not apply Assure II or Poast on days that are unusually hot and humid. With quizalofop, do not apply within 60 days of harvest of dry Southern peas, or within 30 days of harvest of succulent Southern peas
	sethoxydim Poast 1.5 EC	1	1 to 1.5 pt	12	15, 60	With sethoxydim, do not apply within 15 days and 60 days of harvest succulent and dry peas, respectively.
	clethodim Arrow, Clethodim, Intensity, Select 2 EC Select Max, Intensity One 1 EC	1	6 to 8 oz 9 to 16 oz	24	30	Minimum 14 days between applications. Do not apply more than 2 applications per year. Do not exceed 16 fl oz per acre per year. Use of Poast or clethodim with crop oil may increase the likelihood of crop injury at high air. Temperatures and high humidity.

* Mode of action (MOA) code developed by the Weed Science Society of America with the Herbicide Resistance Action Committee (HRAC)
** REI – Restricted Entry Interval; PHI- Pre-Harvest Interval

TABLE 4-19. CHEMICAL WEED CONTROL IN PEPPERS

Weed	Herbicide and Formulation	MOA*	Amount of Formulation Per Acre	Intervals** REI (Hours)	Intervals** PHI (Days)	Precautions and Remarks
PEPPERS, Preplant						
Suppression or control of most annual grasses and broadleaf weeds, full rate required for nutsedge control	metam sodium Vapam HL 42%	N/A	37.5 to 75 gal	120	N/A	Rates are dependent on soil type and weeds present. For nutsedge control, apply 75-gal per acre. Apply when soil moisture is at field capacity (100 to 125%). Apply through soil injection using a rotary tiller or inject with knives no more than 4 in. apart; follow immediately with a roller to smooth and compact the soil surface or with mulch. May apply through drip irrigation prior to planting a second crop on mulch however adhere to label guidelines on crop plant-back interval. Plant back interval is often 14 to 21 days and can be 30 days in some environments. See label for all restrictions and additional information.
Contact kill of all green foliage, stale bed application	paraquat Parazone 3 SL Gramoxone 2 SL	22	1.3 to 2.7 pt 2 to 4 pt	24	N/A	Apply in a minimum of 10-gal of spray mix per acre to emerged weeds before transplanting as a broadcast or band treatment over a preformed row. Row should be formed several days ahead of planting and treating to allow maximum weed emergence. Use a nonionic surfactant at a rate of 16 to 32 oz per 100-gal spray mix or 1-gal approved crop oil concentrate per 100-gal spray mix. Paraquat product labels require applicators to take an EPA-approved training every 3 years to mix, load, and apply paraquat.
Most broadleaf weeds less than 4 in. tall or rosettes less than 3 in. in diameter; does not control grasses	carfentrazone-ethyl Aim 1.9 EW or 2 EC	14	Up to 2 oz	12	N/A	**Transplanted crop.** Apply no later than 1 day before transplanting crop. **Seeded crop.** Apply no later than 7 days before planting seeded crop. Use a nonionic surfactant or crop oil. See label for rate. Coverage of weed is essential for good weed control. Can be tank mixed with other registered burndown herbicides.

* Mode of action (MOA) code developed by the Weed Science Society of America with the Herbicide Resistance Action Committee (HRAC)
** REI – Restricted Entry Interval; PHI- Pre-Harvest Interval

TABLE 4-19. CHEMICAL WEED CONTROL IN PEPPERS (cont'd)

Weed	Herbicide and Formulation	MOA*	Amount of Formulation Per Acre	Intervals** REI (Hours)	Intervals** PHI (Days)	Precautions and Remarks
PEPPERS, Preplant (cont'd)						
Annual and perennial grass and broadleaf weeds, stale bed application	glyphosate 4SL 5.5SL 5.88SL	9	 16 to 32 fl oz 11 to 22 fl oz 10 to 21 fl oz	4	N/A	Apply to emerged weeds at least 3 days before seeding or transplanting. When applying Roundup before transplanting crops into plastic mulch, care must be taken to remove residues of this product from the plastic prior to transplanting. To prevent crop injury, residues can be removed by 0.5 in. rainfall or by applying water via a sprinkler system. Perennial weeds may require higher rates of glyphosate. Consult the manufacturer's label for specific weeds. Certain glyphosate formulations require the addition of a surfactant. Adding nonionic surfactant to glyphosate formulated with nonionic surfactant may result in reduced weed control.
Emerged broadleaf weeds and some annual grasses	caprylic acid + capric acid Homeplate, Suppress	26	 3 to 9% v/v	24	0	May be applied prior to planting as a burndown treatment for emerged weeds, as a preemergence application after seeding but before emergence, as a directed or shielded application between rows or as a postharvest application. Use higher spray volumes for high weed density and weeds larger than 5 in. Coverage is important for acceptable weed control. May be tank mixed with other herbicides. See label for further instructions.
Broadleaf weeds including Carolina geranium and cutleaf evening primrose and a few annual grasses	oxyfluorfen Galigan 2E, Goal 2XL GoalTender 4 F	14	 up to 2 pt up to 1 pt	24	45	**Plasticulture only.** Apply to soil surface of pre-formed beds at least 30 days prior to transplanting crop. While incorporation is not necessary, it may result in less crop injury. Plastic mulch can be applied any time after application, but best results are likely if applied soon after application. Label is state restricted- FL, GA, NC, SC, and VA only.
Palmer amaranth, redroot pigweed, smooth pigweed, *Galinsoga* spp., black nightshade, Eastern black nightshade, common purslane, partial control of yellow nutsedge	fomesafen Reflex 2 EC	14	 1 to 1.5 pt	24	60	**This is a Section 24(c) Special Local Need Label** for transplanted pepper. Growers must obtain the label prior to applications. See label for further instructions. **Plasticulture In-row Application for Transplanted Pepper.** Apply after final bed formation and the drip tape is laid but prior to laying plastic mulch. Avoid soil disturbance after application. Unless restricted by other products such as fumigants, pepper may be transplanted immediately following the application of Reflex and the application of the mulch. **Bareground for Transplanted Pepper.** Apply pretransplant up to 7 days prior to transplanting pepper. Weed control will be reduced if soil is disturbed after application. During the transplanting operation, make sure the soil in the transplant hole settles flush or above the surrounding soil surface. Avoid cultural practices that may concentrate Reflex-treated soil around the transplant root ball. An overhead irrigation or rainfall event between Reflex herbicide application and transplanting will ensure herbicide activation and will likely reduce the potential for crop injury due to splashing. **Plasticulture Row Middle Application.** Apply to row middles with a hooded or shielded sprayer. Avoid drift of herbicide on mulch. If drift occurs, 0.5 in. of rain or irrigation must occur prior to transplanting. Carryover is a large concern; see label for more information.
PEPPERS, Preplant and Preemergence						
Annual grasses and small-seeded broadleaf	clomazone Command 3 ME	13	 0.67 to 2.7 pt	12	N/A	**Not labeled for banana pepper.** Apply preplant before transplanting. Weak on pigweed. See label for instructions on use.
	napropamide Devrinol 50 DF Devrinol 2 EC	15	 2 to 4 lb 2 to 4 qt	24	N/A	**Bareground:** Apply preplant and incorporate into the soil 1 to 2 in. as soon as possible with a rototiller or tandem disk. Can be used on direct-seeded or transplanted peppers. See label for instructions on use. **Plasticulture:** Apply to a weed-free soil before laying plastic mulch. Soil should be well worked yet moist enough to permit a thorough incorporation to a depth of 2 in. Mechanically incorporate or irrigate within 24 hours after application. If weed pressure is from small-seeded annuals, apply to the surface of the bed immediately in front of the laying of plastic mulch. If soil is dry, water or sprinkle irrigate with sufficient water to wet to a depth of 2 to 4 in. before covering with plastic mulch. **Between rows:** Apply to a weed free soil surface between the rows (bare ground or plastic mulch). Mechanically incorporate or irrigate Devrinol into the soil to a depth of 1 to 2 in. within 24 hours of application. See XT labels for information regarding delay in irrigation event.
	pendimethalin Prowl H2O 3.8 AS	3	 1 to 3 pt	24	70	May be applied in chili pepper, cooking pepper, pimento, jalapeno, and sweet pepper. Do not apply more than 3 pt per acre per season. See label for specific use rate for your soil type. Avoid direct contact with pepper foliage or stems. See label for further instructions and precautions. **Between rows.** Can be applied as a post-directed spray on the soil at the base of the plant beneath plants and between rows. **In-row.** May be applied as a broadcast preplant incorporated surface application prior to transplanting peppers.

* Mode of action (MOA) code developed by the Weed Science Society of America with the Herbicide Resistance Action Committee (HRAC)
** REI – Restricted Entry Interval; PHI- Pre-Harvest Interval

TABLE 4-19. CHEMICAL WEED CONTROL IN PEPPERS (cont'd)

Weed	Herbicide and Formulation	MOA*	Amount of Formulation Per Acre	Intervals** REI (Hours)	Intervals** PHI (Days)	Precautions and Remarks
PEPPERS, Preplant and Preemergence (cont'd)						
Annual grasses and small-seeded broadleaf (cont'd)	trifluralin Treflan 4 L, Treflan HFP 4 EC Treflan 10 G	3	 1 to 2 pt 5 to 10 lb	12	N/A	Apply pre-transplant and incorporate to a depth of 2 to 3 in. within 8 hr with a rototiller or tandem disk. Use lower rates on coarse soil types or soils with low organic matter.
	bensulide Prefar 4 EC	15	 5 to 6 qt	12	N/A	Apply preplant incorporated (1 in. incorporation is optimum) or preemergence. With preemergence application, irrigate immediately after application. See label for directions.
Annual grass and broadleaf weeds, yellow nutsedge suppression	S-metolachlor Dual Magnum 7.62 EC	15	 8 to 16 oz	24	60	**Bell pepper transplants only. This is a Section 24(c) Special Local Need Label.** May be applied to formed beds, prior to laying plastic. May be applied 12 oz per acre overtop of bell pepper between 1 and 3 weeks after planting. Does not control emerged weeds. Limited data available for NC. May be applied as a row middle application. **Bareground Production:** Apply conservative rate overtop of transplants 2 days after both transplanting and sealing soil around root ball. Dual also may be applied pre-transplant without disturbance but some stunting is expected.
PEPPERS, Postemergence						
Broadleaf, grass (suppression only), and yellow nutsedge	imazosulfuron League 75 DF	2	 4 to 6.4 oz	12	21	**Pepper (Bell and non-bell).** Apply to pepper plants that are well established and at least 10 in. tall. Apply directed to the base of the plants stem, no higher than 2 in. from the soil surface and do not contact fruit. Consult label for approved surfactants and crop rotation restrictions.
Annual and perennial grasses only	clethodim Arrow, Clethodim, Intensity, Select 2 EC Select Max, Intensity One 1 EC	1	 6 to 8 oz 9 to 16 oz	24	20	Apply postemergence to control grasses. See label for adjuvant and rate. Adding crop oil may increase the likelihood of crop injury at high air temperatures and high humidity. Highly effective in controlling annual blue grass. Apply to actively growing grasses not under drought stress.
	sethoxydim Poast 1.5 EC	1	 1 to 1.5 pt	12	7	Apply to emerged grasses. Consult manufacturer's label for specific rates and best times to treat and adjuvant and rate. Adding crop oil to Poast may increase the likelihood of crop injury at high air temperatures and high humidity.
PEPPERS, Row Middles						
Most broadleaf weeds less than 4 in. tall or rosettes less than 3 in. in diameter; does not control grasses	carfentrazone-ethyl Aim 1.9 EW or 2 EC	14	 Up to 2 oz	12	0	Apply post-directed using hooded sprayer for control of emerged weeds. If crop is contacted, burning of contacted area will occur. Use a nonionic surfactant or crop oil with Aim. See label for rate. Coverage is essential for good weed control. Can be tank mixed with other registered herbicides.
Most emerged weeds	glyphosate 4SL 5.5SL 5.88SL	9	 16 to 32 fl oz 11 to 22 fl oz 10 to 21 fl oz	4	14	Apply as a hooded spray in row middles, as shielded spray in row middles, as wiper applications in row middles, or postharvest. To avoid severe injury to crop, do not allow herbicide to contact foliage, green shoots, stems, exposed, roots, or fruit of crop.
Yellow and purple nutsedge and broadleaf weeds	halosulfuron-methyl Profine 75, Sandea 75 DF	2	 0.5 to 1 oz	12	30	Apply to row middles as a postemergence spray. In plasticulture, do not allow spray to contact plastic. Early season application will give postemergence and preemergence control. For postemergence applications, use nonionic surfactant at 1 qt per 100-gal of spray solution.
Broadleaf, grass (suppression only), and yellow nutsedge	imazosulfuron League 75 DF	2	 4 to 6.4 oz	12	21	**Pepper (Bell and non-bell).** Apply to pepper plants that are well established and at least 10 in. tall. Avoid contact with the pepper crop. When application is being made to peppers grown in plastic mulch culture, equipment must be adjusted to prevent the spray from contacting the plastic. Consult label for approved surfactants and crop rotation restrictions.
Contact kill of all green foliage	paraquat Parazone 3 SL Gramoxone 2 SL	22	 1.3 pt 2 pt	24	N/A	Apply in a minimum of 20-gal spray mix per acre as a shielded spray to emerged weeds between rows of peppers. Use a nonionic surfactant at a rate of 16 oz per 100-gal spray mix. Do not apply more than 3 applications per season. Paraquat product labels require applicators to take an EPA-approved training every 3 years to mix, load, and apply paraquat.

* Mode of action (MOA) code developed by the Weed Science Society of America with the Herbicide Resistance Action Committee (HRAC)
** REI – Restricted Entry Interval; PHI- Pre-Harvest Interval

TABLE 4-20. CHEMICAL WEED CONTROL IN POTATOES, IRISH

Weed	Herbicide and Formulation	MOA*	Amount of Formulation Per Acre	Intervals** REI (Hours)	Intervals** PHI (Days)	Precautions and Remarks
POTATOES, IRISH, Preplant and Preemergence						
Contact kill of all green foliage, stale bed application	paraquat Parazone 3 SL Gramoxone 2 SL	22	 1.3 pt 2 pt	24	N/A	Apply in a minimum of 20-gal spray mix per acre to emerged weeds up to ground cracking before crop emergence. May be used instead of the drag-off operation to kill emerged weeds before the application of preemergence herbicides. Use a nonionic surfactant at a rate of 16 to 32 oz per 100-gal spray mix or 1-gal approved crop oil concentrate per 100-gal spray mix. Paraquat product labels require applicators to take an EPA-approved training every 3 years to mix, load, and apply paraquat.
Most broadleaf weeds less than 4 in. tall or rosettes less than 3 in. in diameter; does not control grasses	carfentrazone-ethyl Aim 1.9 EW or 2 EC	14	 Up to 2 oz	12	7	Apply prior to planting, or up to 1 day after planting crop. Use a nonionic surfactant or crop oil with Aim. See label for rate. Coverage of weed is essential for good weed control. Can be tank mixed with other registered burndown herbicides.
Emerged broadleaf weeds and some annual grasses	caprylic acid + capric acid Homeplate, Suppress	26	 3 to 9% v/v	24	0	May be applied prior to planting as a burndown treatment for emerged weeds, as a preemergence application after seeding but before emergence, as a directed or shielded application between rows, as a harvest aid and desiccant, or as a postharvest application. Use higher spray volumes for high weed density and weeds larger than 5 in. Coverage is important for acceptable weed control. May be tank mixed with other herbicides. OMRI approved material. See label for further instructions.
Annual and perennial grass and broadleaf weeds, stale bed application	glyphosate 4 SL 5.5 SL 5.88 SL	9	 16 to 32 fl oz 11 to 22 fl oz 10 to 21 fl oz	4	N/A	Apply to emerged weeds before crop emergence. Do not feed crop residue to livestock for 8 weeks following treatment. Perennial weeds may require higher rates of glyphosate. Consult the manufacturer's label for rates for specific weeds. Certain glyphosate formulations require the addition of a surfactant. Adding nonionic surfactant to glyphosate formulated with nonionic surfactant may result in reduced weed control.
Annual grasses and small-seeded broadleaf weeds	pendimethalin Prowl H2O 3.8 AS Prowl 3.3 EC	3	 1.5 to 3.0 pt 1.8 to 3.6 pt	24	N/A	Apply just after planting or drag-off to weed-free soil before crop emerges or from emergence until crop reaches 6 in. tall.
Annual grasses and small-seeded broadleaf weeds, plus, yellow nutsedge suppression	S-metolachlor Brawl, Dual Magnum, Medal 7.62 EC Brawl II, Dual II Magnum, Medal II 7.64 EC	15	 1 to 2 pt 1 to 2 pt	24	40, 60	Apply just after planting or drag-off to weed-free soil before crop emerges. Dual Magnum can also be applied at lay-by for control of late season weeds. Do not harvest within 60 days after preplant applications or within 40 days after a lay-by application. See label for further instruction.
Annual broadleaf weeds and some grasses	dimethenamid-P Outlook 6 EC	15	 12 to 21 oz	12	40	Apply just after planting or drag-off to weed-free soil before crop emerges. See label for further instructions.
Annual grasses, most broadleaf weeds, plus yellow and purple nutsedge suppression	EPTC Eptam 7 EC	15	 3.5 to 9.0 pt	12	30	Apply preplant and incorporate into the soil 2 to 3 in. with a rototiller or tandem disk. The variety Superior has been shown to be sensitive to Eptam. See label for specific methods of incorporation. For late season preemergence nutsedge control, apply and incorporate as a directed spray to the soil on both sides of the crop row. See label for more detail.
Most annual broadleaf weeds and some annual grasses	linuron Lorox DF 50 WDG Linex 4L	7	 1.5 to 3 lb 1.5 to 3 pt	24	N/A	Apply just after planting or drag-off or hilling but before crop emerges. If emerged weeds are present, add 1 pt surfactant for each 25-gal spray mixture. Weeds may be up to 3 in. tall at time of application.
	metribuzin TriCor DF & other trade 75 WDG	5	 0.33 to 0.67 lb	12	60	Apply just after planting or drag-off but before crop emerges. Weeds may be emerged at time of application. On sand soils or sensitive varieties, do not exceed 0.67 lb per acre. See label for list of sensitive varieties.
	rimsulfuron Matrix, Pruvin 25 WDG	2	 1 to 1.5 oz	4	60	Apply after drag-off or hilling but before potatoes and weeds emerge. If emerged weeds are present, add surfactant. See label for rate. Can be tank mixed with Eptam, Prowl, Sencor, Lorox, or Dual Magnum. See label for further instructions.
Broadleaf, grass and nutsedge weeds	fomesafen Reflex 2 EC	14	 1 pt	24	70	Apply preemergence after planting but prior to potato emergence. Do not apply as a preplant incorporated application or to emerged potato or severe crop injury may occur. Do not exceed rate of 1 pt per acre per season.
Broadleaf, grass (suppression) and yellow nutsedge	imazosulfuron League 75 DF	2	 4 to 6.4 oz	12	45	Apply as a preemergence application after crop has been planted but prior to emergence or immediately after hilling. Postemergence application (3.2 to 4 oz per acre) may be made after crop has emerged if weeds are less than 3 in. in height. Do not apply more than 6.4 oz per acre per season. Consult label for sequential application program and crop rotation restrictions.

* Mode of action (MOA) code developed by the Weed Science Society of America with the Herbicide Resistance Action Committee (HRAC)
** REI – Restricted Entry Interval; PHI- Pre-Harvest Interval

TABLE 4-20. CHEMICAL WEED CONTROL IN POTATOES, IRISH (cont'd)

Weed	Herbicide and Formulation	MOA*	Amount of Formulation Per Acre	REI (Hours)	PHI (Days)	Precautions and Remarks
POTATOES, IRISH, Postemergence						
Most annual broadleaf weeds and some annual grasses	**metribuzin** TriCor DF & other trade 75 WDG	5	0.33 to 0.67 lb	12	60	Do not use on early maturing smooth-skinned white or red-skinned varieties. Apply only if there have been at least three successive days of sunny weather before application. Treat before weeds are 1 in. tall. Treatment may cause some chlorosis or minor necrosis.
	rimsulfuron Matrix, Pruvin 25 WDG	2	1 to 1.5 oz	4	60	Apply to young actively growing weeds after crop emergence. More effective on small weeds. Add nonionic surfactant at 1 to 2 pt per 100-gal water. Can be tank mixed with Eptam or Sencor or some foliar fungicides. See label for further instructions.
Most broadleaf weeds less than 4 in. tall or rosettes less than 3 in. in diameter; does not control grasses	**carfentrazone-ethyl** Aim 1.9 EW or 2 EC	14	Up to 2 oz	12	7	Apply post-directed using hooded sprayers for control of emerged weeds. If crop is contacted, burning of contacted area will occur. Use a nonionic surfactant or crop oil with Aim. See label for rate. Coverage of weed is essential for good weed control. Can be tank mixed with other registered herbicides.
Most emerged weeds	**glyphosate** 4SL 5.5SL 5.88SL	9	16 to 32 fl oz 11 to 22 fl oz 10 to 21 fl oz	4	14	**Row middles only.** Apply as a hooded spray in row middles, as shielded spray in row middles, as wiper applications in row middles, or postharvest. To avoid severe injury to crop, do not allow herbicide to contact foliage, green shoots, stems, exposed, roots, or fruit of crop.
Annual and perennial grasses only	**clethodim** Arrow, Clethodim, Intensity, Select 2 EC Select Max, Intensity One 1 EC	1	6 to 16 oz 9 to 32 oz	24	30	Apply postemergence for control of grasses. See label for adjuvant and rate. Adding crop oil may increase the likelihood of crop injury at high air temperatures and high humidity. Highly effective in controlling annual bluegrass. Apply to actively growing grasses not under drought stress.
	sethoxydim Poast 1.5 EC	1	1 to 1.5 pt	12	30	Apply to emerged grasses. Consult manufacturer's label for specific rates and best times to treat and adjuvant and rate. Adding crop oil to Poast may increase the likelihood of crop injury at high air temperatures and high humidity.
Annual grasses and small-seeded broadleaf weeds, plus, yellow nutsedge suppression	**S-metolachlor** Brawl, Dual Magnum, Medal 7.62 EC Brawl II, Dual II Magnum, Medal II 7.64 EC	15	1 to 2 pt 1.67 pt	24	40	Apply as interrow or interhill application. Leave a 1 ft. untreated area over the seeded row (6 in. on either side of the row). **Application made as a broadcast spray over the planted row or hill or directly to crop foliage will increase the risk of injury to the crop.** Apply before weeds emerge. See label for further instructions.

* Mode of action (MOA) code developed by the Weed Science Society of America with the Herbicide Resistance Action Committee (HRAC)
** REI – Restricted Entry Interval; PHI- Pre-Harvest Interval

TABLE 4-21. CHEMICAL WEED CONTROL IN PUMPKINS

Weed	Herbicide and Formulation	MOA*	Amount of Formulation Per Acre	REI (Hours)	PHI (Days)	Precautions and Remarks
PUMPKINS, Preplant						
Contact kill of all green foliage, stale bed application	**paraquat** Parazone 3 SL Gramoxone 2 SL	22	1.3 to 2.7 pt 2 to 4 pt	24	N/A	Apply in a minimum of 20-gal spray mix per acre to emerged weeds before crop emergence or transplanting as a band or broadcast treatment over a preformed row. Use sufficient water to give thorough coverage. Row should be formed several days ahead of planting or treating to allow maximum weed emergence. Use a nonionic surfactant at a rate of 16 to 32 oz per 100-gal spray solution or 1-gal approved crop oil concentrate per 100-gal spray mix. Paraquat product labels require applicators to take an EPA-approved training every 3 years to mix, load, and apply paraquat.
Most broadleaf weeds less than 4 in. tall or rosettes less than 3 in. in diameter; does not control grasses	**carfentrazone-ethyl** Aim 1.9 EW or 2 EC	14	Up to 2 oz	12	N/A	**Not registered for use on seeded crop.** Apply prior to transplanting crop. Use a nonionic surfactant or crop oil with Aim. See label for rate. Coverage is essential for good weed control. Can be tank mixed with other registered burndown herbicides.
Annual and perennial grass and broadleaf weeds, stale bed application	**glyphosate** 4SL 5.5SL 5.88SL	9	16 to 32 fl oz 11 to 22 fl oz 10 to 21 fl oz	4	N/A	Apply to emerged weeds at least 3 days before seeding or transplanting. Perennial weeds may require higher rates of glyphosate. Consult the manufacturer's label for rates for specific weeds. Certain glyphosate formulations require the addition of a surfactant. Adding nonionic surfactant to glyphosate formulated with nonionic surfactant may result in reduced weed control.

* Mode of action (MOA) code developed by the Weed Science Society of America with the Herbicide Resistance Action Committee (HRAC)
** REI – Restricted Entry Interval; PHI- Pre-Harvest Interval

TABLE 4-21. CHEMICAL WEED CONTROL IN PUMPKINS (cont'd)

Weed	Herbicide and Formulation	MOA*	Amount of Formulation Per Acre	REI (Hours)	PHI (Days)	Precautions and Remarks
PUMPKINS, Preplant (cont'd)						
Palmer amaranth, redroot pigweed, smooth pigweed, *Galinsoga* spp., black nightshade, Eastern black nightshade, common purslane, partial control of yellow nutsedge	fomesafen Reflex 2 EC	14	8 to 10 oz	24	32	**This is a Section 24(c) Local Need Label** and supplemental label must be obtained prior to this use. **Bare Ground Seeded:** Apply Reflex Herbicide at 8 to 10 fl oz/A within 24 hours after planting. **Bareground transplants.** Prepare land for planting; apply Reflex; lightly irrigate to activate herbicide and move it into soil; and then prepare plant holes and plant.
PUMPKINS, Preplant and Preemergence						
Annual grasses and some small-seeded broadleaf weeds	bensulide Prefar 4 EC	15	5 to 6 qt	12	N/A	Registered for cucurbit vegetable group (Crop grouping 9). Apply preplant and incorporate into the soil 1 to 2 in. (1 in. incorporation is optimum) with a rototiller or tandem disk or apply to the soil surface after seeding and follow with irrigation. Check re-plant restrictions for small grains on label.
	ethalfluralin Curbit 3 EC	3	3 to 4.5 pt	24	N/A	Apply postplanting, to seeded crop prior to crop emergence, or as a banded spray between rows after crop emergence or transplanting. See label for timing. Shallow cultivation, irrigation, or rainfall within 5 days is needed for good weed control. Do not use under mulches, row covers, or hot caps. Under conditions of unusually cold or wet soil and air temperatures, crop stunting or injury may occur. Crop injury can occur if seeding depth is too shallow.
	ethalfluralin + clomazone Strategy 2.1 L	3 + 13	2 to 6 pt	24	N/A	Apply to the soil surface immediately after crop seeding for preemergence control of weeds. **D NOT APPLY PRIOR TO PLANTING THE CROP. DO NOT SOIL INCORPORATE. Do not use under mulch.** May also be used as a **banded** treatment **between** rows after crop emergence or transplanting.
Yellow and purple nutsedge suppression, non-ALS resistant pigweed, wild radish, and ragweed	halosulfuron-methyl Profine 75, Sandea 75 DF	2	0.5 to 1 oz	12	30	**Direct-seeded pumpkin or winter squash.** Apply after seeding but prior to soil cracking. **Transplanted pumpkin and winter squash.** Apply 7 days prior to transplanting. See label for specific rate. See label for crop rotational restrictions and other information. **Post-transplant in pumpkin and winter squash.** Can be applied as an over-the-top application, a directed spray application, or with crop shields. Apply to transplants that are established, actively growing and in the 3- to 5- true leaf stage or no sooner than 14 days after transplanting unless local conditions demonstrate safety at an earlier interval, but before first female flowers appear. Row middle/furrow applications in direct-seeded and transplant pumpkin and winter squash. Apply between rows of direct-seeded or transplanted crop while avoiding contact of the herbicide with the planted crop. If plastic is used on the planted row, adjust equipment to keep the application off the plastic. Rate can be increased to 1 oz per acre if needed for row middle/furrow applications.
Annual broadleaf, grass and yellow nutsedge	S-metolachlor Brawl, Dual Magnum, Medal 7.62 EC Brawl II, Dual II Magnum 7.64 EC	15	1 to 1.33 pt 1 to 1.33 pt	24	30	Apply as interrow or interhill application. Leave a 1-foot untreated area over the seeded row (6 in. on either side of the row). Application made as a broadcast spray over the planted row or hill or directly to crop foliage will increase the risk of injury to the crop. Apply before weeds emerge. See label for further instructions.
PUMPKINS, Postemergence						
Yellow and purple nutsedge suppression, non-ALS resistant pigweed, wild radish, and ragweed	halosulfuron-methyl Profine 75, Sandea 75 DF	2	0.5 to 1 oz	12	30	**Direct-seeded pumpkin and winter squash.** Apply after crop has reached the 2 to 5 true leaf stage, preferably 4 to 5 true leaves, but before first female flowers appear. **Post-transplant in pumpkin and winter squash.** Can be applied as an over-the-top application, a directed spray application, or with crop shields. Apply to transplants that are established, actively growing and in the 3 to 5 true leaf stages or no sooner than 14 days after transplanting unless local conditions demonstrate safety at an earlier interval, but before first female flowers appear. **Row middle/furrow applications in direct-seeded and transplant pumpkin or winter squash.** Apply between rows of direct-seeded or transplanted crop while avoiding contact of the herbicide with the planted crop. If plastic is used on the planted row, adjust equipment to keep the application off the plastic. Rate can be increased to 1 oz per acre if needed for row middle/ furrow applications.
PUMPKINS, Row Middles						
Annual and perennial grasses only	clethodim Arrow, Clethodim, Intensity, Select 2 EC Select Max, Intensity One 1 EC	1	6 to 8 oz 9 to 16 oz	24	14	Apply postemergence for control of grasses. See label for adjuvant and rate. Adding crop oil concentrate may increase the likelihood of crop injury at high air temperatures and high humidity. Highly effective in controlling annual bluegrass. Apply to actively growing grasses not under drought stress.

* Mode of action (MOA) code developed by the Weed Science Society of America with the Herbicide Resistance Action Committee (HRAC)

** REI – Restricted Entry Interval; PHI- Pre-Harvest Interval

TABLE 4-21. CHEMICAL WEED CONTROL IN PUMPKINS (cont'd)

Weed	Herbicide and Formulation	MOA*	Amount of Formulation Per Acre	REI (Hours)	PHI (Days)	Precautions and Remarks
PUMPKINS, Row Middles (cont'd)						
	sethoxydim Poast 1.5 EC	1	1 to 1.5 pt	12	14	Apply to emerged grasses. Consult manufacturer's label for specific rates and best times to treat and adjuvant and rate. Crop oil may increase the likelihood of crop injury at high temperatures and high humidity.
Annual grasses and some small-seeded broadleaf weeds	trifluralin Treflan 4 L, Treflan HFP 4 EC Treflan 10 G	3	1 to 2 pt 5 to 10 lb	12	30	**Row middles only.** To improve preemergence control of late emerging weeds. Apply after emergence when crop plants have reached the three to four true leaf stage of growth. Apply as a directed spray to soil between the rows. Avoid contacting foliage as slight crop injury may occur. Set incorporation equipment to move treated soil around base of crop plants.
Most broadleaf weeds less than 4 in. tall or rosettes less than 3 in. in diameter; does not control grasses	carfentrazone-ethyl Aim 1.9 EW or 2 EC	14	Up to 2 oz	12	0	Apply post-directed using hooded sprayers for control of emerged weeds. If crop is contacted, burning of contacted area will occur. Use a nonionic surfactant or crop oil with Aim. See label for rate. Coverage is essential for good weed control. Can be tank mixed with other registered herbicides.
Most emerged weeds	glyphosate 4SL 5.5SL 5.88SL	9	16 to 32 fl oz 11 to 22 fl oz 10 to 21 fl oz	4	14	**Row middles only.** Apply as a hooded spray in row middles, as shielded spray in row middles, as wiper applications in row middles, or postharvest. To avoid severe injury to crop, do not allow herbicide to contact foliage, green shoots, stems, exposed, roots, or fruit of crop.
Yellow and purple nutsedge and broadleaf weeds	halosulfuron-methyl Profine 75, Sandea 75 DF	2	0.5 to 1 oz	12	30	**Row middles only.** Apply to row middles as a postemergence spray. In plasticulture, do not allow spray to contact plastic. Early season application will give postemergence and preemergence control. For postemergence applications, use nonionic surfactant at 1 qt per 100-gal of spray solution.

* Mode of action (MOA) code developed by the Weed Science Society of America with the Herbicide Resistance Action Committee (HRAC)
** REI – Restricted Entry Interval; PHI- Pre-Harvest Interval

TABLE 4-22. CHEMICAL WEED CONTROL IN RADISH

Weed	Herbicide and Formulation	MOA*	Amount of Formulation Per Acre	REI (Hours)	PHI (Days)	Precautions and Remarks
RADISH, Preplant						
Annual and perennial grass and broadleaf weeds	glyphosate 4SL 5.5SL 5.88SL	9	16 to 32 fl oz 11 to 22 fl oz 10 to 21 fl oz	4	N/A	Apply to emerged weeds before planting. Perennial weeds may require higher rates of glyphosate. Consult the manufacturer's label for rates for specific weeds. Certain glyphosate formulations may require addition of a surfactant. Adding nonionic surfactant to glyphosate formulated with non-ionic surfactant may result in reduced weed control.
Emerged broadleaf weeds and some annual grasses	caprylic acid + capric acid Homeplate, Suppress	26	3 to 9% v/v	24	0	May be applied prior to planting as a burndown treatment for emerged weeds, as a preemergence application after seeding but before emergence, as a directed or shielded application between rows, as a harvest aid or desiccant, or as a post-harvest application. Use higher spray volumes for high weed density and weeds larger than 5 in. Coverage is important for acceptable weed control. May be tank mixed with other herbicides. See label for further instructions.
Annual grasses and broadleaf weeds	trifluralin Treflan 4 L, Treflan HFP 4 EC	3	1 to 1.5 pt	12	N/A	Apply preplant and incorporate immediately after application for preemergence weed control. Low rate should be used on coarse-textured soil.
RADISH, Postemergence						
Most broadleaf weeds less than 4 in. tall or rosettes less than 3 in. in diameter; does not control grasses	carfentrazone-ethyl Aim 1.9 EW or 2 EC	14	Up to 2 oz	12	0	Apply post-directed using hooded sprayers for control of emerged weeds. If crop is contacted, burning of contacted area will occur. Use a nonionic surfactant or crop oil with Aim. See label for rate. Coverage is essential for good weed control. Can be tank mixed with other registered herbicides.
Annual and perennial grasses	clethodim Arrow, Clethodim, Intensity, Select 2 EC Select Max, Intensity One 1 EC	1	6 to 8 oz 9 to 16 oz	24	15	Apply postemergence to emerged grasses. See label for rates for specific grasses. See label for adjuvant and rate.

* Mode of action (MOA) code developed by the Weed Science Society of America with the Herbicide Resistance Action Committee (HRAC)
** REI – Restricted Entry Interval; PHI- Pre-Harvest Interval

TABLE 4-23. CHEMICAL WEED CONTROL IN SPINACH

Weed	Herbicide and Formulation	MOA*	Amount of Formulation Per Acre	Intervals** REI (Hours)	Intervals** PHI (Days)	Precautions and Remarks
SPINACH, Preemergence						
Annual and perennial grass and broadleaf weeds, stale bed application	glyphosate 4SL 5.5SL 5.88SL	9	 16 to 32 fl oz 11 to 22 fl oz 10 to 21 fl oz	4	N/A	Apply to emerged weeds before crop emergence. Do not feed residue to livestock for 8 weeks. Perennial weeds may require higher rates of glyphosate. Consult the manufacturer's label for rates for specific weeds. Certain glyphosate formulations require the addition of a surfactant. Adding nonionic surfactant to glyphosate formulated with nonionic surfactant may result in reduced weed control.
Annual grasses (crabgrass spp., foxtail spp., barnyardgrass, annual ryegrass, annual bluegrass) and broadleaf weeds (*Lamium* spp., lambsquarters, common purslane, redroot pigweed, shepherds' purse)	cycloate Ro-Neet 6 EC	15	 2 qt	48	N/A	Use on sandy mineral soils only. Read label for further instructions.
Annual grass and broadleaf weeds, Palmer amaranth, yellow nutsedge suppression	S-metolachlor Dual Magnum 7.62 EC	15	 0.3 to 0.67 pt	24	50	**This is a section 24(c) Special Local Need Label.** Growers must obtain label prior to application. Do not apply preplant. Do not incorporate after application. Injury potential is greatest when applied to sands or loamy sands especially if a heavy rainfall event occurs following application. See label for further information.
SPINACH, Postemergence						
Broadleaf weeds including sowthistle clover, cocklebur, jimsonweed, and ragweed	clopyralid Stinger 3 EC	4	 0.25 to 0.33 pt	12	21	Apply to spinach in the 2- to 5-leaf stage when weeds are small and actively growing. Will control most legumes. See label for more precautions.
Broadleaf weeds	phenmedipham Spin-aid 1.3 EC	5	 3 to 6 pt	12	21	**For processing spinach only.** Do not use when expected high temperatures are expected to be above 75°F. For best results, spray when weeds are in the two-leaf stage. Use the 6 pt rate only on well-established crops that are not under stress. Spinach plants must have more than six true leaves.
Most broadleaf weeds less than 4 in. tall or rosettes less than 3 in. in diameter; does not control grasses	carfentrazone-ethyl Aim 1.9 EW or 2 EC	14	 Up to 2 oz	12	0	Apply post-directed using hooded sprayers for control of emerged weeds. If crop is contacted, burning of contacted area will occur. Use a nonionic surfactant or crop oil with Aim. See label for rate. Coverage is essential for good weed control. Can be tank mixed with other registered herbicides.
Most emerged weeds	glyphosate 4SL 5.5SL 5.88SL	9	 16 to 32 fl oz 11 to 22 fl oz 10 to 21 fl oz	4	14	**Row middles only.** Apply as a hooded spray in row middles, as shielded spray in row middles, as wiper applications in row middles, or postharvest. To avoid severe injury to crop, do not allow herbicide to contact foliage, green shoots, stems, exposed roots, or fruit of crop.
Annual and perennial grasses only	sethoxydim Poast 1.5 EC clethodim Arrow, Clethodim, Intensity, Select 2 EC Select Max, Intensity One 1 EC	1 1	 1 to 1.5 pt 6 to 8 oz 9 to 16 oz	12 24	14 14	Apply to emerged grasses. Consult manufacturer's label for specific rates and best times to treat. See label for adjuvant and rate. Adding crop oil to Poast or Select may increase the likelihood of crop injury at high air temperatures and high humidity.

* Mode of action (MOA) code developed by the Weed Science Society of America with the Herbicide Resistance Action Committee (HRAC)

** REI – Restricted Entry Interval; PHI- Pre-Harvest Interval

TABLE 4-24. CHEMICAL WEED CONTROL IN SQUASH

Weed	Herbicide and Formulation	MOA*	Amount of Formulation Per Acre	Intervals** REI (Hours)	Intervals** PHI (Days)	Precautions and Remarks
SQUASH, Preplant						
Suppression or control of most annual grasses and broadleaf weeds, full rate required for nutsedge control	metam sodium Vapam HL 42%	N/A	37.5 to 75 gal	120	N/A	Rates are dependent on soil type and weeds present. For nutsedge control, use 75-gal per acre. Apply when soil moisture is at field capacity (100 to 125%). Apply through soil injection using a rotary tiller or inject with knives no more than 4 in. apart; follow immediately with a roller to smooth and compact the soil surface or with mulch. May apply through drip irrigation prior to planting a second crop on mulch. Plant back interval is often 14 to 21 days and can be 30 days in some environments. See label for all restrictions and additional information.
Emerged broadleaf and grass weeds	pelargonic acid Scythe 4.2 EC	26	1 to 10% v/v	12	N/A	Apply before crop emergence to control emerged weeds. There is no residual activity. May be tank mixed with soil residual herbicides. See label for more instructions. May also be used as a banded spray between row middles. Use a shielded sprayer directed to the row middles to reduce drift to the crop.
Contact kill of all green foliage, stale bed application	paraquat Parazone 3 SL Gramoxone 2 SL	22	1.3 to 2.7 pt 2 to 4 pt	24	N/A	Apply in a minimum of 10-gal spray mix per acre to emerged weeds before transplanting or crop emergence as a band or broadcast treatment over a preformed row. Use sufficient water to give thorough coverage. Row should be formed several days ahead of planting or treating to allow maximum weed emergence. Use a nonionic surfactant at a rate of 16 to 32 oz per 100-gal spray mix or 1-gal approved crop oil concentrate per 100-gal spray mix. Paraquat product labels require applicators to take an EPA-approved training every 3 years to mix, load, and apply paraquat.
Most broadleaf weeds less than 4 in. tall or rosettes less than 3 in. in diameter; does not control grasses	carfentrazone-ethyl Aim 1.9 EW or 2 EC	14	Up to 2 oz	12	0	Not registered for seeded crop. Apply prior to transplanting crop. Use a nonionic surfactant or crop oil with Aim. See label for rate. Coverage is essential for good weed control. Can tank mixed with other registered burndown herbicides.
Annual and perennial grass and broadleaf weeds, stale bed application	glyphosate 4SL 5.5SL 5.88SL	9	16 to 32 fl oz 11 to 22 fl oz 10 to 21 fl oz	4	N/A	Apply to emerged weeds at least 3 days before seeding or transplanting. When applying Roundup before transplanting crops into plastic mulch, care must be taken to remove residues of this product from the plastic prior to transplanting. To prevent crop injury, residues can be removed by 0.5 in. rainfall or by applying water via a sprinkler system. Perennial weeds may require higher rates of glyphosate. Consult the manufacturer's label for rates for specific weeds. Certain glyphosate formulations require the addition of a surfactant. Adding nonionic surfactant to glyphosate formulated with nonionic surfactant may result in reduced weed control.
Emerged broadleaf weeds and some annual grasses	caprylic acid + capric acid Homeplate, Suppress	26	3 to 9% v/v	24	0	May be applied prior to planting as a burndown treatment for emerged weeds, as a preemergence application after seeding but before emergence, as a directed or shielded application between rows or as a post-harvest application. Use higher spray volumes for high weed density and weeds larger than 5 in. Coverage is important for acceptable weed control. May be tank mixed with other herbicides. See label for further instructions.
Annual grasses and small-seeded broadleaf weeds	bensulide Prefar 4 EC	15	5 to 6 qt	12	N/A	Registered for cucurbit vegetable group (Crop grouping 9). Apply preplant and incorporate into the soil 1 to 2 in. (1 in. incorporation is optimum) with a rototiller or tandem disk or apply to the soil surface after seeding and follow by irrigation. Check re-plant restrictions for small grains on label.
	Bareground ethalfluralin Curbit 3 EC	3	1.5 to 2 pt	24	N/A	**For squash grown on bareground only.** Apply to the soil surface immediately after seeding. Seed must be covered with soil to prevent crop injury. For coarse-textured soils, use lowest rate of rate range. Shallow cultivation, irrigation, or rainfall within 5 days is needed for good weed control. If weather is unusually cold or soil wet and cold, crop stunting or injury may occur. Crop injury can also occur if seeding depth is too shallow. See label for further precautions and instruction.
	Plasticulture ethalfluralin Curbit 3 EC	3	3 to 4.5 pt	24	N/A	**For squash grown on plastic only.** Apply to soil surface between the rows of black plastic immediately after seeding or transplanting. **Do not use under mulches, row covers, or hot caps.** Do not apply prior to planting or over plastic. See label for further instruction.
Annual grasses and broadleaf weeds	ethalfluralin + clomazone Strategy 2.1 L	3 + 13	2 to 6 pt	24	45	Apply to the soil surface immediately after crop seeding for preemergence control of weeds. **DO NOT APPLY PRIOR TO PLANTING CROP. DO NOT SOIL INCOPORATE.** May also be used as a **banded** treatment **between** rows after crop emergence or transplanting.
SQUASH, Preplant and Preemergence						
Suppression of annual grasses and broadleaf weeds; weak on pigweed and morningglory	clomazone Command 3 ME	13	0.67 to 1.3 pt	12	45	Apply immediately after seeding or prior to transplanting. Seeds and roots of transplants must be below the chemical barriers when planting. Command should only be applied between rows when squash is grown on plastic. Some cultivars may be sensitive to Command (see label). Use lower rates on coarse soils. Higher rates can be used on winter squashes. See label about rotation restrictions.

* Mode of action (MOA) code developed by the Weed Science Society of America with the Herbicide Resistance Action Committee (HRAC)
** REI – Restricted Entry Interval; PHI- Pre-Harvest Interval

TABLE 4-24. CHEMICAL WEED CONTROL IN SQUASH (cont'd)

Weed	Herbicide and Formulation	MOA*	Amount of Formulation Per Acre	Intervals** REI (Hours)	Intervals** PHI (Days)	Precautions and Remarks
SQUASH, Preplant and Preemergence (cont'd)						
Yellow and purple nutsedge, non-ALS-resistant pigweed, broadleaf weeds	halosulfuron-methyl Profine 75, Sandea 75 DF	2	0.5 to 1 oz	12	30	**Row middles only.** Apply to row middles as preemergence spray. In plasticulture, do not allow spray to contact plastic. Early season application will give postemergence and preemergence control. For postemergence applications, use nonionic surfactant at 1 qt per 100-gal of spray solution.
SQUASH, Postemergence						
Most emerged weeds	glyphosate 4SL 5.5SL 5.88SL	9	16 to 32 fl oz 11 to 22 fl oz 10 to 21 fl oz	4	14	**Row middles only.** Apply as a hooded spray in row middles, as shielded spray in row middles, as wiper applications in row middles, or postharvest. To avoid severe injury to crop, do not allow herbicide to contact foliage, green shoots, stems, exposed roots, or fruit of crop.
Yellow and purple nutsedge, non-ALS-resistant pigweed, broadleaf weeds	halosulfuron-methyl Proline, Sandea 75 DG	2	0.5 to 1 oz	12	30	**Row middles only.** Apply to row middles as postemergence spray. In plasticulture, do not allow spray to contact plastic. Early season application will give postemergence and preemergence control. For postemergence applications, use nonionic surfactant at 1 qt per 100-gal of spray solution.
Annual and perennial grasses only	clethodim Arrow, Clethodim, Intensity, Select 2 EC Select Max, Intensity One 1 EC	1	6 to 8 oz 9 to 16 oz	24	14	Apply postemergence for control of grasses. Adding crop oil may increase likelihood of crop injury at high air temperatures and high humidity. Highly effective control of annual bluegrass. Apply to actively growing grasses not under drought stress.
	sethoxydim Poast 1.5 EC	1	1 to 1.5 pt	12	30	Apply to emerged grasses. Consult manufacturer's label for specific rates and best times to treat and adjuvant and rate. Adding crop oil to Poast may increase the likelihood of crop injury at high air temperatures and high humidity

* Mode of action (MOA) code developed by the Weed Science Society of America with the Herbicide Resistance Action Committee (HRAC)
** REI – Restricted Entry Interval; PHI- Pre-Harvest Interval

TABLE 4-25. CHEMICAL WEED CONTROL IN SWEETPOTATOES

Weed	Herbicide and Formulation	MOA*	Amount of Formulation Per Acre	Intervals** REI (Hours)	Intervals** PHI (Days)	Precautions and Remarks
SWEETPOTATO, Preplant						
Annual and perennial grass and broadleaf weeds, stale seed bed application	glyphosate 4SL 5.5SL 5.88SL	9	16 to 32 fl oz 11 to 22 fl oz 10 to 21 fl oz	4	N/A	Apply to emerged weeds before transplanting. Perennial weeds may require higher glyphosate rates. Consult label for rates for specific weeds. Certain glyphosate formulations may require the addition of a surfactant. Adding nonionic surfactant to glyphosate formulated with nonionic surfactant may result in reduced weed control.
Broadleaf weeds and some annual grasses	caprylic acid + capric acid Homeplate, Suppress	26	3 to 9% v/v	24	0	May be applied prior to planting as a burndown treatment for emerged weeds, as a preemergence application after seeding but before emergence, as a directed or shielded application between rows, as a harvest aid or desiccant, or as a postharvest application. Use higher spray volumes for high weed density and weeds larger than 5 in. Coverage is important for acceptable weed control. May be tank mixed with other herbicides. See label for further instructions.
Contact kill of all green foliage, stale bed application	paraquat Parazone 3 SL Gramoxone 2 SL	22	0.7 to 1.3 pt 1 to 2 pt	24	N/A	Apply to emerged weeds prior to transplanting sweetpotato. Paraquat product labels require applicators to take an EPA-approved training every 3 years to mix, load, and apply paraquat.

* Mode of action (MOA) code developed by the Weed Science Society of America with the Herbicide Resistance Action Committee (HRAC)
** REI – Restricted Entry Interval; PHI- Pre-Harvest Interval

TABLE 4-25. CHEMICAL WEED CONTROL IN SWEETPOTATOES (cont'd)

Weed	Herbicide and Formulation	MOA*	Amount of Formulation Per Acre	Intervals** REI (Hours)	Intervals** PHI (Days)	Precautions and Remarks
SWEETPOTATO, Preplant (cont'd)						
Suppression or control of most annual grasses and broadleaf weeds, full rate required for nutsedge control	metam sodium Vapam HL 42%	N/A	37.5 to 75 gal	120	N/A	Rates are dependent on soil type and weeds present. For nutsedge, use 75-gal per acre. Plant back interval is often 14 to 21 days and can be 30 days in some environments. See label for all restrictions and additional information.
Pigweed species, *Galinsoga* spp., black nightshade, Eastern black nightshade, common purslane, partial control of yellow nutsedge	fomesafen Reflex 2 EC	14	1 pt	24	70	**This is a Section 24(c) special local needs label.** Check to make sure Reflex is registered for use in your state prior to making an application. See label for further instructions. See label for potential carryover to rotational crops. Apply prior to transplanting for preemergence control. May be tank-mixed with other herbicides registered for preplant application however do not tank mix with flumioxazin.
Annual broadleaf weeds including Palmer amaranth and other pigweeds, smartweed, morning glory, wild mustard, wild radish, purslane spp., eclipta, common lambsquarters	flumioxazin Valor SX 51 WDG	14	3 oz	12	N/A	Apply prior to transplanting crop. Use on "Beauregard" only unless user has tested on another variety and found the crop tolerance to be acceptable. Movement of soil during transplanting should not occur or reduced weed control may result. Do not use on greenhouse-grown transplants. Do not apply postemergence or serious crop injury will occur. Do not use on transplant propagation beds. See label for further instructions.
SWEETPOTATO, Preemergence						
Annual grass and broadleaf weeds, Palmer amaranth, yellow nutsedge suppression	*S*-metolachlor Dual Magnum 7.62 EC	15	0.75 pt	24	60	**This is a Section 24(c) Special Local Need Label.** Check to make sure Dual Magnum is registered for use in your state. Apply over top of sweetpotatoes after transplanting but prior to weed emergence. Do not apply preplant. Do not incorporate after application. See label for further information.
Annual grasses and broadleaf weeds including velvetleaf, purslane, prickly sida	clomazone Command 3 ME	13	up to 2 pt	12	N/A	Apply preplant or after transplanting prior to weed emergence for preemergence control. Weak on pigweed. The label allows up to 4 pt per acre. See label for other instructions and precautions.
Annual grasses including crabgrass, foxtail, goosegrass, fall panicum and broadleaf weeds including pigweed, Florida pusley, purslane	napropamide Devrinol DF, Devrinol DF-XT 50 DF Devrinol 2-XT 2 EC	15	2 to 4 lb 2 to 4 qt	24	N/A	**Plant beds.** Apply to the soil surface after sweetpotato roots are covered with soil but prior to soil cracking and sweetpotato plant emergence. Does not control emerged weeds. **Production fields.** Apply to the soil surface immediately after transplanting. If rainfall does not occur within 24 hr, shallow incorporate or irrigate with sufficient water to wet the soil to a depth of 2 to 4 in. See XT labels for information regarding delay in irrigation
SWEETPOTATO, Postemergence						
Annual and perennial grasses only	clethodim Arrow, Clethodim, Intensity, Select 2 EC Select Max, Intensity One 1 EC	1	6 to 8 oz 9 to 16 oz	24	30	Apply to actively growing grasses not under drought stress. See label for adjuvant and rate. Adding crop oil may increase the likelihood of crop injury at high air temperatures and high humidity. Highly effective in controlling annual bluegrass.
	fluazifop Fusilade DX 2 EC	1	6 to 12 oz	12	14	Apply to actively growing grasses not under drought stress. Consult manufacturer's label for specific rates and best times to treat and adjuvant and rate. Do not apply Fusilade on days that are unusually hot and humid.
	sethoxydim Poast 1.5 EC	1	1 to 1.5 pt	12	30	Apply to actively growing grasses not under drought stress. Adding crop oil to Poast may increase the likelihood of crop injury at high air temperatures and high humidity.
SWEETPOTATO, Row Middles						
Most broadleaf weeds less than 4 in. tall or rosettes less than 3 in. in diameter; does not control grasses	carfentrazone-ethyl Aim 1.9 EW or 2 EC	14	Up to 2 oz	12	0	Apply post-directed using hooded sprayers for control of emerged weeds. If crop is contacted, burning of contacted area will occur. Use a nonionic surfactant or crop oil with Aim. Coverage is essential for good weed control. Can be tank mixed with other registered herbicides.
Most emerged weeds	glyphosate 4SL 5.5SL 5.88SL	9	16 to 32 fl oz 11 to 22 fl oz 10 to 21 fl oz	4	14	Apply as a hooded spray in row middles, as shielded spray in row middles, as wiper applications in row middles, or postharvest. To avoid severe injury to crop, do not allow herbicide to contact foliage, green shoots, stems, exposed, roots, or fruit of crop. May cause cracking of sweetpotato storage roots if spray solution is exposed to sweetpotato foliage.

* Mode of action (MOA) code developed by the Weed Science Society of America with the Herbicide Resistance Action Committee (HRAC)

** REI – Restricted Entry Interval; PHI- Pre-Harvest Interval

TABLE 4-26. CHEMICAL WEED CONTROL IN TOMATOES

Weed	Herbicide and Formulation	MOA*	Amount of Formulation Per Acre	Intervals** REI (Hours)	Intervals** PHI (Days)	Precautions and Remarks
TOMATO, Preplant						
Suppression or control of most annual grasses and broadleaf weeds (full rate required for nutsedge control)	metam sodium Vapam HL 42%	N/A	37.5 to 75 gal	120	N/A	Rates are dependent on soil type and weeds present. For nutsedge, use 75-gal per acre. Apply when soil moisture is at field capacity (100 to 125%). Apply through soil injection using a rotary tiller or inject with knives no more than 4 in. apart; follow immediately with a roller to smooth and compact the soil surface or with mulch. May apply through drip irrigation prior to planting a second crop on mulch; however, adhere to label guidelines on crop plant back interval. Plant back interval is often 14 to 21 days and can be 30 days in some environments. See label for all restrictions and additional information. Chloropicrin (150 lb per acre broadcast) will also be needed when laying first crop mulch to control nutsedge.
Most broadleaf weeds less than 4 in. tall or rosettes less than 3 in. in diameter; does not control grasses	carfentrazone-ethyl Aim 1.9 EW or 2 EC	14	Up to 2 oz	12	N/A	**Transplanted crop.** Apply no later than 1 day before transplanting. **Seeded crop (Aim 2 EC only).** Apply no later than 7 days before planting seeded crop. Use a nonionic surfactant or crop oil with Aim. See label for rate. Coverage is essential for good weed control. Can be tank mixed with other registered burndown herbicides.
Contact kill of all green foliage, stale bed application	paraquat Parazone 3 SL Gramoxone 2 SL	22	1.3 to 2.7 pt 2 to 4 pt	24	N/A	Apply to emerged weeds in a minimum of 20-gal spray mix per acre before crop emergence as a broadcast or band treatment over a pre-formed row. Row should be formed several days ahead of planting and treating to allow maximum weed emergence. Use a nonionic surfactant at a rate of 16 to 32 oz per 100-gal spray mix or 1-gal approved crop oil concentrate per 100-gal spray mix. Paraquat product labels require applicators to take an EPA-approved training every 3 years to mix, load, and apply paraquat.
Broadleaf weeds and some annual grasses	caprylic acid + capric acid Homeplate, Suppress	26	3 to 9% v/v	24	0	May be applied prior to planting as a burndown treatment for emerged weeds, as a preemergence application after seeding but before emergence, as a directed or shielded application between rows or as a postharvest application. Use higher spray volumes for high weed density and weeds larger than 5 in. Coverage is important for acceptable weed control. May be tank mixed with other herbicides. See label for further instructions.
Broadleaf weeds including Carolina geranium and cutleaf eveningprimrose and a few annual grasses	oxyfluorfen Galigan 2E, Goal 2XL GoalTender 4 F	14	up to 2 pt up to 1 pt	24	45	**Plasticulture only.** Apply to soil surface of pre-formed beds at least 30 days prior to transplanting crop. While incorporation is not necessary, it may result in less crop injury. Plastic mulch can be applied any time after application, but best results are likely if applied soon after application. **Label is state restricted: FL, GA, NC, SC, and VA only.**
Annual grasses and small-seeded broadleaf weeds including common lambsquarters, pigweed, carpetweed, and common purslane	napropamide Devrinol DF, Devrinol DF-XT 50 DF Devrinol 2-XT 2 EC	15	2 to 4 lb 2 to 4 qt	24	N/A	**Baregound:** Apply preplant and incorporate into the soil 1 to 2 in. as soon as possible with a rototiller or tandem disk. Can be used on direct-seeded or transplanted tomatoes. See label for instructions on use. **Plasticulture:** Apply to a weed-free soil before laying plastic mulch. Soil should be well worked yet moist enough to permit a thorough incorporation to a depth of 2 in. Mechanically incorporate or irrigate within 24 hours after application. If weed pressure is from small-seeded annuals, apply to the surface of the bed immediately in front of the laying of plastic mulch. If soil is dry, water or sprinkle irrigate with sufficient water to wet to a depth of 2 to 4 in. before covering with plastic mulch. **Between rows:** Apply to a weed free soil surface between the rows (bareground or plastic mulch). Mechanically incorporate or irrigate Devrinol into the soil to a depth of 1 to 2 in. within 24 hours of application. See XT labels for information regarding delay in irrigation event.
	pendimethalin Prowl H2O 3.8 AS	3	1 to 3 pt	24	21	**Plasticulture In-row.** May be applied as a preplant surface application or a preplant incorporated application prior to transplanting tomato. **Baregound In-row.** May be applied as a broadcast preplant surface application or preplant incorporated application prior to transplanting tomato. **Post-directed spray.** May be applied as a post-directed spray on the soil at the base of the plant, beneath plants, and between rows. Avoid direct contact with tomato foliage or stems. Do not apply over the top of tomato. PHI=21 days. Do not apply more than 3 pt per acre per season. See label for specific use rate for your soil type. Emerged weeds will not be controlled. See label for further instructions and precautions.
	trifluralin Treflan 4 L, Treflan HFP 4 EC Treflan TR-10 G	3	1 to 1.5 pt 5 to 10 lb	12	N/A	**Transplant tomato.** Apply pretransplant and incorporate into the soil 2 to 3 in. within 8 hours using a rototiller or tandem disk. Can be applied postplanting as a directed spray to soil between the rows and beneath plants and then incorporated. Cold soil temperatures may exacerbate injury from trifluralin.

* Mode of action (MOA) code developed by the Weed Science Society of America with the Herbicide Resistance Action Committee (HRAC)
** REI – Restricted Entry Interval; PHI- Pre-Harvest Interval

TABLE 4-26. CHEMICAL WEED CONTROL IN TOMATOES (cont'd)

Weed	Herbicide and Formulation	MOA*	Amount of Formulation Per Acre	REI (Hours)	PHI (Days)	Precautions and Remarks
TOMATO, Preplant (cont'd)						
Yellow and purple nutsedge and broadleaf weeds including pigweed, wild radish, common ragweed, suppression of purslane	halosulfuron-methyl Profine 75, Sandea 75 DF	2	0.5 to 1 oz	12	30	For pretransplant application under plastic mulch, apply to pre-formed bed just prior to plastic mulch application and delay transplanting at least 7 days. Can be applied for pretransplant application in bareground tomato. Early season application will give postemergence and preemergence control. The 1 oz rate is for preemergence and postemergence control in row middles only. For postemergence applications, use nonionic surfactant at 1 qt per 100-gal of spray solution.
Yellow nutsedge, annual grasses, and broadleaf weeds including pigweed, Palmer amaranth, Florida pusley, Hairy galinsoga, Eastern black nightshade, and carpetweed	S-metolachlor Brawl, Dual Magnum, Medal 7.62 EC Brawl II, Dual II Magnum, Medal II 7.64 EC	15	1 to 2 pt 1 to 2 pt	24	30, 90	Apply preplant or postdirected to transplants after the first settling rain or irrigation. In plasticulture, apply to preformed beds just prior to applying plastic mulch. Lower rates of rate range for S-metolachlor are safest to tomato. May also be used to treat row middles in bedded tomato. Minimize contact with crop. Also registered for use in row middles, and in seeded crop. See label for further instructions. If cumulative rate ≤ 1.33 pt/acre, PHI = 30 days. If annual rate exceed 1.33 pt/acre, PHI = 90 days.
Palmer amaranth, redroot pigweed, smooth pigweed, *Galinsoga* spp., black nightshade, Eastern black nightshade, common purslane, partial control of yellow nutsedge	fomesafen Reflex 2 EC	14	1 to 1.5 pt	24	70	**This is a Section 24(c) Special Local Need Label** for transplanted tomato. Growers must obtain the label prior to making an application of Reflex. See label for further instructions. Carryover is a large concern. **Plasticulture In-row Application for Transplanted Tomato.** Apply after final bed formation and the drip tape is laid but prior to laying plastic mulch. Avoid soil disturbance after application. Unless restricted by other products such as fumigants, tomato may be transplanted immediately following the application of Reflex and the application of the mulch. **Bareground for Transplanted Tomato.** Apply pretransplant up to 7 days prior to transplanting tomato. Weed control will be reduced if soil is disturbed after application. During the transplanting operation, make sure the soil in the transplant hole settles flush or above the surrounding soil surface. Avoid cultural practices that may concentrate Reflex-treated soil around the transplant root ball. An overhead irrigation or rainfall event between Reflex herbicide application and transplanting will ensure herbicide activation and will likely reduce the potential for crop injury due to splashing. **Plasticulture Row Middle Application.** Apply to row middles with a hooded or shielded sprayer. Avoid drift of herbicide on mulch. If drift occurs, 0.5 in. of rain or irrigation must occur prior to transplanting. Carryover is a large concern; see label for more information.
Annual grasses and broadleaf weeds including jimsonweed, common ragweed, smartweed, and velvetleaf	metribuzin TriCor DF, Dimetric 75 WDG Dimetric 3 F	5	0.33 to 0.67 lb 0.67 to 1 pt	12	7	Apply to soil surface and incorporate 2 to 4 in. deep before transplanting. See label for instructions.
Broadleaf weeds including Carolina geranium and cutleaf eveningprimrose and a few annual grasses	oxyfluorfen Galigan 2E, Goal 2XL GoalTender 4 F	14	Up to 2 pt Up to 1 pt	24	45	**Plasticulture (fallow beds) only.** Apply to soil surface of pre-formed beds at least 30 days prior to transplanting crop. While incorporation is not necessary, it may result in less crop injury. Plastic mulch can be applied any time after application, but best results are likely if applied soon after application. **Label is state restricted: FL, GA, NC, SC, and VA only.**
Broadleaf, grass (suppression), yellow nutsedge (PRE or POST), purple nutsedge (POST only)	imazosulfuron League 75 DF	2	4 to 6.4 oz	12	21	Apply to planting beds before plastic is laid. Tomato may be transplanted 1 day after application. Refer to label for further application instructions. Consult label for approved surfactants and crop rotation restrictions.
TOMATO, Postemergence						
Yellow and purple nutsedge and broadleaf weeds	halosulfuron-methyl Profine 75, Sandea 75 DF	2	0.5 to 1 oz	12	30	Apply no sooner than 14 days after transplanting. For postemergence applications, use nonionic surfactant at 1 qt per 100-gal of spray solution. Some weeds, such as nutsedge, may require two applications of Sandea; if a second application is needed, spot-treat only weed-infested areas. See label for further instructions.
Annual grasses and broadleaf weeds, yellow nutsedge, and morningglory	metribuzin TriCor DF 75 WDG	5	0.33 to 1.33 lb	12	7	Use either as a broadcast or directed spray but do not exceed 0.5 lb a.i. with a broadcast spray. Do not apply within 7 days of harvest. Do not exceed 1 lb a.i. per year. Do not apply as a broadcast spray unless 3 sunny days precede application.

* Mode of action (MOA) code developed by the Weed Science Society of America with the Herbicide Resistance Action Committee (HRAC)

** REI – Restricted Entry Interval; PHI- Pre-Harvest Interval

TABLE 4-26. CHEMICAL WEED CONTROL IN TOMATOES (cont'd)

Weed	Herbicide and Formulation	MOA*	Amount of Formulation Per Acre	REI (Hours)	PHI (Days)	Precautions and Remarks
TOMATO, Postemergence (cont'd)						
Most broadleaf weeds including wild radish, common purslane, redroot and smooth pigweed	rimsulfuron Matrix 25 WDG, Pruvin 25 WDG	2	1 to 2 oz	4	45	Apply in tomatoes after the crop has at least two true leaves and weeds are small (1 in. or less) and actively growing. Add nonionic surfactant at 1 qt per 100-gal of spray solution. See label for further instruction.
Yellow nutsedge, morningglory and other broadleaf weeds	trifloxysulfuron-sodium Envoke 75 DG	2	0.1 to 0.2 oz	12	45	Apply post-directed to tomato grown on plastic for control of nutsedge and certain broadleaf weeds. Crop should be transplanted at least 14 days prior to application. The application should be made prior to fruit set and at least 45 days prior to harvest. Use nonionic surfactant at 1 qt per 100-gal spray solution with all applications.
Annual and perennial grasses only	clethodim Arrow, Clethodim, Intensity, Select 2 EC Select Max, Intensity One 1 EC	1	6 to 8 oz 9 to 16 oz	24	20	Apply to actively growing grasses not suffering from drought stress. Adding crop oil may increase the likelihood of crop injury at high air temperatures and high humidity. Highly effective in controlling annual bluegrass.
	sethoxydim Poast 1.5 EC	1	1 to 1.5 pt	12	20	Apply to actively growing grasses not under drought stress. See label for adjuvant and rate. Adding crop oil to Poast may increase the likelihood of crop injury at high air temperatures and high humidity.
TOMATO, Row Middles						
Yellow nutsedge, morningglory and other broadleaf weeds	trifloxysulfuron-sodium Envoke 75 DG	2	0.1 to 0.2 oz	12	45	Crop should be transplanted at least 14 days prior to application. Use nonionic surfactant at 1 qt per 100-gal spray solution with all applications. See label for information on registered tank mixes. Tank mixtures with Select or Poast may reduce grass control. See label for more information.
Yellow and purple nutsedge and broadleaf weeds	halosulfuron-methyl Profine 75, Sandea 75 DF	2	0.5 to 1 oz	12	30	For postemergence applications, use nonionic surfactant at 1 qt per 100-gal of spray solution. Some weeds, such as nutsedge, may require two applications of Sandea. If a second application is needed, spot-treat only weed-infested areas. See label for further instructions.
Most broadleaf weeds less than 4 in. tall or rosettes less than 3 in. in diameter; does not control grasses	carfentrazone-ethyl Aim 1.9 EW or 2 EC	14	Up to 2 oz	12	0	Apply post-directed using hooded sprayers for control of emerged weeds. If crop is contacted, burning of contacted area will occur. Use a nonionic surfactant or crop oil with Aim. See label for rate. Coverage is essential for good weed control. Can be tank mixed with other registered herbicides.
Annual grasses and small-seeded broadleaf weeds	napropamide Devrinol DF, Devrinol DF-XT 50 DF Devrinol 2-XT 2 EC	15	 2 to 4 lb 2 to 4 qt	24	N/A	**Plasticulture.** Apply to a weed-free soil surface. Apply within 24 hours of rainfall, or mechanically incorporate or irrigate into the soil to a depth of 1 to 2 in.
	pendimethalin Prowl H2O 3.8 AS	3	1 to 3 pt	24	21	Post-directed spray on the soil at the base of the plant, beneath plants and between rows. Avoid direct contact with tomato foliage or stems. Do not apply more than 3 pt per acre per season. See label for specific use rate for your soil type. Emerged weeds will not be controlled. Avoid direct contact with tomato foliage or stems. See label for further instructions and precautions.
Contact kill of all green foliage	paraquat Parazone 3 SL Gramoxone 2 SL	22	 1.3 pt 2 pt	24	30	Apply for control of emerged weeds between rows of tomatoes. Do not allow spray to contract crop or injury will occur. Do not make more than 3 applications per season. Paraquat product labels require applicators to take an EPA-approved training every 3 years to mix, load, and apply paraquat.

* Mode of action (MOA) code developed by the Weed Science Society of America with the Herbicide Resistance Action Committee (HRAC)
** REI – Restricted Entry Interval; PHI- Pre-Harvest Interval

TABLE 4-27. CHEMICAL WEED CONTROL IN WATERMELON

Weed	Herbicide and Formulation	MOA*	Amount of Formulation Per Acre	Intervals** REI (Hours)	Intervals** PHI (Days)	Precautions and Remarks
WATERMELON, Preplant						
Suppression or control of most annual grasses and broadleaf weeds, full rate required for nutsedge control	metam sodium Vapam HL 42%	N/A	37.5 to 75 gal	120	N/A	Rates are dependent on soil type and weeds present. For nutsedge, apply 75- gal per acre. Apply when soil moisture is at field capacity (100 to 125%). Apply through soil injection using a rotary tiller or inject with knives no more than 4 in. apart; follow immediately with a roller to smooth and compact the soil surface or with mulch. May apply through drip irrigation prior to planting a second crop on mulch. Plant back interval is often 14 to 21 days and can be 30 days in some environments. See label for all restrictions and additional information.
Contact kill of all green foliage, stale bed application	paraquat Parazone 3 SL Gramoxone 2 SL	22	 1.3 to 2.7 pt 2 to 4 pt	24	N/A	Apply in a minimum of 10-gal spray mix per acre to emerged weeds before crop emergence or transplanting as a broadcast or band treatment over a preformed row. Row should be formed several days ahead of planting and treating to allow maximum weed emergence. Plant with a minimum of soil movement for best results. Use a nonionic surfactant at a rate of 16 to 32 oz per 100-gal spray mix or 1-gal approved crop oil concentrate per 100-gal spray mix. Paraquat product labels require applicators to take an EPA-approved training every 3 years to mix, load, and apply paraquat.
Morningglory and small pigweed	pyraflufen ethyl ET Herbicide 0.208 L	14	1 to 2 oz	12	N/A	**Bareground.** Wait 1 day following preplant burndown application before planting. **Plastic Mulch Production.** May apply over mulch; however, a single 0.5 in. irrigation/rain event plus a 7-day waiting period must occur before transplanting. Apply ET with a crop oil concentrated at 1% v/v to sensitive weeds that are less than 3 in.
Emerged broadleaf and grass weeds	pelargonic acid Scythe 4.2 EC	26	1 to 10% v/v	12	N/A	Apply as a preplant burndown treatment, or prior to crop emergence from seed, or as a directed or shielded spray between beds. Avoid contact of watermelon foliage.
Most broadleaf weeds less than 4 in. tall or rosettes less than 3 in. in diameter; does not control grasses	carfentrazone-ethyl Aim 1.9 EW or 2 EC	14	Up to 2 oz	12	7	**Transplants only.** Apply prior to transplanting of crop. Use a nonionic surfactant or crop oil with Aim. See label for rate. Cover age is essential for good weed control. Can be tank mixed with other registered herbicides.
Annual and perennial grass and broadleaf weeds, stale bed application	glyphosate 4SL 5.5SL 5.88SL	9	 16 to 32 fl oz 11 to 22 fl oz 10 to 21 fl oz	4	N/A	Apply to emerged weeds at least 3 days before seeding or transplanting. When applying Roundup before transplanting crops into plastic mulch, care must be taken to remove residues of this product from the plastic prior to transplanting. To prevent crop injury, residues can be removed by 0.5 in. rainfall or by applying water via a sprinkler system. Perennial weeds may require higher rates of glyphosate. Consult the manufacturer's label for rates for specific weeds. Certain glyphosate formulations require the addition of a surfactant. Adding a nonionic surfactant to glyphosate formulated with nonionic surfactant may result in reduced weed control.
Emerged broadleaf weeds and some annual grasses	caprylic acid + capric acid Homeplate, Suppress	26	3 to 9% v/v	24	0	May be applied prior to planting as a burndown treatment for emerged weeds, as a preemergence application after seeding but before emergence, as a directed or shielded application between rows or as a postharvest application. Use higher spray volumes for high weed density and weeds larger than 5 in. Coverage is important for acceptable weed control. May be tank mixed with other herbicides. See label for further instructions.
Annual grasses, some small-seeded broadleaves	bensulide Prefar 4 EC	15	5 to 6 qt	12	N/A	Registered for cucurbit vegetable group (Crop grouping 9). Apply preplant and incorporate into the soil 1 to 2 in. (1 in. incorporation is optimum) with a rototiller or tandem disk or apply to the soil surface after seeding and follow by irrigation. Check re-plant restrictions for small grains on label.
WATERMELON, Preemergence						
Palmer amaranth, redroot pigweed, smooth pigweed, *Galinsoga* spp., black nightshade, Eastern black nightshade, common purslane, partial control of yellow nutsedge	fomesafen Reflex 2 EC	14	10 to 12 oz	24	35	**This is a Section 24(c) Special Local Needs Label.** Growers must check to make sure fomesafen is registered in their state and to obtain the label prior to making an application. See label for further instructions. **Plasticulture transplants or seeds.** May apply under plastic mulch if plastic laying process does not disturb treated soil; thus, do not apply prior to laying drip or forming bed. May apply over plastic mulch if the mulch is washed with 0.5 in. rainfall/irrigation in a single event prior to punching holes and planting; bed formation must allow herbicide to wash off the mulch and not concentrate in low areas on the mulch. **Bareground transplant.** Prepare land for planting; apply Reflex; lightly irrigate to activate herbicide and move it into soil; and then prepare plant holes and plant. **Bareground seeded.** Apply within 1 day of planting; lightly irrigate after application but at least 36 hours prior to emergence. **Row middles.** Must apply prior to crop emergence or transplanting. May use up to 16 oz per acre in watermelon. See label for potential carryover to rotational crop.
Annual grasses and broadleaf weeds	clomazone Command 3 ME	13	0.4 to 0.67 pt	12	N/A	Apply immediately after seeding, or just prior to transplanting. Roots of transplants must be below the chemical barrier when planting. Offers weak control of pigweed. See label for further instructions.
	bicyclopyrone Optogen	27	3.5 fl oz	24	14	Apply before transplanting. Minimize soil disturbance to maintain weed control. Do not make more than one application per crop per year. **Transplants only.**

* Mode of action (MOA) code developed by the Weed Science Society of America with the Herbicide Resistance Action Committee (HRAC)
** REI – Restricted Entry Interval; PHI- Pre-Harvest Interval

TABLE 4-27. CHEMICAL WEED CONTROL IN WATERMELON (cont'd)

Weed	Herbicide and Formulation	MOA*	Amount of Formulation Per Acre	Intervals** REI (Hours)	Intervals** PHI (Days)	Precautions and Remarks
WATERMELON, Preemergence (cont'd)						
Annual grasses and some small-seeded broadleaf weeds	ethalfluralin Curbit 3 EC	3	3 to 4.5 pt	24	N/A	Apply post immediately after seeding crop prior to crop emergence, or as a banded spray between rows after crop emergence or transplanting. May also be used as a banded spray between rows of plastic mulch. See label for timing. Shallow cultivation, irrigation, or rainfall within 5 days is needed for good weed control. Do not use under mulches, row covers, or hot caps. Under conditions of unusual cold or wet soil and air temperatures, crop stunting or injury may occur. Crop injury can occur if seeding depth is too shallow.
Annual grasses and broadleaf weeds	ethalfluralin + clomazone Strategy 2.1 L	3 + 13	2 to 6 pt	24	N/A	Apply to the soil surface immediately after crop seeding for preemergence control of weeds. **DO NOT APPLY PRIOR TO PLANTING. DO NOT INCORPORATE. DO NOT APPLY UNDER MULCH.** May also be used as a banded treatment between rows after crop emergence or transplanting.
Broadleaf weeds	terbacil Sinbar 80 WP	5	2 to 4 oz	12	70	Apply after seeding but before crop emerges, or prior to transplanting crop. With plasticulture, Sinbar may be applied preemergence under plastic mulch or to row middles. Maybe applied over plastic mulch prior to transplanting, or prior to punching holes into the plastic mulch for transplanting. Sinbar must be washed off the surface of the plastic mulch with a minimum of 0.5 in. of rainfall or irrigation prior to punching transplant holes or transplanting watermelon. See label for further instructions.
Yellow and purple nutsedge suppression, non-ALS-resistant pigweed, and non-ALS-resistant ragweed control	halosulfuron-methyl Profine 75, Sandea 75 DF	2	0.5 to 1 oz	12	30	**Bareground.** Apply after seeding but before cracking or prior to transplanting crop. **Plasticulture.** Application may be made to preformed beds prior to laying plastic. If application is made prior to planting, wait 7 days after application to seed or transplant. Stunting may occur but should be short-lived with no negative effects on yield or maturity in favorable growing conditions. **SEE LABEL FOR INFORMATION ON ROTATION RESTRICTIONS AND OTHER RESTRICTIONS.**
WATERMELON, Postemergence						
Annual and perennial grasses only	clethodim Arrow, Clethodim, Intensity, Select 2 EC Select Max, Intensity One 1 EC	1	6 to 8 oz 9 to 16 oz	24	14	Apply postemergence for control of grasses not under drought stress. See label for adjuvant and rate. Adding crop oil may increase the likelihood of crop injury at high air temperatures and high humidity. Highly effective in controlling annual bluegrass.
	sethoxydim Poast 1.5 EC	1	1 to 1.5 pt	12	14	Apply to emerged grasses. See label for adjuvant and rate. Adding crop oil may increase the likelihood of crop injury at high air temperatures and high humidity.
WATERMELON, Row Middles						
Annual grasses and some small-seeded broadleaf weeds	trifluralin Treflan 4 L, Treflan HFP 4 EC Treflan 10 G	3	1 to 2 pt 5 to 10 lb	12	60	To improve preemergence control of late emerging weeds. Apply after emergence when crop plants have reached the three to four true leaf stage of growth. Apply as a directed spray to soil between the rows. Avoid contacting foliage as slight crop injury may occur. Set incorporation equipment to move treated soil around base of crop plants. Will not control emerged weeds.
	pendimethalin Prowl H2O 3.8 AS	3	Up to 2.1 pt	24	35	May be applied sequentially in bareground and plasticulture production systems at a minimum of 21 days apart. Refer to label for specific instructions.
Broadleaf weeds	terbacil Sinbar 80 WP	5	2 to 4 oz	12	70	With plasticulture, Sinbar may be applied to row middles. See label for further instructions.
Most broadleaf weeds less than 4 in. tall or rosettes less than 3 in. in diameter; does not control grasses	carfentrazone-ethyl Aim 1.9 EW or 2 EC	14	Up to 2 oz	12	0	Apply post-directed using hooded sprayers for control of emerged weeds. Apply before fruit is present. If crop is contacted, burning of contacted area will occur. Use a nonionic surfactant or crop oil with Aim. See label for rate. Coverage is essential for good weed control. Can be tank mixed with other registered herbicides.
Most emerged weeds	glyphosate 4SL 5.5SL 5.88SL	9	16 to 32 fl oz 11 to 22 fl oz 10 to 21 fl oz	4	14	Apply as a hooded spray in row middles, as shielded spray in row middles, as wiper applications in row middles, or postharvest. To avoid severe injury to crop, do not allow herbicide to contact foliage, green shoots, stems, exposed, roots, or fruit of crop.
Annual grasses and broadleaf weeds	bicyclopyrone Optogen	27	2.6 to 3.5 fl oz	24	14	Apply to row middles. Avoid contacting watermelon foliage or injury will occur. Using a hooded sprayer will minimize crop injury during application to middles. Add NIS (0.25% v/v) or COC (1% v/v). Do not make more than 1 application per crop per year.
Yellow and purple nutsedge and broadleaf weeds	halosulfuron-methyl Profine 75, Sandea 75 DF	2	0.5 to 1 oz	12	57	Apply to row middles as a postemergence spray. In plasticulture, do not allow spray to contact plastic. Early season application will give postemergence and preemergence control. For postemergence applications, use nonionic surfactant at 1 qt per 100-gal of spray solution.

* Mode of action (MOA) code developed by the Weed Science Society of America with the Herbicide Resistance Action Committee (HRAC)
** REI – Restricted Entry Interval; PHI- Pre-Harvest Interval

TABLE 4-27. CHEMICAL WEED CONTROL IN WATERMELON (cont'd)

Weed	Herbicide and Formulation	MOA*	Amount of Formulation Per Acre	Intervals** REI (Hours)	Intervals** PHI (Days)	Precautions and Remarks
WATERMELON, Row Middles (cont'd)						
Many broadleaf weeds less than 4 in. tall	caprylic acid + capric acid Homeplate, Suppress, Suppress	26	3 to 9% v/v	24	0	May be applied prior to planting as a burndown treatment for emerged weeds, as a preemergence application after seeding but before emergence, as a directed or shielded application between rows or as a postharvest application. Use higher spray volumes for high weed density and weeds larger than 5 in. Coverage is important for acceptable weed control. May be tank mixed with other herbicides. See label for further instructions.

* Mode of action (MOA) code developed by the Weed Science Society of America with the Herbicide Resistance Action Committee (HRAC)

** REI – Restricted Entry Interval; PHI- Pre-Harvest Interval

TABLE 4-28. WEED RESPONSE TO HERBICIDES USED IN VEGETABLE CROPS

HERBICIDE TIME OF APPLICATION	Atrazine PRE	Command PRE	Curbit PRE	Dual Magnum PRE	Devrinol PRE	Goal PRE	Metribuzin PRE	Reflex PRE	Sandea PRE	Valor/Chateau PRE	Aim POST	Atrazine POST	Envoke POST	Fusilade POST	Glyphosate POST/HOOD	Paraquat POST/HOOD	Poast POST	Pursuit POST	Select POST	Sandea POST
PERENNIAL WEEDS																				
Johnsongrass (rhizome)	P	P	P	P	P	P	P	N	P	P	P	P	P	G-E	G-E	P	G	P	G	P
SEDGES																				
Annual sedge	–	P	P	G	F	G	P	–	G	N	P	P	–	G-E	P-F	P	G	P	P	E
Purple nutsedge	P	P	P	P	P-G	P	P	P-F	F	N	P	P	F-G	P	F-G	P-F	P	G	P	E
Yellow nutsedge	P	P	P	F	G	P	P	G-E	F	N	P	P	G	P	F	P-F	P	F	P	E
ANNUAL GRASSES																				
Barnyardgrass	F	G-E	G-E	E	G-E	P	G	–	P	P	–	P	E	E	G	E	F	E	G-E	P
crabgrass, Large	G	E	G-E	E	G-E	P	G	F-G	P	P	P	G	P	E	E	F	G-E	P-F	G-E	P
Crowfootgrass	G	E	G-E	E	G-E	P	G	–	P	P	P	G	P	–	E	G	F-G	P-F	G	P
Fall panicum	P	G-E	G	G-E	G	P	E	–	P	P	P	G	P	P	E	G-E	P-F	G-E	P	
Foxtails	F	E	G-E	E	G-E	P	E	–	P	P	P	–	P	–	E	–	–	–	P	
Goosegrass	F-G	E	G	E	G	P	E	–	P	P	F-G	P	P	E	E	G	G-E	P-F	G-E	P
Johnsongrass (seedling)	P	F-G	G	P-F	G	P	F	–	P	P	P	F	P	–	E	E	F	E	P	
signalgrass, Broadleaf	P	E	G	G	G	P	P	F-G	P	P	P-F	P	P	E	G-E	E	P-F	E	P	
Texas panicum	P	F-G	F-G	P-F	G	P	P	F-G	P	P	P	P-F	P	–	E	G	E	P-F	E	P
ANNUAL BROADLEAF WEEDS																				
Cocklebur	G-E	P-F	P	P	P	–	G	G	G	F-G	E	G-E	N	P	G	P	G-E	P	E	
Florida beggarweed	E	F-G	P	F	P-F	–	–	P	F	G	F	G	G-E	N	E	E	P	P	P	P-F
Florida pusley	E	F-G	G-E	G	G	–	–	P	F	G-E	F	G	P	N	P-G	P-F	P	F	P	P-F
Jimsonweed	E	F-G	P	P	P	–	E	–	G	G	F	E	P	N	E	P	P	G	P	F
lambsquarters, Common	E	G-E	G	F-G	G	F	E	E	F-G	G	G	E	G	N	P	F	P	P-F	P	P
Morningglory	G	P-F*	P	P	P	–	E	P-G**	P-F	F	G-E	E	G-E	N	F-G	F-G	P	F-G	P	P-F
nightshade, Eastern black	G	–	P	–	P	–	F	–	P	–	G	–	–	N	G	P	–	P	P	
Pigweeds	E	P	G-E	G	F-G	G-E	P	E	G-E	E	F-G	E	F****	N	E***	G	P	E****	P	G****
Prickly sida	E	E	P	P-F	F	–	G	–	–	F-G	F	E	P	N	F-G	P-F	P	P-F	P	–
Purslane	E	G-E	G-E	G	G-E	G-E	E	G	F	E	F	E	–	N	G	G	P	F-G	P	P
ragweed, Common	E	F-G	P	P	G	F	G	G	G	F-G	P	E	G	N	E	F	P	P-F	P	G-E
Sicklepod	G	P	P	P	P	P	F	P	P	P	P	E	E	N	E	G-E	P	P	P	P
Smartweed	E	G	F	F	G	E	G	–	G	–	–	–	–	N	–	G	N	–	N	F
WINTER ANNUALS																				
Annual ryegrass	G	–	–	F-G	–	P	G	–	P	F	P	F	P	–	G	F-G	E	–	G-E	P
Shepherdspurse	G	F	P	–	F-G	G-E	G	–	–	–	–	–	–	N	G	F	P	P-F	P	–
Swinecress	G	–	–	–	–	G-E	G	–	E	F	–	–	–	N	G	P	P	–	P	–
Wild mustard	G	–	P	P	P	G-E	G-E	E	G-E	E	F	–	–	N	F-G	F-G	P	G-E	P	E
Wild radish	G	–	P	P	P	G-E	G-E	E	G-E	E	–	F-G	–	N	F-G	F-G	P	G-E	P	E

Modified from Commercial Vegetables - Weed Control Publication by A. Stanley Culpepper, Extension Agronomist Weed Science, University of Georgia

* Command provides fair control of pitted morningglory but poor control of other morningglory species

** Reflex provides "P-F" control on Ipomoea morningglory and "G" control of smallflower morningglory

*** Glyphosate will not control glyphosate-resistant weeds

**** Will not control ALS-resistant pigweed

Emergency Numbers by State

POISON CONTROL CENTERS
Poison Centers maintain a 24-hour consultant service in diagnosis and treatment of human illness resulting from toxic substances. Make sure that your physician knows the Poison Center's telephone number and do not hesitate to call in case of an emergency. Always provide the pesticide label to your physician or emergency personnel to ensure proper treatment in the event of a poisoning.

Alabama Poison Control Center
1-800-222-1222

Arkansas Poison Control Center
1-800-222-1222

Florida Poison Control Center
1-800-222-1222

Georgia Poison Control Center
1-800-222-1222

Kentucky Poison Control Center
1-800-222-1222

Louisiana Poison Control Center
1-800-222-1222

Mississippi Poison Control Center
1-800-222-1222

North Carolina – Carolinas Poison Center
1-800-222-1222

South Carolina – Palmetto Poison Center
1-800-222-1222

Tennessee Poison Control Center
1-800-222-1222

Texas Poison Control Network
1-800-222-1222

Virginia Poison Control Center
1-800-222-1222

PESTICIDE SPILLS

Alabama – CHEMTREC
1-800-424-9300 (24 hours)

Arkansas Department of Emergency Management
1-800-322-4012

Florida – CHEMTREC
1-800-424-9300 (24 hours)

Georgia – Environmental Protection Division
1-800-241-4113 (24 hours)

Kentucky – CHEMTREC
1-800-424-9300 (24 hours)

Louisiana – Louisiana Department of Ag & Forestry
1-855-452-5323

Mississippi – Mississippi Emergency Management Agency (MEMA)
1-800-222-6362 (24 hr emergency line)

North Carolina
1-800-262-8200 (24 hours)

South Carolina – SCDHEC
1-888-481-0125 (24 hours)

Tennessee – CHEMTREC
1-800-424-9300 (24 hours)

Texas – CHEMTREC
1-800-424-9300 (24 hours)

Virginia – CHEMTREC
1-800-424-9300 (24 hours)

HAZARDOUS MATERIAL CLEANUP

Alabama
(334) 260-2700 after 5 pm (334) 242-4378

Arkansas Department of Emergency Management
1-800-322-4012

Florida – Florida Highway Patrol
(850) 617-2000

Georgia – Georgia Highway Patrol
*GSP (*477)

Kentucky
1-800-928-2380

Louisiana – Louisiana Department of Ag & Forestry
1-855-452-5323

Mississippi – Mississippi Emergency Management Agency (MEMA)
1-800-222-6362 (24 hr emergency line)

North Carolina – NC Highway Patrol
1-800-662-7956

South Carolina – SCDHEC
1-888-481-0125 (24 Hours)

Tennessee
(615) 741-0001

Texas
1-800-424-9300 (24 hours)

Virginia
1-800-424-8802

PESTICIDE CONTAINER RECYCLING

Alabama
(334) 240-7100

Arkansas
(501) 225-1598   See note below*

Florida
(352) 392-4721

Georgia
(404) 656-4958

Kentucky
1-800-205-6543

Louisiana – Louisiana Department of Ag & Forestry
1-855-452-5323

Mississippi – USAg Recycling, Inc.
1-800-654-3145

North Carolina
(919) 733-3556

South Carolina
1-800-654-3145

Tennessee
1-800-654-3145

Texas
See note below*

Virginia
Contact your local Extension agent for info

MISUSE OF PESTICIDES
It is a violation of law to use any pesticide in a manner not permitted by its labeling. To protect yourself, never apply any pesticide in a manner or for a purpose other than as instructed on the label or in labeling accompanying the pesticide product that you purchase. Do not ignore the instructions for use of protective clothing and devices and for storage and disposal of pesticide wastes, including containers. All recommendations for pesticide use included in this manual were legal at the time of publication, but the status of registration and use patterns are subject to change by actions of state and federal regulatory agencies.

* In Arkansas, Texas, and Virginia, pesticide container recycling is not required as according to state law "properly rinsed agricultural chemical containers are not classified as hazardous waste."

Recommendations for the use of agricultural chemicals and other products are included in this publication as a convenience to the reader. The use of brand names and any mention or listing of commercial products or services in this publication does not imply endorsement by Alabama A&M University, Auburn University/Alabama Cooperative Extension System, Clemson University, Florida A&M University, Louisiana State University AgCenter, Mississippi State University, North Carolina State University, Oklahoma State University, Texas A&M, Tuskegee University, University of Arkansas System Division of Agriculture, University of Georgia, University of Kentucky, University of Tennessee, and Virginia Tech nor discrimination against similar products or services not mentioned. Recommendations and labels will vary from state to state, and we have made every attempt to assure that these exceptions are noted. However, individuals who use agricultural chemicals are responsible for ensuring that the intended use complies with current regulations and conforms to the product label in their respective home state. Be sure to obtain current information about usage regulations and examine a current product label before applying any chemical. For assistance, contact your county Cooperative Extension Service agent.

Notes

Notes

Notes

Copyright © 2026 Southeastern Vegetable Extension Workers Group.
All rights reserved.

ISBN 978-1-4696-9667-6 (paperback)

Visit www.vegcrophandbook.com to view the information in this handbook in an open access format.

For product safety concerns under the European Union's General Product Safety Regulation (EU GPSR), please contact gpsr@mare-nostrum.co.uk or write to the University of North Carolina Press and Mare Nostrum Group B.V., Doelen 72, 4831 GR Breda, The Netherlands.

Published by the Southeastern Vegetable Extension Workers Group

Distributed by the University of North Carolina Press